EIGHTH EDITION

Estelle Levetin
The University of Tulsa

Karen McMahon
The University of Tulsa

Plants & Society

PLANTS & SOCIETY, EIGHTH EDITION

Published by McGraw-Hill Education, 2 Penn Plaza, New York, NY 10121. Copyright © 2020 by McGraw-Hill Education. All rights reserved. Printed in the United States of America. Previous editions © 2016, 2012, and 2008. No part of this publication may be reproduced or distributed in any form or by any means, or stored in a database or retrieval system, without the prior written consent of McGraw-Hill Education, including, but not limited to, in any network or other electronic storage or transmission, or broadcast for distance learning.

Some ancillaries, including electronic and print components, may not be available to customers outside the United States.

This book is printed on acid-free paper.

2 3 4 5 6 7 8 9 LKV 21 20

ISBN 978-1-259-88004-9 (bound edition)
MHID 1-259-88004-4 (bound edition)
ISBN 978-1-260-81260-2 (loose-leaf edition)
MHID 1-260-81260-X (loose-leaf edition)

Product Developers: *Lora Neyens, Christine Scheid*
Marketing Manager: *Kelly Brown*
Content Project Managers: *Becca Gill, Jeni McAtee*
Buyer: *Laura Fuller*
Designer: *Matt Diamond*
Content Licensing Specialist: *Melissa Homer*
Cover Image: *©Rodrigo A Torres/Glow Images*
Compositor: *Lumina Datamatics, Inc.*

All credits appearing on page or at the end of the book are considered to be an extension of the copyright page.

Library of Congress Cataloging-in-Publication Data

Names: Levetin, Estelle, author. | McMahon, Karen, editor.
Title: Plants & society / Estelle Levetin, The University of Tulsa, Karen
   McMahon, The University of Tulsa.
Other titles: Plants and society
Description: Eighth edition. | New York, NY : McGraw-Hill Education, [2020] |
   Includes index.
Identifiers: LCCN 2019001477| ISBN 9781259880049 (alk. paper) | ISBN
   9781260812602 (loose-leaf edition)
Subjects: LCSH: Botany. | Botany, Economic. | Plants—Social aspects.
Classification: LCC QK47 .L48 2020 | DDC 580—dc23
LC record available at https://lccn.loc.gov/2019001477

The Internet addresses listed in the text were accurate at the time of publication. The inclusion of a website does not indicate an endorsement by the authors or McGraw-Hill Education, and McGraw-Hill Education does not guarantee the accuracy of the information presented at these sites.

mheducation.com/highered

In loving memory of our mothers

Pauline Levetin

& *Dorothy Sink McMahon*

Contents in Brief

UNIT I   Plants and Society: The Botanical Connections to Our Lives

         1   Plants in Our Lives   1

UNIT II   Introduction to Plant Life: Botanical Principles

         2   The Plant Cell   16
         3   The Plant Body   28
         4   Plant Physiology   47
         5   Plant Life Cycle: Flowers   69
         6   Plant Life Cycle: Fruits and Seeds   85
         7   Genetics   101
         8   Plant Systematics and Evolution   120
         9   Diversity of Plant Life   136

UNIT III   Plants as a Source of Food

         10   Human Nutrition   152
         11   Origins of Agriculture   175
         12   The Grasses   187
         13   Legumes   210
         14   Starchy Staples   224
         15   Feeding a Hungry World   241

UNIT IV   Commercial Products Derived from Plants

         16   Stimulating Beverages   271
         17   Herbs and Spices   289
         18   Materials: Cloth, Wood, and Paper   308

UNIT V   Plants and Human Health

         19   Medicinal Plants   337
         20   Psychoactive Plants   358
         21   Poisonous and Allergy Plants   378

UNIT VI   Algae and Fungi: The Impact of Algae and Fungi on Human Affairs

         22   The Algae   400
         23   Fungi in the Natural Environment   419
         24   Beverages and Foods from Fungi   446
         25   Fungi That Affect Human Health   469

UNIT VII   Plants and the Environment

         26   Plant Ecology   486

Contents

Preface xii
List of Boxed Readings xv

UNIT I   Plants and Society: The Botanical Connections to Our Lives

1   Plants in Our Lives   1

Plants and Human Society 2
    The Flowering Plants 2
    The Non-Flowering Plants 4
    The Algae 4
    The Fungi 4
    Plant Sciences 4
    Scientific Method 5

A CLOSER LOOK 1.1 Biological Mimics 6

Fundamental Properties of Life 6
Molecules of Life 8
    Carbohydrates 8
    Proteins 9
    Lipids 11
    Nucleic Acids 12

A CLOSER LOOK 1.2 Perfumes to Poisons: Plants as Chemical Factories 14

UNIT II   Introduction to Plant Life: Botanical Principles

2   The Plant Cell   16

Early Studies of Cells 17
The Cell Wall 19
The Protoplast 19
    Membranes 19
    Moving Into and Out of Cells 19
    Organelles 20

A CLOSER LOOK 2.1 Origin of Chloroplasts and Mitochondria 22

    The Nucleus 23
Cell Division 24
    The Cell Cycle 24
    Prophase 24
    Metaphase 25
    Anaphase 27
    Telophase 27
    Cytokinesis 27

3   The Plant Body   28

Plant Tissues 29
    Meristems 29
    Dermal Tissue 29
    Ground Tissue 31
    Vascular Tissue 32
Plant Organs 33
    Stems 33
    Roots 35
    Leaves 37

A CLOSER LOOK 3.1 Studying Ancient Tree Rings 37

Vegetables: Edible Plant Organs 41
    Carrots 41

A CLOSER LOOK 3.2 Plants That Trap Animals 42

    Lettuce 44
    Radishes 44

A CLOSER LOOK 3.3 Supermarket Botany 44

    Asparagus 45

4   Plant Physiology   47

Plant Transport Systems 48
    Transpiration 49
    Absorption of Water from the Soil 49

A CLOSER LOOK 4.1 Mineral Nutrition and the Green Clean 50

    Water Movement in Plants 51
    Translocation of Sugar 51

A CLOSER LOOK 4.2 Sugar and Slavery 53

Metabolism 55
    Energy 55
    Redox Reactions 55
    Phosphorylation 55
    Enzymes 56
Photosynthesis 56
    Energy from the Sun 56
    Chloroplasts and Light-Absorbing Pigments 57
    Overview 59
    The Light Reactions 59
    The Calvin Cycle 61
    Variation to Carbon Fixation 63
Cellular Respiration 63
    Glycolysis 64
    The Krebs Cycle 64
    The Electron Transport System 64
    Aerobic vs. Anaerobic Respiration 67

5 Plant Life Cycle: Flowers 69

The Flower 70
Floral Organs 70

A CLOSER LOOK 5.1 Mad about Tulips 71

Modified Flowers 72

Meiosis 75
Stages of Meiosis 75
Meiosis in Flowering Plants 77
Male Gametophyte Development 77

A CLOSER LOOK 5.2 Pollen Is More Than Something to Sneeze At 79

Female Gametophyte Development 79

Pollination and Fertilization 79
Animal Pollination 80

A CLOSER LOOK 5.3 Alluring Scents 82

Wind Pollination 83
Double Fertilization 83

6 Plant Life Cycle: Fruits and Seeds 85

Fruit Types 86
Simple Fleshy Fruits 86
Dry Dehiscent Fruits 86
Dry Indehiscent Fruits 86
Aggregate and Multiple Fruits 86

Seed Structure and Germination 88
Dicot Seeds 88
Monocot Seeds 88
Seed Germination and Development 88

Representative Edible Fruits 88
Tomatoes 90

A CLOSER LOOK 6.1 The Influence of Hormones on Plant Reproductive Cycles 92

Apples 94
Oranges and Grapefruits 95
Chestnuts 97
Exotic Fruits 98

7 Genetics 101

Mendelian Genetics 102
Gregor Mendel and the Garden Pea 102
Monohybrid Cross 103
Dihybrid Cross 105

Beyond Mendelian Genetics 106
Incomplete Dominance 106

A CLOSER LOOK 7.1 Solving Genetics Problems 107

Multiple Alleles 109
Polygenic Inheritance 109
Linkage 109

Molecular Genetics 110
DNA—The Genetic Material 111

A CLOSER LOOK 7.2 Try These Genes On for Size 112

Genes Control Proteins 114
Transcription and the Genetic Code 114
Translation 116
Other Roles of RNA 116
Mutations 117
Epigenetics 118
Recombinant DNA 118

8 Plant Systematics and Evolution 120

Early History of Classification 121
Carolus Linnaeus 121

How Plants Are Named 123
Common Names 123

A CLOSER LOOK 8.1 The Language of Flowers 125

Scientific Names 126

Taxonomic Hierarchy 127
Higher Taxa 127
What Is a Species? 129

Phylocode 130

A CLOSER LOOK 8.2 Saving Species through Systematics 131

Barcoding Species 132
The Influence of Darwin's Theory of Evolution 132
The Voyage of the HMS Beagle 133
Natural Selection 134

9 Diversity of Plant Life 136

The Three-Domain System 137
Survey of the Plant Kingdom 137

A CLOSER LOOK 9.1 Alternation of Generations 139

Liverworts, Mosses, and Hornworts 141
Lycophytes and Ferns 143
Gymnosperms 146

A CLOSER LOOK 9.2 Amber: A Glimpse into the Past 148

Angiosperms 151

UNIT III Plants as a Source of Food

10 Human Nutrition 152

Macronutrients 153
Sugars and Complex Carbohydrates 153

A CLOSER LOOK 10.1 Famine or Feast 154

Fiber in the Diet 156
Proteins and Essential Amino Acids 156
Gluten and Celiac Disease 158
Fats and Cholesterol 159
Micronutrients 162
Vitamins 162
Minerals 167
Dietary Guidelines 169
Balancing Nutritional Requirements 169
Healthier Dietary Guidelines 170
Glycemic Index 171
Meatless Alternatives 172

A CLOSER LOOK 10.2 Eat Broccoli for Cancer Prevention 172

11 Origins of Agriculture 175

Foraging Societies and Their Diets 176
Early Foragers 176
Modern Foragers 177
The Paleo Diet 177

A CLOSER LOOK 11.1 Forensic Botany 178

Agriculture: Revolution or Evolution? 179
Latitudinal Spread 180
Early Sites of Agriculture 180
The Near East 181
The Far East 182
The New World 183
Characteristics of Domesticated Plants 184
Centers of Plant Domestication 184

12 The Grasses 187

Characteristics of the Grass Family 188
Vegetative Characteristics 188
The Flower 188
The Grain 188
Wheat: The Staff of Life 189
Origin and Evolution of Wheat 190
Modern Cultivars 191
Wheat Genome 191

A CLOSER LOOK 12.1 The Rise of Bread 192

Nutrition 193
Corn: Indian Maize 193
An Unusual Cereal 194
Types of Corn 194

A CLOSER LOOK 12.2 Barbara McClintock and Jumping Genes in Corn 196

Hybrid Corn 197
Ancestry of Corn 197

Corn Genome 199
Value of Corn 199
Rice: Food for Billions 200
A Plant for Flooded Fields 200
Varieties 201
Rice Genome 202
Flood-Tolerant Rice 202
Other Important Grains 202
Rye and Triticale 202
Oats 204
Barley and Tritordeum 204
Sorghum and Millets 204
Other Grasses 205
Forage Grasses 205
Lawn Grasses 205
Bioethanol: Grass to Gas 206
Corn 206
Sugarcane 207
Cellulosic Ethanol 207

13 Legumes 210

Characteristics of the Legume Family 211

A CLOSER LOOK 13.1 The Nitrogen Cycle 212

Important Legume Food Crops 214
Beans and Peas 214
Peanuts 215
Soybeans 217

A CLOSER LOOK 13.2 Harvesting Oil 219

Other Legumes of Interest 221
A Supertree for Forestry 221
Forage Crops 221
Beans of the Future 222

14 Starchy Staples 224

Storage Organs 225
Modified Stems 225
Storage Roots 226

A CLOSER LOOK 14.1 Banana Republics: The Story of the Starchy Fruit 226

White Potato 228
South American Origins 228
The Irish Famine 228
Continental Europe 229
The Potato in the United States 229
Solanum tuberosum 230
Pathogens and Pests 231
Potato Genome 232
Modern Cultivars 232
Sweet Potato 232
Origin and Spread 233
Cultivation 233

Cassava 234
    Origin and Spread 234
    Botany and Cultivation 234
    Processing 235

A CLOSER LOOK 14.2 Starch: In Our Collars and in Our Colas 237

Other Underground Crops 238
    Yams 238
    Taro 238
    Jerusalem Artichoke 238

15 Feeding a Hungry World 241

Breeding for Crop Improvement 242
The Green Revolution 243
    High-Yield Varieties 243
    Disease-Resistant Varieties 244
    History of the Green Revolution 245
    Problems with the Green Revolution: What Went Wrong? 245
    Solutions? 246
Genetic Diversity 247
    Monoculture 247
    Sustainable Agriculture 248
    Genetic Erosion 249
    Seed Banks 249
    Heirloom Varieties 250
    Germplasm Treaty 250
    Crops and Global Warming 251
Alternative Crops: The Search for New Foods 252
    Quinoa 252
    Amaranth and Chia 252
    Tarwi 253
    Tamarillo and Naranjilla 254
    Oca 254
Biotechnology 254

A CLOSER LOOK 15.1 Mutiny on the HMS *Bounty*: The Story of Breadfruit 255

    Cell and Tissue Culture 256
    Molecular Plant Breeding 257
Genetic Engineering and Transgenic Plants 257
    Herbicide Resistance 258
    Against the Grain 260
    Insect Resistance 260
    Bt Corn and Controversy 261
    The Promise of Golden Rice 262
    Other Genetically Engineered Foods 262
    Disease Resistance 263
    Farming Pharmaceuticals 264
    Nonfood Crops 265
    Genetically Modified Trees 265
    Regulatory Issues 266
    Environmental and Safety Considerations 266
Gene Editing: CRISPR/Cas9 268

UNIT IV Commercial Products Derived from Plants

16 Stimulating Beverages 271

Physiological Effects of Caffeine 272
    Medical Benefits of Caffeine 273
Coffee 273
    An Arabian Drink 273
    Plantations, Cultivation, and Processing 274

A CLOSER LOOK 16.1 Climate Change and the Future of Coffee 276

    From Bean to Brew 277
    Varieties 277
    Decaffeination 278
    Variations on a Bean 279
    Shade Coffee vs. Sun Coffee 279
    Fair Trade Coffee 279
Tea 279
    Oriental Origins 279
    Cultivation and Processing 280

A CLOSER LOOK 16.2 Tea Time: Ceremonies and Customs around the World 280

    The Flavor and Health Effects of Tea 281
    History 282
Chocolate 283
    Food of the Gods 283
    Quakers and Cocoa 283

A CLOSER LOOK 16.3 Candy Bars: For the Love of Chocolate 284

    Cultivation and Processing 285
    The High Price of Chocolate 286
Coca-Cola: An "All-American" Drink 287
Other Caffeine Beverages 287

17 Herbs and Spices 289

Essential Oils 290
History of Spices 290
    Ancient Trade 290

A CLOSER LOOK 17.1 Aromatherapy: The Healing Power of Scents 291

    Marco Polo 291
    Age of Exploration 292
    Imperialism 293
    New World Discoveries 293
Spices 293
    Cinnamon: The Fragrant Bark 293
    Black and White Pepper 294

Cloves 295
Nutmeg and Mace 295
Ginger and Turmeric 296
Saffron 297
Hot Chilies and Other Capsicum Peppers 297
Vanilla 299
Allspice 300
Herbs 300
The Aromatic Mint 301
The Parsley Family 303
The Mustard Family 303
The Pungent Alliums 304

A CLOSER LOOK 17.2 Sweet Talk 306

18 Materials: Cloth, Wood, and Paper 308

Fibers 309
Types 309

A CLOSER LOOK 18.1 A Tisket, a Tasket: There Are Many Types of Baskets 310

King Cotton 312
Linen: An Ancient Fabric 315
Other Bast Fibers 316
Miscellaneous Fibers 317
Rayon: "Artificial Silk" 319
Bark Cloth 319

A CLOSER LOOK 18.2 Herbs to Dye For 321

Wood and Wood Products 323

A CLOSER LOOK 18.3 Good Vibrations 324

Hardwoods and Softwoods 326
Lumber, Veneer, and Plywood 327
Fuel 329
Other Products from Trees 329
Wood Pulp 330
Paper 331
Early Writing Surfaces 331
The Art of Papermaking 333
Alternatives to Wood Pulp 333
Bamboo 334

UNIT V   Plants and Human Health

19 Medicinal Plants 337

History of Plants in Medicine 338
Early Greeks and Romans 338
Age of Herbals 338

A CLOSER LOOK 19.1 Native American Medicine 340

Modern Prescription Drugs 340
Herbal Medicine Today 342
Active Principles in Plants 343
Alkaloids 343
Glycosides 343

Medicinal Plants 343
Foxglove and the Control of Heart Disease 343
Aspirin: From Willow Bark to Bayer 345
Malaria, the Fever Bark Tree, and Sweet Wormwood 346
Diabetes, French Lilac, and Metformin 349
Snakeroot, Schizophrenia, and Hypertension 350
The Burn Plant 351
Ephedrine 351
Cancer Therapy 352
Herbal Remedies: Promise and Problems 354

20 Psychoactive Plants 358

Psychoactive Drugs 359
The Opium Poppy 360
An Ancient Curse 360
The Opium Wars 360
Opium Alkaloids 361
Heroin 361
Withdrawal 362
Marijuana 362
Early History in China and India 362
Spread to the West 363
THC and Its Psychoactive Effects 364
Medical Marijuana 365
Legal Issues 365
Cocaine 366
South American Origins 366
Freud, Holmes, and Coca-Cola 366

A CLOSER LOOK 20.1 The Tropane Alkaloids and Witchcraft 367

Coke and Crack 369
Medical Uses 369
A Deadly Drug 369
Tobacco 370
A New World Habit 370
Cultivation Practices 371
Health Risks 371
Peyote 374
Mescal Buttons 374
Native American Church 375
Kava—The Drug of Choice in the Pacific 375
Preparation of the Beverage 375
Active Components in Kava 375
Lesser Known Psychoactive Plants 376

21 Poisonous and Allergy Plants 378

Notable Poisonous Plants 379
Poisonous Plants in the Wild 379
Poisonous Plants in the Backyard 382

A CLOSER LOOK 21.1 Allelopathy—Chemical Warfare in Plants 383

Poisonous Plants in the Home 386
Plants Poisonous to Livestock 386
Plants That Cause Mechanical Injury 387
Insecticides from Plants 388
Allergy Plants 389
Allergy and the Immune System 390
Respiratory Allergies 390
Hay Fever Plants 391
Climate Change and Allergy Plants 394
Allergy Control 394
Contact Dermatitis 395
Food Allergies 397
Latex Allergy 398

UNIT VI   Algae and Fungi: The Impact of Algae and Fungi on Human Affairs

22 The Algae 400

Characteristics of the Algae 401
Classification of the Algae 401
Cyanobacteria 402
Euglenoids 403
Dinoflagellates 403
Diatoms 404
Brown Algae 404
Red Algae 405
Green Algae 405
Algae in Our Diet 409
Seaweeds 409
Biofuels from Algae 410
Other Economic Uses of Algae 411
Toxic and Harmful Algae 411
Toxic Cyanobacteria 412

A CLOSER LOOK 22.1 Drugs from the Sea 413

Red Tides 414
Pfiesteria 414

A CLOSER LOOK 22.2 Killer Alga—Story of a Deadly Invader 415

Other Toxic Algae 417

23 Fungi in the Natural Environment 419

Fungi and Fungal-Like Organisms 420
Fungal-Like Protists 420
Slime Molds 421
The Kingdom Fungi 422
Characteristics of Fungi 422
Classification of Fungi 424
Role of Fungi in the Environment 433
Decomposers—Nature's Recyclers 433

A CLOSER LOOK 23.1 Lichens: Algal-Fungal Partnership 434

Mycorrhizae: Root-Fungus Partnership 435

A CLOSER LOOK 23.2 Dry Rot and Other Wood Decay Fungi 436

Plant Diseases with Major Impact on Humans 437
Late Blight of Potato 438
Rusts—Threat to the World's Breadbasket 439
Corn Smut—Blight or Delight 441
Dutch Elm Disease—Destruction in the Urban Landscape 441
Sudden Oak Death—Destruction in the Forest 443

24 Beverages and Foods from Fungi 446

Making Wine 447
The Wine Grape 448
Harvest 449
Red or White 449
Fermentation 449

A CLOSER LOOK 24.1 Disaster in the French Vineyards 450

Clarification 452
Aging and Bottling 452
"Drinking Stars" 453
Fortified and Dessert Wines 454
Climate Change and the Wine Grape 454
The Brewing of Beer 454
Barley Malt 455
Mash, Hops, and Wort 455
Fermentation and Lagering 455
Sake, a Rice "Beer" 457
Distillation 457
The Still 458
Distilled Spirits 458
The Whiskey Rebellion 458
Tequila from Agave 459
Hard Cider 460
A Victorian Drink Revisited 461
Absinthe 461
The Green Hour 461
Active Principles 461
Absinthism 462

A CLOSER LOOK 24.2 Alcohol and Health 463

Fungi as Food 465
Edible Mushrooms 465
Fermented Foods 466
Quorn Mycoprotein 467

25 Fungi That Affect Human Health 469

Antibiotics and Other Wonder Drugs 470
   Fleming's Discovery of Penicillin 470
   Manufacture of Penicillin 471
   Other Antibiotics and Antifungals 471
Fungal Poisons and Toxins 472
   Mycotoxins 472

A CLOSER LOOK 25.1 The New Wonder Drugs 473

   Ergot of Rye and Ergotism 475
   The Destroying Angel and Toadstools 477
   Soma and Hallucinogenic Fungi 478
Human Pathogens and Allergies 480
   Dermatophytes 480
   Systemic Mycoses 481
   Allergies 482
   Indoor Fungi and Toxic Molds 482

UNIT VII Plants and the Environment

26 Plant Ecology 486

The Ecosystem 487
   Ecological Niche 487
   Food Chains and Food Webs 488
   Energy Flow and Ecological Pyramids 489
   Biological Magnification 491

Biogeochemical Cycling 492
   The Carbon Cycle 492
   Source or Sink? 492
   The Greenhouse Effect 494
   Species Shift 496
   Global Warming Pact 496
Ecological Succession 497
The Anthropocene Epoch 498
The Green World: Biomes 500
   Deserts 501
   Crops from the Desert 501
   Chaparral 504
   Grasslands 504
   Forests 505
   Harvesting Trees 506

A CLOSER LOOK 26.1 Buying Time for the Rain Forest 508

Appendix A Atoms, Molecules, and Chemical Bonds 513
Appendix B Classification of Plants and Those Organisms Traditionally Classified as Plants Discussed in *Plants and Society* 517
Appendix C Metric System: Scientific Measurement 519
Glossary 521
Index 533

Preface

In the twenty-first century, plant science is once again assuming a prominent role in research. Renewed emphasis on developing medicinal products from native plants has encouraged ethnobotanical endeavors. The destruction of the rain forests has made the timing for this research imperative and has spurred efforts to catalog the plant biodiversity in these environments. Efforts to feed the growing populations in developing nations have also positioned plant scientists at the cutting edge of genetic engineering with the creation of transgenic crops. However, in recent decades botany courses have seen a decline in enrollment, and some courses have even disappeared from the curriculum in many universities. We have written *Plants and Society* in an effort to offset this trend. By taking a multidisciplinary approach to studying the relationship between plants and people, we hope to stimulate interest in plant science and encourage students to further study. Also, by exposing students to society's historical connection to plants, we hope to instill a greater appreciation for the botanical world.

AUDIENCE

Recently, general botany courses have emphasized the impact of plants on society. In addition, many institutions have developed plants and society courses devoted exclusively to this topic. This emphasis has transformed the traditional economic botany from a dry statistical treatment of "bushels per acre" to an exciting discussion of "botanical marvels" that have influenced our past and will change our future. *Plants and Society* is intended for use in this type of course, which is usually one semester or one quarter in length. There are no prerequisites because it is an introductory course. The course covers basic principles of botany and places a strong emphasis on the economic aspects and social implications of plants and fungi.

Students usually take a course of this nature in their freshman or sophomore year to satisfy a science requirement in the general education curriculum. Typically, they are not biology majors. Although most students enroll to satisfy the science requirement, many become enthusiastic about the subject matter. Students, even those with a limited science background, should not encounter any problems with the level of scientific detail in this text.

As indicated, the primary market for this text is a plants and society course; however, it would certainly be suitable for an introductory general botany course as well.

ORGANIZATION

We feel that *Plants and Society* is a textbook with a great deal of flexibility for course design. It offers a unique balanced approach between basic botany and the applied or economic aspects of plant science. Other texts emphasize either the basic or applied material, making it difficult for instructors who wish to provide better balance in an introductory course. Another distinctive feature is the unit on algae and fungi. While other texts cover certain aspects of this topic, we have an expanded coverage of algae and fungi and their impact on society.

Plants and Society is organized into 26 chapters that are grouped into seven units. The first nine chapters cover the basic botany found in an introductory course. However, even in these chapters we have included many applied topics, some in the boxed essays but others directly in the chapter text.

UNIT I Plants and Society: The Botanical Connections to Our Lives. Chapter 1 stresses the overall importance of plants in everyday life. The properties of life, molecules of life, flowering and non flowering plants, algae, and fungi are introduced. The scientific method is explained as the process used by scientists to study and expand our knowledge of the natural world. The diversity and applications of phytochemicals are also presented.

UNIT II Introduction to Plant Life: Botanical Principles. This unit addresses basic botany. Chapters cover plant structure from the cellular level to the mature plant. Reproduction, including mitosis and meiosis and the life cycle of flowering plants, is discussed in two chapters. Other chapters cover genetics, evolution, plant physiology, plant systematics, and plant diversity. Some of the economic aspects of plants discussed in this unit are the importance of vegetables and fruits, the connection between sugar and slavery, plant essential oils and perfumes, phytoremediation, the applications of palynology, and species conservation.

UNIT III Plants as a Source of Food. This unit describes the major food crops. It begins with a chapter on the requirements for human nutrition and continues with a chapter on the origins of agriculture. Other chapters cover the grasses, the legumes, and starchy staples. The unit ends with a chapter on the Green Revolution, the loss of genetic diversity, the search for alternative crops, and the controversial development of transgenic crops.

UNIT IV Commercial Products Derived from Plants. This unit covers other crops that provide us with consumable products, such as beverages, herbs and spices, and materials such as cloth, wood, and paper. The historical origin and societal impact of these crops are explored.

UNIT V Plants and Human Health. This unit introduces students to the historical foundations of Western medicine, the practice of herbal medicine, and the chemistry of secondary plant products. Descriptions of the plants that provide us with medicinal products and psychoactive drugs are discussed. The unit also covers the common poisonous and allergy plants that are found in the environment.

UNIT VI Algae and Fungi: The Impact of Algae and Fungi on Human Affairs. This unit describes the economic importance of the algae and fungi, including their biology and crucial roles in the environment. The algae are recognized as key producers in aquatic environments and as sources of human food, devastating blooms, and industrial products. Fermented beverages and foods from fungi are discussed, as is the medical importance of fungi as sources of antibiotics, toxins, and diseases affecting crops and people.

UNIT VII Plants and the Environment. Chapter 26 is an introduction to the principles of ecology: the ecosystem, niches, food chains, biogeochemical cycles, and ecological succession. The major biomes of the world are discussed, with an emphasis on the economic value of certain desert plants and the strategy of extractive reserves in the rain forest. The problems associated with rising levels of the greenhouse gas CO_2 and the environmental consequences of global warming are addressed.

APPROACH

This textbook is written at the introductory level suitable for students with little or no background in biology. Like any introductory book, this book uses a broad-brush treatment. The nature of the course dictates an applied approach, with the impact of plants on society as the integrating theme, but the theoretical aspects of basic botany are thoroughly covered.

LEARNING AIDS

In addition to the textual material, each chapter begins with a chapter outline and key concepts. Important terms are in boldface type throughout the text, and each chapter ends with a summary and review questions. Thinking Critically questions are inserted in the text to draw the attention of the students as they read the chapter. The questions begin with either a summary of the preceding text or an introduction to new information that is complementary to the chapter. The questions that follow are designed not only to test comprehension but also, in many instances, to promote critical thinking by asking students to apply their knowledge to real-life situations. Thinking Critically questions may also be assigned by instructors or used to initiate in-class discussions. Three appendices and a glossary conclude the text. The classification of plants and other organisms discussed in the text and a review of metric units are located for quick access on the inside front and back covers, respectively.

NEW TO THE EIGHTH EDITION

The eighth edition of *Plants and Society* been updated to spotlight exciting discoveries and update major advancements in the science of plants, algae, and fungi, with special emphasis on how these organisms impact humanity. These include:

- Expanded information on the regulation of cell division (Chapter 2)
- Subsidiary cells in grasses enable the greater inflation of guard cells to bring in more carbon dioxide for photosynthesis as well as facilitating guard cells to respond rapidly to changing environmental conditions (Chapter 3).
- Sun pitchers (*Heliamphora*), carnivorous pitcher plants native to South America, exhibit remarkable adaptations: a nectar spoon to lure prey and a drainage pipe to siphon off excess water from the pitcher (Chapter 3).
- The section on orchids has been updated to include significantly more detail on the unique modifications of orchid flowers (Chapter 5).
- There are updates on the latest research surrounding colony collapse disorder and the search for alternative pollinators (Chapter 5).
- There is additional coverage of the ways in which researchers are using genomic analysis to discover how to enhance the flavors of mass-market tomatoes to better resemble heirloom varieties (Chapter 6).
- Information on CRISPR, a powerful technique in the gene editing of organisms, is presented (Chapters 7 and 15).
- Fiber in the diet may promote health as the food to support microbiota beneficial to the digestive tract (Chapter 10).
- The paleo diet, the so-called original diet of ancient humans, is evaluated (Chapter 11).
- A new update on the International Wheat Genome Sequencing Consortium's work to sequence the bread wheat genome is featured (Chapter 12).
- A Closer Look 13.1 has been updated to include new information on biofertilizers and A Closer Look 13.2 on Harvesting Oil also has data updates (Chapter 13).
- A Closer Look 14.1 features new information on the recent resurgence of Panama disease and its effect on bananas (Chapter 14).
- New evidence on the disjunct distribution of sweet potatoes is detailed (Chapter 14).
- A section on the effects of global warming on crops has been included (Chapter 15).

- Information on genetically modified rice and corn has been added (Chapter 15).
- A new Closer Look on Climate Change and the Future of Coffee has been added (Chapter 16).
- Update on the legal status of hemp cultivation (Chapter 18).
- There is a new content on the increased use of plant biomass (such as wood pellets) to supplement or replace coal (Chapter 18).
- A new section highlights the remarkable history of French lilac in the development of the medicine Metformin, which is used to treat both diabetes and polycystic ovarian syndrome (Chapter 19).
- Updated content is included on the antimalarial properties of certain plants used in Chinese medicine (Chapter 19).
- Updates on the legal status of marijuana has been included (Chapter 20).
- Giant Hogweed, an oversized invasive weed from Asia, produces a toxic sap that causes debilitating blisters and burns to the skin upon exposure to sunlight (Chapter 21).
- There are updates to gluten and nonceliac gluten sensitivity (Chapter 21).
- There is information on new research suggesting a potential connection between algae blooms and global warming (Chapter 22).
- A 2018 research on *Batrachochytrium dendrobatidis* fungus highlights how the pathogen traveled from the Korean peninsula and has been responsible for the extinction or decline of frog and salamander populations around the world (Chapter 23).
- A new section covers the history and cultivation of apple ciders, the alcoholic beverages made by fermenting apple juice (Chapter 24).
- With the rise of MRSA infections worldwide, there has been renewed interest in the therapeutic use of fusidic acid, covered now in additional detail (Chapter 25).
- The global, indelible impact of human activity on Earth has initiated a demand for the creation of a new geologic epoch, the Anthropocene (Chapter 26).
- The Paris Climate Agreement has revised its long-term goal of limiting global warming from an increase of 2.0°C above preindustrial temperatures down to 1.5°C (Chapter 26).
- Over 50 new photographs and several new or revised figures and tables have been added to the eighth edition.

ACKNOWLEDGMENTS

From our first introduction to botany as college students, we became irrevocably fixated on the lives of plants. We can remember the fascination we felt when we read about the plant explorers who discovered *extinct* ginkgo trees alive in China and how a trichome of the stinging nettle was the inspiration for the invention of the hypodermic needle. It is our hope that *Plants and Society* will present the world of plants and how they sustain humanity in a way that will inspire students to have a lifelong appreciation of plants.

We wish to thank the editorial staff at McGraw-Hill Higher Education for their editorial expertise and their endless patience during the publication of the eighth edition of *Plants and Society*. We especially want to acknowledge our Portfolio Manager, Michael Ivanov; Senior Product Developer, Chipper Scheid; Freelance Product Developer, Jen Thomas; and Content Project Manager, Becca Gill.

REVIEWERS

We are indebted to our colleagues who have taken time from demanding schedules to meticulously review *Plants and Society* for errors, inconsistencies, or omissions and to offer constructive feedback and suggestions. We thank you for making the eighth edition of *Plants and Society* the best edition.

Karen Amisi
    Grand Valley State University
Elise C. Hollister
    Grand Valley State University
Meshagae Hunte-Brown
    Drexel University
Arthur Kneeland
    University of Wisconsin–Stout
Diane M. Lahaise
    Georgia State University–Perimeter College
Elizabeth Martin
    Lewis-Clark State College
Michael Sundue
    University of Vermont
Jacob Thompson
    College of Coastal Georgia
Alexander Wait
    Missouri State University

List of Boxed Readings

1. Biological Mimics (Chapter 1)
2. Perfumes to Poisons: Plants as Chemical Factories (Chapter 1)
3. Origin of Chloroplasts and Mitochondria (Chapter 2)
4. Studying Ancient Tree Rings (Chapter 3)
5. Plants That Trap Animals (Chapter 3)
6. Supermarket Botany (Chapter 3)
7. Mineral Nutrition and the Green Clean (Chapter 4)
8. Sugar and Slavery (Chapter 4)
9. Mad about Tulips (Chapter 5)
10. Pollen Is More Than Something to Sneeze At (Chapter 5)
11. Alluring Scents (Chapter 5)
12. The Influence of Hormones on Plant Reproductive Cycles (Chapter 6)
13. Solving Genetics Problems (Chapter 7)
14. Try These Genes On for Size (Chapter 7)
15. The Language of Flowers (Chapter 8)
16. Saving Species through Systematics (Chapter 8)
17. Alternation of Generations (Chapter 9)
18. Amber: A Glimpse into the Past (Chapter 9)
19. Famine or Feast (Chapter 10)
20. Eat Broccoli for Cancer Prevention (Chapter 10)
21. Forensic Botany (Chapter 11)
22. The Rise of Bread (Chapter 12)
23. Barbara McClintock and Jumping Genes in Corn (Chapter 12)
24. The Nitrogen Cycle (Chapter 13)
25. Harvesting Oil (Chapter 13)
26. Banana Republics: The Story of the Starchy Fruit (Chapter 14)
27. Starch: In Our Collars and in Our Colas (Chapter 14)
28. Mutiny on the HMS *Bounty:* The Story of Breadfruit (Chapter 15)
29. Climate Change and the Future of Coffee (Chapter 16)
30. Tea Time: Ceremonies and Customs around the World (Chapter 16)
31. Candy Bars: For the Love of Chocolate (Chapter 16)
32. Aromatherapy: The Healing Power of Scents (Chapter 17)
33. Sweet Talk (Chapter 17)
34. A Tisket, a Tasket: There Are Many Types of Baskets (Chapter 18)
35. Herbs to Dye For (Chapter 18)
36. Good Vibrations (Chapter 18)
37. Native American Medicine (Chapter 19)
38. The Tropane Alkaloids and Witchcraft (Chapter 20)
39. Allelopathy—Chemical Warfare in Plants (Chapter 21)
40. Drugs from the Sea (Chapter 22)
41. Killer Alga—Story of a Deadly Invader (Chapter 22)
42. Lichens: Algal-Fungal Partnership (Chapter 23)
43. Dry Rot and Other Wood Decay Fungi (Chapter 23)
44. Disaster in the French Vineyards (Chapter 24)
45. Alcohol and Health (Chapter 24)
46. The New Wonder Drugs (Chapter 25)
47. Buying Time for the Rain Forest (Chapter 26)

Supplements

PLANTS AND SOCIETY COMPANION WEBSITE

The companion website to accompany *Plants and Society* offers a variety of additional resources for instructors and students. Instructors will appreciate full-color PowerPoint image slides that contain illustrations and photos from the text, along with suggested activities. A comprehensive bank of test questions, aligned with each chapter of the text, is also available along with access to TestGen. TestGen. allows instructors to create paper and online tests or quizzes in one easy-to-use program. Students will find multiple-choice quizzes, short-answer concepts, and further resources to aid in their study. Also included is a listing of useful and poisonous plants, as well as tips for growing houseplants and home gardening.
www.mhhe.com/levetin8e

 Craft your teaching resources to match the way you teach! With McGraw-Hill Create™, www.mcgrawhillcreate.com, you can easily rearrange chapters, combine material from other content sources, and quickly upload content you have written like your course syllabus or teaching notes. Find the content you need in Create by searching through thousands of leading McGraw-Hill textbooks. Arrange your book to fit your teaching style. Create even allows you to personalize your book's appearance by selecting the cover and adding your name, school, and course information. Order a Create book, and you'll receive a complimentary print review copy in 3–5 business days or a complimentary electronic review copy (eComp) via email in minutes. Go to www.mcgrawhillcreate.com today and register to experience how McGraw-Hill Create™ empowers you to teach your students your way.

THE *LABORATORY MANUAL FOR APPLIED BOTANY* BY LEVETIN, MCMAHON, AND REINSVOLD

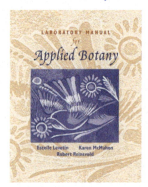 The lab manual features 18 exercises that focus on examining plants and plant products that have sustained or affected human society. Although the manual includes standard information on plant cells and tissues, there is a practical approach to the investigations. Students extract plant dyes, make paper from plant fibers, and study starch grains used in archeology. Several laboratory topics are devoted exclusively to economically important crops—grasses, legumes, starchy staples, and spices. Four additional appendixes—titled Science as a Process, A Field Trip to a Health Food Store, A Taster's Sampler of Caffeine Beverages and Foods, and Notes for Instructors—provide additional information for each of the labs.

©Estelle Levetin

UNIT I

CHAPTER 1

Plants in Our Lives

The botanical connections to our lives are many: food, medicines, materials, and beverages are just a few of the ways plants serve humanity.

KEY CONCEPTS

1. Green plants, especially flowering plants, are more than just landscaping for the planet, since they supply humanity with all the essentials of life: food and oxygen as well as other products that have shaped modern society.

2. The algae are an extremely diverse group of photosynthetic organisms that are key producers in aquatic food chains, a valuable source of human food, and the base for a number of commercial and industrial products.

3. Fungi are also an economically and ecologically important group of organisms that impact society in numerous ways, from fermentation in the brewing process to the use of antibiotics in medicine to their role as decomposers in the environment and as the cause of many plant and animal diseases.

4. All living organisms share certain characteristics: growth and reproduction, ability to respond, ability to evolve and adapt, metabolism, organized structure, and organic composition.

5. The processes of life are based on the chemical nature and interactions of carbohydrates, lipids, proteins, and nucleic acids.

CHAPTER OUTLINE

Plants and Human Society 2
    The Flowering Plants 2
    The Non-Flowering Plants 4
    The Algae 4
    The Fungi 4
    Plant Sciences 4
    Scientific Method 5
Fundamental Properties of Life 6
A CLOSER LOOK 1.1 Biological Mimics 6
Molecules of Life 8
    Carbohydrates 8
    Proteins 9
    Lipids 11
    Nucleic Acids 12
A CLOSER LOOK 1.2 Perfumes to Poisons: Plants as Chemical Factories 14

Chapter Summary 13
Review Questions 15

Much of modern society is estranged from the natural world; people living in large cities often spend over 90% of their time indoors and have little contact with nature. Urbanized society is far removed from the source of many of the products that make civilization possible: most food is purchased in large supermarkets, most medicines are purchased at pharmacies, and most building supplies are purchased at lumber yards. Society's dependence on nature, especially plants, is forgotten (table 1.1).

In less urbanized environments, lifestyles are more attuned to nature. The farmer's existence is dependent on crop survival, and the farmer's work cycle is timed to the growing season of the crops. The few hunter-gatherer cultures that remain in isolated areas of the world are even more dependent on nature as they forage for wild plants and hunt wild animals. These foragers know that without grains there would be no flour or bread; without plant fibers there would be no cloth, baskets, or rope; without medicinal herbs there would be no relief from pain; without wood there would be no shelter; without firewood there would be no fuel for cooking or heat; and without vegetation there would be no wild game.

PLANTS AND HUMAN SOCIETY

Whether forager, farmer, or city dweller, humans have four great necessities in life: food, clothing, shelter, and fuel. Of the four, an adequate food supply is the most pressing need, and, directly or indirectly, plants and algae are the source of virtually all food through the process of **photosynthesis.** Through photosynthesis, plants and algae use solar energy to convert carbon dioxide and water into sugars and, as such, are the **producers** in the **food chain.** They are the base of most food chains, whether eaten by humans directly as **primary consumers** or indirectly as **secondary consumers** when eating beef (which comes from grain-fed or pasture-fed cattle). In addition to the food produced by photosynthesis, the oxygen given off as a by-product is Earth's only continuous supply of oxygen. As sources of food, oxygen, lumber, fuel, paper, rope, fabrics, beverages, medicines, and spices, plants support and enhance life on the planet.

Thinking Critically

Plants are crucial to the existence of many organisms, including human beings.

Could life on Earth exist without plants? Explain.

The Flowering Plants

The word *plant* means different things to different people: to an ecologist, a plant is a producer; to a forester, it is a tree; to a home gardener, a vegetable; and to an apartment dweller, a houseplant. Although there are many different types of plants, the most abundant and diverse plants in the environment are the flowering plants, or **angiosperms.** These are also the most economically important members of the Plant Kingdom and are the primary focus of this book. From the more than 350,000 known **species*** of angiosperms, an overwhelming diversity of products has been obtained and utilized by society. The food staples of civilization—wheat, rice, and corn—are all angiosperms; in fact,

---

*Each kind of organism, or species, has a two-part scientific name consisting of a genus name and a specific epithet; for example, white oak is known scientifically as *Quercus alba*. After the first mention of a scientific name, the genus name can be abbreviated, *Q. alba*. When referring to oaks in general, it is acceptable to use the genus name, *Quercus*, alone. Sometimes an abbreviation for species, "sp." or plural "spp.," stands in for the specific epithet—for example, *Quercus* sp. or *Quercus* spp. Both common and scientific names are used throughout this book; details on this topic are found in Chapter 8.

Table 1.1
How Much Do Plants Affect Society?

_____	1.	True or False—Plants provide most of the calories and protein for the human diet.
_____	2.	True or False—Today plant extracts are widely used in herbal remedies and alternative medicine, but they are no longer important in prescription drugs.
_____	3.	True or False—The search for cinnamon led to the discovery of North America.
_____	4.	True or False—New varieties of plants are being created through genetic engineering; these provide enormous profits for large agrotechnology companies but have no practical value.
_____	5.	True or False—The introduction of the potato to Europe in the sixteenth century initiated events that led to a devastating famine in Ireland.
_____	6.	True or False—Trees are the only source of pulp for papermaking.
_____	7.	True or False—The estimated number of genes in *Arabidopsis thaliana*, the first plant genome sequenced, has about one-fourth the number of genes estimated for the human genome.
_____	8.	True or False—The Salem witchcraft trials in the 1690s might have resulted from a case of fungal poisoning.
_____	9.	True or False—Tomatoes were once considered to be an aphrodisiac.
_____	10.	True or False—A poisonous plant is one of the most important dietary staples in the tropics.

(continued)

Table 1.1 continued		
Answers		
1. True		In nations such as the United States and those in Western Europe, approximately 65% of the total caloric intake and 35% of the protein are obtained directly from plants, while in developing nations close to 90% of the calories and over 80% of the protein are from plants (Chapters 10, 15).
2. False		Approximately 25% of all prescription drugs in Western society contain ingredients derived from plants; however, 80% of the world's population does not use prescription drugs but relies exclusively on herbal medicine (Chapter 19).
3. True		Columbus was one of many explorers trying to find a sea route to the rich spicelands of the Orient. Cinnamon and other spices were so valued in the fifteenth century that a new, faster route to the East would bring untold wealth to the explorer and his country (Chapter 17).
4. False		Transgenic crops, containing one or more genes from another organism, are being planted throughout the world. Some of these crops have been engineered to be more nutritious, disease resistant, or insect resistant and have been found to be beneficial to people and the environment (Chapter 15).
5. True		The potato, native to South America, became a staple food for the poor in many European countries, especially Ireland. The widespread dependence on a single crop led to massive starvation when a fungal disease, late blight of potato, destroyed potato fields in the 1840s. Over 1 million Irish died from starvation or subsequent diseases; another 1.5 million emigrated (Chapter 14).
6. False		While trees provide a sizeable percentage of pulp for the world's paper, many types of plant material can be used. Historically, cotton, hemp, linen, rice straw, and bamboo have been used as sources of pulp. Also, recycling paper helps decrease our dependence on trees for pulp. Recycling a 1.2-meter (4-foot) stack of newspapers would save a 12-meter (40-foot) tree (Chapter 18).
7. False		*Arabidopsis* is estimated to have over 27,000 genes, more than that of the fruit fly and even more than the estimate of 20,500 genes for the human genome (Chapter 7).
8. True		Searching for the cause of the hysteria that led to the accusations of witchcraft in Salem, Massachusetts, some historians have suggested ergot poisoning. Caused by a fungal disease of rye plants, an ergot forms in place of a normal grain and produces hallucinogenic toxins. Consumption of contaminated rye flour can lead to hallucinations, neurological symptoms, or even death (Chapter 25).
9. True		When tomato plants were first introduced to Europe, they were viewed with suspicion by many people, since poisonous relatives of the tomato were known. It took centuries for the tomato, neither poisonous nor an aphrodisiac, to fully overcome its undeserved reputation (Chapters 6, 20).
10. True		Bitter varieties of cassava (*Manihot esculenta*) contain deadly quantities of hydrocyanic acid (HCN), which can cause death by cyanide poisoning. Cultures in South America, Africa, and Indonesia have developed various processing methods to remove HCN and render the cassava edible (Chapter 14).

with minor exceptions, all food crops are angiosperms. The list of other products from angiosperms is considerable and includes cloth, hardwood, herbs and spices, beverages, many drugs, perfumes, vegetable oils, gums, and rubber.

All angiosperms are characterized by flowers and fruits. A typical angiosperm flower consists of four whorls of parts: **sepals, petals, stamens**, and one or more **carpels** (fig. 1.1). The stamens and carpels are the sexual reproductive structures. It is from the carpels that the fruit and its seeds will develop. The angiosperms traditionally have been divided into two groups, the **monocots** and the **dicots**, on the basis of structural and anatomical differences. Among the most familiar monocots are lilies, grasses, palms, and orchids. A few common dicots are geraniums, roses, tomatoes, dandelions, and most broad-leaved trees. The structure and reproduction of the angiosperms will be described in detail in later chapters.

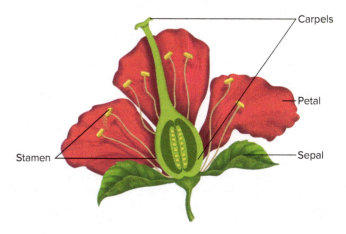

Figure 1.1 A flower, one of the defining characteristics of angiosperms.

The Non-Flowering Plants

In the Plant Kingdom, several distinct groups of non-flowering plants can be found; these range from green algae (fig. 1.2a) to mosses and ferns to giant redwood trees, which are the largest organisms on Earth. Redwoods belong to a group of plants called gymnosperms. Like angiosperms, gymnosperms are seed-bearing plants, but the seeds are not formed in fruits. Gymnosperm seeds are generally produced in cones. One group of gymnosperms consists of conifers, such as pines, cedars, and redwoods. Among the non-flowering plants, the conifers have the greatest impact on society as a source of wood for construction, fuel, and paper. Non-flowering land plants are presented in Chapter 9, and additional material on conifer wood is presented in Chapter 18.

The Algae

Algae are a diverse group of photosynthetic organisms that are found in marine and freshwater habitats where they serve as the base of food chains. They range from microscopic organisms to large seaweeds such as those found in the intertidal zone (fig 1.2a) and giant kelp that form extensive underwater forests. All algae were once considered the most primitive members of the Plant Kingdom, but today most types of algae are classified in separate kingdoms along with other simple organisms. Only the green algae are considered part of the Plant Kingdom. Many species of algae are recognized as important and nutritious food for people throughout the world; however, the widespread uses of algal extracts for industrial applications and as food additives generally go unrecognized.

A negative aspect of the algae is related to environmental damage caused by algal blooms, which are sudden population explosions of certain algal species. In recent years, the occurrence of algal blooms has increased throughout the world. Although these blooms sometimes occur naturally, the increase is believed to be related to nutrient pollution, especially from agricultural runoff, human sewage, and animal wastes. Blooms are particularly dangerous when the algae are capable of producing toxins that can cause massive fish kills or human poisoning. The algae and their connections to society will be examined in Chapter 22.

The Fungi

One other group of organisms that has had a significant impact on society is the fungi, including the molds, mildews, yeast, and mushrooms (fig. 1.2b). Although biologists once considered the fungi a type of simple plant, today they classify them as neither plants nor animals but put them in other kingdoms. The fungi are of major economic importance as they provide many beneficial items, such as penicillin, edible mushrooms, and, through the process of fermentation, beer, wine, cheese, and leavened bread. A negative aspect of their economic importance is the impact of fungal disease and spoilage. The most serious diseases of our crop plants are caused by fungi, resulting in billions of dollars in crop losses each year.

Fungi generally have a threadlike body, the **mycelium**, and propagate by reproductive structures called **spores**. Fungi

(a)

(b)

Figure 1.2 (a) Close-up view of *Fucus* (sometimes called rockweed or bladder wrack), a genus of brown algae commonly found in the intertidal zones of rocky shorelines in most parts of the world. (b) Cluster of deer mushrooms, *Pluteus cervinus*, growing on mulch in an Oklahoma garden.

are nonphotosynthetic organisms, obtaining their nourishment from decaying organic matter as **saprobes** or as **parasites** of living hosts. Ecologically, the fungi play an essential role as **decomposers,** recycling nutrients in the environment. Many fungi are also involved in symbiotic relationships with other organisms. The best known of these relationships are lichens, which are composite organisms formed by a fungus and an alga living together. Because of their traditional ties to botany (the study of plants), the fungi and their impact on humanity will be considered in this book and are presented in Chapters 23–25.

Plant Sciences

When humans began investigating the uses of plants for food, bedding, medicines, and fuel, the beginnings of plant science were evident. Early peoples were skilled regional botanists and passed on their knowledge to succeeding generations. This

Table 1.2 Subdisciplines of Botany

Bryology	Study of mosses and liverworts
Economic botany	Study of the utilization of plants by humans
Ethnobotany	Study of the use of plants by indigenous peoples
Forestry	Study of forest management and utilization of forest products
Horticulture	Study of ornamental plants, vegetables, and fruit trees
Mycology	Study of fungi
Paleobotany	Study of fossil plants
Palynology	Study of pollen and spores
Phycology	Study of algae
Plant anatomy	Study of plant cells and tissues
Plant ecology	Study of the role of plants in the environment
Plant biotechnology	Study and manipulation of genes between and within species
Plant genetics	Study of inheritance in plants
Plant morphology	Study of plant form and life cycles
Plant pathology	Study of plant disease
Plant physiology	Study of plant function and development
Plant systematics	Study of the classification and naming of plants

folk botany gradually amassed a great body of knowledge, laying the foundation for scientific botany, which began in ancient Greece. As the body of knowledge expanded over the centuries, areas of specialization developed within botany, and today many of them are recognized as disciplines in their own right (table 1.2).

Scientific Method

Like other biologists, botanists make advances through a process called the scientific method. This process is the tool that scientists use to study nature and develop an understanding of the natural world. Although the exact steps vary depending on the scientific discipline, generally the scientific method includes careful observation of some natural phenomenon, the development of a hypothesis (tentative explanation for the observation), the use of the hypothesis to make predictions, and experimentation to test the hypothesis. It is often necessary to modify the hypothesis based on the results of the experiments. This, too, is part of the scientific method.

Observation

Scientific study often begins with an observation. It may be something seen repeatedly, such as the blooming of tulips only in the spring. Another type of observation might be the realization that you and others in your family have allergy problems only in September and early October. Observations lead to speculations and questions. You might wonder what causes your September hay fever. With some research on the subject, you might learn that ragweed pollen in the air is the leading cause of fall hay fever. A visit to an allergist confirms the fact that you are allergic to ragweed pollen; however, you cannot find any ragweed plants in your neighborhood. This may lead you to ask the following questions: "Is there ragweed pollen in the air even though there are no plants near my home? Could this be causing my hay fever symptoms?"

Hypothesis

A hypothesis is a possible explanation or working assumption for the original observations. It comes directly from your observations and questions. In the example given here, you might form the following hypothesis: "Airborne ragweed pollen causes my hay fever symptoms every September and October." You may even make some predictions that your symptoms will increase when the airborne pollen level is high.

Hypothesis Testing

Once you have stated your hypothesis, you can find ways to test the hypothesis through experimentation. First, you must decide the type of evidence you will need. You find out information about air sampling and pollen identification and decide to conduct air sampling from July to October during the coming year and determine the types of pollen in the air. You also decide to keep a daily diary of hay fever symptoms during this time and to search for ragweed plants in other locations. Your field work shows that ragweed plants are abundant in an abandoned field about 1 mile south of your neighborhood and along the banks of the river running through your town. Your air sampling data show that ragweed pollen first appeared in the air in late August and increased during the first 2 weeks of September, with the peak on September 12. The pollen levels then began decreasing and were gone from the air by late October. Your symptom chart showed a similar pattern, and, with the help of a friend who is studying statistics, you find a significant correlation between the pollen level and symptoms. The occurrence of ragweed pollen in the air, your symptom diary, and the presence of ragweed plants in town allow you to accept the hypothesis as correct.

Through the use of the scientific method, the body of knowledge increases, allowing scientists to expand their understanding of the workings of the natural world and leading to the development of scientific theories. In science, a theory is an accepted explanation for natural phenomena that is supported by extensive and varied experimental evidence. This definition is very different from the common usage of the word *theory*, which often means a guess. The scientific meaning of theory will always be used

A CLOSER LOOK 1.1

Biological Mimics

The architecture of nature far surpasses any design developed by modern technology. In fact, engineers and inventors often appropriate their best ideas directly from the natural world. Both Velcro™ and barbed wire duplicate the designs found in certain plants.

Today, Velcro™ has hundreds of uses in diapers, running shoes, and space suits, even in sealing the chambers of artificial hearts. But it started from observations during the tedious task of removing cockleburs from clothing. In 1948, a Swiss hiker, George de Mestral, observed the manner in which cockleburs clung to clothing and thought that a fastener could be designed using the pattern. Cockleburs (box fig. 1.1a) have up to several hundred curved prickles that function in seed dispersal. These tiny prickles tenaciously hook onto clothing or the fur of animals and are thus transported to new areas. De Mestral envisioned a fastener with thousands of tiny hooks, mimicking the cocklebur prickles on one side and thousands of tiny eyes for the hooks to lock onto on the other side (box fig. 1.1b). It took 10 years to perfect the original concept of the "locking tape" that has become Velcro™, so common in modern life.

Osage orange (*Maclura pomifera*) is a tree in the mulberry family native to the south-central region of the United States in the area common to Oklahoma, Missouri, Texas, and Arkansas. It has several notable features. Female trees bear large, yellow-green fruits, nicknamed hedge apples (see fig. 8.5a). Hedge apples apparently contain chemicals that repel many insects, and they have been collected for that purpose. The bark is brown with a definite orange tint and becomes more furrowed and shaggier with age. The wood, which is bright orange, very dense, and resistant to rot and termites, was used by several Native American tribes to make war clubs and bows. This usage prompted the French to call the tree *bois d'arc,* meaning wood of the bow. But this story concentrates on the thorns. They are quite formidable. About an inch (2.54 centimeters) long, they alternate in spiral fashion along the length of the branch (box fig. 1.1c). It is these thorny branches that made the osage orange so valuable in the settling of the western plains in North America.

Osage orange was in great demand for its use as a living fence in the vast, treeless plains of the West. The trees were planted close and pruned aggressively to promote a bushy and thorny hedge. Because osage orange is a quick-growing tree, a fence of osage orange took only 4 or 5 years to fill out and could survive for more than a hundred years. Cuttings and seeds of osage orange were collected and sent to farmers throughout America to establish a thorny hedgerow to corral livestock and protect crops. In 1850, a single bushel of osage orange seed cost $50—a fantastic sum in those days. In 1860 alone, 10,000 bushels of seeds were sold, enough to produce 60,000 miles of hedge. In abandoned fields, you can still come across some of these old osage orange hedges or their descendants.

The osage orange hedge did have some drawbacks. A living fence could house insects and other vermin, rob the soil of nutrients and water, and produce shade that could interfere with crop growth. Also, they were not readily movable. What was needed was a new and improved hedgerow, and Michael Kelly was the first, in 1868, to patent a thorny fence made of wire that mimicked the branches of osage orange. It consisted of a single strand of wire with fitted, diamond-shaped sheet metal "barbs" at 6-inch intervals. Kelly established the

---

in this book. The Cell Theory and Darwin's theory of evolution through natural selection are two of the theories that will be described.

FUNDAMENTAL PROPERTIES OF LIFE

Although living organisms can be as different as oak trees, elephants, and bacteria, they share certain fundamental properties. These properties include the following:

1. **Growth and Reproduction** Living organisms have the capacity to grow and reproduce. *Growth* is defined as an irreversible increase in size and should not be confused with simple expansion. Although balloons and crystals can enlarge, this enlargement is not true growth. The ability to reproduce, or produce new individuals, is common to all life. Reproduction can be **sexual,** involving the fusion of **sperm** and **egg** to form a **zygote,** or **asexual,** in which the offspring are genetic clones of a single parent.

2. **Ability to Respond** The environment is never static; it is always changing, and living organisms have the capabilities to respond to these changes. These responses can be obvious, such as a stem turning toward the light (fig. 1.3) or an animal hibernating for the winter. Sometimes, however, the responses are subtle, such as

(a)

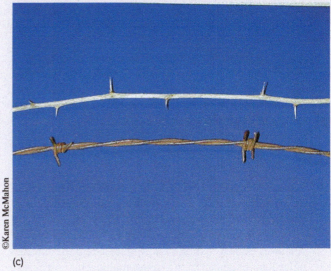
(c)

(b)

Box Figure 1.1 Biological mimicry. (a) The prickles on cocklebur. (b) Hooks on Velcro™. (c) Barbed wire is a design based on the thorny branches of osage orange.

Thorn Wire Hedge Company in 1876 to manufacture his invention. Ultimately, he was bested by his competitors, who had similar ideas and a more successful design. Soon, the vast, open plains of the West were crisscrossed with fences of barbed wire (box fig. 1.1c). In a real sense, the Wild West was tamed by the thorny branches of osage orange.

Figure 1.3 A field of sunflowers with their heads all facing the sun.

changes in the chemical composition of leaves in trees under attack by insects. The chemical composition of intact leaves is altered, making the leaves unpalatable to the insects.

3. **Ability to Evolve and Adapt** All life constantly changes, or evolves. This process has been going on for billions of years, as evidenced by the fossil record. Sometimes changes promote survival because the altered species is better adapted to its environment. Many desert plants have evolved water-storing tissue, an adaptation that helps them survive in their arid environment.

4. **Metabolism** Metabolism is the sum total of all chemical reactions occurring in living organisms. Two of the most important metabolic reactions are **cellular respiration** and

photosynthesis. Respiration is a metabolic process in which food is chemically broken down to release energy. All life requires energy to run chemical reactions, and respiration occurs in all living organisms. Photosynthesis occurs in green plants, algae, and some bacteria. It is the process that links the energy of the sun with life on Earth. In this process, photosynthetic organisms utilize solar energy to manufacture sugars.

5. **Organized Structure** All living organisms are composed of one or more cells; the cell is the basic structure of life. The unique structures encountered in living organisms are often the inspiration for manufactured items, as seen in A Closer Look 1.1: Biological Mimics. From the smallest unicellular organism to the largest multicellular organism, all show a high degree of organization and coordination. The simplest level of organization is seen in bacteria, which are **prokaryotic cells** (fig. 1.4a). These are the most primitive types of cells known. All other organisms are composed of **eukaryotic cells.** In a eukaryotic cell, the **nucleus,** containing the hereditary material, is clearly visible (fig. 1.4b), and different metabolic activities are compartmentalized into specialized membrane-bound structures called **organelles.** Prokaryotic cells lack a discernible internal organization. Prokaryotes have no organized nucleus or other obvious membrane-bound structures, but they have hereditary material and carry out all the activities of life.

6. **Organic Composition** All living organisms are composed mainly of four types of compounds: **carbohydrates, proteins, lipids,** and **nucleic acids.** These are the molecules of life.

MOLECULES OF LIFE

The chemical composition of life is based on the element carbon and the classes of carbon compounds known as carbohydrates, lipids, proteins, and nucleic acids. Carbon is covalently bonded to other carbon atoms to create carbon chains that form the skeletons of these molecules. These four classes of compounds are the most important molecules in living organisms and often exist as large, complex **macromolecules;** however, other compounds also occur (see A Closer Look 1.2: Perfumes to Poisons). Carbohydrates, lipids, and proteins also constitute the major nutrients in the human diet and are discussed in detail in Chapter 10.

Carbohydrates

Carbohydrates, which include **sugars** and **starches** as well as **cellulose,** are composed of carbon, hydrogen, and oxygen (fig. 1.5). Many carbohydrates, especially **glucose,** are sources of energy for cells, while other carbohydrates, such as cellulose, are structural materials. The smallest carbohydrates are the **monosaccharides,** or the simple sugars. These contain only one sugar molecule; the most familiar examples of monosaccharides are glucose and **fructose.** The general formula for monosaccharides is $C_nH_{2n}O_n$, with n equal to 3, 4, 5, 6, or 7. Glucose and fructose have the same general formula, $C_6H_{12}O_6$, but they have different arrangements of the atoms and react differently.

Two sugar molecules chemically bonded together are known as a **disaccharide.** Common table sugar, **sucrose,** is a disaccharide composed of one glucose molecule and one

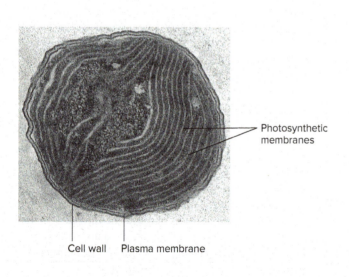

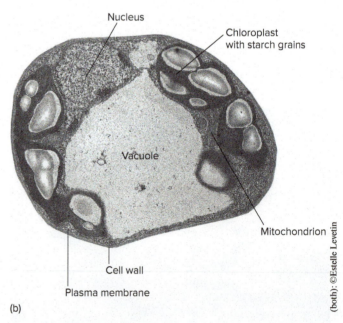

(a)   (b)

Figure 1.4 Cellular organization. (a) Prokaryotic cell. Although this cyanobacterial cell does contain internal photosynthetic membranes, there are no membrane-bound organelles or nucleus. (b) Eukaryotic cell. This *Elodea* leaf cell shows a nucleus and such membrane-bound organelles as chloroplasts, mitochondria, and a vacuole.

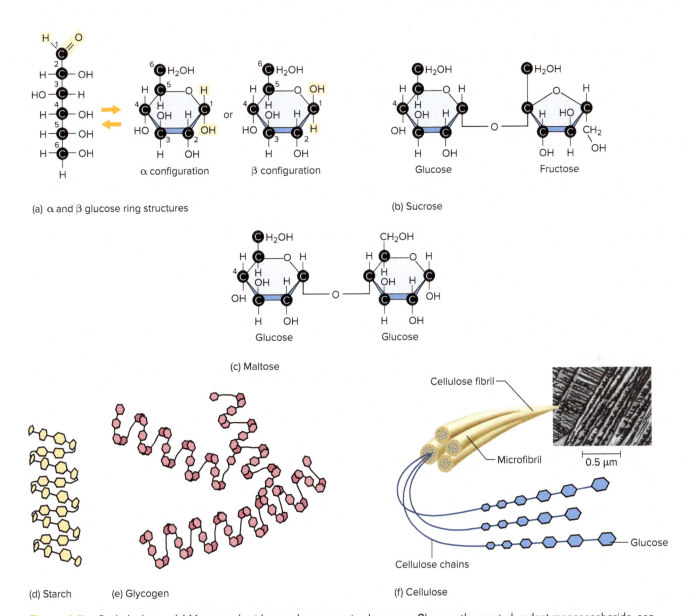

Figure 1.5 Carbohydrates. (a) Monosaccharides are known as simple sugars. Glucose, the most abundant monosaccharide, can exist in a straight chain or ring configuration. (b) Sucrose, a disaccharide, is composed of a molecule of glucose and a molecule of fructose bonded together. (c) Maltose, another disaccharide, forms from two glucose molecules. (d)–(f) Polysaccharides. All three molecules are made from thousands of glucose molecules, but they have different bonding arrangements. (d) Starch found as a storage molecule in green plants. (e) Glycogen found as a storage molecule in animals, bacteria, and fungi. (f) Cellulose, a structural component of plant cell walls, scanning electron micrograph.

fructose molecule. Although most plants transport carbohydrates from one part of the plant to another in the form of sucrose, only a few plants actually store this molecule (see A Closer Look 4.2: Sugar and Slavery). Most sucrose for table use comes from either sugarcane or sugar beet (fig. 1.5). Maltose, another disaccharide, contains two glucose molecules. This sugar is seldom found free in plants but is a breakdown product of starch and an important ingredient in the brewing of beer.

Polysaccharides consist of many thousands of sugar molecules bonded together. The three most common polysaccharides are starch, glycogen, and cellulose. These three are all composed of repeating glucose molecules, but they have different chemical bonding and arrangements (fig. 1.5). Both starch and glycogen are storage molecules; starch occurs in green plants, while glycogen is found in fungi, bacteria, and animals. Starch stored in plant stems, roots, seeds, and fruit is a major source of food for the human population (Chapters 6, 12, and 14). Cellulose is a structural component of plant cell walls, while chitin, a more complex molecule, is the major structural component in fungal cell walls.

Proteins

Proteins are large, complex macromolecules composed of smaller molecules known as **amino acids.** Carbon, hydrogen, oxygen, nitrogen, and sulfur are the elements found in

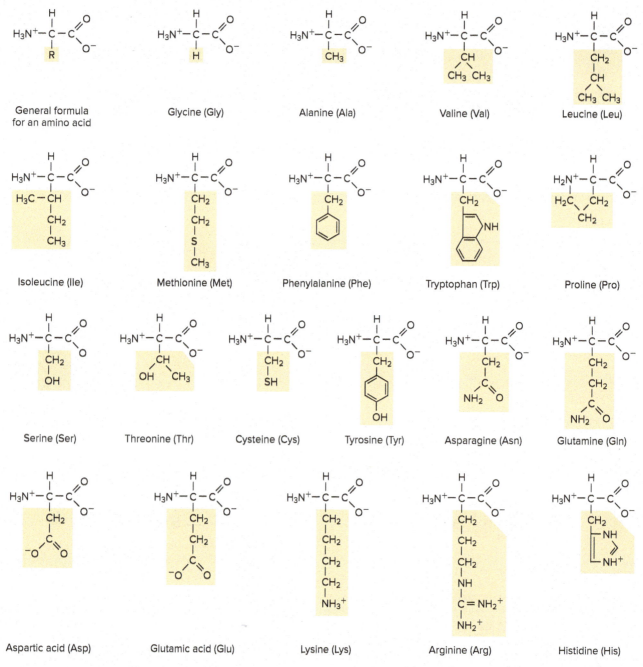

Figure 1.6 Amino acids. General formula for an amino acid and the 20 naturally occurring amino acids. All have the same backbone (N-C-C) and differ only in the side group (R-group) attached to the center carbon.

proteins. There are 20 different amino acids that are common to all life forms. All amino acids have a common backbone with a nitrogen atom and two carbon atoms (N-C-C) and differ only in the side group, called an R-group, attached to the central carbon atom (fig. 1.6). The number and arrangement of these 20 amino acids result in an infinite variety of proteins. Amino acids are attached to each other by a special covalent bond called a peptide bond, and long chains of amino acids are called **polypeptide** chains. In the complete protein structure, the polypeptide chain is twisted and folded into a specific, three-dimensional shape (fig. 1.7). Proteins have many functions; they can serve as enzymes (biological catalysts), structural materials, regulatory molecules, or transport molecules, to name a few of their many roles. Proteins produced by plants, especially legumes, are an important source of nutrients for the human diet (Chapter 13).

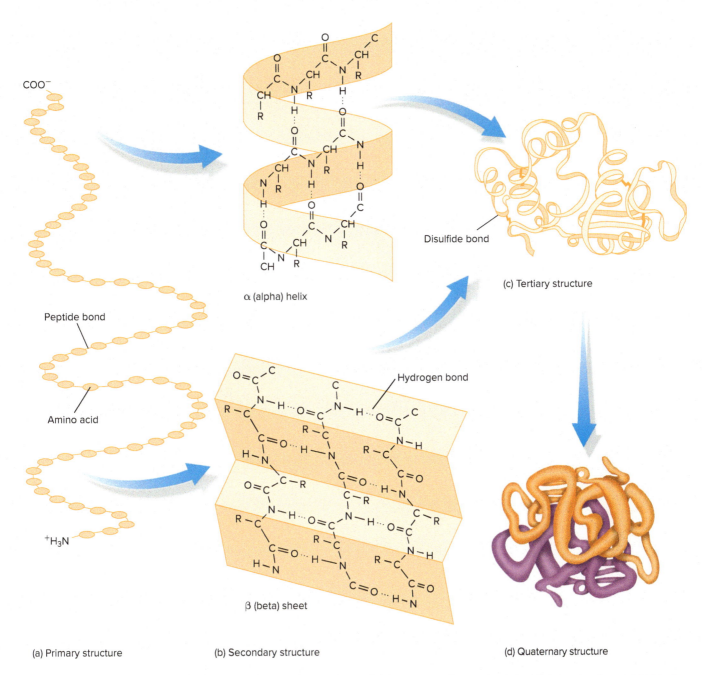

Figure 1.7 Protein structure. (a) Primary protein structure consists of the sequence of amino acids bonded together by peptide bonds to make a polypeptide chain. (b) Secondary protein structure consists of a helix, or pleated sheet, that spirals or folds the polypeptide chain. This is stabilized by hydrogen bonds. (c) Tertiary structure is a twisting and folding of the molecule. (d) Quaternary structure contains more than one polypeptide chain, each with its own tertiary structure.

Thinking Critically

The four major groups of macromolecules in living organisms are carbohydrates, proteins, lipids, and nucleic acids.

What are the primary roles of each of the molecules?

Lipids

Lipids are a diverse group of substances largely composed of only carbon and hydrogen. Small amounts of oxygen may occur in some lipids. There are many different types of lipids; what they have in common is that they are insoluble in water. Lipids include such compounds as **triglcerides, phospholipids, steroids** (fig. 1.8), and waxes. Different types of lipids have different functions. They can be important as sources of energy

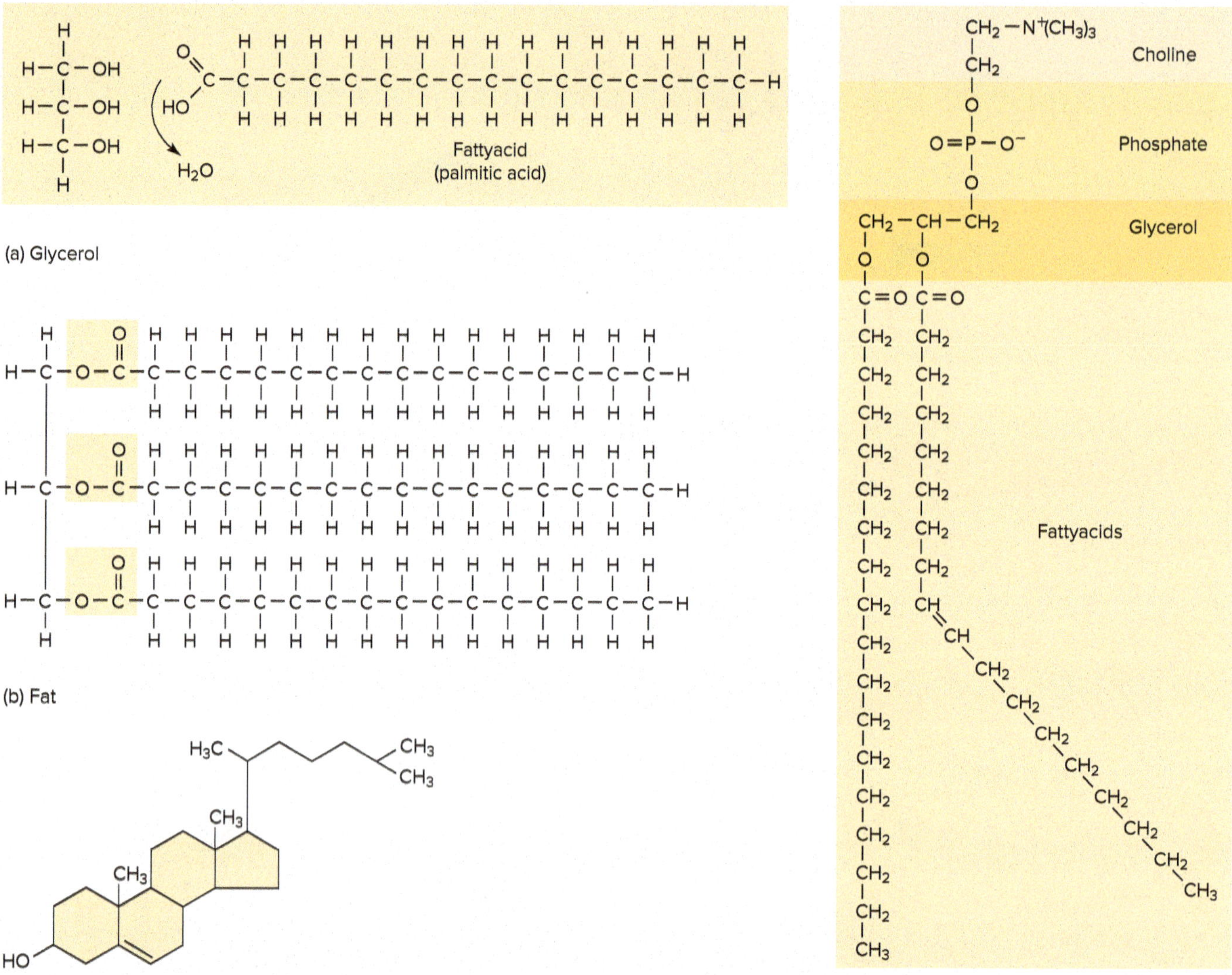

Figure 1.8 Lipids. (a) The building blocks of fats and oils consist of glycerol and fatty acids. (b) A fat, or triglyceride, formed from glycerol and three fatty acids. (c) A phospholipid is formed from glycerol, two fatty acids, a phosphate group, and choline. (d) Cholesterol, one of many steroids, is a more complex lipid. The four-ring steroid backbone is shaded.

(triglycerides), as structural components of cell membranes (phospholipids and cholesterol), or as hormones (steroids). Triglycerides, better known as fats and oils, function as food reserves in many organisms. Fats are the usual energy reserves in animals, while seeds and fruits of certain plants store appreciable amounts of oil, which has been used by humans for thousands of years (Chapter 13).

Nucleic Acids

Nucleic acids contain carbon, hydrogen, oxygen, nitrogen, and phosphorus. They are composed of repeating units called **nucleotides,** which consist of a sugar (either **ribose** or **deoxyribose**), a phosphate group $(PO_4)^{-3}$, and a nitrogenous base (either a **purine** or a **pyrimidine** base) (fig. 1.9). Five different types of nucleotides occur, depending on the type of base. There are two purine bases, **adenine** and **guanine,** and three pyrimidine bases, **thymine, cytosine,** and **uracil.** **Deoxyribonucleic acid (DNA)** and **ribonucleic acid (RNA)** are the two types of nucleic acids. Nucleotides containing adenine, guanine, and cytosine occur in both DNA and RNA. Thymine nucleotides occur only in DNA; uracil replaces thymine in RNA. Thus, both DNA and RNA contain four types of nucleotides, two purines and two pyrimidines.

It is the sequence of these nucleotide bases in the DNA molecule that is the essence of the genetic code. DNA is the hereditary material of life, unique in its ability to replicate itself and thus pass on the genetic code from one generation to the next. DNA, often called the **double helix** (fig. 1.9), exists as a double-stranded molecule that is twisted into a helix.

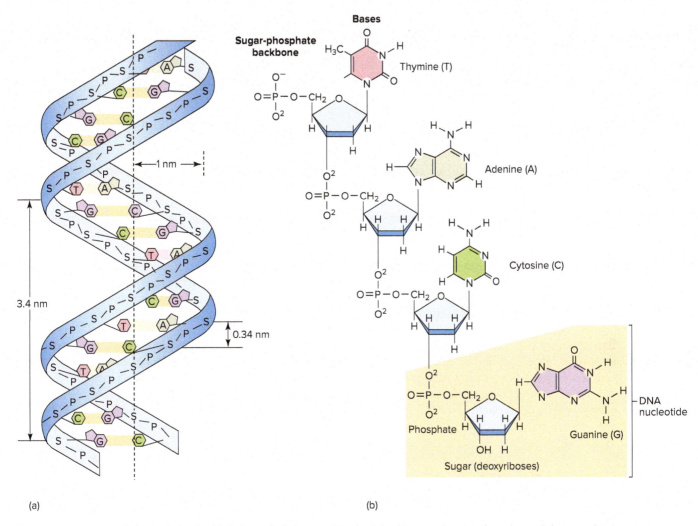

Figure 1.9 DNA molecule. (a) The double helix with the sugar-phosphate backbone making up the sides and the paired nitrogenous bases the interior. (b) Structures of the nucleotides (sugar, phosphate, and base) that make up the DNA molecule.

The sides of the helix are made of alternating sugar (deoxyribose) and phosphate groups, and the nitrogenous bases are found as purine-pyrimidine pairs (adenine always pairs with thymine and guanine with cytosine) in the interior of the helix.

Unlike DNA, RNA usually consists of a single strand, with ribose as part of the sugar-phosphate backbone. RNA is involved in the manufacture of proteins, using the instructions coded on the DNA molecule. The sequence of bases in DNA makes up a **gene,** and each gene codes for the formation of a specific product (see Chapter 7).

CHAPTER SUMMARY

1. Angiosperms, also called flowering plants, supply humanity with the essentials of life. The food staples of civilization—wheat, rice, and corn—are all angiosperms, as are almost all other food crops. Other angiosperm products that have shaped modern society include cloth, hardwood, herbs and spices, beverages, drugs, perfumes, vegetable oils, gums, and rubber.

2. The algae are aquatic, photosynthetic organisms that show a great diversity of form, ranging in size from the microscopic unicellular algae to gigantic seaweeds. They are important as components of aquatic food chains, contributors to the global photosynthetic rate, and sources of a number of economically important products. However, in the case of algal blooms, they can be detrimental to both the environment and the economy.

3. Fungi are also an economically important group of organisms. They include molds, mildews, yeast, and mushrooms. Fungi provide many beneficial items, such as penicillin, beer, wine, edible mushrooms, and leavened bread. A negative aspect of their economic importance is the impact of fungal disease and spoilage.

A CLOSER LOOK 1.2

Perfumes to Poisons: Plants as Chemical Factories

Carbohydrates, lipids, proteins, and nucleic acids are essential to life and are termed *primary metabolites* because they occur in the major, or primary, metabolic pathways. Many plants and fungi also produce other chemical compounds that are produced along secondary pathways and are referred to as secondary compounds (or secondary metabolites). They include four large classes of chemicals: terpenes, phenolics, glycosides, and alkaloids. Many secondary compounds are actually derived from primary metabolites, such as lipids, carbohydrates, or amino acids, and may even be combinations of these. At one time, it was thought that these compounds were by-products of metabolism; however, it is now known that these compounds have diverse functions in plants. They may attract pollinators, inhibit bacterial and fungal pathogens, deter grazing animals, deter insects, or inhibit the growth of competing plants. Humans have discovered that many of these compounds have other applications, and this discovery has made certain plants of tremendous value to society. Some are medicinal; some impart interesting flavors to food; some produce useful products; and some are highly toxic.

Terpenes are hydrocarbons, which are compounds containing only carbon and hydrogen atoms; they range greatly in size and structure and include essential oils, resins, and polyterpenes. Essential oils provide the flavor and aroma of many herbs and spices as well as the scents used in perfumes and incense. Resins are used in the production of pharmaceuticals, dental adhesives, varnishes, insecticides, chewing gum, turpentine, rosin, perfumes, and oil-based paints. Polyterpenes include the elastic compounds found in latex, which is a milky sap produced by many plants. The most important of these compounds is natural rubber from *Hevea brasiliensis*, the rubber tree. Other terpenes include the carotenoid pigments, which are the red, orange, and yellow pigments found in plants; however, these are usually classified as primary metabolites. Taxol is a terpene from the bark of Pacific yew trees that is important in chemotherapy for treating ovarian and breast cancer.

Phenolics are a large and diverse category of compounds, which all contain one or more aromatic benzene rings (a ring of six carbon atoms with six hydrogen atoms attached) with one or more hydroxyl (OH) groups. They range from small molecules to large, complex macromolecules and include flavonoids, tannins, and lignin. The natural browning in cut surfaces of apples and potatoes is caused by the interaction of phenolics with oxygen in the air. Although many essential oils are terpenes, others are phenolic compounds; examples are clove oil and bergamot oil, which is used to flavor Earl Grey tea. Flavonoids include water-soluble plant pigments known as anthocyanins, which are found in red cabbage and in many flower petals. These pigments, along with other phenolics, are important sources of natural dyes. Tannins occur in many plants and have been traditionally used to tan animal skins to form leather. Also, tannins in tea, red wine, and some fruits are important components of the flavors. The flavonoids and possibly the tannins found in red grapes (and red wine) are believed to reduce the risk of heart disease. Urushiol is the phenolic compound in poison ivy and poison oak that causes the blistering, itchy rash that comes from contact with the plants. Tetrahydrocannabinol (THC), a phenolic resin, is the active ingredient in marijuana plants. Finally, mention should be made of lignin, a primary metabolite composed of thousands of phenolic molecules. Lignin is found in the cell wall of certain plants and is the substance that gives wood its hardness and strength.

Glycosides are compounds containing glucose (or a different sugar) combined with another, nonsugar molecule. Typically, these are combinations of glucose and a terpene, a steroid, or a phenolic compound. Three common categories of glycosides are saponins, cardioactive glycosides, and cyanogenic glycosides. Saponins, which form a soapy lather when vigorously mixed with water, consist of a combination of a sugar and a steroid. Saponins are bitter tasting and can cause gastric upsets. Plants rich in saponins have

4. All living organisms have the capacity to grow and reproduce, the ability to respond, the ability to evolve and adapt, a metabolism, an organized structure, and an organic composition.

5. The chemical nature of living matter is based on the element carbon and its ability to covalently bond to other carbon atoms to form the skeletons of carbohydrates, lipids, proteins, and nucleic acids—the molecules of life. Monosaccharides, especially glucose molecules, serve as sources of energy for cells; polysaccharides have storage and structural functions. Proteins, composed of long chains of amino acids, have many functions as enzymes, structural molecules, regulatory molecules, and transport molecules. Lipids are a diverse group of compounds that are insoluble in water. Some serve as energy reserves, others as structural materials or hormones. DNA serves as the hereditary material of life by encoding information in the sequences of bases. RNA functions in the manufacture of proteins using information encoded in the DNA molecule.

been used in detergents, shampoos, and other products. Saponins from yams are the source of steroids that have been used in the manufacture of human sex hormones and cortisone. Cardioactive glycosides are similar in structure to the saponins; however, the steroid portion of the molecule is modified. As the name implies, these compounds affect the heart; in fact, they are fatal if consumed in enough quantity. However, at the proper dosage, digitoxin, a cardiac glycoside from foxglove, is one of the most important treatments for congestive heart failure. Cyanogenic glycosides release hydrogen cyanide (HCN) when metabolized. HCN is a deadly compound, yet cassava, which contains cyanogenic glycosides, is a dietary staple in many tropical countries. Proper processing of cassava removes these toxic metabolites. Another category of glycosides are glucosinolates, which are mainly found in the Brassicaceae (the mustard and cabbage family). Glucosinolates constitute a large group of important flavor molecules in broccoli, cabbage, and other vegetables and also impart the sharp biting taste of mustard and horseradish. Research suggests that the glucosinolates in broccoli and similar vegetables have anticancer properties and help protect people from colon and rectal cancer.

Alkaloids are a large group of nitrogen-containing secondary metabolites that are synthesized from various amino acids and found in many plants and fungi. These compounds are well known for their effects on mammalian physiology, especially on the central nervous system. Many alkaloids are structurally similar to neurotransmitters found in the brain, and most are considered psychoactive. Some alkaloids, such as caffeine and cocaine, are stimulants; others, such as morphine and codeine, are depressants. Still others—such as mescaline, the tropane alkaloids, and the ergot alkaloids—are hallucinogenic. In high doses, some alkaloids are deadly poisons; these include nicotine, the alkaloid in tobacco. Many have important medicinal applications, but the addictive properties of other alkaloids have caused widespread problems throughout the world.

Throughout this book are numerous examples of how terpenes, phenolics, glycosides, and alkaloids have been used by people and even examples of how they have altered the course of civilization (table 1.A).

Table 1.A Commonly Occurring Secondary Products

Class of Compound	Examples	Use by Humans	Chapter
Terpenes	Essential oils	Herbs and spices/flavor	Chapter 17
	Essential oils	Perfumes and incense	Chapter 5
	Taxol	Chemotherapy	Chapter 19
Phenolics	THC	Hallucinogen/glaucoma treatment	Chapter 20
	Urushiol	Allergen	Chapter 21
Glycosides	Cassava—cyanogenic glycosides	Starchy staple	Chapter 14
	Yam—saponin	Starchy staple/source of steroids	Chapter 14
	Digitoxin	Heart medication	Chapter 19
Alkaloids	Caffeine	Stimulant	Chapter 16
	Ephedrine	Stimulant/decongestant	Chapter 19
	Quinine	Treatment for malaria	Chapter 19
	Morphine	Pain relief, psychoactive	Chapter 20
	Cocaine	Anesthetic/psychoactive	Chapter 20
	Mescaline	Hallucinogen	Chapter 20

REVIEW QUESTIONS

1. What are the characteristics of angiosperms? of fungi? of algae?
2. Crystals can increase in size and seemingly grow. Would you consider crystals to be living? Why or why not?
3. Describe the levels of protein structure.
4. What are the differences between monosaccharides, disaccharides, and polysaccharides?
5. How do triglycerides and phospholipids differ in structure and function?
6. Define the following terms: *simple sugar, starch, amino acid,* and *polypeptide.*
7. Investigate the similarity of design between the hairs of the stinging nettle and the hypodermic syringe (Chapter 21).

UNIT II

CHAPTER 2

The Plant Cell

©Steven P. Lynch

Bulblets often replace flowers in the wild onion; the bulblets drop off to the ground and vegetatively produce clones of the parent plant. Mitosis is the underlying cell division for vegetative, or asexual, reproduction.

KEY CONCEPTS

1. The Cell Theory establishes that the cell is the basic unit of life, that all living organisms are composed of cells, and that cells arise from preexisting cells.
2. Plant cells are eukaryotic, having an organized nucleus and membrane-bound organelles.
3. Substances can move into and out of cells by diffusion and osmosis.
4. Mitosis, followed by cytokinesis, results in two genetically identical daughter cells. Growth, replacement of cells, and asexual reproduction all depend on the process of cell division.

CHAPTER OUTLINE

Early Studies of Cells 17
The Cell Wall 19
The Protoplast 19
    Membranes 19
    Moving Into and Out of Cells 20
    Organelles 20

A CLOSER LOOK 2.1 Origin of Chloroplasts and Mitochondria 22
    The Nucleus 23

Cell Division 24
    The Cell Cycle 24
    Prophase 24
    Metaphase 25
    Anaphase 27
    Telophase 27
    Cytokinesis 27
Chapter Summary 27
Review Questions 27

CHAPTER 2 The Plant Cell

All plants (and every other living organism) are composed of cells. In some algae and fungi, the whole organism consists of a single cell, but angiosperms are complex, multicellular organisms composed of many different types of cells. Plant cells are microscopic and typically range from 10 to 100 µm in length. This means that there would be between 254 and 2,540 of these cells to an inch (fig. 2.1). In Chapter 3, we will be looking at the variety of cells, but in this chapter we will focus on a composite angiosperm plant cell.

EARLY STUDIES OF CELLS

The first person to describe cells was the Englishman Robert Hooke in 1665. Hooke was examining the structure of cork with a primitive microscope (fig. 2.2) and noticed that it was organized into small units that resembled the cubicles in monasteries where monks slept. These rooms were called "cells." He gave that name to each of the little compartments in cork, and the term was eventually applied to mean the basic unit of life. Although the cork was not living, Hooke later looked at living plants and identified cells there also.

Other scientists in the late seventeenth and eighteenth centuries continued the microscopic examination and study of a variety of organisms. It was not until the mid-nineteenth century, however, that Matthias Schleiden and Theodor Schwann, and later Rudolf Virchow, firmly established the **Cell Theory,** which recognizes the cell as the basic unit of life. The Cell Theory further states that all organisms are composed of cells and all cells arise from preexisting cells. This theory is one of the major principles in biology.

Although these early scientists were unable to identify many structures within a cell, today it is possible to magnify extremely small details of the cell using an electron microscope. Use of the electron microscope has greatly expanded our knowledge of cellular structure and function. The structures in a eukaryotic plant cell that are visible with an electron microscope are illustrated in Figure 2.3.

Figure 2.2 Robert Hooke's microscope.

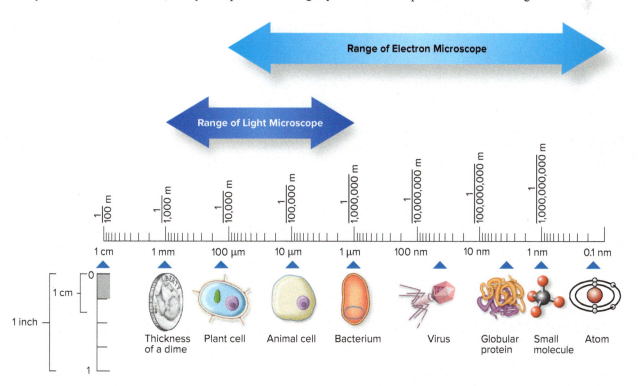

Figure 2.1 Biological measurements. The scale ranges from 1 centimeter (0.01 meter) down to 0.1 nanometer (0.0000000001 meter).

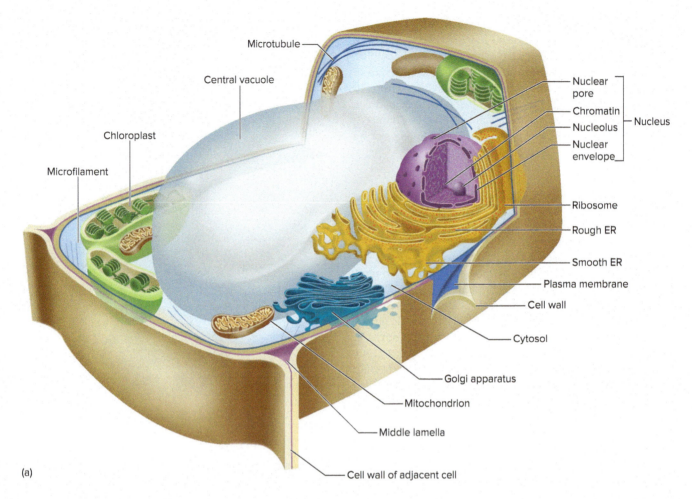

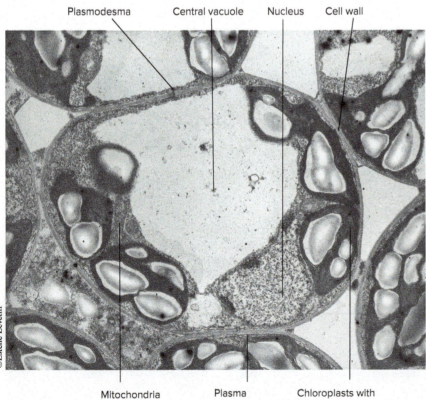

Figure 2.3 Plant cell structure. (a) Diagram of a generalized plant cell as seen under an electron microscope. (b) Electron micrograph of a plant cell.

THE CELL WALL

The **cell wall** encloses all other parts of the plant cell, collectively called the **protoplast.** The cell wall material is formed by the protoplast. Plant cell walls may consist of one or two layers. The first layer, or the **primary wall,** is formed early in the life of a plant cell. It is composed of a number of polysaccharides, principally cellulose. The cellulose is in the form of fibrils, extremely fine fibers (see fig. 1.5f). These fibrils are embedded in a matrix of other polysaccharides.

The **secondary wall** is laid down internal to the primary wall. In cells with secondary walls, **lignin,** a very complex organic molecule, is a major component of the walls, in addition to the cellulose and other polysaccharides. Considering all the plant material on Earth, it is not surprising that cellulose is the most abundant organic compound, with lignin not far behind.

Only certain types of plant cells have secondary walls, usually just those specialized for support, protection, or water conduction. Lignin is known for its toughness; it gives wood its characteristic strength and provides protection against attack by **pathogens** (disease-causing agents) and consumption by herbivores (although certain species of wood-rotting fungi have the ability to break down lignin—see A Closer Look 23.2: Dry Rot and Other Wood Decay Fungi). To compare the characteristics of primary and secondary walls, imagine a chair made of lettuce leaves instead of wood!

Although the cell wall is one or two layers thick, it is not a solid structure. Minute pores, or **pits,** exist; most of these are large enough to be seen with a light microscope. Pits allow for the transfer of materials through cell walls. Cytoplasmic connections between adjacent plant cells often occur. These are called **plasmodesmata** and pass through the pits in the cell wall. These allow for the movement of materials from cell to cell (fig. 2.4).

A sticky layer called the **middle lamella** (fig. 2.3) can be found between the walls of adjacent plant cells. This acts as a cellular cement, gluing cells together. It is composed of **pectins,** the additive often used in making fruit jellies.

THE PROTOPLAST

The protoplast is defined as all of the plant cell enclosed by the cell wall. It is composed of the nucleus plus the **cytoplasm.** The cytoplasm consists of various **organelles** (cellular structures) distributed in the **cytosol,** a matrix consisting of large amounts of water (in some cells, up to 90%), proteins, other organic molecules, and ions. Also found in the cytoplasm is a network of proteinaceous **microtubules** and **microfilaments** that make up the **cytoskeleton,** a cellular scaffolding that helps support and shape the cell and is involved in all aspects of cell movement (fig. 2.3).

Membranes

The outermost layer of the protoplast is the plasma membrane, which is composed of phospholipids and proteins. The **fluid mosaic model,** the currently accepted idea of membrane structure, is shown in Figure 2.5. This model consists of a double layer of phospholipids with scattered proteins. Some of the proteins go through the lipid bilayer **(integral proteins),** while others are on either the inner or the outer surface **(peripheral proteins).** Some of the membrane proteins and lipids have carbohydrates attached; they are called **glycoproteins** and **glycolipids,** respectively. The carbohydrates are usually short chains of about five to seven monosaccharides. Some have described this membrane model as "protein icebergs in a sea of lipids." The plasma membrane serves as a permeability barrier, allowing some molecules (such as water), but not others, to pass through.

Moving Into and Out of Cells

Cells constantly exchange materials with their environment. One way this exchange occurs is by **diffusion.** Diffusion is the spontaneous movement of particles or molecules from areas of higher concentration to areas of lower concentration. Examples of diffusion occur everywhere. Open a bottle of perfume; soon the scent spreads throughout the room. Try the same thing with a bottle of ammonia. In both cases, the

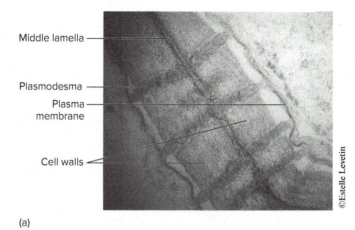

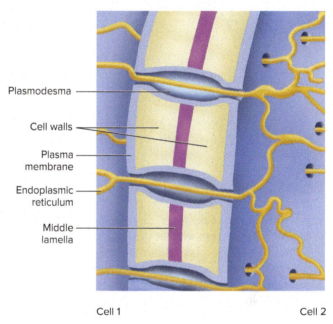

Figure 2.4 Plasmodesmata permit the passage of materials from cell to cell. (a) Electron micrograph. (b) Drawing.

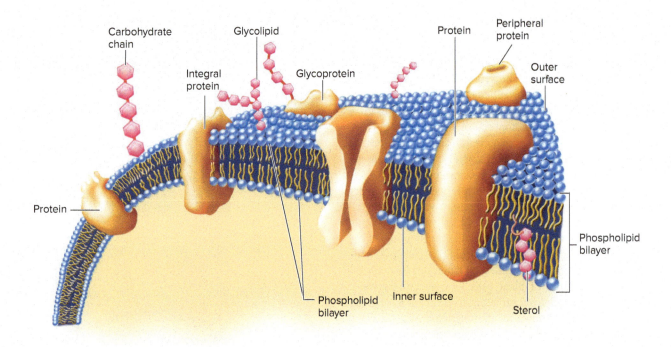

Figure 2.5 Fluid mosaic model of the plasma membrane. The plasma membrane is composed of a phospholipid bilayer with embedded proteins.

molecules diffuse from where they were most concentrated. Diffusion can also be easily demonstrated in liquids. Place a sugar cube in a cup of hot tea; eventually, the sugar will diffuse and be distributed, even without stirring.

Diffusion also occurs within living organisms, but the membranes present barriers to this movement of molecules. Membranes such as the plasma membrane can be described as **differentially permeable.** They permit the diffusion of some molecules but present a barrier to the passage of other molecules. Many molecules are simply too large to diffuse through membranes.

The diffusion of water across cell membranes is called **osmosis.** Water can move freely through membranes. The direction the water molecules move is dependent on the relative concentrations of substances on either side of the membrane. If you place a cell in a highly concentrated solution of salt or sugar, water will leave the cell. The water is attracted to the solute molecules and associates with them. More water will remain on the side of the differentially permeable member that has the higher solute concentration. On the other hand, if you place a cell in distilled (pure) water, water will enter the cell where the concentration of solutes is higher (fig. 2.6).

If a plant cell is left in a highly concentrated, or **hypertonic,** solution for any length of time, so much water will leave that the protoplast actually shrinks away from the cell wall. When this happens the cell is said to be **plasmolyzed.** In a wilted leaf, many of the cells are plasmolyzed (fig. 2.6).

When a plant cell is in pure water or a very weak, **hypotonic,** solution, water will enter until the vacuole is fully extended, pushing the cytoplasm up against the cell wall. Such cells look plump, or **turgid.** This is the normal appearance of cells in a well-watered plant. The crispness and crunch of fresh celery are due to its turgid cells. When the cell is placed in a solution of the same concentration, **isotonic,** there is no net movement of water, and the cell is not turgid (fig. 2.6).

Diffusion and osmosis take place when molecules move along a **concentration gradient,** from higher to lower concentrations. However, cells can also move substances against a concentration gradient; sometimes sugars are accumulated this way. This type of movement is called **active transport** and requires the expenditure of energy by the cell. Membrane proteins are involved in transporting these substances across the membrane.

Organelles

A variety of organelles can be found in the plant cell (fig. 2.3). Most of these are membrane-bound, with the membrane being similar in structure and function to the plasma membrane. In leaf cells, the most distinctive organelles are the disk-shaped **chloroplasts,** which, in fact, are double membrane-bound. These organelles contain several pigments; the most abundant pigments are the chlorophylls, making leaves green. Carotenes and xanthophylls are other pigments present; these orange and yellow pigments are normally masked by the more abundant chlorophylls but become visible in autumn when the chlorophylls break down before the leaves are shed. The pigments are located within **thylakoids,** the internal membranes of the chloroplasts, and are most concentrated in membranous stacks called **grana** (sing., **granum**). The individual grana are interconnected and embedded in the **stroma,** a protein-rich environment. Although chloroplasts are easily seen with a light microscope, the internal organization, or ultrastructure, is visible only with an electron microscope. Photosynthesis occurs in the chloroplasts; this process allows plants to manufacture

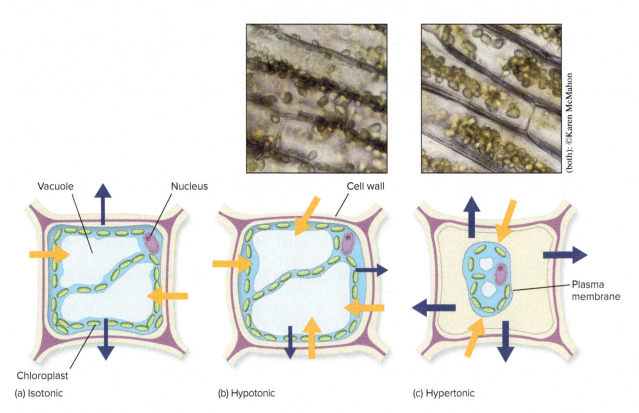

Figure 2.6 Osmosis in plant cells. The direction of the arrows indicates the direction of the water movement; the size of the arrow indicates the relative amount of water moving into or out of the cell. (a) In an isotonic solution, the cell neither gains nor loses water; water flows equally both into and out of the cell. (b) In a hypotonic solution, the cell gains water because more water enters the cell than leaves. (c) In a hypertonic solution, the cell loses water because more water leaves the cell than enters.

food from carbon dioxide and water using the energy of sunlight. More details on the ultrastructure of chloroplasts and the photosynthetic process are covered in Chapter 4.

Two other organelles that may be found in plant cells are **leucoplasts** and **chromoplasts**. Chloroplasts, leucoplasts, and chromoplasts are collectively called **plastids**. Leucoplasts are colorless organelles that can store various materials, especially starch. The starch grains filling the cells of a potato are found in a type of leucoplast called an **amyloplast** (fig. 2.7). Chromoplasts contain orange, red, or yellow pigments and are abundant in colored plant parts, such as petals and fruits. The orange of carrots, the red of tomatoes, and the yellow of marigolds result from pigments stored in chromoplasts.

These pigments are called carotenoids; although they can be part of the color in animals—the yellow in egg yolks or the pink in the feather of a flamingo—it was thought that animals received these carotenoids directly from the plants, algae, or fungi in their diet, but animals did not possess the genes to produce carotenoids themselves. However, it was discovered that the pea aphid (*Acyrthosiphon pisum*), which lives off the sugary sap in phloem and can come in green, yellow, and red forms, actually manufactures its own carotenoids. Some time ago, the gene coding for the synthesis of carotenoids was transferred from a fungus to an aphid and has been passed down from aphid to aphid over the generations. Apparently, the yellow aphids have one copy of the carotenoid gene and the red aphids two copies, while the green pea aphids lack the carotenoid gene altogether.

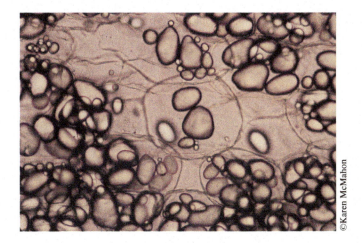

Figure 2.7 Amyloplasts in white potato.

Another organelle bound by a double membrane is the **mitochondrion** (pl., **mitochondria**), which is the site of many of the reactions of cellular respiration in all eukaryotic cells (fig. 2.3). Recall that cellular respiration is the metabolic process in which glucose is chemically broken down to release energy in a usable form, ATP. Mitochondria are not easily studied with a light microscope; the electron microscope has made the study of their ultrastructure possible. The size, shape, and numbers of mitochondria vary among different types of cells, but all mitochondria have a smooth outer membrane and an inner membrane with numerous infoldings called **cristae**. The compartment

A CLOSER LOOK 2.1
Origin of Chloroplasts and Mitochondria

As stated in Chapter 1, prokaryotes were the first organisms on Earth. Evidence indicates that prokaryotes first appeared approximately 3.5 billion years ago, while eukaryotes appeared only around 1.5 billion years ago. One question that has intrigued biologists for many years is, How did the eukaryotic cell evolve? The Endosymbiont Theory attempts to answer how eukaryotic cells evolved from prokaryotic cells. This theory states that the organelles of eukaryotic cells are the descendants of once free-living prokaryotes that took up residence in a larger cell, establishing a symbiotic relationship (symbiosis: two or more organisms living together). This association evolved into the well-studied eukaryotic cell.

Chloroplasts and mitochondria provide the best examples of this theory. Both organelles resemble free-living prokaryotes. In fact, as long ago as the 1880s, some biologists observed that chloroplasts of eukaryotic cells resembled cyanobacteria (then called blue-green algae). Both chloroplasts and mitochondria have structures that are associated with free-living cells. For example, they contain both DNA and ribosomes, which are bacterial in size and nature, allowing them to synthesize some of their own proteins. Both chloroplasts and mitochondria can divide to produce new chloroplasts and mitochondria in a manner very similar to prokaryotic cell division. The inner membranes of both organelles closely resemble the plasma membrane of prokaryotes. These features, as well as additional biochemical similarities, provide support for the validity of the Endosymbiont Theory.

Recent research has discovered certain bacteria that appear to be in the process of evolving into organelles as predicted by the Endosymbiont Theory. Approximately 10% of insect species house bacterial endosymbionts. Some of the best studied are bacteria that live inside specialized gut cells of sap-sucking pests. The sugary sap of plants is deficient in amino acids, and apparently the bacterial endosymbionts produce needed amino acids and other essential nutrients for their insect hosts. In return, bacterial endosymbionts have been passed from generation to generation in insect hosts for over hundreds of millions of years. During this time, the bacterial endosymbionts have lost most of the genes that are necessary for bacteria to be self-sufficient. They no longer possess the genes to make the outer plasma membrane, to metabolize lipids and nucleotides, to transport materials into a cell, or for cell division. There is evidence that some of these bacterial genes may have been transferred to the nucleus of the host cell that now supports the endosymbiont.

Carsonella ruddii, an endosymbiont found in the gut cells of psyllids, a type of agricultural pest also known as jumping plant lice, has the smallest genome known for any bacterium, with only 160,000 base pairs of DNA. Its genome size is similar to that of the mitochondria (<600,000 base pairs) and chloroplasts (220,000 base pairs) found in terrestrial plants. Perhaps this endosymbiont will one day evolve into an organelle.

---

enclosed by the inner membrane is called the **matrix;** the matrix contains enzymes that are used in cellular respiration, while the cristae are the sites of ATP formation. Chapter 4 contains additional information on the role of mitochondria in cellular respiration, and A Closer Look 2.1: Origin of Chloroplasts and Mitochondria details how these organelles may have evolved.

Most mature plant cells (fig. 2.3) are characterized by a large **central vacuole** that is separated from the rest of the cytoplasm by its own membrane. In some cells, the vacuole takes up 90% of the cell volume, pushing the cytoplasm into a thin layer against the plasma membrane. The vacuole contains the cell sap, a watery solution of sugars, salts, amino acids, proteins, and crystals, all separated from the cytoplasm by the vacuolar membrane. The cell sap is often acidic; the tartness of lemons and limes is due to their very acidic cell sap. Some of the substances in the vacuole are waste products; others can be drawn upon when needed by the cell. The concentrations of these materials in the vacuole may become so great that they precipitate out as crystals. The leaves of the common houseplant dumb cane (*Dieffenbachia* spp.) are poisonous because of the presence of large amounts of calcium oxalate crystals (see Chapter 21).

If consumed, the crystals can injure the tissues of the mouth and throat, causing a temporary inability to speak—hence the common name dumb cane. Pigments can also be found in the vacuole; these are called **anthocyanin** and are responsible for the deep red, blue, and purple colors of many plant organs, such as red onions and red cabbage. Unlike the pigments of the chloroplasts and chromoplasts, the anthocyanins are water soluble and are distributed uniformly in the cell sap. Anthocyanins have also been utilized for millenia as dyes for fabrics as discussed in A Closer Look 18.2: Herbs to Dye For.

An internal membrane system also occurs in plant cells (figs. 2.3 and 2.8). This consists of the **endoplasmic reticulum (ER),** Golgi apparatus, and microbodies. These structures are all involved in the synthesizing, packaging, and transporting of materials within the cell. The ER is a network of membranous channels throughout the cytoplasm. In some places the cytoplasmic side of the ER is studded with minute bodies called **ribosomes.** Ribosomes, composed of RNA and protein, are not membrane-bound and are the sites of protein synthesis. Portions of the ER with ribosomes attached are referred to as **rough ER.** Owing to the presence of ribosomes, rough ER

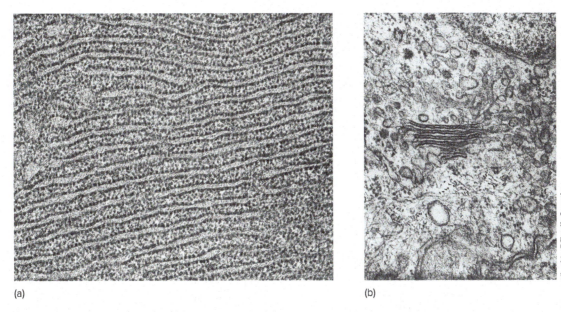

Figure 2.8 The internal membrane system of plant cells. (a) Rough endoplasmic reticulum. (b) Golgi apparatus.

is active in protein synthesis. Ribosomes are also found free in the cytoplasm. Portions of the ER without ribosomes are called **smooth ER,** which functions in the transport and packaging of proteins and the synthesis of lipids.

The **Golgi apparatus** is a stack of flattened, hollow sacs with distended edges; small vesicles are pinched off the edges of these sacs (figs. 2.3 and 2.8). The Golgi apparatus functions in the storage, modification, and packaging of proteins that are produced by the ER. Once the proteins are transported to the Golgi sacs, they are modified in various ways to form complex biological molecules. Often, carbohydrates are added to proteins to form glycoproteins. The vesicles that are pinched off contain products that will be secreted from the cell. Some of the polysaccharides (not cellulose) found in the cell wall are also secreted by these Golgi vesicles.

Microbodies are small, spherical organelles in which various enzymatic reactions occur. Plant cells can contain two types of microbodies: **peroxisomes,** which are found in leaves and play a limited role in photosynthesis under certain conditions, and **glyoxysomes,** which are involved in the conversion of stored fats to sugars in some seeds.

Proteasomes are tunnel-shaped complexes of proteases, protein-degrading enzymes. There can be as many as 30,000 proteasomes in a cell. Proteasomes disassemble proteins tagged by an identifier called ubiquitin for destruction. Proteins targeted for destruction may be misfolded or otherwise abnormal and would not be able to function properly. The activity of proteasomes is critical to regulating metabolism, controlling cell reproduction, and understanding the mechanism for certain diseases. For example, destruction of a regulatory protein that is an enzyme in a biochemical pathway will stop the reaction, while destroying a regulator that inhibits cell division will promote reproduction. Diseases such as Parkinson's or Alzheimer's accumulate anomalous proteins in nerve cells; a drug that could stimulate proteasome activity might alleviate symptoms. Drugs that inhibit proteasome activity could be used to destroy mutated cells like those in cancer.

The **tannosome** is a newly discovered organelle unique to plants. As the name implies, these plastids contain **tannins,** compounds produced by a wide variety of plants, especially woody plants, to ward off herbivory by insects and other animals. Tannins also afford plants protection from UV radiation. Naturally dark brown in color, tannins have a tart taste and contribute to the flavor and color of black tea and red wine (Chapters 16 and 24). Tannins from oaks have also been used in the processing (accordingly called *tanning*) of animal skins into leather. Leather is more durable because tannins deactivate proteins, making the skins less subject to microbial decomposition. Tannins have also been used in the dyeing of fabrics from natural sources (Chapter 18).

Tannosomes arise from the thylakoids, the internal membrane system in chloroplasts, which are arranged in stacks called grana. At first, the thylakoid membranes become loose and swell. Next, small sections of the membrane bud off and form tiny spheres. The newly constructed tannosomes are then collected into larger membrane-bound transport structures called shuttles. Shuttles convey the tannosomes out of the chloroplasts through the cytoplasm to be released into the large central vacuole. Along this journey, tannins are constructed from subunits in the tannosomes. As the concentration of tannins rises, the tannosomes change color from chlorophyll green to the characteristic brown hue. Eventually the vacuole will completely fill up with tannosomes.

The Nucleus

One of the most important and conspicuous structures in the cell is the **nucleus,** the center of control and hereditary information (fig. 2.3). The nucleus is surrounded by a double membrane with small openings called **nuclear pores,** which lead to the cytoplasm. In places, the nuclear membrane is connected to the ER. Contained within the nucleus is granular-appearing **chromatin,** which consists of DNA (the hereditary material), RNA, and proteins. Another structure within the nucleus is the **nucleolus;** one or more dark-staining nucleoli are always

Table 2.1
Plant Cell Structures and Their Functions

Structure	Description	Function
Cell wall	Cellulose fibrils	Support and protection
Plasma membrane	Lipid bilayer with embedded proteins	Regulates passage of materials into and out of cell
Central vacuole	Fluid-filled sac	Storage of various substances
Nucleus	Bounded by **nuclear envelope**; contains chromatin	Control center of cell; directs protein synthesis and cell reproduction
Nucleolus	Concentrated area of RNA and protein within the nucleus	Ribosome formation
Ribosomes	Assembly of protein and RNA	Protein synthesis
Endoplasmic reticulum	Membranous channels	Transport and protein synthesis (rough ER)
Golgi apparatus	Stack of flattened, membranous sacs	Processing and packaging of proteins; secretion
Chloroplast	Double membrane-bound; contains chlorophyll	Photosynthesis
Leucoplast	Colorless plastid	Storage of various materials, especially starch
Chromoplast	Pigmented plastid	Imparts color
Mitochondrion	Double membrane-bound	Cellular respiration
Microbodies	Vesicles	Various metabolic reactions
Cytoskeleton	Microtubules and microfilaments	Cell support and shape
Plasmodesmata	Cytoplasmic bridges	Movement of materials between cells
Proteasome	Tunnel containing proteases	Disassembly of proteins targeted for destruction
Tannosome	Membrane-bound sphere	Production of tannins

present. The nucleolus is not membrane-bound and is roughly spherical; it is involved in the formation of ribosomes. Table 2.1 is a summary of the functions of the cellular components.

Thinking Critically
Researchers synthesized a bacterial chromosome and transplanted it into another bacterial cell. The artificial chromosome replaced the native DNA, and the cell soon began replicating and making proteins according to the instructions from the synthetic genes.

Have these scientists created a living cell? Why or why not?

CELL DIVISION

The cell, with its organelles just described, is not a static structure but dynamic, continually growing, metabolizing, and reproducing. Inherent in all cells are the instructions for cell reproduction or **cell division**, the process by which one cell divides into two.

The Cell Cycle

The life of an actively dividing cell can be described in terms of a cycle, which is the time from the beginning of one division to the beginning of the next (fig. 2.9). Most of the cycle is spent in the nondividing, or **interphase**, stage. This metabolically active stage consists of three phases: G_1, S, and G_2. The G_1 (first gap) phase is a period of intense biochemical activity; the cell is actively growing; enzymes and other proteins are rapidly synthesized; and organelles are increasing in size and number. The S, or synthesis, phase is crucial to cell division, for this is the time when DNA is duplicated; other chromosomal components such as proteins are also synthesized in this phase. After the S phase, the cell enters the G_2, or second gap, phase, during which protein synthesis increases and the final preparations for cell division take place. The G_2 phase ends as the cell begins division. Regulatory proteins, called **kinases** and **cyclins,** signal when a cell is to begin a phase of the cell cycle (G_1, S, G_2 or mitosis). Cyclins are proteins whose concentrations fluctuate during the cell cycle. **Cyclin–dependent kinases (CdK)** are enzymes activated when attached to a particular cyclin to initiate a specific stage in the cell cycle. Conversely if the kinase is dissociated from the cyclin, the stage is arrested.

During cell division, two exact copies of the nucleus result from a process known as **mitosis. Cytokinesis,** the division of the cytoplasm, usually occurs during the later stages of mitosis.

Chromatin, consisting of DNA and protein, is prominent in the nucleus of a nondividing, or interphase, cell. Although the chromatin appears granular when viewed through a microscope, it is actually somewhat threadlike (fig. 2.10). The chromatin has already been duplicated during the S phase prior to mitosis. The events of mitosis are described as four intergrading stages: **prophase, metaphase, anaphase,** and **telophase** (fig. 2.10).

Prophase

During prophase, the appearance of the nucleus changes dramatically. The chromatin begins to condense and thicken,

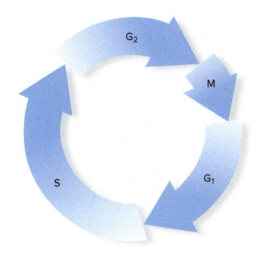

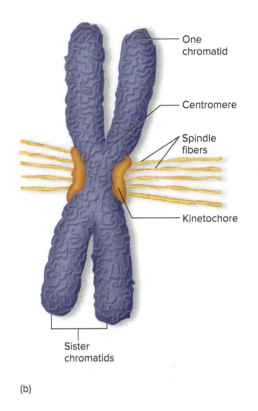

Figure 2.9 (a) The cell cycle consists of four stages (G_1, S, G_2, and M [mitosis]). The events that occur in each stage and the length of each stage, using the broad bean (*Vicia faba*) as an example, are depicted. (b) A duplicated chromosome consists of two sister chromatids held together at the centromere.

coiling up into bodies referred to as chromosomes. Each **chromosome** is double, composed of two identical **chromatids**, which represent the condensed duplicated strands of chromatin. The chromatids are joined at a constriction known as the **centromere** (fig. 2.9). By the end of prophase, the chromosomes are fully formed. Also during prophase, the nuclear membrane and the nucleoli disperse into the cytoplasm and are no longer visible. This leaves the chromosomes free in the cytoplasm.

Telomeres are repeated sequences of DNA that are found at the ends of chromosomes. Telomeres work like caps at the end of shoelaces, protecting the chromosomes from being shortened or otherwise damaged during cell divisions. Shortening of the telomeres happens each time cells divide and, consequently, the genetic content becomes more susceptible to damage. In 2009, three scientists from the United States (Elizabeth Blackburn, Jack Szostak, and Carol Greider) who worked on identifying and elucidating the role of telomeres received the Nobel Prize in Physiology or Medicine.

Factors in the environment can be either protective or destructive of telomere length. One study compared the length of telomeres between women who took multiple vitamin supplements and those who did not. Women who took multivitamins had telomeres that were an average of 5% longer, with approximately an extra 273 DNA base pairs. Women who had been taking multivitamin supplements for 5 years or more had telomeres that were 8% longer. The women who did not take multivitamins had telomeres that appeared to be about 10 years older because telomeres shorten by approximately 28 base pairs each year as a person ages. Not all supplements are equal in protecting telomeres. Taking an iron supplement alone results in telomeres that are an average of 9% shorter. Iron is associated with oxidation reactions in the cell, which are known to be linked to cell stress and damage. On the other hand, supplements rich in the antioxidant vitamins (C, D, and E) result in even longer telomeres.

Metaphase

The chromosomes arrange themselves across the center of the cell during metaphase, the second stage of mitosis. The **spindle,** which begins forming in prophase, is evident during this stage. Spindle fibers, composed of microtubules, stretch from each end, or pole, of the cell to the **kinetochore** of each chromatid. Kinetochores are formed during late prophase; they are specialized regions on the centromere that attach each chromatid to the spindle. Other spindle fibers stretch from each pole to the equator (fig. 2.10).

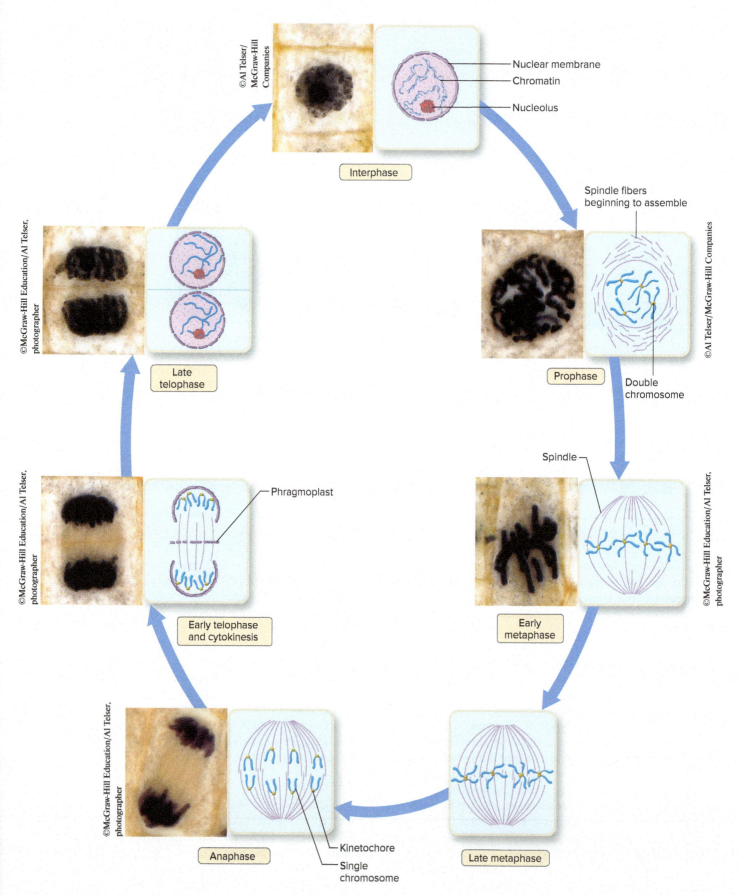

Figure 2.10 Mitosis in a plant cell.

Anaphase

In anaphase, chromatids of each chromosome separate, pulled by the spindle fibers to opposite ends of the cell. This step effectively divides the genetic material into two identical sets, each with the same number of single chromosomes. At the end of anaphase, the spindle is less apparent (fig. 2.10).

Telophase

During telophase, the chromatin appears again as the chromatids, at each end of the cell, begin to unwind and lengthen. At each pole, a nuclear membrane reappears around the chromatin. Now two distinct nuclei are evident. Within each nucleus, nucleoli become visible (fig. 2.10).

Cytokinesis

Cytokinesis, the division of the cytoplasm, separates the two identical daughter nuclei into two cells. Cytokinesis begins during the latter part of anaphase and is completed by the end of telophase. The **phragmoplast,** which consists of vesicles, microtubules, and portions of ER, accumulates across the center of the dividing cell. These coalesce to form the **cell plate,** which becomes the cell wall separating the newly formed daughter cells (fig. 2.10).

The production of new cells through cell division enables plants to grow, repair wounds, and regenerate lost cells. Cell division can even lead to the production of new, genetically identical individuals, or **clones**. This type of reproduction is known as asexual, or vegetative. When you make a leaf cutting of an African violet and a whole new plant develops from the cutting, you are facilitating asexual reproduction and seeing the results of cell division on a large scale. Many of our crops are actually propagated through asexual methods that will be discussed in detail in future chapters. Also, in Chapter 5, you will learn of another type of cell division called meiosis, which is involved in sexual reproduction.

Thinking Critically

Cloning based on cell division has become a major issue for discussion since researchers have cloned sheep, cattle, pigs, goats, cats, dogs, horses, primates, and other animals. *Is there a biological difference between cloning plants and cloning animals? an ethical difference?*

CHAPTER SUMMARY

1. All life on Earth, including plant life, has a cellular organization. The plant cell shares many characteristics with other eukaryotic cells. The plant protoplast includes the cytoplasm with the embedded organelles and nucleus. Within the nucleus is DNA, the genetic blueprint of all cells.

2. The plasma membrane, composed of phospholipids and proteins according to the fluid mosaic model, regulates the passage of materials into and out of the cell. Numerous mitochondria can be found within the cytoplasm; they are the sites of cellular respiration. The endoplasmic reticulum, Golgi apparatus, and microbodies make up an internal membrane system that functions in the synthesizing, packaging, and transporting of materials.

3. Some features of a plant cell are unique. The primary cell wall containing cellulose surrounds a plant protoplast, providing protection and support. In certain specialized plant cells, a secondary cell wall, impregnated with the toughening agent lignin, imparts extra strength. Chloroplasts are the site for photosynthesis; they are one of several types of plastids. Other plastids are the food-storing leucoplasts and the pigment-containing chromoplasts. Tannosomes store tannins responsible for the dark brown color and tartness of many plant products. A large central vacuole may take up approximately 90% of the mature plant cell and act as a storage site for many substances.

4. The life of a cell can be described in terms of a cycle. Most cells spend the majority of the time in interphase, a nondividing stage. But at certain times in its life a cell may undergo division whereby one cell divides into two. Mitosis is the duplication of the nucleus into two exact copies. There are four intergrading stages in mitosis: prophase, metaphase, anaphase, and telophase. The division process is complete when, in the process of cytokinesis, the cytoplasm is split and two identical daughter cells are formed.

REVIEW QUESTIONS

1. What is the significance of the Cell Theory to biology?
2. List the parts of a plant cell, and for each part describe its structure and function.
3. Describe the events occurring during the G_1, S, and G_2 stages of interphase.
4. Describe the stages of mitosis.
5. Describe the similarities and differences between chloroplasts and mitochondria.
6. Differentiate between osmosis and diffusion.
7. Cancer cells are abnormal cells undergoing repeated cell divisions. Vincristine is a drug obtained from the Madagascar periwinkle that has been highly effective in treating certain cancers. Vincristine disrupts microtubules, preventing spindle formation. Explain the success of vincristine on the cellular level.
8. Plant cells are compartmentalized into organelles, each with a specialized function. Which organelles would be abundant in the following: leaf cells of a spinach plant, cells of a potato tuber, yellow petals of a tulip?

CHAPTER

3

The Plant Body

©Karen McMahon

The cytoskeleton of a leaf reveals the extensive network of the xylem vascular tissue supplying water and minerals to plant cells.

KEY CONCEPTS

1. Tissues are groups of cells that perform a common function and have a common origin and structure.
2. Flowering plants are made up of three basic tissue types: dermal, ground, and vascular.
3. These tissues make up the vegetative organs of higher plants: roots, stems, and leaves.

CHAPTER OUTLINE

Plant Tissues 29
    Meristems 29
    Dermal Tissue 29
    Ground Tissue 31
    Vascular Tissue 32
Plant Organs 33
    Stems 33
    Roots 35
A CLOSER LOOK 3.1 Studying Ancient Tree Rings 36
    Leaves 36
Vegetables: Edible Plant Organs 41
    Carrots 41
A CLOSER LOOK 3.2 Plants That Trap Animals 42
A CLOSER LOOK 3.3 Supermarket Botany 44
    Lettuce 44
    Radishes 44
    Asparagus 45
Chapter Summary 46
Review Questions 46

CHAPTER 3　The Plant Body　29

The earliest life forms were unicellular, and that single cell was capable of carrying out all the necessary functions of life. When multicellular organisms evolved, certain cells became specialized in structure and function, leading to a division of labor. Groups of specialized cells performing specific functions are usually referred to as **tissues**. In flowering plants, various tissues compose the familiar organs: roots, stems, and leaves.

PLANT TISSUES

Meristems

All flowering plants are multicellular, with the cells all originating from regions of active cell division. These regions are known as **meristems** (fig. 3.1). Plant growth is localized in meristems. The cells originating from meristems give rise to the various tissue types that make up a plant, such as the cells of the epidermis that form the protective layer in a plant. The three basic tissue types in higher plants are **dermal, ground,** and **vascular** (fig. 3.2).

Apical meristems are located at the tips of all roots and stems and contribute to the increase in length of the plant. Tissues that develop from these apical meristems are part of the **primary growth** of the plant and give rise to the leaves and nonwoody stems and roots. Some plants have additional meristematic tissues that contribute to increases in diameter. These are the **vascular cambium** and **cork cambium**. Tissues developing from them are considered part of the plant's **secondary growth** (fig. 3.2).

Dermal Tissue

Dermal tissues are the outermost layers in a plant. In young plants and nonwoody plant parts, the outermost surface is the **epidermis** (fig. 3.3a). It is usually a single layer of flattened cells. Epidermal cells in leaves and stems secrete **cutin,** a waxlike

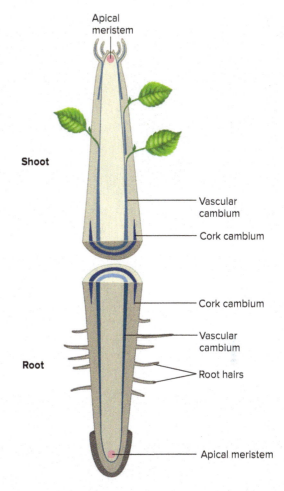

Figure 3.1 Plant meristematic tissues in a diagram of a shoot tip and root tip. Apical meristems contribute to increases in the length of the plant, or primary growth. Vascular cambium and cork cambium are present in plants that have secondary growth, an increase in the girth of the plant body.

Figure 3.2 Organization of plant tissues.

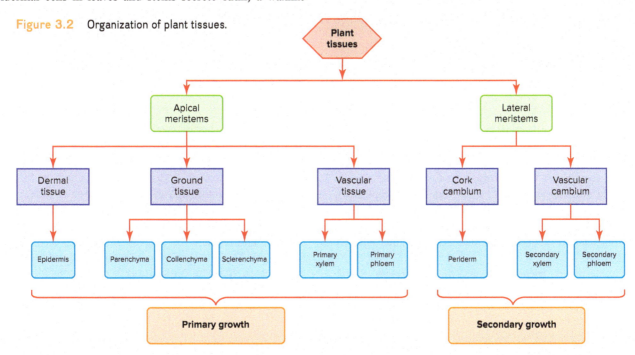

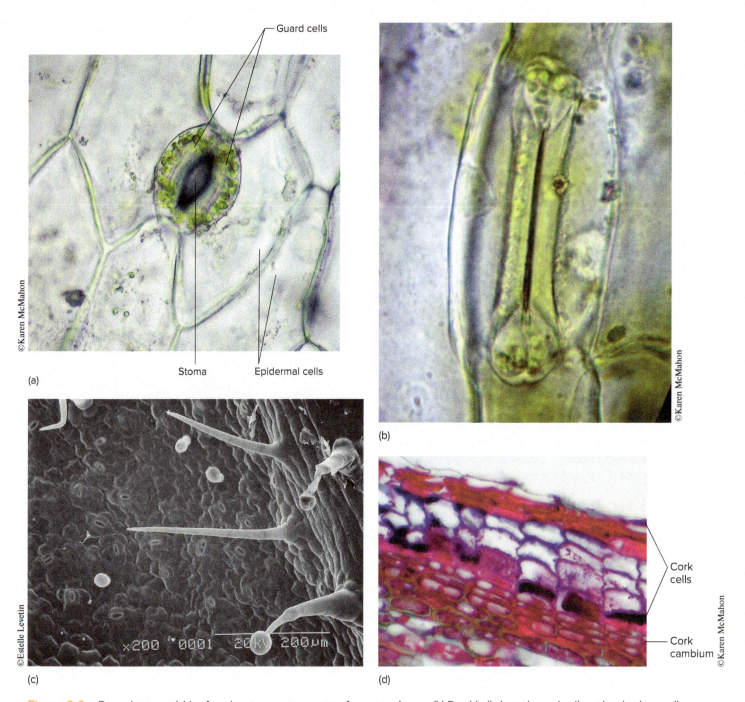

Figure 3.3 Dermal tissues. (a) Leaf epidermis contains stomata for gas exchange. (b) Dumbbell-shaped guard cells with subsidiary cells in a grass leaf. (c) Trichomes. (d) Periderm is a complex tissue consisting of a thick outer layer of cork cells that arise from the cork cambium.

substance that makes up the **cuticle** on the external surface. The cuticle prevents evaporative water loss from the plant by acting as a waterproof barrier. In many leaves, the cuticle is so thick that the leaf has a shiny surface; this is especially true in succulents, such as the jade plant, and tropical plants, such as philodendron.

In some plants, **trichomes** (hairs) may be present on the epidermis (fig. 3.3c). Although usually microscopic, they may be abundant enough to give a fuzzy appearance and texture to leaves or stems. Trichomes may also be glandular, often imparting an aroma when they are brushed, as you can experience by rubbing a geranium or tomato leaf.

Scattered through the leaf epidermis are pores known as **stomata** (sing., **stoma**). Gases, such as carbon dioxide, oxygen, and water vapor, are exchanged through these stomata. A pair of kidney bean shaped cells, **guard cells**, occur on each side of the pore and regulate the opening and closing of each stoma (fig. 3.3a). The guard cells are the only epidermal cells with chloroplasts. Stomata and guard cells can also be found in the epidermis of some stems.

Guard cells of grasses differ in that their shape resembles a pair of dumbbells. Additionally, grass guard cells are linked to two adjacent cells called **subsidiary cells** (fig 3.3b). Recently, research has shown that it is the flexibility of the subsidiary cells that enable guard cells in grasses to open the stomata wider to bring in more carbon dioxide for photosynthesis and to close down the stomata quickly when changing environmental conditions, such as the sudden onset of strong winds, accelerate water loss. As grass guard cells expand to open stomata, the subsidiary cells deflate providing room for the greater inflation of the guard cells and, consequently, the widening the stomata. The role of subsidiary cells in making grass stomata more responsive to environmental conditions may explain why grasses can be found in a variety of habitats and are one of the most widely distributed plant families.

In plant parts that become woody, the epidermis cracks and is replaced by a new surface layer, the **periderm,** which is continuously produced by the cork cambium as the tree increases in girth. The periderm, which consists of cork cells, the cork cambium, and sometimes other cells, makes up the outer bark seen on mature trees (fig. 3.3d). In fact, the cork in wine bottles is the periderm from *Quercus suber,* the cork oak tree native to the western Mediterranean. Cork is principally made up of dead cells whose walls contain **suberin,** another waterproofing fatty substance. It prevents water loss and protects underlying tissues (see Chapter 18).

Ground Tissue

Ground tissues make up the bulk of nonwoody plant organs and perform a variety of functions. The three categories of ground tissue are **parenchyma, collenchyma,** and **sclerenchyma.**

The most versatile of these is parenchyma. Although often described as a thin-walled, 14-sided polygon, parenchyma cells can be almost any shape or size. Usually parenchyma tissue is loosely arranged, with many intercellular spaces. Parenchyma cells are capable of performing many different functions (fig. 3.4a). They are the photosynthetic cells in leaves and green stems and the storage cells in all plant organs. The starch in potato tubers, the water in cactus stems, and the sugar in sugar beet roots are all stored in parenchyma cells.

Collenchyma cells are the primary support tissue in young plant organs. They can be found in stems, leaves, and petals. Collenchyma cells are elongated cells with unevenly thickened primary cell walls, often with the walls thickest at the corners (fig. 3.4b). They are found tightly packed together just below the epidermis. The tough strings in celery are actually strands of collenchyma cells.

Sclerenchyma tissue has two cell types: **fibers** and **sclereids.** Like collenchyma cells, the fibers are elongated cells that function in support. Unlike collenchyma, they are nonliving at maturity and have thickened secondary walls (fig. 3.4c). For centuries, people have used leaf and stem fibers from many plants in the making of cloth and rope (see Chapter 18). Sclereids have many shapes but are seldom elongate like fibers. The major function of these cells is to provide mechanical support and protection. The extremely thick secondary walls of sclereids account for the hardness in walnut shells and the grit of pear fruit.

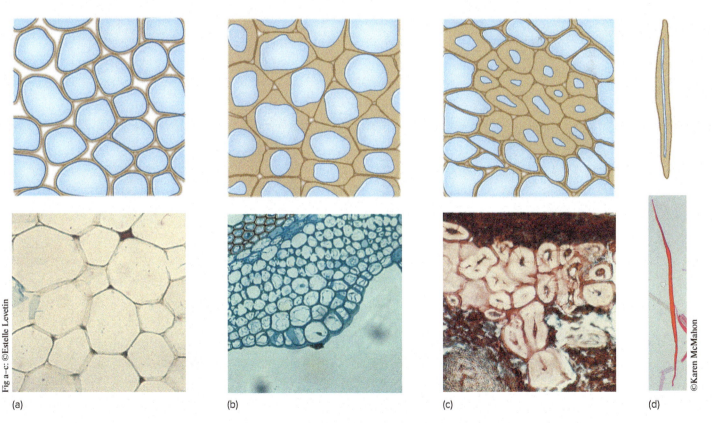

Figure 3.4 Ground tissues. (a) Parenchyma cells are the most abundant plant tissue type and have characteristically thin cell walls. (b) Collenchyma cells have primary cell walls that are thickest at the corners. (c) Sclerenchyma cells have very thick secondary cell walls and are nonliving. (d) Fiber is type of sclerenchyma cell that is elongated and narrow.

Vascular Tissue

Vascular tissues are the conducting tissues in plants. You can readily see the vascular tissues in a leaf; they are the **veins.** The vascular tissues form a continuum throughout the plant, allowing the unrestricted movement of materials. There are two types: **xylem,** which conducts water and minerals from the roots upward, and **phloem,** which transports organic materials synthesized by the plant. Both xylem and phloem are complex tissues composed of several cell types.

Tracheids and **vessel elements** are the water-conducting cells in the xylem. Both cell types have secondary walls, and at maturity these cells are dead and consist only of cell walls. Tracheids are long, thin cells with tapering ends and numerous pits in the walls; these cells also function in support. **Pits** are depressions in plant cell walls where the wall is thinner because the primary wall is not covered by secondary wall. Vessel elements are usually shorter and wider and often have horizontal end walls with large openings. Like the tracheids, the side walls have numerous pits. Vessel elements are attached end to end to form a long, pipelike **vessel** (fig. 3.5a).

Tracheids and vessel elements are found in angiosperms, but only tracheids occur in other plants with vascular tissue. Fibers are present in the xylem, where they provide

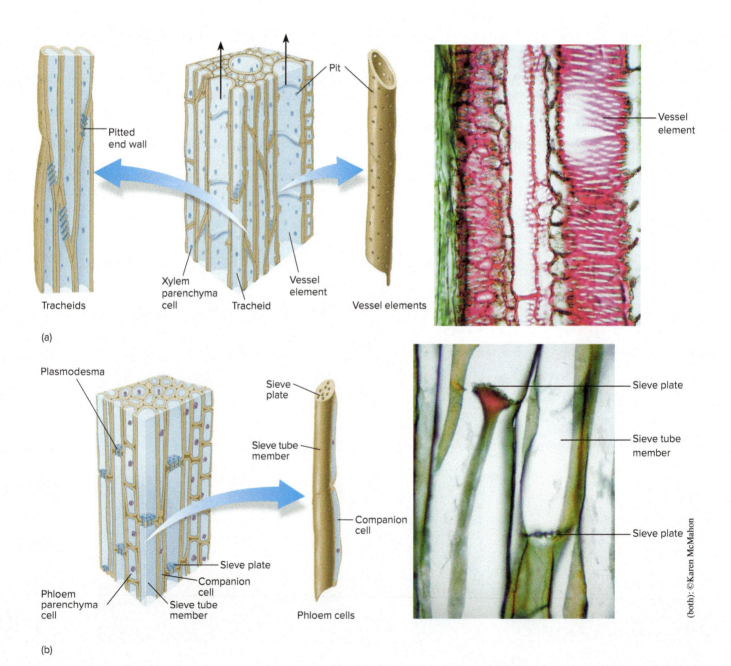

Figure 3.5 (a) Xylem. The conducting cells of xylem are tracheids and vessel elements. (b) Phloem. Sugars are loaded by companion cells into sieve tube members for transport.

additional support. Parenchyma cells, which also occur in the xylem, are the only living and metabolically active cells in this tissue.

Xylem can be either primary or secondary; primary xylem originates from the apical meristem, while secondary xylem comes from the vascular cambium. In trees, secondary xylem is very extensive; it is what we call **wood.**

The cells involved in the transport of organic materials in the phloem are the **sieve tube members.** Unlike the conducting cells in the xylem, the sieve tube members are living cells with only primary walls. But they are unusual living cells because the nucleus and some organelles degenerate as the sieve tube member matures. The end walls of these cells have several to many large pores and are called **sieve plates.** They allow plasmodesmata, cytoplasmic connections, to occur between adjacent sieve tube members and provide channels for conduction. The column of connected sieve tube members is referred to as a **sieve tube** (fig. 3.5b).

Adjoining each sieve tube member is a **companion cell,** which is physiologically and developmentally related to its sieve tube member. The smaller companion cell has a large nucleus that controls the adjacent sieve tube member through the numerous plasmodesmata that connect the two cells. The companion cells are involved in the loading and unloading of organic materials for transport. As in the xylem, both fibers and parenchyma cells are found in the phloem. Both primary and secondary phloem occur; again, the primary phloem is produced by the apical meristem and the secondary by the vascular cambium. Table 3.1 is a summary of these plant tissues.

Thinking Critically

Xylem and phloem are the vascular, or conducting, tissues in plants. Xylem conducts water and dissolved minerals, while phloem conducts organic materials.

How do the conducting cells of xylem (vessel elements and tracheids) differ from the sieve tube members and companion cells in phloem?

PLANT ORGANS

The principal vegetative organs of flowering plants are stems, roots, and leaves. Roots anchor the plant and absorb water and nutrients from the soil; stems support the plant and transport both water and organic materials; and leaves are the main photosynthetic structures.

Stems

Recall that angiosperms are divided into two classes of plants, the dicots and the monocots. Although the major differences between these classes are in the flower and seed, anatomical differences can also be seen in stems, roots, and leaves.

A monocot stem is best exemplified by examining a cross section of a corn stem. The outermost tissue is a single layer of epidermis. Beneath the epidermis are two to three layers of sclerenchyma for support. **Vascular bundles** are scattered throughout the stem. These vascular bundles are composed of both xylem and phloem and are usually surrounded by a **bundle sheath** of fibers. Parenchyma fills in the rest of the stem (figs. 3.6a and b).

Dicot stems can be either **herbaceous** (nonwoody) or **woody.** In herbaceous dicots, the vascular tissue occurs as a ring of separate vascular bundles. Again, each vascular bundle contains both xylem and phloem, with the xylem toward the center of the stem and the phloem toward the outside. This ring of vascular bundles surrounds the **pith,** a central area of ground tissue composed of parenchyma cells. On the other side of the ring of vascular bundles, toward the outside of the stem, is the **cortex,** another region of ground tissue. Although the cortex consists mainly of parenchyma cells, fibers often occur in this region. Between the vascular bundles, the ground tissue of the pith and cortex is continuous. The outermost layer of the stem is the epidermis. In some plants, support tissue, either sclerenchyma or collenchyma, can be found beneath the epidermis (figs. 3.6c and d).

In woody dicots, the vascular tissue, especially the xylem, is much more extensive and makes up the bulk of the stem. As it does in the herbaceous dicots, the pith occupies the center of the stem. Surrounding the pith are rings of secondary xylem. Each ring represents the xylem formed by the vascular cambium during one growing season and is called an

Table 3.1
Plant Tissues

Tissue Type	Cell Types	Function
Dermal		
Epidermis	Epidermal cells	Protection
Periderm	Cork cells	Protection
Ground		
Parenchyma	Parenchyma cells	Storage, photosynthesis
Collenchyma	Collenchyma cells	Support
Sclerenchyma	Sclereids, fibers	Support, protection
Vascular		
Xylem	Tracheids, vessel elements, fibers, parenchyma	Water conduction, support
Phloem	Sieve tube members, companion cells, fibers, parenchyma	Food transport

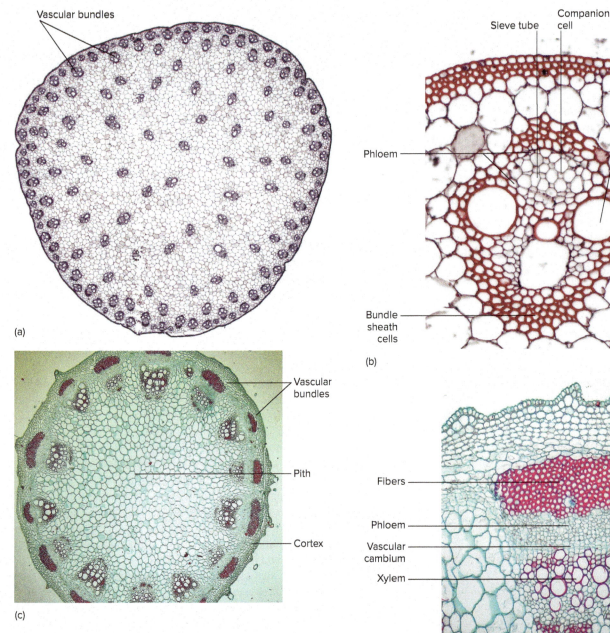

Figure 3.6 Herbaceous stems. (a) In monocot stems, the vascular bundles are scattered. (b) Close-up of a vascular bundle in monocot stem. (c) Dicot stems have vascular bundles in a ring. (d) Close-up of vascular bundles in a dicot stem.

annual ring. The rings, which are easily visible to the naked eye, are due to the different sizes of cells formed through the growing season. Wood produced in the spring, when water is more abundant, is called **springwood** or **early wood** and consists of cells noticeably larger than those found in **summerwood**, or **late wood** produced during the late summer. The portion of each ring with springwood appears lighter than the area with the smaller, densely packed cells of summerwood. Since each ring typically represents one growing season, in temperate regions the age of the tree can be determined by counting the annual rings (fig. 3.7 and see A Closer Look 3.1: Studying Ancient Tree Rings).

Surrounding the outermost ring of xylem is the vascular cambium, the meristematic tissue that produces both secondary xylem toward the inside and secondary phloem toward the outside. The amount of secondary phloem produced each year is very small when compared with the xylem. No annual rings are evident in the phloem, although bands of fibers occur in some plants (fig. 3.7).

Vascular rays, resembling spokes of a wheel, are seen crossing both xylem and phloem. Composed of parenchyma cells, these rays are involved in radial transport of materials.

A small band of cortex can be found outside the phloem. In older trees, however, the cortex is completely replaced by

Roots

Two major types of root systems can be found in flowering plants: **taproots** and **fibrous roots.** Taproots have one large main root with small lateral, or branch, roots. Taproots can be enlarged for storage, as evident in carrots, turnips, and beets (see also A Closer Look 4.2: Sugar and Slavery). Fibrous roots are highly branched and lack a central main root, as in many grasses (fig. 3.8).

At the tips of all main roots or branch roots are thimble-shaped **root caps,** which protect the root meristems as the roots grow through the soil. The meristem **(zone of cell division)** accounts for primary growth in roots. Just behind the meristem the newly formed cells elongate **(zone of elongation)** considerably before they begin to differentiate into the various tissues that constitute the root (fig. 3.9a).

A cross section of a dicot root in the region where cells have differentiated **(zone of maturation)** is seen in Figure 3.9b. The vascular tissue is found in the center of the root, making up the **stele,** or **vascular cylinder.** In the very center of the stele is the xylem, usually in a star-shaped configuration. The number of arms of this star is variable, with bundles of phloem found between the arms of xylem. In monocot roots, a pith is present and is encircled by alternating bundles of xylem and phloem (fig. 3.10a). The outermost layer of cells in the stele is known as the **pericycle,** which is a meristematic layer that can give rise to branch roots (figs. 3.9 and 3.10).

Surrounding the stele is the cortex, composed of parenchyma cells, which are sites of storage. The innermost layer of the cortex (just outside the pericycle) is known as the **endodermis.** Endodermal cells are characterized by the presence of a **Casparian strip,** a waxy material ringing each endodermal cell. The faces of the cell wall next to the cortex and stele do not have a Casparian strip. Because of this strip, water and minerals must pass through the endodermal cells, not between them (see Chapter 4).

The cortex is usually quite large, making up the bulk of the root. The outermost layer of cells is the epidermis. Extensions of the epidermal cell are called **root hairs;** these greatly increase the surface area and are the sites of maximum water and mineral absorption.

Root structure can be further modified as a result of adaptations to the environment. Native to North America, the bald cypress tree (*Taxodium distichum*) is a conifer commonly found in freshwater swamps. Under these conditions, the tree produces aerial outgrowths, called cypress knees or **pneumatophores,** from its submerged root system. These conical roots project about the water surface and contain **aerenchyma** tissue (a type of parenchyma in which there are numerous intercellular spaces). Pneumatophores provide a passageway for oxygen to pass from the air to the submerged

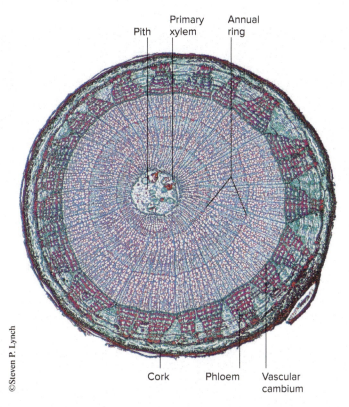

Figure 3.7 Anatomy of a woody stem.

Figure 3.8 Root systems. (a) The fibrous root system of barley. (b) The taproot of a dandelion.

the periderm, or cork (fig. 3.7). In fact, even the older, outermost layers of phloem are replaced by the periderm. The thickness and texture of the periderm depend on the type of tree, and the periderm varies from thin and papery in cherry or paper birch to extremely thick in cork oak.

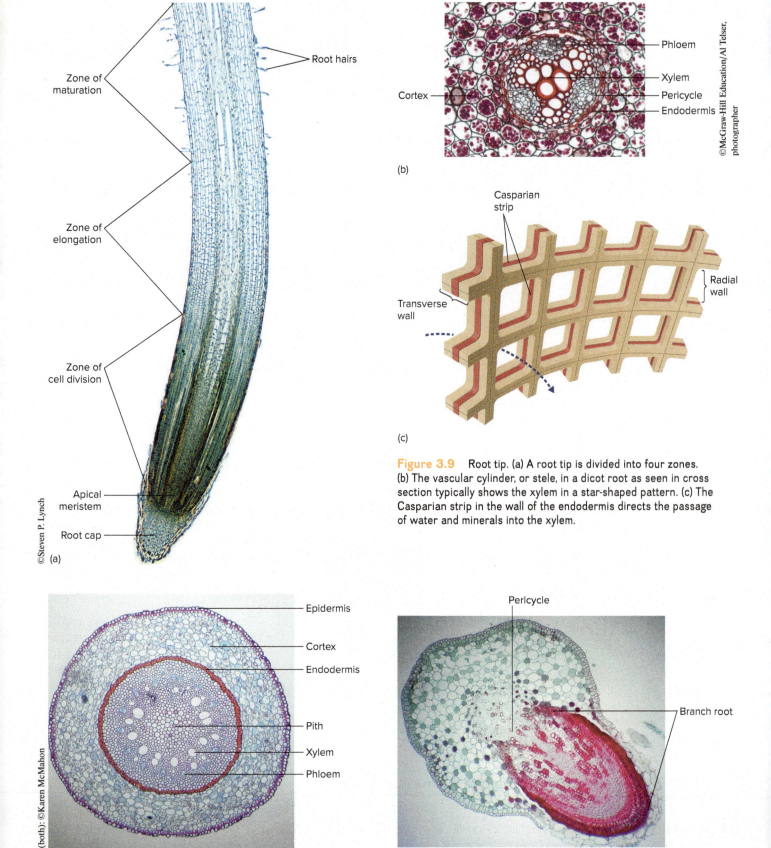

Figure 3.9 Root tip. (a) A root tip is divided into four zones. (b) The vascular cylinder, or stele, in a dicot root as seen in cross section typically shows the xylem in a star-shaped pattern. (c) The Casparian strip in the wall of the endodermis directs the passage of water and minerals into the xylem.

Figure 3.10 (a) The vascular cylinder of monocot roots typically contains a pith. (b) Branch roots originate from the pericycle.

A CLOSER LOOK 3.1

Studying Ancient Tree Rings

The study of tree rings is known as **dendrochronology** and is of value to fields as diverse as astronomy, ecology, and anthropology. The science began in the early twentieth century by Andrew Douglass in Arizona. Douglass, an astronomer, frequently visited logging camps to study the annual ring patterns on tree stumps. The size of a ring can indicate climatic conditions that existed when the ring was formed (box fig. 3.1). A very narrow ring may indicate a year of low rainfall or drought, while a wide ring may indicate abundant rainfall. Douglass wondered if the climatic changes brought about by the 11-year cycle of sunspots was evident in tree-ring patterns. Although Douglass did not find the answer to the sunspot question, he did see that tree-ring patterns from different areas throughout northern Arizona showed the same patterns of wide and narrow rings.

Douglass continued the study of tree rings for many years. By matching patterns from living trees, remains of fallen trees, and wood samples from Pueblo ruins, Douglass was able to date all the ancient pueblos throughout the Southwest. In 1937, Douglass founded the Laboratory of Tree Ring Research at the University of Arizona. Today, this laboratory is still a major world center for tree-ring study.

Conditions in Arizona are ideal for this type of study. Since rainfall is always limiting for tree growth, a small change in the weather has a great effect on the width of the tree ring. Also, the arid climate prevents the decay of dead trees and wooden artifacts. In fact, in the Southwest, scientists have been able to construct a chronology of tree rings going back approximately 9,000 years.

By contrast, in other areas of the United States and in Europe, tree-ring analysis is more difficult because more favorable growing conditions that are relatively consistent result in more uniform tree rings. Also, when trees die they decay in the moister environment.

Another aspect of tree-ring research is **dendroclimatology.** By studying the annual rings of very old trees, scientists have been able to reconstruct major climatic changes of the past. Tree-ring specialists are trying to determine if droughts occur in a cyclic pattern. Others are looking at the effects of pollution, pests, forest fires, volcanoes, and earthquakes on tree rings.

Recently, tree-ring data have provided insight into the high mortality of the first Jamestown colonists and the disappearance of the Lost Colony of Roanoke Island. Taking cores from 800-year-old bald cypress trees (*Taxodium distichum*) from Virginia, the Tree-Ring Laboratory at the University of Arkansas was able to reconstruct the precipitation and temperature patterns in the region from A.D. 1185 to 1984. They discovered that the last sighting of the settlers at Roanoke Island off the North Carolina coast in August 1587 coincided with the beginning of an extreme drought (1587–1589), the driest period in 800 years. Similarly, the Jamestown colonists had the misfortune to begin their settlement in April 1607 during the driest 7-year period (1606–1612) in over 770 years. These studies suggest that the disappearance of the Lost Colony at Roanoke Island and the 80% mortality of colonists during the establishment of Jamestown were in part due to the drought. Both colonies had planned to live off the land and barter for additional supplies from indigenous peoples. This strategy failed as the lack of rainfall caused crops and livestock to die, affecting the food supply not only of the colonists but also of the native peoples. The extreme

---

roots. Without pneumatophores, the bald cypress roots would not have the oxygen necessary for cellular respiration and would not survive under flood conditions.

Many tropical rain forest orchids are **epiphytes,** plants that physically lodge on other plants. Tropical rain forest trees can be festooned with epiphytic orchids on their upper limbs in the full sunlight of the tree canopy. Epiphytic orchids have aerial roots, which are at first green and photosynthetic but later become covered with **velamen,** a white outgrowth of the epidermis, which waterproofs the aerial root against water loss. **Prop roots** sprout from the nodes of stems but then grow down into the soil. Prop roots provide aboveground support and belowground anchorage plus conduction of water and dissolved nutrients to the stem. Examples of plants with prop roots are corn (*Zea mays*) and the screw pine (*Pandanus* spp.).

Contractile roots can be found in the common dandelion (*Taraxacum officinale*) and other plants with short stems. Cells in the upper cortex of the root change shape, expanding in width but losing more than half their height. This shape change pulls the stem down, either keeping it at soil level or slightly below. Storage adaptations in roots are discussed in Chapter 14.

Plants that have woody stems have extensive secondary xylem and annual rings in roots as well as in stems. One major difference between a woody root and a woody stem is that the woody root has no pith.

Leaves

Leaves have often been called the photosynthetic factories of the plant, since photosynthesis is their major function. (Some plants have leaves that are modified for other functions, such as trapping insects. See A Closer Look 3.2: Plants That Trap Animals.) The flat, expanded **blade** of the leaf is ideally suited

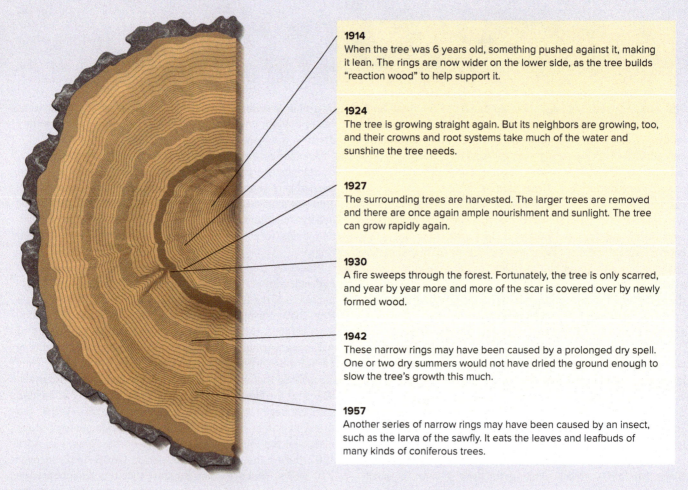

Box Figure 3.1 The pattern of annual rings is correlated with events in the life of this tree.
Source: Adapted from St. Regis Paper Company, 1966.

climatic conditions of 1587–1589 and of 1606–1612, determined by the deciphering of tree-ring data, can explain the fate of the early colonists in Roanoke and Jamestown.

Overall, we know that trees are living histories. Contained within the tissues of the tree is the history of the environment for the year in which a ring was formed.

for the photosynthetic process. The **petiole,** or leaf stalk, connects the leaf blade to the stem and transports materials to and from the blade. Some leaves have no petiole; in those cases, the blade is attached directly to the stem. Small, paired appendages called **stipules** may be present at the base of the leaf (fig. 3.11a). Stipules are varied in form: in some plants, they are leaflike; in others, they are thornlike.

The place where the petiole is attached to the stem is called the **node.** The areas of the stem between adjacent nodes are **internodes.** There are three patterns of leaf arrangement on stems. If only one leaf is present at a node, the arrangement is known as **alternate.** If two leaves occur at a node, the arrangement is **opposite;** with three or more, the arrangement is **whorled** (fig. 3.11b).

There is a great variety of leaf forms and shapes, ranging from small, **simple leaves,** as in elm, whose blade is undivided, to large, **compound leaves,** as in pecan and buckeye, whose blades are divided into **leaflets.** When the leaflets occur in a featherlike pattern, it is called **pinnately compound,** whereas it is called **palmately compound** when the leaflets have a common attachment. It may be difficult to determine whether you are looking at a leaf or a leaflet. One reliable indicator is the position of the **axillary bud.** The upper angle that forms between the top surface of a leaf and the stem is called the **axil,** and it is here that a **bud** (embryonic shoot) is located. Axillary buds are found only at the base of leaves (fig. 3.11a), so if you see an axillary bud you are looking at a leaf. Figure 3.11a illustrates the varieties of simple and compound leaves.

The vascular tissues of leaves make up the venation patterns usually visible to the naked eye. Monocot leaves usually have **parallel venation** because the vascular bundles are arranged in parallel lines running the length of the blade. In contrast, dicots have **net,** or **reticulate, venation,** in which the vascular tissue is highly branched, forming a network

Figure 3.11 Leaf morphology. (a) Leaf composition. Leaves may be simple, consisting of a single undivided blade, or compound, in which the blade is subdivided into leaflets. (b) Leaf arrangement. Alternate, opposite, or whorled indicates the number of leaves coming off a node. (c) Leaf venation. The venation pattern is commonly parallel in monocot leaves and net in dicot leaves.

Thinking Critically

Plant leaves may be simple, with the blade undivided, or compound, with a blade dissected into leaflets.

How can you distinguish a leaf from a leaflet?

throughout the blade (fig. 3.11c). In late fall or winter, you may encounter decaying leaves that have only the vascular tissue remaining as a lacy network.

A cross section of a blade reveals epidermis covering both the upper and lower leaf surfaces (fig. 3.12). Recall that the epidermal cells are covered by a waxy cuticle of variable thickness. Guard cells and stomata are distributed throughout the epidermis. Although thousands of stomata occur on both the upper and lower surfaces, their number and distribution vary considerably (table 3.2). The number of stomata in fossil leaves can also be used to deduce information about the paleoenvironment. Jennifer McElwain of the Field Museum in Chicago examined the density of stomata in leaves of the California black oak (*Quercus kelloggii*) at different altitudes, from near sea level to elevations of 2,500 meters (8,125 feet). She developed an equation that uses stomatal density to calculate the altitude at which the tree lives and was able to apply this to fossil leaves of the California black oak. At higher altitudes, the number of stomata per leaf increased to compensate for the lower concentrations of carbon dioxide in the thinner air. Examining fossil leaves from other species will further

Table 3.2 Distribution of Stomata on Leaves of Various Species

Species	Average Number of Stomata per cm^2	
	Upper Epidermis	Lower Epidermis
Apple (*Malus pumila*)	0	29,400
Black oak (*Quercus velutina*)	0	58,000
Cabbage (*Brassica oleracea*)	14,100	22,600
Corn (*Zea mays*)	5,200	6,800
Geranium (*Pelargonium domesticum*)	1,900	5,900
Mulberry (*Morus alba*)	0	48,000
Pea (*Pisum sativum*)	10,100	21,600
Scarlet oak (*Quercus coccinea*)	0	103,800
Sunflower (*Helianthus annuus*)	8,500	15,600
Wheat (*Triticum aestivum*)	3,300	1,400

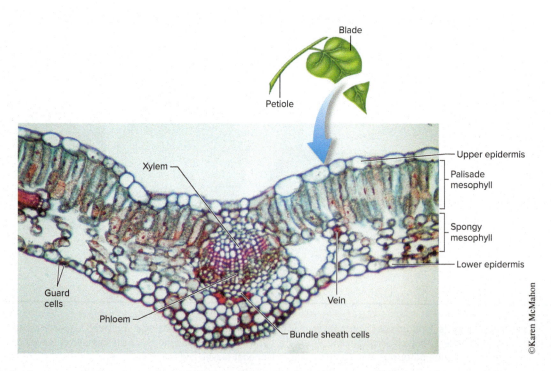

Figure 3.12 Leaf anatomy. Cross section of a leaf illustrates that palisade and spongy cells make up the mesophyll.

increase knowledge about the range of species and chronicle the formation of mountains.

The middle of the leaf, or **mesophyll,** is composed mainly of photosynthetic parenchyma cells that may be of two types: **palisade** and **spongy** (fig. 3.12). The palisade parenchyma cells are tightly packed, columnar cells lying just beneath the upper epidermis. Spongy parenchyma are loosely packed, spherical cells with many large intercellular spaces. Scattered within the mesophyll are the vascular bundles that bring water up to the leaf and carry away the sugars produced by the mesophyll cells.

The topography of a leaf epidermis is the inspiration for a new category of biological mimics (see A Closer Look 1.1: Biological Mimics), self-cleaning materials. In 1982, Wilhelm Barthlott, a botanist at the University of Bonn in Germany, noticed that the leaves of a water lily-like plant, the sacred lotus (*Nelumbo nucifera*), did not have to be cleaned of dirt particles when being prepared for viewing with the scanning electron microscope. Although the sacred lotus grows in muddy ponds, its emergent leaves are radiantly clean because rainwater thoroughly washes dirt from its leaf surface apparently better than in any other plant. Barthlott discovered that the self-cleaning property is due to both the leaf cuticle and the three-dimensional shape of the leaf epidermal cells. The surface of the leaf epidermal cells is microscopically bumpy, covered with many minute projections just a few micrometers in height. Combined with an especially thick waxy cuticle, the leaf surface of the sacred lotus is a super water-repellent or **hydrophobic** (water-hating) surface. Rain falling on the leaf surface will form a spherical droplet and be restricted to the crests of the epidermal cells. (In contrast, water on a **hydrophilic** or water-loving surface, would spread across the surface in a large, flatter smear.) Water droplets perched on the crests of the leaf epidermal surface picked up dirt particles as they rolled off the epidermal peaks, and, in this manner, cleaned the leaf. Barthlott realized that if he could re-create the Lotus Effect™, he could produce a novel group of self-cleaning products. Barthlott's first invention was the honey spoon, which has a silicone surface roughened with miniscule projections that makes a super hydrophobic surface to which honey cannot stick. Lotusan® paint for the exterior of buildings was introduced in 1999, and other self-cleaning products are becoming available. SLIPS is a different approach to designing a repellent surface inspired by the slippery surface of a pitcher plant (see A Closer Look 3.2: Plants That Trap Animals).

VEGETABLES: EDIBLE PLANT ORGANS

Most of us should have more than a passing interest in plant organs because we consume many of them daily; they are the vegetables in our diet. Technically, vegetables are edible parts of the vegetative plant body: stems, roots, leaves, or even flower parts. This definition excludes fruits, which develop from the ovary of a flower and contain seeds. See A Closer Look 3.3: Supermarket Botany to test your botanical powers of recognition. Interestingly, the legal definitions of fruits and vegetables differ from these botanical definitions. In 1893, the U.S. Supreme Court ruled that tomatoes and similar fruits are legally vegetables, since fruits are generally thought of as sweet, while vegetables are not (see Chapter 6). The debate continues, but we will give the botanical definition priority. Next, we will consider a few examples of some common vegetables, examining their history and folklore.

Carrots

Carrots (*Daucus carota*) are one of the most popular vegetables in the American diet; we each consume about 10 pounds of this root each year. Cutting through a carrot clearly shows the organization of the root, with the stele more deeply pigmented than the surrounding cortex.

Carrots are **biennial** plants; *biennial* means that it takes 2 years for the plant to complete its life cycle. During the first year, the plant stores a large amount of food in an enlarged taproot, which we call the carrot. During the second summer, the plant uses the stored food in the taproot to produce a flowering stalk. Normally, we harvest carrots after the first growing season, but we can see the flowering stalk in the wild carrot, a beautiful summer weed also called Queen Anne's lace. Carrots were first introduced into North America by the early colonists, and Queen Anne's lace is the wild descendant of those first carrots (fig. 3.13).

Today, we associate the color orange with carrots, but carrots were not always orange. Originally, carrots were purple and branched. It was not until the sixteenth century that a pale yellow variety appeared in western Europe. In the seventeenth century, Dutch plant breeders developed a deep orange carrot that is the ancestor of all orange carrots grown today.

Figure 3.13   Queen Anne's lace, the flowers of wild carrot.

A CLOSER LOOK 3.2

Plants That Trap Animals

Usually, the natural order is for animals to eat plants but, with some plants, tables are turned. These are the **carnivorous plants**. Although Audrey II from the *Little Shop of Horrors* had a taste for human flesh, most carnivorous plants "snack" on insects. Out of more than 350,000 known species of angiosperms, 400 have developed the carnivorous habit. Most of these plants are found in nutrient-poor soils, such as acid freshwater bogs, and it is believed that this carnivorous trait supplies nutrients lacking in the soil. The plants have been reported to grow with renewed vigor after a "meal."

Since plants are stationary and animals are motile, the plants have evolved elaborate traps to lure their prey. These traps are all modified leaves that offer various incentives, such as nectar or color, to attract insects. Once the insects are ensnared, digestive enzymes are released, and soon only the empty shell of the insect remains. There are various types of traps; we will consider three of the best known.

Venus Flytrap A native to the North Carolina coastal region, *Dionaea muscipula* has a trap that imprisons its victims (box fig. 3.2a). Each leaf is a two-sided trap with trigger hairs on each side. When the trigger hairs are touched, the trap snaps shut tightly around the insect. Once the trap has closed, digestive enzymes begin their job. After a few days the trap opens, ready for another unsuspecting insect.

Sundew Sundews, *Drosera* spp., are small plants that use flypaper-like leaves to trap insects (box fig. 3.2b). Glandular hairs on the leaf surface produce an adhesive that is the "superglue of the plant kingdom." An insect that has been lured to the plant for its nectar or by its coloration sticks tight and is soon digested away.

Pitcher Plants In pitcher plants, *Sarracenia* spp. and other genera, the leaf has evolved into a vase or pitcher shape that acts as a pitfall trap (box fig. 3.2c). Once lured to the pitcher, the insects slip into the pool of rainwater that has collected at the base. The pool also contains digestive juices that eat away the soft parts of the insect. At the end of a season, a pitcher may be filled with the indigestible shells of its many victims. *Nepenthes* (box fig. 3.2d) are vines found in the Asian tropics that produce juglike pitchers. *Nepenthes rajah* is found on only two mountains in Borneo and has the largest pitchers in the world, more than a foot (30 cm) deep. Uncharacteristically, *Nepenthes ampullaria* does not produce aerial pitchers but instead produces pitchers in large mats on the forest floor. The pitchers of *N. ampullaria* are unusual in other ways in that they have small nectar glands and lack a slippery waxy layer on the inner pitcher wall. The pitcher itself has a wide opening and lacks a covering lid. These adaptations indicate that *N. ampullaria* traps not animals but leaves and flowers that fall from the forest canopy into the mouthlike pitchers. Another pitcher (*Nepenthes lowii*) sprouts aerial pitchers, which have broad, flat openings. Investigators rarely find insects in the traps but do find another source of animal nitrogen—feces. Apparently, these pitchers look like toilet bowls to tree shrews, which are attracted to a sweet secretion exuded from the lids. While licking the sweet concoction, the tree shrew defecates into the pitchers. Isotope analysis has shown that 57%–100% of the nitrogen of the pitcher originates as tree-shrew manure.

Recently *Nepenthes* pitcher plants have been the inspiration for a material called SLIPS (slippery liquid-infused porous surfaces), a nonstick surface that can be used in a variety of applications. Researchers at Harvard University observed that when the lip of the pitcher was dry, ants could navigate the pitcher easily, but if the lip was wet from rain or nectar, the surface became slippery and the ants could not keep their footing and slid into the pitcher to be devoured. Closer examination showed that the dry surface of the pitcher is covered with microscopic ridges that provide traction as the ants cross the surface; however, if the lip becomes wet, the liquid fills in the spaces between the ridges, creating a slippery surface in which the ants cannot maintain a foothold.

SLIPS technology involves applying a liquid, usually an oil, to a porous solid etched with minute ridges. The liquid has a chemical attraction to the porous solid, which keeps the liquid in place. Once the liquid fills in the texture, the treated surface becomes nonstick. It can be applied to a variety of surfaces (metal, plastic, glass) and can be used to make walls graffiti-proof or to create windshields that remain ice-free because the water rolls off before it has time to freeze.

Sun pitchers (*Heliamphora*), more than 20 species of carnivorous plants native to South America, have been the subject of recent investigations. These carnivorous plants are located on tepuis, the mesas that arise abruptly 1,000 to 3,000 meters (3,300 to 9,800 feet) from the rainforest floor in Venezuela, northern Brazil, and western Guyana. As with other pitcher plants, the traps of the sun pitchers are formed from a single leaf folded into a tube but differ in lacking lids. Instead, a small nectar-secreting projection, a nectar spoon, is found on the upper margin of the trap above the opening. The nectar is the lure for insects, typically ants, and downward pointing hairs in the trap prevent the prey from escaping. Another feature unique to the sun pitchers is a channel in the trap to drain off excess rainwater from accumulating in the pitcher, an adaptation to compensate for the frequent

Box Figure 3.2 Carnivorous plants. (a) Venus flytrap has guard hairs to prevent prey from escaping. (b) Glandular hairs of the sundew produce a sticky glue. (c) Pitcher plants. (d) Pitcher of *Nepenthes*.

rainfall on the tepuis. Surprisingly, investigations have so far revealed that only one species of sun pitchers produces its own digestive enzymes. Rather, the sun pitchers rely on microbes living within the trap to produce and release a variety of enzymes that degrade the prey into nutrients which the sun pitchers then absorb.

A CLOSER LOOK 3.3

Supermarket Botany

A trip through the produce section of your neighborhood market is a chance to test your knowledge of plant structure (box fig. 3.3). To play "supermarket botany," determine what plant organs are represented by the following vegetables:

1. Beet _____
2. Celery _____
3. Cabbage _____
4. Potato _____
5. Water chestnut _____
6. Onion _____
7. Rhubarb _____
8. Pumpkin _____
9. Rutabaga _____
10. Brussels sprouts _____

Box Figure 3.3 The produce section of a market offers a chance to study plant anatomy up close.

The orange color is due to the pigment beta-carotene, an important dietary nutrient. When we eat carrots, the pigment is converted into vitamin A, a vitamin with many functions in the body. One important role of this vitamin is in night vision. The old wives' tale may be true: eating carrots may help you see better in the dark. Although vitamin A is necessary for healthy skin, eating too many carrots can turn the skin yellow. This condition is known as carotenemia, a harmless condition that will disappear a few weeks after the person stops eating carrots. Recently, scientific research suggested an additional benefit to eating foods rich in vitamin A; they may lower the risk of developing cancers of the larynx, esophagus, and lung. Current research is also investigating the value of vitamin A as an antioxidant (see Chapter 10).

Carrots also contain sizable amounts of potassium, calcium, phosphorus, and sugar. Their high sugar content has made them the key ingredient in many desserts, such as the familiar carrot cake.

Lettuce

Contemporary Americans seem to have a love affair with lettuce (*Lactuca sativa*). It is the national favorite cold vegetable. This passion is something we share with the ancient Romans, who traditionally began their feasts with a salad of lettuce. The Romans also believed that lettuce had medicinal values as a soporific (sleep inducer). Lettuce juice and lettuce teas were used for their sedative effects in colonial America; in fact, extracts from wild lettuce were used for this purpose until World War II.

Lettuce was first introduced into the New World by Columbus, who planted some in the West Indies in 1493. By the 1880s, there were over 100 cultivars (cultivated varieties) available in the United States, and many of our modern varieties can be found in seed catalogs from that time.

Nutritionally, the lettuce leaf is mainly water with some vitamin A, calcium, and vitamin C. There are three basic types of this leafy vegetable: head lettuce, loose-leaf, and cos. Head lettuce forms a dense, tightly packed head of leaves, as in iceberg lettuce, which was introduced in 1894. Iceberg lettuce dominates the U.S. market because of its ease in transport and storage, even though it is the least tasty and least nutritious variety. Other popular head lettuces are Boston and Bibb. Many loose-leaf varieties of lettuce can be found in today's supermarket. Among the most popular are redleaf, oakleaf, and green leaf. These lettuces are nonheading, with their ruffled leaves forming loose clusters. Cos varieties form an upright, cylindrical head composed of long leaves. The heads of cos varieties are not compact; romaine lettuce is the most popular. Both the names *cos* and *romaine* reflect the ancestry of this type. Romans originally obtained the lettuce from the Greek island of Cos.

Radishes

Today the radish, *Raphanus sativus*, is considered just a peppery garnish for salads, but in the past radishes were enormous, mild-tasting vegetables that were usually cooked. These cooking, or winter radishes, were valued because they could be easily stored in root cellars through the winter. Our present-day radishes, known as summer radishes, were first developed in the eighteenth century. Containing only potassium and some iron, the summer radish is not particularly nutritious, but it makes an attractive and tasty addition to salad greens.

Answers

1. A taproot; almost all root vegetables are taproots
2. A petiole, although the petiole is greatly enlarged and you can see the remains of the blade
3. Leaves, of course, but actually a whole stem with shortened internodes and tightly packed leaves
4. An underground stem, actually; even buds are present as the "eyes"; see Chapter 14
5. Another underground stem; see Chapter 14
6. Underground leaves and stem, but the stem is usually too tough to eat
7. A petiole again; in this case, knowledge of plant anatomy may save your life, since all other parts of the plant are poisonous
8. Botanically, not a vegetable at all but a fruit; see Chapter 6
9. A taproot again, the result of a medieval cross between cabbage and turnip
10. Axillary, or lateral, buds that look like miniature cabbages

Botanically, the radish is a composite vegetable consisting of both root and hypocotyl (the base of the stem). Like the carrot, the radish is a biennial that has developed this underground storage organ to fuel the second year's growth.

Although Americans favor the small, red globose varieties, radishes come in all shapes, sizes, and colors. The Japanese daikon is scarcely recognizable as a radish. It is white, carrot-shaped, and approximately 46 cm (18 in.) long. This vegetable is a staple of Asian cuisine; the Japanese have at least 100 ways of cooking daikon.

An unusual use of radishes is seen in the festival La Noche de Rábanos (The Night of the Radishes), which is held each year on December 23 in Oaxaca City in southern Mexico. Radishes grown in the rocky soils of this region are often grotesquely misshapen and may resemble human or animal forms. In 1889, these bizarre vegetables were part of displays in an agricultural exhibit. This event developed into a yearly festival. Over the years, the displays became more and more elaborate. Today, the radishes are carved into detailed figures and arranged into dioramas that depict historical and religious themes, such as the Mexican Revolution or the Nativity.

Asparagus

Asparagus is an aerial stem vegetable with a royal reputation. This monocot was a favorite of France's King Louis XIV. Native to southern Europe and the eastern Mediterranean region, asparagus was a popular vegetable of the ancient Greeks and Romans. The Puritans valued asparagus as a spring tonic and brought "sparagus" seeds with them to North America.

Asparagus is a perennial plant that, once established, can be harvested for many years. The species of commerce is *Asparagus officinialis*, although several wild species have also been harvested for their stalks. Asparagus is dioecious, having separate male and female plants; however, newer, all-male varieties have been developed. These varieties are disease resistant, have greater longevity, and produce larger stems, since no energy is diverted to seed production. The female plants produce small, whitish-green flowers and, later, red to purple berries.

California leads the United States in the cultivation of asparagus, with most of the domestic market favoring the familiar green variety. A less common purple variety is gaining popularity, and a white variety is most often found in European cuisine. Mounding soil around the growing stem or covering the growing shoots under black plastic tunnels produces the white variety. The tender young stem tips, or spears, are cut by hand in the early spring (making asparagus one of the more expensive vegetables) while still under the surface. Both methods prevent sunlight from reaching the stems, inhibiting the production of chlorophyll, the green pigment seen in the normally photosynthetic asparagus stem. Fresh asparagus is highly perishable; the season lasts for only a few weeks.

If the asparagus stalk is not harvested but left to grow, the stem is said to "fern out" and takes on a delicate, leafy appearance (fig. 3.14). Several species of asparagus "ferns" are cultivated as house or garden plants or used in cut flower arrangements. The delicate, leafy structures are not leaves at all but modified stems called **cladophylls** that mimic leaves in appearance and function.

Asparagus is rich in vitamins A, C, and folic acid, one of the B vitamins. A deficiency of folic acid raises the risk of

Figure 3.14 Cladophylls, leaflike stem branches, can be seen on some of the emerging asparagus shoots. Cladophylls.

heart disease and, during pregnancy, raises the risk of spina bifida and other neural tube birth defects (see Chapter 10).

For some people, eating asparagus has an unpleasant aftereffect. Sulfur compounds in the asparagus are voided in the urine, imparting an offensive odor. Why some people can detect these malodorous compounds but others cannot is a question that researchers are debating. Some researchers suggest that the foul-smelling compounds are produced through an interaction of the chemicals in asparagus with a particular type of personal biochemistry that is genetically determined. Other researchers believe that the compounds in asparagus taint everyone's urine but only some people have the capacity to smell the odor-causing compounds.

In a few areas of the world, such as Australia, asparagus species are considered noxious weeds, crowding out native species (see A Closer Look 22.2: Killer Alga—Story of a Deadly Invader for more information on invasive species). Some pets have been known to develop dermatitis from contact with asparagus and digestive upset from eating the berries.

Delving into the backgrounds of other common vegetables will reveal other interesting facts and folklore; there's more to vegetables than just good nutrition.

CHAPTER SUMMARY

1. Meristems are regions of active cell division and are the source of cells for the various tissue types in the plant body. Growth at apical meristems is primary growth. Increase in the girth in woody plants is called secondary growth.

2. There are three basic tissue types in plants: dermal, ground, and vascular. Dermal tissues include the epidermis and periderm and cover the surface of a plant. Ground tissues include parenchyma, collenchyma, and sclerenchyma. Parenchyma is the most abundant and versatile plant tissue, functioning in storage and photosynthesis. Both collenchyma and sclerenchyma are tissues of support. There are two types of sclerenchyma: fibers and sclereids.

3. Vascular tissues are the conducting tissue in plants. Xylem conducts water and minerals from the soil upward; phloem moves organic solutes throughout the plant body. Tracheids and vessel elements are the water-conducting cells of xylem. Sieve tube members, assisted by companion cells, conduct organic solutes through the phloem.

4. The vegetative organs of the plant body include roots, stems, and leaves. Roots anchor the plant and absorb water and minerals from the soil. The two types of root systems are taproot and fibrous root. At the tip of any root are four distinct regions, or zones: root cap, zone of cell division, zone of cell elongation, and zone of maturation. Vascular tissue in a nonwoody root is organized into a stele; the pattern of xylem and phloem in the stele varies in monocot and dicot roots. The stele is surrounded by a cortex and an outer layer of epidermis. Extending from the epidermal cells are root hairs, through which most of the water and minerals that enter a root are absorbed.

5. Stems support leaves to maximize light absorption and are part of the conduit for the transport of water, minerals, and organic solutes. Leaves are the main photosynthetic structures in most plants. Unlike roots, the vascular tissue in both stems and leaves is organized into vascular bundles. In stems of herbaceous dicots, the vascular bundles are arranged in a ring around a pith; in monocots, the vascular bundles are scattered. In woody dicots, the discrete vascular bundles are replaced by continuous rings of xylem that correspond to the xylem produced during a single growing season.

6. A leaf may have three parts: the blade, the petiole, and a pair of stipules. If the blade is undivided, the leaf is said to be simple; if the blade is divided into separate leaflets, the leaf is compound. According to the pattern of the leaflets, compound leaves may be pinnately or palmately compound. Leaf venation patterns are either parallel (most monocots) or net (most dicots). The entire leaf surface is covered by epidermis; the epidermis secretes a waxy layer, the cuticle. Guard cells are found in the leaf epidermis. They regulate the entry and exit of gases through the stomata. The mesophyll of the leaf is composed of two types of photosynthetic parenchyma cells: palisade and spongy cells.

7. Carnivorous plants have modified leaves that trap animals for nourishment. The carnivorous habit has evolved in plants that inhabit environments that are characteristically deficient in soil nutrients.

8. An examination of the produce section of a supermarket reveals that many of our common vegetables can be identified as being one of the organs of a plant. Many vegetables have a value beyond the culinary as sources of medicine and folklore.

REVIEW QUESTIONS

1. What is the role of meristematic tissues?
2. Describe the organization of a typical herbaceous dicot or monocot stem.
3. Describe the anatomical differences between monocots and dicots.
4. What cell types and tissues are involved in support?
5. What are the functions of roots, stems, and leaves?
6. How can scientists date wood artifacts from archeological sites?
7. Describe the trapping mechanisms of some common carnivorous plants.
8. Investigate the origins and folklore of the following vegetables: cabbage, turnips, and beets.
9. How old is the woody stem pictured in Figure. 3.7? Was it cut down in the spring or summer?
10. In A Closer Look 3.3: Supermarket Botany, you learned that the potato, mistaken by most to be a root, is actually an underground stem. What must be the anatomical features of the dicot potato tuber?

©SAK_PD/Shutterstock

CHAPTER 4

Plant Physiology

Products of photosynthesis are translocated in the phloem and stored in various plant organs. The stalks of these sugarcane plants provide the raw material for sugar for consumption as well as for biofuels.

KEY CONCEPTS

1. The movement of water in xylem is a passive phenomenon dependent on the pull of transpiration and the cohesion of water molecules, whereas the translocation of sugars in the phloem is best described by the Pressure Flow Hypothesis.

2. Plants are dynamic metabolic systems with hundreds of biochemical reactions occurring each second, which enable plants to live, grow, and respond to their environment.

3. Life on Earth is dependent on the flow of energy from the sun, and photosynthesis is the process during which plants convert carbon dioxide and water into sugars using this solar energy with oxygen as a by-product.

4. In cellular respiration, the chemical-bond energy in sugars is converted into an energy-rich compound, ATP, which can then be used for other metabolic reactions.

CHAPTER OUTLINE

Plant Transport Systems 48
    Transpiration 49
    Absorption of Water from the Soil 49

A CLOSER LOOK 4.1 Mineral Nutrition and the Green Clean 50

    Water Movement in Plants 51
    Translocation of Sugar 51

A CLOSER LOOK 4.2 Sugar and Slavery 53

Metabolism 55
    Energy 55
    Redox Reactions 55
    Phosphorylation 55
    Enzymes 56

Photosynthesis 56
    Energy from the Sun 56
    Chloroplasts and Light-Absorbing Pigments 57
    Overview 59
    The Light Reactions 59
    The Calvin Cycle 61
    Variation to Carbon Fixation 63

Cellular Respiration 63
    Glycolysis 64
    The Krebs Cycle 64
    The Electron Transport System 64
    Aerobic vs. Anaerobic Respiration 67

Chapter Summary 68
Review Questions 68

Although plants lack mobility and appear static to the casual observer, they are nonetheless active organisms with many dynamic processes occurring within each **part of the plant.** Materials are transported through specialized conducting systems, energy is harnessed from the sun, storage products are manufactured, stored foods are broken down to yield chemical energy, and a multitude of products are synthesized. Plants synthesize hormones that control growth, reproduction, and development. They also respond to environmental cues that influence these processes, as well as to pests, pathogens, and environmental stress. Put simply, plants are bustling with activity. This chapter considers some of the major transport and metabolic pathways in higher plants. The role of plant hormones in reproduction will be discussed in A Closer Look 6.1: The Influence of Hormones on Plant Reproductive Cycles.

PLANT TRANSPORT SYSTEMS

As described in Chapter 3, there are two conducting, or vascular, tissues in higher plants, the xylem and the phloem, each with component cell types. Water and mineral transport in the xylem will be described first. Tracheids and vessel elements, which consist of only cell walls after the cytoplasm degenerates, are the actual conducting components in xylem.

The source of water for land plants is the soil. Even when the soil appears dry, there is often abundant soil moisture below the surface. Roots of plants have ready access to this soil water; leaves, however, are far removed from this water source and are normally surrounded by the relatively drier air. The basic challenge is moving water from the soil up to the leaves across tremendous distances, sometimes up to 100 meters (300 feet). This challenge is, in fact, met when water moves through the xylem. There are three components to this movement: transpiration from the leaves, the uptake of water from the soil, and the conduction in the xylem (fig. 4.1).

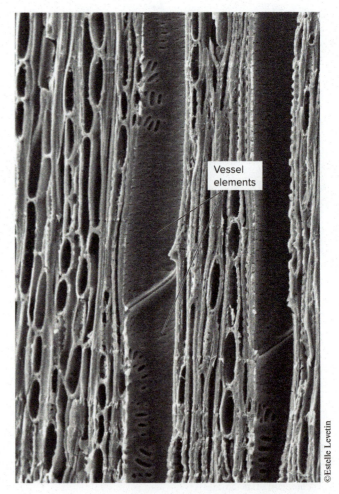

(a) Xylem transport

(b)

Figure 4.1 Transpiration-Cohesion Theory of xylem transport. (a) As transpiration occurs in the leaf, it creates a cohesive pull on the whole water column downward to the roots, where water is absorbed from the soil. (b) Vessel elements join to form a long vessel that may reach from the roots to the stem tip.

Transpiration

Transpiration, the loss of water vapor from leaves, is the force behind the movement of water in xylem. This evaporative water loss occurs mainly through the stomata (90%) and to a lesser extent through the cuticle (10%). When stomata are open, gas exchange occurs freely between the leaf and the atmosphere. Water vapor and oxygen (from photosynthesis) diffuse out of the leaf while carbon dioxide diffuses into the leaf (fig. 4.2a). The amount of water vapor that is transpired is astounding, with estimates of 2 liters (0.5 gallons) of water per day for a single corn plant, 5 liters (1.3 gallons) for a sunflower plant, 200 liters (52 gallons) for a large maple tree, and 450 liters (117 gallons) for a date palm. Imagine the quantities of water lost each day from the acres of corn and wheat planted in the farm belt of the United States! Clearly, transpiration by plants is a major force in the global cycling of water.

It is the action of the guard cells that regulates the rate of water lost through transpiration and, at the same time, regulates the rate of photosynthesis by controlling the CO_2 uptake. Each stoma is surrounded by a pair of guard cells, which have unevenly thickened walls. The walls of the guard cells that border the stoma are thicker than the outer walls. When guard cells become turgid they can only expand outward owing to the radial orientation of cellulose fibrils; this outward expansion of the guard cells opens the stoma. Stomata are generally open during daylight and closed at night. As long as the stomata are open, both transpiration and photosynthesis occur, but when water loss exceeds uptake, the guard cells lose turgor and close the stomata (fig. 4.2b). (See A Closer Look 6.1: The Influence of Hormones on Plant Reproductive Cycles.) On hot, dry, windy days the high rate of transpiration frequently causes the stomata to close early, resulting in a near shutdown of photosynthesis as well as transpiration. A fine balance must be struck in this photosynthesis–transpiration dilemma to allow enough CO_2 for photosynthesis while preventing excessive water loss. Some plants have evolved an alternate pathway for CO_2 uptake at night when rates of transpiration are lower (see CAM Pathway later in this chapter). Other plants have morphological or anatomical adaptations that reduce rates of transpiration while keeping the stomata open. These physiological and anatomical adaptations are most common in **xerophytes,** plants occurring in arid environments, and are of interest to scientists attempting to develop drought-tolerant crops (Chapter 15).

The basis of transpiration is the diffusion of water molecules from an area of high concentration within the leaf to an area of lower concentration in the atmosphere. Unless the atmospheric relative humidity is 100%, the air is relatively dry compared with the interior of a leaf, where the intercellular spaces are saturated with water vapor. As long as stomata are open, a continuous stream of water vapor transpires from the leaf, creating a pull on the water column that extends from the leaf through the plant to the soil.

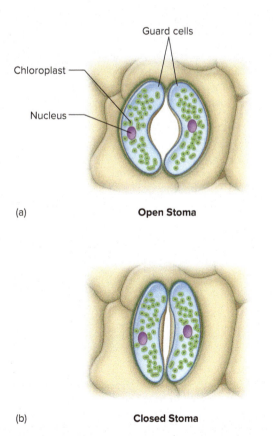

Figure 4.2 Transpiration is the basic driving force behind water movement in the xylem. (a) When stomata are open, both transpiration and photosynthesis occur as H_2O molecules diffuse out of the leaves and CO_2 molecules diffuse in. When guard cells are turgid, stomata are open, and (b) when guard cells are flaccid, stomata are closed.

Thinking Critically

Xerophytes are plants that are able to grow in arid environments.

Explain how the following adaptations of xerophytes would reduce transpiration rates and enhance these plants' survival in arid regions:

Thick cuticle

Sunken stomata (stomata are found in cavities)

Leaf surface covered with dense mat of trichomes (hairs)

(Hint: See Chapter 3.)

Absorption of Water from the Soil

Water and dissolved minerals enter a plant through the root hairs and can follow two paths, via either the **symplast** or the **apoplast.** Water molecules can diffuse through the plasma membrane into the cytoplasm of a root hair cell and continue on this intracellular movement through the cytoplasm of cells in the cortex. Recall that molecules will move from an area of high concentration to one of low concentration. The water

A CLOSER LOOK 4.1

Mineral Nutrition and the Green Clean

Research has shown that certain minerals are required by plants for normal growth and development. These are included in the essential elements listed in Appendix Table 1. The soil is the source of these minerals, which are absorbed by the plant with the soil water. Even nitrogen, which is a gas in its elemental state, is normally absorbed from the soil in the form of nitrate ions (NO_3^-). Some soils are notoriously deficient in micronutrients and are therefore unable to support most plant life. Serpentine soils, for example, are deficient in calcium, and only plants able to tolerate the low levels of this mineral can survive. In modern agriculture, mineral depletion of soils is a major concern, since harvesting crops interrupts the natural recycling of nutrients back to the soil.

Mineral deficiencies can often be detected by specific symptoms, such as chlorosis (loss of chlorophyll resulting in yellow or white leaf tissue), necrosis (isolated dead patches), anthocyanin formation (development of deep red pigmentation of leaves or stem), stunted growth, and development of woody tissue in an herbaceous plant. Soils are most commonly deficient in nitrogen and phosphorus. Nitrogen-deficient plants exhibit many of the symptoms just described. Leaves develop chlorosis (box fig. 4.1); stems are short and slender; and anthocyanin discoloration occurs on stems, petioles, and lower leaf surfaces. Phosphorus-deficient plants are often stunted, with leaves turning a characteristic dark green, often with the accumulation of anthocyanin. Typically, older leaves are affected first as the phosphorus is mobilized to young growing tissue. Iron deficiency is characterized by chlorosis between veins in young leaves.

Much of the research on nutrient deficiencies is based on growing plants hydroponically, using soilless nutrient solutions. This technique allows researchers to create solutions that selectively omit certain nutrients and then observe the resulting effects on the plants. **Hydroponics** has applications beyond basic research, since it facilitates the growing of greenhouse vegetables during winter. Aeroponics, a technique in which plants are suspended and the roots misted with a nutrient solution, is another method for growing plants in soilless culture.

While mineral deficiencies can limit the growth of plants, an overabundance of certain minerals can be toxic and can also limit growth. Saline soils, which have high concentrations of sodium chloride and other salts, also limit plant growth, and research continues to focus on developing salt-tolerant varieties of agricultural crops. Research has focused on the toxic effects of heavy metals, such as lead, cadmium, mercury, and aluminum; however, even copper and zinc, which are essential elements, can become toxic in high concentrations. Although most plants cannot survive in these soils, certain plants have the ability to tolerate high levels of these minerals.

molecules move from the dilute soil solution and enter the more concentrated cytoplasm of the root cells. The cytoplasm of all cells is interconnected through plasmodesmata and is referred to as the symplast. Thus, this pathway follows the symplast from a root hair cell into the stele (fig. 4.3).

A second path is the diffusion of water through the cell walls and intercellular spaces from the root hair through the cortex (fig. 4.3). The intercellular spaces and the spaces between the cellulose fibrils in the cell walls constitute the apoplast of a plant; thus, the water molecules move unimpeded through the apoplast until they reach the endodermis. The innermost layer of the cortex consists of a specialized cylinder of cells known as the endodermis. The presence of a **Casparian strip** on the walls of the endodermal cells regulates the movement of water and minerals into the stele. The Casparian strip is a layer of suberin (and in some instances lignin as well) on the radial and transverse walls (top, bottom, and sides) that prevents the apoplastic movement of water into the stele (see Chapter 3, fig. 3.9c). The movement of water through the differentially permeable plasma membrane is therefore directed to the tangential walls of the endodermis and into the cytoplasm of these cells and, thus, the symplast. By forcing the water and minerals through the symplastic

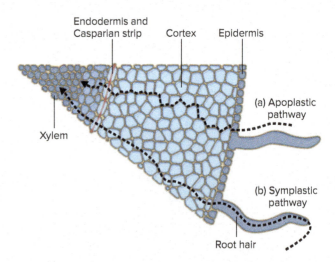

Figure 4.3 Water and minerals can follow one of two pathways across the cortex into the vascular cylinder: (a) the apoplastic pathway, in which water diffuses through the cell walls and intercellular spaces, or (b) the symplastic pathway, in which water diffuses into the cytoplasm of a root hair cell and continues moving through the cytoplasm from one cell to the next. In both pathways, water must move through the symplast of the endodermal cells.

Box Figure 4.1 Chlorosis caused by nitrogen deficiency is evident in this azalea leaf.

Scientists have known for some time that certain plants, called **hyperaccumulators,** can concentrate minerals at levels 100-fold or greater than normal. Certain minerals are more likely to be hyperaccumulated than others. A survey of known hyperaccumulators identified that 75% of them amassed nickel. Cobalt, copper, zinc, manganese, lead, and cadmium are other minerals of choice.

Hyperaccumulators run the gamut of the plant world. They may be herbs, shrubs, or trees. Many members of the Brassicaceae (mustard family), Euphorbiaceae (spurge family), Fabaceae (legume family), and Poaceae (grass family) are top hyperaccumulators. Many are found in tropical and subtropical areas of the world where accumulation of high concentrations of metals may afford some protection against plant-eating insects and microbial pathogens. Hyperaccumulators are even found among non-flowering plants; the fern, *Pteris vittata,* has been shown to accumulate arsenic from contaminated sites.

Only recently have investigators considered using these plants to clean up soil and waste sites that have been contaminated by toxic levels of heavy metals—an environmentally friendly approach known as **phytoremediation.** A green clean scenario begins with the planting of hyperaccumulating species in the target area, such as an abandoned mine or an irrigation pond contaminated by runoff. Toxic minerals would first be absorbed by roots but later translocated to the stem and leaves. A harvest of the shoots would remove the toxic compounds off-site to be ashed or composted to recover the metal for industrial uses. After several years of cultivation and harvest, the site would be restored at a cost much lower than the price of excavation and reburial, the standard practice for remediation of contaminated soils.

In field trials, alpine pennycress (*Thlaspi caerulescens*) removed zinc and cadmium from soils near the site of a zinc smelter. Indian mustard (*Brassica juncea*) native to Pakistan and India has been effective in reducing the level of selenium salts by 50% in contaminated soils. Much interest has focused on Indian mustard, since it has also been shown to concentrate lead, chromium, cadmium, nickel, zinc, and copper in the laboratory. The aquatic weed parrot feather (*Myriophyllum aquaticum*) shows promise in restoring contaminated waterways. In addition to these hyperaccumulators, other plants, including poplars (*Populus* spp.) and willows (*Salix* spp.), have been used in the cleanup of various organic pollutants. Research is ongoing as the search continues for the plants best suited to purify polluted sites quickly and cheaply.

---

pathway, some control over the uptake of minerals is exerted. Some minerals are prevented from entering the stele, while others are selectively absorbed by active transport. (See A Closer Look 4.1: Mineral Nutrition and the Green Clean.) Once inside the cytoplasm of the endodermal cells, water moves symplastically into the living cells of the pericycle, the outermost layer of the stele. The water moves into the conducting cells of the xylem, drawn by the pull of transpiration.

Water Movement in Plants

Once water is in the xylem of the stele, its movement upward in the plant is driven by the pull of transpiration as well as certain properties of the water molecule itself, **cohesion** and **adhesion.** The polarity of water molecules creates hydrogen bonds between adjacent molecules (see Appendix). These hydrogen bonds may form between water molecules themselves (cohesion) or between water molecules and the molecules in the walls of vessel elements and tracheids (adhesion). The presence of these water molecules adhering to the cell walls provides a continuous source of water that can evaporate into the intercellular spaces of the leaf and transpire through the stomata. The cohesive force is so strong that any force or pull on one water molecule acts on all of them as well, resulting in the bulk flow of water within the plant. Bulk flow is usually defined as the movement of a fluid because of pressure differences at two locations. In the xylem, the pull of transpiration is the force causing the bulk flow. As transpiration occurs in the leaf, it creates a cohesive pull on the whole water column downward from the leaf through the xylem to the root, where water uptake occurs to replace the water lost through transpiration (fig. 4.1). This mechanism of water movement in plants is known as the **Transpiration-Cohesion Theory** and has been used to explain rates of water movement as high as 44 meters (145 feet) per hour in angiosperm trees.

Translocation of Sugar

Organic materials are translocated by the sieve tube members of the phloem. In contrast to the xylem, where the conducting elements function when the cells are dead, the sieve tube members of the phloem are living but highly specialized cells (see Chapter 3). While water movement in the xylem is upward from the soil, phloem translocation moves in the direction from **source** (supply area) to **sink** (area of metabolism or storage). In late winter, the source may be an underground storage organ translocating sugars to apical meristems (the sink) in the branches of a tree. In summer, the source may be photosynthetic leaves sending sugars for storage to sinks such as roots or developing fruits (fig. 4.4). In most plants, the primary material translocated in phloem is sucrose in a watery solution that also may include small amounts of amino acids, minerals, and other organic compounds. Translocation in the phloem is quite rapid and has been timed at speeds averaging 1 meter (3.3 feet) per hour. The amount of material translocated is also quite impressive. In a growing pumpkin, which reaches a size of 5.5 kg (11 lb) in 33 days, approximately 8 g (0.3 oz) of solution are translocated each hour.

Each fall at state and local fairs all across the United States and other countries, prize-winning pumpkins routinely weigh well over 454 kg (1,000 lb). Although giant pumpkin competitions had been going on for decades, the first pumpkin to pass the 1,000-lb barrier was grown in 1996. Since then, new world records have been set almost every year and the size has increased dramatically. In 2012, the first pumpkin to break the one-ton (909 kg or 2,000 lb) barrier was grown by Ron Wallace from Green, Rhode Island; it weighed in at 911.3 kg (2,009 lb). This amazing bulk was surpassed in 2013 when a pumpkin (fig. 4.4c) grown by Tim Mathison from Napa, California, weighed in at 921.7 kg (2,032 lb). In 2014, new records were again set; Swiss gardener Beni Meier produced three giant pumpkins, each of which broke world records, with the largest weighing in at 1,054 kg (2,323.7 lb). In 2016, another world record was set with a 1,190 kg (2,623 lb) pumpkin grown by Mathias Willemijns from Belgium. Although no new world record was set in 2017, 15 pumpkins that weighed over one ton were entered in giant pumpkin competitions in North American and Europe. Consider the amount of phloem translocation occurring during the growth of these giants!

The hypothesis currently accepted to explain translocation in the phloem is the **Pressure Flow** (or Mass Flow) **Hypothesis.** This is a modified version of a hypothesis first proposed by Ernst Münch in 1926. According to this hypothesis, there is a bulk flow of solutes from source to sink (fig. 4.4). At the source, phloem loading takes place as sugar molecules are first actively transported into companion cells and then move symplastically into sieve tube members through plasmodesmata. This highly concentrated solution in the sieve tube members causes water to enter by osmosis from nearby xylem elements, resulting in a buildup of pressure. When pressure starts to build in these cells, the solute-rich phloem sap is pushed through the pores in the sieve plate into the adjacent sieve tube member and so on down to the sink. This movement of material en masse is known as mass flow. At the sink, companion cells function in active phloem unloading, which reduces the concentration of sugars and allows water to diffuse out of these cells. Sugars unloaded at the sink are taken up by nearby cells and either stored as starch or metabolized. (See A Closer Look 4.2: Sugar and Slavery.) The loading and unloading of sugars by active transport are energy-requiring steps.

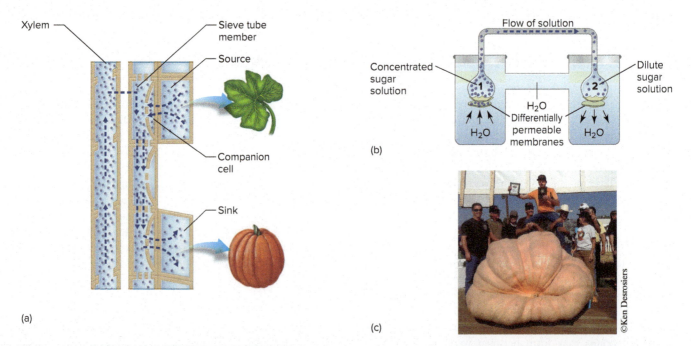

Figure 4.4 Translocation of sugar. (a) Products of photosynthesis move from source to sink. At the source, sugar molecules are loaded into the phloem. As the sugar concentration increases, water moves in from adjacent xylem, pressure builds up, and the sugar solution is forced through the plant to the sink, where sugar is unloaded from the phloem. (b) A physical model can be used to demonstrate the Pressure Flow Hypothesis. The concentrated sugar solution on the left is comparable to the source, and the dilute solution on the right is comparable to the sink. In bulb 1, water moves into the concentrated sugar solution from the surroundings, pressure builds up, and the sugar solution flows through the tube to bulb 2, where the sugars begin to accumulate, and the pressure causes the water to move out. Recall the discussion of osmosis and diffusion in Chapter 2. (c) Photograph of the 2013 world record giant pumpkin that weighed 921.7 kg (2,032 lb).

A CLOSER LOOK 4.2
Sugar and Slavery

Products of photosynthesis are typically transported to growing fruits, storage organs, and other sinks throughout the plant. After being unloaded from the phloem, sugars are usually converted to starch or other storage carbohydrates. Although the disaccharide sucrose is the material translocated in the phloem of most plants, very few species store significant amounts of this sugar. Only two plants, sugarcane and sugar beet, are commercially important sources of sucrose, commonly known as table sugar. Sugarcane is the more important crop and, in terms of sheer tonnage, leads the global crop production list (see fig. 12.1).

Sugarcane is native to the islands of the South Pacific and has been grown in India since antiquity. Small amounts of sugarcane reached the ancient civilizations in the Near East and Mediterranean countries through Arab trading routes, but it was not grown in those regions until the seventh century. Even after cultivation was established, honey remained the principal sweetener in Europe until the fifteenth century. During the Middle Ages, sugar was an expensive luxury that found its greatest use in medicinal compounds to disguise the bitter taste of many herbal remedies. Early in the fifteenth century, sugar plantations were established on islands in the eastern Atlantic on the Canary Islands by Spain as well as on Madeira and the Azores by Portugal.

Columbus introduced sugarcane to the Caribbean Islands on his second voyage in 1493. By 1509, sugarcane was harvested in Santo Domingo and Hispaniola and soon spread to other islands. In fact, many Caribbean Islands were eventually denuded of native forests and planted with sugarcane. The Portuguese saw the opportunities in South America and started sugar plantations in Brazil in 1521. Although late to enter the West Indies, the British established colonies in the early seventeenth century, and by the 1640s, sugar plantations were thriving on Barbados. The first sugarcane grown in the continental United States was in the French colony of Louisiana in 1753.

The growing of sugarcane was responsible for the establishment of slavery in the Americas, and early in the sixteenth century, sugar and the slave trade became interdependent. Decimation of the native populations led to the need for new workers on the sugar plantations. The first suggestion to use African slaves was made in 1517. Within a short time, the importation of slaves was a reality in both the Spanish and Portuguese colonies, with the greatest number of slaves imported into Brazil. The introduction of African slaves to these colonies was an outgrowth of the slave trade in Spain and Portugal that had begun in the 1440s. Initially, Spain exported the slaves, but by 1530 slaves were sent directly from Africa to the Caribbean.

The sugar production in the Caribbean came at a time when supplies of honey in Europe were decreasing. The Catholic monasteries were the traditional source of honey. Beehives were kept, principally, to produce beeswax for church candles. During the Protestant Reformation, Catholic monasteries were suppressed, and the sources of honey fell short of demand. Also, in the late seventeenth century, the introduction and growing popularity of coffee, tea, and cocoa in Europe accelerated the demand for sugar, since Europeans generally disliked the naturally bitter taste of these beverages. Sugar became the most important commodity traded in the world, and eventually England became the dominant force in this enterprise. The Triangular Trade was the source of many fortunes. The first leg of the triangle was from England to West Africa, where trinkets, cloth, firearms, salt, and other commodities were bartered for slaves. The second leg brought slaves to the Caribbean Islands, where they were sold. The final leg of the journey carried rum, molasses, and sugar back to England (box fig. 4.2a). A second triangle became important in the mid-eighteenth century, linking the West Indies, New England, and West Africa. The use of slaves on sugar plantations continued until the early nineteenth century, when the slave trade was abolished. It has been estimated that during this period 10 million to 20 million African slaves were brought to the New World. Approximately 40% of the slaves brought to the New World went to the Caribbean Islands. During the seventeenth century, the British sugar plantations in these islands created

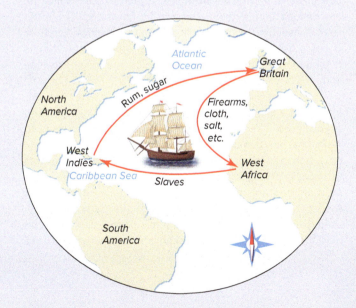

Box Figure 4.2a The Triangular Trade.

Box Figure 4.2b Sugarcane harvest near Luxor, Egypt.

Box Figure 4.2c Sugar beet ready for harvest.

one of the harshest systems of slavery in history. It was physically crueler and more demanding than slave conditions on the North American continent. Although the life of a slave was never easy, it was especially arduous on the sugar plantations. Few survived the hard labor; consequently, the slave population had to be constantly replenished.

Sugarcane, *Saccharum officinarum,* is a perennial member of the grass family, Poaceae (box fig. 4.2b). Several species of *Saccharum* are known to exist in the tropics, and it is believed that *S. officinarum* originated as a hybrid of several species. The species owes its importance to the sucrose stored in the cells of the stem. Sugarcane, which uses the C_4 Pathway for photosynthesis, is considered one of the most efficient converters of solar energy into chemical energy. Canes are often 5 to 6 meters (15 to 20 feet) tall, with individual stalks up to 10 cm (4 in.) in diameter. Plants are grown from stem cuttings, which are laid horizontally, bud upward, in shallow trenches. Generally, 12–18 months are needed before the canes can be harvested. On fertile land, subsequent crops develop from the rhizome for 2 or 3 years before replanting is necessary. Sugarcane thrives in moist lowland tropics and subtropics and today provides about 60% of the world's sugar supply.

Canes generally contain 12%–15% sucrose. After harvesting, canes are crushed by heavy steel rollers to extract the sugary juice. The fibrous residue (bagasse) can be used to make fiberboard, paper, and other products or used as compost. Successive boilings concentrate the sucrose; impurities are usually precipitated by adding lime water (calcium oxide solution) and are removed by filtering. The solution is then evaporated to form a syrup from which the sugar is crystallized. Centrifuges separate the thick, brown, liquid portion from the crystals. The liquid portion is molasses, which is used in foods or is fermented to make rum, ethyl alcohol, or vinegar. The crystallized sugar (about 96%–97% pure sucrose) is further refined to free it from any additional impurities. Sugarcane is also an important source of ethanol biofuel, especially in Brazil, the world's leading sugarcane producer. Brazil's bioethanol industry, considered the most successful alternative fuel, produced 30.2 billion liters (8 billion gallons) of sugarcane ethanol in 2015/2016 (see Chapter 12).

Sugar beet, *Beta vulgaris,* a member of the amaranth family, Amaranthaceae (formerly in the Chenopodiaceae), is not closely related to sugarcane (box fig. 4.2c). It is actually the same species as red beets, which are native to the Mediterranean region and have been consumed since the time of the ancient Romans. In the mid-eighteenth century, a German chemist, Andreas Marggraf, discovered that the roots contain sugar that is chemically identical to that from cane. The rise of the sugar beet industry can be tied to the emperor Napoleon I. When a British naval blockade cut off the sugar imports to France, Napoleon realized the value of a domestic source of sugar and encouraged scientific research on the sugar beet. After 1815, sugarcane imports were restored, halting the developing sugar beet industry. By the early twentieth century, the sugar beet industry had been revived in both Europe and North America, and today sugar beets provide close to 40% of the world's supply of table sugar.

Sugar beet is a biennial plant, but it is harvested at the end of the first year, when the sucrose content is greatest. Selective breeding gradually raised the percentage of sugar in the root from 2% to approximately 20%. After harvesting, roots are shredded, steeped in hot water, and then pressed to extract the sucrose. Further processing is similar to sugarcane processing and produces an identical final product.

METABOLISM

Metabolism is the sum total of all chemical reactions occurring in living organisms. Metabolic reactions that synthesize compounds are referred to as **anabolic** reactions and are generally **endergonic,** requiring an input of energy. In contrast, **catabolic** reactions, which break down compounds, are usually **exergonic** reactions, which release energy. Many of these reactions also involve the conversion of energy from one form to another.

Energy

All life processes are driven by energy, and consequently, a cell or an organism deprived of an energy source will soon die. **Energy** is defined by physicists as the ability to do work and is governed by certain physical principles, such as the laws of thermodynamics.

The first law of thermodynamics states that energy can neither be created nor destroyed, but it can be converted from one form to another.

Among the forms of energy are radiant (light), thermal (heat), chemical, mechanical (motion), and electrical. One focus of this chapter is photosynthesis, the process that converts radiant energy from the sun into the chemical energy of a sugar molecule.

The second law of thermodynamics states that in any transfer of energy there is always a loss of useful energy to the system, usually in the form of heat.

When gasoline is burned as fuel to drive an automobile engine, chemical energy is converted into mechanical energy, but the conversion is not very efficient. Some of the energy is lost as heat to the surroundings.

All forms of energy can exist as either **potential energy** or **kinetic energy.** Potential energy is stored energy that has the capacity to do work; kinetic energy actually is doing work or is energy in action. For example, a boulder at the top of a hill has a tremendous amount of potential energy. If it rolls down the hillside, the potential energy is changed into kinetic energy, an exergonic process (fig. 4.5). To push it back up to the top of the hill would be an endergonic process requiring considerable input of energy. Transformations from potential to kinetic and vice versa occur constantly in biological systems and are part of the underlying principles of both photosynthesis and respiration.

Redox Reactions

Many energy transformations in cells involve the transfer of electrons or hydrogen atoms. When a molecule gains an electron or a hydrogen atom, the molecule is said to be **reduced**, and the molecule that gives up the electron is said to be **oxidized**. A molecule that has been reduced has gained energy; likewise, the oxidized molecule has lost energy. Oxidation and reduction reactions are usually coupled (sometimes called **redox reactions**); as one molecule is oxidized, the other is simultaneously reduced.

$$AH_2 + B \longrightarrow A + BH_2$$
(A-reduced) (B-oxidized) (A-oxidized) (B-reduced)

Figure 4.5 Potential and kinetic energy. (a) The boulder at the top of a hill has a tremendous amount of potential energy. (b) If the boulder rolls down the hill, the potential energy is converted to kinetic energy. To push the boulder back up to the top would require a considerable input of energy.

In many oxidation-reduction reactions, an intermediate is used to transport electrons from one reactant to another. One such electron intermediate is **NAD** (nicotinamide adenine dinucleotide), which can exist in both oxidized and reduced states (NAD^+ = **oxidized form** and $NADH + H^+$ = **reduced form**). Similarly, **NADP** (nicotinamide adenine dinucleotide phosphate) and **FAD** (flavin adenine dinucleotide) also can exist as $NADP^+/NADPH + H^+$ and $FAD/FADH_2$, respectively. NAD and FAD are common electron carriers in respiration; NADP serves the same function in photosynthesis.

Phosphorylation

Other energy transformations involve the transfer of a phosphate group. When a phosphate group is added to a molecule, the resulting product is said to be **phosphorylated** and has a higher energy level than the original molecule. These phosphorylated compounds may also lose the high-energy phosphate group and thereby release energy. The energy currency of cells, **ATP** (adenosine triphosphate), is constantly recycled in this way. When a phosphate group is removed from ATP,

ADP (adenosine diphosphate) is formed, and energy is released in this exergonic reaction.

$$ATP \rightarrow ADP + PO_4 + energy$$

Re-creating ATP requires the addition of a phosphate group to ADP (an endergonic reaction) with the appropriate input of energy.

$$ADP + PO_4 + energy \rightarrow ATP$$

Enzymes

Proteins that act as catalysts for chemical reactions in living organisms are known as **enzymes. Catalysts** speed up the rate of a chemical reaction without being used up or changed during the reaction. The majority of chemical reactions in living organisms require enzymes in order to occur at biological temperatures. Enzymes are highly specific for certain reactants; the compound acted upon by the enzyme is known as the **substrate.** The names of enzymes most commonly end in the suffix **-ase,** which is sometimes appended to the name of the substrate or the type of reaction. For example, protease enzymes degrade proteins, lipases split fats (lipids), and cellulase enzymes break down cellulose. Some enzymes function properly only in the presence of **cofactors** or **coenzymes.** Cofactors are inorganic, often metallic, ions (such as Mg^{++} and Mn^{++}), while organic molecules (such as NAD, NADP, and some vitamins) are coenzymes. Both cofactors and coenzymes are loosely associated with enzymes; however, **prosthetic groups** are nonprotein molecules that are attached to some enzymes and are necessary for enzyme action.

The first enzyme that was discovered was amylase, identified from germinating grain. Amylase breaks down starch into maltose, which contains two glucose molecules (see Chapter 1). In addition to plants, many organisms produce amylase including humans. Amylase in saliva begins digesting starch in a person's mouth. Other enzymes break down the maltose into glucose, which is used by organisms as an energy source (see Chapter 10).

Although enzymes are synthesized within cells, there are many plant and fungal enzymes that are useful commercially. One of the most familiar is papain, which is a protease enzyme from papaya fruit. Papain breaks the peptide bonds between specific amino acids in proteins. It occurs in greatest concentration in the latex (a milky sap produced by many plants) from unripe papaya fruit. A similar enzyme, bromelain, is found in pineapple fruits and stems. Both papain and bromelain are widely used in commercial meat tenderizers found in grocery stores. These are sprinkled on meat for a short time before cooking; the enzymes break down muscle fiber and connective tissues and are particularly useful for less expensive cuts of meat. Fresh papaya and pineapple as well as kiwifruit and figs, which also contain protease enzymes, can be used in marinades to produce the same tenderizing effects. Careful timing for all these applications is a must to avoid having the meat become mushy.

Lactase is an enzyme that breaks down lactose, a disaccharide found in dairy products. In people lactose is produced in the small intestine and during digestion the enzyme splits lactose into two monosaccharides; however, many adults lose the ability to synthesize lactase. They become lactose–intolerant and experience digestive distress when they consume dairy products. Several fungal species also produce lactase. These fungi are the source of commercial lactase, which is added to milk and other dairy products to make them lactose-free and safe for lactose-intolerant individuals.

Pectinases are enzymes that degrade pectin, the complex polysaccharide found in the primary walls and middle lamella between adjacent plant cells. In plants, pectinases are produced during the process of fruit ripening. It is one of the many enzymes that change fruits from hard and unpalatable to fleshy, sweet, and juicy. Pectinases are also made by numerous fungi as well as bacteria, and commercial pectinases are primarily produced by the fungus *Aspergillus niger*. These enzymes are used commercially in wide variety of industries that process plant material. For example, in the preparation of fruit juices and wines, pectinases speed up the release of juice from the fruit pulp; the enzymes also help clarify juices such as apple juice, which are naturally cloudy. In the textile industry, pectinase enzymes are used to separate and process stem and phloem fibers for manufacturing fabric and rope (see Chapter 18).

These are just a few illustrations of how enzymes have become important commercial products. In the remainder of this chapter, the enzymes that are described are involved in the essential cellular processes of photosynthesis and respiration.

PHOTOSYNTHESIS

Photosynthesis is the process that transforms the vast energy of the sun into chemical energy and is the basis for most food chains on Earth. The overwhelming majority of life depends on the photosynthetic ability of green plants and algae, and without these producers, life as is known today could not survive.

Energy from the Sun

The sun is basically a thermonuclear reactor producing tremendous quantities of **electromagnetic radiation,** which bathes Earth. Visible light is only a small portion of the electromagnetic spectrum, which includes radio waves, microwaves, infrared radiation, ultraviolet radiation, X rays, and gamma rays (fig. 4.6). This radiant energy, or light, has a dual nature consisting of both particles and waves. The particles are known as **photons** and have a fixed quantity of energy. It is believed that the photons travel in waves and thus display characteristic wavelengths. Wavelengths vary from radio waves, which may be over 1 kilometer long, to gamma rays, which are a fraction of a nanometer (nm) long. The energy content also varies and is inversely proportional to the wavelength—that is, the longer the wavelength, the lower the energy.

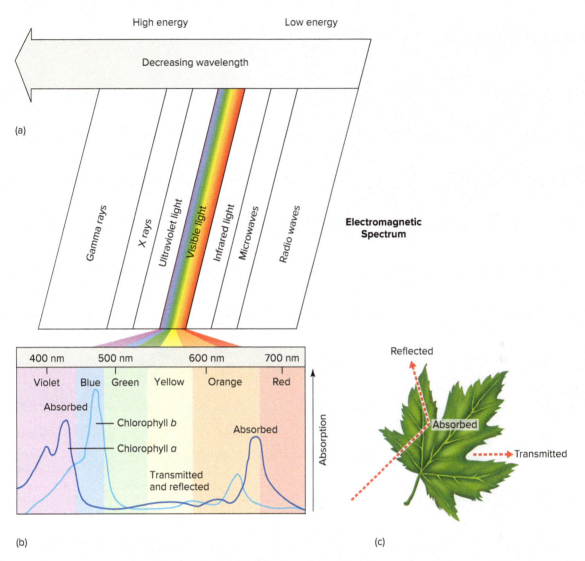

Figure 4.6 Energy from the sun drives the process of photosynthesis. (a) Visible light is only a small portion of the electromagnetic spectrum. (b) If visible light is passed through a prism, the component colors are apparent. The chlorophyll pigments in leaves absorb the blue-violet and orange-red portions of the spectrum. (c) Leaves reflect the green and yellow portions of the spectrum.

Approximately 40% of the radiant energy reaching Earth is in the form of **visible light.** If visible light passes through a prism, the component colors become apparent; these range from red at one end of the visible band to blue-violet at the other end. The wavelengths of visible light range from 380 nm (violet) to 760 nm (red) and are the wavelengths most important to living organisms (fig. 4.6). In fact, the wavelengths within this range are the ones absorbed by the chlorophylls and other photosynthetic pigments in green plants and algae.

Chloroplasts and Light-Absorbing Pigments

When light strikes an object, the light can pass through the object (be transmitted), be reflected from the surface, or be absorbed. For light to be absorbed, pigments must be present. Pigments absorb light selectively, with different pigments absorbing different wavelengths and reflecting others. Each pigment has a characteristic **absorption spectrum**, which depicts the absorption at each wavelength. If all visible wavelengths are absorbed, the object appears black; however, if all wavelengths are reflected, the object appears white. Green leaves appear green because these wavelengths are reflected.

In higher plants, the major organ of photosynthesis is the leaf, and the green chloroplasts within the mesophyll are the actual sites of this process. Recall from Chapter 2 that chloroplasts are double membrane-bound organelles with extensive internal membranes that are easily seen in the electron microscope. Starch grains can often be found within chloroplasts as well. The major photosynthetic pigments are the green **chlorophylls,** which are located on the **thylakoid membranes** of the chloroplasts (fig. 4.7). Thylakoids can be

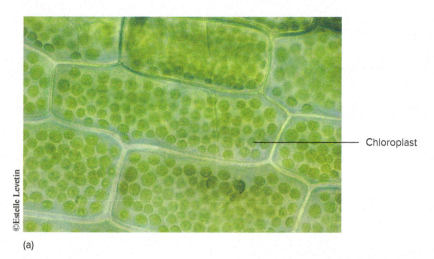
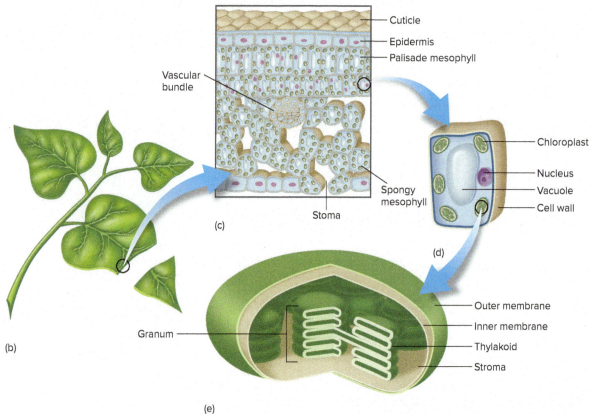
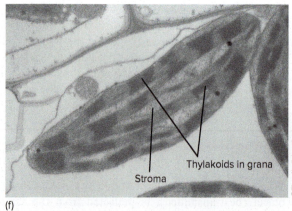

Figure 4.7 The major organ of photosynthesis is the leaf, and the actual site of photosynthesis is the chloroplast. (a)–(d) Leaf cells with chloroplasts; (e)–(f) internal structure of the chloroplast. Thylakoid sacs comprise the grana, sites of the light reactions. The stroma contains the enzymes that carry out the Calvin Cycle.

found in stacks, known as **grana,** as well as individually in the **stroma,** the enzyme-rich ground substance of the chloroplast. Chloroplasts may have 50 to 80 grana, each with about 10 to 30 thylakoids. The chlorophylls can be located on the **stroma thylakoids** as well as in the grana, and it is the abundance of these pigments that makes leaves appear green.

In green plants there are two forms of chlorophyll, *a* and *b*, which differ slightly in chemical makeup. Most chloroplasts have three times more chlorophyll *a* than *b*. The absorption spectra of the chlorophylls show peak absorbances in the red and blue-violet regions, with much of the yellow and green light reflected (fig. 4.6). Other forms of chlorophyll and other photosynthetic pigments occur in the algae (Chapter 22).

In addition to chlorophylls, chloroplasts also contain **carotenoids.** These include the orange **carotenes** and yellow **xanthophylls,** which absorb light in the violet, blue, and blue-green regions of the spectrum. Carotenoids are called accessory pigments, and the light energy absorbed by these pigments is transferred to chlorophyll for photosynthesis. Although present in all leaves, carotenoids are normally masked by the chlorophylls. Recall that these carotenoids become apparent in autumn in temperate latitudes, when chlorophyll degrades.

Overview

Photosynthesis consists of two major phases, the **light reactions** and the **Calvin Cycle.** The light reactions constitute the photochemical phase of photosynthesis, during which radiant energy is converted into chemical energy. During the light reactions, water molecules are split, releasing oxygen and providing electrons for the reduction of $NADP^+$ to $NADPH + H^+$. The light reactions also provide the energy for the synthesis of ATP. The Calvin Cycle constitutes the biochemical phase and involves the fixation and reduction of CO_2 to form sugars using the ATP and NADPH produced in the light reactions (fig. 4.8).

The Light Reactions

The light reactions are composed of two cooperating photosystems, **Photosystems I and II,** and take place on the thylakoid membranes within the chloroplasts. Each photosystem is a complex of several hundred chlorophyll and carotenoid molecules (known as **light-harvesting antennae**) and associated membrane proteins. Countless units of these photosystems are arrayed on the thylakoid membranes throughout the chloroplast. When light strikes a pigment molecule in either photosystem, the energy is funneled into a **reaction center,**

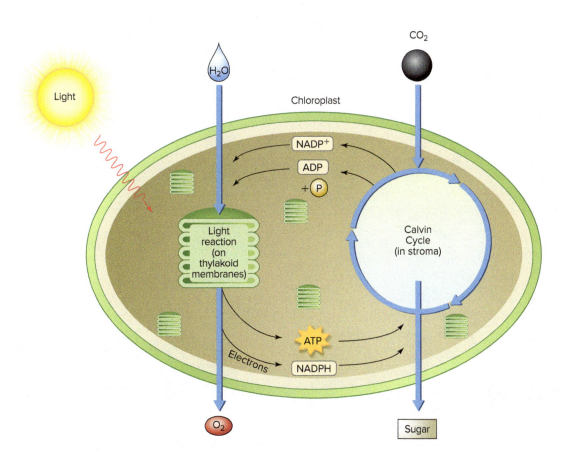

Figure 4.8   Photosynthesis consists of two major phases, the light reactions and the Calvin Cycle.

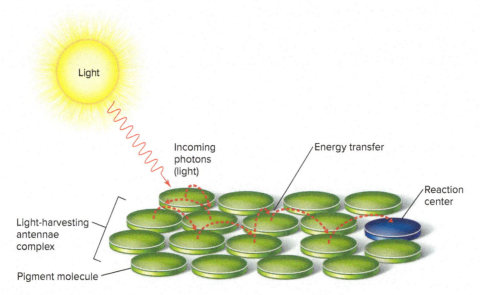

Figure 4.9 Photosystem. Each photosystem is composed of several hundred chlorophyll and carotenoid molecules that make up light-harvesting antennae within the thylakoid membranes of the chloroplast. When light strikes a pigment molecule, the energy is transferred and funneled into a reaction center.

which consists of a chlorophyll *a* molecule bound to a membrane protein (fig. 4.9). The reaction center for Photosystem I is known as **P_{700}**, which indicates the wavelength of maximum light absorption in the red region of the spectrum; the reaction center for Photosystem II is **P_{680}**, again indicating the peak absorbance. Associated with the photosystems are various enzymes and coenzymes that function as electron carriers and are components of the thylakoid membranes.

When a photon of light strikes a pigment molecule in the light-harvesting antennae of Photosystem I, the energy is funneled to P_{700} (fig. 4.10). When P_{700} absorbs this energy, an electron is excited and ejected, leaving P_{700} in an oxidized state. The ejected electron is picked up by a primary electron acceptor, which then passes the electrons on to ferredoxin (Fd), another electron intermediate, and eventually to $NADP^+$, reducing it to $NADPH + H^+$, one of the products of the light reactions.

Another photon of light absorbed by a chlorophyll molecule in Photosystem II will transfer its energy to the reaction center P_{680} (fig. 4.10). When P_{680} absorbs this energy, an excited

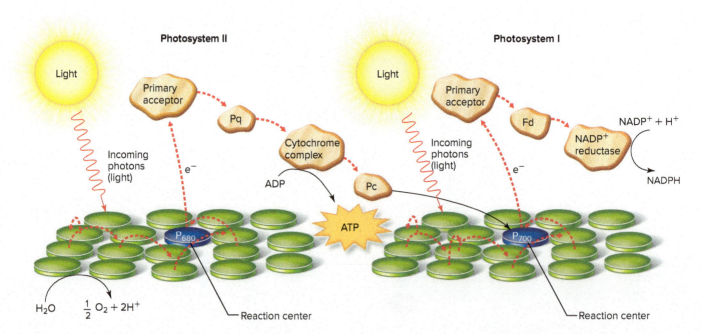

Figure 4.10 Two cooperating photosystems work together to transfer electrons from water to **NADPH**. The passage of electrons from Photosystem II to Photosystem I also drives the formation of ATP, a process known as noncyclic or linear photophosphorylation.

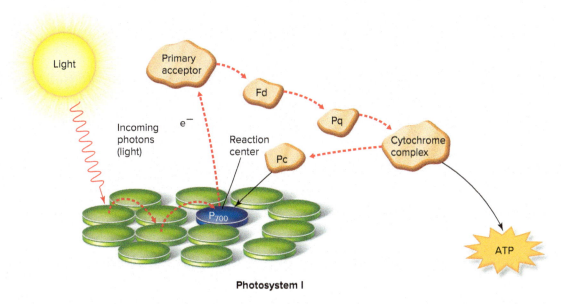

Figure 4.11 Photosystem I can also function in a cyclic fashion. Instead of reducing NADP, electrons are passed back to P_{700}. This process results in the generation of ATP by cyclic or linear photophosphorylation.

electron is ejected and passed on to another primary electron acceptor, leaving P_{680} in an oxidized state. The electron lost by P_{680} is replaced by an electron from water, in a reaction that is not fully understood, and catalyzed by an enzyme on the thylakoid membrane that requires manganese atoms. In this reaction, water molecules are split into oxygen and hydrogen; the hydrogen is a source of both electrons and protons.

The primary electron acceptor passes the electron on to a series of thylakoid membrane-bound electron carriers that include plastoquinone (Pq), cytochrome complex, plastocyanin (Pc), and others. The electron is eventually passed to the oxidized P_{700} in Photosystem I. During the transfer of electrons, ATP is synthesized as protons are passed from the thylakoid lumen into the stroma by an **ATP synthase** in the membrane. It is actually the passage of protons through this enzyme that drives the production of ATP; however, the mechanism for this reaction is still not completely understood. This synthesis of ATP is known as **photophosphorylation**, since the energy that drives the whole process is from sunlight.

In the process just described, the two photosystems are joined together by the one-way transfer of electrons from Photosystem II to Photosystem I. Water is the ultimate source of these electrons, continually replenishing electrons lost from P_{680}. The photophosphorylation that occurs during this process is referred to as **noncyclic or linear photophosphorylation,** since the electron transfer is linear, with the reduction of NADP as the final step.

Photosystem I is also capable of functioning independently, transferring electrons in a cyclic fashion. The electrons, instead of being passed to NADP from ferredoxin, may be passed to the cytochrome complex and then back to P_{700} (fig. 4.11). ATP, but not NADPH, may be generated during this process, which is known as **cyclic photophosphorylation,** since the flow of electrons begins and ends with P_{700}.

As just described, when water is split, oxygen is released. The oxygen eventually diffuses out of the leaves into the atmosphere and is Earth's only constant supply of this gas. The current 20% oxygen content in the atmosphere is the result of 3.5 billion years of photosynthesis. The atmosphere of early Earth did not contain this gas; oxygen began to accumulate only after the evolution of the first oxygen-producing photosynthetic organisms, the cyanobacteria. Today, the vast majority of living organisms depend on oxygen for cellular respiration and, therefore, the energy that maintains life.

The overall light reactions proceed with breathtaking speed as a constant flow of electrons moves from water to NADPH, powered by the vast energy of the sun. The ATP and NADPH that result from the light reactions are needed to drive the biochemical reactions in the Calvin Cycle.

The Calvin Cycle

The source of carbon used in the photosynthetic manufacture of sugars is carbon dioxide from the atmosphere. This gas makes up just a tiny fraction, approximately 0.04%, of Earth's atmosphere and enters the leaf by diffusing through the stomata. The reactions that involve the fixation and reduction of CO_2 to form sugars are known as the Calvin Cycle and are sometimes referred to as the **C_3 Pathway.** These reactions utilize the ATP and NADPH produced in the light reactions but do not involve the direct participation of light and hence are sometimes referred to as light-independent reactions, or dark reactions. The Calvin Cycle takes place in the stroma of the chloroplasts, which contains the enzymes that catalyze the many reactions in the cycle. This pathway was worked out by Melvin Calvin, in association with Andrew Benson and James Bassham, during the late 1940s and early 1950s. The pathway is named in honor of Calvin, who received a Nobel Prize for his work in 1961.

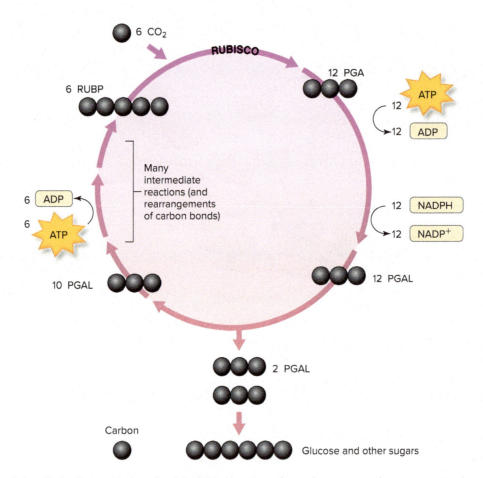

Figure 4.12 The Calvin Cycle. For every six molecules of CO_2 that enter the cycle, one six-carbon sugar is produced. The ATP and NADPH required by this cycle are generated by the light reactions.

The following discussion will be limited to the main events of the Calvin Cycle, which are depicted in Figure 4.12. The end product of this pathway is the synthesis of a six-carbon sugar; this requires the input of carbon dioxide. Six turns of the cycle are needed to incorporate six molecules of CO_2 into a single molecule of a six-carbon sugar. The initial event is the fixation or addition of CO_2 to ribulose-1,5-bisphosphate (RUBP), a five-carbon sugar with two phosphate groups. This carboxylation reaction is catalyzed by the enzyme **ribulose bisphosphate carboxylase (RUBISCO).** In addition to its obvious importance to photosynthesis, RUBISCO appears to be the most abundant protein on Earth, since it constitutes 12.5%–25% of total leaf protein. The product of the carboxylation is an unstable six-carbon intermediate that immediately splits into two molecules of a three-carbon compound, with one phosphate group called phosphoglyceric acid (PGA), or phosphoglycerate. Six turns of the cycle would yield 12 molecules of PGA.

The 12 PGA molecules are converted into 12 molecules of **glyceraldehyde phosphate,** or phosphoglyceraldehyde (PGAL). This step requires the input of 12 NADPH + H$^+$ and 12 ATP (both generated during the light reactions), which supply the energy for this reaction. Ten of the 12 glyceraldehyde phosphate molecules are used to regenerate the six molecules of RUBP in a complex series of interconversions that require six more ATP and allow the cycle to continue. Two molecules of glyceraldehyde phosphate are the net gain from six turns of the Calvin Cycle; these are converted into one molecule of fructose-1,6-bisphosphate, which is soon converted to glucose. The glucose produced is never stored as such but is converted into starch, sucrose, or a variety of other products, thus completing the conversion of solar energy into chemical energy. Almost all plants store starch within chloroplasts during daylight and then break this starch down later that night. This temporary starch storage is referred to as transitory starch, and transitory starch grains are often visible within chloroplasts in electron micrographs (fig. 4.7f). Transitory starch allows for continuous export of sugars from leaves during the night when photosynthesis is not taking place.

The complex steps of photosynthesis can be summarized in the following simple equation, which considers only the raw materials and end products of the process:

$$6CO_2 + 12H_2O + \text{Sunlight} \xrightarrow{\text{CHLOROPHYLL}} C_6H_{12}O_6 + 6O_2 + 6H_2O$$

Thinking Critically

Photosynthesis consists of two major phases: the light reactions, in which light energy is converted into chemical energy, and the Calvin Cycle, in which the fixation and reduction of carbon dioxide to form sugars take place.

How is each phase dependent on the other?

Variation to Carbon Fixation

Many plants utilize a variation of carbon fixation that consists of a prefixation of CO_2 before the Calvin Cycle. There are two pathways in which this prefixation occurs, the **C_4 Pathway** and the **CAM Pathway**. The C_4 Pathway occurs in several thousand species of tropical and subtropical plants, including the economically important crops of corn, sugarcane, and sorghum. This pathway, occurring in mesophyll cells, involves incorporating CO_2 into organic acids, resulting in a four-carbon compound and, hence, the name of the pathway. This compound is soon broken down to release CO_2 to the Calvin Cycle, which occurs in cells surrounding the vascular bundle. The C_4 Pathway ensures a more efficient delivery of CO_2 for fixation and greater photosynthetic rates under conditions of high light intensity, high temperature, and low CO_2 concentrations.

The same steps are part of the **CAM (Crassulacean Acid Metabolism)** Pathway, which functions in a number of cacti and succulents, plants of desert environments. This pathway was initially described among members of the plant family Crassulaceae. CAM plants are unusual in that their stomata are closed during the daytime but open at night. Thus, they fix CO_2 during the nighttime hours, incorporating it into four-carbon organic acids. During the daylight hours, these compounds are broken down to release CO_2 to continue into the Calvin Cycle. This alternate pathway allows carbon fixation to occur at night when transpiration rates are low, an obvious advantage in hot, dry desert environments.

CELLULAR RESPIRATION

As previously discussed, photosynthesis converts solar energy into chemical energy, stored in a variety of organic compounds. Starch and sucrose are common storage compounds in plants and, as such, are the energy reserves for the plants themselves and the animals that feed on them. Ultimately, the survival of all organisms on Earth is dependent on the release of this chemical energy through the catabolic process of cellular respiration. All living organisms require energy to maintain the processes of life. Even at the cellular level, life is a highly dynamic system, requiring continuous input of energy, which is used in the processes of growth, repair, transport, synthesis, motility, cell division, and reproduction. Cellular respiration occurs continuously, every hour of every day, in all living cells; the need for energy is nonstop.

Cellular respiration is a step-by-step breakdown of the chemical bonds in glucose, involving many enzymatic reactions, and results in the release of usable energy in the form of ATP. The overall process is the complete oxidation of glucose, resulting in CO_2 and H_2O and the formation of ATP.

$$C_6H_{12}O_6 + 6O_2 \rightarrow 6CO_2 + 6H_2O + 36ATP$$

This equation of cellular respiration is merely a summary of a complex, step-by-step process that has three major stages or pathways: **glycolysis**, the **Krebs Cycle**, and the **Electron Transport System**.

Glycolysis is a series of reactions that occur in the cytoplasm and result in the breakdown of glucose into two molecules of a three-carbon compound. Along the way, NAD is reduced and some ATPs are produced. The Krebs Cycle continues the breakdown of the three-carbon compounds in the matrix of the mitochondria and results in the release of CO_2. Additional ATP, NADH, and $FADH_2$ are also generated during these steps. The final stage of respiration, the Electron Transport System, occurs on the cristae, the inner membrane of the mitochondria, and consists of a series of redox reactions during which significant amounts of ATP are synthesized. A comparison of photosynthesis and cellular respiration is presented in Table 4.1.

Table 4.1
Comparison of Photosynthesis and Cellular Respiration

	Photosynthesis	Cellular Respiration
Overall reaction	$6CO_2 + 12H_2O + \text{Sunlight} \rightarrow C_6H_{12}O_6 + 6O_2 + 6H_2O$	$C_6H_{12}O_6 + 6O_2 \rightarrow 6CO_2 + 6H_2O + 36ATP$
Reactants	Carbon dioxide, water, sunlight	Glucose, oxygen
Products	Glucose	Energy
By-products	Oxygen	Carbon dioxide + water
Cellular location	Chloroplasts	Cytoplasm, mitochondria
Energetics	Requires energy	Releases energy as ATP
Chemical pathways	Light reactions, Calvin Cycle	Glycolysis, Krebs Cycle, Electron Transport System
Summary	Sugar synthesized using the energy of the sun	Energy released from the breakdown of sugar

Glycolysis

The word *glycolysis* means the splitting of sugar. It starts with glucose, which arises from the breakdown of polysaccharides, most commonly either starch or glycogen (in animals and fungi), or the conversion from other substances, especially other sugars. The first few steps in glycolysis add energy to the molecule in the form of phosphate groups (fig. 4.13). These phosphorylations are at the expense of two molecules of ATP. In addition, the glucose molecule undergoes a rearrangement that converts it to fructose-1,6-bisphosphate. These steps prime the molecule for the later oxidation. The next step splits fructose-1,6-bisphosphate into glyceraldehyde phosphate and dihydroxyacetone phosphate, but the latter is converted into a second molecule of glyceraldehyde phosphate. Both glyceraldehyde phosphate molecules continue on in the glycolytic pathway so that each of the remaining steps actually occurs twice. Glyceraldehyde phosphate is phosphorylated and oxidized in the next step, which also reduces NAD^+ to $NADH + H^+$. The resulting organic acids, with two phosphate groups, give up both phosphates in the remaining steps of glycolysis, yielding two molecules of pyruvate (pyruvic acid), plus 4 ATP and $2 NADH + H^+$. Note that during Steps 7 and 10 a total of 4 ATP are produced; however, 2 ATP molecules are used during the initial steps, resulting in a net gain of only 2 ATP.

The Krebs Cycle

The remainder of cellular respiration occurs in the mitochondria of the cell (fig. 4.14). Recall from Chapter 2 that mitochondria are organelles with a double membrane. Although the outer membrane is smooth, the inner membrane is invaginated; these folds are referred to as **cristae.** The area between the outer and inner membranes is the intermembrane space; the enzyme-rich area enclosed by the inner membrane is known as the **matrix.** The enzymes in the matrix catalyze each step in the Krebs Cycle.

Once inside the mitochondrial matrix, each of the two pyruvate molecules from glycolysis undergoes several changes before it enters the Krebs Cycle. The molecule is oxidized and decarboxylated, losing a CO_2, with the remaining two-carbon compound joining to coenzyme A to form a complex known as acetyl-CoA. During this step, NAD^+ is also reduced to $NADH + H^+$. Acetyl-CoA enters the Krebs Cycle by combining with a four-carbon organic acid known as oxaloacetate (oxaloacetic acid) to form a six-carbon compound known as citrate (citric acid). (The Krebs Cycle, named in honor of Hans Krebs, who worked out the steps in this pathway in 1937 and later received a Nobel Prize for this work, is alternatively known as the **Citric Acid Cycle.**)

The steps in the cycle consist of a series of reactions during which two more decarboxylations (going from a six-carbon to a five-carbon and then to a four-carbon compound) and several oxidations occur (fig. 4.15). During these steps, three more molecules of NAD^+ are reduced to $NADH + H^+$, a molecule of FAD is reduced to $FADH_2$, and one molecule of ATP is formed. At the end of these steps, the four-carbon oxaloacetate is regenerated, allowing the cycle to begin anew. For each molecule of pyruvate that entered the mitochondrion, three molecules of CO_2 are released and 1 ATP, $4 NADH + H^+$, and 1 $FADH_2$ are produced. Since two pyruvate molecules are formed from each glucose molecule, the cycle turns twice, resulting in 6 CO_2 released and yielding 2 ATP, $8 NADH + H^+$, and 2 $FADH_2$ as energy-rich products. At this point, the entire glucose molecule has been totally degraded; a portion of its energy has been harvested in these Krebs Cycle products as well as the 2 ATP and $2 NADH + H^+$ produced in glycolysis.

The Electron Transport System

The third and final stage of cellular respiration, the Electron Transport System, occurs on the inner membranes of the mitochondria and involves a series of enzymes and coenzymes, including several iron-containing **cytochromes** that are embedded in this layer and function as electron carriers. (This series of electron carriers is similar to the ones described in the light reactions of photosynthesis.) During this stage, electrons and hydrogen ions are passed from the $NADH + H^+$ and $FADH_2$ molecules formed in glycolysis and the Krebs Cycle down a series of redox reactions and are finally accepted by oxygen-forming water in the process (fig. 4.16).

The Electron Transport System is a highly exergonic process and is coupled to the formation of ATP. This method of ATP synthesis is referred to as **oxidative phosphorylation.** When the electron flow begins from NADH produced within the mitochondria, enough energy is available to produce three molecules of ATP from each NADH for a total of 24 ATP from the 8 NADH. Two molecules of ATP are also synthesized during the flow of electrons from each $FADH_2$ produced in the Krebs Cycle (4 ATP) and each NADH from glycolysis (4 ATP). During the Electron Transport System, then, a total of 32 ATP are generated. This number is added to the net yield of 2 ATP from glycolysis and the 2 ATP produced in the Krebs Cycle, for a grand total of 36 ATP for each glucose molecule that completes cellular respiration (table 4.2). These ATP molecules are then transported out of the mitochondria and are available for use within the cell. It should be noted, however, that this production of 36 ATP harnesses only a fraction, 39%, of the original chemical energy of the glucose molecule; the remainder is lost as heat. (Although this 39% efficiency seems low, it is actually much higher than energy conversions in mechanical systems.)

The formation of ATP during the transport of electrons is believed to occur by the same mechanism described for photophosphorylation during photosynthesis. During the transfer of electrons, protons pass from the intermembrane space into the

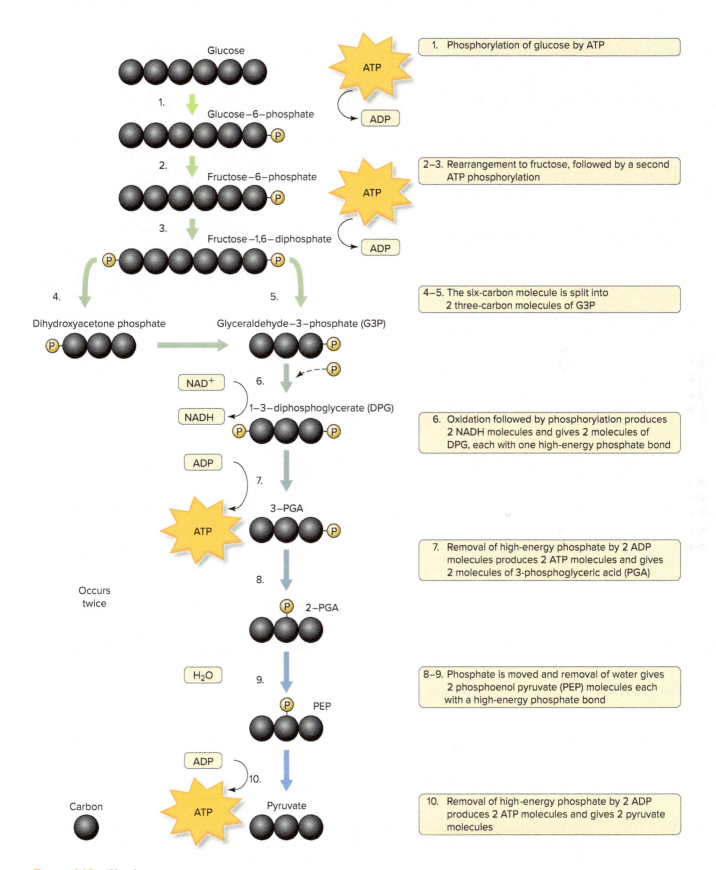

Figure 4.13 Glycolysis.

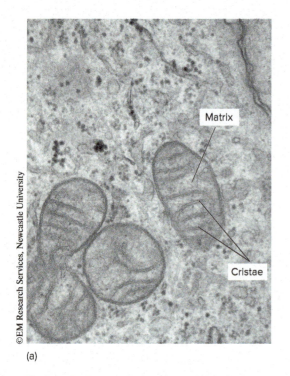
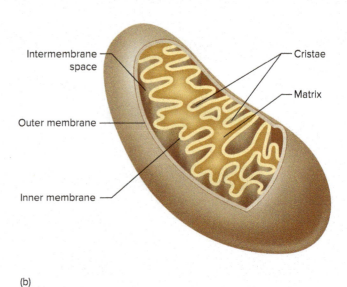

(a)   (b)

Figure 4.14 Mitochondrial structure. The inner mitochondrial membrane has numerous infoldings known as cristae. The enzymes and coenzymes of the Electron Transport System occur on these membranes. The matrix contains the enzymes that carry out the Krebs Cycle. (a) Electron micrograph showing four mitochondria. (b) Cutaway diagram of mitochondrion to show internal organization.

matrix through an ATPase in the membrane (fig. 4.17). It is the passage of protons through this ATP synthase enzyme that somehow drives the production of ATP. This model for ATP synthesis is known as **chemiosmosis** and was first proposed by Peter Mitchell during the early 1960s. Mitchell received a Nobel Prize for this theory, which applies to ATP synthesis in both respiration and photosynthesis.

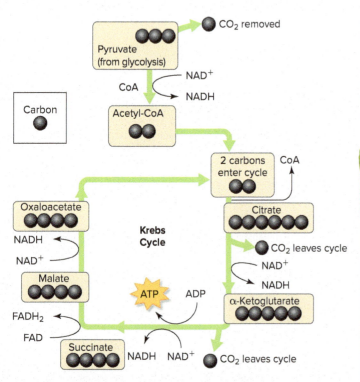

Figure 4.15 Krebs Cycle. Before the cycle begins, pyruvate is converted to acetyl-CoA with the loss of CO_2. The two-carbon acetyl group combines with oxaloacetate to form the six-carbon citrate. Two decarboxylations and several redox reactions regenerate oxaloacetate. For every pyruvate that enters the mitochondrion, 4 NADH, + $FADH_2$, and 1 ATP are produced, and 3 CO_2 are released.

Table 4.2 Tally of ATP Produced from the Breakdown of Glucose during Cellular Respiration

Pathway	Net ATP Yield*
Glycolysis	
2 ATP	2 ATP
2 NADH	4 ATP
Acetyl-CoA formation (2 turns)	
1 NADH × 2	6 ATP
Krebs Cycle (2 turns)	
1 ATP × 2	2 ATP
3 NADH × 2	18 ATP
1 $FADH_2$ × 2	4 ATP
TOTAL	36 ATP

*Each NADH produced in the mitochondrion yields 3 ATP, while each NADH produced during glycolysis has a net yield of 2 ATP. Each $FADH_2$ also yields 2 ATP.

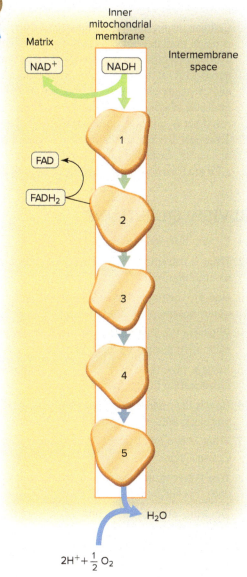

Figure 4.16 Electron Transport System. Electrons from NADH and $FADH_2$ are passed along electron-carrier molecules (numbers 1 to 5), including several cytochromes, and finally are accepted by oxygen. This process drives the formation of ATP by chemiosmosis.

Thinking Critically

All living organisms carry out cellular respiration. In plants, the sugars that are broken down during cellular respiration were produced by the plant during photosynthesis.

What is the immediate source of compounds broken down during cellular respiration in humans? What is the ultimate source of these compounds?

Aerobic vs. Anaerobic Respiration

The complete oxidation of glucose requires the presence of oxygen and is therefore known as aerobic respiration. As indicated, it is the last step of respiration and involves the direct participation of oxygen as the final electron acceptor. Without oxygen, this last step cannot occur, since no other compound can serve as the ultimate electron acceptor. In fact, both the Electron Transport System and the Krebs Cycle are dependent on the availability of oxygen and cannot operate in its absence.

Some organisms, however, have metabolic pathways that allow respiration to proceed in the absence of oxygen. This type of respiration is known as **anaerobic respiration,** or **fermentation,** and is found in some yeast (a unicellular fungus), in bacteria, and even in muscle tissue. The most familiar example is alcoholic fermentation found in certain types of yeast and utilized in the production of beer and wine (see Chapter 24). When oxygen is not available, the yeast cells can switch to a pathway that can convert pyruvic acid to ethanol and CO_2. In the process, NADH is oxidized back to NAD^+, allowing this coenzyme to recycle back to glycolysis. Recycling of the

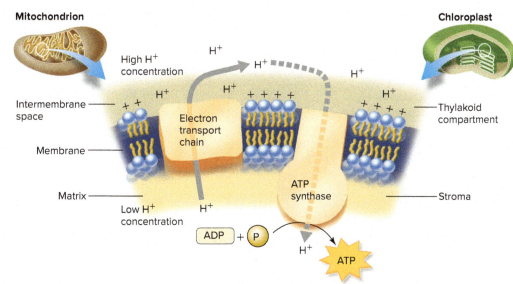

Figure 4.17 Chemiosmosis in mitochondria and chloroplasts. In the mitochondria, protons (H^+) are translocated to the intermembrane space during the transfer of electrons down the Electron Transport System. The proton gradient drives ATP synthesis as protons move through the ATP synthase complex back to the matrix. In the chloroplast, protons are translocated into the thylakoid compartment. As protons move through the ATP synthase complex back to the stroma, ATP is synthesized.

coenzyme allows glycolysis to continue and thus supply the energy needs of the yeast, at least in a limited way. The only energy yield from this alcoholic fermentation is the 2 ATP produced during glycolysis (compared with 36 ATP during aerobic respiration). The still energy-rich alcohol is merely a by-product of the oxidation of NADH. If oxygen becomes available, yeast can switch back to aerobic respiration, with its higher energy yield. Other anaerobic pathways also exist in bacteria and muscle cells, in which the by-products are different from alcohol but the yield of NAD^+ and ATP is similar.

CHAPTER SUMMARY

1. Plants obtain water from the soil, moving it up within the xylem through the entire plant. This movement of water is a passive phenomenon dependent on the pull of transpiration and the cohesion of water molecules. Minerals are also obtained from the soil and transported in the xylem. The translocation of sugars occurs in the phloem, moving from source (photosynthetic leaves or storage organs) to sink (growing organs or developing storage tissue) through mass flow within sieve tube members.

2. Sucrose is the usual sugar transported in the phloem; however, very few plants actually store this economically valuable carbohydrate. Sugarcane and sugar beet are the major sucrose-supplying crops. The early development of sugarcane plantations in North America greatly influenced the course of history by introducing the slave trade to the continent. Today, sugarcane is also an important source of ethanol for biofuels.

3. Plants are dynamic metabolic systems with hundreds of reactions occurring each second to enable plants to live, grow, and respond to their environment. All life processes are driven by energy, with some metabolic reactions being endergonic and others exergonic. Energy transformations occur constantly in biological systems and are part of the underlying principles of both photosynthesis and cellular respiration.

4. Photosynthesis takes place in chloroplasts of green plants and algae and results in the conversion of radiant energy into chemical energy (linking the energy of the sun with life on Earth). Using the raw materials carbon dioxide and water, along with chlorophyll and sunlight, plants are able to manufacture sugars. In the light reactions of photosynthesis, energy from the sun is harnessed, forming molecules of ATP and NADPH. During this process, water molecules are split, releasing oxygen to the atmosphere as a by-product. During the Calvin Cycle, carbon dioxide molecules are fixed and reduced to form sugars, using the energy provided by the ATP and NADPH from the light reactions.

5. Cellular respiration is the means by which stored energy is made available for the energy requirements of the cell. Through respiration, the energy of carbohydrates is transferred to ATP molecules, which are then available for the energy needs of the cell. During aerobic respiration, each molecule of glucose is completely oxidized during the many reactions of glycolysis, the Krebs Cycle, and the Electron Transport System, resulting in the formation of 36 ATP molecules. In anaerobic respiration, only 2 molecules of ATP are formed from each glucose molecule.

REVIEW QUESTIONS

1. Explain how water enters a root.
2. What is transpiration, and how does it affect water movement in plants?
3. How are the properties of water important to the theory of water movement in plants?
4. What are the advantages and disadvantages to having stomata open during the daylight hours?
5. Explain how the Pressure Flow Hypothesis accounts for translocation in the phloem.
6. How is light harnessed during the light reactions of photosynthesis, and what pigments are involved?
7. What is carbon fixation? How is carbon fixed during the Calvin Cycle?
8. Why is glycolysis important to living organisms, and where does it occur?
9. Describe mitochondria and the respiratory events that occur in them.
10. Few plants can survive the saline soils of deserts or coastal areas, where mineral salts, such as sodium chloride, accumulate in extremely high concentrations. Why?
11. If a 500-kg (1,100-lb) pumpkin develops during a 4-month growing season, determine how much photosynthate (products of photosynthesis) is transported into the growing fruit each hour.
12. How are the processes of transpiration and photosynthesis interrelated?
13. In what way is life on Earth dependent on the energy of the sun?

©Estelle Levetin

CHAPTER

5

Plant Life Cycle: Flowers

This purple coneflower (Echinacea purpurea) *is a member of the Asteraceae (sunflower family). What appears as a single flower is actually a type of inflorescence known as a head containing two types of flowers. Ray flowers with pink to purple petals surround the small disk flowers.*

KEY CONCEPTS

1. Angiosperms are unique among plants in that they have their sexual reproductive structures contained in a flower.
2. Meiosis is a type of cell division that reduces the number of chromosomes from the diploid to the haploid number and is an integral part of sexual reproduction.
3. Pollination is the transfer of pollen from the anther to the stigma and largely occurs through the action of wind or animals.
4. In angiosperms, reproduction is accomplished through the process of double fertilization.

CHAPTER OUTLINE

The Flower 70
   Floral Organs 70
A CLOSER LOOK 5.1 Mad about Tulips 71
   Modified Flowers 72
Meiosis 75
   Stages of Meiosis 75
   Meiosis in Flowering Plants 77
A CLOSER LOOK 5.2 Pollen Is More Than Something to Sneeze At 79
   Male Gametophyte Development 77
   Female Gametophyte Development 79
Pollination and Fertilization 79
   Animal Pollination 80
A CLOSER LOOK 5.3 Alluring Scents 82
   Wind Pollination 83
   Double Fertilization 83
Chapter Summary 84
Review Questions 84

The natural beauty of flowers has always been a source of inspiration, and the appearance of the first flower of spring lightens the heart of anyone weary of winter. (See A Closer Look 5.1: Mad about Tulips.) But what role do flowers play in the lives of plants? Their beauty notwithstanding, flowers play a pivotal role in the life cycle of angiosperms, since they are the sites of sexual reproduction. The events leading to flowering are very complex and may include internal factors, such as plant hormones (see A Closer Look 6.1: The Influence of Hormones on Plant Reproductive Cycles) and biological clocks (internal rhythms that regulate the timing of biological functions) as well as external factors, such as temperature and photoperiod (length of light and darkness in a 24-hour period). The interconnection between these internal and external features allows plants to coordinate their reproduction with the environment. This chapter emphasizes the reproductive role of flowers once they have developed.

THE FLOWER

Flowers, unique to angiosperms, are essentially modified branches bearing four sets of specialized appendages or floral organs. These appendages are grouped in whorls and consist of the following from outermost to innermost whorls: sepals, petals, stamens, and carpels. They are inserted into the **receptacle**, the expanded top of the **pedicel** or **peduncle** (flower stalk) (fig. 5.1).

Floral Organs

The outermost whorl consists of the sepals, leafy structures that cover the unopened flower bud; they are usually green and photosynthetic. The whole whorl of sepals of a single flower is called the **calyx.** The petals that make up the next whorl of flower parts are collectively called the **corolla.** Often brightly colored and conspicuous, the petals function by attracting animal pollinators. Together, the calyx and corolla constitute the **perianth.**

In the center of the flower, the male and female structures can be found. The **androecium,** the whorl of male structures, is composed of stamens, each of which consists of a pollen-producing **anther** supported on a stalk, the **filament.** Each anther houses four chambers, where **pollen** develops. The pollen chambers can be seen in the cross section of the anther, Figure 5.1. The **gynoecium** is the collective term for the female structures, or carpels, which are located in the middle of the flower. Flowers can have one to many carpels. (The old term *pistil*, which referred to one or more carpels, will not be used in this book.) A gynoecium with just one carpel is illustrated

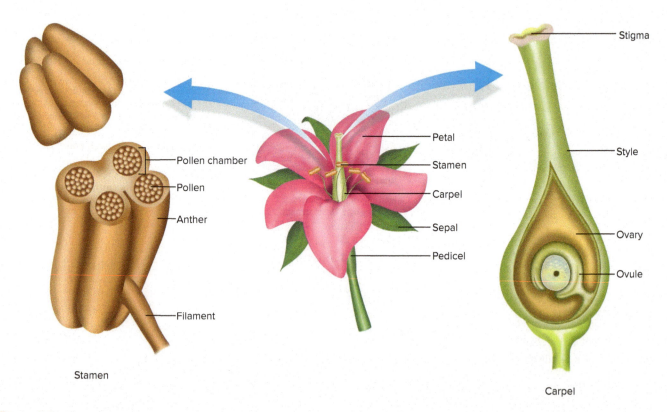

Figure 5.1 Flower structure.

A CLOSER LOOK 5.1

Mad about Tulips

The flower, the crowning characteristic of the angiosperms, technically is a modified branch bearing specialized leaves that are integral to sexual reproduction. But flowers have meaning beyond this technical definition. Few of us could imagine or want a world without flowers. We are attracted to them because of their beauty of color, form, and fragrance. Perhaps we appreciate them all the more because their beauty is so delicate and ephemeral. They have been praised in song and poem and have decorated our homes and persons. Certain flowers are so revered that they become representations of human emotions or nations, yet in a bizarre episode in history, the desire for a beautiful flower created a frenzy that brought a nation and its people to economic ruin.

> To see a World in a Grain of Sand,
> And a Heaven in a Wild Flower,
> Hold Infinity in the palm of your hand,
> And Eternity in an hour.
>
> —William Blake (1757–1827),
> "Auguries of Innocence"

The story begins in 1554 in the Ottoman court of Suleiman the Magnificent in Constantinople. Some years earlier, the Turks had been the first to bring into cultivation a wildflower whose beauty had captivated the court and all who saw it. Ogier Ghislain de Busbecq, the ambassador for the Austrian-Hapsburg Empire, was so entranced by the floral beauties that he had some of the flowers sent back to Vienna. The Turks called the flower *lale*, but Busbecq called it *tulipam*, a corruption of *dulban*, meaning turban. The Ottoman court favored tulips with elongated, pointed petals and, through selected crosses, had achieved this effect. Perhaps Busbecq thought the form of the flower resembled a turban. In any case, the first tulips arrived in Europe in 1554.

Wild tulips (*Tulipa* spp.) originated in Central Asia and the Caucasus Mountains. Tulips are monocots belonging to the lily family, the Liliaceae. There are about 120 species of tulip, and typically a single flower is borne on a stalk. Most tulips have three sepals and three petals apiece (all similarly vibrantly colored), six stamens, and a central gynoecium of three carpels. The shape of the flower and petals varies; some have a narrow, elongated perianth with the flower resembling a star. The familiar tulips of horticulture have rounded, broad sepals and petals with a bowl-shaped flower. The tulip can produce seeds, but it takes 7 years before a tulip grown from seed will flower. A quicker way is to grow the tulips from bulbs. A bulb, such as the familiar onion, is actually an underground stem with fleshy storage leaves (see Chapter 14). As the bulb grows, two to four new bulbs develop within the layers, or skirts, of the original bulb. The bulbs can be divided and planted and will flower within the few months of a single growing season. In addition, since propagation by bulbs is a method of asexual reproduction, the integrity of the flower, its form and color, remains true to the parent plant, whereas tulip flowers grown from seeds, as the products of sexual reproduction, may be quite variable.

In 1593, Carolus Clusius, who had been the director of the Imperial Medicinal Garden in Vienna, was persuaded to go to the new University of Leiden, in the Netherlands, and establish a physick, or medicinal, garden. Naturally, he brought many exotic plants with him, including his collection of tulips. Although the tulip was known in Holland, it was still a rarity and, as such, a status symbol for the wealthy. Clusius was loath to part with any of his bulbs. Under the darkness of night, thieves broke into the garden and stole most of his tulip collection. The thieves lost no time in propagating and selling the tulips.

As tulips became more available, their popularity spread across Europe, but especially in Holland. Part of the fascination was due to a phenomenon called *breaking*. A small number of tulips, perhaps just one or two in 100, were prone to spontaneous, unpredictable eruptions of color, with stripes or feathers of bold color against a contrasting background. Today, the variety of tulips with contrasting perianth colors are called Rembrandts because the Dutch painter painted some of the most famous tulips of the time. We now know that the breaks are caused by the tulip breaking virus spread by the peach-potato aphid (*Myzus persicae*), an insect that sucks the sap of plants. Since peach trees were practically a staple in every Dutch garden, there was a ready source of aphids to spread the disease.

The color of tulips is due to the presence of two types of pigments. The base color of all tulips is either white or yellow. On top of this there are anthocyanin pigments of reds and blues. The base and anthocyanin pigments combine to produce the final color. The tulip-breaking virus causes the suppression of the anthocyanins in infected cells of the perianth; in these cells, the base color *breaks* through in striking contrast with the more vibrant color of the uninfected cells. Although the flower palette may be spectacular, the broken tulips are, in fact, diseased. The virus eventually weakens the bulb, and fewer and fewer offshoots are produced with each generation. Thus, the very nature of breaking ensures rarity.

Since the cause of breaking at the time was unknown and appeared to be unpredictable, many superstitious practices arose in a vain attempt to encourage breaking. One suggested sprinkling pigments on a tulip field. When it rained, the pigments would dissolve and be absorbed by the roots. When the pigments were transported to the flower, its color would be transformed!

Tulips that had a particularly pleasing color pattern with symmetrical breaks were the most desired. One of the most famous broken tulips was *Semper augustus* (box fig. 5.1), with its carmine red feathers against a white background. A single bulb commanded a price of what today would be $4,600!

By 1634, tulips were no longer being bought to grow in a personal garden but for resale at a profit. Tulipomania had begun. People sold their homes and possessions to invest in the tulip trade for a sure profit. Tulip clubs were formed to buy and sell the hottest properties. The Dutch government established a Tulip Notary, which dealt exclusively with the tulip trade. At first, sales took place between the end of the growing season in June and September when the bulbs were ready for replanting. Later, the sales took place throughout the year because people were buying and selling ownership to future bulbs. They bought and sold papers for bulbs that never left the ground.

On February 2, 1637, there were the first signs that the tulip business was souring. Tulips were up for sale but could no longer command the expected price. Sellers got worried and tried to sell their bulbs at lower and lower prices. Eventually, confidence in the market dropped when there were more sellers than buyers, and the prices of bulbs plummeted. The madness for tulips that had inspired wild speculation ended in an economic crisis, but the tulip and the Netherlands are forever linked. A 2017 movie called *Tulip Fever* was set in seventeenth-century Holland during this period of wild speculation. The main theme of the movie was

Box Figure 5.1 *Semper augustus* tulip, the most expensive tulip sold during tulipomania. Artist Johannes Simon Holtzbecher (1610–1671), Statens Museum for Kunst/National Gallery of Denmark.

a love triangle set against the backdrop of the rise and then crash of tulip futures. Although the movie was not a huge box office success, it accurately depicted the hopefulness and disappointment of tulipomania.

in Figure 5.2a. If many carpels are present, they may either be fused together (fig. 5.2b) or remain separate. Carpels, whether individual or fused, consist of a **stigma,** a **style,** and an **ovary** (fig. 5.1). Contained within the basal ovary are one to many **ovules** (structures that will eventually become seeds); rising from the top of the ovary is a slender column called the style. The expanded tip of the style is the stigma, which functions in receiving pollen. Flowers containing all four floral appendages are known as **complete flowers.**

Although flowers have been described in terms of only four floral appendages, some flowers may have additional floral structures called **bracts,** which are found outside the calyx. Bracts may appear leaflike or petal-like and be of various sizes. The showy red "petals" of poinsettia are actually bracts.

In Chapter 3, the vegetative differences between monocots and dicots (eudicots, see Chapter 8) were described; there are also easily recognizable differences in the floral structures (fig. 5.3). Monocots generally have their floral parts in threes or multiples of three; for example, lilies have three sepals, three petals, six stamens, and a three-part ovary (formed from the fusion of three carpels). On the other hand, dicots (eudicots) generally have a numerical plan of four or five or multiples; a wild geranium flower contains five sepals, five petals, ten stamens, and five fused carpels with separate stigmas.

Modified Flowers

The basic pattern of flower structure is often modified. In flowers such as tulips and lilies, the sepals are brightly colored and identical to the petals. In such flowers, the petals and sepals are often referred to as **tepals.** Other modifications of the perianth occur when the petals are partly or completely fused together forming a corolla tube. Whole or partial fusion of the sepals may also occur, forming a cuplike structure

Figure 5.2 Examples of gynoecia. (a) Gynoecium composed of a single carpel. (b) Gynoecium composed of three carpels. Multiple ovules are visible in the carpels of each gynoecium.

Figure 5.3 Monocot and dicot flowers. (a) Coast rose gentian, *Sabatia arenicola*, illustrates a dicot (eudicot) flower with flower parts in multiples of five. (b) Tiger lily, *Lilium lancifolium*, a monocot, shows flower parts in multiples of three.

around the petals called a calyx tube. In contrast, flowers of the grasses possess neither sepals nor petals; these flowers are incomplete (fig. 5.4a). In fact, flowers lacking any of the four floral structures are known as **incomplete flowers.** Flowers with both stamens and carpels are called **perfect flowers** even if sepals or petals are lacking. Some flowers, such as squash and holly, are unisexual; they are either **staminate** or **carpellate.** Incomplete flowers lacking either stamens or carpels are **imperfect.** A single plant may have both staminate and carpellate flowers; this plant is said to be **monoecious.** Alternatively, **dioecious** plants have only unisexual flowers on a single individual. Corn, squash, and pecans are familiar monoecious species (see fig. 5.10), while spinach, date palms, and some hollies are dioecious.

One feature that is important in the classification of flowers is the position of the ovary in relation to the other floral parts. If the sepals, petals, and stamens are inserted beneath the ovary, this arrangement is referred to as a **superior ovary.** The ovary is **inferior** if the sepals, petals, and stamens are inserted above it. Corresponding terms that refer to these arrangements are **hypogynous, epigynous,** and **perigynous.** Hypogynous (below the gynoecium) flowers, such as buttercup, have flower parts inserted beneath a superior ovary; epigynous (on the gynoecium) flowers, such as apple, have flower parts inserted above an inferior ovary; and perigynous (around the gynoecium) flowers, such as cherry, have the bases of the flower parts fused into a cuplike structure surrounding a superior ovary (fig. 5.4b).

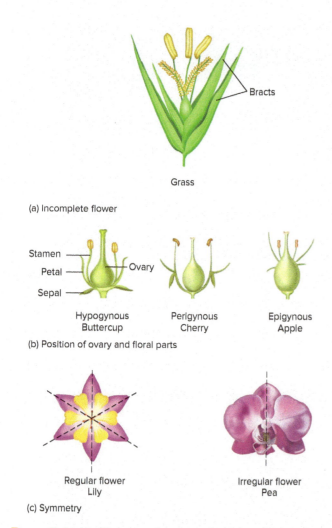

Figure 5.5 *Oncidium* sp. orchid flower.

Figure 5.4 Modifications of the basic floral design result in diverse flower types. (a) Incomplete flowers lack one or more of the four floral organs. Grass flowers lack both sepals and petals. (b) Various positions of the floral whorls in relation to the ovary are possible. (c) Regular flowers can be bisected along many planes, but irregular flowers can be bisected along only one.

Flowers can also be described by their pattern of symmetry. **Regular flowers (actinomorphic),** such as the tulip, lily, rose, and daffodil, display radial symmetry; they can be dissected into mirror-image halves along many lines. **Irregular flowers (zygomorphic),** such as the orchid, iris, snapdragon, and pea, display bilateral symmetry. They can be dissected into mirror-image halves along only one line (fig. 5.4c).

In addition to being irregular, orchid flowers have some fascinating modifications. Orchids typically have three sepals and three petals; the sepals may be colored the same as the petals (fig 5.5). Two petals occur on either side of the flower, while the third petal, called a lip or labellum, is found at the base of the flower. This third petal is often highly modified and showy; it is the most diverse and distinctive part of orchids, functioning to attract pollinators. The lip may be elaborately adorned with frills, fringes, or other ornamentation and may have distinctive color patterns that are quite different from the other petals. The reproductive portions of orchids are also modified. Opposite the lip is the column that develops from the fusion of the stamens to the stigma and style of the carpel. In most orchids there is only a single stamen and the anther appears as an anther cap at the top of the column. Pollen grains produced by the anther are held together by sticky material and packaged into one or two pollinia (sing., pollinium). The sticky pollinia are easily transferred to the body of visiting insects.

Some flowers (termed solitary flowers) are borne singly on a stalk, but in many cases, flowers are grouped in clusters called **inflorescences.** Sometimes what is commonly called a single flower is actually an inflorescence, as in the case of sunflowers, daisies, and chrysanthemums. The dogwood flower is also an inflorescence, but here the pink or white "petals" are bracts surrounding a cluster of small flowers. The arrangement of flowers in the cluster determines the type of inflorescence, with many patterns possible: **spike, raceme, panicle, umbel, head,** and **catkin** (fig. 5.6a-g). Often the type of inflorescence is an important characteristic in classification. An unusual type of inflorescence is found in *Ficus* (the fig genus). The fig inflorescence is called a syconium, which has an urn-shaped receptacle bearing numerous (up to several thousand) small unisexual flowers on its inner surface (fig. 5.6h). As a result, none of the flowers are visible from outside. In some *Ficus* species, both carpellate (female) and staminate (male) flowers occur in the same syconium. Each syconium has a small opening known as an ostiole at the apex. Pollination is accomplished by tiny fig wasps, which enter through the ostiole. A specialized species of wasp carries out the pollen transfer for each species of *Ficus*. This relationship between the *Ficus* species and its wasp pollinator is an excellent example of coevolution (see Animal Pollination below).

To understand the flower's role in sexual reproduction, it is necessary to learn about a special form of cell division, **meiosis,** that occurs within stamens and carpels.

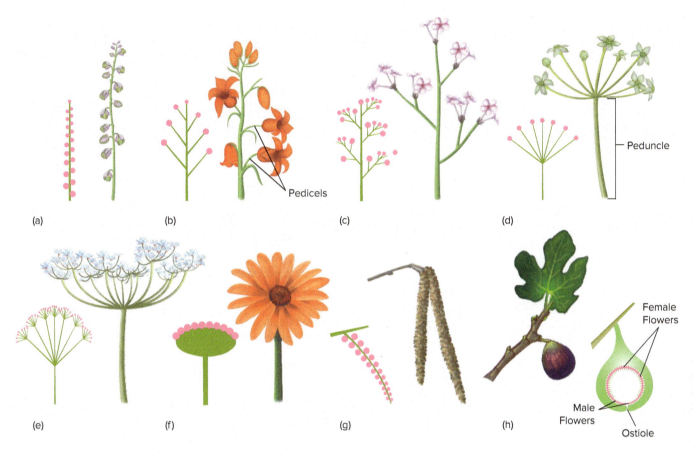

Figure 5.6 Inflorescence types: (a) spike, (b) raceme, (c) panicle, (d) umbel, (e) compound umbel, (f) head often with small disk flowers in the center and showy ray flowers at the margin, (g) catkin, which is a unisexual inflorescence, and (h) syconium.

MEIOSIS

Sexual reproduction, whether in a plant or an animal, is basically the fusion of male and female **gametes**, sperm and egg, to produce a **zygote**, which will develop into a new individual. When the egg is fertilized by the sperm, the zygote receives an equal number of chromosomes from each gamete. Gametes are different from most other cells in angiosperms because they are **haploid** (containing only one set of chromosomes), whereas other body cells are **diploid** (containing two complete sets of chromosomes). During the process of fertilization, the diploid number is restored in the zygote. When the chromosomes in a diploid cell are examined microscopically, it can be seen that there are two of each kind of chromosome. These pairs of chromosomes are known as **homologous chromosomes;** the members of a pair are derived from the contributing haploid gametes. The homologous chromosomes not only look alike but also carry genes for the same traits.

Meiosis has a major role in all sexually reproducing organisms because it is the process that reduces the number of chromosomes from the diploid to the haploid number. This reduction compensates for the doubling that occurs during fertilization. Without meiosis, the number of chromosomes would double with each generation.

In animals, gametes are produced directly by meiosis; however, in plants the products of meiosis are haploid **spores.** Spores are reproductive units formed in a sporangium. The diploid plant that undergoes meiosis to form these spores is known as a **sporophyte.** Spores develop into haploid **gametophytes** that produce gametes, egg or sperm (fig. 5.7). The sporophyte and gametophyte are the two stages in the life cycle of each plant. (Later in this chapter, the gametophytes of flowering plants will be studied.) The process of fertilization brings together egg and sperm to produce a genetically unique zygote. Sexual reproduction in this way introduces variation into a population, whereas offspring produced by asexual reproduction, such as cuttings or bulbs, are genetic clones.

Stages of Meiosis

Meiosis is a specialized form of cell division that consists of two consecutive divisions and results in the formation of four haploid cells (fig. 5.8). Both the first and second divisions of meiosis are divided into four stages: prophase, metaphase, anaphase, and telophase. Recall that these are the same names used for the stages of mitosis.

During the first meiotic division, the chromosome number is halved; in fact, it is often called the reduction division. The most significant events occur during prophase I.

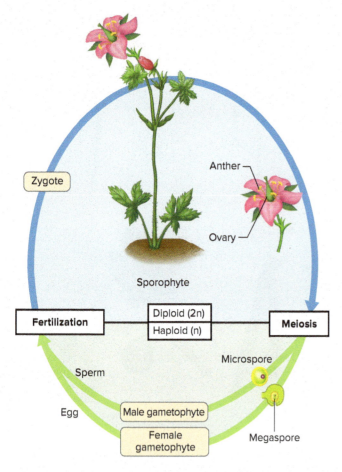

Figure 5.7 Alternation of diploid (sporophyte) and haploid (gametophyte) generations in a flowering plant.

Metaphase I

During the next stage, metaphase I (fig. 5.8), the homologous chromosome pairs line up at the equatorial plane (across the center of the cell). Spindle fibers that actually begin to appear in late prophase attach to the centromeres of each homologous pair. Two types of spindle fibers occur: those that run from pole to pole and those that run from one pole to a centromere. Recall that spindle fibers are composed of microtubules.

Anaphase I

Homologous chromosomes separate during anaphase I (fig. 5.8); they are pulled by the spindle fibers to opposite poles of the cell. During metaphase I, the orientation of the homologous chromosome pairs occurred by chance. As a result, chromosomes from each parent are mixed randomly into a number of possible combinations during this separation process. In contrast to the events in mitotic anaphase, in meiotic anaphase the chromatids of each chromosome are still united; it is only the homologous pairs that are separating. By the end of this stage, the chromosome number has been halved.

Telophase I

Telophase I (fig. 5.8) is similar to telophase of mitosis in that the spindle disappears, the chromosomes become less distinct, and the nuclear membrane may re-form. Cytokinesis generally follows, dividing the cell into two daughter cells, each with half the number of chromosomes of the original parent cell.

Second Meiotic Division

In some organisms, an interphase occurs between the two meiotic divisions; in other cases, the cells proceed directly from telophase I to prophase II. The second meiotic division is essentially similar to mitosis; the chromatids, which are still joined together, finally separate. Prophase II is identical to mitotic prophase; in each cell, the chromosomes become evident and the nuclear membrane breaks down (fig. 5.8). During metaphase II, the chromosomes line up at the equatorial plane of each cell and spindles appear, with the spindle fibers stretching from pole to pole and pole to centromere. During anaphase II, the chromatids separate, pulled to the poles by the spindle fibers. Cytokinesis occurs in telophase II, and the nuclear membranes and nucleoli reappear as the single-stranded chromosomes become threadlike chromatin. By the end of telophase II, four haploid cells are produced. Because of crossing over and the random associations of parental chromosomes, the four cells contain unique genetic combinations that differ from each other and the parent cell from which they originated. This result contrasts with the process of mitosis, in which the two daughter cells are genetically identical to the parent cell (see Chapter 2, Cell Division).

Prophase I

At the beginning of prophase I, the chromosomes appear threadlike. As in mitosis, the DNA is duplicated during the S phase of the preceding interphase, so that each chromosome actually consists of two chromatids. As the chromosomes continue to condense and coil, the homologous chromosomes pair up gene for gene in a process called **synapsis**. Since each chromosome is doubled, the synapsed homologous chromosomes actually consist of four chromatids. As the synapsed chromosomes continue to condense, breaks and exchanges of genetic material can occur between the chromatids in an event called **crossing over**. This results in chromatids that are complete but have new genetic combinations. Soon synapsis starts to break down, and the homologous chromosomes repel each other; however, they are held together at points where crossing over occurred. These places are referred to as **chiasmata** (sing., **chiasma**). While these chromosome events are occurring, the nucleolus and nuclear membrane break down, leaving the chromosomes free in the cytoplasm. Prophase I is the longest and most complex stage of meiosis (figs. 5.8 and 7.6).

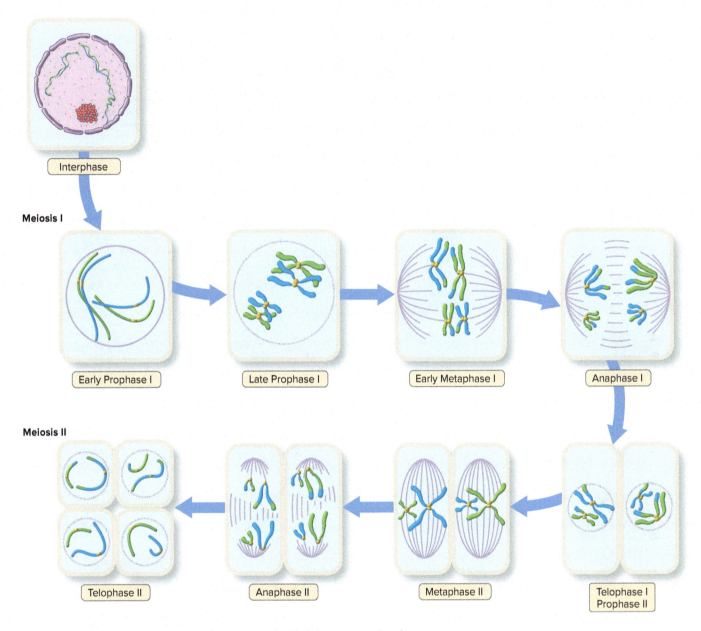

Figure 5.8 Stages of meiosis. See the text for a detailed description of each stage.

Meiosis in Flowering Plants

Within the flower, meiosis occurs during the formation of pollen in the anthers of the stamen and the formation of ovules within the ovary of the carpel (fig. 5.9).

Male Gametophyte Development

During the development of the stamen, certain cells in the pollen chambers of the anther become distinct as **microspore mother cells;** in fact, the pollen chambers are technically referred to as **microsporangia** (sing., **microsporangium**). Each microspore mother cell undergoes meiosis to produce four **microspores** (male spores). Initially, the four spores stay together as a **tetrad,** but eventually they separate and each will develop into a pollen grain, an immature male gametophyte. In the development of the pollen grain, the microspore undergoes a mitotic division to produce two cells, a small **generative cell** and a large **vegetative cell** (or tube cell). Also, the wall of the microspore becomes chemically and structurally modified into the pollen wall. The pollen wall consists of an inner layer, the **intine,** and an outer layer, the **exine,** which may be ornamented with spines, ridges, or pores. When the pollen grains are fully developed, they are released as the anthers open, or dehisce. (See A Closer Look 5.2: Pollen Is More Than Something to Sneeze At for further discussion on pollen.)

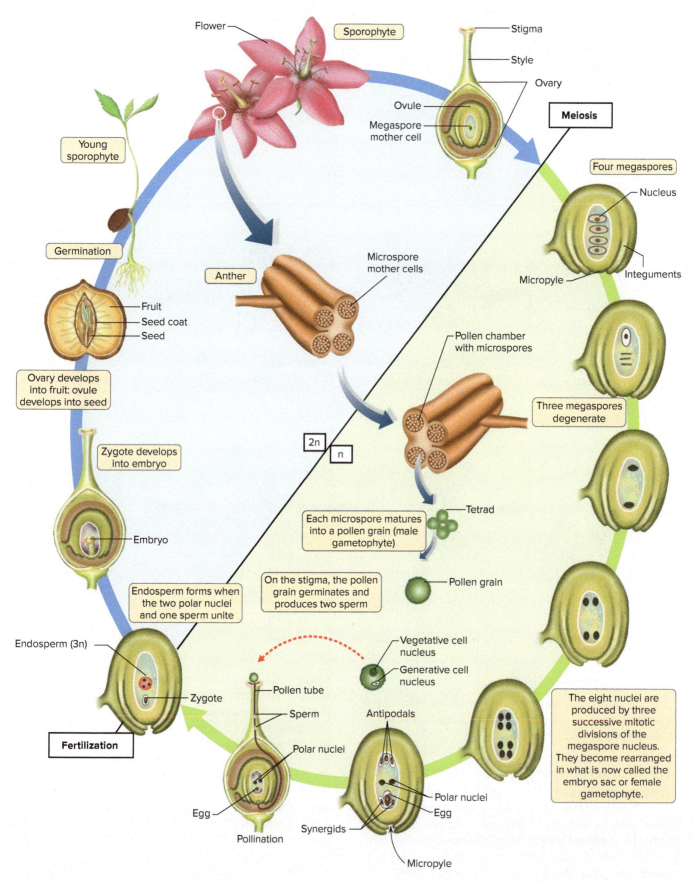

Figure 5.9 Reproductive cycle of an angiosperm.

A CLOSER LOOK 5.2

Pollen Is More Than Something to Sneeze At

The essential role that pollen plays in the life cycle of seed plants is well documented. Less well known is the significance that palynology (the study of pollen) has had in many diverse fields: petroleum geology, archeology, criminology, anthropology, aerobiology, and the study of allergy.

When pollen is released by wind-pollinated plants (either gymnosperms or angiosperms), only a tiny percentage reaches the receptive female organ. At the proper season, pollen is so abundant that clouds of it can be seen emanating from vegetation disturbed by wind or shaking (box fig. 5.2a). Most of it is carried by the wind and eventually settles back to the ground. It is this excess pollen that is the focus of study.

The distinctive ornamentation on the outer wall of a pollen grain allows for the identification of most types of pollen to the genus level (box figs. 5.2b and c). Under certain conditions, pollen can be preserved, leaving a record of area vegetation. Some fossil pollen dates back over 200 million years and has revealed information about the changing vegetation patterns over evolutionary time.

Palynology is essential to the petroleum industry; an examination of fossil pollen from core samples can determine if an area is likely to be a rich source of oil. Certain fossil species in a particular region are known to be associated with oil deposits; palynologists look for the pollen of these indicator species in core samples.

Archeologists have sought the help of palynologists in determining when agriculture originated in certain areas and what plants were consumed by ancient peoples. Examination of the fossil pollen can pinpoint the shift from gathering native vegetation to cultivating cereal grains. Pollen residues found in storage vessels and coprolites (fossilized feces) give direct evidence of the diet of prehistoric groups; for example, the Viking recipe for mead was determined by examining pollen scrapings in drinking horns.

Pollen has proved instrumental in solving many criminal cases. The scene of a crime or the whereabouts of a suspect at the time of the crime can often be determined by analyzing pollen clinging to the victim's body or to the shoes and clothing of a suspect.

Box Figure 5.2a A cloud of pollen can be seen wafting from the pollen cones when a cedar branch is disturbed.

Female Gametophyte Development

Within the ovary, one or more ovules develop; an ovule consists of a **megasporangium** enveloped by one or two layers of tissue called **integuments.** The integuments completely surround the megasporangium except for an opening called the **micropyle.** The ovule first appears as a bulge in the ovary wall. During the development of the ovule, one cell becomes distinct as a **megaspore mother cell;** it is surrounded by tissue called the **nucellus.** The megaspore mother cell undergoes meiosis to produce four **megaspores;** generally, three of these degenerate, leaving one surviving megaspore. This megaspore then undergoes a series of mitotic divisions, eventually producing a mature female gametophyte, which is often called the **embryo sac.** In the typical pattern of development, a series of three mitotic divisions produces eight nuclei within the greatly enlarged megaspore. These eight nuclei are distributed with three (the **egg apparatus**) near the micropyle end of the ovule, three **antipodals** at the opposite end, and two **polar nuclei** in the center. The egg apparatus consists of two **synergids** and one **egg**. Cell walls soon develop around the egg, synergids, and antipodals; at this stage, the female gametophyte is mature (fig. 5.9).

POLLINATION AND FERTILIZATION

Pollination involves the transfer of pollen from the anther to the stigma. Pollen transfer within the same individual plant is known as **self-pollination**. **Cross-pollination** involves the transfer of pollen from one plant to another. It is obvious that cross-pollination prevents the potentially detrimental effects of inbreeding, and most perfect flowers have physiological mechanisms to prevent self-fertilization. Pollination can be accomplished by various methods. Large, showy flowers usually attract animal pollinators, whereas small, inconspicuous flowers are often wind-pollinated.

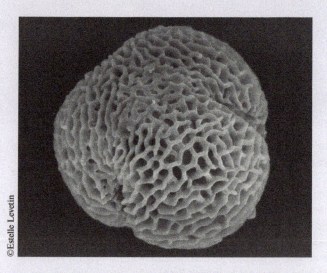

Box Figure 5.2b *Geranium* (geranium) pollen grain with netlike pattern of ornamentation on the exine.

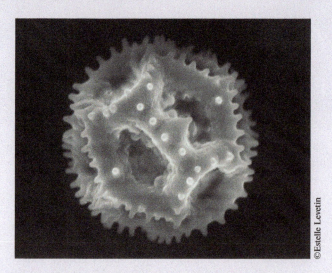

Box Figure 5.2c *Taraxicum* (dandelion) pollen grain with elaborate ornamentation consisting of spine-topped ridges that divide the exine into angular lacunae (depressions).

Anthropologists have learned that pollen has symbolic meaning to several Native American tribes in the Southwest. Among the Navajos, pollen is revered as a symbol of life and fulfillment; it is used in sacred ceremonies and chants throughout the stages of life from birth to death. The mystique of pollen has even been adopted by current health food faddists who claim that bee pollen (pollen collected by bees) is a power food that cures ailments, prevents disease, and promotes fitness. A few athletes take daily bee pollen supplements to maintain a winning edge, but most nutritionists discount these claims and even express concern about allergic reactions.

Airborne pollen is well known to trigger hay fever, asthma, and other allergic reactions in sensitized individuals. Despite its name, hay fever is not caused by hay but is due to pollen from inconspicuous flowers of wind-pollinated trees, grasses, and weeds. This pollen is responsible for the misery of at least the 20% of the U.S. population identified as allergy sufferers. Aerobiologists study airborne pollen to document the species responsible and the factors influencing pollen abundance and distribution. Their findings suggest that there should be greater care in the selection of landscaping plants, so that hay fever-causing plants are avoided. Allergies and their causes will be discussed further in Chapter 21.

Animal Pollination

Although bees are the most familiar animal pollinators, a host of species are also involved in the transfer of pollen (figs. 5.10a and b). Other insects, such as wasps, flies, ants, butterflies, and moths, are equally important pollinators for many flowers. Even larger animals, such as birds and bats, are efficient pollinators for some species. Pollination is accomplished inadvertently when the animal visitor, dusted with pollen from one flower, visits a second flower of the same species.

Color and scent are what attract animals to flowers. Certain colors are associated with specific pollinators; for example, bee-pollinated flowers are often yellow, blue, purple, or some combination of those colors. Many bird-pollinated flowers, such as columbine and trumpet creeper, are red. In addition, many white or light-colored flowers are pollinated in the evening by night-flying visitors. Various contrasting color patterns (**nectar guides**) seen on petals direct insects toward the nectar. Often, nectar guides cannot be seen by the human eye but are visible in ultraviolet light, which can be perceived by certain insects. To the insect eye, the nectar guides seem like airport lights lining a runway. **Essential oils,** volatile oils that impart a fragrance, attract pollinators by scent. The essential oils of flowers such as the rose, orange, and jasmine have been used in perfumes for hundreds of years (see A Closer Look 5.3: Alluring Scents). Not all scents are appealing to humans; for example, the carrion flower (*Stapelia* spp.), which is fly-pollinated, gives off an aroma of rotting meat.

In most cases, the flower provides a reward of **nectar,** pollen, or both to the animal. Nectar is a sugary liquid produced in glands, called **nectaries,** found in the epidermis of a floral organ. Many children have tasted its sweetness when they sipped the nectar of honeysuckle (*Lonicera japonica*) blossoms. The amount of nectar produced by flowers varies greatly; flowers pollinated by birds generally produce copious amounts of nectar.

Flowering plants and their animal pollinators are a classic example of **coevolution.** Coevolution is a case of reciprocal adaptations as two interacting species modify and adjust to each other over time. Adaptations occur that make a flower more attractive to a specific type of pollinator, thus ensuring

(a)

(b)

Figure 5.10 Flowers and their animal pollinators. (a) Butterfly-pollinated flowers have a broad expanse for the butterfly to land. (b) Honey bees are important pollinators of many crops. This bee is visiting a carpellate watermelon flower.

a greater chance of a successful pollination. The pollinator, in turn, changes in ways that enhance its efficiency in exploiting the nutritional rewards offered by the flower. As the evolutionist Charles Darwin (see Chapter 8) observed,

> Thus, I can understand how a flower and a bee might slowly become either simultaneously or one after the other modified and adapted in the most perfect manner to each other.

Honey Bees and Colony Collapse Disorder

Honey bees (*Apis mellifera*) are pollinators of flowers for many important crops including almonds, apples, apricots, beans, cherries, pumpkins, pears, peaches, plums, and squashes. Approximately 25% of the U.S. diet comes from crops pollinated by honey bees, and it is estimated that the global value of honey bee-pollinated crops is more than $200 billion each year. In large commercial orchards and farms growing these crops, farmers rely on beekeepers to provide bee colonies for pollination. Although the number of honey bees needed varies with the crop, generally one hive per acre is adequate. As a result, commercial beekeeping for pollination is big business, and hives are routinely moved cross-country to provide pollination services. The California almond industry best illustrates the use of honey bees. California is the major almond-growing region in the world and produces over 80% of the world's almonds at an annual value of $5.3 billion. When almond trees are in flower, approximately 1.9 million beehives are brought in for pollination. In fact, close to 75% of the bee colonies in the United States are used for California almond pollination.

Beginning in 2006, beekeepers in the United States started reporting significant losses of honey bees in their hives. Although some loss of bees during winter is expected, these losses were especially severe. Often just a queen bee and a few workers have been found in a hive. Losses of 30% to 50% have been reported by commercial beekeepers. This massive die-off is called **colony collapse disorder,** and the cause is still uncertain. Similar but less severe losses have been reported in many European countries. Among the suggested causes are pests such as *Varroa* mites as well as *Nosema* parasites and at least two viruses. Others feel that the cause may be related to the combination of insecticides, herbicides, and fungicides the bees are exposed to. Although the evidence is not conclusive, one class of insecticides, the neonicotinoids, has been suggested as contributing to colony collapse disorder; however, other studies have suggested that the mixture of multiple chemicals may be to blame. Colony collapse disorder has an enormous economic impact. Not only are the beekeepers hurt financially, but fewer bees could mean the growers have a smaller harvest. This translates into higher prices for all consumers.

A CLOSER LOOK 5.3
Alluring Scents

Since earliest times, the fragrances of certain plants, owing to their essential oils, have been valued as a source of perfumes. It is difficult to pinpoint when people first began using plant fragrances to scent their bodies, but 5,000 years ago the Egyptians were skilled perfumers, producing fragrant oils that were used by both men and women to anoint their hair and bodies. Fragrances were also used as incense to fumigate homes and temples in the belief that these aromas could ward off evil and disease. In fact, our very word *perfume* comes from the Latin *per* meaning through and *fumus* meaning smoke, possibly referring to an early use of perfumes as incense.

Today most perfumes are a mixture of several hundred scents that are carefully blended, using formulas that are highly guarded secrets (table 5.A). Many of these scents are now synthetics that resemble the natural essences from plants, but many costly perfumes still rely on the natural essential oils extracted directly from plants.

Various methods are used to extract the essential oils from plant organs, including distillation, solvent extraction, expression, and enfleurage. The method used depends to a large extent on the location and chemical properties of the essential oil. Distillation, one of the most common methods, exposes the tissue to boiling water, or steam, thereby volatilizing the essential oil, which can then be separated from the condensate (box fig. 5.3). Since this method employs heat, only the most stable essential oils can be extracted by distillation. For solvent extraction, plant material is immersed in an organic solvent at room temperature; later, the essential oil is recovered from the solution. At no time is heat used, thus avoiding damage to temperature-sensitive essential oils. Expression is the simplest method and is mainly used to express the oil from citrus rinds with mechanical pressure. In enfleurage, flower petals are layered on trays containing cold fat that absorbs the essential oils from the blossoms. The petals are continually replaced until the fat is saturated with the floral essence. The essential oil is then extracted from the fat with alcohol. Enfleurage is slow and labor intensive and is used to extract only those delicate essential oils that would be destroyed by other methods. Regardless of the method used, tremendous quantities of plant material are needed to produce even small quantities of the pure essential oil; for example, 60,000 roses are needed for 1 ounce (28 grams) of rose oil.

Once the essential oils have been extracted and blended for the characteristic fragrance of a particular perfume, fixatives are added to retard the evaporation of highly volatile essential oils. The fixatives may be plant or animal oils, such as musk oil from the musk deer. Today, however, most animal oils have been replaced by synthetics.

The final perfume concentrate is then diluted with alcohol and a small amount of distilled water: perfumes generally contain 18%–25% concentrate; eau de parfum, 10%–15%; eau de cologne, 5%–8%; and eau de toilette, 2%–4%.

Table 5.A Commonly Used Plant Materials for Essential Oil Extraction in the Perfume Industry

Plant Organ	Source
Flowers	Roses, carnations, orange blossoms, ylang-ylang, violets, lavender
Leaves and stems	Mints, rosemary, geranium, citronella, lemon grass
Seeds and fruits	Oranges, lemons, nutmeg
Roots	Sassafras
Rhizomes	Ginger
Bark	Cinnamon, cassia
Wood	Cedar, sandalwood, pine
Gums	Balsam, myrrh

Box Figure 5.3 Rose petals undergo distillation to extract rose oil, one of the perfume industry's most valued scents.

While research is still underway on the cause of colony collapse disorder, other researchers are looking at alternatives to honey bees for use as pollinators. Suggestions include bumblebees and mason bees. In particular, the blue orchard bee (*Osmia lignaria*), which is a type of mason bee, has been proposed as a possible substitute or supplementary pollinator. The blue orchard bee is well known as an efficient pollinator of fruit trees including apples, pears, plums, peaches, cherries, and almonds. Unlike honey bees, blue orchard bees are solitary bees; there is no hive, no queen, and no workers. Each female establishes her own nest and tends to her own progeny. Nests are built in natural holes in tree trunks, reeds, and bamboo as well as in artificial structures such as tubes or wooden blocks with holes provided by growers. For managed pollination, artificial nests can be placed in orchards in the spring, when the next generation of bees are about to emerge from the nests. The interest in blue orchard bees among commercial fruit growers and even home gardeners is expanding rapidly. Hopefully, blue orchard bees will become useful alternative pollinators.

Figure 5.11 Staminate catkins of pecan, *Carya illinoensis*, a wind-pollinated species.

Thinking Critically

The coevolution of flowers and animal pollinators is one of the marvels of nature and can greatly enhance the rate of successful pollinations. Few flowers, however, are so specialized that they can be pollinated by only one type of animal.

Why would extreme specialization between pollinator and flower be both an advantage and a drawback?

Wind Pollination

As described, animal-pollinated flowers have a variety of mechanisms used to attract the pollinator; wind-pollinated flowers, on the other hand, have a much simpler structure. Color, nectar, and fragrance, which play an integral role in animal pollination, are usually not prominent in the wind-pollinated flower.

Wind-pollinated flowers are often small and inconspicuous, usually lacking petals and sometimes even sepals; these drab flowers are frequently arranged in inflorescences such as the catkins of oak, pecan (fig. 5.11), and willow and the panicles, racemes, and spikes of the grasses. Although most grasses have perfect flowers, many other wind-pollinated species are imperfect. Stamens and stigmas are also modified for this method of pollination. The filaments are usually long, allowing the anthers to hang freely away from the rest of the flower, thereby enabling the pollen to be caught by the wind. Stigmas are often feathery, increasing the surface area for trapping pollen.

Although individual flowers are small, their small size is offset by the large number of flowers formed and by the production of copious amounts of dry, lightweight pollen. A single stamen of corn contains between 2,000 and 2,500 pollen grains, with the whole plant producing about 14 million pollen grains. One wind-pollinated plant whose pollen causes misery to hay fever sufferers is ragweed (*Ambrosia* spp.); one healthy ragweed plant can release 1 billion pollen grains. It is estimated that 1 million tons of ragweed pollen are produced in the United States each year (see Chapter 21)!

Thinking Critically

In many perfect flowers, the stamens and carpels mature at different times. For instance, the anthers of a flower may release pollen while the stigma is still immature and unreceptive, and by the time the stigma is receptive, the anthers have released all of their pollen and are empty.

What is the advantage of this adaptation?

Double Fertilization

Once pollination has been accomplished, the stage is set for fertilization. Recall that the pollen grain at the time of pollination contains a tube cell and a generative cell. On a compatible stigma, the pollen grain germinates; a **pollen tube** begins growing down into the style toward the ovary. The vegetative nucleus is generally found at the growing end of the pollen tube while behind it the generative nucleus divides mitotically, producing two nonmotile sperm. The pollen tube continues to grow until it reaches and grows into the micropyle of an ovule, penetrating the ovule at one synergid. The vegetative nucleus, synergids, and antipodals usually degenerate during the fertilization process, leaving the two sperm, egg, and polar nuclei as the remaining participants (fig. 5.9).

Both sperm are involved in fertilization. One sperm fertilizes the egg to produce a zygote that will develop into an **embryo.** The zygote produced from the fusion of haploid egg and sperm is diploid; this restores the chromosome number for the sporophyte generation. The second sperm fertilizes or fuses with the two polar nuclei, producing the **primary endosperm nucleus,** which develops into **endosperm,** a nutritive tissue for the developing embryo. The fusion of the haploid sperm and polar nuclei normally produces **triploid** endosperm. Endosperm development generally begins immediately, followed by division of the zygote to produce the embryo. This **double fertilization** is a distinctive feature of angiosperm reproduction.

The value of endosperm as a food source for the human population cannot be overemphasized. The nutritive value of wheat, rice, and corn, the world's major crops, is due to the large endosperm reserves in these grains. The development of early civilizations in various parts of the world is linked to the cultivation of these grains, which provided stable food sources (see Chapters 11 and 12).

After fertilization, changes begin to occur within the whole flower. Sepals, petals, and stamens often wither and drop off as the ovary greatly expands, becoming a **fruit.** Within the ovary, each fertilized ovule becomes a **seed** containing an embryo and nutritive tissue; the integuments of the ovule develop into the **seed coat,** the outer covering of the seed. Occasionally, fruit forms without fertilization. This process is referred to as **parthenocarpy** and, understandably, results in seedless fruit. Parthenocarpy occurs naturally in certain fruits, such as bananas (see Chapter 14) and navel oranges (see Chapter 6). Hormone applications can induce artificial parthenocarpy in other fruits. (See A Closer Look 6.1: The Influence of Hormones on Plant Reproductive Cycles.) The discussion of fruits and seeds continues in Chapter 6.

CHAPTER SUMMARY

1. Flowers, the characteristic reproductive structures of angiosperms, are composed of sepals, petals, stamens, and carpels. Modifications of the basic floral organs are common, often resulting in incomplete and imperfect flowers.

2. Meiosis is a form of cell division that reduces the number of chromosomes from diploid to haploid. The process consists of two consecutive divisions, with the reduction in chromosome number occurring in the first division. The most significant events of meiosis occur in prophase I, when synapsis occurs, and anaphase I, when the homologous chromosome pairs separate.

3. In angiosperms, meiosis occurs before the formation of male and female gametophytes, which are small and relatively short-lived. Ovules, which include the female gametophytes, develop within the carpels; the pollen grains, or male gametophytes, develop in the stamen.

4. Pollen is transferred passively by animals or wind from stamen to stigma. Insect-pollinated flowers typically have bright, showy petals and fragrant aromas and are rich in nectar. Pollen in these flowers is often sticky, adhering to the insect body. Wind-pollinated flowers are usually small and inconspicuous but produce copious amounts of dry, lightweight pollen. Only a small amount of pollen from wind-pollinated plants reaches the female organ. Most pollen grains settle to the ground, where they can leave a lasting record in the sediment.

5. Before fertilization, the pollen tube grows down the style into the ovary and ovule. The generative nucleus gives rise to two sperm. Within the ovule, double fertilization occurs as one sperm fertilizes the egg, producing the zygote, while the second sperm fuses with the polar nuclei, giving rise to the primary endosperm nucleus. After fertilization, the ovary becomes a fruit and each ovule becomes a seed.

REVIEW QUESTIONS

1. Describe the parts of a flower, and indicate some common modifications.

2. Detail the events of meiosis. Why is prophase I of meiosis an important stage?

3. Describe the male and female gametophytes.

4. What is the general appearance of a wind-pollinated flower? of an animal-pollinated flower? Take a walk in a garden or tour a greenhouse and try to determine whether the flower is wind- or animal-pollinated by examining the flower's structure.

5. What is meant by double fertilization?

6. What fields use the study of pollen as a tool? What types of information have been gained from this approach?

7. Self-pollination occurs when the stigma is pollinated by pollen from the same flower or another flower of the same plant. Cross-pollination results when the stigma of one flower is pollinated by the pollen from a different individual. What is the advantage of self-pollination? the disadvantage? What is the advantage of cross-pollination? the disadvantage?

8. A number of flowering plants are adapted to aquatic environments. Investigate how pollen is transferred in eelgrass (*Vallisneria*), a dioecious angiosperm.

©Karen McMahon

CHAPTER

6

Plant Life Cycle: Fruits and Seeds

Each golden, translucent fruit of the soapberry tree (Sapindus saponaria) *contains a single black seed.*

KEY CONCEPTS

1. Fruits are ripened ovaries that are the end products of sexual reproduction in angiosperms and are a major vehicle for the dispersal of their enclosed seeds.
2. Protected by a tough outer coat, seeds are ripened ovules that contain an embryonic plant plus some nutritive tissue and are the starting point for the next generation.
3. Edible fruits of various types play a major role in the human diet.

CHAPTER OUTLINE

Fruit Types 86
    Simple Fleshy Fruits 86
    Dry Dehiscent Fruits 86
    Dry Indehiscent Fruits 86
    Aggregate and Multiple Fruits 86
Seed Structure and Germination 88
    Dicot Seeds 88
    Monocot Seeds 88
    Seed Germination and Development 88
Representative Edible Fruits 88
    Tomatoes 90

A CLOSER LOOK 6.1 The Influence of Hormones on Plant Reproductive Cycles 92
    Apples 94
    Oranges and Grapefruits 95
    Chestnuts 97
    Exotic Fruits 98
Chapter Summary 99
Review Questions 100

Fruits, as are flowers, are unique aspects of sexual reproduction in angiosperms; they protect the enclosed seeds and aid in their dispersal. Not only are fruits essential in the angiosperm life cycle, but they are also widely utilized as significant food sources.

FRUIT TYPES

The fruit wall that develops from the ovary wall is known as the **pericarp** and is composed of three layers: the outer **exocarp,** the middle **mesocarp,** and the inner **endocarp.** The thickness and distinctiveness of these three layers vary among fruit types.

Simple Fleshy Fruits

Simple fruits are derived from the ovary of a single carpel or several fused carpels and are described as **fleshy** or **dry.** When ripe, the pericarp of fleshy fruits is often soft and juicy. Seed dispersal in the fleshy fruits is accomplished when animals eat the fruits. The following are the most common types of fleshy fruits (fig. 6.1):

- A **berry** has a thin exocarp; a soft, fleshy mesocarp; and an endocarp enclosing one to many seeds. Tomatoes, grapes, and blueberries are familiar berries.
- A **hesperidium** is a berry with a tough, leathery rind, such as oranges, lemons, and other citrus fruits.
- A **pepo** is a specialized berry with a tough outer rind (consisting of both receptacle tissue and exocarp); the mesocarp and endocarp are fleshy. All members of the squash family, including pumpkins, melons, and cucumbers, form pepos.
- A **drupe** has a thin exocarp, a fleshy mesocarp, and a hard, stony endocarp that encases the seed; cherries, peaches, and plums are examples.
- Apples and pears are **pomes;** most of the fleshy part of pomes develops from the enlarged base of the perianth that has fused to the ovary wall.

As described for the pepo and pome, some fruits develop from flower parts other than the ovary; fruits of these types are termed **accessory** fruits.

Dry Dehiscent Fruits

The pericarp of dry fruits may be tough and woody or thin and papery; dry fruits fall into two categories, **dehiscent** and **indehiscent.** Dehiscent fruits split open at maturity and so release their seeds. They usually contain more than one seed and often many seeds. When the fruit wall opens, the seeds can be dispersed individually rather than en masse. Wind often aids the dispersal of seeds from dehiscent fruits. Three common types of dehiscent fruit—**follicles, legumes,** and **capsules**—are characterized by the way in which they open. Follicles, as found in magnolia and milkweed, split open along one seam, while legumes (such as bean pods and pea pods) split along two seams (fig. 6.1). The most common dehiscent fruit is a capsule that may open along many pores or slits; cotton and poppy are representative capsules.

Dry Indehiscent Fruits

Indehiscent fruits do not split open. Instead, they use other means of dispersing their seeds. **Achenes, samaras, grains,** and **nuts** are examples of indehiscent fruits. Sunflower "seeds" are, in fact, achenes, one-seeded fruits in which the pericarp is free from the seed (fig. 6.1). Carried by the wind, the winged fruits of maple, elm, and ash trees are familiar types of samaras. Samaras are usually described as modified achenes. The fruits of all our cereal grasses are grains, single-seeded fruits in which the pericarp is fused to the seed coat. Also called a **caryopsis,** this type of fruit is found in wheat, rice, corn, and barley. Botanically, nuts are one-seeded fruits with hard, stony pericarps, such as hazelnuts, chestnuts, and acorns. In common usage, however, the term *nut* has also been applied to seeds of other plants; peanuts, cashews, and almonds are actually seeds, not nuts.

Aggregate and Multiple Fruits

Aggregate fruits develop from a single flower with many separate carpels, all of which ripen at the same time, as in raspberries and blackberries. Strawberries, another aggregate fruit, also contain accessory tissue. The brownish yellow spots on the surface are actually achenes inserted on the enlarged, fleshy, red receptacle (fig. 6.1).

Multiple fruits result from the fusion of ovaries from many separate flowers on an inflorescence. Figs and pineapples are examples of multiple fruits (fig. 6.1).

Thinking Critically

Many large herbivores in North America became extinct around 13,000 years ago. Plants that had coevolved with native animals to distribute their fruits and seeds were left without animal assistance. The Osage orange tree (*Maclura pomifera*) produces a grapefruit-sized fruit with a fibrous texture (see fig. 8.5a). After the presumed loss of its animal distributor, the range of Osage orange, which originally was widespread throughout the eastern half of North America, had shrunk to a few valleys in east Texas. Introduced from Europe, horses (but not cattle) can bite through the tough pericarp and eat the Osage oranges (also known as horse apples) after they fall to the ground.

What features would you look for to discover if animals distributed a fruit? What may the information about European horses reveal about the extinct North American distributor of Osage oranges? What alternatives do plants have for reproduction and survival if they lose their animal dispersal agents to extinction?

Figure 6.1 The berry, hesperidium, pepo, and pome are fruits in which at least part of the pericarp is soft and juicy. Fruits such as the follicle, legume, and capsule are characterized by the way in which they open. Achenes, grains, and nuts are dry fruits that do not split open to disperse the seed. A samara (as in elm or maple) is a winged fruit that uses wind as the dispersal agent. Blackberries and strawberries are collections of fruits that develop from the many separate carpels of a single flower. The pineapple is a multiple fruit that forms when the ovaries of individual flowers in a flower cluster fuse.

SEED STRUCTURE AND GERMINATION

A seed contains the next generation and so completes the life cycle of a flowering plant. The seed develops from the fertilized ovule and includes an embryonic plant and some form of nutritive tissue within a seed coat. Differences between dicots and monocots are apparent within seeds. The very names dicot and monocot refer to the number of seed leaves, or **cotyledons,** present in the seed. Dicot seeds usually have two cotyledons that are attached to and enclose the embryonic plant. The cotyledons, which are often large and fleshy, occupy the greatest part of the dicot seed and have absorbed the nutrients from the endosperm. Thus, the endosperm in many dicot seeds either is lacking entirely or is very much reduced. Monocots have a single, thin cotyledon that transfers food from the endosperm to the embryo. In several monocot families, large amounts of endosperm are apparent. Because of these stored nutrient reserves (in either the cotyledons or the endosperm), many seeds, like many fruits, are valuable foods for humans and other animals.

Dicot Seeds

The garden bean, because of its large size, is a good example of a dicot seed (fig. 6.2a). A thin, membranous **seed coat,** also known as the **testa,** encloses the seed. A **hilum** and **micropyle** are visible on the surface of the testa. The hilum is a scar that results from the separation of the seed from the ovary wall. Recall that the micropyle, seen as a small pore, is the opening in the integument through which the pollen tube enters the ovule. If the seed coat is removed, the two large food-storing cotyledons are easily seen and separated. Sheltered between the cotyledons is the embryo axis consisting of the **epicotyl,** the **hypocotyl,** and the **radicle.** The epicotyl develops into the shoot (stems and leaves) of the seedling and typically bears embryonic leaves within the seed. The hypocotyl is the portion of the embryo axis between the cotyledon attachment and the radicle, the embryonic root.

Monocot Seeds

The corn kernel is a familiar grain that can be used to illustrate the composition of a monocot seed (fig. 6.2b). It is important to remember that a grain is a fruit in which the seed coat of the single seed is fused to the pericarp. One major difference from the garden bean is the presence of extensive endosperm that occupies much of the volume of the seed. The small embryo has only a single cotyledon, called a **scutellum.** Seeds in the grass family (such as corn) have other differences, including the presence of a **coleoptile** (a protective sheath that surrounds the epicotyl) and a **coleorhiza** (a protective covering around the radicle).

Thinking Critically
Monocot seeds have a single, thin cotyledon, whereas dicot seeds usually contain two prominent cotyledons.

How does nourishment of the embryonic plant differ between monocots and dicots?

Seed Germination and Development

With appropriate environmental conditions (adequate moisture and oxygen and appropriate temperature), seeds germinate (fig. 6.2). The first structure to emerge from the seed is the radicle, which continues to grow and produces the primary root. In corn, the radicle first breaks through the coleorhiza. This early establishment of the root system enables the developing seedling to absorb water for continued growth. Next, the shoot emerges. In garden beans, the hypocotyl elongates and breaks through the soil in a characteristic arch that protects the epicotyl tip with its embryonic leaves. In most dicots, the cotyledons are carried aboveground with the expanding hypocotyl, while in others the cotyledons remain belowground. Soon after the tissues of the seedling emerge from underground and are exposed to sunlight, they develop chlorophyll and begin to photosynthesize. The exposure to sunlight also triggers the hypocotyl to straighten into an erect position. The coleoptile of corn emerges from the soil; the epicotyl soon breaks through the coleoptile, and the embryonic leaves begin expanding. Establishment of the seedling is the most critical phase in the life of a plant, and high mortality is common. Seedlings are sensitive to environmental stress and vulnerable to attack by pathogens and predators; established plants have a greater array of defenses.

Thinking Critically
Violets and thousands of other plant species have evolved seeds with a small, fatty appendage, known as an **elaiosome,** rich in oils. Elaisomes attract ants, which carry these seeds back to their nests and remove the elaiosomes to feed them to their larvae.

Speculate on the fate of the seeds in which the elaiosomes have been removed. What is the benefit to the plant in making a seed with an elaiosome? How could you test your hypothesis?

REPRESENTATIVE EDIBLE FRUITS

Of the more than 350,000 known species of angiosperms, only a small percentage produce fruits that have been utilized by humans; however, these have made a significant impact on

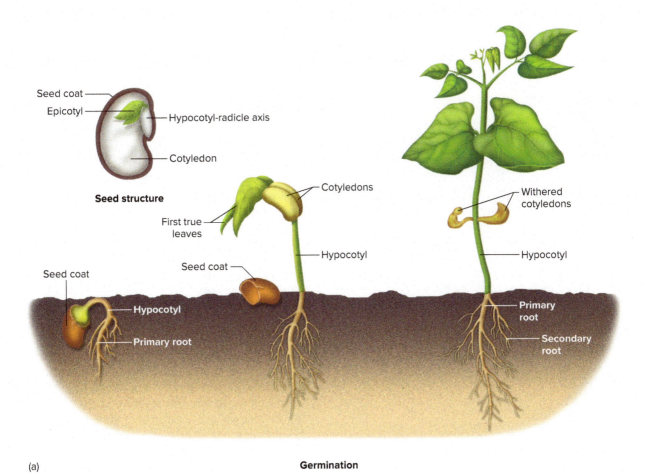

Figure 6.2 Seed structure and germination of (a) a dicot, the garden bean, and (b) a monocot, corn.

Figure 6.3 In 1893, the U.S. Supreme Court decided that the tomato is legally a vegetable.

our diet and economics. Fruits are packed with nutrients and are particularly excellent sources of vitamin C, potassium, and fiber. As this chapter describes, a fruit is a mature or ripened ovary, but this botanical definition has been ignored in the marketplace. Even the U.S. Supreme Court has debated the question, What is a fruit?

It all started in the late nineteenth century when an enterprising New Jersey importer, John Nix, refused to pay the vegetable import tariff on a shipment of tomatoes from the West Indies. He argued that the 10% duty placed on vegetables by the Tariff Act of 1883 was not applicable to tomatoes, since botanically they are fruits, not vegetables. This fruit-vegetable debate eventually reached the U.S. Supreme Court in 1893 (fig. 6.3). Justice Horace Gray wrote the decision, stating,

> Botanically speaking, tomatoes are the fruits of the vine, just as are cucumbers, squashes, beans, and peas. But in the common language of the people, whether sellers or consumers of provisions, all these are vegetables which are grown in kitchen garden, and which, whether eaten cooked or raw, are, like potatoes, carrots, parsnips, turnips, beets, cauliflower, cabbage, celery, and lettuce, usually served at dinner, in, with, or after the soup, fish, or meats, which constitute the principal part of the repast, and not, like fruits, generally as dessert.

Tomatoes were legally declared vegetables and Nix paid the tariff. Despite the legal definition, botanists still consider tomatoes to be berries.

Tomatoes

Tomatoes, *Solanum lycopersicum* (formerly *Lycopersicon esculentum*), are native to Central and South America, in the Andes region of Chile, Colombia, Ecuador, Bolivia, and Peru, and are believed to have been first domesticated in Mexico. The Spanish conquistadors introduced the tomato to Europe, where it was first known as the "Apple of Peru," the first of many names for this fruit. Later it was known as *pomo doro*, golden apple (early varieties were yellow), in Italy and *pomme d'amour*, love apple (it was believed by many to be an aphrodisiac) in France. The common name for the fruit comes from the Aztec's word for it, *tomatl*. The scientific name also reflects another early myth about the tomato; literally translated, *lycopersicon* means wolf peach, a reference to its poisonous relatives, including the deadly nightshade and henbane. Despite its described edibility, it took years for the tomato to live down its poisonous reputation. One bizarre demonstration of its lack of toxicity took place in Salem, New Jersey, in 1820 when Colonel Robert Gibbon Johnson ate a bushel of tomatoes in front of a crowd gathered to witness his certain demise. His obvious survival, without ill effects, finally settled the issue of the tomato's edibility.

Over the years since Johnson's demonstration, the tomato, owing to its attractiveness, taste, and versatility, has become one of the most commonly eaten "vegetables" in the United States; we each consume approximately 36 kilograms (80 pounds) every year. Individually, the tomato is largely water, with only small amounts of vitamins and minerals, but because of the large volume consumed, the tomato leads all fruits and vegetables in supplying these dietary requirements. Its versatility is almost unequaled; what would pizza, ketchup, bloody Marys, salads, and lasagna be without tomatoes?

Cultivars

There are over 500 **cultivars**, cultivated varieties, in the species *Solanum lycopersicum*. The most familiar varieties, such as Big Boy, Mammoth Wonder, and Beefsteak, have large fruits. At the other extreme are the cherry tomatoes, which scientists believe are most similar to the ancestral wild type. Today's large-scale commercial production yields a tomato very different from the home garden varieties. Commercial tomatoes have been bred for efficient mechanical harvesting, transportability, and long shelf life, not necessarily for taste. Another characteristic that has improved the commercial varieties is the determinate habit. This trait originally appeared as a spontaneous mutation in Florida in 1914. Determinate plants are shorter, bushier, and more compact than the indeterminate habit, which has a more sprawling pattern of growth that requires extensive staking or trellising. Although no other species are widely cultivated, scientists are interested in the salt tolerance of *Solanum cheesmanii*, which is native to the Galápagos Islands. This species, unlike *S. lycopersicum,* is able to survive in seawater, with its high salt concentration. Developing salt tolerance in crop plants is one of the aims of plant-breeding programs, since salt-tolerant crops will allow agriculture to expand into areas with saline soils. A new and unusual species of tomato was recently discovered in northern Australia. It is a type of bush tomato (*Solanum ossicruentum*) and produces small fruits, 1.5–2.5 cm in diameter, covered in spikes that latch onto the fur of animals for dispersal. Cutting into the young fruits reveals a white-green flesh, which within minutes upon exposure to air turns blood red and continues to darken to a deep maroon. This "bleeding effect" accounts for its scientific name, *ossicruentum,* which translates as bones and blood!

Heirloom tomatoes are traditional varieties that have become popular selections for both the home gardener and the gourmet market in recent years. Heirlooms offer a wider diversity of tastes, shapes, and colors that can be found in the typical selection of a supermarket, as indicated by a survey of their names: Green Sausage, Black Prince, Red Strawberry, and Marvel Stripe (fig. 6.4). They offer a nostalgic connection to ancestral farmers and are valued for the fact that most are grown on small, family-owned organic farms without the intervention of harmful synthetic pesticides or fertilizers.

Recently, the genome of the domesticated tomato (*S. lycopersicum*) was deciphered after analyzing 7,000 heirloom and modern varieties. Researchers were able to identify 13 species of tomatoes and four closely related species from wild populations. Genomic analysis revealed that the current tomato (*S. pimpinellifolium*) is the closest wild relative of *S. lycopersicum;* they shared a common ancestor about 1.4 million years ago.

Most of the characteristics associated with the domesticated tomato are the result of mutations in just 30 of the 35,000 genes found in the tomato. One of these mutations has been identified as the gene *fasciated,* which doubles fruit size by increasing the number of locules, compartments in the ovary containing the seeds, from only two in the cherry tomato to eight in the beefsteak variety. Other mutations have been identified that changed fruit shape and color. Some species of wild tomatoes in South America produce purple fruits, but the tomatoes are small and often poisonous. Seeds of purple tomatoes were collected in the 1960s and 1970s and then bred to modern varieties. The resulting hybrids are larger, nontoxic, and deep purple, similar in color to an eggplant. Researchers found a gene in the wild purple tomatoes called anthocyanin fruit (*Aft*), which codes for the production of high levels of these pigments (see Chapter 2) and the purple hue. Through breeding experiments, the *Aft* gene in the wild tomato was passed on to the new purple hybrid. The skin of the purple tomatoes still contains lycopene in addition to the anthocyanins. Both types of pigments are powerful antioxidants and are believed to reduce the risk of cancer and heart disease. Purple tomatoes may soon be found in farmers' markets around the country.

Domestication brought about an increase in the number of genes producing sugars and acids, making the domesticated

Figure 6.4 Heirloom tomatoes display a variety of sizes, shapes, and colors.

tomato more flavorful. At the same time, the number of genes that imparted protection against diseases was lost; most heirloom varieties possess only a single gene that confers disease resistance against fungal infections.

Lately, researchers have used genomic analysis to analyze nearly 400 modern, heirloom, and wild tomatoes to identify desirable flavor compounds in the tomato. Many modern varieties have declined in flavor as breeders selected for increased yield, extended shelf life, firmness, appearance, and disease resistance. Thirteen volatile compounds were identified associated with increased sugar content, acidity, and aroma in the tastier heirloom varieties. By pinpointing the flavorful genes, researchers hope to breed back these important compounds missing in tomatoes bred for the mass market.

Growth and Ripening

Tomatoes are one of the most common plants grown in the home garden. Even in large metropolitan areas, many apartment dwellers grow patio varieties in containers. Home gardeners have all watched the stages of growth, flowering, and fruit ripening while waiting for the first harvest. Tomato fruits require 40 to 60 days from flowering to reach maturity. Once

A CLOSER LOOK 6.1

The Influence of Hormones on Plant Reproductive Cycles

Many phases of the plant life cycle, including flowering and fruit development, are influenced by hormones, which act as chemical messengers that are effective at very low concentrations. Hormones are produced in one part of an organism and have their effects on another part of that organism. The major types of plant hormones are **auxins**, **gibberellins**, **cytokinins**, **abscisic acid**, and **ethylene**. In addition, several other hormones have recently been described.

Auxins are the best known of the plant hormones and were the first to be discovered. The research done by Charles Darwin and his son Francis in the late 1870s led to the discovery of this hormone several decades later. The Darwins studied phototropism (growth toward the light) and suggested that this response was due to an "influence" produced in the tip of a coleoptile that then moved to the growing area. Experiments by Dutch physiologist Frits Went in the 1920s demonstrated that the "influence" is actually a chemical compound, which Went named auxin. Auxins are produced in apical meristems and other actively growing plant parts, including young leaves, flowers, fruits, and pollen tubes.

In addition to phototropism, auxins are involved in many stages of growth and development. Auxins promote elongation of young stems and coleoptiles by stimulating cell elongation. They inhibit lateral bud development and thus promote apical dominance, producing a plant with a main stem and limited branching. Auxins stimulate adventitious root initiation and are involved in growth responses to gravity (gravitropism).

Auxins regulate fruit development. They are produced by the pollen tube as it grows through the style and by the embryo and endosperm in developing seeds. Fruit growth depends on these sources of auxin. The application of auxins to the flowers of some plants, such as tomato and cucumber, before the pollen is mature can promote parthenocarpy, leading to the development of seedless fruits (see Chapter 5).

Synthetic auxins have auxinlike activity but do not occur naturally in plants. Often, they are more effective than natural auxins in stimulating plant responses. There are many agricultural and horticultural uses for both naturally occurring and synthetic auxins. In general, auxins have no effect on flowering; the exception is pineapple. Auxins can be applied to pineapple plants to promote uniform flowering; however, this is a secondary effect because the auxins stimulate ethylene formation, and ethylene is the hormone that actually promotes flowering in pineapple.

Maturing fruit of apples, oranges, and grapefruit can be sprayed with auxins to prevent the premature development of abscission layers (separation zones) and the resulting fruit drop. Higher doses of auxins, however, can cause abscission, and this phenomenon can also be used to growers' advantage. Heavy applications of synthetic auxins are used commercially to promote a coordinated abscission of various fruits to facilitate harvesting.

Gibberellins were isolated and chemically identified in 1939 by Japanese botanists. Like auxins, gibberellins are involved in many aspects of plant growth and development. Gibberellins promote stem elongation of dwarf plants by stimulating internode elongation. Dwarf varieties of many species will grow to normal size if supplied with gibberellins. Gibberellins promote seed germination of some plants by substituting for an environmental (cold or light) trigger.

Gibberellins can stimulate flowering in biennials during the first year. Biennials typically produce a tight cluster of leaves (called a rosette) in their first year. This rosette occurs on stems with very short internodes. In the second year, internodes expand greatly and the plants flower; this expansion and flowering is called bolting. The application of gibberellins to the rosette will promote bolting during the first year. This hormone-induced bolting allows growers to harvest seeds after 1 year instead of 2. Commercially, gibberellins are also used to increase the size of seedless grapes and stimulate the germination of barley seeds for beer production.

Cytokinins stimulate cell division and differentiation of plant organs. This hormone, along with auxins, is necessary as an ingredient in media for tissue culture of plant cells (see Chapter 15). Cytokinins also delay senescence in detached plant parts. Treated areas remain green and healthy while the surrounding tissues age and die. For this reason, cytokinins

---

fertilization has occurred, the fruit rapidly increases in size, reaching its mature size in 20 to 30 days. In the latter half of fruit development, color changes reflect internal changes in the acidity, sweetness, and vitamin C content of the fruit. The first hint of ripening is seen when the green fruit lightens as a result of chlorophyll breakdown. As the chlorophyll content continues to decrease, additional carotenoids are synthesized. The carotenoids, beta-carotene (orange) and especially lycopene (red), give the mature fruit its characteristic color. As the red color deepens, the acidity decreases, the sugars and vitamin C increase, the flavor develops, and the fruit softens. These changes coincide with increases in ethylene and a sudden peak in respiration in the fruit. Ethylene is a gaseous plant hormone involved in several developmental stages but is best known for its involvement with fruit ripening. (See A Closer Look 6.1: The Influence of Hormones on Plant Reproductive

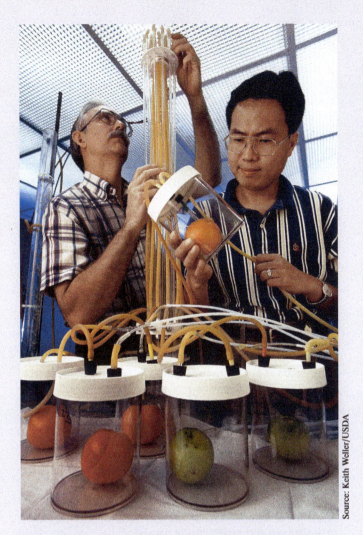

Box Figure 6.1 Researchers measure the respiration and ethylene production of tomatoes in experiments aimed at maintaining the quality of fresh produce.

Ethylene is an unusual plant hormone because it is a gas. This gas can be produced in all parts of the plant but is especially prominent in roots, the shoot apical meristem, senescing (aging) flowers, and ripening fruit. In addition, high levels of auxin stimulate ethylene production.

Ethylene promotes flowering in a few species, such as pineapple. As described, synthetic auxins have been used to trigger flowering on pineapple plantations. The auxins stimulate ethylene formation, which induces uniform flowering. Today, many growers take a direct approach and spray their plants with ethephon, a compound that breaks down to release ethylene.

Possibly the most dramatic effects of ethylene are on fruit ripening. Ethylene stimulates fruit softening, the conversion of starch to sugars, and the production of volatile compounds that impart aromas and flavor in many types of fruit, especially apples, oranges, tomatoes, and bananas. These changes are generally accompanied by a peak in cellular respiration of the fruit (box fig. 6.1). Commercially, ethylene can be used to produce ripe fruits throughout the winter. Apples and other fruits are picked green and stored under conditions that inhibit ethylene synthesis. When fruits are required for sale, they are exposed to small amounts of ethylene, which induce ripening. As the fruits begin to ripen, they produce their own ethylene, which further accelerates the maturing process and provides table-ready fruit throughout the year.

Ethylene is sometimes considered a stress hormone because it is produced in wounded or infected tissues and flooded plants. In addition to ethylene, other plant hormones, salicylic acid (see Chapter 19) brassinosteroids, and jasmonic acid, are also involved in the response to pathogens by regulating plant defenses. The role of these hormones in plant health is a major focus of research; scientists hope this knowledge can help reduce the use of fungicides and pesticides on our crops.

Thinking Critically

Plant hormones influence many phases of plant growth and development, from seed germination to flowering and fruit formation.

Which of the plant hormones can be utilized commercially because of their effects on fruit development or maturation?

---

are sometimes used commercially to maintain the freshness of cut flowers.

Abscisic acid is an inhibitor hormone that promotes dormancy in seeds and buds. It is also involved in regulating water balance in plants by causing stomata to close. This last response is the one that is understood best. In well-watered plants, abscisic acid concentrations in leaves are very low, but levels increase rapidly if plants are exposed to severe drought. The buildup of abscisic acid results in stomatal closure, which causes transpiration rates to decrease dramatically.

Cycles.) When tomatoes are picked green and ripened in storage, they have a lower vitamin C and sugar content and poorer flavor; this explains why most people prefer a vine-ripened tomato.

Many home gardeners have noticed the role that temperature plays in tomato development and ripening. Most varieties do best when air temperatures are between 18° and 27°C (65° and 80°F). When temperatures are either too hot or too cold, fruit set is inhibited. Also, temperatures above 29°C (85°F) inhibit the development of the red pigments.

Spaceflight

During the spring of 1990, millions of elementary school children and high school and college students in the United States and 30 other countries were given the opportunity to participate in the SEEDS project. The project was a cooperative venture between NASA and the George W. Park Seed Company to compare the growth of space-exposed seeds with the growth of control seeds stored on Earth. The space-bound seeds had been launched into orbit aboard the LDEF satellite on April 7, 1984, by the space shuttle *Challenger*. The LDEF (Long Duration Exposure Facility) satellite is a cylindrical structure approximately 9 meters (29.5 feet) long and 4 meters (13 feet) in diameter and built to house many separate experiments that were designed to test the effects of space on various systems. For almost 6 years, 12.5 million Rutgers California Supreme tomato seeds were aboard LDEF, experiencing weightlessness and exposure to cosmic radiation for longer than any previous NASA experiment involving biological tissue. The satellite was recovered by the space shuttle *Columbia* and returned to Earth on January 20, 1990. The first germination test began in late February and continued throughout the spring as seed packages were mailed to teachers and students. The first conclusion reached was that, after nearly 6 years in space, the tomato seeds were still viable. Germination rates were approximately 65% for both the space-exposed and Earth-based seeds, but the space-exposed seeds exhibited an 18%–30% faster germination rate than their Earth-based counterparts. Growth rates of the space seedlings were also accelerated for the first 3 or 4 weeks; after that, the Earth-based seedlings caught up and no significant differences between the Earth and space tomato plants or their fruits were observed. Researchers also observed a greater number of mutant individuals, such as albino plants, stunted individuals, and one plant with no leaves at all, in the space-exposed group. Interestingly, plants of space-exposed seeds had greater levels of chlorophylls and carotenes than did those of the Earth group. Perhaps the most significant conclusion of the SEEDS project is the proof that seeds can survive relatively undamaged in space for long periods of time. Tomatoes are still blasting off into space. In 2007, a dwarf tomato variety, Micro-Tom, was sent along on the space shuttle *Endeavour* to test the feasibility of growing crops for long-duration space flights.

Apples

Apples, *Malus pumila,* have a long history of human use; they were among the first tree fruits to be domesticated in temperate regions. Most of today's cultivated varieties are descendants of apples native to central and western Asia, where the apple has been domesticated for thousands of years. Near Almaty, formerly Alma-Ata in Kazakhstan, there are forests of 17-meter (50-foot) wild apple trees, some 350 years old. *Alma-Ata,* in fact, means father of the apple. Because of its long history, the apple also has a place in the imagery and folklore of many cultures. The following expressions reveal the apple as more than a tasty fruit in American culture: the "Big Apple," "American as apple pie," "apple of your eye," "apple polishing," "apple pie order," and "an apple a day keeps the doctor away."

Most people are familiar with the biblical story of Adam and Eve and the presumed role of the "apple" in the downfall of humankind. In reality, the apple's only involvement was due to a faulty translation from the Latin version of the Old Testament. The confusion starts with the Latin word *mali,* which could refer to *malum,* meaning evil, or to *malus,* meaning apple tree. In Genesis, Adam and Eve were told not to partake of the fruit of the tree of the knowledge of good and evil. The erroneous association with apple trees began with thirteenth- and fourteenth-century artists, whose Latin was poor; they incorrectly translated *mali* to mean apple tree.

The legend of Johnny Appleseed is another story familiar to most Americans. There really was a Johnny Appleseed; his name was John Chapman. He was born in Massachusetts in 1774, but we know nothing of his early life. He appeared in 1797, sowing apple seeds in what was the Northwest Territory in frontier America: western Pennsylvania, Ohio, and eastern Indiana. Chapman probably saw more of America than most of his contemporaries; he traveled hundreds of miles by foot, horseback, and canoe. Chapman was an itinerant orchardist who gave away or sold—and even planted with his own hands—apple seeds and seedlings that gave rise to acres of apple trees throughout the region. Some of the orchards are still in existence today. He continued his mission until his death in 1845 in Fort Wayne, Indiana, a city that still honors his memory every summer with a Johnny Appleseed Festival.

Apple trees are medium-sized trees with a broad, rounded crown and a short trunk. They are generally spreading, long-lived trees that can bear fruit for up to a century. In modern commercial orchards, however, dwarf trees have become the norm. Often almost 2 meters (about 6 feet) tall, these trees are easily pruned and mechanically harvested.

The apple blossoms appear in profusion early in the spring before the leaves develop. The fragrant, pinkish white blossoms are five-merous flowers (containing five sepals, five petals, numerous stamens, and a five-carpeled ovary) that are usually pollinated by bees (fig. 6.5). The apple tree requires cold winter temperatures in order to flower and, therefore, cannot be grown in tropical and subtropical climates. The leading apple-growing areas in the United States are Washington, Oregon, and northern California in the West and Michigan, New York, and Virginia in the East.

Figure 6.5   The apple blossom is being pollinated by a bee.

Although Johnny Appleseed distributed thousands of apple seeds, today's modern orchardist does not grow the trees from seeds. Each seed is a unique combination of traits that are not identical to either parent; in a sense, planting a seed is a genetic experiment. Although some seeds may develop into valuable varieties, most will not. (Interestingly, most of our familiar cultivars did develop as chance or volunteer seedlings from naturally produced seeds.) Also, planting seeds is a long-term experiment, and it would take a number of years to know the results. Today's apple growers need to ensure uniformity in their orchards, not only to produce apples with the desired flavor and taste but also to maximize efficiency at harvest time. As a result, most apple trees are produced by **grafting** (a form of asexual reproduction or cloning) in which stem cuttings or buds from a desirable cultivar are joined to the base of a second tree. The cutting or bud, called the **scion,** will become the upper or top portion of the new tree, while the **rootstock,** or simply **stock,** of the second tree is the root system of the graft combination. Grafting can create thousands of identical copies of a variety that will continue to produce apples with the desired characteristics.

As described earlier, the apple is a pome, a simple fleshy fruit with accessory tissue. The core of the apple is a five-carpeled ovary with seeds; the ovary wall is visible as a fine, brownish line, and the endocarp is prominent as the parchmentlike material around the seeds (fig. 6.6). The skin and flesh of the apple develop from the receptacle and base of the perianth. A ripe apple, although mostly water, contains about 12% sugar, 1% fiber, and negligible amounts of fat and protein. Approximately 50% of the apples harvested are consumed as fresh fruit, with the remainder being processed into applesauce, apple butter, apple cider (see Chapter 24), apple juice, cider vinegar, dried apples, and canned apples.

Apple Varieties

Although there are thousands of varieties of apple, only a few can be found in the modern supermarket. In the past, many more varieties were available to consumers; virtually gone from the market are old staples such as Baldwin, Early Harvest, Fall Pippin, and Gravenstein. Today, MacIntosh, Crispin, Braeburn, Red Delicious, Golden Delicious, and Granny Smith are some of the most popular apples varieties. The Red Delicious apple was first discovered in Iowa in the 1870s, but its popularity can be traced to the growth of the large supermarket chains during the 1950s and 1960s. It became the generic "red apple" to the American consumer. Other commercially important red apples include the Rome, Jonathan, and McIntosh. A popular red-gold variety is Jonagold, considered by many experts as the top-ranking apple. It was created from a cross between Golden Delicious and Jonathan apples. The Golden Delicious, the leading yellow apple, arose by chance on a West Virginia farm in 1910. The Granny Smith is a tart green apple that originated in New Zealand. Growers are constantly developing new varieties. One of the newest is Gala, a red and yellow apple also developed in New Zealand. Fuji was created by Japanese breeders who crossed a Red Delicious with the heirloom Ralls Janet. Its sweetness and crispness have made it a popular variety.

The world's most extensive collection of apple varieties is maintained in Geneva, New York, at the Agricultural Experiment Station run by Cornell University and the Plant Genetic Resources Unit of the U.S. Department of Agriculture (USDA). More than 5,000 apple trees are planted on this 50-acre farm; these consist of 2,500 apple varieties, with two trees of each variety. An additional 500 types are stored as seeds at this repository. The apples come from all over the world and include wild apples, modern cultivars, and obsolete cultivars. In an attempt to preserve the genetic diversity of apples, researchers from Geneva have traveled throughout central Asia searching for unique apples. In addition to cultivating these varieties, the Geneva Repository distributes over 3,000 specimens (as seeds or scions) each year to apple researchers throughout the world. This facility can be considered a modern-day Johnny Appleseed.

Oranges and Grapefruits

The citrus family (Rutaceae) is the source of many edible fruits; sweet oranges (*Citrus sinensis*), grapefruits (*C. paradisi*), tangerines (*C. reticulata*), lemons (*C. limon*), and limes (*C. aurantifolia*) come to mind most immediately. But there are also pummelos (*C. grandis*), citrons and etrogs (*C. medica*), bergamots (*C. bergamia*), sour oranges (*C. aurantium*), and kumquats (*Fortunella japonica*). Most of these species are native to southeastern Asia, where they were undoubtedly cultivated by native peoples. Brought by caravan from the East, the citron was the first citrus fruit

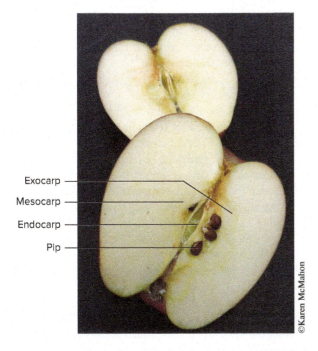

Figure 6.6 In this longitudinal section of the apple fruit, the brown line (exocarp) delineates the true ovary wall from the outer accessory tissue. The endocarp is the brown, papery material surrounding each seed, or pip.

known to Western civilization; it was widely grown throughout the Mediterranean area during Greek and Roman times, being spread by the Jewish people, for whom it had religious significance. Centuries later, Europeans became acquainted with sour oranges, lemons, and limes when Arabic traders during the Middle Ages introduced them from the East. It was not until the sixteenth century, however, that the sweet orange, the most familiar type today, was taken to Europe by Portuguese traders. It is believed that these first sweet oranges came from India, but later oranges were brought from China, which is believed to be the country of origin, as reflected in its scientific name, *Citrus sinensis.*

Spanish and Portuguese explorers introduced citrus to the New World; by 1565, sour orange trees were growing in Florida. After Florida became a state in 1821, the wild groves of sour oranges became the rootstocks for the sweet orange industry. Florida soon became, and remains, the leading orange-producing state. Valencia oranges are the main variety cultivated in Florida; this is a thin-skinned, seeded orange grown primarily for juice. Citrus was taken to California by the Spanish missionaries in the eighteenth century. Today, we associate the navel orange with California, but these oranges have their origins in Bahia, Brazil. In 1870, an American missionary stationed in Bahia was impressed by the appearance and flavor of a local variety and sent 12 saplings to the USDA in Washington, DC. The USDA propagated the trees and offered free plants to anyone wishing to grow them. In 1873, a Riverside, California, resident, Mrs. Luther Tibbets, received two trees that were such a success that they were soon widely propagated. In fact, it is believed that all navel oranges today are descendants of Mrs. Tibbets's trees.

Botanically, a citrus fruit is a hesperidium. The thick rind is impregnated with oil glands, a characteristic of the family. Fragrant essential oils attract animals for fruit and seed dispersal and are important commercially for perfumes and cosmetics. Within the fruit, the individual carpels are filled with many one-celled juice sacs. Although the carpels are always distinct, their ease of separation varies with the type and variety of fruit. Navel oranges are noted for their seedless condition and for the navel, which is actually a small, aborted ovary near the top of the fruit. Although widespread cultivation of navel oranges is a fairly recent phenomenon, Europeans were aware of navel oranges in the seventeenth century. Navel oranges have probably appeared spontaneously many times throughout the history of orange growing.

The color development in oranges is not related to ripening. The orange color associated with the fruit develops only under cool nighttime temperatures; in tropical climates, the fruits stay green. In most areas, a deep orange color is needed for successful marketing, and growers use various methods to achieve the desired color. The most widely used method involves exposing the ripened fruit to ethylene, which promotes the loss of chlorophyll, thereby making the orange carotenes visible. As a group, the citrus fruits are high in vitamin C. Those that are orange colored also provide some beta-carotene (precursor to vitamin A) and calcium.

Varieties of Citrus

The blood, or Maltese, orange is a newly introduced cultivar of the sweet orange to U.S. markets, in which the flesh of the fruit is a deep red. The color is due to the presence of anthocyanins, pigments which are not usually found in citrus fruits. Growing blood oranges under environmental conditions of hot days and cold nights increases the production of anthocyanins and intensifies the color. Blood oranges are usually smaller and thicker skinned than the more familiar varieties. The blood orange is not known in the wild and apparently developed as a mutation during the cultivation of sweet oranges in southern Europe sometime in the mid-nineteenth century. Sicily is famous for its production of blood oranges, but they have only been cultivated in the citrus-growing regions of Florida, California, and Texas within the last decade.

The largest citrus fruits are pummelos (*Citrus maxima*), which can be up to 30 cm (12 in.) in diameter and have a thick, yellow rind. There may be up to 14 carpels, with pulp varying from pale yellow to deep red and from slightly sweet to markedly acidic. Although pummelos are well known throughout Asia, it is only in recent years that they have started appearing in grocery stores in the United States. They are especially popular in cities with a large Chinese population and are a favorite for Chinese New Year festivities.

Pummelos are native to Southeast Asia and have been widely cultivated throughout that area. It has been determined by genetic analysis that the pummelo is one of the five species that gave rise through natural hybridizations to all other species within the *Citrus* genus (the other ancestors are: kunquat, small-flowered papeda, citron, and mandarin). Arab traders introduced pummelos to Spain in the twelfth century. Pummelo seeds were taken to the Western Hemisphere in the seventeenth century by an English sea captain, James Shaddock, on a return voyage from the East Indies; by 1696, the fruit was being cultivated in Barbados and Jamaica. In many areas, pummelos are also known as shaddocks in tribute to this seventeenth-century sea captain.

Pummelos' most enduring claim to fame is as one of the ancestors of the grapefruit. It is generally believed that grapefruits arose as a natural hybrid between pummelos and sweet oranges on the island of Barbados during the eighteenth century. The grapefruit was first described in 1750, when it was called a forbidden fruit. The name *grapefruit,* which refers to the fact that fruits develop in grapelike clusters of three or more, was first used in 1814, and grapefruits were botanically characterized as a separate species and named *Citrus paradisi* in 1837.

The first U.S. grapefruit trees were planted in Florida in the early nineteenth century, and the first commercial groves were established in 1875. By the beginning of the twentieth century, the Florida grapefruit industry was going strong, and today Florida is the world's leading grapefruit producer. Noteworthy grapefruits are the Texas Ruby Reds, which initially developed as a mutation from a pink-fleshed grapefruit tree in 1929. About half of the U.S. grapefruit crop is processed into juice or fruit sections.

Grapefruit juice made medical headlines after the first report of a grapefruit juice–drug interaction in 1991. Since that time, hundreds of studies have been performed to determine the mode of action and which prescription drugs are affected. Among the prescription drugs that show interaction with grapefruit juice are various types of tranquilizers, antihistamines, calcium channel blockers (used to lower blood pressure and to treat heart disease), cholesterol-lowering drugs, and immunosupressants. The grapefruit juice effect centers on the enzyme CYP3A in the intestinal wall that normally metabolizes drugs and thereby regulates their uptake. Grapefruit juice inactivates or inhibits the enzyme, causing a greater uptake of the drugs and elevated drug levels in the bloodstream. In some instances, the drug levels have been dangerously high and have resulted in death. A study looked at compounds called furanocoumarins (see Chapter 21), which are plentiful in grapefruit juice but are found in insignificant concentrations or not at all in most other citrus juices, as the responsible culprits. Researchers processed grapefruit juice by removing its furanocoumarins. Subjects were given a blood pressure-lowering medication with furanocoumarin-free grapefruit juice, normal grapefruit juice, or orange juice. Over a 24-hour period, the medication lingered about twice as long in those who had drunk normal grapefruit juice as in those who had drunk either processed grapefruit juice or orange juice.

Today, many prescription drugs carry warnings to avoid taking them with grapefruit juice; however, research is being conducted on ways to use grapefruit juice therapeutically to enhance drug uptake. Someday, instructions on a bottle of pills may say "take only with grapefruit juice."

Citrus Greening Disease

Citrus greening disease, also known as Huanglongbing or HLB, has become a serious threat to the citrus industry in Florida and California. First identified in Florida in 2008, the disease has spread throughout the state, reducing orange and grapefruit production by 50% in 2009. Citrus greening disease was first identified in California in 2009 and has since spread to Texas.

The disease originated in Asia and is caused by the bacterium *Candidatus liberibacter* and transmitted by the Asian citrus psyllid. Psyllids, also known as jumping lice, are tiny flying insects that feed on the sap of leaves and young shoots of citrus trees. Infected psyllids harbor *C. liberibacter* in their salivary glands and inject the bacteria into the phloem as they feed. The resultant bacterial infection spreads throughout the phloem vascular system and blocks the flow of nutrients from the leaves to the roots. Dying roots are no longer able to absorb water and soil nutrients to send up through the xylem to the leaves and fruits. Symptoms of infection include mottled yellow leaves and green misshapen fruits that taste bitter and drop prematurely. Diseased trees are also less able to withstand drought or other environmental stress and eventually die.

Growers have employed several strategies to try to reduce the spread of the disease. One promising strategy has been to introduce Asian wasps (*Tamarixia radiata*), natural predators of the citrus psyllid. The female wasp lays her egg under the belly of a psyllid; when the egg hatches into a larva, it begins eating the psyllid and does not stop until all that is left is the empty shell. The larva then seals itself inside the shell, weaves a cocoon, and later emerges as an adult by chewing a small hole in the psyllid husk. A wasp can also kill a psyllid by stabbing it with its egg-laying tube and drinking the gushing blood.

Other strategies have been used to save the citrus industry. Protective screens have been erected to cover entire nurseries to protect uninfected trees from psyllid infestation. Foliar spraying has been employed in some citrus plantations to keep infected trees productive for as long as possible. The leaves of infected trees are sprayed with the nutrients that they would normally receive from healthy roots. Antibiotics, such as penicillin, have been injected into the vascular system of the tree trunk to combat the bacterial infection directly.

There has also been a search for survivor trees that appear to be immune to the disease to breed citrus trees that are resistant to the disease. If genes resistant to the bacterium can be identified, these could be transferred to create disease-resistant citrus trees.

Chestnuts

Images of the winter holidays are often filled with the aroma of "chestnuts roasting on an open fire." These flavorful nuts are products of *Castanea* spp., a genus native to temperate regions of eastern North America, southern Europe, northern Africa, and Asia. In North America, *Castanea dentata,* the American chestnut tree, was one of the most useful trees to the native peoples and settlers. The trees, once some of the most abundant in the Eastern forests, were towering specimens, often reaching heights of 36 meters (120 feet) and diameters of 2 meters (7 feet). The wood was highly prized for furniture, fence posts, telegraph poles, shingles, ship masts, and railroad ties. The high tannin content of the wood made the lumber resistant to decay and was a source of tannins for the tanning of leather.

The nuts have long been consumed by both humans and animals; they can be eaten raw but are more commonly boiled or roasted. The chestnuts can also be made into a rich confectionery paste known as *creme de marrons* (chestnut spread) used in French desserts. Nutritionally, chestnuts are high in starch (approximately 78%) and unusually low in fat (4%-5%) for a nut.

Usually, three nuts (one-seeded fruits with a stony pericarp) are borne in a spiny bur or husk that splits open at maturity (fig. 6.7a). Each nut is produced by a single female flower, which is borne in a cluster of three; the cluster is subtended by an involucre, a collection of bracts that develop into the spiny bur as the fruits mature. The trees are monoecious, with the staminate flowers borne in long, slender catkins on the same individual.

Chestnut Blight

The reign of the chestnut trees in American forests began to decline early in the twentieth century because of chestnut blight, a disease caused by the fungus *Cryphonectria parasitica.* The fungus is believed to have been introduced in 1890 from some Asian chestnut trees taken to New York. The first reported case of the disease was in 1904 at the Bronx

Zoological Park. Cankers, localized areas of dead tissue, were noted on several of the trees in the park (fig. 6.7b). Although attempts were made to stop the disease by pruning away diseased branches, the fungus soon spread to all the chestnut trees in the park. By 1950, chestnut blight had spread throughout the natural range of *Castanea dentata* from Maine to Alabama and west to the Mississippi River. Although the trunk of an infected tree dies, the roots are usually not infected. The chestnut's ability to resprout from the roots has saved the species from extinction. These young saplings can reach heights of 4–6 meters (12–20 feet) before they succumb to the blight. Today, intense research efforts are focusing on various techniques to restore the American chestnut to its former glory. Asian chestnut trees show resistance to the blight fungus, and breeding programs in Connecticut have developed hybrids between the Asian and American species that are also resistant. The hybrids are backcrossed to the American chestnut for several generations with the aim that the only foreign genes remaining in the hybrids are those that confer blight resistance. Biological control, another line of ongoing research, uses a virus that infects *Cryphonectria parasitica*. Strains of the fungus with the virus are hypovirulent (less potent) and do not destroy the chestunut trees. Inoculating infected trees has resulted in disease remission. It is hoped that the combination of resistant trees and biological control may one day restore the American chestnut tree.

Not long ago, a stand of mature chestnut trees that had somehow escaped the blight were found in the Appalachian range in Georgia. The oldest of the half-dozen trees is about 40 feet tall and between 20 and 30 years old. Research will determine whether the dry, rocky ridge somehow inhibits the fungus or these trees are naturally resistant. Pollen from the stand may be used in the blight resistance breeding program.

Scientists at the American Chestnut Research and Restoration Project have finally succeeded in genetically engineering a blight–resistant American chestnut tree. Researchers tested more than 30 genes from different plant species for resistance to the blight; the OxO gene from bread wheat (*Triticum aestivum*) was found most effective. The wheat gene produces oxalic oxidase, an enzyme that breaks down oxalate, the compound the fungus produces that results in the deadly cankers on chestnut stems. Degrading oxalate is a common defense strategy against fungal pathogens and is found in many crop plants. By incorporating the OxO gene into the approximately 40,000 genes found in the American chestnut genome, researchers have created a blight-resistant tree that is over 99.999% genetically identical to wild-type trees. The next step in the restoration of the American chestnut to eastern forests, is to seek permission from federal regulatory agencies to plant these blight resistant American chestnuts into the wild and restore the native forest ecology.

Exotic Fruits

Many of the dessert and snack fruits that we commonly enjoy are not native to North America but had their origins

Figure 6.7 (a) The developing bur of the chestnut fruit is covered with sharp spines. (b) Chestnut blight has produced cankers on the trunk of a chestnut tree.

in exotic lands and faraway places. Consider these familiar fruits and their origins: apples (central Asia), oranges (southeastern Asia), peaches (China), bananas (southeastern Asia), watermelons (Africa), and pineapples (Latin America). Today this trend is accelerating, and every year new fruits are introduced to the North American public. The kiwifruit is a good example of a fruit that, within a short time, made the jump from exotic to familiar.

Around 1980 the kiwifruit, *Actinidia chinensis,* began appearing in supermarkets throughout the United States, and soon after that, many people were enjoying this fuzzy, brown, egg-shaped fruit. The distinctive flavor of its emerald-green flesh is reminiscent of a strawberry-banana-pineapple combination. Its current popularity is the result of successful marketing that began in New Zealand. Originally native to China, the plants were introduced into New Zealand in 1904, where kiwifruit was known as the Chinese gooseberry and often was grown as an ornamental vine. Commercial farming began in the 1930s, and the first exports were delivered to England in 1952. Around this time, marketing strategists renamed the Chinese gooseberry; the new name, kiwifruit, fit because its fuzzy rind resembles New Zealand's flightless bird, the kiwi. The name change paid off. Sales and exports increased steadily, and by the late 1980s, it was an American produce staple. In the United States, cultivation of this berry is found mainly in northern California, with a large percentage of the crop exported to Europe, Japan, and Canada.

One exotic fruit that may be seen more frequently in the near future is the cherimoya, *Annona cherimola.* This tree fruit, native to the uplands of Peru and Ecuador (Fig. 6.8), was first cultivated by the ancient Incas. In appearance, the aggregate fruit looks like a leathery green pinecone. Its custardlike flesh can be scooped out and is delicious with cream or orange juice. Today, cherimoya, described by some as the aristocrat of fruits, is widely grown in Chile, Spain, and Israel and is making inroads in North America through California growers. Atemoya, a new hybrid resulting from a cross between the cherimoya and the closely related sugar apple (*Annona squamosa*), has the advantage of being more tolerant of environmental conditions than either parent and can, therefore, be grown in a wide variety of climates. The sweet taste of atemoya makes this a superb fruit for fresh consumption as well as for frozen desserts.

Carambola, *Averrhoa carambola,* is another ancient fruit that has recently been introduced to North American markets. Native to Malaysia, the carambola is also known as star fruit because a series of five-pointed yellow stars results from slicing the elongate, fluted fruit. This tart fruit adds an appealing shape that brightens up seafoods, salads, desserts, and fruit punches. The fruit can also be squeezed to make a refreshing juice or can be picked green, cooked, and eaten as a vegetable. Presently, star fruits are grown throughout the tropics, with the U.S. cultivation centered in Florida.

While cherimoyas are the aristocrats of fruit in South America, durians, *Durio zibethinus,* are considered the king of fruit throughout much of Southeast Asia. The best durians are said to have originated in Malaysia, although Thailand and South Vietnam are the leading producers. Durian trees produce melon-sized fruits that have a thick green rind and are covered with stout spines. Inside the fruit, there are usually five compartments, with smooth, creamy white pulp surrounding the seeds. This pulp, which is typically eaten fresh,

Figure 6.8 Cherimoya, a fruit cultivated by the ancient Incans.

is usually described as having a heavenly flavor. By contrast, the most notorious feature of durian fruits is their incredibly foul smell, which has been described as similar to the smell of rotten eggs. The odor is due to the presence of sulfur compounds; in fact, there are 43 sulfur compounds in durians. Some of these are similar to the compounds in onions and garlic; others are similar to the compounds produced by skunks. The odor is so strong that, in Singapore, fresh durians are banned in buses, subways, taxis, and airlines. Despite the odor, during the harvest season, thousands of tourists from Japan and other Asian countries attend durian festivals and tours in Malaysia. Durian trees require tropical conditions and abundant rainfall; seedlings introduced to Florida survived only a short time. The trees have been introduced successfully to Hawaii, some Caribbean islands, and Honduras, although so far the plantings are not extensive. Fresh durians have a short shelf life and are seldom shipped out of Asia. However, the United States is the largest importer of frozen durians, which can be found in Asian groceries in major U.S. cities. The world's smelliest fruit may soon be arriving at your neighborhood market.

After 20 years of research, a scientist at Thailand's Horticultural Research Institute has created an odorless variety of durian. Without the offensive smell, Chantaburi No. 1 is predicted to become a major export crop for Thailand when commercial production begins in a few years. In the meantime, research is progressing on a durian variety that is both odorless and thornless.

CHAPTER SUMMARY

1. Fruits are unique to the sexual reproduction of angiosperms. They protect the enclosed seeds and aid in seed dissemination. Botanically, a fruit is a ripened ovary, although in the United States the legal definition of a fruit is something that tastes sweet and is eaten as dessert.

2. Fruits can be classified according to the characteristics of the fruit wall, or pericarp. In fleshy fruits, the pericarp is soft and juicy; berry, hesperidium, pepo, drupe, and pome are all examples of fleshy fruits. In dry fruits, the pericarp is often tough or papery. Dry fruits can also be dehiscent, splitting open along one or more seams to release their seeds. Follicles, legumes, and capsules are examples of dehiscent fruits. Dry fruits that do not split open are indehiscent; examples of this fruit type are achenes, samaras, grains, and nuts. Simple fruits are derived from a single ovary. Aggregate fruits develop from the separate ovaries within a single flower; multiple fruits result from the fusion of ovaries from separate flowers in an inflorescence.

3. Seeds are the end products of sexual reproduction in flowering plants. Each seed contains an embryonic plant, nutrient tissue to nourish the embryo, and a tough outer seed coat. Differences exist between monocot and dicot seeds. Monocotyledonous seeds have a single thin cotyledon, whereas dicotyledonous seeds typically have two large cotyledons.

4. Edible fruits have played an important role not only as a significant contribution to the human diet but also in scientific studies and folklore. Once-exotic fruits are becoming commonplace as they are incorporated into the world's marketplace.

REVIEW QUESTIONS

1. What is the function of the fruit in the life cycle of an angiosperm?
2. Give the botanical meaning of the following: berry, nut, legume, and grain.
3. What is a seed? What is a fruit?
4. Compare and contrast monocot and dicot seeds in structure and germination.
5. How did the tomato, fruit of the Americas, become the staple of Italian cookery?
6. What factors contributed to the successful introduction of the kiwifruit into North American markets?
7. Although "chestnuts roasting on an open fire" is an American image, most chestnuts eaten in the United States are from Italy. Why?
8. Research the history, uses, and folklore of the following fruits: pineapple, mango, and papaya.
9. How does fruit structure reflect the method of seed dispersal? Give examples.

Source: Stephen Ausmus/USDA

CHAPTER

7

Genetics

A USDA scientist applies pollen from the anthers of a desired blackberry parent to the stigma of another blackberry flower.

KEY CONCEPTS

1. Gregor Mendel, an Austrian monk, discovered the basic principles of inheritance in the 1860s.
2. Mendelian genetics explains the inheritance of genes that are located on separate chromosomes, but other patterns of inheritance occur that are not governed by Mendel's principles.
3. The chemical basis of inheritance is the gene, a segment of the DNA molecule that contains a specific nucleotide sequence that codes for the formation of a particular product.

CHAPTER OUTLINE

Mendelian Genetics 102
    Gregor Mendel and the Garden
      Pea 102
    Monohybrid Cross 103
    Dihybrid Cross 105
Beyond Mendelian Genetics 106
    Incomplete Dominance 106

A CLOSER LOOK 7.1 Solving Genetics
  Problems 107

    Multiple Alleles 109
    Polygenic Inheritance 109
    Linkage 109

Molecular Genetics 110
    DNA–The Genetic Material 111

A CLOSER LOOK 7.2 Try These Genes
  On for Size 112

    Genes Control Proteins 114
    Transcription and the Genetic Code 114
    Translation 116
    Other Roles of RNA 116
    Mutations 117
    Epigenetics 118
    Recombinant DNA 118
Chapter Summary 119
Review Questions 119
Additional Genetics Problems 119

Genetics is the study of inheritance, the transmission of traits from parent to offspring and the expression of these traits. From earliest times, people have realized that certain traits in both plants and animals are passed on from parents to offspring. Artificial selection was practiced by farmers both consciously and unconsciously in establishing many domesticated plants and animals. It has only been in the twentieth century that science has provided a clear understanding of the nature of **genes** and the scientific basis for selective breeding. The hereditary material, DNA (deoxyribonucleic acid), found in chromosomes is organized into units called genes. Each gene contains a specific sequence of nucleotides that codes for the formation of a particular product. Genes control all phases in the life of an organism, including its metabolism, size, color, development, and reproduction.

A gene is a segment of the DNA molecule, and its location on a specific chromosome is called its **locus.** The discussion of meiosis in Chapter 5 indicated that each gamete contains only one of each homologous chromosome, so diploid offspring receive one set of chromosomes (a haploid set) from each parent. For any genetic character, the offspring, therefore, will have two genes. It will receive one gene on a specific chromosome from one parent and another gene for that character on the **homologous** chromosome from the other parent. Genes often exist in at least two alternate forms, known as **alleles**. For example, a gene that controls flower color may have alleles that specify purple or white flowers. An offspring can receive two identical alleles for a specific character, a condition known as **homozygous**, or two different alleles for that character, a condition called **heterozygous**.

Today, these concepts constitute a clear foundation for understanding the principles of genetics. When Gregor Mendel, considered the founder of genetics, began his studies of inheritance in the middle of the nineteenth century, none of these concepts were known. The Cell Theory was generally accepted among scientists of the time, including the view that cells arose only from preexisting cells. In 1831, Robert Brown had described the nucleus as a fundamental and constant component of the cell, but its importance was not recognized. Nucleic acids and chromosomes were not known, and neither mitosis nor meiosis had been described. This makes Mendel's discoveries all the more spectacular.

MENDELIAN GENETICS

The science of genetics can be traced back to the work of Gregor Mendel (1822–1884), an Augustinian monk who lived and worked in an Austrian monastery in the mid-nineteenth century (fig. 7.1). The monastery at Brunn (now Brno in the Czech Republic) was a center of enlightenment and scientific thought. It had an excellent library, botanical garden, and herbarium. Many of these resources were due to Abbot Cyril Napp, who was a skilled plant breeder as well as a leader in church matters. From 1851 to 1853, Napp sent Mendel to

Figure 7.1 Gregor Mendel (1822–1884) is considered the father of genetics.

study biology, physics, and math at the University of Vienna, where he was influenced by some of the leading scientists of the day. There, Mendel learned to apply a quantitative experimental approach to the study of natural phenomena.

Mendel was interested in the inheritance of traits, and when he returned to the monastery, he began a series of experiments that eventually demonstrated the nature of heredity. In 1865, Mendel summarized the results of his research in a paper he read to the Brunn Society for the Study of Natural Science and published a detailed written account the following year. Although Mendel sent copies of his paper to leading biologists of the day, his work was ignored for over 30 years. In 1900, 16 years after Mendel's death, three scientists independently carried out genetic studies of their own. They reached the same conclusions as Mendel and cited his pioneering work.

Gregor Mendel and the Garden Pea

Mendel chose the garden pea for his experiments, possibly because garden peas were available in many easily distinguishable varieties and were easy to cultivate. Mendel obtained 34 distinct strains of the pea from local farmers and raised generations of the plants for 2 years to determine which plants bred true. (Today, of course, true-breeding plants are called homozygous.) In plants that breed true, the offspring are identical to the parents for the trait in question. In his initial crosses, Mendel used only those plants that bred true. It was possible to maintain these strains because garden peas, unlike many plants, are self-pollinating. The flower does not open fully, so pollen from the anther lands on the stigma within the same flower and fertilizes it. Although this is the normal method of sexual reproduction, experimental crosses can also be made by carefully opening the flower and removing the anthers before the pollen is mature. Pollen from the flower of another pea plant can be transferred to the stigma of the first plant (fig. 7.2). Mendel made many of these crosses. As added insurance, Mendel covered these flowers with small cloth bags to prevent any pollinating insects from visiting the flower and introducing unwanted pollen.

For his experiments, Mendel eventually selected strains with seven clearly contrasting pairs of traits: *Tall* versus *dwarf* plants, *yellow* versus *green seeds*, *smooth* versus *wrinkled seeds*, *green* versus *yellow pods*, *inflated* versus *constricted pods*, *purple* versus *white flowers*, and *terminal* versus *lateral flowers* (fig. 7.3). Mendel studied only one or two of these pairs of traits at a time. He traced and recorded the type and number of all offspring produced from each pair of plants that he cross-fertilized. Mendel also followed the results of each cross for at least two generations.

Monohybrid Cross

Mendel carried out a series of **monohybrid crosses**, mating individuals that differed in only one trait. The members of the first generation of offspring all looked alike and resembled one

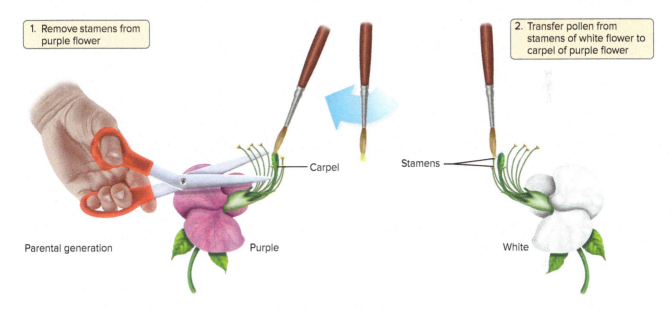

Figure 7.2 The flower of the garden pea. Although the garden pea is normally self-pollinated, experimental crosses can be made by opening the flower and removing the anthers before they mature. Pollen from another pea flower is then transferred.

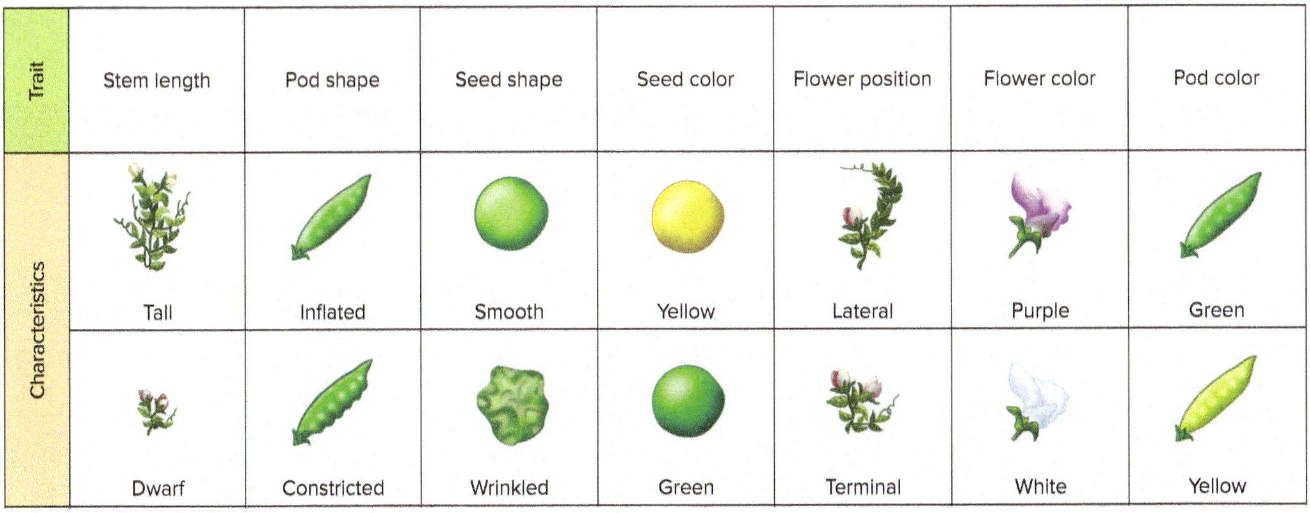

Trait	Stem length	Pod shape	Seed shape	Seed color	Flower position	Flower color	Pod color
Characteristics	Tall	Inflated	Smooth	Yellow	Lateral	Purple	Green
	Dwarf	Constricted	Wrinkled	Green	Terminal	White	Yellow

Figure 7.3 The seven pairs of characteristics studied by Mendel.

of the two parents. In one set of experiments, Mendel crossed a pea plant that produced yellow seeds with a strain that produced green seeds. These were called the P, or parental, generation. The offspring from this cross, called the **F₁ generation**, or first filial generation, produced only yellow seeds. Mendel then allowed the F₁ generation to self-pollinate and produce the **F₂ generation**, or second filial generation, which developed 8,023 seeds. Out of the total, 6,022 were yellow and 2,001 were green or approximately three-fourths yellow and one-fourth green. This represented a 3:1 ratio. The green trait reappeared in the F₂ generation, having been masked in the F₁ generation. The yellow trait had dominated over the green trait in the F₁ generation.

Mendel got similar results from other crosses (table 7.1). From these results he proposed that each kind of inherited character is controlled by two hereditary factors. (Today we call these hereditary factors genes and know that they are located on homologous chromosomes.) For each pair of traits he studied, one allele masked, or was **dominant** over, the other allele in the F₁ generation. This concept is usually referred to as Mendel's principle of dominance. Mendel called the nondominant allele **recessive**. Each recessive trait reappeared in the F₂ generation in the ratio of three dominant to one recessive.

Although nothing was known about meiosis at this time, Mendel proposed that, when gametes are formed, the pairs of hereditary factors (gene pairs) become separated so that each sex cell (egg or sperm) receives only one of each kind of factor or gene. This concept is known as the principle of segregation. Recall from Chapter 5 that in flowering plants it is actually the gametophytes, the ovule and pollen, that produce the gametes.

The actual appearance of a trait is known as the **phenotype**. In the earlier example describing Mendel's research, the phenotypes of the true-breeding parents were yellow seeds for one parent and green seeds for the other. If the symbol *Y* is used to designate yellow and *y* is used to designate green, then the genetic makeup (known as the **genotype**) of one parent is *YY* and the genotype of the other parent is *yy*. One parent was homozygous for yellow seeds, and the other parent was homozygous for green seeds. The gametes produced by the ovule or pollen from the yellow-seeded parent all contained the *Y* allele and those produced by the green-seeded parent all contained the *y* allele. The phenotype of all the F₁ generation was yellow seeds, but the genotype was *Yy*. These plants were heterozygous, but the dominant allele masked the expression of the recessive trait:

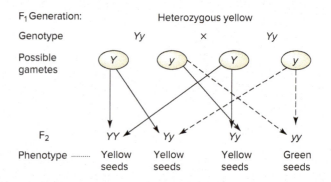

When the F₁ plants were allowed to self-pollinate and produce the F₂ generation, gametes containing either *Y* or *y* were formed. When fertilization took place, a random combination of the gametes produced the following results:

Although three-fourths of the F₂ generation contained the recessive allele *(y)*, only one-fourth expressed the recessive trait, green seeds.

Table 7.1
Mendel's Results from Monohybrid Crosses

Parents					Ratios
Dominant	Recessive	F₁ Generation	F₂ Generation		
Yellow seeds	Green seeds	All yellow seeds	6,022 yellow seeds	2,001 green seeds	3.01:1
Purple flowers	White flowers	All purple flowers	705 purple flowers	224 white flowers	3.15:1
Smooth seeds	Wrinkled seeds	All smooth seeds	5,474 smooth seeds	1,850 wrinkled seeds	2.96:1
Inflated pods	Constricted pods	All inflated pods	822 inflated pods	299 constricted pods	2.95:1
Green pods	Yellow pods	All green pods	428 green pods	152 yellow pods	2.82:1
Tall plants	Dwarf plants	All tall plants	787 tall plants	277 dwarf plants	2.84:1
Lateral flowers	Terminal flowers	All lateral flowers	651 lateral flowers	207 terminal flowers	3.14:1

Another way of displaying the probable recombinants that can occur is the use of a Punnett square. The alleles representing the gametes from one parent are on one side of the square, and those representing the other parent are along the other side. Combining gametes with the indicated alleles fills in all the boxes of the square, thereby displaying all possible combinations of offspring. Using a Punnett square to show the self-pollination of the F_1 generation gives

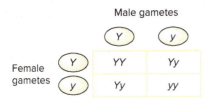

Thinking Critically

Alternate forms of a gene are known as alleles. For each pair of alleles that Mendel studied, he found that one allele was dominant over the other allele.

You have discovered a new species of plant on a remote mountain in Colorado. Some individuals have blue flowers and some have red flowers. How can you determine which allele is dominant?

To test his theory, Mendel allowed the F_2 plants to grow to maturity and self-pollinate and thus produce the F_3 generation. Green-seeded plants gave rise only to green seeds in the third generation. One-third of the yellow-seeded plants (those that were homozygous *YY*) gave rise only to yellow seeds, while two-thirds of the yellow-seeded plants (which were heterozygous *Yy*) gave rise to both yellow and green in the same 3:1 ratio.

Dihybrid Cross

Mendel also analyzed a series of **dihybrid crosses**, matings that involved parents that differed in two independent traits.

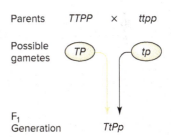

Testcross

Y? Yellow-seeded plant of unknown genotype × yy Green-seeded plant

If the plant being tested is homozygous:

Male gametes: Y Y
Female gametes: y Yy Yy / y Yy Yy

All the seeds are yellow.

If the plant being tested is heterozygous:

Male gametes: Y y
Female gametes: y Yy yy / y Yy yy

Half the seeds are yellow and half are green.

Mendel also used testcrosses to support his hypotheses. A **testcross** involves mating an individual with an unknown genotype to a homozygous recessive individual. For example, a pea plant with yellow seeds can be *Yy* or *YY*. If the unknown were homozygous *YY*, then all the offspring would be yellow-seeded when the unknown was crossed with a green-seeded plant. If the unknown were heterozygous *Yy*, then half the offspring would have yellow seeds *(Yy)* and half would have green seeds *(yy)*.

Testcrosses are used today to determine unknown genotypes, especially in commercial seed production when the breeder is trying to determine if a strain will breed true for a certain trait.

For the parent, Mendel crossed true-breeding tall plants with purple flowers *(TTPP)* and true-breeding dwarf plants with white flowers *(ttpp)*. The entire F_1 generation was tall with purple flowers, heterozygous for both characters *(TtPp)*. The results were predictable, in keeping with the dominance of both the tall and purple alleles.

Mendel allowed the F_1 generation to self-pollinate and observed the phenotypes in the F_2 generation. Four phenotypes appeared in the F_2 generation, including two phenotypic combinations not seen in either parent (tall plants with white flowers and dwarf plants with purple flowers). For this result to occur, the alleles for the two genes (for height and flower color) must have segregated independently from each other during gamete formation. Instead of gametes having only *TP* and *tp* combinations, four kinds of gametes would be produced in equal numbers: *TP, Tp, tP,* and *tp*. When these gametes combined randomly, they would produce nine genotypes and four phenotypes (tall purple, tall white, dwarf purple, and dwarf white) in the F_2 generation. The four phenotypes occurred in a 9:3:3:1 ratio. Not only were the original parental types represented in the F_2 generation but two new recombinant types were also present.

F₁ Generation		TtPp		
Possible gametes	TP	Tp	tP	tp

		Male gametes			
		TP	Tp	tP	tp
Female gametes	TP	TTPP	TTPp	TtPP	TtPp
	Tp	TTPp	TTpp	TtPp	Ttpp
	tP	TtPP	TtPp	ttPP	ttPp
	tp	TtPp	Ttpp	ttPp	ttpp
		Tall purple 9	Tall white 3	Dwarf purple 3	Dwarf white 1

Mendel's actual results were close to the 9:3:3:1 ratio, not only for this cross but also with several other dihybrid crosses. On the basis of these results, Mendel postulated the principle of independent assortment, which states that members of one gene pair segregate independently from other gene pairs during gamete formation. In other words, different characters are inherited independently. Each gamete receives one gene for each character, but the alleles of different genes are assorted at random with respect to each other when forming gametes. Today, it is known that this principle applies to genes that occur on separate chromosomes, not to those that lie close together on the same chromosome. To learn how genetic principles can be applied, read A Closer Look 7.1: Solving Genetics Problems.

BEYOND MENDELIAN GENETICS

The patterns of heredity explained by Mendel's principles, or laws of inheritance, are often called Mendelian genetics. The characters and traits that Mendel chose to study were lucky choices because they showed up very clearly. For each of the characters, one allele was dominant over the other, and both phenotypes were easy to recognize. Sometimes, however, the outcome is not that obvious because of variations in the usual dominant-recessive relationship.

Incomplete Dominance

When homozygous red snapdragons *(RR)* are crossed with homozygous white snapdragons *(rr)*, the flowers produced by the F₁ generation are pink *(Rr)*. When the F₁ generation is self-pollinated, the resulting F₂ plants are produced in a ratio of 1 red: 2 pink: 1 white (fig. 7.4). When a pink snapdragon is crossed with a white one, one-half of the offspring are pink and one-half are white. The pink-flowered plants are heterozygous, but neither the red allele nor the white allele is completely dominant.

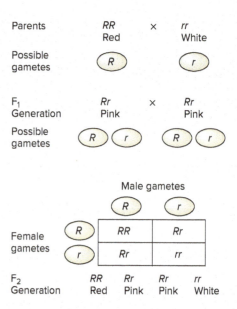

The *R* allele codes for a protein that is involved in the synthesis of red pigment, while the *r* allele codes for an altered protein that does not lead to pigment production. Plants that are homozygous *rr* cannot synthesize pigment at all, and the flowers are white. Heterozygous plants with one *R* gene produce just enough red pigment that the flowers are pink, while homozygous *RR* plants produce more red pigment. Whenever the heterozygous phenotype is intermediate, the genes are said to show **incomplete dominance**. This phenomenon is not unique to snapdragons; in fact, it is relatively common.

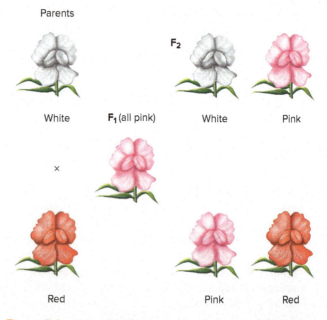

Figure 7.4 Incomplete dominance. In a cross between a red and a white snapdragon, all the offspring in the F₁ generation have pink flowers. In the F₂ generation, there is a 1:2:1 ratio of red to pink to white flowers.

A CLOSER LOOK 7.1
Solving Genetics Problems

Solving genetics problems is like solving a puzzle. They can be challenging and perplexing but rewarding when finally solved. Using the following steps will make solving the problems easier.

Step 1. Establish a key to the alleles for each trait described. Use an uppercase letter to designate the dominant allele and the same letter, but lowercase, for the recessive allele. If the problem does not indicate which trait is dominant, the phenotype of the F_1 generation is often a good clue.

Step 2. Determine the genotypes of the parents. If the problem states the parents are true breeding, they should be considered homozygous. If one parent expresses the recessive phenotype, the genotype can be readily deduced. The phenotype of the offspring can also provide evidence of the parental genotypes.

Step 3. Determine the possible types of gametes formed by the ovules or pollen from each parent. If the problem is a monohybrid cross, the principle of segregation must apply. For example, if the parent is heterozygous *Tt*, two kinds of gametes are produced: *T* and *t*. A homozygous parent forms only one kind of gamete: *TT* can produce only gametes with *T*, and *tt* can produce only gametes with *t*. If the problem involves a dihybrid cross, the principle of segregation and the principle of independent assortment both apply. If a parent is heterozygous for two traits, *TtSs*, and the genes for the two traits occur on separate chromosomes, then four possible types of gametes are formed: *TS*, *Ts*, *tS*, and *ts*. Often, the most difficult part of the problem is solved by determining the gametes.

Step 4. Set up a Punnett square, placing the gametes from one parent across the top and the gametes from the other parent down the left-hand side. Fill in the Punnett square. If a dominant allele is present, always place it first in each pair of genes (i.e., always write the dihybrid as *TtSs* and never *tTsS*). Read off the phenotypes (and genotypes, if requested in the problem) and calculate the phenotypic ratios.

Try the following problems:

Problem 1. In squash, the allele for white color *(W)* is dominant to the allele for yellow color *(w)*. Give the phenotypic ratios for each of the following crosses:

a. *WW* × *ww*
b. *Ww* × *ww*
c. *Ww* × *Ww*

Answers: Solving these crosses step by step, as suggested here, reveals that Steps 1 and 2 are given in the problem. The key to the alleles is provided (white, *W*, is dominant to yellow, *w*), and the genotypes of the parents are provided for each cross. Start with Step 3.

a. Both parents, *WW* and *ww*, are homozygous; therefore, each can form only one kind of gamete. Parent *WW* can produce only gametes with the gene for white color *(W)*, and parent *ww* can form only gametes with the gene for yellow color *(w)*. Step 4 involves setting up a Punnett square. For this simple monohybrid cross, it may not be necessary, since each parent can produce only one kind of gamete:

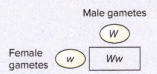

Answer to a. Since all the offspring will have the genotype *Ww*, all will show the white phenotype.

b. One parent is heterozygous, and one is homozygous. The heterozygous parent, *Ww*, can produce two kinds of gametes, either *W* or *w*, while the homozygous parent, *ww*, can form only gametes with the *w* allele. The Punnett square should show the following:

Since one-half of the offspring are heterozygous, they will show the dominant phenotype, white. The other half are homozygous recessive and will be yellow.

Answer to b. There is a 1:1 ratio of white to yellow.

c. Both parents are heterozygous and can therefore form two kinds of gametes, either *W* or *w*. The Punnett square should show the following:

Male gametes

	W	w
W	WW	Ww
w	Ww	ww

Female gametes

Three-fourths of the offspring (homozygous *WW* and the heterozygous *Ww*) will show the dominant phenotype, white. One-fourth of the offspring (homozygous *ww*) will show the recessive phenotype, yellow.

Answer to c. There is a 3:1 ratio of white to yellow.

Problem 2. In garden peas, the allele for tall plants (*T*) is dominant over the allele for dwarf plants (*t*); and the allele for smooth peas (*S*) is dominant over the allele for wrinkled peas (*s*). Determine the phenotypic ratios for each of the following crosses:

  a. *TtSs* × *ttss*
  b. *TtSs* × *TtSs*

Answers: Again, Steps 1 and 2 are provided in the problem, and the solutions can begin with Step 3, determining the gametes.

a. One parent is heterozygous for both traits, *TtSs*. Remember that the principles of segregation and independent assortment both apply. This will result in the formation of four possible types of gamete: *TS*, *Ts*, *tS*, and *ts*. The other parent is homozygous for both traits, *ttss*, so only one type of gamete can form, *ts*. The Punnett square should show the following:

Male gametes

	TS	Ts	tS	ts
ts	TtSs	Ttss	ttSs	ttss

Female gametes

One-fourth of the offspring will show both dominant characters, tall with smooth peas; one-fourth will be tall with wrinkled peas; one-fourth will be dwarf with smooth seeds; and one-fourth will show both recessive characters, dwarf with wrinkled seeds.

Answer to a. There will be a 1:1:1:1 ratio of tall smooth to tall wrinkled to dwarf smooth to dwarf wrinkled.

b. Both parents are heterozygous for both traits; therefore, the Punnett square should show the following:

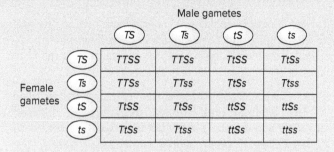

Reading off the phenotypes from the Punnett square shows that nine-sixteenths of the offspring will show both dominant characters, tall plants with smooth seeds; three-sixteenths will be tall with wrinkled seeds; three-sixteenths will be dwarf plants with smooth peas; and one-sixteenth of the offspring will show both recessive characters, dwarf plants with wrinkled peas.

Answer to b. There will be a 9:3:3:1 ratio of tall smooth to tall wrinkled to dwarf smooth to dwarf wrinkled.

Problem 3. In watermelons, the genes for green color and for short shape are dominant over their alleles for striped fruit and long shape. Suppose a plant that has genes for long striped fruit is crossed with a plant heterozygous for both these traits. What would be the phenotypic ratio of offspring from this cross?

Answer: To solve this problem, begin at Step 1. Assign letters to represent the dominant and recessive alleles for each trait. Green fruit (*G*) is dominant to striped fruit (*g*), and short shape (*S*) is dominant to long shape (*s*).

Next, determine the genotypes of the parents. The plant with long striped fruit will be *ssgg*, and the heterozygous parent will be *SsGg*.

The homozygous recessive parent *ssgg* can form only one type of gamete, *sg*. The heterozygous parent can form four possible types of gametes: *SG*, *Sg*, *sG*, and *sg*. The Punnett square should show the following:

Male gametes

	SG	Sg	sG	sg
sg	SsGg	Ssgg	ssGg	ssgg

Female gametes

Among the offspring, there will be a 1:1:1:1 ratio of short green to short striped to long green to long striped watermelons.

Sometimes when neither allele is dominant, both alleles are expressed independently in the heterozygote. This condition is known as **codominance**. (In incomplete dominance, the heterozygote shows an intermediate phenotype, but in codominance both phenotypes are expressed.) The classic example of codominance is seen in the human AB blood type. Some individuals have type A blood and others have type B; the alleles refer to the presence of A or B antigens on the surface of a red blood cell. Individuals with type AB blood have both types of antigens present on their red blood cells.

Multiple Alleles

Sometimes more than two alleles exist for a given character. Of course, any single individual has only two alleles (on homologous chromosomes). Human blood types also serve as an excellent example of this non-Mendelian feature, known as **multiple alleles**. Three alleles exist for human blood types I^A, I^B, and i. Both I^A and I^B code for the formation of surface antigens, whereas i does not result in antigen production. Different combinations of these alleles result in four different phenotypes (blood types): A, B, AB, and O. Although I^A and I^B are codominant with respect to each other, both are dominant to i and will produce antigens if one allele is present. The homozygous recessive condition, ii, produces blood type O with no surface antigens.

Other well-studied examples of multiple alleles are found in fruit flies and clover leaves. In the fruit fly, a large number of alleles affect eye color by determining the amount of pigment produced. The final eye color depends on which two alleles are inherited. Some leaves of clover are solid green, while other clover plants have patterns (stripes, chevrons, or triangles) of white present on the leaves. Seven different alleles influence this trait. The resulting coloration and pattern depend on which two alleles are inherited.

Polygenic Inheritance

Often, a character is controlled by more than one pair of genes, and each allele has an additive effect on the same character. Examples of **polygenic inheritance** are relatively common, especially for human characteristics such as skin color and height. The inheritance of seed color in wheat, controlled by three sets of alleles, is also an example of polygenic inheritance. When plants producing white seeds are crossed with plants producing dark red seeds, the F_1 generation is intermediate in color. However, if the F_1 generation are allowed to self-pollinate, the second generation shows seven phenotypes ranging from white to dark red. Each dominant allele has a small but additive effect on the pigmentation of the seeds. If A, B, and C represent the dominant alleles that code for pigment production and a, b, and c represent the recessive alleles that lead to no pigment production, the three generations can be described as follows:

Parents:	*AABBCC*	×	*aabbcc*
	Dark red seeds		White seeds
F_1 Generation		*AaBbCc*	
		Medium red seeds	
F_2 Generation	All dominant alleles *(AABBCC)*—dark red seeds		
	Any five dominant alleles—deeper red seeds		
	Any four dominant alleles—deep red seeds		
	Any three dominant alleles—medium red seeds		
	Any two dominant alleles—light red seeds		
	Any one dominant allele—pale red seeds		
	All recessive alleles *(aabbcc)*—white seeds		

Most of the F_2 generation show intermediate phenotypes. On average, only 1 out of 64 would show the dark red seeds of one parental type and 1 out of 64 would show the white seeds of the other parental type (fig. 7.5).

Linkage

When Mendel conducted his experiments and developed his principles of inheritance, the existence of chromosomes was not known. Mendel considered his "hereditary factors" as separate particles, and this idea was integrated into his principle for independent assortment. However, chromosomes are inherited as units, so genes that occur on one chromosome tend to be inherited together, a condition known as **linkage**. Because the genes are on the same chromosome, they move together through meiosis and fertilization.

Linkage was first described by British geneticists William Bateson and Reginald C. Punnett in 1906 while working with sweet peas. In the years from Mendel's work until 1905, much progress had been made studying cells and reproduction. Chromosomes had been described, and so had the events of mitosis and meiosis. Bateson and Punnett knew from previous experiments that purple flowers were dominant to red flowers and that the oval (long) pollen shape was dominant to round. When they crossed a homozygous purple-flowered plant with long pollen (*PPLL*) to a homozygous red-flowered plant with round pollen (*ppll*), the F_1 generation plants (*PpLl*) all resembled the dominant parent, with purple flowers and long pollen. However, when they allowed the F_1 generation to self-pollinate, the F_2 generation did not occur in the 9:3:3:1 ratio that was expected based on Mendel's experiments. Instead, there were 284 purple-flowered plants with long pollen, 21 plants with purple flowers and round pollen, 21 plants with red flowers and long pollen, and 55 plants with red flowers and round pollen. The results seemed to indicate that genes for both flower color and pollen shape occurred on the same chromosome. The linkage of these genes should yield a 3:1 ratio of

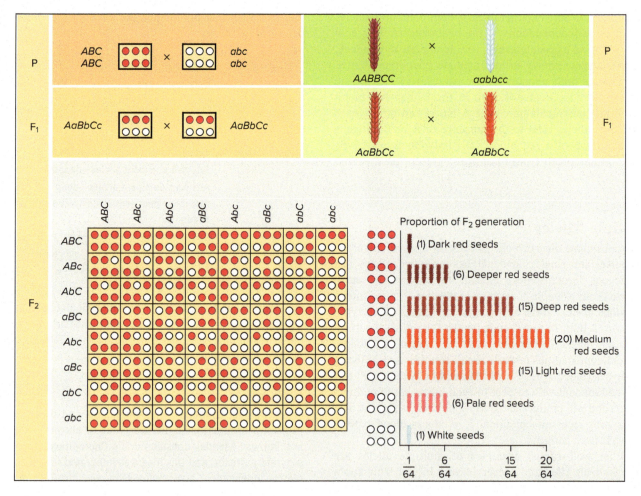

Figure 7.5 The inheritance of seed color in wheat is controlled by three sets of alleles, each of which has an additive effect on the phenotype.

purple flowers with long pollen to red flowers with round pollen. The linking of these genes is indicated as follows:

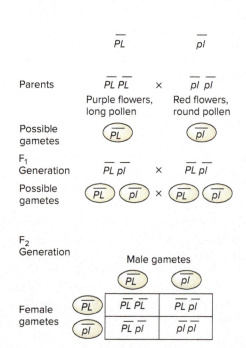

Although Bateson and Punnett described the linkage of these genes, they could not explain the 21 purple-flowered plants with round pollen and the 21 red-flowered plants with long pollen that occurred in the F₂ generation. It was the work of Thomas Hunt Morgan that explained the phenotype of these plants. In the early 1900s, Morgan was conducting genetics studies on the fruit fly *Drosophila melanogaster* at Columbia University. Morgan's work identified linked genes and showed that a small number of offspring received new combinations of alleles. Morgan proposed that an occasional exchange of segments between homologous chromosomes takes place during meiosis, thereby breaking the linkage between genes. Recall from Chapter 5 that this exchange of segments, called **crossing over**, occurs during prophase I of meiosis. Nonsister chromatids break at corresponding places, but the fragments reunite with the opposite chromatid. Thus, crossing over accounts for the recombination of linked genes (fig. 7.6).

MOLECULAR GENETICS

Probably the greatest achievement in biology in the twentieth century was the development of the field of molecular genetics. Because of the advances in this field, scientists now

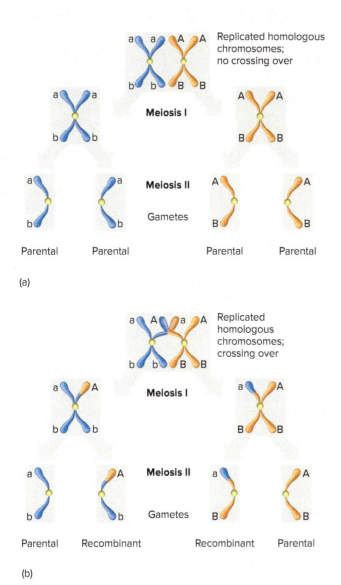

Figure 7.6 Crossing over during prophase 1 of meiosis can result in recombination of linked genes. In (a), no crossing over has occurred, so only two types of gametes are formed. In (b), crossing over occurred, resulting in four possible types of gametes that include two recombinant types.

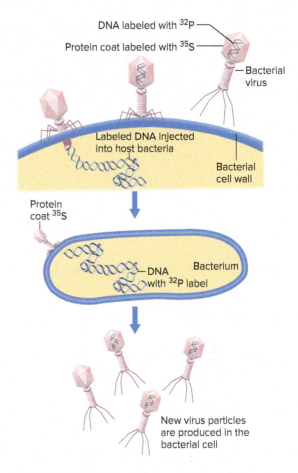

Figure 7.7 The Hershey and Chase experiment that proved conclusively that DNA is the genetic material. The two researchers used bacterial viruses with radioactively labeled protein (with ^{35}S) and DNA (with ^{32}P). They found that only the viral DNA entered the bacterial cell and provided the information to produce new viral particles.

have a precise understanding of the chemical nature of the gene, how it functions, how mutations can occur, how to transfer genes from one organism to another, and how to sequence genes (see A Closer Look 7.2: Try These Genes On for Size).

DNA—The Genetic Material

While genetics studies continued through the first four decades of the twentieth century, the exact nature of the gene remained unknown. Scientists sought to learn what genes were made of and how they worked. Although it was known that genes occurred on chromosomes and that chromosomes were composed of DNA and protein, it was widely assumed that proteins were the genetic material. The proof that DNA is the genetic material was established through the efforts of several researchers in the 1940s and early 1950s. The final proof was provided by Alfred Hershey and Martha Chase, who showed that, when bacterial cells are infected with viruses, only the viral DNA enters the cell. The protein coat remains on the outside while the viral DNA provides all the instructions for the manufacture of new virus particles (fig. 7.7).

In 1953, about the same time that Hershey and Chase were proving that DNA was the genetic material, James Watson and Francis Crick described the structure of the DNA molecule. Recall from Chapter 1 that DNA is a double helix, with alternating sugar molecules (**deoxyribose**) and phosphate groups making up sides of the helix and nitrogenous bases forming the cross-rungs of the helix. Watson and Crick suggested that the nitrogenous bases always pair up in a specific pattern, with one **purine** base (**adenine** or **guanine**) hydrogen bonding to one **pyrimidine** base (**thymine** or **cytosine**). Adenine always pairs with thymine and guanine with cytosine; this unvarying pattern is referred to as complementary base pairing. Watson and Crick also proposed that complementary base pairing provided a mechanism for the replication of the hereditary material. The model suggested that the double helix unwound, and each strand served as a template for manufacturing a

A CLOSER LOOK 7.2
Try These Genes On for Size

The Human Genome Project began in 1990 to determine the base sequence of the entire human genome as well as the genomes of select organisms representative of such major groups as the bacteria, protists, fungi, flowering plants, insects, and rodents. In mid-February 2001, the first draft of a human sequence map of all 24 chromosomes (1–22 plus the X and Y sex chromosomes) was completed by two competing, at times cooperative, groups: the International Human Genome Sequencing Consortium (Human Genome Project in the United States and sequencing centers in Great Britain, Japan, France, Germany, and China), supported by public funds (such as the National Institutes of Health and the Department of Energy in the United States), and Celera Genomics, a private biotechnology company in Rockville, Maryland. In April 2003, the International Human Genome Sequencing Consortium announced that it had finished the complete sequencing of the human genome.

DNA sequencing is the process of determining the exact order of the nucleotide bases (A, T, C, and G) in a chromosome. With automated DNA sequencing, the order of nucleotide bases in DNA can be determined in a few hours (box fig. 7.2). The Human Genome Project researchers collected blood (female) or sperm (male) samples from a large number of donors and selected a few samples from this collection for sequencing. Celera collected DNA samples from five donors who represented Hispanic, Asian, Caucasian, and African American ethnic groups.

Much has been learned about the human genome. The human genome contains about 3 billion bases. Almost all (99.9%) of the nucleotide bases are identical in all people. The areas of the genome that have a high density of genes are predominantly composed of the nucleotide bases G–C (guanine and cytosine), whereas the gene-poor regions are A–T bases (adenosine and thymine).

Not all chromosomes have the same number of genes. Chromosome 1 has the most genes at 3,168 genes, and the Y chromosome has the fewest, only 344. Many genes involved in disease have been located. The gene linked to cystic fibrosis is located on chromosome 7, and BRCA 1, one of the genes implicated in some breast cancers, is found on chromosome 17. So far, about 3 million locations have been identified on the human genome in which a single base of DNA differs among individuals. These are called **single nucleotide polymorphisms**, or **SNPs**. These differences can be used to find the location of disease-associated sequences or to understand why one drug for type 2 diabetes is helpful to some people but toxic to others.

The size of an average gene is 3,000 bases, but the largest human gene known consists of 2.4 million bases. One big surprise was the small number of genes, estimated to be between 19,000 to 20,500 genes, in the human genome. Much earlier predictions based on gene-rich areas had speculated that the human genome would contain as many as 100,000 genes. Although the number of genes in the human genome is about the same as found in a worm (table 7.A), there is evidence that human genes are capable of doing more. Instead of a gene coding for just one protein, each human gene, through a process called alternative splicing, can code an average of three. Another major surprise is how little of the human genome actually consists of genes, DNA that codes for proteins. Only about 2% of the genome consists of coding regions of DNA. Also, at least 50% of the genome consists of repeat sequences of DNA bases that do not code for proteins.

When the human genome is compared with previously sequenced genomes, some similarities and differences are evident. Genes in the human genome are randomly distributed, whereas genes in the genomes of other organisms examined are evenly spaced throughout the genome. About 223 human genes are nearly identical to bacterial genes and are not found in fungi, insects, or worms.

Now that the human genome is complete, there are several new initiatives. The Joint Genome Institute plans to focus sequencing efforts on more nonmammalian genomes. The National Human Genome Research Institute plans to investigate haplotypes, the individual and group DNA variations among people. Another project called ENCODE will concentrate on deciphering the function of genes.

Shortly before the publication of the human genome, the first flowering plant genome was completely sequenced in December 2000.

Although plant genomes are usually larger and more complicated than those of many animals, *Arabidopsis thaliana* (commonly called mouse-ear cress), a tiny weed in the commercially important mustard family (Brassicaceae) has

Table 7.A Genomes of Representative Organisms

Organism	Number of Base Pairs	Estimated Number of Genes
Virus	5,386	11
Agrobacterium	4,674,062	5,419
Yeast	12,495,682	5,770
Roundworm	100,258,171	21,733
Arabidopsis	135,000,000	27,000
Fruit fly	139,500,000	15,682
Mouse	2,800,000,000	23,000
Human	3,300,000,000	19,000–20,500

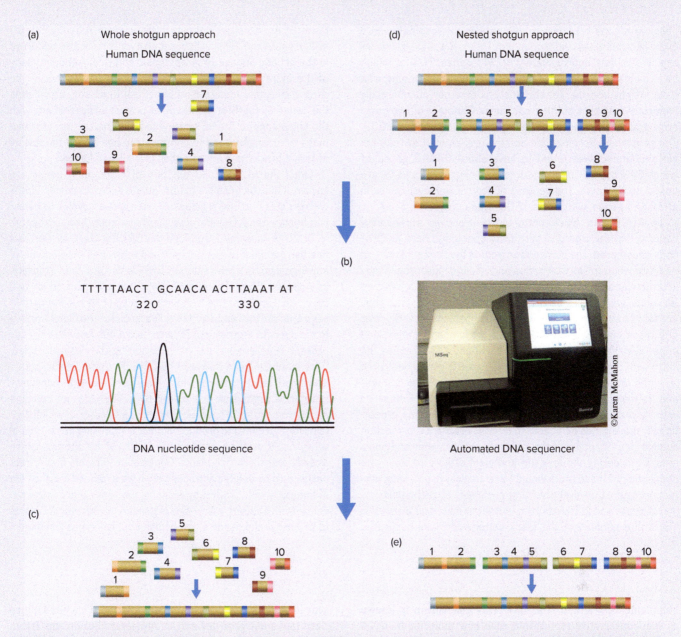

Box Figure 7.2 Genome sequencing techniques. In the whole shotgun approach, the strategy taken by Celera Genomics: (a) first the whole genome is shredded into fragments: (b) then the DNA sequence of each fragment is read by automated DNA sequencers to reveal the nucleotide base pattern; and (c) the fragments are reassembled in order using computers to see where the sequences overlap. The Human Genome Project used the nested shotgun approach to sequence the genome. First the 23 pairs of chromosomes are separated out and (d) fragments of the DNA of each chromosome are cut into smaller and smaller sizes and kept in relative order; (b) then each fragment is sequenced with automatic DNA sequencers to reveal the nucleotide base pattern, and (e) the sequenced fragments are reassembled according to the known order.

been sequenced. With 135 million bases and an estimate of 27,407 genes, this wild mustard exceeds the gene number for animals: the *Drosophila* fruit fly and the *Caenorhabditis elegans* roundworm. In fact, its gene number is even larger than that estimated for the human genome (table 7.A). Interestingly, about 60% of its genes are duplicates.

Now that the DNA sequences of many genes have been identified, the next step is to figure out what the genes do. One way to identify gene function is to use a BLAST (Basic Local Alignment Search Tool) search. The base sequence of a gene is entered into a database, such as GenBank, that contains sequences of previously identified genes. Next, there is a search of the database for matches. If the new gene matches the DNA sequence of a known gene whose function is also known, then it can be assumed that the new gene probably has a similar function. The functions of many of the genes have been assigned using this comparative genomic approach. About 100 of the genes in *Arabidopsis* are nearly identical to those found in the human genome. Some genes that humans and the mouse-ear cress share include those that have been implicated in human diseases, such as cystic fibrosis and breast cancer.

The complete sequencing of *Arabidopsis* has made it easier to identify genes in other flowering plant species. The publication of the first draft of the rice genome in 2002 made it possible for the first time to compare sequences between a dicot (*Arabidopsis*) and a monocot (rice). Rice (*Oryza sativa* ssp. *indica*) has approximately 40,745 genes, 70% of which have also been found in *Arabidopsis*. The rice genome also gave insight into the genetic analysis of other commercially important members of the grass family. Genomes of two other crop grasses, sorghum and maize, were completed in 2009. The genome of *Sorghum bicolor*, the first plant of African origin sequenced, was found to have about 34,496 genes, of which 24% appear to be characteristic of grasses and 7% are unique to sorghum. The genome of corn (*Zea mays*) comes in at 39,475 genes, with about 1,600 genes unique to maize.

In 2006, the black cottonwood (*Populus trichocarpa*) became the first tree and third plant species to have its DNA code deciphered. The black cottonwood is the largest hardwood native to western North America and is found most prominently in bottomlands of streams and rivers. Cottonwoods are named for white, wispy hairs on the seeds that catch the wind. They are fast-growing trees, reaching sexual maturity in 4–6 years. The black cottonwood is harvested for lumber, veneers, and paper pulp. With a diploid number of 38 and 41,335 genes, the cottonwood genome includes more than 90 genes dedicated to wood development, such as the synthesis of the cell wall and its major components of cellulose and lignin. Compared with *Arabidopsis*, the *Populus* genome also contains proportionally more genes dedicated to disease resistance, protection against herbivorous insects, and transport of water and nutrients. Considering the long life and giant stature of trees, the dominance of these traits would be necessary in providing protection against an ever-changing onslaught of pathogens and insect pests and in transporting nutrients over long distances. The entire genome of the black cottonwood was duplicated three times in its evolutionary history, more recently 60–65 million years ago. Similar duplication events have also occurred in the genetic lineage of *Arabidopsis* and rice. Although many of the extra gene copies are deleted over time, some of the gene duplicates acquire new beneficial functions. Elucidating the genes of the first woody plant provides researchers with the knowledge to create trees with taller stature, greater biomass, or stronger wood to improve tree health and the quality of forest products, such as paper, biofuel, and lumber.

Once the identity and function of more plant genes are known, it may be possible to manipulate genes to design plants with greater nutritional value or faster growth cycles to better suit human need. Comparative plant genomics has just begun.

In 2013, Addgene introduced the **CRISPR-Cas9** system that has the potential to alter, delete, and rearrange the genome of any organism with precision. CRISPR-Cas9 (Clustered Regularly Interspaced Short Palindromic Repeats) is composed of two parts: the Cas9 enzyme, a molecular scalpel, to cut out certain segments of DNA and an RNA guide to direct the Cas9 enzyme to the specific sequence of DNA to be cut out. Enzymes already present in cells mend the cut by inserting the desired nucleotides that have been delivered by CRISPR. Unlike GMOs (genetically modified organisms) which have been met with consumer resistance, organisms engineered by CRISPR do not combine genes from different species. Golden rice, GMO crop which contains foreign genes to produce vitamin A improving its nutritional value, could now be created with CRISPR by altering genes already present in rice plants. Genes in crop plants that attract pests could be deleted and scientists have extended the shelf life of tomatoes with CRISPR by silencing the genes that control ripening. With CRISPR, a new age of genetic modification of plants to better suit the needs of the human population has just begun.

---

complementary strand (fig. 7.8). This mechanism is known as semiconservative replication; each new molecule of DNA contains one old strand and one new strand. Today, we know that the process is more complicated than originally proposed, but the basics of the process are the same.

Genes Control Proteins

Recall that proteins are composed of one or more long chains of amino acids called polypeptide chains (see Chapter 1). Many types of proteins occur in living organisms, including enzymes (see Chapter 4), structural proteins, regulatory proteins, and transport proteins. The number of amino acids and their sequence are distinctive for each protein.

As early as 1908, Archibald Garrod, a British physician, suggested that genes and enzymes were somehow related. In studying the inheritance of genetic diseases, he suggested that the absence of a certain enzyme was associated with a specific gene. Little work was done in this area until early 1940, when George Beadle and Edward Tatum found mutants of the fungus *Neurospora crassa* that interfered with known metabolic pathways. Each mutant strain was shown to have a mutation in only one gene. Beadle and Tatum found that, for each individual gene identified, only one enzyme was affected. This was summarized into a *one gene-one enzyme hypothesis*. Beadle and Tatum received the 1958 Nobel Prize in Physiology or Medicine for their findings. It was soon realized, however, that many genes code for proteins other than enzymes and that many proteins contain more than one polypeptide chain. The Beadle and Tatum gene hypothesis was later modified to state that *each gene codes for one polypeptide chain*. Nevertheless, it was not until the early 1960s that scientists understood how cells are able to convert the DNA information into the polypeptide chains.

Transcription and the Genetic Code

The sequence of bases in the DNA molecule determines the sequence of amino acids in proteins, but the information in the DNA is not used directly. A molecule of **messenger RNA (mRNA)** is made as a complementary copy of a gene, a portion of one strand of the double helix. The process of RNA synthesis from DNA is known as **transcription**. Messenger RNA forms as a single strand by a mechanism that is similar to that of DNA replication. Enzymes cause a portion of the DNA molecule to unwind, and bases of RNA nucleotides bond with the exposed bases on one strand of the DNA. The enzyme **RNA**

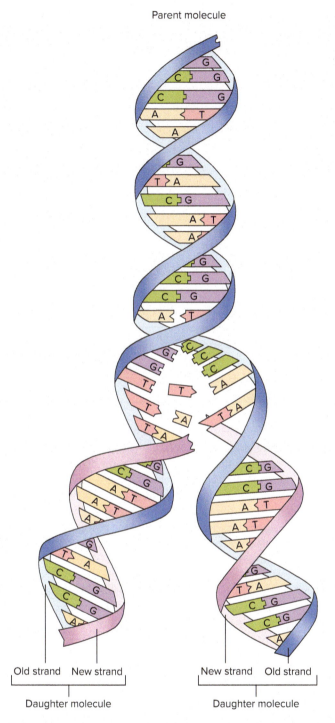

Figure 7.8 Semiconservative replication of DNA. Each strand serves as a template for the formation of a new strand of DNA. Each of the daughter molecules produced is composed of one old strand and one new strand.

polymerase is responsible for attaching nucleotides together in the sequence specified by the DNA (fig. 7.9). For example, if the sequence of bases on one strand of the DNA molecule is ATTAGCAT, the synthesized sequence on the RNA strand is UAAUCGUA. (Remember that the pyrimidine base uracil replaces thymine in RNA.) This molecule of mRNA represents a gene, and each gene in an organism is represented by a different RNA molecule. Each mRNA contains in its sequence

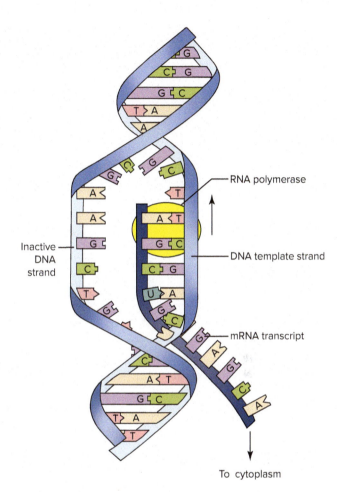

Figure 7.9 Transcription. During transcription, a molecule of messenger RNA is formed as a complementary copy of a region on one strand of the DNA molecule.

of bases information that will be **translated** into the sequence of amino acids that constitute a specific protein. Once the mRNA has been formed, other RNA molecules that function as enzymes within the nucleus process the molecule, modifying and removing segments called **introns**, which are not specifically involved in the coding for the polypeptide. The remaining segments, **exons** (or expressed segments), are spliced together to form the final mRNA molecule, which leaves the nucleus and goes into the cytoplasm ready for translation.

Each of three consecutive bases on the mRNA molecule constitutes a code word, or **codon**, that specifies a particular amino acid. This **genetic code** contains a total of 64 codons, with 61 of them coding for amino acids (table 7.2). Three codons do not code for amino acids but act to signal termination (UAA, UAG, and UGA). One codon, AUG, which codes for the amino acid methionine, acts as a signal to start translation. The genetic code is redundant because two or more codons that differ in the third base code for the same amino acid (table 7.2 and fig. 1.6). With very few exceptions, the genetic code is a universal code among living organisms. Codons specify the same amino acid in all organisms. This means that bacteria can translate genetic information from potato cells or human cells, or vice versa, and the eukaryotic cells can translate the genetic information of the bacterial cell (see Chapter 15).

Table 7.2
The Genetic Code

First Base	Second Base: U		Second Base: C		Second Base: A		Second Base: G		Third Base
U	UUU UUC	} Phe (phenylalanine)	UCU UCC	} Ser (serine)	UAU UAC	} Tyr (tyrosine)	UGU UGC	} Cys (cysteine)	U C
U	UUA UUG	} Leu (leucine)	UCA UCG	}	UAA-Stop UAG-Stop		UGA-Stop UGG-Trp (tryptophan)		A G
C	CUU CUC CUA CUG	} Leu (leucine)	CCU CCC CCA CCG	} Pro (proline)	CAU CAC	} His (histidine)	CGU CGC CGA CGG	} Arg (arginine)	U C A G
C					CAA CAG	} Gln (glutamine)			
A	AUU AUC AUA	} Ile (Isoleucine)	ACU ACC ACA ACG	} Thr (threonine)	AAU AAC	} Asn (asparagine)	AGU AGC	} Ser (serine)	U C A G
A	AUG-Met (methionine) or Start				AAA AAG	} Lys (lysine)	AGA AGG	} Arg (arginine)	
G	GUU GUC GUA GUG	} Val (valine)	GCU GCC GCA GCG	} Ala (alanine)	GAU GAC	} Asp (aspartic acid)	GGU GGC GGA GGG	} Gly (glycine)	U C A G
G					GAA GAG	} Glu (glutamic acid)			

Translation

The translation of the mRNA codons into an amino acid sequence occurs on ribosomes in the cytoplasm of the cell. In addition to mRNA, two other types of RNA function in translation. **Ribosomal RNA (rRNA)** joins with a number of proteins to form ribosomes, the sites of protein synthesis. A ribosome consists of two subunits (small and large), each with rRNA and protein. Subunits are packaged in the nucleolus and float free in the cytoplasm. Two subunits come together when they attach to the end of an mRNA molecule.

Transfer RNA (tRNA) molecules are transport molecules that carry specific amino acids to a ribosome and align the amino acids to form a polypeptide chain. There are binding sites for two tRNA molecules on the ribosome where each tRNA recognizes the correct codon on the mRNA molecule.

There are at least 20 different types of tRNA molecules, one specific for each amino acid. They are all relatively small molecules, containing between 70 and 80 nucleotides. Although tRNA molecules are single stranded, some of the bases hydrogen bond with each other, folding the molecule into a specific shape. The specified amino acid attaches to one end of the tRNA molecule, while the bottom of the molecule attaches to an mRNA codon by base pairing with the anticodon. The **anticodon** is a series of three bases on the tRNA molecule that pairs with the complementary codon on mRNA. For example, the mRNA codon UUU specifies the amino acid phenylalanine. The tRNA molecule that brings phenylalanine to the ribosome has the anticodon AAA and will, therefore, base pair with the codon, inserting phenylalanine in the correct place in the growing polypeptide chain. Once the amino acid becomes part of the polypeptide, the bond with the tRNA is broken, allowing the tRNA to cycle back to the cytoplasm and bind another amino acid molecule (fig. 7.10). The ribosome now moves along the mRNA molecule, reading a new codon, and a tRNA brings in the specified amino acid. Translation continues, with the ribosome sliding over the mRNA, codon by codon, translating the sequence of bases until a termination codon is reached, signaling the end of the process. A summary of the events involved in gene expression, from transcription through translation, is shown in Figure 7.11.

As the discussion of translation indicates, tRNA and rRNA molecules have vital roles in protein synthesis. Like mRNA, these and other types of RNA are produced in the nucleus by transcription; however, only mRNA is translated. Ribosomal RNA and tRNA carry out their functions as RNA molecules. Previously, the definition of a gene was considered to be the segment of DNA that codes for the formation of one polypeptide or one protein. Today, the segments of DNA that code for the formation of rRNA, tRNA, and other RNA molecules are also considered genes. As the knowledge and understanding of gene expression has increased, so has the definition of a gene. Scientists now define a gene as a segment of the DNA molecule that codes for the formation of a specific functional product, either a protein or an RNA molecule.

Other Roles of RNA

In addition to mRNA, rRNA, and tRNA, which are involved in protein synthesis, scientists now recognize many other types of RNA. Recall, from the discussion of transcription, mRNA undergoes modification before it leaves the nucleus. Small RNA molecules, complexed with proteins and referred to as spliceosomes, function as enzymes in intron removal

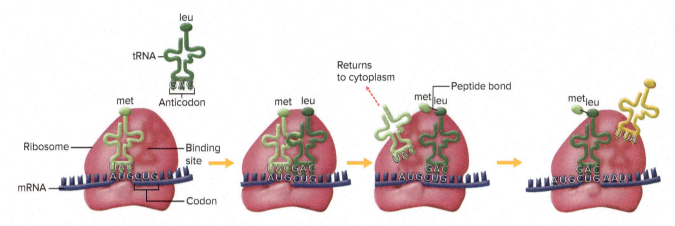

Figure 7.10 Translation. Transfer RNA molecules arrive at the ribosome, bringing amino acids. Codon–anticodon base pairing ensures that the amino acids will be incorporated into the sequence specified on the mRNA.

and splicing exons together. Several other types of RNA have regulatory roles in controlling gene expression. Two of the types involved in gene regulation are **microRNA (miRNA)** and **small interfering RNA (siRNA)**. The first, miRNAs, are small, single-stranded RNA molecules that are 20 to 24 nucleotides in length; thousands of different miRNA molecules have been identified since their initial discovery. In fact, even in a single plant, *Arabidopsis thaliana*, which is used extensively in plant genetic studies (see A Closer Look 7.2: Try These Genes On for Size), there have been over 400 miRNA molecules identified. Small interfering RNA molecules are about the same length as miRNA but are double stranded. Thousands of siRNAs have also been identified. Both miRNA and siRNA molecules have multiple functions but are primarily involved in gene expression, often by binding to complementary sites on messenger RNA and stopping translation. This action is often described as post-transcriptional gene silencing, or RNA interference. Experimental studies have shown that these short segments of RNA play major roles in growth and development, responses to environmental stress, and pathogen attack. Researchers are actively investigating the potential of these small RNA molecules for multiple uses, especially disease prevention in both plants and animals.

Thinking Critically

During protein synthesis, certain genes on the DNA molecule are transcribed into a molecule of messenger RNA. The mRNA is then translated at a ribosome to form a specific protein.

The following sequence of bases is found in the DNA of a strawberry plant. What amino acids are specified by this DNA sequence: GGG-TTA-CAT-TTC-AAA?

Mutations

Mutations are changes in the nucleotide sequence of DNA. Once a DNA sequence has changed, DNA replication copies the altered sequence and passes it along to future generations of that cell line. The smallest mutations are called **point mutations**, caused by a base substitution in which one base is substituted for another. When the gene is transcribed, the altered mRNA codon may insert a different amino acid and possibly alter the shape and function of the final protein. A single base

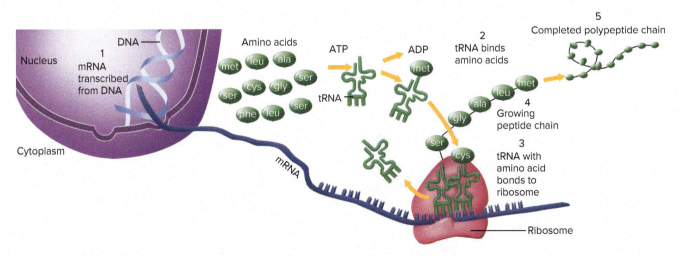

Figure 7.11 Summary of protein synthesis.

substitution is responsible for the altered protein that causes the inherited disease sickle cell anemia. A mutation in which a small segment of the DNA is lost is known as a **deletion**, and a mutation in which a segment is added is called an **insertion**. The segment lost or added may be as small as a single base. This may result in a modification of the mRNA reading frame, known as a **frame-shift mutation**, and all codons downstream of this site will specify different amino acids. An analogy to a frame-shift mutation can be seen in the following sentence, THE RAT SAW THE CAT AND RAN. The deletion of the letter *R* in RAT can create a shift of the reading frame. The resulting sentence, THE ATS AWT HEC ATA NDR AN, is meaningless. Similarly, the deletion of a nucleotide can create a meaningless mRNA message.

Mutations can occur in any cell. If they occur in cells that do not lead to gametes, they are called **somatic mutations**. Somatic mutations can occur in cells of leaves, stems, or roots and are not usually passed on to offspring. However, some plants undergo extensive asexual reproduction. If the organ affected is involved in this process, the resulting clone will also incorporate the somatic mutation. If the mutation occurs during the formation of gametes, it will be passed on to the offspring, possibly creating a new allele for a genetic character. This is one small way in which genetic change can lead to the evolution of new species (see Chapter 8).

Epigenetics

In addition to mutations, other inherited changes can be caused by chemical changes to chromatin. Recall that chromatin consists of DNA and protein. The major proteins in chromatin are histone proteins. Chemical changes to either the nucleotides in the DNA or the histone proteins can cause inherited changes even when there is no change in the DNA sequence. **Epigenetics** is the study of these inherited changes. The result of the epigenetic change is that transcription of the affected DNA segment is prevented. Although scientists have been aware of epigenetics for decades, research in this field has rapidly expanded in recent years. Some of these epigenetic changes are accomplished with the help of siRNA molecules.

A common example of an epigenetic change is the attachment of a methyl group ($-CH_3$) to either a nucleotide or a histone. In particular, methyl groups are frequently attached to cytosine bases in DNA. When the cell divides, the methyl groups on the cytosine bases are accurately inherited, thereby maintaining the altered gene. Other chemical modifications are also passed along to the daughter cells during cell division. Epigenetic changes within an organism are passed on through mitosis; however, some of these changes can also be passed on through meiosis and, thereby, affect the offspring as well. Currently, the inheritance of epigenetic changes by offspring is far better known in plants than in animals.

Epigenetic changes occur in various ways, and some of these are a natural part of development. All somatic cells in an organism carry the same genetic makeup; however, large multicellular organisms have many different cell types. These result from epigenetic changes that occur during development. Changes such as methylation (addition of methyl groups) of DNA allow unnecessary genes to be switched off in certain cell types. For example, in a developing leaf, epigenetic changes help explain how one cell specializes to become a leaf epidermal cell and another specializes to become a palisade mesophyll cell or a phloem sieve tube member.

Chemicals, nutrient deficiencies, and other types of environmental stress, including pests and pathogens, can also cause epigenetic changes to be made or even removed, resulting in altered gene activity. In one set of experiments using genetically identical dandelions, researchers found different patterns of DNA methylation when the plants were grown under different types of environmental stress. Understanding the importance of DNA methylation and other epigenetic changes may soon help plant geneticists develop crops that are more resistant to environmental stress.

Recombinant DNA

By the 1970s, scientists had an understanding of the chemical nature of the gene, how it functioned, and how mutations occurred. Since then, this knowledge has been applied to practical uses in a number of areas that affect society, such as health care, agriculture, plant breeding, and food processing. New applications are occurring so rapidly that it is difficult to predict future directions.

Many of these applications involve **recombinant DNA** technology, which entails the introduction of genes from one organism into the DNA of a second organism. This technique has allowed bacteria to produce human insulin and plants to express a gene for herbicide resistance normally found in bacteria (see Chapter 15). The formation of recombinant DNA makes use of proteins called **restriction enzymes** to cut a gene from its normal location. Restriction enzymes, which normally occur in bacterial cells, are able to cut DNA at specific base sequences.

Transferring the isolated gene to another species requires the use of a vector, usually a **plasmid**, which is a small, circular strand of DNA that also occurs in bacterial cells. The plasmid is cut with the same restriction enzyme, then mixed with the isolated gene. The ends of the plasmid join to the ends of the gene, with the result being a recombinant DNA molecule that is transferred to a cell in another organism (fig. 7.12).

In much of the early work on recombinant DNA, the plasmids were transferred to bacterial cells. When a recombinant bacterial cell divides, so does the plasmid, and within hours a whole colony of cells are produced, all with the transferred, or foreign, gene. Because the genetic code is a universal code, the bacterial cell is able to transcribe and translate the information on the plasmid, producing the desired protein. This technique has been used to manufacture high-quality human insulin, human growth factor, and many other hormones. Recombinant DNA technology is now applied to higher organisms, and both animals and plants that express genes from other organisms have been produced (see Chapter 15).

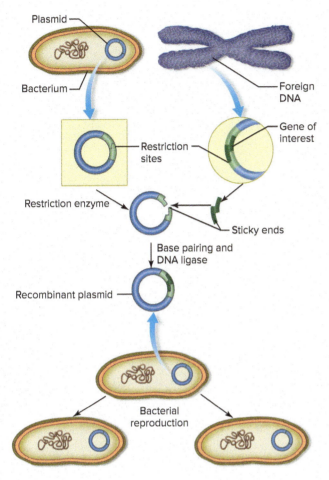

Figure 7.12 Recombinant DNA technology. A gene of interest from one organism is cut from the DNA molecule using a restriction enzyme. The same enzyme is used to cut a bacterial plasmid. The two DNA samples are mixed together and the DNA sealed with the enzyme DNA ligase. The recombinant plasmid is taken up by the bacterial cell, reproducing when the cell divides and expressing the foreign gene.

CHAPTER SUMMARY

1. Genes control all phases of life in an organism, including its metabolism, size, color, development, and reproduction. Genes occur in pairs and are located on homologous chromosomes, which are inherited from parents. Through the use of monohybrid and dihybrid crosses with the garden pea, Gregor Mendel discovered the basic principles that govern the inheritance of genes.

2. Mendelian genetics explains the inheritance of genes that are located on separate chromosomes and have two allelic forms, but other patterns of inheritance occur that are not governed by Mendel's principles. Incomplete dominance and codominance occur when neither allele is completely dominant. Multiple phenotypes can occur when more than two alleles exist for a given trait or when that trait is controlled by more than one pair of genes. Genes that occur on one chromosome are inherited together or linked. Crossing over can result in the recombination of linked genes during meiosis.

3. James Watson and Francis Crick described the structure of the DNA molecule as a double helix of nucleotides. A gene is a segment of the DNA molecule that codes for the formation of a protein or RNA molecule. Through transcription and translation, the information coded in the sequence of bases is expressed as a specific sequence of amino acids. Some mutations arise from changes in the DNA that result in altered proteins with different characteristics. Since the genetic code is a universal code, the genetic information from organisms can be translated by another organism. Recombinant DNA technology is now being used for practical applications.

REVIEW QUESTIONS

1. What can the phenotype tell you about the genotype of an individual? Distinguish between phenotype and genotype.
2. In the commercial development of seeds, how would you determine if a newly established strain will breed true?
3. What do we mean when we say that genes are linked? Why does linkage alter the normal ratio of phenotypes?
4. How are the various types of RNA involved in the process of protein synthesis?
5. What is a gene? How are genes related to chromosomes?
6. What is the relationship between inherited traits and proteins?
7. What is a codon? How many exist? What does a codon represent?
8. Compare and contrast DNA replication with transcription.
9. Other than protein synthesis, what are other roles of RNA molecules?
10. What is epigenetics? What causes epigenetic changes?
11. What are restriction enzymes? Why are they important tools in recombinant DNA research?

ADDITIONAL GENETICS PROBLEMS

1. In garden peas, yellow seeds are dominant to green. What are the phenotypic ratios of offspring in the following crosses?
   a. homozygous yellow × heterozygous yellow
   b. heterozygous yellow × homozygous green
   c. heterozygous yellow × heterozygous yellow
2. Also in garden peas, the tall allele (T) is dominant to the dwarf allele (t). What are the phenotypic ratios of offspring in the following crosses?
   a. $TT \times Tt$ b. $tt \times TT$ c. $Tt \times Tt$
3. If you used the same alleles in a cross of tall green garden peas with dwarf yellow peas, what would be the appearance of the F_1 and F_2 generations?
4. In tomatoes, red fruit is dominant to yellow fruit and tall plants are dominant to short plants. A short plant with yellow fruit is crossed with a plant that is heterozygous for both traits. What would be the phenotypic ratio of the offspring from such a cross?

CHAPTER

8

Plant Systematics and Evolution

©Karen McMahon

Fossils, such as the leaf of the Dawn Redwood (Metasequoia), *were part of the evidence Charles Darwin used to formulate his theory of evolution by means of natural selection.*

KEY CONCEPTS

1. Scientific names are two-word names called binomials that are internationally recognized by the scientific community.
2. Carolus Linnaeus, an eighteenth-century Swedish botanist, started the binomial system and is therefore known as the Father of Taxonomy.
3. With the publication in 1859 of *On the Origin of Species,* Charles Darwin proposed that species are not static entities but are works in progress that evolve in response to environmental pressures.
4. Natural selection favors the survival and reproduction of those individuals in a species that possess traits that better adapt them to a particular environment.

CHAPTER OUTLINE

Early History of Classification 121
   Carolus Linnaeus 121
How Plants Are Named 123
   Common Names 123
A CLOSER LOOK 8.1 The Language of Flowers 125
   Scientific Names 126
Taxonomic Hierarchy 127
   Higher Taxa 127
   What Is a Species? 129
PhyloCode 130
A CLOSER LOOK 8.2 Saving Species through Systematics 131
Barcoding Species 132
The Influence of Darwin's Theory of Evolution 132
   The Voyage of the HMS *Beagle* 133
   Natural Selection 134
Chapter Summary 135
Review Questions 135

Plant systematics is the branch of botany that is concerned with the naming, identification, evolution, and classification (arrangement into groups with common characteristics) of plants. In a strict sense, plant taxonomy is the science of naming and classifying plants; however, in this book the terms *taxonomy* and *systematics* are used interchangeably. The simplest form of classification is a system based on need and use; early humans undoubtedly classified plants into edible, poisonous, medicinal, and hallucinogenic categories.

EARLY HISTORY OF CLASSIFICATION

The earliest known formal classification was proposed by the Greek naturalist Theophrastus (370-285 B.C.). In his botanical writings (*Enquiries into Plants* and *The Causes of Plants*), he described and classified approximately 500 species of plants into herbs, undershrubs, shrubs, and trees. Because his influence extended through the Middle Ages, he is regarded as the Father of Botany.

Two Roman naturalists who also had long-lasting impacts on plant taxonomy were Pliny the Elder (A.D. 23-79) and Dioscorides (first century A.D.). Both described medicinal plants in their writings and Dioscorides's *De Materia Medica* remained the standard medical reference for 1,500 years. From this period through the Middle Ages, little new botanical knowledge was added. Blind adherence to the Greek and Roman classics prevailed, using manuscripts painstakingly copied by hand in monasteries throughout Europe.

The revival of botany after its stagnation in the Middle Ages began early in the Renaissance with the renewed interest in science and other fields of study. The invention of the printing press in the middle of the fifteenth century allowed botanical works to be more easily produced than ever before. These richly illustrated books, known as **herbals,** dealt largely with medicinal plants and their identification, collection, and preparation. The renewal of interest in taxonomy can be traced to the work of several herbalists; in fact, this period of botanical history from the fifteenth through the seventeenth centuries is known as the Age of Herbals. Another factor in the revival of taxonomy was the global exploration by the Europeans during this period, which led to the discoveries of thousands of new plant species. In less than 100 years, more plants were introduced to Europe than in the previous 2,000 years.

Carolus Linnaeus

By the beginning of the eighteenth century, it was common to name plants using a **polynomial** (see fig. 8.4), which included a single word name for the plant (today called the genus name), followed by a lengthy list of descriptive terms, all in Latin. This system had flaws. It was not standardized; different polynomials existed for the same plant; and it was cumbersome to remember some of the longer polynomials, which could be a paragraph in length. This was the state of taxonomy during the time of Linnaeus.

Figure 8.1 Statue of Carolus Linnaeus (1707–1778) holding flowers of Indian blanket (*Gaillardia pulchella*) at the Linnaeus Teaching Garden, Tulsa, OK.

Carolus Linnaeus (fig. 8.1) was born in May 1707 in southern Sweden, the son of a clergyman. He became interested in botany at a very young age through the influence of his father, who was an avid gardener and amateur botanist. It was expected that Linnaeus would also become a clergyman, but in school he did not do well in theological subjects. He did, however, excel in the natural sciences and entered the University of Lund in 1727 to pursue studies in natural science and medicine. (At this time, medical schools were the centers of botanical study because physicians were expected to know the plant sources of medicines in use.) After 1 year he transferred to the University of Uppsala, the most prestigious university in Sweden. It was there that he published his first botanical papers, which laid the foundations for his later works in classification and plant sexuality. In 1732, he undertook a solo expedition to Lapland to catalog the natural history of this relatively unknown area. He later published *Flora Lapponica,* a detailed description of the plants of this area.

Linnaeus received his medical degree in 1735 from the University of Harderwijk in the Netherlands. Soon he came under the patronage of George Clifford, a director of the Dutch East India Company and one of the wealthiest men in Europe. He served as Clifford's personal physician and as curator of his magnificent gardens, which housed specimens from around the world. The 3 years he spent in the Netherlands were the most productive period in his life. During that time, he completed several books and papers, including *Systema Naturae*, *Fundamenta Botanica*, and *Genera Plantarum*, which expanded on his ideas of classification (fig. 8.2).

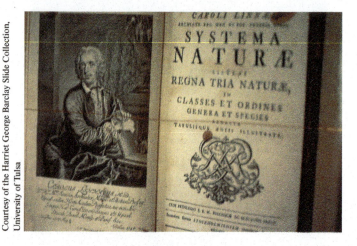

Figure 8.2 Frontispiece of *Systema Naturae*, one of the writings of Linnaeus, in which he expounded on his ideas of classification.

He returned to Sweden in 1738 and soon married Sara Elisabeth Moraea. After setting up a medical practice in Stockholm, he was appointed physician to the Swedish Admiralty, specializing in the treatment of venereal diseases. In 1741, he returned to the University of Uppsala as professor of medicine and botany, a position he retained until retirement in 1775. Linnaeus was a popular teacher who attracted students from all over Europe. Many of his students became famous professors in their own right; others traveled to distant lands collecting unknown specimens for Linnaeus to classify. After suffering several strokes, he died in January 1778.

One of Linnaeus's achievements was his sexual system of plant classification, which did much to popularize the study of botany. This system was based on the number, arrangement, and length of stamens and thus divided flowering plants into 24 classes. Using this system, it was possible for anyone to identify and name unknown plants. At the time, his language was risqué because he compared floral parts to human sexuality, with stamens referred to as husbands and pistils* as wives; for example, "husband and wife have the same bed" meant stamens and pistils in the same flower (fig. 8.3). Dr. Johann

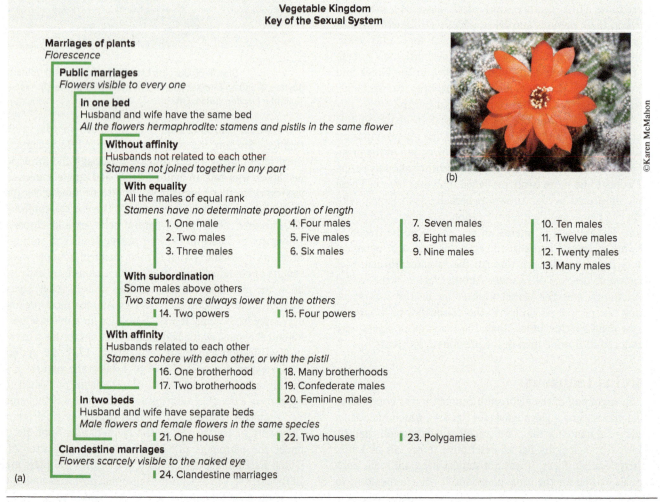

*Note that the older term *pistil*, rather than *carpel* has been used in this key.

Figure 8.3 (a) Linnaeus's sexual system related floral parts to human sexuality. (b) Peanut cactus (*Echinopsis chamaecereus*) in the cactus family (Cactaceae) keys out to *With equality, Many males*.

Siegesbeck, a contemporary of Linnaeus and director of the botanical garden in St. Petersburg, was shocked at the analogies and said such

> loathsome harlotry as several males to one female would not be permitted in the vegetable kingdom by the Creator. . . . Who would have thought that bluebells, lilies, and onions could be up to such immorality?

Despite the opposition of some, the Linnaean sexual method was easy to understand and simple for even the amateur botanist to use. This method, however, was an artificial system grouping together clearly unrelated plants (in his system, cherries and cacti were grouped together); by the early nineteenth century, it had been abandoned in favor of systems that reflected natural relationships among plants.

Linnaeus's greatest accomplishment was his adoption and popularization of a **binomial** system of nomenclature. When he described new plants, he conformed to the current practice of using a polynomial. For convenience, however, he began to add in the margin a single descriptive adjective that would identify unequivocally a particular species (fig. 8.4). He called this adjective the trivial name. This combination later developed into the two-word scientific name, or binomial, described in the next section. Linnaeus used this system consistently in *Species Plantarum,* published in 1753. This work contains descriptions and names of 5,900 plants, all the plants known to Linnaeus. The binomial system simplified scientific names and was soon in wide use. In 1867, a group of botanists at the International Botanical Congress in Paris established rules governing plant nomenclature and classification. They established *Species Plantarum* as the starting point for scientific names. Although the rules (formalized in the *International Code of Botanical Nomenclature*) have been modified over the years, the 1753 date is still valid, and many names first proposed by Linnaeus are still in use today.

Linnaeus's contributions were not limited to botany, since the binomial system is used for all known organisms. He is credited with naming approximately 12,000 plants and animals; for all his contributions to the field of taxonomy, he is known as the Father of Taxonomy.

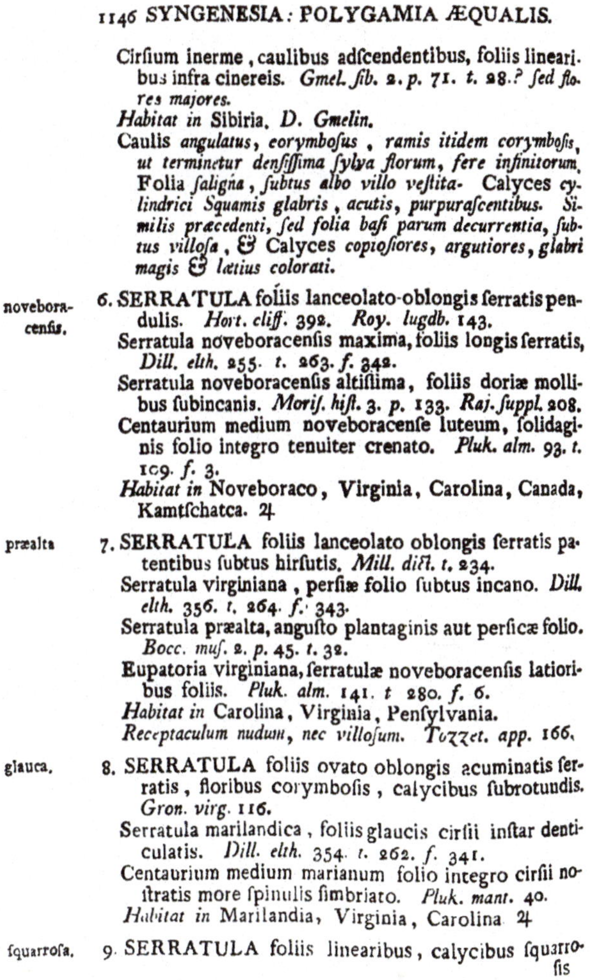
1146 SYNGENESIA: POLYGAMIA ÆQUALIS.

Cirſium inerme, caulibus adſcendentibus, foliis linearibus infra cinereis. *Gmel. ſib.* 2. *p.* 71. *t.* 28.? *ſed flores majores.*
Habitat in Sibiria. *D. Gmelin.*
Caulis *angulatus, corymboſus, ramis itidem corymboſis, ut terminetur denſiſſima ſylva florum, fere infinitorum.* Folia *ſaligna, ſubtus albo villo veſtita.* Calyces *cylindrici Squamis glabris, acutis, purpuraſcentibus. Similis præcedenti, ſed folia baſi parum decurrentia, ſubtus villoſa,* & Calyces *copioſiores, argutiores, glabri magis & lætius colorati.*

noveboracenſis. 6. SERRATULA foliis lanceolato-oblongis ſerratis pendulis. *Hort. cliff.* 392. *Roy. lugdb.* 143.
Serratula noveboracenſis maxima, foliis longis ſerratis, *Dill. elth.* 255. *t.* 263. *f.* 342.
Serratula noveboracenſis altiſſima, foliis doriæ mollibus ſubincanis. *Moriſ. hiſt.* 3. *p.* 133. *Raj. ſuppl.* 208.
Centaurium medium noveboracenſe luteum, ſolidaginis folio integro tenuiter crenato. *Pluk. alm.* 93. *t.* 109. *f.* 3.
Habitat in Noveboraco, Virginia, Carolina, Canada, Kamtſchatca. ♃

præalta 7. SERRATULA foliis lanceolato oblongis ſerratis patentibus ſubtus hirſutis. *Mill. dict. t.* 234.
Serratula virginiana, perſiæ folio ſubtus incano. *Dill. elth.* 356. *t.* 264. *f.* 343.
Serratula præalta, anguſto plantaginis aut perſicæ folio. *Bocc. muſ.* 2. *p.* 45. *t.* 32.
Eupatoria virginiana, ſerratulæ noveboracenſis latioribus foliis. *Pluk. alm.* 141. *t* 280. *f.* 6.
Habitat in Carolina, Virginia, Penſylvania.
Receptaculum nudum, nec villoſum. Tozzet. app. 166.

glauca. 8. SERRATULA foliis ovato oblongis acuminatis ſerratis, floribus corymboſis, calycibus ſubrotundis. *Gron. virg.* 116.
Serratula marilandica, foliis glaucis cirſii inſtar denticulatis. *Dill. elth.* 354. *t.* 262. *f.* 341.
Centaurium medium marianum folio integro cirſii noſtratis more ſpinulis fimbriato. *Pluk. mant.* 40.
Habitat in Marilandia, Virginia, Carolina ♃

ſquarroſa. 9. SERRATULA foliis linearibus, calycibus ſquarroſis

Figure 8.4 A photograph from *Species Plantarum* illustrates the beginning of the binomial system. Note the trivial names in the margin next to the polynomial description for each species. The trivial name was later designated as the species epithet, which, together with the generic, forms the binomial.

HOW PLANTS ARE NAMED

Names are useful because they impart some information about a plant; it may be related to flower color, leaf shape, flavor, medicinal value, season of blooming, or location. Names are necessary for communication; "if you know not the name, knowledge of things is wasted."

This discussion begins with a look at common names, or what plants are called locally, and follows with an examination of internationally recognized scientific names.

Common Names

A close look at common names often reveals a keen sense of observation, a fanciful imagination, or even a sense of humor: trout lily, milkweed, Dutchman's pipe, Texas bluebonnet, ragged sailor, and old maid's nightcap (table 8.1). Sometimes the names even convey feelings or emotions (see A Closer Look 8.1: The Language of Flowers).

Names have evolved over centuries but are sometimes used only in a limited geographical area. Even short distances away, other common names may be used for the same plant. Consider, for example, the many names for the tree that some people call osage orange (*Maclura pomifera*) (fig. 8.5a): bodeck, bodoch, bois d'arc, bow-wood, osage apple tree, hedge, hedge apple, hedge osage, hedge-plant osage, horse apple, mock orange, orange-like maclura, osage apple, and wild orange.

Table 8.1 Some Common Names and Their Meanings

Names in Commemoration	
Douglas fir	David Douglas, plant collector (1798–1834)
Camellia	Georg Josef Kamel, pharmacist (1661–1706)
Gerber daisy	Traugott Gerber, German explorer (?–1743)
Freesia	Friedrich H. T. Freese, German physician (?–1876)

Names That Describe Physical Qualities	
Dusty miller	White, woolly leaves
Dutchman's breeches	Shape of flower
Goldenrod	Shape and color of inflorescence
Indian pipe	Shape of flower with stem
Cattail	Inflorescence of carpellate flowers
Lady's slipper	Shape of this orchid's flower
Milkweed	Milky juice when plant is cut
Skunk cabbage	Fetid odor of inflorescence
Cheeses	Fruit resembles a round head of cheese
Smoke tree	Plumelike pedicels
Shagbark hickory	Shedding bark
Redbud	Color of flower buds
Quaking aspen	Rustling leaves
Bluebell	Color and shape of flower
Crape myrtle	Wavy edges of petals

Scientific Names That Have Become Common Names	
Hydrangea	Abelia
Vanilla	Narcissus
Coreopsis	Gladiolus

Names That Indicate Use	
Daisy fleabane	Gets rid of fleas
Boneset	A tonic from this plant can heal bones
Feverwort	Medicinal property to reduce fever
Kentucky coffee tree	Seeds roasted for coffee substitute
Belladonna	Juice used to beautify by producing pallid skin and dilated, mysterious eyes

Names That Indicate Origin, Location, or Season	
Pacific yew	Grows along northern Pacific coast
Spring beauty	One of the first flowers of spring
Marshmallow	Found in wet, marshy habitat
Daylily	Flowers last only a day
Four-o'clock	Flowers open in late afternoon
Japanese honeysuckle	Country of origin is Japan

(a)

(b)

Figure 8.5 Mock orange is a common name shared by (a) the tree *Maclura pomifera*, with its fruit resembling a grapefruit, and (b) the shrub *Philadelphus lewisii*, with its flowers resembling orange blossoms.

On the other hand, different plants may share the same common name. Although mock orange is one of the common names for osage orange, the name mock orange is usually associated with a completely unrelated group of flowering shrubs (*Philadelphus* spp.) (fig. 8.5b). These examples point out the difficulties with common names; one plant may be known by several different names, and the same name may apply to several different plants. The need to have one universally accepted name is fulfilled with scientific names.

A CLOSER LOOK 8.1

The Language of Flowers

Through traditions, some flowers became symbolic of certain emotions and feelings. This was sometimes even reflected in their common names; two straightforward examples are forget-me-not flowers, which conveyed the sentiment "remember me," and bachelor's buttons, which indicated the single status of the wearer. This symbolism reached its peak during the Victorian era, when almost every flower and plant had a special meaning. In Victorian times, it was possible to construct a bouquet of flowers that imparted a whole message (box fig. 8.1). A Victorian suitor might send a bouquet of jonquils, white roses, and ferns to his intended, which indicated that he desired a return of affection, he was worthy of her love, and he was fascinated by her. This "language" became so popular that dictionaries were printed to interpret floral meanings. One of the most popular dictionaries was *Language of Flowers* (1884) by Kate Greenaway, a well-known illustrator of children's books. Following is a small sampling of some common flowers and plants and what they symbolized:

Amaryllis: pride
Apple blossom: preference
Bachelor's buttons: celibacy
Bluebell: constancy
Buttercup: ingratitude
Yellow chrysanthemum: slighted love
Daffodil: regard
Daisy: innocence
Dogwood: durability
Elm: dignity
Goldenrod: caution
Holly: foresight
Honeysuckle: generous and devoted affection
Ivy: fidelity
Lavender: distrust
Lichen: dejection
Lily of the valley: return of happiness
Live oak: liberty
Magnolia: love of nature
Marigold: grief
Mock orange: counterfeit
Oak leaves: bravery
Palm: victory
Pansy: thoughts
Spring crocus: youthful gladness
Dwarf sunflower: adoration

Box Figure 8.1 Flowers convey a message all their own.

Tall sunflower: haughtiness
Yellow tulip: hopeless love
Blue violet: faithfulness
Wild grape: charity
Zinnia: thought of absent friends

Even today, several plants have well-known symbolic meanings. Red roses convey passionate love; a four-leaf clover means luck; orange blossoms symbolize weddings; and an olive branch indicates peace. Floral colors can also communicate feelings, with red indicating passion; blue, security; yellow, cheer; white, sympathy; and orange, friendship. With some thought, it is possible to find the right flower and color to express the exact message.

Scientific Names

Each kind of organism is known as a **species**, and similar species form a group called a **genus** (pl., **genera**). Each species has a scientific name in Latin that consists of two elements; the first is the genus and the second is the specific epithet. Such a name is a binomial, literally two names, and is always italicized or underlined; for example, *Maclura pomifera* is the scientific name for osage orange. A rough analogy of the binomial concept can be seen in a list of names in a telephone directory, where the surname "Smith" (listed first) represents the genus and the first names (John, Frank, and Mary) define particular species within the genus.

In the binomial, the first name is a noun and is capitalized; the second, written in lowercase, is usually an adjective. After the first mention of a binomial, the genus name can be abbreviated to its first letter, as in *M. pomifera*, but the specific epithet can never be used alone. The genus name, however, can be used alone, especially when referring to several species within a genus; for example, *Philadelphus* refers to over 50 species of mock orange. The specific epithet can be replaced by an abbreviation for species, "sp." (or "spp." plural), when the name of the species is unknown or unnecessary for the discussion. In the previous example, *Philadelphus* sp. refers to one species of mock orange, whereas *Philadelphus* spp. refers to more than one species.

Scientific names may be just as descriptive as common names, and translation of the Latin (or latinized Greek) is informative (table 8.2). Sometimes either the genus name or the specific epithet is commemorative, derived from the name of a botanist or other scientist. Specific epithets are frequently used with more than one genus, and knowledge of their meanings will provide some insight into scientific names encountered later in this text (table 8.3).

A complete scientific name also includes the name or names of the author or authors (often abbreviated) who first described the species or placed it in a particular genus. For example, the complete scientific name for corn is *Zea mays* L.; the "L" indicates that Linnaeus named this species. On the other hand, the complete name for osage orange is *Maclura pomifera* (Raf.) Schneid. This author citation indicates that Rafinesque-Schmaltz first described the species, giving it the specific epithet *pomifera*, but Schneider later put it in the genus *Maclura*. In this text, the author citations are omitted for simplicity.

A scientific name is unique, referring to only one species and universally accepted among scientists. It is the key to unlocking the door to the accumulated knowledge about a plant. Imagine the confusion if only common names were used and a reference were made to mock orange. Would this reference allude to *Maclura pomifera* or to a species of *Philadelphus*?

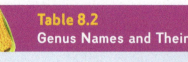

Table 8.2 Genus Names and Their Meanings

Names in Commemoration	
Begonia	Michel Begon, patron of botany (1638–1710)
Forsythia	William Forsyth, gardener at Kensington Palace (1737–1804)
Bougainvillea	Louis Antoine de Bougainville, explorer and scientist (1729–1811)
Fuchsia	Leonhard Fuchs, German physician and herbalist (1501–1566)
Zinnia	Johann Gottfried Zinn (1727–1759)
Wisteria	Caspar Wistar, American professor of plant anatomy (1761–1818)
Names That Describe Physical Qualities	
Myriophyllum	Finely divided leaves
Chlorophytum	Green plant
Lunaria	Moon, refers to appearance of pods
Helianthus	Sunflower
Zebrina	Zebra, refers to striped leaves
Trillium	Floral parts in threes
Tetrastigma	Four-lobed stigma
Ribes	Acid tasting; refers to fruit
Polygonum	Many knees; refers to jointed stems
Zanthoxylum	Yellow wood
Sagittaria	Arrow; refers to arrowhead leaves
Names from Aboriginal or Classic Origins	
Avena	Oats (Latin)
Triticum	Wheat (Latin)
Allium	Garlic (Greek)
Catalpa	Catalpa (North American Aboriginal)
Vitis	Grapevine (Latin)
Ulmus	Elm (Latin)
Pinus	Pine (Latin)
Names That Indicate Use	
Solidago	Make whole or strengthen
Angelica	Angelic medicinal properties
Cimcifuga	Repel bugs
Saponaria	Soap; refers to soap that can be made from the plant
Pulmonaria	Lung; used to treat infections of the lung
Potentilla	Powerful; refers to its potent medicinal properties
Names That Indicate Location	
Elodea	Grows in marshes
Petrocoptis	Break rock; refers to habit of growing in rock crevices

Table 8.3 Common Scientific Epithets and Their Meanings

acidosus, -a, -um	Sour
aestivus, -a, -um	Developing in summer
albus, -a, -um	White
alpinus, -a, -um	Alpine
annus, -a, -um	Annual
arabicus, -a, -um	Of Arabia
arboreus, -a, -um	Treelike
arvensis, -a, -um	Of the field
biennis, -a, -um	Biennial
campester, -tris, -tre	Of the pasture
canadensis, -is, -e	From Canada
carolinianus, -a, -um	From the Carolinas
chinensis, -is, -e	From China
coccineus, -a, -um	Scarlet
deliciosus, -a, -um	Delicious
dentatus, -a, -um	Having teeth
domesticus, -a, -um	Domesticated
edulis, -is, -e	Edible
esculentus, -a, -um	Tasty
europaeus, -a, -um	From Europe
fetidus, -a, -um	Bad smelling
floridus, -a, -um	Flowery
foliatus, -a, -um	Leafy
hirsutus, -a, -um	Hairy
japonicus, -a, -um	From Japan
lacteus, -a, -um	Milky white
littoralis, -is, -e	Growing by the shore
luteus, -a, -um	Yellow
mellitus, -a, -um	Honey-sweet
niger, -ra, -um	Black
occidentalis, -is, -e	Western
odoratus, -a, -um	Fragrant
officinalis, -is, -e	Used medicinally
robustus, -a, -um	Hardy
ruber, -ra, -rum	Red
saccharinus, -a, -um	Sugary
sativus, -a, -um	Cultivated
silvaticus, -a, -um	Of the woods
sinensis, -is, -e	Chinese
speciosus, -a, -um	Showy
tinctorius, -a, -um	Used for dyeing
utilis, -is, -e	Useful
vernalis, -is, -e	Spring flowering
virginianus, -a, -um	From Virginia
vulgaris, -is, -e	Common

TAXONOMIC HIERARCHY

In addition to genus and species, other taxonomic categories exist to conveniently group related organisms. As pointed out, Linnaeus used an artificial system; however, today scientists use a **phylogenetic** system to group plants. In a phylogenetic system, information is gathered from morphology, anatomy, cell structure, biochemistry, genetics, and the fossil record to determine evolutionary relationships and, therefore, natural groupings among plants.

Higher Taxa

Species that have many characteristics in common are grouped into a genus, one of the oldest concepts in taxonomy (fig. 8.6). In almost every society, the concept of genus has developed in colloquial language; in English, the words *oak, maple, pine, lily,* and *rose* represent distinct genera. These intuitive groupings reflect natural relationships based on shared vegetative and reproductive characteristics. Many of the scientific names of genera are directly taken from the ancient Greek and Roman common names for these genera (*Quercus,* old Latin word for oak).

The next higher category, or **taxon** (pl., **taxa**), above the rank of genus is the **family.** Families are composed of related genera that (as in a genus) share combinations of morphological traits. In the angiosperms, floral and fruit features are often used to characterize a family. Ideally, the family represents a natural group with a common evolutionary lineage; some families are very small, while others are very large, but still cohesive groups. A few common angiosperm families that have special economic importance are listed in Table 8.4. According to the *International Code of Botanical Nomenclature,* each family is assigned one name, which is always capitalized and ends in the suffix *-aceae.* The old established names of several well-known families present exceptions to this rule. Both the traditional and standardized names are used for these families in Table 8.5. The taxa above the rank of family and their appropriate endings are presented in Table 8.6. The higher the taxonomic category, the more inclusive the grouping (fig. 8.7). Families are grouped into **orders,** orders into **classes,** classes into **divisions** (phyla),* and divisions into **kingdoms.** A domain is above the kingdom level and is the most inclusive taxonomic category. The complete classification of a familiar species is also illustrated in Table 8.6. In addition to the categories already described, biologists also recognize intermediate categories with the "sub" for any rank; for example, divisions may be divided into subdivisions, and species may be divided into

---

*Either *division* or *phylum* (sing.; *phyla,* plural) may be used to indicate the taxonomic rank that is composed of a group of related classes. Traditionally, *division* has been the term preferred by botanists and will be used throughout this textbook.

Table 8.4
Economically Important Angiosperm Families

Scientific Family Name	Common Family Name	Economic Importance
Apiaceae	Carrot	Edibles (carrot, celery), herbs (dill), poisonous (poison hemlock)
Arecaceae	Palm	Edibles (coconut), fiber oils and waxes, furniture (rattan)
Asteraceae	Sunflower	Edibles (lettuce), oils (sunflower oil), ornamentals (daisy)
Brassicaceae	Mustard	Edibles (cabbage, broccoli)
Cactaceae	Cactus	Ornamentals, psychoactive plants (peyote)
Cannabaceae	Hemp	Psychoactive (marijuana), fiber plants (hemp)
Cucurbitaceae	Gourd	Edibles (melons, squashes)
Euphorbiaceae	Spurge	Rubber, medicinals (castor oil), edibles (cassava), ornamentals (poinsettia)
Fabaceae	Bean	Edibles (beans, peas), oil, dyes, forage, ornamentals
Fagaceae	Beech	Lumber (oak), dyes (tannins), ornamentals
Iridaceae	Iris	Ornamentals
Juglandaceae	Walnut	Lumber, edibles (walnut, pecan)
Lamiaceae	Mint	Aromatic herbs (sage, basil)
Lauraceae	Laurel	Aromatic oils (bay leaves), lumber
Liliaceae	Lily	Ornamentals, poisonous plants
Magnoliaceae	Magnolia	Ornamentals, lumber
Malvaceae	Mallow	Fiber (cotton), seed oil, edibles (okra), ornamentals
Musaceae	Banana	Edibles (bananas), fibers
Myrtaceae	Myrtle	Timber, medicinals (eucalyptus), spices (cloves)
Oleaceae	Olive	Lumber (ash), edible oil and fruits (olive)
Orchidaceae	Orchid	Ornamentals, spice (vanilla)
Papaveraceae	Poppy	Medicinal and psychoactive plants (opium poppy)
Piperaceae	Pepper	Black pepper, houseplants
Poaceae	Grass	Cereals, forage, ornamentals
Ranunculaceae	Buttercup	Ornamentals, medicinal and poisonous plants
Rosaceae	Rose	Fruits (apple, cherry), ornamentals (roses)
Rubiaceae	Coffee	Beverage (coffee), medicinals (quinine)
Rutaceae	Citrus	Edible fruits (orange, lemon)
Salicaceae	Willow	Ornamentals, furniture (wicker), medicines (aspirin)
Sapindaceae	Soapberry	Lumber (ash, maple), maple sugar
Solanaceae	Nightshade	Edible (tomato, potato), psychoactive, poisonous (tobacco, mandrake)
Theaceae	Tea	Beverage (tea)
Vitaceae	Grape	Fruits (grapes), wine

(a) *Quercus phellos*
(willow oak)

(b) *Quercus rubra*
(red oak)

Figure 8.6 A genus is a group of species that share many characteristics. Although willow oak (*Quercus phellos*) and red oak (*Quercus rubra*) are clearly distinct species, they are both recognizable as belonging to the oak (*Quercus*) genus by the presence of acorns.

Table 8.5 Traditional and Standardized Names for Some Common Families

Family Name	Traditional Name	Standardized Name
Sunflower	Compositae	Asteraceae
Mustard	Cruciferae	Brassicaceae
Grass	Gramineae	Poaceae
Mint	Labiatae	Lamiaceae
Pea	Leguminosae	Fabaceae
Palm	Palmae	Arecaceae
Carrot	Umbelliferae	Apiaceae

Table 8.6 The Taxonomic Hierarchy and Standard Endings

Rank	Standard Ending	Example
Division (Phylum)	-phyta	Magnoliophyta
Class	-opsida	Liliopsida
Order	-ales	Liliales
Family	-aceae	Liliaceae
Genus		*Lilium*
Species		*Lilium superbum* L.

subspecies (varieties and forms are also categories below the rank of species).

Although the *International Code of Botanical Nomenclature* has rules that govern the assignment of names and define the taxonomic hierarchy, it does not set forth any particular classification system. As a result, there are several organizational schemes that have supporters. These systems differ in the numbers of classes, divisions, kingdoms, and even domains and how they are related to one another. Presently, most biologists use a three-domain, six-kingdom system, which will be described fully in Chapter 9. There is general agreement about the use of a three-domain, six-kingdom system, but biologists still debate the definition of a species.

What Is a Species?

As indicated previously, each kind of organism is known as a species. Although this intuitive definition, based on morphological similarities, works fairly well in many circumstances, it

Figure 8.7 Major ranks in the taxonomic hierarchy. Note that the higher the ranking, the broader the defining characteristics and the more inclusive the group.

is limited; scientists have given much thought to the biological basis of a species. Many accept the **biological species concept** first proposed by Ernst Mayr in 1942, which defines a species as "a group of interbreeding populations reproductively isolated from any other such group of populations."

This definition presents problems when defining plant species. Many closely related plant species that are distinct morphologically are, in fact, able to interbreed; this is true for many species of oaks and sycamores. By contrast, a single plant species may have diploid and **polyploid** (more than the diploid number of chromosomes) individuals that may be reproductively isolated from each other. It is estimated that as many as 40% of flowering plants may be polyploids, with the evening primrose group a thoroughly studied example; an even higher percentage of polyploid species occurs in ferns. Because of these limitations, alternatives to the biological species concept have been suggested. The **ecological species concept** recognizes a species through its role in the biological community as defined by the set of unique adaptations within a particular species to its environment. The availablity of molecular sequence data for nucleic acids and proteins had led to the development of the **genealogical species concept.** Proponents utilize the distinct genetic history of organisms to differentiate species. Despite the lack of an all-inclusive botanical definition, the concept of "species" facilitates the naming, describing, and classifying of plants in a uniform manner. An inventory of the world's species is the first step in preserving biodiversity, as discussed in A Closer Look 8.2: Saving Species through Systematics.

PHYLOCODE

The Linnaean system of nomenclature and the hierarchy of classification that have been presented in this chapter were created more than 250 years ago, before Charles Darwin and Alfred R. Wallace proposed their evolutionary theory by means of natural selection. Linnaean nomenclature is an artificial system based on the appearance of organisms that often does not reflect their evolutionary relationships, or **phylogeny.** Currently, there is a movement to reject this pre-evolutionary taxonomy and replace it with a new system of nomenclature, called PhyloCode, that is truly phylogenetic.

PhyloCode is based on the work of the twentieth-century German entomologist Willi Hennig, who proposed that only shared derived characteristics should be used to define a group of related organisms. (A derived characteristic is a trait that differs from the ancestral trait; taxa which share these traits are believed to share a common evolutionary history and be closely related.) He further proposed that each group constructed should be monophyletic, or composed of only those organisms that can trace their descent from a common ancestor. These natural groupings are known as **clades.**

A CLOSER LOOK 8.2

Saving Species through Systematics

Earth is rich with a tremendous variety of living organisms. About 1.4 million living microbes, fungi, plants, and animals have been identified by systematics. The number of yet undescribed species is much greater, with estimates ranging between 10 million and 100 million. **Biodiversity** is an inventory of the number and variety of organisms that inhabit Earth. We are currently in the midst of a biodiversity crisis; the variety of living species is declining owing to an accelerated extinction rate. Human activities are responsible for this terrible loss. Over 7.6 billion people at present inhabit Earth, and this number is expected to increase to over 9.7 billion by 2050. Population pressures cause natural areas to be cleared for agriculture or expanding urbanization. More people also results in more pollution that fouls the land, sea, and air. All of these human-induced changes translate into a death toll upon the world's biodiversity.

A recent study predicted that if global temperature rises by 2 °C, the global extinction rate for species will also increase from its current rate of 2.8% to 5.2%. The extinction rate per continent is expected to vary greatly with the lowest species extinction rates of 5% and 6% in North America and Europe, respectively, 23% in South America, 14% in Australia, 13% in Africa, and 8% in Asia. However, if the global temperature warms to 3 °C, the overall extinction rate also rises to 8.5% with subsequent increases in all regions.

Why should we care about biodiversity? Biodiversity is the basis for the necessary essentials to human existence: food, fiber, fuel, and shelter. Of the estimated more than 350,000 species of angiosperms, nearly 20,000 have been used at one time or another as food for humans. Advances in agriculture are dependent upon the interaction between systematics and biodiversity. Since the 1960s, world crop yields have increased two- to four-fold. Part of this increase is due to the creation of improved crop varieties through breeding programs and more recently through genetic engineering. Locating and identifying relatives of crop species have been of critical importance to agricultural research in breeding for desirable characteristics. With the advent of genetic engineering, nearly any plant species is a potential source of genes for transfer to agricultural crops. Ironically, the conversion of native ecosystems to agricultural lands in an attempt to accommodate the food demands of an exponentially growing human population may eliminate the very organisms on which agriculture depends for its future. Fertile soil, obviously essential for the vitality of agricultural crops, is also a by-product of biodiversity because it is formed through the interactions of a number of soil organisms: fungi, earthworms, bacteria, plant roots, and burrowing mammals. Species loss could result in soils unable to support vegetation.

Not all of the world's supply of food comes from cultivated sources. There is still a substantial harvest of wild plants and animals. Commercial fishing is, in essence, the hunting of wild fish populations. Blueberries and maple syrup are just two examples of foods gleaned from nature in the United States. Wood and wood pulp are other products harvested from biodiversity resources. Biodiversity in itself is a major economic force, as evidenced by the increasing popularity of ecotourism. Sport fishing, hunting, and bird-watching are other examples of economically profitable activities that depend on the preservation of biodiversity. Last, nearly half of the medicinals now in use originated from a wild plant, and it has been estimated that between 35,000 and 70,000 species of plants are used directly as medicines worldwide.

Knowledge of systematics has many practical applications. There has been a movement to reduce our dependence on chemical pesticides and instead rely more heavily on biological controls to manage nuisance organisms. Biological control methods depend on proper identification of a pest, knowledge of its life cycle, and correct identification of its predators and susceptibility to disease. Misidentification can be costly. Mealybugs are noxious pests that can cause massive damage to crops. A species of mealybug was identified as the culprit in the devastation of coffee plantations in Kenya. Biological control methods were employed using the natural enemies of the identified species of mealybug but were ineffective. Further investigation revealed that the mealybug had been misidentified. Once the correct species was assigned, natural pests of the mealybug were brought in from its native habitat in Uganda, and the mealybug infestation was soon brought under control. Knowledge of systematics can be used to predict economic uses of little known but related species. Researchers identified anticancer compounds from Kenyan populations of *Maytenus buchananii*. There was a problem, however; the species was rare in this locality. Knowledge of systematics suggested that closely related species would probably possess the same chemical compounds. This proved to be the case when a population of the same genus but different species was collected from India.

Clearly, the preservation of biodiversity should be of utmost importance to everyone. Systematic research is fundamental to learning about the characteristics and dimensions of biodiversity. Systematics is necessary to identify localities of high species diversity or rare species. Baseline data must be collected to ascertain which species are declining in numbers or those whose range is becoming limited. Knowledge of systematics will determine if exotic pests are moving into new areas and threatening native species. Without scientific identification and mapping, valuable habitats and the species found there will be lost. In fact, Systematics Agenda 2020 is an ongoing global initiative by the scientific community to discover, describe, and classify the world's species in an effort to understand and conserve biodiversity.

First introduced in 1983, the PhyloCode abandons the Linnaean ranks of the taxonomic hierarchy. In this system, as new information that may change a group's ranking accumulates, names are not changed, as they would be with the Linnaean system, which associates different suffixes with different ranks. Instead of ranks, clades are the only groups recognized. Opponents fear that a complete abandonment of all ranks will result in a loss of comparative information and encourage a proliferation of names that, without any context, will only confuse the nomenclature.

Released in 1991, the APG (Angiosperm Phylogeny Group) system compared the sequence data of select genes to classify the flowering plants. The highest formal rank in this classification system is the order; higher categories are only identified as clades. As more data accumulated, APG IV, an update of the classification, became available in 2016. In this system, angiosperms are recognized as a clade, sharing several distinct characteristics, such as ovules enclosed in a carpel and double fertilization. Within the angiosperms, all monocots appear to belong to a distinct clade, but molecular data indicate that the traditional dicots represent several evolutionary lineages. Most of the dicots do constitute a clade and are now called the **eudicots,** or *true dicots*. Approximately 75% of all angiosperm species are now classified as eudicots. Traditional dicots excluded from the clade eudicot are called the **paleodicots** (literally, *old dicots*) by some authorities and include several ancient lineages in the evolution of angiosperms.

BARCODING SPECIES

A new way to identify species relies not on an expert for a particular taxonomic group but on a genetic barcode, a short segment of DNA that is standard for a particular group of organisms. Identifying a universal barcode for every species is a quick and clear-cut way to discriminate among species and is a necessary first step in the conservation and protection of endangered species, as well as in biodiversity studies. It will also reliably identify larval or juvenile forms whose morphology differs strikingly from the adult animal or root or leaf samples from angiosperms in which the flower is not available.

For animal and algal groups, the nucleotide sequence of one gene, *cox1* or CO1, in the mitochondrial genome has been identified as the standard for distinguishing among species. The *cox1* gene codes for the enzyme cytochrome *c* oxidase subunit 1, an enzyme that has a role in cellular respiration. It has only 648 base pairs, which makes it relatively easy to sequence the DNA. Analysis of the *cox1* gene has proven to be a reliable DNA barcode for animal and algal groups because the difference in its nucleotide sequences is greater between species (interspecific differences) than within a species (intraspecific differences). The human sequence of *cox1*, when compared with that of chimpanzees and gorillas, differs by only 60 and 70 sites, respectively.

Sequencing CO1 has already confirmed that a collection of skipper butterflies in Costa Rica that look remarkably similar as adults and were once considered part of a single species but differ with respect to their caterpillars in appearance, food choice, and habitat are, indeed, separate species.

The *cox1* gene has not worked as an identifiable genetic DNA segment for land plants and fungi for several reasons. In these groups, the mitochondrial genes evolve too slowly; in other words, different species have similar sequences for this gene and therefore it cannot be used to discriminate between species. Additionally, in fungi this gene has undergone many duplications. For these reasons, the nuclear ribosomal DNA segment has been identified as the fungal universal barcode and, in land plants, the gene *rbcL* found in the chloroplast has been used. The *rbcL* gene codes for the large subunit of RUBISCO, ribulose bisphosphate carboxylase/oxidase, an enzyme necessary for the fixation of carbon dioxide into glucose during the dark reaction of photosynthesis (see Chapter 4). Another gene, *matK*, which codes for a maturase enzyme in chloroplasts, an enzyme that splices out introns (see Chapter 7), has also being considered as a plant barcode. The problem in fungi and land plants is that there are not enough differences in these genes between closely related species, making identification to the species level less certain. However, if the analysis is restricted only to the species found in a particular region under investigation, the likelihood of correctly identifying the species is almost 100%.

Organellar DNA is preferred over nuclear DNA because it is less likely to occur in multiple copies and it is much easier to sequence; however, in the case of plant hybrids, the use of organellar DNA ignores the paternal contribution, as only the maternal DNA will be analyzed in sequencing the plastid gene. In the future, researchers are looking at developing nuclear DNA genes that have a low copy number (less than 10 copies) to solve this problem. Low-copy nuclear DNA may also be useful in slowly evolving organisms like fungi which have not yet accumulated discriminating differences in many mitochondrial genes.

Several organizations have been created to support and disseminate information about barcoding species. The Barcode of Life Data (BOLD) system collects barcodes that have been sequenced for various organisms. The Consortium for the Barcode of Life (CBOL) has the support of a number of scientific institutions and universities that are sponsoring the development of barcoding as the standard method for the identification of species.

THE INFLUENCE OF DARWIN'S THEORY OF EVOLUTION

The theory of evolution by means of natural selection was to irrevocably change the way biologists view species. Instead of unchanging organisms and generations created all alike, it was realized that species are dynamic and variable, continually evolving through the mechanism of natural selection in which adaptions are refined to a changing environment.

The Voyage of the HMS *Beagle*

Charles Robert Darwin (fig. 8.8) was born in England in 1809 to a family of distinguished naturalists and physicians. His grandfather was Erasmus Darwin, a well-known poet and physician, and his father, Robert Darwin, was a successful country doctor.

At 15 years of age, Charles was sent to the University of Edinburgh Medical School to study medicine. Not finding it to his liking, he transferred after 2 years to Cambridge University to study theology. While at Cambridge, he spent much of his free time with the students and professors of natural history. This association later proved invaluable.

In 1831, at the age of 22, Darwin graduated from Cambridge with the intention of pursing a degree in theology. Shortly thereafter, he was recommended by John Henslow, one of the natural history professors at Cambridge, for an unpaid position, as a gentleman companion for Captain FitzRoy, who was an accomplished naturalist himself and wanted a similarly trained person to assist in the collection and identification of specimens during the voyage. The ship in question was the HMS *Beagle*, commissioned by King William IV to undertake a voyage around the world for the purpose of charting coastlines, particularly that of South America, for the British navy.

The voyage of the *Beagle* began on December 27, 1831, and would last 5 years (fig. 8.9). During his time on the *Beagle*, Darwin collected thousands of plants and other specimens from South America, the Galápagos Islands (off the coast of

Figure 8.8 Charles Darwin (1809–1882) published *On the Origin of Species* in 1859.

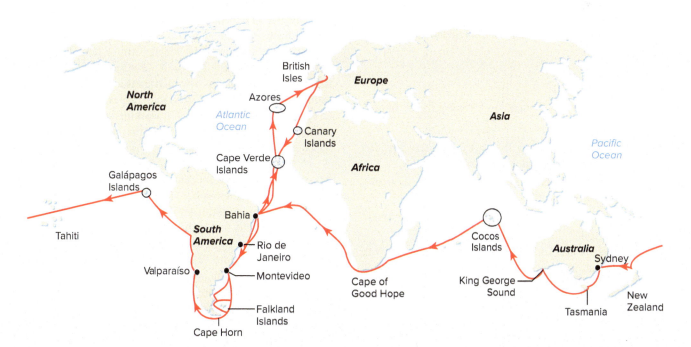

Figure 8.9 The 5-year voyage of the HMS *Beagle*. Darwin's observations on the geology and distributions of plants and animals in South America and the Galápagos Islands were the groundwork for the development of the theory of evolution by means of natural selection.

Ecuador), Australia, and New Zealand. He studied geological formations and noted fossil forms of extinct species. He found that some fossils of extinct species bore a striking resemblance to extant species, as though the former had given rise to the latter. Darwin spent some time studying the species found on the Galápagos Islands. He noted that animals and plants found in the Galápagos were obviously similar to species found in South America, but there were distinct differences. These observations led Darwin to question the fixity of species concept. According to this concept, widely held at the time of Darwin, species were acts of Divine Creation, unchanging over time.

When the *Beagle* returned to England in 1836, Darwin married his cousin, Emma Wedgwood (of the famous Wedgwood china family), and settled, at age 27, in the English countryside. He continued his work in natural history, conducting experiments, writing papers, and corresponding with other naturalists. Among his works was a four-volume treatise on the classification and natural history of barnacles.

In 1842, he began putting his thoughts together on what was to become his theory of **evolution** by natural selection. Darwin continued to expand and fine-tune his thoughts over the next 16 years. In June of 1858, he received a manuscript from Alfred Russel Wallace (1823-1913), a young British naturalist working in Malaysia. Wallace's work was entitled *On the Tendency of Varieties to Depart Indefinitely from the Original Type;* Wallace had independently arrived at the concept of natural selection. Wallace and Darwin jointly presented their ideas on July 1, 1858, at a meeting of the Linnean Society in London. During the next few months, Darwin completed writing what was to become one of the most influential texts of all time. With the publication on November 24, 1859, of *On the Origin of Species by Means of Natural Selection, or the Preservation of Favoured Races in the Struggle for Life* by Charles Darwin, biological thought was changed forever.

Natural Selection

There are four underlying premises to Darwin's theory of evolution by natural selection:

1. *Variation:* Members within a species exhibit individual differences, and these differences are heritable.
2. *Overproduction:* Natural populations increase geometrically, producing more offspring than will survive.
3. *Competition:* Individuals compete for limited resources, what Darwin called "a struggle for existence."
4. *Survival to reproduce:* Only those individuals that are better suited to the environment survive and reproduce (survival of the fittest), passing on to a proportion of their offspring the advantageous characteristics.

Offspring that inherit the advantageous traits are selected for survival and many will live to reproductive age, passing on the desirable attributes. Those that do not inherit these traits are not likely to survive or reproduce. Gradually, the species evolves, or changes, as more and more individuals carry these traits. Darwin gave this example:

> If the number of individuals of a species with plumed seeds could be increased by greater powers of dissemination within its own area (that is, if the checks to increase fell chiefly on the seeds), those seeds which were provided with ever so little more down, would in the long run be most disseminated; hence a greater number of seeds thus formed would germinate, and would tend to produce plants inheriting the slightly better-adapted down.

Thinking Critically

Darwin identified four conditions that are necessary if evolution is to occur: genetic variation, overproduction of offspring, competition for limited resources, and reproduction of the fittest.

Imagine a plant population that reproduces entirely by asexual methods, such as spreading by underground stems. Although there are many individual plants in the population, they are essentially a single plant genetically; that is, they are clones. Can natural selection act on a population of clones? Is this population capable of evolving? Explain.

In addition to natural selection, humans have long used **artificial selection,** selective breeding as practiced by humans (see Chapter 11), to shape the characteristics of crop plants to suit the needs of humanity. The most serious flaw in Darwin's theory of evolution was the mechanism of heredity. Darwin had not worked out the source of variation in species, nor did he understand the means by which traits are passed down from generation to generation. It would take an Austrian monk, Gregor Mendel (see Chapter 7), working in relative obscurity with pea plants, to come up with the answers to Darwin's questions about inheritance.

A well-known example of natural selection is the case of heavy-metal tolerance in bent grass, *Agrostis tenuis.* Certain populations of bent grass were found growing near the tailings, or soil heaps, excavated from lead mines in Wales despite the fact that mine soils had high concentrations of lead and other heavy metals (copper, zinc, and nickel). When mine plants were transplanted into uncontaminated pasture soil, all survived but were small and slow growing. A nearby population of bent grass from uncontaminated pasture soil exhibited no such tolerance when transplanted into mine soil; in fact, most (57 out of 60) of the pasture plants died in the lead-contaminated soil. The survival of the three pasture plants in mine soil is significant; undoubtedly, these three possessed an advantageous trait, the ability to tolerate heavy-metal soil. A trait that promotes the survival and reproductive success of an organism in a particular environment is an **adaptation.** The mine plants had descended from bent grass plants that possessed the adaptation that conferred tolerance to the mine soil; over time (less than 100 years, in this case) populations of *Agrostis* tolerant to heavy metal evolved from those few tolerant individuals.

Although Darwin's theory of natural selection is the foundation of modern evolutionary concepts, biologists today are still learning about the forces that shape evolution.

Thinking Critically

Natural selection favors the survivorship of those individuals in a population that possess characteristics crucial for survival.

You observe that trees in a part of a forest in which deer are plentiful have higher branches than the trees in a fenced-off part of the forest. Explain the different selective forces at work in these two environments.

CHAPTER SUMMARY

1. Plant systematics has its origins in the classical works of Theophrastus of ancient Greece, who is generally regarded as the Father of Botany. The study of plants, as did many other intellectual endeavors, went into a decline during the Dark Ages of Europe but was later revived because of renewed interest in herbalism during the fifteenth to seventeenth centuries.
2. Linnaeus, a Swedish botanist of the eighteenth century, is credited with the creation of the binomial, or scientific, name. Although common names are often informative and readily accessible, scientific names have the advantage of being recognized the world over and unique to a single species.
3. The taxonomic hierarchy includes the major ranks: domain, kingdom, division (phylum), class, order, family, genus, and species.
4. Biologists have wrestled with the concept of the species; the biological concept describes a species as a group of interbreeding populations, reproductively isolated from other populations.
5. Charles Darwin and his theory of evolution by natural selection irrevocably changed the way biologists viewed species. Natural selection favors those individuals that possess traits that better enable them to survive in the environment. These individuals survive to reproduce, and many of their offspring will tend to have these adaptations and pass them on to future generations. In this way, populations change over time. The four underlying conditions of Darwin's theory of evolution by natural selection are variation, overproduction of offspring, competition, and survival to reproduce.

REVIEW QUESTIONS

1. List the common names of some of the wildflowers in your area. Determine the type of information each name imparts.
2. Using a plant dictionary, look up the scientific names and their meanings for common houseplants and landscape plants in your area.
3. Briefly describe the concept of evolution by natural selection.
4. Why are only inherited traits important in the evolutionary process?
5. How do mutations (see Chapter 7) lead to the evolution of new species?
6. What was the lasting contribution of Linnaeus? How was the binomial system an improvement over polynomials?
7. In what ways can systematics preserve biodiversity?
8. What are genetic barcodes? What are some of the challenges in identifying universal genetic barcodes in plants and fungi?

CHAPTER

9

Diversity of Plant Life

Close up of leaves and cones of Wollemi pine (Wollemia nobilis) *considered by many botanists as a "living fossil."*

KEY CONCEPTS

1. Living organisms are classified into three domains: Archaea, Bacteria, and Eukarya.
2. Land plants occur in the kingdom Plantae of the domain Eukarya and include a diverse group of organisms that impacts our lives in many economically and ecologically important ways.
3. All plants have an alternation of sporophyte and gametophyte generations, although the structure of each generation differs in the various divisions.

©Suzanne Long/Alamy Stock Photo

CHAPTER OUTLINE

The Three-Domain System 137
Survey of the Plant Kingdom 137

A CLOSER LOOK 9.1 Alternation of Generations 139

    Liverworts, Mosses, and Hornworts 141
    Lycophytes and Ferns 143
    Gymnosperms 146

A CLOSER LOOK 9.2 Amber: A Glimpse into the Past 148

    Angiosperms 151
Chapter Summary 151
Review Questions 151

With the overwhelming diversity of life on Earth, scientists have long sought to categorize these organisms into a meaningful system. For many years, organisms were classified as either plants or animals. As our knowledge increased, it became evident that many organisms could not be conveniently classified as either. Other kingdoms have been suggested to solve the problems with the two-kingdom system. One system proposed by Robert Whitaker in 1969 classified organisms into five kingdoms: Animalia, Plantae, Fungi, Protista, and Monera. In this system, the kingdom Monera contained organisms with prokaryotic cells while the other four kingdoms had eukaryotic cells. Recall that a prokaryotic cell lacks a nucleus and membrane-bound organelles that occur in eukaryotes. Research during the 1970s and 1980s changed the scientific understanding about relationships among prokaryotic organisms; two distinctly different types of prokaryotic organisms were recognized. This research led to the current classification of living organisms into three domains. A domain is a taxonomic category above the rank of kingdom.

THE THREE-DOMAIN SYSTEM

In 1990 microbiologist Carl Woese proposed the reorganization of life into the domains **Eukarya, Archaea,** and **Bacteria** (Eubacteria). The domain Eukarya includes all the eukaryotic kingdoms, while Archaea and Bacteria are prokaryotic domains. Although archaea look like bacteria, they represent a distinct evolutionary line.

The majority of prokaryotic organisms are in the domain Bacteria. The Archaea include methane producers as well as organisms that live in hot springs and environments with high salt content. It has been suggested that these extreme environments may be similar to conditions that existed during early Earth history. The two domains of prokaryotic organisms have differences in ribosomal RNA as well as several other molecular and biochemical characteristics. In fact, for some of the molecular characteristics, archaea are similar to eukaryotic cells.

The domain Eukarya includes the kingdoms Plantae, Fungi, Animalia, as well as a large and varied group of organisms often called the kingdom Protista. Organisms in the kingdom Protista include unicellular and simple multicellular organisms that represent many diverse lineages with some species showing plantlike, animal-like, or fungi-like characteristics. As a result, a number of biologists have abandoned the kingdom Protista and have split the organisms into several smaller kingdoms. However, the evolutionary relationships within this group are still uncertain, and other biologists retain the kingdom Protista only as a convenient tool to discuss these organisms. We will follow this practice in this book. The remaining three kingdoms can be distinguished by their modes of nutrition. Members of the kingdom Plantae are **autotrophic,** capable of manufacturing their own food through photosynthesis. The organisms in the kingdoms Animalia and Fungi cannot make their own food and rely on external sources of nutrition. They are, therefore, considered **heterotrophic.** Animals, from primitive sponges to highly evolved mammals, are ingestive heterotrophs, engulfing their food and digesting it internally. The fungi, from molds to mushrooms, are **absorptive heterotrophs,** secreting into their surroundings enzymes that break down food, which is then absorbed. Although historically fungi were considered members of the plant kingdom, recent molecular evidence suggests a closer evolutionary relationship between fungi and animals.

Other organisms that were once regarded as plants in the old two-kingdom system now are included in two eukaryotic kingdoms and the domain Bacteria. In particular, many algae are not included in the kingdom Plantae. The algae consist of a diverse grouping of photosynthetic organisms that have been classified according to pigment types, storage products, and ultrastructural features. They range from prokaryotic microscopic forms to giant kelps and can be found in marine and freshwater habitats where they form the base of the food chains. The cyanobacteria, which were previously known as blue-green algae, are in the domain Bacteria, the green algae are in the kingdom Plantae, and all the other groups of algae are in the kingdom Protista. The algae will be discussed in detail in Chapter 22. Organisms traditionally called fungi are also included in two kingdoms. In the kingdom Protista are several groups of fungus-like organisms that evolved along separate evolutionary pathways. The majority of fungi are classified in the kingdom Fungi. The fungi will be covered in Chapters 23-25.

SURVEY OF THE PLANT KINGDOM

The kingdom Plantae includes a diverse group of complex photosynthetic organisms ranging from green algae to flowering plants. Recall from Chapter 8 that large groupings of similar organisms are called divisions (or phyla). This kingdom includes 12 divisions with living representatives; however, other groups are extinct and known only from the fossil record. Two divisions of green algae are considered part of the kingdom Plantae; however, these will be considered in Chapter 22 along with other groups of algae. The remaining 10 divisions are often referred to as land plants to distinguish them from the algae. One of the features of all land plants is the retention of the embryo. After fertilization, the zygote develops into a multicellular embryo while still enclosed in the female gametangium (reproductive structure). Land plants are often called **embryophytes** to reflect this trait. Although some algae are found on land (damp soil), none retain an embryo; therefore, the term *land plants* will be used only to refer to plants that retain an embryo. The terms *land plant* and *embryophyte* will be used interchangeably in this text.

The fossil record indicates that land plants first evolved more than 470 million years ago. Evidence from biochemistry, ultrastructure, and molecular biology indicates that these plants evolved from freshwater green algae along a common evolutionary path. The 10 divisions of land plants

Figure 9.1 Phylogeny of embryophytes (land plants) showing a hypothesis regarding the evolutionary relationships between plant groups.

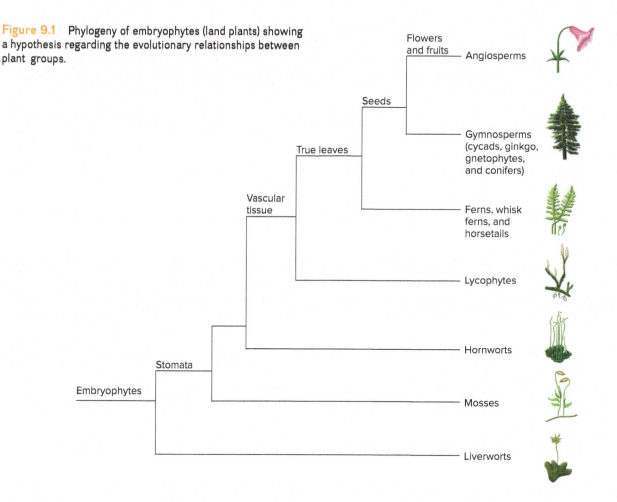

can be easily distinguished from each other by various traits (fig. 9.1). Plants in the divisions Hepatophyta, Bryophyta, and Anthocerophyta lack vascular tissue, while the other seven divisions contain both xylem and phloem (table 9.1). Although all land plants produce spores and gametes at different stages of their life cycles, five divisions of plants also produce seeds. Recall from Chapter 6 that seeds *contain an embryo* and a food supply enclosed by a seed coat. Lastly, members of one division of seed plants, the Magnoliophyta, produce flowers. The life cycles of organisms in several of these divisions are described in A Closer Look 9.1: Alternation of Generations.

Table 9.1
Embryophyte Divisions in the Plant Kingdom

Division	Common Name	Vascular Tissue	Reproduction
Hepatophyta	Liverworts	Absent	No seeds
Bryophyta	Mosses	Absent	No seeds
Anthocerophyta	Hornworts	Absent	No seeds
Lycophyta	Lycophytes	Present	No seeds
Pterophyta	Ferns, whisk ferns, and horsetails	Present	No seeds
Cycadophyta	Cycads	Present	Naked seeds
Ginkgophyta	Ginkgo	Present	Naked seeds
Coniferophyta	Conifers	Present	Naked seeds
Gnetophyta	Gnetophytes	Present	Naked seeds
Magnoliophyta	Flowering plants	Present	Seeds in a fruit

A CLOSER LOOK 9.1

Alternation of Generations

One characteristic of land plants is a life cycle with an alternation of gametophyte and sporophyte generations. Each cell in the sporophyte generation has two sets of chromosomes (diploid), while the cells in the gametophyte generation have only one set (haploid). The gametophyte gives rise to gametes; the sporophyte produces spores. The links between the two generations are the processes of fertilization and meiosis. During fertilization, haploid gametes fuse to form a diploid zygote, which develops into an embryo and then the mature sporophyte. In time, the diploid sporophyte undergoes meiosis to produce haploid spores, which begin the gametophyte generation (box fig. 9.1a). The prominence of each generation is variable among the different divisions. For example, the gametophyte is highly reduced and dependent on the sporophyte in angiosperms, but in some non-flowering plants the gametophyte generation may be prominent and is often independent.

One division in which the gametophyte is conspicuous is the Bryophyta, the mosses. In fact, in this group the gametophyte is the dominant form; the mossy carpet often seen growing on a rock or tree trunk consists of gametophytes. At certain times of the year, the moss undergoes sexual reproduction, resulting in a sporophyte that is attached to and dependent on the gametophyte (box fig 9.1b.1). The gametophytes of mosses are usually either male or female. The male gametophyte develops structures known as **antheridia** (sing., **antheridium**), where the sperm form. The female gametophyte produces flask-shaped **archegonia** (sing., archegonium); each contains a single egg cell. When mature, the flagellated sperm will leave the antheridium and, if sufficient moisture is present, will swim to the archegonium of a female gametophyte. The sperm will fertilize the egg, forming a zygote, which establishes the diploid sporophyte. The sporophyte consists of a stalk and a **capsule** (**sporangium**); the stalk is embedded in the female gametophyte and receives both water and nutrients from it (box fig. 9.1b.2). Within the sporangium (capsule) meiosis occurs, generating haploid spores. If the spores released from the capsule land on a suitable environment, each will germinate into a threadlike structure called a protonema. From the protonema, the mature moss gametophyte develops. The moss life cycle is shown in Box Figure 9.1b.1; similar life cycles exist for liverworts and hornworts.

The vascular plants in the divisions Lycophyta and Pterophyta all share a similar life cycle, with a dominant sporophyte and a small but free-living gametophyte. The ferns will be used as a model for this group (box fig. 9.1c). Unlike the mosses, it is the sporophyte stage of ferns that is the dominant form. In the fern sporophyte, sporangia generally form on the underside of the leafy frond. Sporangia are often clustered into **sori** (sing., sorus), which are visible with the naked eye; meiosis occurs in these sporangia, producing haploid spores. In a hospitable environment, spores give rise to small, flat, often heart-shaped gametophytes that bear both archegonia and antheridia. Flagellated sperm from the antheridia swim to the archegonia, fertilizing the eggs; however, only one sporophyte will emerge from each gametophyte. Although initially attached to the gametophyte, the sporophyte soon grows and becomes independent, while the gametophyte dies.

Like the angiosperms (see Chapter 5), the gymnosperms have an extremely reduced, dependent gametophyte generation, with the sporophyte generation being the dominant and familiar form. A pine in the division Coniferophyta will be used as an example of the gymnosperm life cycle (box fig. 9.1d). The characteristic pollen and seed (or ovulate) cones are the reproductive structures of a pine. The pollen cone consists of sporophylls, which are modified leaves that bear microsporangia; two occur on each sporophyll. Microspores are produced following meiosis in these microsporangia. The microspore develops into the pollen grain that constitutes the male gametophyte generation. At the time the pollen grain is released from the cone, it consists of four cells: two body cells, one generative cell, and one tube cell. Pines are wind-pollinated; air bladders in the

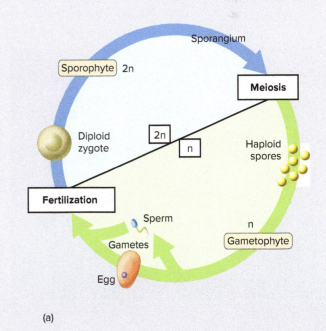

Box Figure 9.1 (a) Alternation of generations. Plants display an alternation of gametophyte and sporophyte phases in their life cycles.

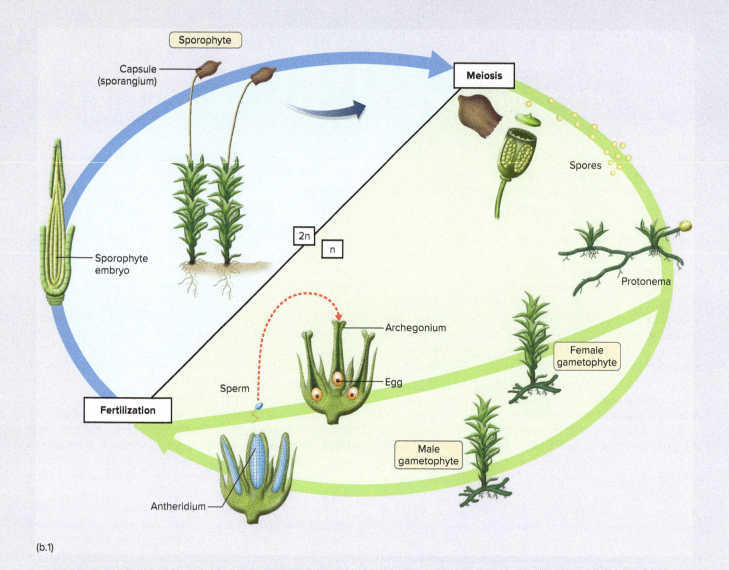

Box Figure 9.1 Continued (b.1) Moss life cycle. (b.2) Moss sporophytes arising from the gametophytes.

walls of the pollen grain contribute to their buoyancy. The pollen grains are carried to the seed cone, where they are trapped by a sticky fluid. The pollen grain germinates with the tube cell, producing a pollen tube that will eventually carry nonflagellated sperm to the egg.

The seed cone is larger and structurally more complex than the pollen cone. Ovules develop on the upper surface of cone scales, which are arranged around a central axis. Normally, two ovules are borne on each cone scale. Recall from the life cycle of angiosperms that the ovule is a megasporangium surrounded by integuments, with an opening, the micropyle, facing the central axis. Meiosis produces four megaspores within each ovule, but three degenerate. The one surviving megaspore develops into the female gametophyte. At maturity the female gametophyte usually contains two archegonia, each with a single egg. While the female gametophyte is developing, the pollen tube is slowly growing through the surrounding tissues. During this time, the generative cell within the pollen tube divides to form two sperm. Although both nonmotile sperm are carried to the archegonium, only one sperm fertilizes the egg, giving rise to the new sporophyte generation. Within the ovule, only one embryo is produced, with the female gametophyte as its nutrient tissue. As the fertilized ovule develops into the seed, the integuments develop into the seed coat. When the seeds are mature, the seed cone opens and sheds the seeds, which are wind dispersed.

The three life cycles described here illustrate the variation of the basic alternation of a gametophyte generation with a sporophyte generation found in the plant kingdom. Other types of life cycles can be found in the algae (see Chapter 22) and the fungi (see Chapter 23).

Thinking Critically

In the majority of land plants, the sporophyte is the dominant stage while the gametophyte is reduced, becoming completely dependent on the diploid phase in the most advanced divisions.

Suggest a reason why natural selection favored the dominance of the diploid sporophyte over the gametophyte.

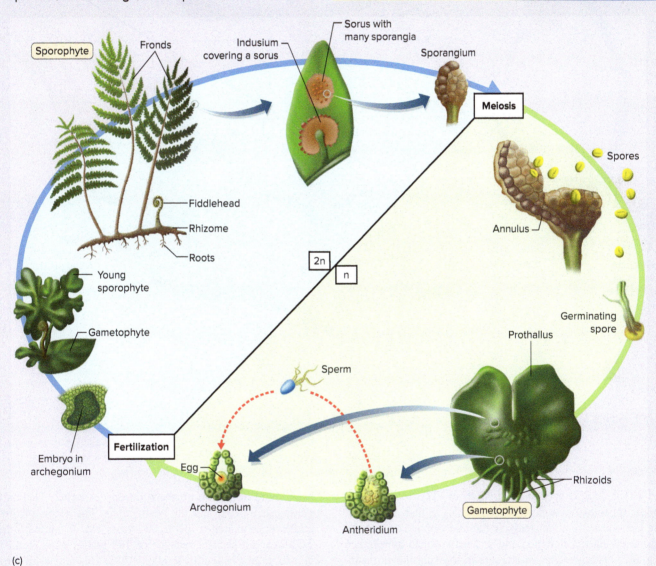

Box Figure 9.1 Continued (c) Fern life cycle.

Liverworts, Mosses, and Hornworts

The first land plants to be considered are the liverworts, mosses, and hornworts. Traditionally, these plants had been grouped together in a single division. However, distinct differences in these three groups have led to their reclassification into three separate divisions: the Hepatophyta or Marchantiophyta (liverworts), the Bryophyta (mosses), and the Anthocerophyta (hornworts). Today the term *bryophyte* is still used as a collective term to refer to plants in any or all of these divisions. Bryophytes are small plants generally restricted to moist environments, although many bryophytes can withstand extended dry periods. It is estimated that there

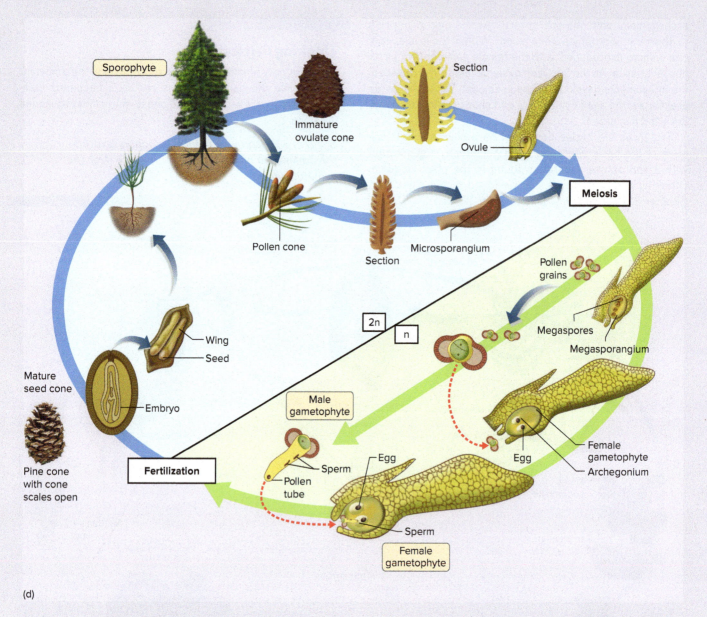

Box Figure 9.1 Continued (d) Pine life cycle.

are about 25,000 species of bryophytes. The mosses have small appendages, appearing somewhat similar to leaves; the liverworts are either leafy or flat and ribbonlike; and the hornworts have a flattened and somewhat lobed appearance (fig. 9.2). The presence of stomata in the sporophyte distinguishes mosses and hornworts from liverworts. A distinctive feature found in hornworts is a meristem near the base of the sporophyte that permits continued growth of the sporophyte under favorable conditions.

The well-known peat moss, *Sphagnum*, is unusual because it grows in acid water, making the water even more acidic and often converting the pond into a bog. The acidity restricts the growth of other plants and prevents the growth of microorganisms that cause decay. This creates a perfect environment for preservation. Important archeological finds have been made in peat bogs; several hundred well-preserved bodies have been found in bogs in northern Europe, especially Denmark, Germany, Ireland, and the Netherlands. The oldest of these bodies is estimated to be 10,000 years old. In addition, the 7,000-year-old skeletal remains of 168 bodies were found in Windover Bog in south Florida in 1982. This remarkable find was a burial site of ancient Americans who lived in the area. Although most of the flesh was decayed, many of the brains were preserved, and were recently used for DNA testing. The yield of DNA from the brain tissue was low; however, preliminary DNA analysis from five brains found that the DNA did not match that of known Native American tribes. The DNA appeared to match DNA markers from Europeans, suggesting these bog people were of European descent.

There are several economic uses for *Sphagnum* as well. Dried peat moss is commonly used as a potting and bedding material by gardeners, since it has a great water-holding

Figure 9.2 Bryophytes include mosses, liverworts, and hornworts. (a) *Sphagnum* moss with sporophytes arising from the gametophyte. (b) *Marchantia*. Gametes are produced in the umbrella-like structures shown arising from the flattened liverwort. (c) Hornwort. *Anthoceros* with erect sporophytes arising from the lobed gametophyte. (d) *Sphagnum* moss. A peat bog being drained and peat harvested.

capacity. For centuries, the peat bogs of northern Europe, especially those in Ireland, have been the source of a home heating fuel. The peat from these bogs, which is partially decomposed *Sphagnum*, is harvested, dried (fig. 9.2d), and burned in large quantities. As late as World War I, *Sphagnum* also was used as a dressing for wounds because of its antiseptic properties. In contrast, most other bryophytes are of limited economic importance; however, all bryophytes are of great ecological importance. They are among the first organisms to colonize barren areas and help retain soil and minerals in these areas. Mosses are even known to colonize mountaintops and tundra.

Lycophytes and Ferns

The remaining divisions in the plant kingdom are all vascular plants; that is, they all contain vascular tissues that conduct both water and food throughout the plant body. The evolution of vascular tissue allowed the establishment of land plants in areas where freestanding water was limited. Some of the oldest well-studied vascular plant fossils are dated approximately 420 million years ago. The first two divisions of vascular plants are nonseed plants and include the lycophytes and ferns.

The division Lycophyta consists of club mosses, spike mosses, and quillworts. Today these plants range from small, prostrate forms commonly found on the forest floor to larger, epiphytic forms in the tropics. (**Epiphytes** are plants hanging from other, larger forms of vegetation.) Superficially, these plants resemble mosses because their stems are covered with small, overlapping, scalelike leaves; however, they are not mosses but vascular plants with conducting tissue in the roots, stems, and leaves. The leaves of club mosses, called microphylls, are very simple with a single unbranched vascular bundle. Two to three hundred million years ago, there were treelike species of lycophytes that formed extensive forests. These prehistoric forests are the basis for many of our coal deposits today (fig. 9.3).

Figure 9.3 *Lepidodendron* (left) and *Calamites* (right) fossils. These tree-sized plants were often important components of the Carboniferous period forests (360 to 286 million years ago). Modern descendents of *Lepidodendron* include the club mosses, and horsetails are descendents of *Calamites*. *Lepidodendron* fossils are characterized by diamond-shaped leaf scars, while *Calamites* fossils are recognized by ribbed, jointed stems similar to modern horsetails.

In many species of *Lycopodium*, a member of the division Lycophyta, the sporangia are arranged in compact clusters at the ends of erect or hanging stems. The clusters of sporangia often have a clublike appearance (fig. 9.4a). Creeping jenny and ground pine are two well-known club mosses often used in Christmas wreaths. In the early days of photography, *Lycopodium* spores were the original flash powder. *Lycopodium* powder was also used in the past as a type of talcum powder, as a coating for pills, to stop bleeding from wounds, and for other medicinal purposes. Another common plant in this division is the resurrection plant, *Selaginella lepidophylla*, a native of arid regions, which appears as a dried brown clump when water is scarce yet quickly becomes green and photosynthetic when water is available. This spike moss is often sold as a novelty in shops.

The division Pterophyta (also known as Monilophyta) consists of horsetails, whisk ferns, and ferns. Until recently the horsetails and whisk ferns were placed in two separate divisions; however, molecular analysis of these two groups shows that these plants are closely related to ferns and evolved along a single evolutionary path. Electron microscopy of sperm cells from these groups shows this same relationship. As a result, in modern classification systems, the horsetails and whisk ferns are considered members of the division Pterophyta.

Today the horsetails are a small group of widely distributed plants in the genus *Equisetum*. These plants have ribbed, jointed, photosynthetic stems with whorls of tiny leaves that soon become brown and nonphotosynthetic. Sporangia are grouped into conelike structures at the tips of some stems. Although most horsetails are small, some species are 2 meters (6 feet) or more, but have very slender stems; some species have whorls of lateral branches. (fig. 9.4b). As in the division

(a)

(b)

Figure 9.4 Club mosses and horsetails are modern-day relatives of plants that were common in the Carboniferous period forests. (a) *Lycopodium*, a club moss. (b) *Equisetum*, a horsetail.

Figure 9.5 Branching stems of *Psilotum* with numerous sporangia.

(a)

(b)

(c)

Figure 9.6 Ferns. (a) Fiddleheads unrolling in springtime. (b) *Dryopteris* sp., wood fern, growing in a cloister garden. (c) Sporangia grouped into sori on the underside of a fern frond.

Lycophyta, tree-sized species existed 200 to 300 million years ago and also contributed to coal deposits that are mined today (fig. 9.3). One interesting feature of the horsetails is the presence of silica in their cell walls. The silica in the walls makes the stems abrasive, and for this reason, pioneers in North America used these plants as a primitive scouring pad to clean pots and pans. People in developing nations still use them for this purpose. This feature explains another common name for these organisms—the scouring rushes. Musicians also use the scouring rushes to sand the reeds on wind instruments.

The whisk ferns are a small group of primitive-looking vascular plants with only two genera *Psilotum* and *Tmesipteris*. *Psilotum* consists of green, branching stems with tiny, scale-like appendages and globose yellow sporangia (fig. 9.5). Traditionally, many botanists considered these plants related to the earliest vascular land plants that existed over 400 million years ago; however, no fossil evidence supports this link.

The ferns are the largest group of nonseed vascular plants (fig. 9.6). Ferns range from small, aquatic types less than 1.25 cm (0.5 in.) in size to large tropical tree ferns that grow to 20 meters (60 feet) in height. The ferns that we are most familiar with, from temperate forests and as houseplants, have large, divided leaves called fronds that arise from a horizontal underground stem, a **rhizome,** which also gives rise to roots. The sporangia of some ferns are borne on the underside of the fronds; the distribution of sporangia on the frond is a characteristic that can be used as an aid in identification. Young fern leaves first appear as tightly coiled fiddleheads, which gradually unroll as they grow. The fiddleheads of some ferns are considered a gourmet delicacy. Although there are about 12,000 species of ferns in the world, the ferns were more abundant during past geological times and contributed to our coal deposits.

Various ferns have a long history of use in folk medicine in different cultures. They have also been exploited as sources of dyes, fiber, and thatching materials for roofs. *Azolla*, a small, aquatic fern, is considered a source of green manure, especially

Figure 9.7 (a) Gnarled trunk of a bristlecone pine tree (*Pinus aristata*). This is one of three species called bristlecone pines, which included some of the oldest living trees. Individual bristlecone pines may live 3,000 to 5,000 years. (b) The General Sherman tree, *Sequoiadendron giganteum*. As a group the giant sequoias are considered the world's largest (most massive) trees. The largest on record is this General Sherman tree, which is reported to be 83.9 m (274.9 ft) tall with a circumference of 31.3 m (102.6 ft).

in rice fields (see Chapter 12 and fig. 12.12). In 2001, the brake fern, *Pteris vittata,* was found to accumulate arsenic from contaminated soils. This was the first plant shown to hyperaccumulate arsenic, which is a growing environmental problem; *P. vittata* is also the first fern found to be effective in phytoremediation. Since the initial discovery, other *Pteris* species have also been described as effective arsenic accumulators (see A Closer Look 4.1: Mineral Nutrition and the Green Clean).

Gymnosperms

The remaining five divisions in the plant kingdom are all **seed plants.** The seed plants are the dominant vegetation in the world. Recall that a seed contains a small, embryonic plant with a food supply encased in a tough, protective coat. Seeds are found in either cones or fruits. In cones, the seeds are exposed at maturity and said to be "naked." Plants with naked seeds are called **gymnosperms**, while plants with seeds enclosed or hidden within fruits are called **angiosperms.**

There are four divisions of gymnosperms: Coniferophyta, Cycadophyta, Ginkgophyta, and Gnetophyta. Conifers, members of the Coniferophyta (also known as the Pinophyta), are the most familiar gymnosperms and include pines, spruces, yews, firs, cedars, redwoods, and larches. The largest, tallest, and oldest trees in the world are all conifers (fig. 9.7). Conifers are cone bearers, typically producing both pollen and seed cones. The pollen cones are usually small and inconspicuous and produce pollen, while the seed cones (or ovulate cones), which produce seeds, are larger and often take several years to mature (fig. 9.8). The pollen and seed cones may be on the same or different trees.

The conifers, like all gymnosperms, are an ancient group of plants that reached their greatest diversity during past geological periods. Periodically, scientists have discovered certain gymnosperm species alive that were previously known only from the fossil record. The most recent discovery of a "living fossil" is the Wollemi pine (*Wollemia nobilis*). Forty specimens of this tree were discovered in August 1994 in a secluded rain forest gully of Wollemi National Park about 200 kilometers (125 miles) west of Sydney, Australia. Today approximately 100 trees have been found from three close locations. Genetic testing has shown that these trees are genetically indistinguishable, suggesting that at some point the population became so small that all genetic variability was lost. Wollemi pines

Figure 9.8 (a) Seed cone of *Pinus monticola*, western white pine. (b) Seed cones of *Juniperus ashei*, mountain cedar. The small, berrylike cones of *Juniperus* are a sharp contrast to the seed cones of *Pinus*.

have been classified in the family Araucariaceae along with Norfolk Island pines, but their closest relatives are fossils from the Jurassic period (200–136 million years ago) and the Cretaceous period (136–65 million years ago). Although the oldest trees in the stand at Wollemi National Park are only about 150 years old, the survival of this species is considered remarkable and may be one of the most significant botanical discoveries of the twentieth century. It has been described as equivalent to finding a small dinosaur alive. Intense study of these trees is underway. The National Parks and Wildlife Service in Australia has been carefully protecting the grove of trees from tourists and is working on efforts to propagate the species and establish other populations. The propagation program has provided Wollemi pines for botanical gardens around the world. In addition, seedlings are now available for sale in many countries. They grow well in a variety of locations and climates, so you, too, can have this living fossil in your garden.

The conifers are the dominant trees in the northernmost forests of the world. They are important sources of lumber, paper pulp, and products such as turpentine, rosin, and pitch (see Chapter 18 and A Closer Look 9.2: Amber: A Glimpse into the Past). Although conifers are not usually considered as food, pine nuts, seeds of the pinyon pine, are enjoyed by many. Also, some types of cedar, *Juniperus* spp., are essential in the production of gin, a popular alcoholic beverage. The seed cones of these cedars, or junipers, are unusual because they are small and fleshy, resembling waxy blueberries (fig. 9.8b). These "berries" impart the flavoring to gin. Next time you see a *Juniperus*, pick a few "berries," crush them, and sniff the aroma of gin. *Taxus* species, commonly called yews, are noted as the source of taxol used in cancer chemotherapy (see Chapter 19) and also the source of poisonous alkaloids (see Chapter 21).

The cycads (Cycadophyta) are a small group of short shrubs to moderate-sized, long-lived trees native to tropical and subtropical regions. These trees are often mistaken for palms because the leaves are large and palmlike (fig. 9.9a). The sago palm, a cycad commonly used for landscaping in subtropical to tropical climates, is also a popular houseplant. Like the conifers, most cycads bear cones. The seed cones of certain species can be quite large; some weigh up to 100 pounds (45.36 kilograms)! Cycads are dioecious, with seed cones and pollen cones on separate individuals. Although the conifers are wind-pollinated, recent research has shown that cycads are pollinated by weevils and beetles. Some botanists believe that cycads may be the earliest insect-pollinated plants.

Today the cycads are mere remnants of plants that were the dominant features of the Mesozoic landscape. Although the Mesozoic era (240–65 million years ago) is usually known as the Age of Dinosaurs, botanists consider this time the Age of Cycads. A real Jurassic Park would have had cycad forests. Only about 250 species of cycads in 11 genera are found today, with the greatest number of species occurring in the Southern Hemisphere. Many species are threatened with extinction because they live in tropical forests, which are endangered habitats, while other species are threatened by extensive collecting. Cycad breeding programs are underway

A CLOSER LOOK 9.2

Amber: A Glimpse into the Past

Many economically useful products are derived from conifers; much of the lumber and paper industry is based on pine, spruce, and other coniferous species. Resin is another product that is usually obtained from conifers; it is the material often seen oozing from the trunk of a pine tree. Resins are sticky secretions that harden as they dry. Wounding a plant initiates the release of resins that promote healing by sealing off a wound against water loss, pathogens, and insects. Resins are best known as the source of turpentine and rosin (see Chapter 18).

If conditions are right and the resin is buried before it can oxidize, it fossilizes. Amber is fossilized tree resin. Although most amber is from conifers, a few angiosperm families, including the Fabaceae (bean or legume family), can also produce resins that fossilize and become amber. Also, not all amber is chemically identical. Recent chemical analyses have shown that resins from different tree species have slightly different chemical signatures. These chemical methods are being applied to ancient amber as well to identify the species of trees producing amber in specific locations.

Translucent and typically golden brown, amber has been valued as a semiprecious stone for millennia. Stone Age figurines of amber have been found in the Baltic area; they may have been talismans. Roman soldiers wore amber-studded armor for good luck, and the emperor Nero established trade routes with the Germanic tribes of the north to obtain it. The pinnacle of amber artistry was the famed Amber Room in the Catherine Palace of Russia. A gift from the Prussian king Frederick Wilhelm I to the Russian czar Peter the Great, intricately carved pieces of amber were fit together on panels that lined the walls of the room. The panels were disassembled and stolen by the Nazis during World War II. The reassembled panels were later put on display in Königsberg, Germany, but disappeared during the last days of the war. The fate of the panels remains a mystery. However, in 2003, after decades of work, a full-scale reconstruction of the Amber Room was completed in the Catherine Palace. The new carvings were based on old black-and-white photographs of the original room. In 2017, another reconstruction of the famed Amber Room was completed and opened at the residence of the Russian Ambassador to France in Paris.

Amber has been gathered for centuries on the shores of the Baltic Sea, an area well known for its rich deposits. As waves erode away the shoreline, amber is washed free and can be found floating in the water. The largest amber-producing mines are located in the town of Palmnicken on the Baltic Sea. Approximately 30 million years ago, this area was subtropical in climate and had forests of

Box Figure 9.2 A leaf of a tree in the legume family is preserved in amber from the Dominican Republic. An insect trapped in the amber is visible near the bottom of the photo.

at botanical gardens in various parts of the world to preserve these primitive seed plants.

Cycads are rich in starch, which is found in the roots, stems, and seeds, and they have a long history as a food source as well as a medicinal plant. However, cycads contain toxic compounds, which must be removed by processing the starch before the flour is used. Cycasin and BMAA (beta-methylamino-alanine) are distinct toxins found in cycads. Cycasin is a carcinogenic and neurotoxic glycoside; BMAA is an unusual amino acid that is also neurotoxic. BMAA is believed to block the function of receptors that enable nerve cells to communicate with each other in the brain. Most detoxification methods focus on soaking the crushed seeds in water; soaking slowly dissolves the toxins out of the plant material over several days to weeks. The resulting paste is baked into breads. Toxins also occur in the leaves of many cycads, and these have caused poisonings in cattle and sheep grazing on the leaves.

Although gastrointestinal and liver problems can also occur, cycad toxins have been implicated in a number of neurological conditions, including Guam disease. Following World War II, physicians found that ALS (amyotrophic lateral sclerosis, also known as Lou Gehrig's disease) was about 100 times greater among members of the Chamorro tribe, the native population in Guam, than the rest of the world. Later studies found that the incidence of Parkinson's disease and Alzheimer's-like dementia were also much higher among the Chamorros. Although many conifers. The resins of these conifers are the source of the 700 tons of amber mined each year. Amber-rich beds are found 40 meters (120 feet) below the surface; as much as 4.05 kilograms per cubic meter (7.7 pounds per cubic yard) can be found. The earth is strip-mined to the amber beds; the sediments are then washed with water from huge, pressurized hoses and the amber is separated out. Only 13% of the amber is usable for jewelry; the rest is used in paint thinners, varnishes, and polishes.

There are other amber deposits in many parts of the world, dating from the late Carboniferous to the Pleistocene (300–1.5 million years ago). Some of the oldest amber hails from the central Appalachian region in the eastern United States. Amber from the Dominican Republic holds the record for the greatest number of inclusions, encasing insects, spiders, bits of wood, flowers, seeds and leaves (box fig. 9.2), and even feathers. For every 100 pieces of amber from the Dominican Republic, one piece contains an insect fossil; in Baltic amber, the ratio drops to a mere one in 1,000. Excavations in the 1990s of a large deposit of Cretaceous amber from New Jersey have been the source of several exciting finds. What was once an ancient marsh in coastal New Jersey and is now a sandy coastal pit is the site of charcoalized flowers so well preserved that petals, pollen grains, and even tiny ovules are discernible. More than 200 species of angiosperms have been identified, with many of the flowers related to hydrangeas, carnations, azaleas, pitcher plants, oaks, and the tropical mangosteen. The origin of flowering plants as well as many insect lineages dates back to the Cretaceous period. In the New Jersey amber, 90-million-year-old flowers with scent glands and nectaries have been uncovered that were obviously insect-pollinated. The association between flowering plants and their insect pollinators extends further back in time than previously thought. The same Cretaceous amber field also yielded the oldest intact mushrooms ever found. An even older fossil of a coral fungus (related to mushrooms) was found in amber from Myanmar (Burma) in 2003. The reproductive structures were clearly visible in the amber, which is dated to the Cretaceous period and is approximately 100 million years old. In 2006, scientists identified the oldest known bee preserved in amber, also from Myanmar. This 100-million-year-old specimen is approximately 40 million years older than other bee fossils.

Amber and the biological specimens trapped within have been the focus of much interest and controversy because of the possibility of extracting and deciphering the entrapped fossil DNA. Amber is especially good at preserving ancient DNA because the resin contains compounds that dry and fix living tissue and, at the same time, inhibit bacterial decomposition. Forty-million-year-old fungus gnats were so perfectly preserved in amber that organelles, such as mitochondria, ribosomes, and cell membranes, could be clearly seen in the abdominal cells. The first extraction of amber DNA was from a species of extinct bee. Since that first extraction, the genetic material of a 120-million-year-old weevil and that of a 30-million-year-old extinct species of termite have also been reported.

One problem with the technique is contamination with extant DNA. One research team thought they had isolated fossil DNA only to find out it was contaminated with the DNA from modern species. In fact, some scientists have questioned the authenticity of all of these fossil DNA finds. The method of extraction presents another problem. The amber is cracked and destroyed in order to scrape out the specimen's DNA. Less destructive methods are being investigated, such as drilling a tiny hole in which a small needle is inserted, so that only a portion of the specimen is extracted and the amber is saved.

Once the DNA has been isolated, it can be amplified to determine the base sequence. This information will be invaluable in tracing evolutionary lineages. It may even be possible someday to reconstruct long-extinct species from a deciphered genetic blueprint. The plot of the science fiction book and movie *Jurassic Park* is based on this possibility.

Right now, reconstructing dinosaur DNA is only a remote possibility, since the most successful isolation of amber-fossilized DNA has sequenced only 200 base pairs. Even a bacterium contains approximately 2 million base pairs, and humans have over 3 billion. Nevertheless, amber has opened the gateway to looking at fossil DNA and examining the evolutionary past of ancestral and fossil organisms.

---

scientists believed that the neurotoxins in cycads were at least partly responsible, it was only in 2003 that the proposed connection between the toxins and the neurological conditions was explained. Initially, it was believed that the direct consumption of a tortilla-like cycad bread led to these conditions, but the concentration of toxin is low in cycad flour and probably would not be sufficient to cause the symptoms. A team of researchers led by Paul Cox and Sandra Banack suggested that biological magnification of the toxin may be the answer. Biological magnification is the increase in concentration of toxins as they are passed along a food chain (see Chapter 26).

The consumption of a type of bat, called a flying fox, by the Chamarro people may be a key link in the toxicity. Cycad seeds are a major food source for flying foxes, and the Chamarro people traditionally ate these animals boiled in coconut milk at weddings and other festive events. Following World War II, firearms became more widely available, and it became easier to hunt bats. As a result, the consumption of bats greatly increased. In fact, hunting led to the extinction of one species of flying fox and the near extinction of the second species native to Guam. Chemical analysis of tissues collected from museum specimens of flying foxes showed levels of the BMAA toxins that were 100 times higher than levels in the cycad seeds.

Although the concentration of BMAA is highest in cycad reproductive structures, it also occurs at low levels in other parts of the plant. Normal roots lack the toxin; however, Cox

Figure 9.9 (a) *Cycas revoluta* with a rosette of young leaves developing at the top. (b) *Ginkgo biloba*, the maidenhair tree, branch with leaves, seeds shown in inset. (c) *Ephedra viridis*, a desert shrub. (d) *Welwitschia mirabilis* in the Namib Desert of Namibia, Africa.

and Banack found BMAA present in modified roots called coralloid roots. These short, wide, lateral roots grow at the soil surface and form a symbiotic relationship with cyanobacteria present within small cavities of the root. (**Symbiosis** is the intimate association of two species living together in a relationship that can be beneficial to one or both organisms.) Investigating further, the researchers found that, when cultured alone, the cyanobacteria synthesized BMAA. Putting all these pieces together, Cox and Banack suggest that the biological magnification of the BMAA starts with the toxin synthesis by cyanobacteria in the roots and continues with its accumulation in cycad seeds, its further accumulation in the tissues of the flying fox, and its final ingestion by the Chamarro people. The incidence of the neurological problems decreased following the population decline of the flying foxes on Guam. Today, different species of flying foxes are imported from other islands in the South Pacific for Chamarro consumption. Since cycads do not occur on these other islands, the toxin is no longer present in the Chamarro food chain.

Ginkgo biloba, the maidenhair tree, is the only living species of the Ginkgophyta. This species is another "living fossil" because it is the only remaining species of a larger group of plants that were abundant in the Mesozoic era over much of the world. Examination of fossils indicates that these plants have changed very little over millions of years. Although ginkgo became extinct over much of its range, it survived in China and has been cultivated for centuries, especially around Buddhist temple gardens and monasteries. It became known to Europeans in the eighteenth century and was first introduced into the United States in 1784. Today ginkgo is a popular tree in city landscaping because it is air pollution tolerant and hardy. It is a moderate-sized tree with unique, fan-shaped leaves that may or may not be notched in the middle (fig. 9.9b).

Ginkgo seeds are partially surrounded by a fleshy coat that smells like rancid butter when the seed is mature. Fortunately, ginkgo is dioecious, with separate male and female trees; therefore, efforts are made to screen for pollen-bearing trees before planting. However, the seed itself is considered a delicacy in China and other Asian countries. The seed, also called a ginkgo nut, is boiled or roasted and included in both sweet and savory dishes. It is frequently included in meals served at Chinese weddings because it is a symbol of good luck. The seed has also been used medicinally to treat coughs, asthma, and other respiratory complaints. Today, extracts of ginkgo leaves are widely used as an herbal remedy to improve short-term memory and enhance concentration (see Chapter 19).

The Gnetophyta is a very small group of gymnosperms with unusual morphological and anatomical features that have intrigued botanists for years. There are three genera in the division, *Ephedra, Gnetum,* and *Welwitschia,* each classified in a separate family. *Ephedra*, commonly called Mormon tea, is an economically important member of this group. This desert shrub produces the alkaloids ephedrine and pseudoephedrine, which are useful in the treatment of bronchial asthma, sinusitis, the common cold, and hay fever (fig. 9.9c). The medicinal uses of *Ephedra* are discussed in Chapter 19. Also the leaves of the tropical vine *Gnetum* are used as a vegetable in parts of central and west Africa. Traditionally the leaves were collected from the wild, cut into strips, and cooked. Overexploitation of this resource has threatened the *Gnetum* populations in some tropical forests. Recently, the Limbe Botanical Garden in Cameroon has developed cultivation methods for two *Gnetum* species to ensure future availability of this nutritious vegetable. Perhaps the most unusual member of the Gnetophyta is *Welwitschia mirabilis.* This species occurs only in the Namib desert of southwest Africa in Angola and Namibia and is the only species in the genus. The plant produces a very long taproot and a short, barrel-shaped stem with two leathery straplike leaves that continue to grow throughout the life of the plant. The plant may live for 1,000 to 2,000 years, and the leaves split into strips as they age (fig. 9.9d).

Thinking Critically

The discovery of the Wollemi pine emphasizes the value of national parks in the protection of natural habitats.

What are some of the other benefits of national parks?

Angiosperms

The last remaining division in the plant kingdom is the Magnoliophyta, the flowering plants, or angiosperms. The angiosperms constitute the most widespread vegetation on Earth today, ranging in form from small herbaceous plants to large trees. This division includes most of our familiar plants, such as lawn grasses, crop plants, vegetables, weeds, and houseplants as well as oaks, elms, maples, and many other trees. Over 350,000 species of angiosperms have been described, and it has been suggested that as many as 1 million undescribed species may exist in tropical forests.

In the previous eight chapters, we focused on the anatomy, reproduction, and physiology of angiosperms. In the chapters that follow, we shall see that most of our economically useful plants are, in fact, angiosperms.

CHAPTER SUMMARY

1. Currently, biologists accept the three-domain system of classification. In this system, living organisms are classified as Archaea, Bacteria, and Eukarya. The kingdom Plantae in the domain Eukarya contains a diversity of photosynthetic organisms from green algae to angiosperms.
2. The land plants in the kingdom Plantae have an alternation of generations and retain a multicellular embryo within the female gametangium.
3. Some plants, such as liverworts, mosses, and hornworts, have a dominant gametophyte generation. These are small plants that are confined to moist environments, since they have no vascular tissue.
4. All vascular plants have a dominant sporophyte. Two divisions of vascular plants do not form seeds. The ferns and lycophytes have a long fossil history, and in past geological ages they dominated the landscape.
5. The seed plants are currently the dominant vegetation in the world. Four divisions of seed plants are gymnosperms. Conifers are the most familiar gymnosperms and include the largest, tallest, and oldest trees in the world. Conifers are of great economic importance as the source of wood, pulp, and chemicals. The angiosperms constitute the most widespread and diverse vegetation today. They provide most of the food for life on the planet as well as economically useful products.

REVIEW QUESTIONS

1. How does the three-domain system compare with the two-kingdom and five-kingdom systems?
2. Organisms traditionally called plants are now assigned to other kingdoms. Describe which groups have been reassigned and why.
3. Describe the divisions of seed plants.
4. What are the basic features of the ferns and lycophytes? What is the economic impact of these groups, both living and fossil?
5. Describe the alternation of generations in vascular plants.
6. Define *sporangium, archegonium, antheridium, sporophyte,* and *gametophyte.*
7. Discuss the economic and ecological value of the gymnosperms.
8. Describe the uses of peat moss.
9. Compare the life cycle of a moss, a fern, a gymnosperm, and an angiosperm. What changes have taken place to the gametophyte generations in seed plants?

UNIT III

CHAPTER 10

Human Nutrition

Nutrition Facts label imparts important nutritional information for the consumer.

KEY CONCEPTS

1. Human nutritional needs are supplied by macronutrients (carbohydrates, proteins, and fats) and micronutrients (vitamins and minerals).
2. If nutritional requirements are not satisfied, deficiency diseases can result that have widespread effects on the bodily systems.
3. Plants can supply the majority of human nutritional requirements, and there is evidence that increasing the proportion of plant foods in the diet can have positive health benefits.

©Karen McMahon

CHAPTER OUTLINE

Macronutrients 153
   Sugars and Complex Carbohydrates 153

A CLOSER LOOK 10.1 Famine or Feast 154

   Fiber in the Diet 156
   Proteins and Essential Amino Acids 156
   Gluten and Celiac Disease 158
   Fats and Cholesterol 159
Micronutrients 162
   Vitamins 162
   Minerals 167

Dietary Guidelines 169
   Balancing Nutritional Requirements 169
   Healthier Dietary Guidelines 170
   Glycemic Index 171

A CLOSER LOOK 10.2 Eat Broccoli for Cancer Prevention 172

   Meatless Alternatives 172
Chapter Summary 173
Review Questions 174

New ideas in nutrition are quickly incorporated by the health-conscious segment of society and advertisers looking for a new marketing gimmick. The benefits of several nutritional concepts, such as fiber, monounsaturated oils, and low-carbohydrate diets, have made headlines and influenced lifestyle changes. All these concepts promise better health and many are, in fact, dependent on a greater consumption of plants in the human diet. This chapter will examine human nutritional needs and how plants can satisfy these needs.

MACRONUTRIENTS

The basic nutritional needs of humans are to supply energy and raw materials for all the various activities and processes that occur in the body. In addition to the need for water, humans require five types of nutrients from their food supply; three of these are required in relatively large amounts and are called **macronutrients,** consisting of **carbohydrates, proteins,** and **fats.** The other two types of nutrients, **vitamins** and **minerals,** are required in small amounts and are known as **micronutrients.** If water were removed, the macronutrients would make up almost all the dry weight of foods.*

Human energy requirements vary with the age, sex, and activity level of the individual, within a wide range of 1,200 to 3,200 kilocalories per day. The current recommendation for Americans is an average daily intake of 2,000 kilocalories or 1,600 kilocalories for women and 2,200 kilocalories for men. (A **calorie** is a measure of energy—technically, the amount of energy needed to raise the temperature of 1 gram of water by 1 degree Celsius.) Food energy is normally measured in **kilocalories** (1,000 calories = 1 kilocalorie), which can be abbreviated as **kcal,** or **Calories** with a capital *C*. Most dietary guides simply use the term *calories,* but this book will use the more accurate Calories or kilocalories. Each gram of carbohydrate or protein can supply 4 kilocalories; for each gram of fat consumed, the amount of energy supplied is more than double, 9 kilocalories. Although all the macronutrients can be used as a source of energy, normally only carbohydrates and fats do so, while proteins provide the raw materials, or building blocks, required for the synthesis of essential metabolites, growth, and tissue maintenance. The consequences of undernutrition, malnutrition, and overnutrition for the world's population are the topics of A Closer Look 10.1: Famine or Feast.

Sugars and Complex Carbohydrates

Although carbohydrates are commonly grouped into **sugars** and starches, recall (see Chapter 1) that these compounds can be chemically classified into **monosaccharides, disaccharides,** and **polysaccharides,** based on the number of sugar units in the molecule.

Monosaccharides

Monosaccharides are the basic building blocks of all carbohydrates, and **glucose** is the most abundant of these sugars. During the process of digestion, many carbohydrates are broken down or converted into glucose, which is then transported by the blood to all the cells in the body. Within cells, the process of cellular respiration metabolizes glucose to produce the energy necessary to sustain life. Other common monosaccharides are **fructose** and **galactose,** which have the same chemical makeup as glucose, $C_6H_{12}O_6$, differing only in the arrangement of the atoms within the molecules. In the body, most of the fructose and galactose is converted into glucose and metabolized as such. In the United States, an inexpensive sweetener for many types of processed foods is high-fructose corn syrup, often preferred because fructose is sweeter-tasting than table sugar. Fructose is commonly found in many fruits, unlike galactose, which does not normally occur free in nature.

Disaccharides

Disaccharides are composed of two monosaccharides chemically joined together. The most common disaccharide is **sucrose,** or table sugar, formed from a molecule of glucose and a molecule of fructose. Other disaccharides are the milk sugar **lactose** (a combination of glucose and galactose) and **maltose** (formed by two glucose molecules), which are largely found in germinating grains. Table sugar, which primarily comes from sugarcane and sugar beet, is at least 97% pure sucrose with little nutritional value, thereby supplying only kilocalories. (See A Closer Look 4.2: Sugar and Slavery.) During digestion, these disaccharides are broken down to yield their component monosaccharides.

Polysaccharides

Polysaccharides, also known as complex carbohydrates, contain hundreds to thousands of individual sugar units, and for the most part, glucose is the only monosaccharide present. The different polysaccharides are distinguished by the way in which the glucose units are joined together, their arrangement, and their number. **Starch** is the storage form of glucose found in plants; it occurs abundantly in seeds, some fruits, tubers, and taproots. The presence of starch in foods can be traced directly to its plant origin; the starch in white bread and pasta was originally stored in the grain of a wheat plant. The major grain crops (wheat, rice, and corn), the major underground crops (potato, sweet potato, and cassava), and the major legumes (beans and peas) supply the majority of starch in the human diet. In the body, starch is broken down into glucose by enzymes in saliva and the small intestine and is transported by the bloodstream to body cells.

Glycogen is the body's storage form of glucose, found in the liver and skeletal muscles. When the levels of glucose in the blood are higher than the demands of the cells, the excess is used for the synthesis of glycogen in liver and muscle cells. Only a limited amount of glycogen can be stored as a reserve—no more than a day's worth of energy needs. Excess glucose beyond this amount is generally converted to fat. During

---

*All organisms have nutrient needs requiring some compounds in large amounts (macronutrients) and other compounds in smaller amounts (micronutrients). The elements that plants require, *macronutrients* and *micronutrients* refer to the mineral requirements. When referring to human nutritional needs, these terms take on a different meaning, as discussed here.

A CLOSER LOOK 10.1

Famine or Feast

In 2018, estimates of the world population size were placed at 7.7 billion, with a projection of 9.8 billion by the year 2050 (box fig. 10.1a). This unprecedented population growth presents many problems related to the production and distribution of sufficient food to meet human nutritional needs. Although global food production has increased during recent decades, chronic hunger and malnutrition are ever-present problems in many developing nations, especially in sub-Saharan Africa and Asia. It is estimated that 815 billion people suffer from undernutrition. In addition, many are malnourished and deficient in vitamin A, iron, iodine, or zinc. **Undernutrition** is defined as an insufficient number of kilocalories to maintain daily energy requirements, while **malnutrition** is a quality deficiency in which one or more essential nutrients is lacking, even though caloric intake may be sufficient. The majority of starvation-related deaths are among children, with about half of those dying under the age of five. Starvation is usually the underlying cause of death, but most die of diseases such as diarrhea or measles, which would not be fatal in a properly nourished individual.

Two devastating conditions specifically related to undernourishment and malnourishment are **kwashiorkor** and **marasmus** (box figs. 10.1b and c). Kwashiorkor occurs when the diet is deficient in protein but has sufficient kilocalories. It is particularly prevalent after weaning, when a child no longer receives the protein-rich breast milk and is switched to a starchy diet low in protein content or quality or both. Symptoms of kwashiorkor include puffy skin and swollen belly due to edema, a fatty liver, a reddish orange cast to the hair, dermatitis, and listlessness.

Marasmus results from starvation when the diet is low in both kilocalories and protein; other nutrients are probably deficient as well. Sufferers from marasmus are extremely thin and shriveled—literally, skin and bones as the muscles of the body, even the heart muscle, are wasted away because muscle protein is digested to supply energy needs. The overt symptoms of both marasmus and kwashiorkor can be reversed if treated in time, but especially in infants and young children, mild mental retardation and stunted growth may be permanent results.

The counterpart to the conditions of undernutrition and malnutrition is **overnutrition,** in which an excessive intake of food can result in obesity and chronic disease. Currently, a global epidemic of obesity is the chief concern of many public health officials.

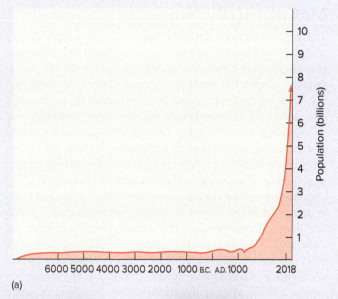

(a)

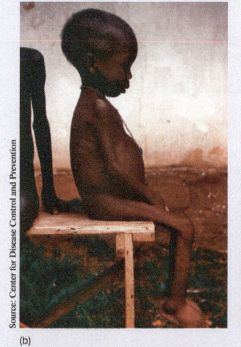

(b)

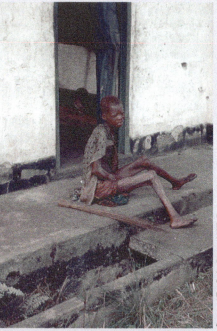

(c)

Box Figure 10.1 (a) The human population has continued to grow rapidly, up to 7.3 billion in 2015, as indicated by the J-shaped curve. (b) A child suffering from kwashiorkor, a protein-deficiency disease. (c) Marasmus victims have a skeletal appearance as the body wastes from starvation.

The WHO (World Health Organization), which first listed obesity as a disease in 1979, has asked the UN member nations to adopt programs in their home countries to reduce the intake of foods high in fats and sugar. It is ironic that developed nations, which have nearly eradicated many nutritional deficiencies, are now faced with a new nutritional threat with the prevalence of overweight and obesity. Developing nations are also experiencing a rise in obesity in certain populations—often at a faster rate than that seen in industrialized nations—at the same time that they are trying to combat the serious problem of undernutrition.

Almost 40% of U.S. adults (20–74 years) in 2015–2016 were overweight or obese, up from 14% in 1971. This increase has been seen for both men and women and across all ethnic, socioeconomic, and age groups. The United States is not alone; obesity has been identified as a growing health problem in the United Kingdom, Australasia, Eastern Europe, the Middle East, and the Pacific Islands—all these areas have seen the incidence of obesity more than triple since 1980. The WHO estimates that globally more than 1.9 billion adults are overweight, with 650 million of them obese. Estimates of obesity vary widely, from below 5% of the population in China and Japan and some African nations to greater than 75% in urban Samoa.

Body mass index (BMI) is the most current method health care professionals and fitness experts use to determine if a person's weight poses a health risk. BMI calculates a person's body weight in kilograms per square meter (BMI = body wt in kg/[ht in m]^2 or body wt in lb/[ht in in.]^2 × 703). Most people are considered overweight if their BMI ranges from 25 up to 30. A BMI of 30 or more indicates obesity; a rating of 40 or greater is indicative of severe obesity. An optimal BMI for most adults is below 25, and a BMI below 18.5 is considered underweight. The average BMI for adults in Africa and Asia ranges from 22 to 23, while a range of 25 to 27 is the norm in North America, Europe, and some Latin American, North African, and Pacific Island nations.

Obesity is associated with a higher risk of many diseases and chronic conditions: type 2 diabetes, hypertension (high blood pressure), cardiovascular disease, stroke, asthma, gallstones, cancers (prostate, breast, and colon), and osteoarthritis. Eighty percent of obese adults suffer from one of these diseases; 40% have two or more. Recently, it has been shown that women who are overweight in their 70s have a higher risk of developing Alzheimer's disease in their late 80s.

The alarming rise in childhood obesity over the past two decades has been documented in many countries: Haiti, Costa Rica, Chile, Brazil, England, Scotland, China, Egypt, Australia, Ghana, Morocco, and others. In 2015–2016, 18.5% of children and adolescents (ages 2–19) in the United States are classified as overweight or obese). In the world, it is estimated that 41 million children under the age of 5 were overweight or obese. Developing countries have a 30% higher rate of increase in childhood overweight and obesity levels than that of developed nations.

The repercussions of childhood obesity can be devastating. As with adult obesity, childhood obesity is linked to numerous complications: hypertension, sleep apnea, asthma, negative self-image, and gallstones, among others. Type 2 diabetes, formerly rare in adolescence, is now on the rise in children, mainly as a consequence of childhood obesity. In some populations, 50% of the newly diagnosed are adolescents. In January 2004, the American Academy of Pediatrics issued the statement that "overweight is the most common medical condition of childhood."

The global epidemic of obesity appears to be directly related to an energy imbalance. In most cases, obesity is caused by an excessive consumption of energy-rich foods, especially those high in sugar and/or saturated fats, coupled with inadequate expenditure of energy due to the limited physical activity associated with a modern, urban lifestyle. In the United States, some researchers suggest that the prevalence of fast foods in the American diet should take much of the blame for the fattening of Americans. Others have advocated a sin tax on sugary snack foods and soft drinks to discourage their consumption.

Responding to the nation's rising obesity rate, the Surgeon General released *The Surgeon General's Call to Action to Prevent and Decrease Overweight and Obesity* in 2001. The report warns that health problems resulting from the current epidemic of obesity could reverse many of the health gains achieved in the United States in recent decades. It details specific steps to educate the American people about obesity-linked health issues and what actions should be taken to reduce the incidence of obesity in the American population. Released in 2002, *Healthier U.S. Initiative* is encouraging Americans to prevent obesity by being physically active every day and making healthier choices in their diets.

Critics have countered that the so-called obesity epidemic was exaggerated by flawed statistical studies and the media. The research behind the obesity epidemic headlines was heavily funded by the weight-loss industry, which may have influenced the results. Newer studies report that there is only a small increase in mortality rates of the mildly obese and, in fact, the underweight, even when smokers are excluded, have a higher death rate when compared with those of normal weight. Critics note that the researchers who identified obesity as a major cause of death in the United States did not consider the impact of recent medical advancements, which have significantly improved the outlook for people with heart disease, diabetes, and high blood pressure, diseases associated with obesity. Another observation ignored is that obesity can have a protective effect in the elderly. Elderly patients who are mildly obese usually outlive their normal-weight cohorts when hospitalized for an extended time. Apparently, a nutritional reserve can be helpful in the recovery process. Colon cancer and postmenopausal breast cancer are slightly elevated in the obese, but lung cancer rates are surprisingly lower. The effects of obesity on health are probably more complex than once thought, and further research is needed.

strenuous exercise, the body's glycogen reserves are called upon; therefore, athletes training for a competition practice a regimen of carbohydrate loading by eating lots of starchy foods to build up muscle glycogen reserves.

Fiber in the Diet

Another important dietary component is **fiber,** which is derived from plant sources. Although not digestible, it does provide bulk and other benefits. There are many types of dietary fiber: cellulose, lignin, hemicellulose, pectin, gums, mucilages, and others. Cellulose, a principal component of plant cell walls, is another polysaccharide composed of glucose; however, humans do not have the enzymatic ability to break the bonds connecting the glucose molecules in cellulose as they do for starch and glycogen, and thus cellulose passes through the digestive tract as roughage, largely unaltered. Other cell wall components considered dietary fiber are lignin, pectins, and hemicelluloses. Lignin, a cell-wall component in plant cells that have secondary walls, is not a polysaccharide but a complex polymer. Pectins and hemicelluloses, which are cell wall polysaccharides, form the matrix in which cellulose fibrils are embedded. Pectins also occur in the middle lamella between adjacent cells. Gums and mucilages are exudates from various plants that are used commercially as thickening agents in prepared foods. Cell wall polysaccharides, refined from some species of red and brown algae, can also be considered dietary fiber. Although not digestible by human enzymes, some fiber, especially some hemicelluloses, can be broken down by intestinal bacteria and the nutrients made available to the body.

Dietary fiber can be conveniently grouped into two types, **soluble** and **insoluble,** relating to their solubility in water. Insoluble fiber includes cellulose, lignin, and some hemicelluloses, while soluble fiber includes other hemicelluloses, pectins, gums, mucilages, and the algal polysaccharides. Soluble fiber is resistant to digestion and absorption in the small intestine, but as soluble fiber enters the large intestine, it is acted upon by naturally occurring bacteria. The bacteria ferment soluble fiber into gases and products beneficial to health. Butyric acid is one of these products. It is known to stabilize blood glucose levels, which decreases the risk of type 2 diabetes. Butyric acid also has been shown to reduce blood levels of cholesterol, reducing the risk of cardiovascular disease, and to raise the acidity of the colon, which prevents cancerous polyps (small, tumorlike growths on the lining of the large intestine) from forming.

Fruits, vegetables, seeds, and whole grains supply most of the fiber in the human diet. Some plants are higher in one or more of these types of fiber, and the beneficial effects of high fiber foods differ depending on which fiber is abundant. For example, the soluble fiber present as gum in oat bran and as pectin in apples is believed to lower cholesterol levels in the blood. Wheat bran, which is largely cellulose, an insoluble fiber, has no particular cholesterol-lowering ability but seems to be most effective in speeding passage through the colon, which may reduce the risk of colon cancer. Psyllium husk from the seed coat of several species of plantain (*Plantago ovata, Plantago arenaria*) is an especially valuable dietary fiber because it is a good source of both soluble and insoluble fiber.

A study, which tracked the diets of more than 700,000 people for over 20 years, found that those eating the least amount of fiber (less than 10 grams per day) were at an increased risk of colorectal cancer. The study also suggested that fiber from cereals and whole grains, but not fruits and vegetables, is best for slightly lowering the risk of rectal cancer. Physicians and nutritionists still advocate the value of fiber in lowering the risk of obesity, heart disease, and diabetes.

The value of fiber in the diet in promoting health has been widely reported, but the exact mechanism by which a high-fiber diet brings about these benefits has been unknown until recently. Studies indicate that fiber plays a critical role in maintaining a healthy microbiota in the gut. Humans have a limited number of enzymes in the digestive tract and cannot make the enzymes needed to breakdown dietary fiber, but certain bacteria in the microbiota that inhabits the human intestines can. In other words, dietary fiber is the food or fuel for these bacteria, and if a diet is high in dietary fiber, the population of these bacteria rises and health benefits accompany this rise. For example, when the bacteria digest fiber, they often produce short-chain fatty acids as waste, which are in turn then consumed by the intestinal cells as an important source of fuel.

Experiments in which mice were switched from feeding on a high-fiber diet to a fiber-free, high fat, and high protein diet show that the microbiota in their guts underwent rapid changes in species composition. Some populations of the microbiota crashed or disappeared completely; on the other hand, the bacterial species that fed on protein and fat increased significantly. Deprived of fiber for fuel, some bacterial species start to feed on the mucus layer of the intestines. Mucus is the layer lining the gut produced by intestinal cells that coats the intestinal lining and protects intestinal cells from digestive enzymes and invasive bacteria. If intestinal cells are deprived of the short-chained fatty acids provided by fiber-eating bacteria, the health of the intestinal cells declines and the result is a thinner mucus layer. With a thinner mucus layer, bacteria can begin to penetrate the intestinal wall triggering the body's immune response to destroy the invaders. The end result is chronic inflammation and may result in conditions such as ulcerative colitis, a painful chronic bowel disease.

Proteins and Essential Amino Acids

Proteins are a group of large, complex molecules that serve as structural components and regulate a large variety of bodily functions (table 10.1). Recall that the constituents of proteins are **amino acids;** there are 20 naturally occurring amino acids, which can be assembled in various combinations and numbers to make thousands of different types of proteins. During digestion, proteins are broken down into their component amino

Table 10.1 Functions of Proteins		
Type of Protein	Function	Examples
Structural	Support	Collagen, keratin
Enzymes	Catalysts	Digestive enzymes
Hormones	Regulation	Insulin
Transport	Transport substance	Hemoglobin
Storage	Storage of amino acids	Ovalbumin in egg white, casein in milk
Contractile	Movement	Actin and myosin in muscles
Defensive	Protection	Antibodies (immunoglobins)

Table 10.2 Essential and Nonessential Amino Acids	
Essential	Nonessential
Histidine	Alanine
Isoleucine	Asparagine
Leucine	Aspartic acid
Lysine	Arginine
Methionine	Cysteine
Phenylalanine	Glutamic acid
Threonine	Glutamine
Tryptophan	Glycine
Valine	Proline
	Serine
	Tyrosine

acids by enzymes in the digestive tract and transported in the bloodstream to the liver and body tissues.

Essential Amino Acids

The necessary role of dietary proteins is to supply amino acids, so that the body can construct human proteins. All 20 amino acids are necessary for protein synthesis, and cells in the human body have the ability to synthesize 11 amino acids from raw materials; the other nine cannot be made by the body. These nine are called the **essential amino acids** (table 10.2 and see fig. 1.6) and must come from the diet. It is important to note that these essential amino acids cannot be stored by the body. For this reason it is critical that the body receive all the essential amino acids in a single day. Persistent lack of these essential amino acids prevents synthesis of necessary proteins and results in protein-deficiency diseases.

Complete Proteins

Complete proteins contain all the essential amino acids and in the right proportions. Almost all proteins derived from animals are complete proteins, whereas proteins derived from plants are usually **incomplete,** deficient in one or more essential amino acids. Although plant proteins are incomplete, the essential amino acid requirements can be met by combining complementary plant proteins. For example, the traditional diet of the native peoples of Mexico, beans and corn, contains complementary protein sources. The beans are low in methionine but adequate in tryptophan and lysine, but corn, which is poor in tryptophan and lysine, contains adequate amounts of methionine.

Although we have an absolute requirement for the essential amino acids, the actual amount of protein required by humans is a small percentage of our nutrient needs. It is recommended that approximately 10% (a range from 8% to 10%) of our total caloric intake be provided by proteins. On the basis of this percentage, individuals on a 2,000-kilocalorie

Table 10.3 Protein Content (in grams) of Some Common Food Items
1 ounce meat (beef, chicken, turkey) 7
1 ounce cheese 7
1 glass milk 8
1/2 cup beans 6
1 slice whole-wheat bread 4
1 egg 8
2 tbsp. peanut butter 8
1 serving oatmeal 5

diet should have 50 grams of protein per day; those on a 1,600-kilocalorie diet, 40 grams of protein per day; and those on a 2,200-kilocalorie diet, 55 grams of protein. The following daily protein intakes have been recommended for specific age groups: infants under 1 year—14 grams; children 1 to 4 years old—16 grams; pregnant women—60 grams; and nursing mothers—65 grams. The protein we require can be obtained from many foods. Table 10.3 contains examples of the protein content of some common foods.

Proteins can be assigned a numerical value that reflects how well they supply the essential amino acids. The protein in eggs has been assigned a biological value of 100, and all other foods are given values using egg protein as the reference standard. Another factor that needs to be considered is the digestibility of a particular protein. Some proteins cannot be broken down completely; that is, the amino acids are not fully released during digestion. This incomplete breakdown reduces the dietary value of the protein. For example, when digestibility is taken into account, even egg protein, considered the perfect protein source, drops to a value of 94. High-quality proteins contain all the essential amino acids in the right proportions and are fully digestible, freeing their amino acids, which are then absorbed into the blood and transported to the body's cells.

> **Thinking Critically**
>
> Humans have a dietary requirement for proteins, although proteins are needed in smaller quantities than carbohydrates and fats.
>
> *On the basis of caloric intake, determine your protein requirement.*

Gluten and Celiac Disease

Eating gluten, the major protein in wheat, or the related proteins secalin and hordein, in rye and barley, respectively, (see Chapter 12), can trigger celiac disease in individuals who have an inherited sensitivity to these proteins. Celiac disease is characterized by destruction of the intestinal lining and recurrent abdominal pain and diarrhea. In the most serious cases, the body is unable to absorb nutrients. Aretaeus, a Greek physician who practiced during the first century A.D., was the first to describe children who, while apparently well fed, were nonetheless exhibiting signs of starvation. In 1887, the English physician Samuel Gee observed that the condition of chronic indigestion could affect people of all ages but was especially prevalent in children of 1-5 years old. Willem-Karel Dicke, a Dutch pediatrician, was the first to link the condition of chronic indigestion to wheat gluten during World War II. He observed that, during times of bread shortages in the Netherlands, children with celiac disease were no longer dying, but the mortality rate returned to the pre-war level of 35% when bread was once more plentiful.

A newly available blood test has enabled physicians to more clearly identify sufferers of celiac disease. When intestinal cells are damaged, an enzyme called transglutaminase is released. Evidence of elevated levels of transglutaminase in the blood is indicative of celiac disease. In 2003, a widespread screening trial revealed that the incidence of celiac disease in North America is about 1 in 133, a ratio that indicates that the affliction is 100 times more common than thought previously. It may seem puzzling that the human body would react in such an abnormal way to the protein in wheat, a staple food for much of humanity, but it is important to realize that wheat and other cereal grains were not part of the original human diet. Our ancestors subsisted primarily by foraging for sweet fruits, starchy tubers, seeds of legumes, and eggs and by hunting animals for the occasional meat meal until agriculture arose approximately 10,000 years ago (see Chapter 11).

Celiac disease is an autoimmune disorder in which the immune system of genetically predisposed individuals reacts abnormally to the presence of gluten. Gluten is made up of large amounts of glutamine and proline, amino acids for which the body lacks the enzymes to digest completely. Instead, the gluten is broken down into peptides, small protein fragments, containing these amino acids. Normally, these peptides are not absorbed by the small intestine but remain inside the gastrointestinal tract and are eventually eliminated in the feces. In patients with celiac disease, gluten penetrates the small intestine, eliciting an immune response, which damages the cells in the lining.

It has been discovered that patients with celiac disease have abnormally high permeability of the intestinal wall. The cells of the intestinal lining are joined together by tight junctions, complexes of proteins that join adjacent cells to form a permeability barrier. Tight junctions prevent most materials from moving through the spaces between the cells and penetrating the lining of the small intestine to reach deeper tissues. However, in the normal functioning of the body, there are circumstances in which this barrier must be breached to allow for the passage of large molecules or cells. Zonulin is the body's protein that, when released, increases intestinal wall permeability. Apparently, the amount of zonulin is abnormally high in patients with celiac disease, and this allows gluten to pass through the gut wall.

The immune system is programmed to react to foreign proteins that may signal the arrival of disease-causing microorganisms. Eating food is one way foreign proteins are introduced into the body. Immune cells are situated beneath the lining of the small intestine, primed to attack any foreign proteins that pass through the intestinal lining. Ninety-five percent of those suffering from celiac disease have certain genetic markers on their cells, either HLA-DQ8 or HLA-DQ2 or both. These protein molecules pick up the peptide fragments of gluten and present them to T cells, a type of immune cell. These T cells are now trained to attack and destroy the gluten fragments whenever they are encountered. As the T cells destroy the gluten fragments, cells in the intestinal lining are damaged, causing the gastrointestinal symptoms of celiac disease. Additionally, it has been discovered that sufferers of celiac disease also release immune-stimulating chemicals that intensify the immune response.

Patients who suffer from celiac disease apparently also have certain abnormalities in their small intestine. Repeated exposure to gluten in patients with celiac disease causes the villi of the small intestine to become chronically inflamed. Villi, fingerlike projections in the lining of the small intestine, absorb digested nutrients from the gut and pass them into the circulatory system for distribution. In celiac disease, the villi become damaged and flattened, which decreases the surface area of the small intestine and its ability to absorb nutrients. Undiagnosed patients are often thought to be suffering from other conditions, such as osteoporosis or chronic fatigue, but the underlying cause is impaired nutrient absorption because of celiac disease. When patients with celiac disease eliminate gluten from their diet, the villi resume their original shape and function normally.

The primary treatment for celiac disease is to remove the gluten trigger by eating a gluten-free diet. Another promising strategy is to administer enzymes that would break down completely any gluten present in the small intestine. Other potential strategies are to depress the immune response or desensitize the immune system by introducing small, repeated exposures to gluten. Some researchers recommend that infants born to a family with an incidence of celiac disease have a gluten-free diet during their first year. By preventing exposure during the period of rapid development of the immune system, gluten sensitivity might be avoided. Another promising tactic is to develop drugs that are zonulin inhibitors to reduce the permeability of the intestinal lining in celiac patients.

Researchers are also working to develop gluten-free varieties of wheat that would be safe to eat for people who suffer from celiac disease. Recently a mutant form of barley was isolated that completely lacked the gliadin-like protein found in gluten but was high in lysine, one of the essential amino acid which is usually deficient in wheat flour. Gliadin is the primary culprit in celiac disease and researchers are hopeful that the gliadin-free barley will be the start of developing gluten-free wheat varieties that will retain all of the flavor and baking qualities associated with wheat flour. Research has shown previously that eliminating gliadins from gluten does not affect the bread-making qualities of wheat flour. It is also known that heirloom varieties of wheat such as einkorn, emmer, and spelt (see Chapter 12) are much lower in gluten content and may also be used to develop gluten-free wheat varieties.

Thinking Critically

Some people apparently develop celiac disease not as infants but much later in life. It has been suggested that the delay of celiac disease in genetically predisposed individuals is due to changes in the population of bacteria that normally inhabit the gut.

What might the bacteria population be supplying to the host that has forestalled the activation of the immune system to attack gluten? Does this suggest another possible treatment for sufferers of celiac disease?

Fats and Cholesterol

Fats are usually considered culprits in the diet because they are associated with cardiovascular disease, but some fat is necessary because it serves several vital functions and some fats are heart-healthy. Fats and related compounds belong to a larger category of organic molecules called **lipids.** Although a diverse group of compounds, all lipids share the characteristic of insolubility in water (table 10.4).

Triglycerides

Ninety-five percent of the lipids in foods are fats and oils; both these compounds are chemically classified as **triglycerides,** which are formed from glycerol and three **fatty acids** (see fig. 1.12). Fatty acids themselves are the simplest type of lipid and serve as building blocks for triglycerides and phospholipids. A glycerol backbone is common to all triglycerides, but many

Table 10.4 Functions of Lipids

Type of Lipid	Function	Examples
Triglyceride	Energy, storage	Animal fat, vegetable oils
	Insulation	Subcutaneous fat
Steroid	Structure	Cholesterol in membranes
	Hormonal regulation	Cortisol, estrogen, testosterone
Phospholipid	Structure	Phosphatidylcholine in cell membranes

types of fatty acids can occur. It is the nature of the fatty acids that determines the chemical and physical properties of the triglyceride. Each fatty acid contains a carbon chain with hydrogen attached to the carbon atoms; different fatty acids vary in the number of carbon and hydrogen atoms.

During digestion in the small intestine, triglycerides are first acted on by bile, which is made by the liver and stored and released by the gallbladder. Bile contains a complex mixture of lipids, bile salts, and pigments; prominent among the lipid components are cholesterol and lecithin. Bile acts as an emulsifier, breaking up the triglycerides into smaller droplets that can be acted on by enzymes. Enzymes from the pancreas and intestinal cells split these smaller droplets of triglycerides into monoglycerides and two fatty acids or into glycerol and three fatty acids. These end products are absorbed into the intestinal cells, where they are resynthesized into new triglycerides that enter the lymphatic system and eventually the bloodstream. High blood triglyceride level is a risk factor for coronary heart disease.

Essential Fatty Acids

The body is capable of synthesizing most fatty acids, but three must be supplied in the diet. Linoleic, linolenic, and arachidonic acids are designated essential fatty acids, but few adults suffer deficiency symptoms because these three fatty acids are widely found in foods, especially vegetable oils. Even if an adult were consuming a totally fat-free diet, as little as 1 teaspoon of corn oil, as an ingredient in foods, would supply the essential fatty acids. Deficiency symptoms, such as poor growth and skin irritation, have been seen in infants fed a formula lacking these essential nutrients.

Saturated and Unsaturated Fats

Fatty acids can be separated into two types, **saturated** and **unsaturated.** Saturated fatty acids contain all single bonds between the carbon atoms and have the maximum number of hydrogen atoms (it is said to be saturated with hydrogen). Unsaturated fatty acids have one or more double bonds between carbon atoms and consequently fewer hydrogen

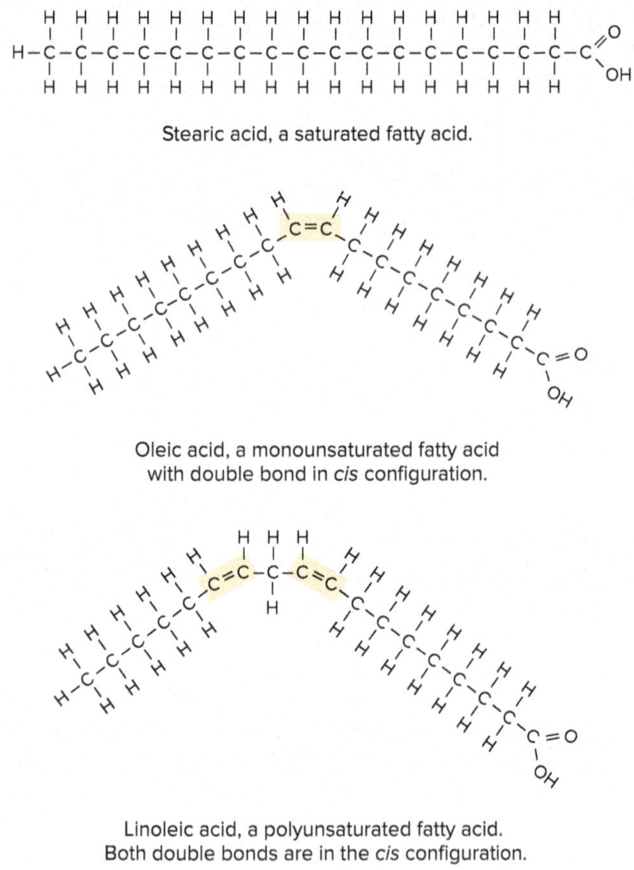

Figure 10.1 Structure of a saturated, monounsaturated, and polyunsaturated fatty acid.

atoms. Each carbon atom can form only four bonds, so if a double bond occurs between two carbons, then less than the full complement of hydrogen atoms can be attached. A fatty acid with one double bond is called **monounsaturated** and lacks two hydrogen atoms; a **polyunsaturated** fatty acid has two or more double bonds and lacks four or more hydrogen atoms (fig. 10.1).

All food **fats** contain a mixture of both saturated and unsaturated fatty acids (fig. 10.2). Saturated fats contain mostly saturated fatty acids and are solid at room temperature; animal fats, such as lard, butter, and beef fat, are familiar examples. Vegetable oils are generally composed of unsaturated fatty acids and are liquid at room temperature. **Oils** containing mostly monounsaturated fatty acids are olive oil, peanut oil, and canola oil; other vegetable oils, such as corn oil, soybean oil, and walnut oil, contain mostly polyunsaturated fatty acids. Coconut oil, palm and palm kernel oils, and cocoa butter are exceptions to the rule. Although they are of plant origin, they consist mostly of saturated fatty acids. On the other hand, certain fish oils are actually unsaturated. The oils from fish such as salmon, tuna, and herring are polyunsaturated and contain omega-3 fatty acids. Omega-3 fatty acids lower the tendency of blood platelets to stick together and form blood clots. Lowering the risk of clot formation reduces the incidence of blocked blood flow to the heart and the onset of coronary heart disease.

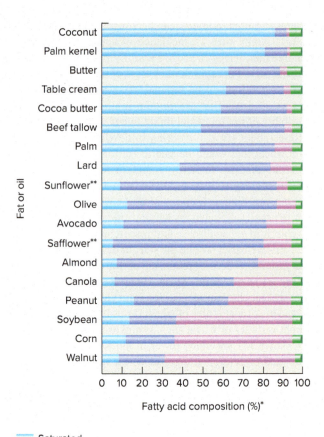

Figure 10.2 Fatty acid composition in common fats and oils.
*Calculated as a percent of total lipids.
**Newer varieties of sunflower and safflower produce oils which are 70% oleic acid, a monounsaturated fatty acid.
Source: Data from USDA National Nutrient Database for Standard Reference.

The health implications of saturated versus unsaturated fats have been intensely studied by the scientific community. Because saturated fats increase blood cholesterol levels, they are linked to cardiovascular diseases. Unsaturated fats, on the other hand, lower the risk of cardiovascular disease by lowering blood cholesterol levels.

Cholesterol

Cholesterol belongs to a subcategory of lipids known as steroids, which are compounds containing four carbon rings (see Chapter 1). Several steroids, including cholesterol, also have a hydrocarbon tail and an —OH group, making them sterols. Cholesterol is a vital constituent of cells; it is part of the lipid component of cell membranes and is used in the synthesis of sex hormones and several other hormones.

Cholesterol is synthesized in the liver from saturated fatty acids and is absorbed by intestinal cells from animal foods, especially eggs, butter, cheese, and meat. If the diet is high in saturated fats, even if it is low in cholesterol, the liver responds by increased cholesterol synthesis. Because cholesterol, like all lipids, is insoluble in the watery medium of the blood, it is transported by a special complex that consists of a cholesterol center with a coating of lipids and water-soluble proteins. These transport molecules, known as lipoproteins, exist in several forms. Two of the most significant are the **low-density lipoproteins (LDLs)** and **high-density lipoproteins (HDLs)**. LDLs transport cholesterol to all the body cells, while HDLs remove excess cholesterol from the body's tissues and carry it to the liver for degradation and elimination. In the popular press, the LDLs are considered the "bad" cholesterol because they can be taken up by the cells that line the arteries. The resulting deposition of cholesterol blocks the arteries, restricting the blood flow and is known as atherosclerosis. This condition can lead to heart attacks if the coronary arteries are blocked and strokes if the arteries delivering blood to the brain are blocked. HDLs are considered "good" cholesterol because they can prevent atherosclerosis by preventing the buildup of cholesterol deposits on the lining of the arteries.

Diets high in cholesterol, saturated fat, or both contribute to high blood cholesterol levels, especially in the form of LDLs. Plant sources do not contribute dietary cholesterol directly, although, as pointed out previously, some are high in saturated fats. On the other hand, plant oils are generally rich in unsaturated fats, which are known to lower blood cholesterol levels. However, monounsaturated and polyunsaturated fats act differently. Polyunsaturated fats tend to lower all cholesterol levels, including the protective HDLs, whereas monounsaturated fats raise the HDLs while lowering total and LDL levels. LDL blood cholesterol level is a better gauge of the risk of heart disease than is total blood cholesterol; the lower the LDL level, the lower the risk of heart disease. Cholesterol ratio, the ratio of total cholesterol to HDL, is another measure for the risk of heart disease. The ratio should be less than 5 to 1; the recommended ratio is 3.5 to 1 or lower. Although hereditary factors, cigarette smoking, and exercise play a role in the LDL-HDL balance, for the majority of people diet is the single most important factor in controlling cholesterol levels and the inherent disease risks.

The HDL hypothesis suggests that raising HDL and lowering LDL levels should lower the risk of cardiovascular disease, but recent findings contest this idea. For example, people with certain genetic mutations that raise HDL levels were studied and were found not to have the expected reduced risk of cardiovascular disease. Also, people who have a different mutation that lowered the level of HDL to about half of what is seen in the general population were found to have a lower, not higher, risk of cardiovascular disease as compared to the general population. In addition, during testing of a proposed drug that raised HDL levels from 30% to 100%, the risk of cardiovascular disease was not lessened but actually increased in some individuals.

The problem may be that not all HDL function similarly. It has been suggested that only young HDL can remove cholesterol that has built up in artery walls and deliver it to the liver for elimination, a process known as reverse cholesterol transport, and thereby reduce the risk of cardiovascular disease. Older HDL particles may no longer be capable of reverse cholesterol transport. Newer strategies aim to develop medications that raise only the young, functional HDL particles.

Trans Fatty Acids

Trans fatty acids that occur naturally in foods such as beef, lamb, whole milk, cream, and butter are derived from ruminant (cud-chewing) animals. Conjugated linoleic acid, a natural trans fatty acid, may be beneficial as an anticarcinogen and in strengthening the immune system. However, most trans fatty acids are made when manufacturers convert liquid oils into solid fats. In this practice, known as **hydrogenation,** vegetable oils are partially or fully hydrogenated to make margarine, vegetable shortening, peanut butter, and salad dressing. **Trans fats** are present in many processed foods, such as crackers, cookies, baked goods, snack foods, and practically any food made with or fried in partially hydrogenated oils.

Trans fats act somewhat like saturated fats, but chemically, trans fats are unsaturated. The carbon-carbon double bonds found in unsaturated vegetable oils are in a *cis* configuration in which the molecular groups attached to each carbon in the double bonds bend in the same direction, either both up or both down. During the hydrogenation process, some of the double bonds are broken and hydrogen is added to the carbons. More significantly, most of the double bonds change configuration from *cis* to the *trans* position. In the *trans* configuration, the molecules attached to each carbon of the double bonds bend in opposite directions or in a zigzag fashion. Thus, trans fatty acids can stack closer together, and the transformed oil readily takes on a solid, spreadable form that many consumers prefer over the natural liquid state.

A study of elderly men in the Netherlands indicated that a diet high in trans fat raised blood LDL cholesterol levels while lowering HDL levels. Coronary heart disease was higher in those men who ate more trans fats, and the risk of developing coronary heart disease within 10 years increased by 25% for each additional 2% of trans fats consumed in their diet. After 10 years, the percentage of trans fats in the diets of these men dropped from a high of over 4% to just less than 2%. It is estimated that trans fats make up about 2%-4% of the typical U.S. diet.

In the Nurses' Health Study, conducted by the Harvard School of Public Health, nearly 90,000 women filled out detailed questionnaires about their diet every few years for 14 years. The results found no correlation between the onset of type 2 diabetes and the total fat, saturated fat, or monounsaturated fat consumed. There was, however, a significant occurrence of this most common type of diabetes when the women had diets high in trans fats. It was also discovered that a diet high in polyunsaturated oils lowered the risk of type 2 diabetes.

Since 1983, the FDA has required that saturated fat and dietary cholesterol be listed on food labels. Due to the apparent association between high trans fats in the diet and the higher risk of cardiovascular disease and, in females, type 2 diabetes, the content of trans fats must be listed on the Nutrition Facts panel of food labels in the United States.

Manufacturers have responded to health concerns and consumer demand by developing new methods of hydrogenating vegetable oils. Using supercritical carbon dioxide as a solvent and under conditions of higher pressure and lower temperatures, USDA researchers produced a hydrogenated soybean oil with only 10% trans fats. This percentage is lower than the usual 10%–30% trans fats found in most hydrogenated vegetable oils.

MICRONUTRIENTS

Like macronutrients, micronutrients are essential for proper nutrition, but they are required in much smaller amounts. While macronutrients make up the bulk dry weight of food, micronutrients constitute only 1%–2% of the dry weight. There are two categories of micronutrients, the organic compounds known as vitamins and the inorganic compounds, the minerals.

Vitamins

Many vitamins play roles as **coenzymes** (molecules that are required for the proper functioning of certain enzymes) in many metabolic pathways in the body; others are directly involved in the synthesis of indispensable compounds. Vitamins are classified according to their solubility, with four **fat-soluble vitamins (A, D, E, and K)** and nine **water-soluble vitamins (eight B-complex vitamins and C)**. The water-soluble vitamins are not readily stored in the body, and any excess is eliminated in the urine; therefore, these are unlikely to become toxic. On the other hand, the fat-soluble vitamins are easily stored in the fatty tissues of the body, and excessive intake can lead to toxicity symptoms. The dietary sources and deficiencies of these vitamins are described in Tables 10.5 and 10.6. The following discussion is limited to vitamins A, D, and C and the B vitamins thiamine, niacin, and B_{12}.

Vitamin A

Vitamin A has many roles in the body. One of the best known involves the formation of vision pigments (rhodopsin and others) present in the retina of the eye. Each pigment is composed of a molecule of retinal (a form of vitamin A) and a protein molecule, called an opsin, that differs from pigment to pigment. The pigments are contained in two types of photoreceptor cells, rods and cones, located deep in the retina of the eye. All light stimulates the rods, providing only black-and-white vision, while the cones are selectively stimulated by different colors, providing the full spectrum of color vision. In dim light or at night, only the rods are used for vision; therefore, a shortage of retinal has especially pronounced effects, leading to one of the earliest signs of vitamin A deficiency, **night blindness.**

Vitamin A is also necessary for the maintenance of epithelial tissues that line both internal and external body surfaces, an

Table 10.5 Fat-Soluble Vitamins

Vitamin	Dietary Source	Results of Deficiency
A	Yellow, orange, and dark green vegetables and fruits; dairy products	Night blindness, xerophthalmia
D	Eggs and enriched dairy products	Rickets
E	Seeds, leafy green vegetables	Unknown
K	Leafy green vegetables	Poor blood clotting

Table 10.6 Water-Soluble Vitamins

Vitamin	Dietary Source	Results of Deficiency
B_1 (thiamine)	Whole grains, legumes, seeds, nuts	Beriberi
B_2 (riboflavin)	Dairy products, whole grains, leafy green vegetables, poultry	Mouth sores, lesions of eyes
Niacin	Meat, eggs, seeds, legumes	Pellagra
B_6 (pyridoxine)	Dried fruits, seeds, poultry, leafy green vegetables	Irritability, muscle weakness, skin disorders
Pantothenic acid	Dried fruits, seeds, poultry, leafy green vegetables, nuts	Insomnia, weakness
Folic acid (folate)	Legumes, whole grains, green vegetables	Anemia, diarrhea, neural tube defects
Biotin	Legumes, vegetables, meat, egg yolks	Fatigue, dermatitis
B_{12} (cobalamin)	Meat, eggs, dairy products	Pernicious anemia
C	Fresh fruits and vegetables	Scurvy

area roughly equivalent to one-fourth of a football field. When vitamin A is lacking, these tissues fail to secrete their protective mucus covering, producing instead a protein called keratin (normally found in hair and nails), which results in the tissues' becoming dried and hardened. This aspect of vitamin A deficiency can therefore affect many different areas in the body. One of the most tragic consequences is a type of blindness known as **xerophthalmia,** in which severe vitamin A deficiency results in irreversible drying and degeneration of the cornea. This permanent blindness, easily preventable with proper nutrition, is found most frequently among malnourished children in developing nations. In the skin, keratinization results in rough, dry, scaly, and cracked skin, often with an accumulation of hard material around a hair follicle that looks like a permanent goose bump. Other epithelial tissues, such as those in the mouth, gastrointestinal tract, and respiratory system, are also affected by the decrease in mucus production, becoming progressively drier and subject to infection. Vitamin A is additionally involved in dozens of other roles in the body, including normal bone and tooth development and hormone production in the adrenal and thyroid glands.

In food derived from animal sources, especially liver, vitamin A occurs primarily as retinol, which is readily absorbed by the body and converted to retinal. In plant sources, no retinol is present, but a vitamin A precursor, **beta-carotene** (first isolated from carrots) occurs abundantly in many yellow, orange, and dark green fruits and vegetables (see Chapter 3) and can be split into two molecules of retinol in the body.

Antioxidants

The importance of beta-carotene as an antioxidant has been investigated. Antioxidants may protect the body against the destructive action of reactive ions called free radicals. Destructive free radicals form when electrons escape from the Electron Transport System during ATP production and combine with oxygen available in mitochondria (see Chapter 4). These highly reactive ions can damage the metabolic machinery of the cell, impairing energy production. They may also damage DNA itself, inducing mutations, and they have been implicated in cancer, heart disease, and even the signs of aging.

Several recent animal studies are challenging the prevailing concept that the build-up of free radicals, destructive ions that produce oxidative damage in cells, accounts for many of the deleterious effects associated with aging. There is also recent evidence that the practice of taking antioxidant supplements to neutralize the effects of free radicals can be harmful. Researchers bred roundworms (*Caenorhabditis elegans*) to overproduce a specific free radical (superoxide) with the expectation that the worms would die prematurely compared to worms with lower levels of this free radical. Surprisingly, the worms which had higher levels of the free radical, lived 32% longer. More astonishing, treating the worms with high levels of free radicals with the antioxidant vitamin C reversed the longer life span.

In another group of experiments with roundworms, normal worms were exposed to an herbicide that is known to induce free radical production in animals. The herbicide treated worms, with the higher level of free radicals, lived almost 60 percent longer than untreated worms; and once again feeding antioxidants to the worms prevented this longevity. Researchers then genetically engineered roundworms without the ability to produce certain natural enzymes that act as antioxidants. As anticipated, the genetically engineered worms had higher levels of free radicals and more potentially damaging oxidative reactions throughout their bodies but unexpectedly the worms did not age prematurely but instead lived just as long as worms with normal levels of antioxidants.

Other model organisms have showed similar results. Some strains of mice were genetically engineered to produce higher than normal levels of antioxidant enzymes, whereas other strains were created that produced lower than normal levels of antioxidant enzymes. It was expected that the mice with higher levels of antioxidant enzymes in their bodies would live longer that those with lower levels but there was no difference in life spans between the strains. Observations from nature also support the experimental data. The naked mole rat with lower levels of natural antioxidants and evidence of more oxidative damage to their bodies lives 25–30 years, about eight times longer, than another rodent, the common mouse.

As antioxidants have been shown to not always be beneficial, new research is indicating that free radicals are not always harmful. At lower levels, free radicals appear to signal the body to turn on genes to repair oxidative cellular damage and prevent further injury. Free radicals have been shown to activate gene *HIF-1* which itself turns on genes that are involved in cellular repair. Current research is suggesting that at lower levels free radicals are not necessarily destructive agents in themselves but are instead signals to the body to turn on genes to repair oxidative cellular damage and prevent further injury. Only, in larger amounts, free radicals destructive.

Vitamin D

The primary function of vitamin D is the regulation of calcium and phosphorus levels, especially for normal bone development. Vitamin D helps control the blood levels of these minerals in three ways:

1. the absorption of calcium and phosphorus from food in the gastrointestinal tract,
2. the removal of these minerals from bones to maintain the concentration in the blood, and
3. the retention of calcium by the kidneys.

Vitamin D is unique in that it can be synthesized by the human body on exposure to sunlight; in fact, it has long been called the sunshine vitamin. The precursor (a cholesterol derivative) is manufactured by the liver and transported to the skin, where exposure to the sun's ultraviolet rays converts it to provitamin D; the final steps in the manufacture of active vitamin D occur in the liver and kidneys. The amount of pigmentation in the skin affects the synthesis of vitamin D because pigment blocks ultraviolet absorption. Darker skin requires longer exposure to sunlight to produce adequate amounts of vitamin D. Thirty minutes of sunlight is adequate for light skin, while darker skin may require up to 3 hours. This exposure to sunlight must be achieved with caution, since overexposure to the ultraviolet rays in sunlight is linked to higher risk of skin cancer. The Environmental Protection Agency has estimated that the average American is indoors 93% of the time, in transit 5% (cars, buses, trains), and outdoors only 2% of the time daily. This limited exposure to sunlight makes it essential that the diet contain vitamin D to avoid deficiency symptoms. Vitamin D is not naturally abundant in any food.

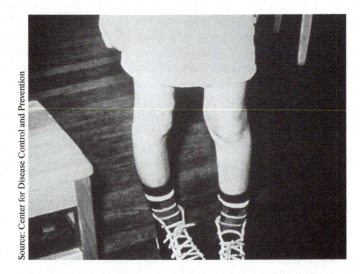

Figure 10.3 Characteristic bowing of the legs and knees in rickets, a disease caused by insufficient vitamin D.

None occurs in plant sources, but vitamin D does occur in limited amounts in animal sources, such as egg yolks, liver, cream, some fish, and butter. Because there is concern about meeting the nutritional needs for vitamin D, especially for children, milk, which does not contain adequate vitamin D, is routinely fortified with this vitamin.

Because of the role of vitamin D in calcium regulation, the deficiency symptoms are most evident in bone formation. The effects of vitamin D deficiency are most pronounced in children and result in a characteristic malformation of the skeleton known as **rickets.** The bowing of the legs so commonly associated with the condition is just one of many abnormalities of the skeletal system (fig. 10.3). Dark-skinned children who live in northern, smoggy cities are especially vulnerable to developing this deficiency. Once believed to have been eradicated in developed countries, rickets has reemerged in the United States and the United Kingdom. The problem seems to be that breastfed children are not receiving enough vitamin D to absorb the amount of calcium necessary to build strong bones and teeth. Pediatricians are now recommending vitamin D supplements for breastfed infants. Adult rickets, also known as **osteomalacia,** is rare but does occur in women who have undergone repeated pregnancies, have low calcium intake, and have inadequate exposure to sunlight. However, a large scale study in 2018 found no benefit in the taking of high–dose vitamin D supplements in either increasing bone density or in the prevention of fractures from falls.

Currently, some researchers in nutrition are advocating that the daily recommended dose of vitamin D be raised to 1,000 IU* (**International Units**) per day for adults not only to promote healthy bones but also to prevent certain forms of cancer. Previous studies have shown that colon cancers are more prevalent in the northern United States where exposure to sunlight is the lowest. Recent studies have correlated blood levels of vitamin D in women to colon cancer. Those with the lowest blood levels of the vitamin had double the risk of colon cancer compared with those women who had the highest levels of vitamin D. Preliminary research shows that this trend may also be repeated for type 2 diabetes and other diseases.

Excess vitamin D causes abnormally high levels of calcium in the blood; this condition often leads to calcium deposits in soft tissues, such as kidneys and blood vessels. If not caught in time, this may result in irreversible damage to the cardiovascular system, kidney failure, and even death. In excess, vitamin D is the most toxic of all the vitamins. However, this toxicity cannot occur from sunlight or food; it results only when megadoses of the vitamin are administered without medical supervision.

Vitamin C

Fresh fruits and vegetables are the richest sources of vitamin C, **ascorbic acid,** with organ meats the only significant animal sources. The most important role of vitamin C in the body is in the synthesis of collagen, a connective tissue protein that serves as a "cellular cement," holding cells and tissues together. Collagen, the most abundant protein in the body, is found in the matrix of bones, teeth, and cartilage and provides the elasticity of blood vessels and skin. Vitamin C also functions as an antioxidant in the body, preventing other molecules from being oxidized (losing electrons). Because vitamin C prevents oxidation, it is sometimes added to packaged foods to extend the shelf life; in a similar manner, orange juice and lemon juice, with their high vitamin C content, prevent the oxidation (browning) of sliced apples and bananas. Vitamin C is additionally involved in promoting iron absorption through the intestines. When foods containing vitamin C are consumed with foods containing iron, absorption is enhanced. Finally, vitamin C is involved in a number of other metabolic reactions, including the production of various hormones.

For centuries, sailors on long ocean voyages faced the possibility of developing **scurvy,** a disease that could cause bleeding of the gums, pinpoint hemorrhages under the skin, severe fatigue, poor healing of wounds, brittle bones, and even sudden death due to massive internal bleeding. It was not uncommon for half to two-thirds of the ship's company to die of scurvy on a long voyage. The first cure for this disease was identified in 1747 by Dr. James Lind, who experimented with 12 sailors afflicted with scurvy. Lind tried various dietary supplements and found that sailors who were given either oranges or lemons for 6 days improved rapidly. Interestingly, it was almost 50 years after these findings that the British admiralty took measures to prevent scurvy by dictating that all sailors receive lemon or lime juice daily. (British sailors were soon nicknamed "limeys" because of this practice, a name that is still heard today.) Now it is well known that the vitamin C in citrus juice prevents scurvy. The typical symptoms of scurvy can be traced directly to the inability of the body to make collagen. The scurvy-vitamin C connection is reflected in the name *ascorbic acid,* which literally means "without scurvy."

---

*International Unit is a unit of measurement for biologically active substances (such as vitamins, hormones, and vaccines) that produces a measured biological effect agreed upon as an international standard.

Since the 1970s, the daily intake of vitamin C has been the focal point of a heated debate, after the publication of *Vitamin C and the Common Cold* by Linus Pauling, a Nobel Prize–winning chemist. The RDI (reference daily intake)** for vitamin C is 60 milligrams (approximately 10 milligrams daily can prevent scurvy), but Pauling recommended megadoses as high as 2,000 to 10,000 milligrams for optimum health. Pauling maintained that large doses of vitamin C can prevent colds and other viral infections and that vitamin C is bactericidal and even cures cancer. Recent research has shown that high doses of vitamin C is effective in killing certain types of colon cancer cells. However, some authorities note that tissues become saturated with vitamin C at levels of 80 to 100 milligrams per day and that intakes above these levels are generally excreted in the urine. Controversy also exists regarding the toxicity of vitamin C. Some scientists maintain that high doses are nontoxic; others feel that toxicity can occur. The most common toxicity symptoms are nausea, abdominal cramps, and diarrhea. Another possible complication of megadosing is rebound scurvy, a condition that may occur after an abrupt cessation of these high doses. Symptoms mimic scurvy, even though vitamin C intake is not deficient; these symptoms do not occur if the megadoses are decreased gradually.

Vitamin B Complex

The vitamin B complex includes a group of eight vitamins that are often found in foods together and have similar roles in the body; that is, they function as coenzymes, involved in thousands of metabolic reactions. They are found in each cell of the body and must be present for normal cell functioning. For each of these vitamins, specific deficiency symptoms occur when the vitamin is lacking in the diet, but in general, no toxicity symptoms have been reported because excess is excreted in the urine. As is vitamin C, the B vitamins are water soluble and can be leached out during food preparation when excess water is used and discarded. In addition, some of the B vitamins may be destroyed by high temperatures during cooking. **Thiamine, niacin,** and **B$_{12}$** will be considered because these vitamins may be deficient in plant sources.

Thiamine Thiamine, also known as **vitamin B$_1$**, is part of the coenzyme thiamine pyrophosphate, which is involved in the metabolic breakdown of carbohydrates just before the Citric Acid Cycle. Because of thiamine's central role in metabolism, the symptoms of thiamine deficiency are profound: fatigue; depression; mental confusion; cramping, burning, and numbness in the legs; edema; enlarged heart; and eventually death from cardiac failure. This thiamine deficiency is known as **beriberi** and was found mainly in Asia, where diets were based mainly on polished or white rice rather than on the whole-grain brown rice that still has the outer bran (husk) intact. The thiamine that occurs in the outer layer of the rice is removed during the polishing process. Beriberi became more prevalent when improved techniques for polishing rice were developed that removed more of the bran and, inadvertently, more of the thiamine. In the 1880s, thiamine deficiency in the Japanese navy was particularly widespread, with 25%–40% of the sailors developing beriberi. A Japanese physician, Dr. K. Takaki, observed that few sailors developed the disease when milk, meat, and eggs were added to the normal staple of white rice. Unfortunately, he did not realize that the enriched diet was supplying a nutrient missing in the white rice diet. A few years later, a Dutch physician, Dr. Christiaan Eijkman, studied beriberi in the East Indies; he showed that, in diets based almost exclusively on rice, the consumption of brown rice instead of white rice prevents the appearance of beriberi. It was not until the twentieth century that thiamine deficiency was actually identified as the cause of the disease. Good dietary sources of thiamine include meat, especially pork and liver, whole grains, seeds and nuts, and legumes.

Niacin Niacin is the collective term for two compounds, **nicotinic acid** and **nicotinamide,** either one of which is used to form the coenzymes NAD$^+$ and NADP$^+$. Recall the importance of these coenzymes for oxidation-reduction reactions in many energy-yielding metabolic pathways (see Chapter 4). Without these coenzymes, the release of energy from the breakdown of foods cannot occur and cellular death results. Niacin can be supplied directly through foods rich in niacin itself or foods rich in the essential amino acid tryptophan because the body can synthesize niacin from the amino acid. Niacin deficiency, therefore, is coupled with a low-protein diet.

A lack of niacin severely affects every organ of the body, and a severe deficiency disease, **pellagra,** develops. The symptoms of pellagra are referred to as the 4 Ds: dermatitis (skin disorders), dementia (mental confusion), diarrhea, and eventually death if niacin is not supplied. The dermatitis (fig. 10.4) is characterized by rough, reddened skin with lesions developing

Figure 10.4 Pellagra, caused by a lack of niacin, is characterized by dermatitis of the hands.

**Daily values are determined by the U.S. FDA (Food and Drug Administration) and indicate the percentage amount of a nutrient that is provided by a single serving of a particular food, based on the current recommendations for a 2,000-kilocalorie diet. There are two categories of daily values: daily reference values (DRVs) and reference daily intakes (RDIs). DRVs have been determined for total fat, saturated fat, cholesterol, total carbohydrate, dietary fiber, sodium, potassium, and protein. RDIs are established for 19 vitamins and minerals and have replaced the older term *recommended daily allowances (RDAs).*

in exposed areas; in fact, *pellagra* means "rough skin." The central nervous system is affected, and confusion, memory loss, dizziness, and hallucinations occur. The gastrointestinal system is also involved, and diarrhea, along with abdominal discomfort, nausea, and vomiting, is common. Another characteristic of the condition is a bright red, or strawberry, tongue.

Pellagra is especially common in areas where corn is the dietary staple. Outbreaks have occurred in southern Europe, particularly Italy, parts of southern India, and the rural South in the United States, where the disease was epidemic early in the twentieth century. It was estimated that 10,000 people died and another 200,000 were afflicted each year. During this period, about half the patients in mental hospitals in the South were suffering from the dementia caused by pellagra.

Although corn does contain some niacin, it is in a form that makes it unavailable; furthermore, corn is also deficient in tryptophan. However, pellagra was not a problem in the traditional diet of the natives of Mexico, Central America, and parts of South America for two reasons. Lime (calcium oxide from wood ash or shells), used in the preparation of corn meal, is able to release the bound niacin. In addition, the beans, squash, tomatoes, and peppers commonly eaten with the corn also supplied niacin.

The addition of milk and meat to the diet was recommended to prevent pellagra long before the vitamin was identified. Although meat contains niacin, milk is low in the vitamin but does supply tryptophan. Sources rich in niacin include meat, poultry, fish, eggs, nuts, seeds, and legumes. One way the pellagra in the South could have been prevented was by the consumption of a handful of peanuts every other day, because peanuts are an excellent source of niacin.

There has been an increased interest in the therapeutic value of megadoses of niacin (the nicotinic acid form) for reducing blood cholesterol levels. Unfortunately, the megadoses can cause some toxicity symptoms, the most common of which is a niacin flush. It produces a temporary warm flush of the skin, with a tingling or stinging sensation. Intestinal irritation and liver damage have also been reported. The other form, nicotinamide, does not produce those toxicity symptoms but is not at all effective in lowering blood cholesterol levels.

Vitamin B_{12} Vitamin B_{12} **(cobalamin)** is unique in that it does not occur naturally in any foods of plant origin but occurs only in animal sources, where it is widely available. Those who completely eliminate meat, dairy products, and eggs from their diets are at risk of developing a B_{12} deficiency unless they take vitamin supplements or eat fortified foods. Soy milk, breakfast cereals, and meat substitutes are often fortified with B_{12}.

The absorption of vitamin B_{12} in the small intestine requires the presence of a substance secreted by the stomach called an intrinsic factor. Poor absorption of B_{12} has been reported in people who, because of a genetic defect, do not produce the intrinsic factor. This defect most often shows up after the age of 60, when intrinsic factor production becomes impaired. In this case, B_{12} deficiency symptoms show up even though there are sufficient quantities in the diet, and the vitamin must be received by injection.

The most common result of B_{12} deficiency is **pernicious anemia,** characterized by the production of improperly formed red blood cells. The associated symptoms include fatigue and weakness because the delivery of oxygen to the body's tissues is impaired. A more serious consequence of B_{12} deficiency is nerve damage that begins as a creeping numbness of the lower extremities.

In general, vitamin B_{12} is involved in nucleic acid synthesis and interacts with **folic acid** (another B vitamin) in this function. Because blood cells are constantly being formed in the bone marrow, this site of rapid cell division is one of the first affected by impaired synthesis of DNA due to a deficiency of either vitamin. These vitamins are, therefore, involved in the normal development of red blood cells, and a deficiency of either vitamin causes anemia. The anemia can be treated with either B_{12} or folic acid supplements. However, folic acid supplements have no effect on the nerve damage caused by a B_{12} deficiency, and the administration of folic acid for the anemia can mask a true B_{12} deficiency and result in permanent neurological degeneration. In this regard, vitamin B_{12} functions in maintaining the sheath surrounding nerve fibers, which is necessary for the transmission of nerve impulses.

Under directions from the FDA and the Department of Health and Human Services, U.S. manufacturers fortify most enriched breads, flours, corn meals, rice, and other grain products with folic acid. This action was taken because it was determined that insufficient levels of folic acid contribute to spina bifida (the backbone does not form properly, leaving the spinal cord exposed) and other neural tube birth defects. Because more than half of all pregnancies are unplanned and these defects of the spine and brain occur in the developing fetus before most women realize they are pregnant, it is important that all women of childbearing age consume 0.4 milligram of folic acid daily. Since 1998, when folic acid fortification of foods began, the number of neural tube birth defects in the United States has dropped by one-third.

Adequate levels of folic acid may also afford protection from early heart disease. In patients with atherosclerosis,

Thinking Critically

Vitamins are classified according to their solubility. The water-soluble vitamins include C and the B complex. A, D, E, and K are the fat-soluble vitamins.

Many people subscribe to the notion of vitamin megadosing to ensure better health and consume many times the RDI for particular vitamins. What are the practical effects of vitamin solubility on this practice?

high levels of plasma homocysteine are commonly found. Homocysteine is an intermediate in amino acid metabolism. It is toxic to the lining of blood vessels, bringing about changes that lead to cardiovascular disease. It may also promote clotting factors in the blood. Normally, homocysteine is not found in high levels in the bloodstream because it is broken down by enzymatic activity. Certain vitamins—folic acid is one—are cofactors for these enzymes. Many patients with early coronary artery disease have low levels of folic acid and correspondingly high levels of homocysteine. Increasing the nutritional intake of folic acid will decrease homocysteine levels as well as the risk of heart attack or stroke.

Minerals

Minerals are inorganic compounds that exist in the body as ions (charged atoms) or as part of complex molecules. At least 18 minerals are required for normal metabolic activities (table 10.7). In 2014, bromine was identified as another essential element necessary for proper human nutrition. Animal studies have shown that if bromine, active in the body in the form of the bromide ion, is eliminated from the diet, abnormal basement membranes form. Basement membranes are extracellular structures found in connective tissue which form scaffolds that support tissues. Bromide ion is a necessary cofactor for the enzyme peroxidasin; this enzyme catalyzes the formation of sulfilimine (sulfur-nitrogen) bonds between protein ropes of collagen that make up the basement membrane. This research is of tremendous interest because it had been shown previously that cells in contact with abnormal basement membranes can become cancerous. Also it is known that thiocyanate, one of the chemicals in tobacco smoke, inhibits the enzyme peroxidasin, which leads to the breakdown of basement membranes in the lungs of smokers.

Minerals are subdivided into two categories, the **major minerals,** needed in amounts greater than 100 milligrams per day, and the **trace minerals,** needed in amounts no more than a few milligrams per day. The following discussion will be limited to **calcium,** a major mineral whose RDI has been recently revised, and two trace minerals that have been extensively studied, **iron** and **iodine.**

Calcium

Calcium is the most abundant mineral in the body, with the average adult containing 800 to 1,300 grams of the element. Ninety-nine percent of the body's calcium is found in the bones and teeth; the other 1% is in the blood and tissues. The concentration of calcium is under the control of several hormones and vitamin D. If the amount in the blood gets too low, calcium reserves in the bone are drawn upon to restore levels to the normal range. If the amount of calcium in the blood is too high, more calcium is deposited in the bone and more is excreted by the kidneys. Excess calcium intakes (12,000 milligrams per day and above are considered toxic) have been associated with increased risk of kidney stone formation. In addition to forming the matrix of bones and teeth, calcium in the body fluids is involved in many important functions: nerve impulse transmission, muscle action (including heartbeat), blood clotting, cell membrane integrity, intracellular communication, and as a **cofactor** for enzymes (cofactors are mineral ions and, like coenzymes, are necessary for the proper functioning of certain enzymes).

Calcium deficiency may lead to **osteoporosis,** a degenerative bone disease that may strike older individuals without warning. In osteoporosis, the bone density is greatly reduced (*osteoporosis* literally means porous bone), resulting in bones that fracture readily. This condition can result from years of low dietary calcium intake or poor absorption of calcium from the intestines (caused by lack of vitamin D or other factors). To maintain blood calcium levels, the reserves in the bone are dangerously depleted. Postmenopausal women are particularly at risk of developing osteoporosis because bone loss is accelerated at this time. Estrogen replacement therapy appears to retard this bone loss, but is associated with a higher risk of breast cancer, heart attacks, and strokes. Adequate dietary calcium and regular exercise also prevent osteoporosis; however,

Table 10.7 Dietary Mineral Requirements

Mineral	Function
Major Minerals	
Calcium	Bone and tooth formation, blood clotting, nerve impulse transmission, muscle contraction
Phosphorus	Nucleic acids, bone and tooth formation, cell membranes, ATP formation
Sulfur	Protein formation
Potassium	Muscle contraction, nerve impulse transmission, electrolyte balance
Chlorine	Gastric juice
Sodium	Nerve impulse transmission, body water balance
Magnesium	Protein formation, enzyme cofactor
Trace Minerals	
Iron	Hemoglobin
Zinc	Component of many enzymes and insulin, wound healing
Iodine	Component of thyroid hormones
Fluorine	Bone and tooth formation
Copper	Enzyme component, red blood cell formation
Selenium	Antioxidant
Cobalt	Component of vitamin B_{12}
Chromium	Normal glucose metabolism
Manganese	Enzyme cofactor
Molybdenum	Enzyme cofactor
Bromine	Enzyme cofactor

there are many interacting factors (both genetic and environmental), and much more research is needed in this area.

Milk and milk products are among the best sources of calcium, but the element is also present in dark green leafy vegetables, many seeds, and other foods. Unfortunately, in some vegetables, the presence of oxalic acid inhibits the absorption of calcium. Recently, the RDI for dietary calcium for adults over 50 has been increased to 1,200 milligrams to prevent the development of osteoporosis. New evidence has also shown that calcium, together with vitamin D, may provide protection against colon cancer; the amount of calcium required for this beneficial action is 1,500 milligrams per day.

Iron

Although most of the trace minerals are usually found in adequate amounts in a well-balanced diet, iron and iodine present special problems. Iron deficiency is common in women and children, and care must be taken to ensure that the diet supplies sufficient quantities of the element. Meat, especially liver and other organ meats, shellfish, fish, and poultry are excellent iron sources. Many foods from plants are also rich in iron, including dark green leafy vegetables, dried fruits, legumes, whole grains, and enriched breads and cereal products. Overall, only about 10% of dietary iron is actually absorbed by the body, with the absorption dependent on the type of iron compound present; the rest is eliminated in the feces. The iron from animal sources may be present as **heme iron** (40%) or **nonheme iron** (60%), whereas the iron in plant sources is nonheme. Heme iron is more readily absorbed by the body, but the absorption of nonheme iron can be improved by the presence of vitamin C.

The most important role of iron is as a component of hemoglobin, the molecule that carries oxygen in red blood cells; in fact, it is the iron that imparts the red color to these cells. In addition, iron occurs in myoglobin, a molecule similar to hemoglobin, the oxygen carrier in muscle cells; in several storage proteins in the liver, bone marrow, and spleen; and in enzymes present in each cell.

Because the majority of iron is found in hemoglobin, iron deficiency has its greatest impact on red blood cells. When iron reserves in the body are low, not enough hemoglobin can be synthesized for newly formed red blood cells. These cells are smaller, paler, and less efficient in oxygen transport than are normal red blood cells and are characteristic of **iron-deficiency anemia,** the most common dietary deficiency disease in the world. The symptoms of iron-deficiency anemia include fatigue, inability to concentrate, pale coloration, weakness, and listlessness.

The greatest risk of iron toxicity comes from overdosing on iron supplements, which can result in damage to the liver and pancreas and even sudden death in young children.

Iodine

The presence of iodine in food is dependent on the availability of iodine in the natural environment where the plant or animal developed. Foods from the ocean are reliable sources of iodine because this element is plentiful in seawater. In general, inland areas, especially mountainous regions, are likely to have iodine-deficient soils. It is in these areas where people may develop the iodine-deficiency disease **endemic, or simple, goiter.** In the United States, the area around the Great Lakes was formerly known as the "goiter belt" because of the high incidence of goiter. The most obvious symptom of goiter is a swelling of the neck caused by an enlargement of the thyroid gland, which straddles the trachea.

Iodine is required for the formation of thyroid hormones, which regulate metabolism in all cells of the body and control body temperature, growth, development, and reproduction. When the amount of iodine is low, hormone production is impaired. The thyroid enlarges in an attempt to produce more of the needed hormones; this attempt is futile without the necessary iodine. A person suffering from simple goiter exhibits a lack of energy, decreased blood pressure, sensitivity to cold temperatures, and weight gain.

Goiter has been a recognized ailment since ancient times, and various treatments have been suggested as a cure. The earliest known treatment is recorded in a Chinese source from 5,000 years ago that recommended eating seaweed and burned marine sponge. Today it is known, of course, that organisms from the ocean are naturally rich in iodine. The relationship between iodine and goiter was confirmed in 1820 by the French physician Jean-Francois Coindet, who reported the treatment of goiter using doses of iodine salts. Today goiter is rare in the United States and Europe because of iodized table salt, first introduced in 1924. However, in other areas of the world, almost 200 million people still suffer from goiter, an easily preventable disorder.

More than half of the salt sold in the United States is iodized, and a single teaspoon of this salt supplies almost twice the RDI of 0.15 milligram. However, overconsumption of iodine-containing substances can also be a problem because iodine can be toxic. As little as 2.0 milligrams per day is considered toxic, resulting also in an enlargement of the thyroid gland.

In addition to a lack (or even an excess) of dietary iodine, goiter can result from the overconsumption of goitrogenic compounds. Certain medications, including some of the sulfa drugs, and vegetables in the cabbage family are known to contain compounds that block the utilization of iodine in the thyroid. In a varied diet, these compounds are harmless, but they may be a problem in diets restricted solely to those vegetables.

Thinking Critically

The macronutrients (carbohydrates, proteins, and fats) are required in relatively large amounts for proper nutrition, whereas the micronutrients, the vitamins and minerals, are needed in smaller quantities.

Since micronutrients are required in much smaller amounts than macronutrients, are they any less important to the human diet? Explain.

DIETARY GUIDELINES

Research has shown that many significant diseases are influenced by nutrition, and beneficial changes in diet can, therefore, reduce the risk of developing these conditions. Diseases linked to nutrition are some of the major causes of death in the United States: cardiovascular diseases, hypertension, some forms of cancer, and type 2 diabetes. These diseases may arise in part from excess consumption of fat (especially saturated fat), cholesterol, refined sugar, and salt. In light of these findings, government agencies and health professionals have recommended dietary guidelines for better health and the prevention of disease.

Balancing Nutritional Requirements

The U.S. Senate Select Committee on Nutrition and Human Needs issued the first *Dietary Goals for the United States* in 1977. This was followed in 1980 by the publication of *Nutrition and Your Health: Dietary Guidelines for Americans,* issued jointly by the U.S. Department of Health and Human Services (HHS) and the U.S. Department of Agriculture (USDA). Revisions of the *Dietary Guidelines* have followed every 5 years. The latest revision of the USDA recommendations, *Dietary Guidelines for Americans, 2010,* featured MyPlate (fig. 10.5), which replaced the Food Pyramid. Some of its recommendations are as follows:

1. Make half your plate fruits and vegetables; especially red, orange, and dark green-vegetables. Eat fruit, vegetables, and unsalted nuts as snacks.
2. Drink fat-free or low-fat (1%) milk. Try calcium-fortified soy products as a substitute for dairy foods.
3. Make at least half of your grains whole by choosing whole-grain cereals, breads, and pasta.
4. Vary protein choices by eating seafood twice a week and eating beans. Eat smaller and lean portions of meat and poultry.
5. Cut back on foods and drinks high in saturated fats, added sugars, and salt. Replace sugary drinks such as soda with water. Choose low-sodium versions of soups, bread, and frozen meals.
6. Balance Calories by eating less and avoiding oversized portions. Be physically active.

The dietary guidelines originally suggested that complex carbohydrates should make up about 60% of the daily caloric intake, proteins about 8%–10%, and fats no more than 30%, with 10% each from saturated, monounsaturated, and polyunsaturated fat sources. Since the first dietary guidelines were issued, some healthy trends have developed among the American people. Fresh fruits, vegetables, and whole-grain products have undergone a resurgence in popularity, owing to expanded selections available in the supermarkets and increased nutritional awareness of the value of fiber.

Although the guidelines call for a reduction in the consumption of refined and processed sugars, consumption has continued to rise (a trend that began early in the twentieth century). At least part of the rise is attributable to the increased consumption of soft drinks, which contain high-fructose corn syrup.

Americans are eating less red meat, eggs, and whole dairy products than ever before, which is a good trend for reducing saturated fats and cholesterol. Consumption of poultry and fish has increased as consumption of beef has declined, and low-fat dairy products have become the preferred choice for many people. Although the intake of saturated fat has decreased (dramatically for butter and lard), more plant oils are being consumed. This trend has had a significant agricultural impact, as farmers have increased the acreage devoted to growing oil crops, such as sunflower, safflower, rapeseed

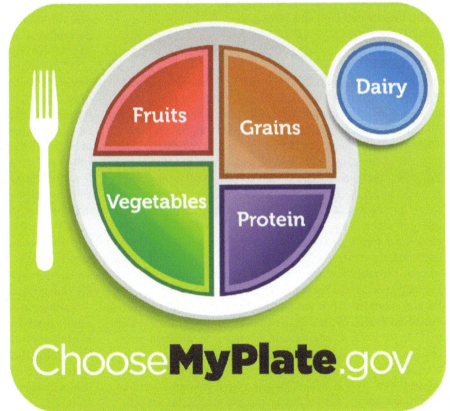

Figure 10.5 MyPlate illustrates the 2010 *Dietary Guidelines for Americans*, which offer the best scientific advice on how to eat for health issued by the USDA.
Source: http://www.choosemyplate.gov/food-groups/downloads/MyPlate/MyPlateGraphicsStandards.pdf.
Source: USDA

(canola oil), corn, and soybean. Margarine, shortening, salad oils, and cooking oils account for the expanded use of plant oils, but remember, if the plant oils are hydrogenated, they have reduced health value.

Since the early 1980s, there has been a heightened awareness about the dangers of high blood cholesterol levels, and Americans have responded with changes in diet and exercise. A report on several thousand men and women showed a decrease in cholesterol levels: for men, the 1980-1982 average was 205 milligrams, lowered to 200 milligrams in the 1985-1987 study; likewise, women showed a drop from 201 milligrams to 195 milligrams in the same period. This trend indicates that many Americans are attempting to keep blood cholesterol at or below the recommended level of 200 milligrams.

Although sodium is one of the major minerals, most Americans ingest more sodium (in the form of sodium chloride—table salt) than required; high sodium intake is related to hypertension (high blood pressure). Excessive salt ions in the bloodstream draw water from the tissues, thereby raising the fluid pressure in the blood vessels. A low-salt diet may be effective in lowering blood pressure in sodium-sensitive individuals. Some foods naturally contain sodium, but much of the sodium in our diet comes from the table salt added during cooking, during the meal, or in prepared foods. In fact, two-thirds of dietary salt is actually "hidden" in commercially prepared food and beverages.

Healthier Dietary Guidelines

In February 2004, the CDC (Centers for Disease Control and Prevention) released the results of a study on the prevalence of obesity and the changes to the American diet over the past 30 years. The incidence of obese individuals in the total U.S. adult population increased from 14.5% in 1971 to nearly 31% in 2000 as the average amount of kilocalories per day rose. For men, the average energy intake increased from 2,450 in 1971 to 2,618. The rise in energy intake for women was nearly double that at 335 kilocalories. As Americans ate more and got fatter, what they ate changed, too. Carbohydrates made up a greater percentage of the American diet. For men, carbohydrate intake increased from about 42% to 49%; for women, carbohydrates rose from approximately 45% to 52% of the daily diet. At the same time, the percentage of both total fat and saturated fat in the diet declined. For men, total fat decreased from 37% to 33%, and saturated fat fell from 14% to 11%; for women, total fat and saturated fat fell from 36% to 33% and 13% to 11%, respectively. Protein levels for both men and women declined slightly from an average of approximately 17% to just over 15%. As many low-carbohydrate dieters suggest, the rise in obesity in the United States appears to be correlated with a rise in carbohydrates, especially refined starches and sugars.

This trend is especially disheartening because earlier versions of the USDA *Dietary Guidelines* advised Americans to minimize consumption of fats to just 30% per day and increase complex carbohydrates, such as breads, cereals, rice, and pasta, to 60% of the daily diet. Other directives were to limit dairy products and protein sources (meat, eggs, fish, poultry, and beans) to two to three servings for each group per day and increase the number of servings of fruits and vegetables eaten daily.

The most controversial aspect of previous *Dietary Guidelines* is the advice about fats and carbohydrates. Many researchers are coming to the conclusion that it is just too simplistic to advocate eating all types of complex carbohydrates and avoid all types of fats. Increasingly, it has been shown that there is no nutritional evidence that a diet high in complex carbohydrates and low in all fats is beneficial to health. In fact, the reverse may be true.

Although it is known that a diet high in saturated fats raises total blood cholesterol levels and that high cholesterol is associated with an increased risk of coronary heart disease, certain fats, as stated previously, are beneficial to health. Monounsaturated fats, such as olive oil, can reduce the risk of cardiovascular disease by lowering LDL cholesterol levels and raising HDL cholesterol. It has also been shown that a diet of polyunsaturated fats can reduce total blood cholesterol levels.

For example, the so-called Mediterranean diet common in Italy, France, and Greece, in which fats make up 40% of the total kilocalories, is associated with a low rate of heart disease. Apparently, the type of fats—mainly monounsaturated olive oil and polyunsaturated fish oils with omega-3 fatty acids—not the percentage, is the determining factor in lowering the incidence of heart disease.

The original intent of the earlier USDA *Dietary Guidelines* was to influence the American public to decrease consumption of saturated fats. At the time, it was thought to be too difficult for the public to distinguish saturated fats from other types of fats, so the message was simplified to decrease consumption of all fats to 30% of total kilocalories, down from the 40% typical of the American diet. To compensate for the kilocalories lost by decreased fat consumption, the percentage of complex carbohydrates was raised from 45% to 60%. The recommended daily percentage of protein stayed about the same at 10%-15% because there was concern that, if increased protein consumption was recommended, people would eat more red meat as a protein source without realizing that red meat is usually associated with saturated fat. "Fats are bad" was the rallying cry of the earlier *Dietary Guidelines*.

Unfortunately, this message was the wrong one. Nutritional studies from the early 1990s have shown that, if people replace kilocalories from saturated fat with an equal amount from carbohydrates, their LDL and total cholesterol levels do fall, but so does their level of HDL. Similarly, if people eat a diet high in monounsaturated or polyunsaturated fats but then switch to an equivalent amount of kilocalories from carbohydrates, their LDL levels rise and HDL levels decline.

Trans fatty acids found in many dietary substitutes for foods rich in saturated fats are uniquely bad because they raise LDL and triglycerides while reducing HDL. Eating trans fatty acids greatly increases the risk of cardiovascular disease.

In contrast, consuming saturated fats increases the risk only slightly because consumption of saturated fats increases LDL but also HDL.

Another common misconception is that eating a greater percentage of fats in the daily diet is linked to obesity because fats have more than double the kilocalories of equal portions of proteins or carbohydrates. The standard advice has been to avoid obesity by eating a low-fat diet. The only way to avoid obesity is to lower total kilocalories, not just fat kilocalories, and increase energy expenditure through physical activity.

No nutritional studies have definitively linked the consumption of fat with a higher risk of breast or colon cancer. A diet high in red meat has been associated with a higher risk of colon cancer, but this link is probably due to carcinogens produced during cooking and the type of chemicals found in processed meat. A low-fat diet has not been shown to reduce the risk of cancer. A 2006 study on women's health compared women who reduced fat consumption from 8% to 10% over a period of 8 years with a control group. In that study, the rates for colon and heart disease were not statistically different, but the rate of breast cancer showed a small decline.

Incorporating weight control and physical activity appear to be the key factors in decreasing the risk of many cancers. Unfortunately, recent data did find that sedentary, overweight women who ate diets high in refined carbohydrates had a high incidence of both pancreatic and breast cancer. Increasingly, evidence points out that a diet high in certain carbohydrates is not synonymous with healthy eating.

Glycemic Index

Some carbohydrates are readily digested and quickly metabolized by the body into glucose. A rapid increase in blood glucose stimulates the release of insulin from the pancreas. Insulin facilitates the uptake of glucose into body cells, such as muscle and fat cells, and quickly lowers blood sugar levels. **Glycemic index (GI)** measures the effect that a particular food has on blood glucose levels. The faster the food is converted to glucose, the higher the GI number. White bread, white rice, and white potatoes are examples of carbohydrates with relatively high GI values. Whole grains, and high-fiber fruits and vegetables have lower GI ratings. A more practical measure that is often used is **glycemic load,** which takes the GI value for a food and multiplies it by the number of carbohydrate grams the food contains. Thus, many fruits and vegetables that have a relatively high GI value have a much lower glycemic load. Fiber slows the rate of digestion, and high-carbohydrate foods, such as many beans, that also are high in fiber have a lower GI.

There are difficulties in accurately determining the GI for a particular food because the calculation is based on the average blood glucose response of 10 subjects after they have eaten a particular food. Different people have different responses to the same food, and the same person may have a different response on a different day. Eating different combinations of foods and changing the method of cooking can also alter the GI value.

A high GI diet is associated with a greater risk of heart disease, type 2 diabetes, and obesity. A diet high in GI carbohydrates raises blood triglyceride levels and lowers HDL cholesterol, increasing the risk of coronary heart disease. Another consequence of a high GI diet is insulin resistance, common in many overweight, sedentary people. One of the main indicators of insulin resistance, also known as Syndrome X, is a sustained high blood glucose level after the ingestion of high GI foods. Although insulin is produced in affected individuals, it is not as effective in moving glucose from the bloodstream into body cells. Because the insulin is not as effective, more and more insulin is released to do the job of moving glucose out of the blood and into cells. Eventually, the insulin-producing cells of the pancreas become overtaxed by higher production levels and give out. People who exhibit insulin resistance have a higher risk of hypertension, heart disease, and most significantly type 2 diabetes. Foods with the highest GI numbers correspond with greater insulin release, and the higher the insulin spike, the lower the blood glucose sinks. Low blood glucose stimulates hunger and a craving for more high GI foods, which may lead to overeating and weight gain. In contrast, low GI foods are digested more slowly, glucose levels in the blood drop gradually, and it takes longer for hunger to return.

Another problem with the previous USDA *Dietary Guidelines* was that they did not distinguish between different sources of protein. They lumped red meat, poultry, fish, legumes, nuts, and eggs into a single group, despite the evidence that a diet high in red meat is high in saturated fats and cholesterol and is associated with an increased risk of coronary heart disease and type 2 diabetes. Poultry and fish possess fewer saturated fats, and fish are a source of beneficial omega-3 fatty acids. Certain nuts, such as walnuts, are also rich in beneficial omega-3 fatty acids but regrettably high in Calories. What's more, the pyramid promoted overconsumption of dairy products, a recommendation that was initially made to ensure significant calcium content in the diet to prevent osteoporosis. The problem is that most dairy products are associated with saturated fat and high caloric values. The best advice for those concerned about obtaining sufficient calcium for preventing osteoporosis would be to take calcium supplements. The *Dietary Guidelines* do advocate eating plenty of fruits and vegetables, but consumption of the white potato and other starchy vegetables should be limited.

An improved version of the *Dietary Guidelines,* as suggested by Walter Willett of the Harvard School of Public Health, emphasizes keeping one's weight under control through daily exercising and avoiding an excessive total intake of kilocalories. It recommends that the majority of one's daily diet should be the consumption of the healthy monounsaturated and polyunsaturated fats and healthy carbohydrates, whole-grain and unrefined. A variety of fruits and nonstarchy vegetables with protective phytochemicals should be eaten in abundance. Healthy protein sources, such as nuts, legumes, fish, poultry, and eggs, should be eaten in moderation. Dairy

A CLOSER LOOK 10.2

Eat Broccoli for Cancer Prevention

Cancer has been a dreaded disease for centuries, and it continues to plague humanity today. The term *cancer* actually refers to over 100 forms of a disease that can strike just about every tissue and organ in the body that shares several basic processes. Cancer cells are abnormal cells that proliferate uncontrollably, forming masses called tumors. Cancer cells also possess the ability to migrate, or metastasize, from the original site, forming tumors in other parts of the body. It is the interference of malignant tumors with normal body functioning that makes cancer lethal.

A number of agents have been identified as **carcinogens** (cancer-causing agents)—certain microbes, ultraviolet radiation, chemicals (such as PCBs, arsenic, and benzene), radon, and various gene mutations—but more than half of all cancers in the United States are related to tobacco smoke and diet. In addition to tars, a number of known carcinogens are found in tobacco smoke (see Chapter 20). Not surprisingly, smoking—especially cigarette smoking—is associated with several cancers: lung, upper respiratory tract, esophageal, bladder, and pancreatic. Tobacco smoke is also implicated in stomach, liver, kidney, and pancreas cancers and in leukemia.

Diet is second only to smoking as a major cause of cancer in the United States. Red meat has been identified as an associative cause of colon, rectal, and prostate cancers. Obesity in adults is linked to cancers of the uterus, breast, colon, kidney, and gallbladder. Abusive drinking of alcoholic beverages enhances the risk of cancers of the upper respiratory tract, digestive tract, and liver. Eating heavily salted foods and drinking extremely hot beverages have been linked to cancers of the stomach and esophagus, respectively, in countries outside the United States where these dietary habits are customary.

Ironically, the latest research suggests that what is missing in our diets may be far more important in causing cancer than what we actually eat or drink. Components called **phytochemicals,** which occur naturally in vegetables, fruits, grains, and seeds, have been investigated for their protective action in the prevention of cancer and other diseases.

Cruciferous vegetables (broccoli, cauliflower, cabbage, and so on) are excellent sources of one class of chemopreventive chemicals: the dithiolthiones. In laboratory animals treated with a synthetic version of one of the dithiolthiones, tumors of the lung, colon, mammary glands, and bladder were inhibited. One of the most potent dithiolthiones is sulforaphane, found in broccoli, which has been shown to inhibit breast cancer in rats. Both of these phytochemicals apparently work by activating liver enzymes that destroy carcinogens in the body.

Genistein, a compound found in soy products (derived from the soybean), prevents the formation of breast tumors in rats in a different manner. It inhibits the formation and growth of blood vessels to a growing tumor. Without these sources of nutrients and oxygen, the tumor cannot grow.

---

products should be limited to one or two servings daily. The consumption of red meat, saturated fats, refined grains, and starchy fruits and vegetables should be restricted. Trans fatty acids should be avoided completely. A daily multiple vitamin and mineral supplement should be taken. Alcohol consumption of beer, wine, or distilled spirits in moderation is acceptable because of evidence of its benefit to the cardiovascular system (see Chapter 24).

Recently, Willett conducted a study on the health of men and women who followed his dietary suggestions. He found that they had reduced the risk of cardiovascular disease by 30% for women and 40% for men. These findings and others have resulted in major revisions within the *USDA Dietary Guidelines, 2010* and illustrated in the release of MyPlate.

Meatless Alternatives

With the awareness about the dangers of saturated fat and cholesterol inherent in animal products, many Americans are incorporating a greater percentage of vegetables, fruits, grains, and legumes into their diets. Some even choose a totally **vegetarian** lifestyle. There are many different forms of vegetarianism: **lacto vegetarians, lacto-ovo vegetarians,** and **vegans.** Some vegetarians, such as the lacto and the lacto-ovo, consume dairy products or dairy products and eggs but do not consume animal flesh. Vegans are pure vegetarians, consuming no animal products at all. Some other vegetarians stretch the concept by consuming fish and poultry, avoiding only red meat.

There are several health benefits to increasing the consumption of plant products while decreasing the consumption of animal products. Vegetarians are less likely to suffer from the chronic diseases that afflict many Americans whose diets are high in animal products. Blood cholesterol and triglyceride levels usually reflect the amount of animal fat in the diet and are lowest in vegans, who, consequently, have a lower incidence of heart disease. Those cancers linked to red meat and dairy consumption (colon and prostate) are less common in vegetarians. High fiber in a vegetarian diet also plays a role in reducing risks of type 2 diabetes and lowering cholesterol levels. (See A Closer Look 10.2: Eat Broccoli for Cancer Prevention.) A further benefit of high-fiber diets is in weight control; the filling effect of fiber suppresses overeating.

Studying populations that drink quantities of green tea (made from unfermented tea leaves, Chapter 16) has revealed a lower incidence of many cancers, especially those of the breast and prostate. Green tea contains a high percentage of chemical agents (flavonoids) known as catechins, and in particular, epigallocatechin gallate. For a cancer to metastasize, certain enzymes are needed. One of these crucial enzymes is urokinase, and it appears that catechins inhibit it and thus prevent the invasion and spread of cancer cells to distant sites. In the processing of black tea, the type of tea most commonly drunk by Americans, epigallocatechin gallate is destroyed and thus does not afford the same protective benefits associated with the drinking of green tea. In 2010, it was shown that epigallocatechin gallate (EGCG) can interfere with certain anticancer drugs, such as bortezomib. Bortezomib is given to patients who are suffering from multiple myeloma, an incurable but treatable form of cancer of blood cells in the bone marrow. Bortezomib acts by deactivating proteasomes (see Chapter 2); however, in the presence of EGCG, its anticancer activity is inhibited. It was observed, however, that only proteasome deactivators that contain boron, as does bortezomib, are inhibited by EGCG. Green tea does not interfere with the proteasome inhibition of other anticancer drugs that lack boron. Patients undergoing treatment with bortezomib should refrain from drinking green tea or taking supplements of green tea extracts.

Phytochemicals appear to be most effective if eaten in foods rich in these cancer-preventive agents rather than administered as supplements. In fact, the National Cancer Institute initiated the 5-A-Day Program in 1991 to encourage the public to eat five or more servings of vegetables and fruits every day. In 2007, the CDC and the Produce for Better Health Foundation (PBH) released Fruits and Veggies—More Matters™ which calculates the servings of fruits and vegetables on the basis of age, gender and activity level. Several studies appear to support this belief. Beta-carotene, the yellow-orange pigment associated most commonly with carrots, was given to people at risk of developing lung cancer as a supplement in a chemoprevention trial sponsored by the National Cancer Institute. Surprisingly, the group given beta-carotene had a slightly higher rate of lung cancer than the group given a placebo. A 2007 review of nearly 70 studies also found that adults who take supplements of beta-carotene, vitamin A, and/or vitamin E had higher mortality rates than those who did not or who were given a placebo. Adults who took supplements of vitamin C or selenium showed no difference in death rates than the control group.

Recent large, long-term studies looking at a diet high in fruits and vegetables to reduce the risk of breast cancer have failed to substantiate the promise of earlier laboratory work. Those who consumed the greatest amount of fruits and vegetables did, however, show a 25% reduction in the risk of cardiovascular disease. Critics point out flaws in questionnaires and the difficulties that participants have in accurately recording and quantifying their diets. Also, some suggest that the protective effect may be conferred during a critical period in childhood, but no study has examined the connection between a girl's diet and breast cancer. Another study found that women who had the highest blood concentration of carotenoids, such as beta-carotene, showed the lowest incidence of breast cancer. But the same study showed no difference in the incidence of breast cancer with the amount of vegetables and fruits consumed, suggesting that individual differences in metabolism to harvest protective phytochemicals may be more important. The work continues.

Vegetarians, especially vegans, must be knowledgeable about all the nutritional requirements and use care in selecting plant sources that meet those requirements. Special attention must be given to ensure that iron, the B vitamins (especially B_{12}), calcium, vitamin D, and the essential amino acids are supplied. Complementing the essential amino acids is of prime importance because all plant proteins are incomplete. This can be easily accomplished by eating legumes and grains that, in combination, provide an excellent source of protein for the diet. In fact, one study indicates that proteins from plant foods may be preferable to animal protein because consumption of excessive animal protein itself, not just animal fat, may be linked to heart disease and cancer. The suggestion is that meat should no longer be the centerpiece of a meal but instead relegated to a side dish for a vegetarian entrée. Perhaps vegetarians have the right idea.

CHAPTER SUMMARY

1. The nutritional needs of the human diet can be categorized into the macronutrients (carbohydrates, proteins, lipids) and micronutrients (vitamins and minerals).

2. Carbohydrates include simple sugars, such as the monosaccharides fructose and glucose and the disaccharides sucrose and lactose. Complex carbohydrates, or polysaccharides, include starch and glycogen and act as the body's fuel. Dietary fiber includes both soluble and insoluble forms that provide beneficial health effects, acting as roughage, promoting regularity, and lowering blood cholesterol levels.

3. All plant proteins are incomplete, deficient in one or more essential amino acids. But complementing incomplete plant proteins, such as combining legumes and cereals in a single meal, can overcome this deficiency. Gluten in wheat can trigger celiac disease in individuals who have an inherited sensitivity to this protein.

4. Lipids include fats and oils, and most of the dietary lipids are classified as triglycerides. A diet high in saturated fats raises the risk of cardiovascular disease. Unsaturated fats, whether monounsaturated or polyunsaturated, lower the risk of cardiovascular disease by lowering LDL cholesterol levels. Foods of animal origin generally are high in saturated fats and cholesterol, but foods of plant origin lack cholesterol and contain mainly unsaturated fats.

5. Deficiency diseases result if a diet lacks any of the essential vitamins and minerals. Fat-soluble vitamins (A, D, E, and K) can be stored by the body but can build up to toxic levels if excessive amounts are taken. Water-soluble vitamins (C and B complex) cannot be stored in appreciable quantities by the body. Most vitamins and minerals (the exception is vitamin B_{12}) can be found in foods of plant origin. The category of minerals (major or trace) is defined by the quantity needed in the body.

6. Americans should modify their diets for a healthier lifestyle. Some fats, such as monounsaturated and polyunsaturated fats, are heart-healthy and should become a regular part of the diet. Saturated fats, however, should be limited and trans fats avoided completely. Limit the consumption of white bread, white rice, white potato, and other high glycemic index carbohydrates to reduce the risk of type 2 diabetes. Limit dairy products that are high in kilocalories and saturated fat. Eat plenty of nonstarchy fruits and vegetables, and take a daily vitamin and mineral supplement. The most important factor for a healthier life is to keep weight under control by limiting total caloric consumption and exercising regularly.

REVIEW QUESTIONS

1. If fiber is largely indigestible, why is it required for a healthy diet?
2. What are the essential amino acids? Why is it important that they be consumed?
3. What is the role of gluten in celiac disease?
4. What is the role of lipids in the body? What is the dietary significance of saturated and unsaturated fats?
5. Describe the following deficiency diseases: scurvy, rickets, marasmus, beriberi, osteoporosis.
6. How closely do you follow the recommended dietary guidelines? Evaluate your diet in terms of a healthy lifestyle.
7. What are the dietary causes of anemia?
8. What differentiates simple carbohydrates from complex carbohydrates?
9. What dietary factors influence the development of cardiovascular diseases or cancer?
10. Estimate the amount of protein you consume in a typical day. What portion is from animal food sources, and what portion is from plant foods?
11. Your friend insists that a vegetarian diet is unhealthy because proteins found in plants are never complete. How would you respond?
12. Concerned about your risk of cardiovascular disease, you undergo a blood test for cholesterol at a health fair at the local supermarket. The test results indicate your total cholesterol count is 190. Is further testing necessary?

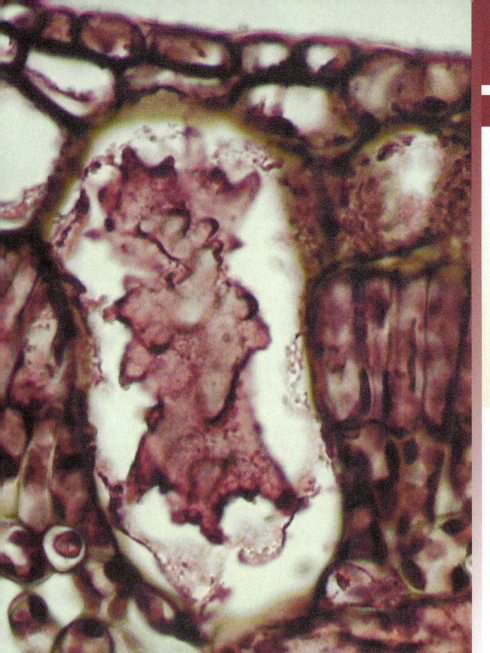

©Karen McMahon

CHAPTER

11

Origins of Agriculture

Plant crystals such as this one found in the leaf of a rubber tree (Ficus) *can be used by forensic botanists to identify species.*

KEY CONCEPTS

1. Early human societies were based on a foraging lifestyle in which wild plants were gathered and animals were hunted.
2. Agriculture evolved independently in several areas of the world.
3. The earliest evidence of agriculture dates back at least 10,000 years in the Near East, the Far East, and Mesoamerica.
4. Domesticated plants are genetically different from their wild counterparts, and many can no longer survive without human intervention.

CHAPTER OUTLINE

Foraging Societies and Their Diets 176
    Early Foragers 176
    Modern Foragers 177
    The Paleo Diet 177

A CLOSER LOOK 11.1 Forensic Botany 178

Agriculture: Revolution or Evolution? 179
    Latitudinal Spread 180

Early Sites of Agriculture 180
    The Near East 181
    The Far East 182
    The New World 183
Characteristics of Domesticated Plants 184
Centers of Plant Domestication 184

Chapter Summary 185
Review Questions 186

The human species, known as *Homo sapiens*, has existed for about 200,000–300,000 years. For most of that time, humans survived as **foragers** or **hunter-gatherers,** gathering wild plants and hunting animals in their natural environment (fig. 11.1). Around 10,000 years ago in many areas of the world, there was a shift in human endeavor from foraging to farming. Most authorities agree that agriculture arose independently in different areas over several thousand years. Why this shift occurred can only be theorized, but the development of agriculture formed the basis of advanced civilization in both the Old and the New Worlds. Over the centuries, agricultural societies spread into those environments that could be easily adapted to agriculture, and foragers gradually became restricted to marginal areas. By the late twentieth century, foraging societies had largely disappeared, constituting only a tiny percentage of the human population and limited to a few tropical rain forests, deserts, savannas, tundras, and boreal forests.

FORAGING SOCIETIES AND THEIR DIETS

Foraging societies are by no means all alike in the types of food they eat. Some groups, such as the Arctic Inuit, subsist almost entirely on meat; at the other extreme, the Hadza of Tanzania are largely vegetarian, rarely hunting for meat. Other groups, such as the !Kung (the exclamation point is pronounced as a click) of southern Africa, however, have had a more varied diet, largely plant-based but supplemented by eggs, insects, fish, small animals, and meat from the hunt.

Many hunter-gatherers have utilized plants by gathering seeds, flowers, roots, and tubers and have a fairly thorough knowledge of the botany in the area. From experience, they know which plants are edible, which are poisonous, which have medicinal properties, which are sources of dyes, which could be used for weaving or building materials, and even which have psychoactive properties. By looking for certain visible clues, such as flowering on so-called calendar plants, foragers know if tubers are ready to be dug up or if turtles are laying their eggs. They have developed remarkable methods to prepare edible foods, even from plants with toxins, such as cassava, which contains poisonous hydrocyanic acid.

Early Foragers

Archeological investigations have supplied knowledge about the diet of early humans from many sources, and radiocarbon dating of artifacts can provide an estimated time frame. Fossilized remains of both plants and animals have been found in early settlements (fig. 11.2). Plants in the diet have been identified from charred seeds and preserved fruits or other plant parts, while bones, teeth, feathers, scales, fur, and shells indicate the animals in the diet. Microscopic remains include plant fibers, plant crystals, and pollen, with crystals and pollen especially useful in identification. **Coprolites,** fossilized fecal materials, provide direct evidence of the diet because some plant materials, especially seeds and pollen, can

(a)

(b)

Figure 11.1 Native North Americans of the Pacific Northwest foraging in the early twentieth century. (a) A Hupa man hunts salmon with a spear. (b) A Pomo woman uses a beater to gather seeds into a basket.

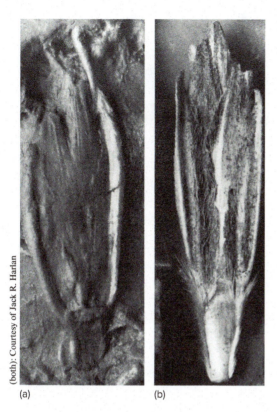

Figure 11.2 (a) A clay impression fossil of wheat as compared with (b) a present-day species of wild wheat.

Figure 11.3 Cave painting depicting a hunt.

pass through the digestive tract largely intact. Often middens, or dump sites, from a human encampment provide a concentrated source of plant and animal remains. (See A Closer Look 11.1: Forensic Botany.)

Grinding stones, sickles, and digging implements provide information on the food in the diet but do not indicate if the plants harvested were wild or domesticated. Tools of the hunt also provide some information on the size of the animals hunted and the method of the kill. Depictions in cave paintings (of animals as well as hunting and gathering activities), pottery fragments, and clay figurines are other windows to the past.

In some excavations at Wadi Kubbaniya in the Nile Valley of Upper Egypt, archeologists have dated plant remains of a hunter-gatherer settlement from 17,000 to 18,000 years ago. The charred remains of fruits, seeds, and tubers from 25 plant species have been found; interestingly, contemporary foraging societies have utilized the same plants or closely related species. The most abundant plant remains found were the tubers of wild nut grass, a type of sedge. Analysis of modern samples of these tubers indicates that they are high in carbohydrates and fiber but low in protein; however, the protein is of good quality because it is high in lysine, one of the essential amino acids. Mature nut grass tubers also contain toxins, but these could be easily removed by various methods still in use today. It is believed that this tuber served as one of the dietary staples, along with acacia seeds, cattail rhizomes, and palm fruits. Evidence from this and other **Paleolithic** (Old Stone Age or preagricultural societies) sites indicates that early foraging groups had a remarkably varied plant diet.

Modern Foragers

Much of our knowledge about the foraging way of life has been documented by studying modern foraging societies, such as the !Kung San of the Kalahari Desert of southern Africa. The !Kung live in the tropical savannas that border the Kalahari in what is now southeastern Angola, northeastern Namibia, and northwestern Botswana. The !Kung have foraged in this area for at least 10,000 years, continuing the hunter-gatherer way of life until recent times (fig. 11.3). Extensive studies of the !Kung during the 1960s revealed that they utilized over 100 species of plants and 50 animal species; two-thirds of their diet was plant-based. The plants included a mixture of fruits, nuts, berries, melons, roots, and greenery; one nut, the mongongo nut, was a very important high-protein component of their diet. The !Kung diet was very nutritious; they consumed an average of 2,355 kilocalories per person per day, with 96 grams of protein and adequate vitamins and minerals. This diet more than meets the nutritional requirements established for people of the !Kung stature and physical activity. As in most foraging societies, the division of labor was along gender lines, with women doing most of the gathering and men the hunting. Surprisingly, the amount of time spent on foraging activities averaged only about 2.5 days per week, which left plenty of time for leisure and socializing.

The Paleo Diet

The Paleo Diet, or Caveman's Diet, is a recent nutritional trend that promotes a return to the eating habits of our human ancestors during the Paleolithic period when humans lived as hunter-gatherers. Followers of this dietary plan do not eat dairy or refined grains, processed sugars, or legumes (such as peanuts, lentils, beans, and peas) because these foods were not available in the original diet. Instead, practitioners of the Paleo Diet eat large quantities of meats, usually cooked in animal fat, fruits, nuts, root vegetables, and an occasional treat of honey. In sum, they eat the foods that could be obtained by hunting animals for meat, digging up tubers, and picking fruits. Their reasons for promoting the Paleo Diet is that it is the original diet of humans; our bodies are adapted to this diet; and many diseases

A CLOSER LOOK 11.1

Forensic Botany

Archeologists have made extensive use of plant remains in reconstructing the lifestyles of ancient foraging and early agricultural peoples. The painstaking task of botanical detection begins with the recovery and processing of plant remains. Not all parts of a plant are equally well preserved: seeds, wood, pollen, phytoliths, starch grains, and fibers are among the most informative recoverable remains. The lignified walls of wood and fibers are more resistant to decay than purely herbaceous tissue because lignin can be degraded only by a few types of fungi. Likewise, the lignin in the sclerified cells of the seed coat, or testa, makes seeds among the most common dietary material recovered from archeological sites. Buried pollen grains are virtually indestructible, owing to the chemical properties of the exine (outer pollen wall). **Phytoliths** (literally, "plant stones") are crystals formed and found in many plants (box fig. 11.1a and b). They are often created within epidermal cells or between plant cells. The crystals may be composed of calcium oxalate or, rarely, calcium carbonate. Another common component of phytoliths is opaline silica (silicon dioxide). After a plant decays, the phytoliths are released into the soil and can remain intact for thousands of years. Phytoliths are so distinctive in shape and size that they can be used to identify the presence of a particular plant family, genus, or even species. In fact, phytoliths can be used to separate wild species from domesticated ones and have proven especially useful in dating the origin of agriculture in many sites.

Phytoliths of wild barley husks were used to identify structures as ancient granaries built by prehistoric hunter-gatherers over 11,000 years ago to store wild cereal grains. Researchers believe that the hunter-gatherers sowed wild cereal grains in what is today Jordan and stored any surplus in the granaries. With a stored food supply, this society of hunter-gatherers was able to establish permanent settlements 1,000 to 2,000 years before the domestication of plants for agriculture.

The same principles of plant detection are applicable to forensic science, in which the identification of plants can be used as incriminating evidence. The first criminal case that used botanical information was the famous 1935 trial of Bruno Hauptmann, who was accused and convicted of kidnapping and murdering the young son of Charles and Anne Morrow Lindbergh. The botanical evidence centered on a homemade wooden ladder used during the kidnapping and left at the scene of the crime. After extensive investigation, plant anatomist Arthur Koehler showed that parts of the ladder were made from wooden planks from Hauptmann's attic floor.

In forensics, even herbaceous plant parts can be useful for identification. In one case, botanical evidence disputed the testimony of an accused rapist. Fragments of tree leaves and bark in his pants cuff indicated that the accused had climbed

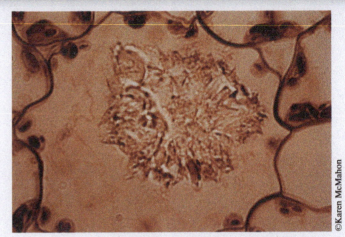

(a)

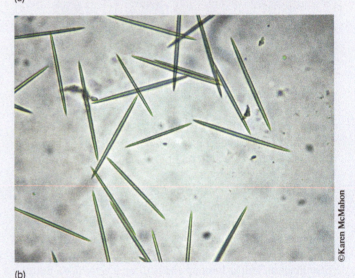

(b)

Box Figure 11.1 (a) Calcium oxalate crystal in the vacuole of a leaf cell, ×270. (b) Needlelike crystals are recognizable in pineapple pulp.

a tree to get into a window of the victim's home, rather than being admitted through the front door, as he claimed. In cases of suspected plant poisonings, identification can be made from leaves or fruits of intact plants or even from analysis of stomach contents. In this type of situation, proper medical treatment depends on accurate identification of the plant or mushroom. Trained botanists and mycologists (scientists who study fungi) are routinely called to hospital emergency rooms for this purpose. Botanical analyses of stomach contents have also played roles in other types of cases: a hunting guide killed a grizzly bear he claimed was eating his supply of alfalfa hay. Botanical evidence showed no alfalfa in the bear's stomach, only native vegetation. The guide was fined and imprisoned for killing a threatened species.

such as obesity, diabetes, cancer, and cardiovascular disease that afflict contemporary society are a result of incompatibility of the contemporary diet with our Stone Age human anatomy.

The Paleo Diet does have some benefits; it cuts down on processed foods such as refined flours that have less protein, fiber, and vitamins and prohibits others foods that may be packed with sodium and artificial preservatives. However, it also eliminates eating dairy products that are rich in calcium; whole grains that are high in fiber and vitamins; and legumes that are abundant in protein.

The Paleo Diet is based on the fallacy that humans have not evolved since the Stone Age; however, there are well-known examples of evolution in human nutrition. Originally the gene producing lactase, the enzyme that breaks down the milk sugar lactose, was turned off after infancy. But several thousands of years ago, a mutation arose in some groups of people around the world, in which lactase production continued throughout adulthood. This allowed humans to eat dairy products made from their herding animal without digestive upset. Also some diseases that plague humans in modern times, such as atherosclerosis, are age-old afflictions as evidenced by signs of this disease in ancient mummies from Egypt, Peru, the southwestern United States, and the Aleutian Islands.

Another mistake of the Paleo Diet is that there is not one type of hunter-gatherer diet. The Inuit of the Arctic region consume a diet that is 95% sea mammals and fish with only 5% fruits and vegetables, whereas the Hadza of Tanzania in Africa have a diet of about 58% meats and fish, 12% fruits and vegetables, 30% roots. The variety in the diets of modern day hunter-gatherers reflects the diversity of the human diet. Humans have the digestive flexibility to consume diets that are almost all animal to all vegetarian and everything in between and this versatility accounts for the fact that humans have thrived in almost every ecosystem on earth.

Thinking Critically

Humans survived as foragers of wild plants and animals for hundreds of thousands of years. Agriculture arose relatively recently, roughly 10,000 years ago, in the course of human history.

Explain the success of foraging as a survival strategy for early humans.

AGRICULTURE: REVOLUTION OR EVOLUTION?

Archeological evidence indicates that about 10,000 years ago human cultures began the practice of agriculture in several areas of the world. Over the next few thousand years in the Near East, the Far East, and Mesoamerica, agriculture flourished. The question that has puzzled archeologists and other scientists is "Why, after thousands of years of foraging, did hunter-gatherers switch to agriculture?"

Many theories have been proposed to answer this question. Some state that agriculture was the discovery of a brilliant sage who, with a flash of insight, realized that, if you sow seeds, the crop will grow. There were many variations of this theme. Some held that the brilliant sage realized that plants growing at middens, or dump sites, were growing from discarded seeds. Others held that the sage observed that seeds buried with the dead (as food for the afterlife) gave rise to plants at grave sites. These theories viewed agriculture as a revolution; in fact, the term *Agricultural Revolution* was used to describe the transition from foraging to agriculture. It was suggested that this revolution spread quickly because agriculture was thought to be an improvement over the hunter-gatherer lifestyle in that a dependable food source could be easily grown rather than collected from the wild.

Beginning in 1960, archeologists questioned that view, suggesting instead that the origin of agriculture was not a revolution but the result of a gradual cultural evolution. They reasoned that hunter-gatherers knew the wild plants, knew how they grew, and would incorporate farming along with foraging as part of an overall food-collection strategy when necessary. For example, certain aboriginal groups in coastal Peru abandoned their farming practices whenever fish became plentiful.

Many archeologists believe that there was a transitional stage between simple foraging, in which small, nomadic bands followed the wild plants and animals, and agricultural societies with their sedentary lifestyle. During this transitional stage, foraging groups formed settlements but sent out members to hunt and gather. This more complex strategy resulted in changes in the social organization of the groups and permitted populations to increase. This transitional stage lasted for several thousand years in some locations until resource stress or environmental change led to the switch to agriculture. Archeologists believe that in the Near East, for example, the climatic dry period around 11,000 years ago brought about a change in the distribution of cereal grains (wheat and barley). Applying their botanical knowledge, these foragers gradually changed from collecting these wild cereals to cultivating them.

Genomic analysis of ancient farmers is shedding light on the switch from hunting-gathering societies to farming during the Neolithic period (New Stone Age) in Europe. After agriculture originated more than 10,000 years ago in the Near East, farming spread slowly to what are now the modern nations of Turkey, Greece, and Bulgaria. However, 7,500 years ago, the expansion of agriculture into central Europe accelerated and spread rapidly both east and west until the activity of farming was the principal lifestyle from France to the Ukraine. Two opposing theories have been proposed to explain the agricultural explosion across Neolithic Europe. As stated previously in this section, the theory of cultural diffusion proposes that the adoption of agriculture spread from culture to culture because agriculture was recognized as superior to the hunter-gatherer way of life. The opposing theory believes that agricultural expansion is the result of a massive migration of farmers who brought agricultural practices as they migrated across Europe.

A study compared DNA from ancient European hunter-gatherers and early farmers. Mitochondrial DNA was fully sequenced from 20 hunter-gatherers and 25 farmers who lived 15,000 to 4,300 years ago from skeletal remains collected in sites across central Europe. A comparison of the mitochondrial DNA revealed that the hunter-gatherers and early farmers were so genetically distinct that it eliminated any possibility that the two groups were related. The ancient mitochondrial DNA was also compared with that of nearly 500 modern Europeans, and the results suggested that modern Europeans are the descendants of the migrant farmers and not directly related to the indigenous hunter-gatherers. Researchers concluded that the first farmers in central Europe were immigrants who spread agriculture throughout the region. The next goal of the researchers is to identify the original home of the immigrant farmers.

Latitudinal Spread

Latitudinal Spread, recently proposed by Jared Diamond, is another theory on why agriculture developed and spread in certain areas of the world and not in others. Diamond noticed that continents such as Eurasia, in which the main orientation of the continent is east to west, favor the development and expansion of domesticated crops in contrast to continents such as the Americas, and to a certain extent Africa, in which the main axis is north to south. Historically, the diffusion of domesticated crops and livestock can be traced from the Fertile Crescent in southwest Asia westward to Europe and east to Central Asia. Another example of a route for the transfer of crops and livestock has been northward from Mexico to America. Diamond calculated that along the east-west axis from southwest Asia, domesticated crops and livestock spread west to Europe and North Africa and east to Central Asia at a rate of 0.7 miles per year. In contrast, crop expansion along the north-south axis from Mexico to the American Southwest was 0.5 miles per year; the spread of corn and beans from Mexico to the East Coast took place at 0.3 miles per year; and the llama spread from Peru north to Ecuador at a rate of 0.2 miles per year. His conclusion was that domesticated species spread more rapidly along an east-west axis than a south-north axis.

Diamond noted that the completeness of the transfer of crops and livestock also varied according to axis orientation. Crops and livestock transferred mainly wholesale from southwest Asia to Europe, but along the north-south axis in the Americas the transfer was often incomplete. Livestock from the Andes of South America, such as the llama, alpaca, and guinea pig, never reached Mesoamerica, although some of the crops such as manioc, sweet potato, and peanuts did. He argues that evidence of this preemptive domestication can be uncovered by examining whether the transferred crop shows the same transforming mutation across its range. A transforming mutation is one that changed the crop from its wild ancestor into a domesticated crop. If there is just a single identical mutation in the domesticated plant, this would indicate that the crop was domesticated once and spread to other areas. If the crop shows alternate mutations across its range, it would indicate that the crop was domesticated independently in several areas.

Diamond found that many of the ancient crops in southwest Asia exhibit just one transforming mutation, suggesting that all modern varieties of the crop stem from a single domestication event and that the domesticated crop spread quickly to other areas. All variants of the domesticated pea along this western spread share the same recessive allele that prevents pods from splitting open, which indicates that the variants are descendants from a common domesticated ancestor. It is important to note that even if wild relatives of a domesticated crop exist elsewhere in the range, they were never domesticated despite possessing desirable traits because the domesticated crop spread so rapidly that there was no need to domesticate another wild relative. For example, *Pisum fulvum* is a wild pea species found along the pathway of domestication from southwest Asia. It has all the qualities that would make it a candidate for domestication: good-tasting seeds that can be eaten both in dried and fresh forms. However, *P. fulvum* was never domesticated because *Pisum sativum,* domesticated in the Fertile Crescent area, expanded rapidly throughout the range. The founder crops from southwest Asia were never domesticated again after their initial domestication.

In the Americas, however, multiple domestication events are evident. Lima beans, common beans, and chili peppers have all been shown to have been domesticated on at least two separate occasions; once in Mesoamerica and again in South America. Likewise, squash and goosefoot were also domesticated independently at least twice, once in Mexico and once in what is now the eastern United States.

These differences in the pattern of domestication can be explained by the differences in axis orientation between Eurasia and the Americas. In the dominant east-west orientation of Eurasia, in which a large area shares the same latitudes, domesticated crops can spread rapidly because crops grown at the same latitude share exactly the same daylength and other environmental conditions such as temperature, rainfall, and habitat. Domesticated forms from the Fertile Crescent expanded rapidly because they were already adapted to the same type of climate along Eurasia's wide band of land at the same latitude. But in the Americas, the south-north axis dominates; the lowlands of Central America will stop the spread of crops adapted to the cool highlands of Peru from transferring to the cool highlands of Mexico. Likewise, sunflowers adapted to the cool conditions along the northeastern coast of North America would have thrived in the Peruvian highlands but never made it through the tropical climates of the south.

EARLY SITES OF AGRICULTURE

Archeological excavations have documented many sites of early agriculture in both the Old and New Worlds (fig. 11.4). The evidence indicates that the earliest sites were in the Far East, dating back approximately 11,500 years.

Figure 11.4 Evidence of the beginnings of agriculture has been found in eastern North America, the Tehuacan and Oaxaca valleys in Mexico (Mesoamerica), the South American highlands, the Fertile Crescent of the Near East, the Yangtze and Yellow River valleys of China (Far East), and the New Guinea highlands.

The Near East

Some of the oldest sites of agriculture are in southwestern Asia, in the foothills around the area known as the **Fertile Crescent,** which today includes parts of Iran, Iraq, Turkey, Syria, Lebanon, and Israel (fig. 11.4). From sites such as Jarmo, Jericho, Ali Kosh, Cayonu, and others, remains of both plants and animals date back 9,000 to 14,000 years. Early plant domesticates include einkorn wheat, emmer wheat, barley, pea, lentil, and vetch, while cattle, goats, and sheep were among the domesticated animals. Research suggests that the animals were domesticated several thousand years before the plants. Evidence indicates that barley may have been the first crop domesticated in the Near East, approximately 10,000 years ago.

One of the sites that had been particularly well studied was Jarmo in northeastern Iraq, in the foothills of the Zagros Mountains. This area was inhabited approximately 9,000 years ago; it was a permanent farming village with about 24 mud-walled houses and a population of about 150 people. The charred grains of wheat and barley found there were a domesticated type already changed from the wild type; for this genetic change to have taken place, initial cultivation must have predated the age of the remains. In addition to the domesticated plants and animals, the inhabitants of Jarmo also continued foraging, as evidenced by the bones of wild animals, snail shells, acorns, and pistachio nuts. Artifacts uncovered at the site include flint sickles and grinding stones for harvesting and milling cereal grains as well as clay figurines, woven baskets, and rugs.

The domestication of cats apparently took place earlier than previously thought. Most authorities had placed their domestication in Egypt around 4,000 years ago. A cat skeleton discovered on the island of Cyprus indicates that people there cherished the domesticated cat. The 8-month-old cat (*Felis silvestris*) was buried in a small grave next to a large grave containing a human skeleton. The joint burial suggests a special association between human and cat. Cats were most likely valued in agricultural societies to protect grain stores from mice.

Recent reports suggest that central China, not the Near East, may be the site of the domesticated cat based on analysis of the bones of cat fossils dating to 5,300 years ago. The dominant crop grown in the region at this time was millet, which belongs to a group of plants that utilize the C_4 pathway in photosynthesis (see Chapter 4), resulting in a signature form of carbon isotope in the glucose produced. The bones of wild animals do not show this unusual form of carbon isotope, but the bones of domesticated cats, which fed on the rodents that ate the millet, do. It is believed that the cats were domesticated to control the rodent population attacking the millet granaries.

A genomic comparison of domesticated and wild cat species has identified 13 genes associated with the so-called *domestication syndrome*. The domestication syndrome refers to the observation that domesticated animals across species share characteristics (white spotted coats, floppy ears, juvenile appearance, and reduced fear of humans). Five of the 13 genes are involved in embryonic development and affect pigmentation and the tissues of bone and cartilage that shape faces. Also some of these genes result in smaller adrenal glands in the domesticated cat species. Adrenal glands produce and release stress hormones. Reducing the size of adrenal glands would lower the level of stress hormones in animals and may account for the reduced fear of humans, a necessary step in the domestication of wild species.

Several sites have been postulated as the site for the domestication of the dog from wolf populations. In one study, researchers examined 48,000 SNPs (or single nucleotide polymorphisms) (see A Closer Look 7.2: Try These Genes On for Size) from 85 different breeds of dogs as well as 11 different populations of grey wolves from across the world. From the analysis, it was discovered that all dog breeds are more closely genetically related to wolf populations from the Middle East than to wolves from anywhere else in the world.

Other researchers have debated these data and point out new findings that doglike animals were present in Europe and Siberia at least 30,000 years ago. There is also new evidence that suggests that the ancestral wolf species that gave rise to the domesticated dog has long been extinct and is not traceable to any existing species of wolf.

Some researchers question several of the archeological findings at some ancient sites of early agriculture. Nicole Miller, an expert in archaeobotany at the University of Pennsylvania Museum of Archaeology and Anthropology, studies sites in the Near East for evidence of plants used by people during ancient times. She documented a 5,000-year-old site in southwestern Iran that showed an increasing presence of charred wild seeds. In previous analyses, the presence of wild seeds at the sites of ancient cook fires was interpreted as spilled food and hence indicated that the plants were consumed by ancient people. Miller has another view. She explains that, as the surrounding forests were cleared for fuel and pastureland, the inhabitants turned to the dung of animals as cooking fuel, and thus, the presence of charred wild seeds reveals more about the human impact on the environment than their diet. Similarly, she disagrees with the conclusion that the diversity of wild seeds points to a varied diet for the inhabitants of an 11,000-year-old site in Syria. She deduced that the cooking fuel used was the dung of gazelles and that the wild seeds found were a food source for an animal that was plentiful at this location in the past.

The Far East

Excavations of dozens of sites in Asia indicate that agriculture arose at several locations in the Far East. These sites include the Yellow River and Yangtze River valleys in China (fig. 11.4). Among the earliest plants domesticated in the Far East were rice, foxtail millet, broomcorn millet, rape, and hemp, with evidence of domesticated cattle, pigs, and poultry.

Current archeological studies in China indicate that rice cultivation began approximately 11,500 years ago along the middle reaches of the Yangtze River. From there, agriculture spread both upstream and downstream. Among the artifacts examined were samples of rice grains, husks, plant remains, and impressions of rice grains in pottery. If these dates were confirmed, this would mean that the domestication of rice predates the domestication of crops in the Near East by about 1,500 years. In settlements to the north, along the Yellow River, foxtail millet was domesticated approximately 8,000 years ago and became a dietary staple; it remained the dominant crop in North China until the historical period. Broomcorn millet was also cultivated but not as extensively. Tilling tools, harvesting tools, and grain-processing tools made of stone, bone, shell, or wood have also been recovered from these sites along with bones of domesticated animals.

Archeological evidence pinpoints the Far East, specifically China, as the site for the domestication of an insect, the silkworm, more than 5,000 years ago. For millennia, the Chinese strictly prohibited the export of silkworms to other countries. Sequencing the genome of the domesticated silkworm (*Bombyx mori*) and comparing it with 20 other domesticated silkworm varieties, as well as 11 populations of wild silkworms (*Bombyx mandarina*), has revealed valuable information about the domestication process. The domesticated silkworm is genetically distinct from the wild species and, although there is some genetic variability between the domesticated varieties of silkworm, there is much less than among the wild populations. It appears that 354 genes may be associated with the domestication process in the silkworm. Domesticated silkworms have been bred for several traits: increased cocoon size, faster growth and reproduction rates, improved digestive efficiency (eating mulberry leaves for silk production), tolerance of human handling, crowding, loss of predator avoidance behavior, and the inability to fly in the moth stage. These traits have rendered the domesticated silkworm unable to survive in the wild.

A new study identifies Central Asia as the site of horse domestication more than 5,000 years ago. The Botai were hunter-gatherers who lived on the plains of Central Asia in what is present-day Kazakhstan. Researchers have long thought that the Botai domesticated horses, which they rode when hunting; however, the Botai did not grow crops or domesticate any other animals. An analysis of Botai pottery identified residues of animal fat which were identified as mares' milk. Modern villagers in the region milk horses to make koumiss, a fermented alcoholic beverage. Additional evidence of domestication comes from the skeletal remains of horses obtained at several Botai sites, showing tooth and cheek damage indicative of horses that had been driven or ridden with a bit and bridle.

Additional evidence strongly indicates that agriculture may have begun independently in the Papua New Guinea highlands nearly 7,000 years ago. The Kuk Swamp site in the Waghi Valley of central New Guinea shows several indications of early agriculture. Analysis of soil sediments for pollen and phytolith microfossils (see A Closer Look 11.1: Forensic Botany) indicates that the area was subjected to accelerated forest clearing. Deforestation led to the establishment of a grass-sedge swampland. Archeological remains of stakeholes and postholes were also found and indicate planting, digging, and supporting posts, all clearly associated with cultivation. Vestiges of soil mounding and deep channels have been interpreted as practices to improve drainage in the grassy wetland. Evidence of taro and banana, historically two of the most important tropical food crops, has also been found at Kuk Swamp. The percentage of banana phytoliths is higher than would be expected in the area and probably indicates their cultivation. The researchers conclude that this species of banana, previously thought to have been domesticated in Southeast Asia, may have instead been domesticated first in New Guinea and later dispersed to Southeast Asia. Microfossils of taro, including starch grains and needlelike calcium oxalate crystals (see A Closer Look 11.1: Forensic Botany), were recovered from stone tools. Finding a lowland crop, such as taro, in the highlands confirms that this crop was deliberately introduced into this site. All evidence points to New Guinea as one of the primary centers of agriculture.

The New World

In contrast to the Old World, the **Neolithic** (New Stone Age, or agricultural society) cultures of the New World had domesticated an impressive array of plants but comparatively few animals. Among the earliest crops domesticated in the New World were squash, corn, chili peppers, amaranth, avocado, gourds, beans, and both white and sweet potatoes, with only dogs, turkeys, llamas, alpacas, guinea pigs, and Muscovy ducks as domesticated animals. Most of the initial archeological evidence for early agriculture in the New World has been obtained from the highlands of Mexico and Peru.

Probably the most thoroughly documented site is a group of caves in the Tehuacan Valley of central Mexico. Research in this area was initiated to obtain information on the ancestry of corn. Working with ancient corncobs, archeologists determined that corn had been domesticated in this region by 5,500 years ago (fig. 11.5). Originally, it was thought that these corncobs were much older (7,000 years old), but newer dating techniques advanced the time frame. The Tehuacan Valley is one of the few sites where the transition from foraging to farming can be thoroughly documented. For thousands of years, people inhabited the caves seasonally as they foraged for plants and hunted animals. At first, there was a shift from hunting to a more intensified foraging of plants, along with the domestication of squash and avocado. Later, the list of domesticated plants expanded to include corn, bottle gourd, two species of squash, amaranth, three species of bean, and chili peppers; however, people still relied on foraging for the majority of their diet. Over the next few thousand years, agriculture became even more important as additional plants (tomato, peanut, and guava), the dog, and later the turkey were domesticated. Artifacts such as stone tools, textiles, and pottery were also found. From this evidence, archeologists have been able to reconstruct a picture of the lifestyles of the inhabitants of these caves over a 12,000-year period. At yet another cave in southern Mexico, Guilá Naquitz in Oaxaca, seeds and the fruit rind of a domesticated squash have been dated to between 10,000 and 8,000 years ago. This date of domestication is much earlier than what has been obtained for maize, beans, or any other New World domesticate. These findings indicate that farming appeared in the Americas at more or less the same time that it did in the Fertile Crescent and Asia.

Dolores Piperno of the Smithsonian Tropical Research Institute has been a leading investigator into the origins of agriculture in lowland tropical soils of Central and South America. Using phytolith data (see A Closer Look 11.1: Forensic Botany), she has discovered that local squash species were independently domesticated in coastal Ecuador between 9,000 and 10,000 years ago, perhaps slightly predating squash domestication in the upland regions of Mexico.

Recently, evidence from microfossils has revealed that chili peppers were domesticated more than 6,000 years ago in South America. Starch grains of domesticated chili peppers differ in appearance from those of wild species. The starch grains of domesticated chili peppers have been preserved on grinding stones, on ceramic shards of cooking vessels, and in soil sediments from archeological sites in Central America, South America, and the Bahamas, with the oldest sites in Ecuador.

Much attention has been focused on the early site of agriculture in the eastern forests of North America. On the basis of the latest evidence, the eastern half of what is now the United States and Canada was another New World center of plant domestication, developing independently at least four domesticated seed crops: sunflower, marsh elder, goosefoot, and wild gourd. These farming societies flourished for some 2,000 years before the arrival of maize and other domesticated crops from the Southwest in A.D. 1–200 (fig. 11.4).

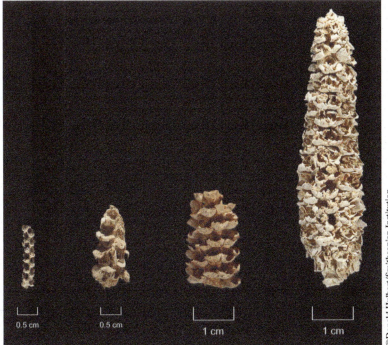

Figure 11.5 Preserved corncobs from the Tehuacan Valley, Mexico. The oldest cob, on the left, is 2.8 cm (approximately 1 in.) long.

CHARACTERISTICS OF DOMESTICATED PLANTS

Plants that have been **domesticated** are genetically distinct from their wild progenitors. Through the process of **natural selection** (see Chapter 8), wild plants have evolved mechanisms that ensure their survival in the environment, but once a plant has been domesticated, traits are **artificially selected** to suit human needs and do not necessarily have a survival value. In fact, some of these traits might be detrimental to survival in the natural environment. For example, modern corn, with its ensheathing husks, cannot disperse its seeds; also, domesticated wheat and other cereals have fruiting heads that are nonshattering, a trait that limits seed dissemination.

Most wild grasses have **shattering** fruiting heads, which will break apart at a slight touch or breeze and scatter their seeds over a wide area. A recessive gene is responsible for a tough spike with a **nonshattering** head. It would be natural for early foragers to gather those seeds attached to the tougher spikes. When agriculture began, the seeds most easily gathered would be planted and so pass on the nonshattering trait. In 2006, researchers pinpointed the exact gene mutation in domesticated rice that codes for the nonshattering fruiting head. By comparing the DNA sequences of domesticated rice with its wild relatives with the shattering trait, researchers found a single gene on chromosome 4 with one difference in the nucleotide bases. The wild rice species had a G (guanine) that had been replaced by a T (thymine) in domesticated rice. After the researchers genetically engineered domesticated rice and restored the G, the shattering trait was also restored.

Likewise, early foragers selected for larger seeds, fruits, or tubers, and over time the domesticated varieties became larger than their wild counterparts. For example, wild barley has two rows of grains, while the domesticated varieties have six rows (fig. 11.6). In fact, archeological evidence of six-row barley is indicative of agriculture at that excavation. Loss of seed dormancy is another general characteristic of domesticated plants. Seeds of most wild plants are dormant. For example, wild seeds formed at the end of autumn typically do not germinate immediately. Instead, germination is delayed via chemical means (hormonal control) or mechanical means (a tough seed coat) until the next spring, when the environmental conditions are favorable for seedling growth and survival.

> **Thinking Critically**
>
> Domesticated plants are genetically different from their wild relatives.
>
> *The fruits of wild cherry trees are sour tasting, yet the domesticated varieties derived from wild populations produce deliciously sweet cherries. Explain the difference between wild and domesticated cherries by explaining the difference between artificial and natural selection.*

Figure 11.6 Two-row wild barley (left) is contrasted with six-row domesticated barley (right).

Other changes commonly seen in domesticated plants when compared to their wild relatives include a change in the plant's life cycle from perennial to annual and a reduction in protective structures (thorns, spines, or trichomes) and chemical defenses (toxins). Additionally, environmental conditions for flowering such as daylength (photoperiod) have been removed and many domesticated plants will both flower and fruit during the same season unlike their wild relatives. Plant breeders continue to select for desired traits today, using traditional as well as more sophisticated genetic manipulations (see Chapter 15).

CENTERS OF PLANT DOMESTICATION

Within each area of the world where agriculture evolved, the native peoples developed indigenous crops for a staple food supply. Crops that were particularly suitable for agriculture slowly spread to surrounding regions as people traded with others or migrated to new areas, taking their crops with them. This diffusion led to the emergence of principle crops associated with major centers of the world. In the Near East, wheat and barley were the dietary staples; in the Far East, rice; in Africa, sorghum and millet; in Mesoamerica, corn; and in South America, the potato and other root crops. As civilization continued to develop, trading and migration expanded the range of crops far from their origin, and today many crops are even more successful outside their native range. Potatoes, which became associated with the Irish, are actually native to the highlands of Peru; coffee, actually native to the mountains of Ethiopia, is most frequently linked to Colombia and Brazil; and the tomato, so essential to Italian cuisine, was first domesticated in the New

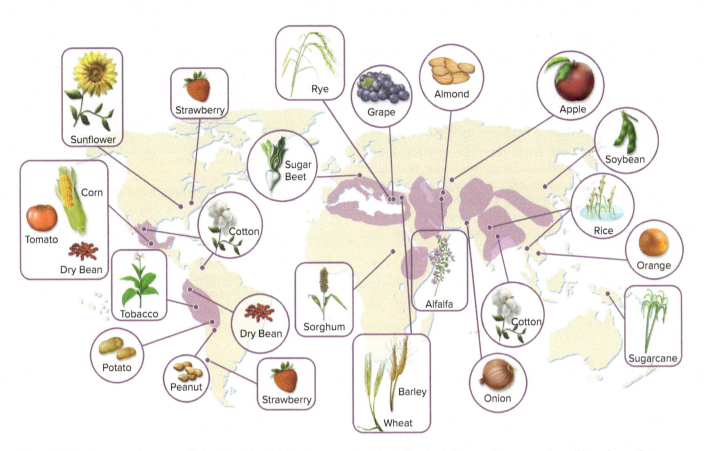

Figure 11.7 Centers of origin as first defined by N. I. Vavilov are indicated by the shaded areas. More recent work has shown that some crops were domesticated outside these areas or were domesticated independently in several regions.

World. The agricultural harvest of the United States would be meager if limited to commercial crops of native origin, such as blueberries, cranberries, sunflower, pecan, and maple syrup.

Pinpointing the exact origin of important crops has intrigued scientists for many years. The name most frequently associated with this endeavor is Nikolai I. Vavilov, a Russian botanist. Vavilov directed plant-collecting expeditions around the world and examined thousands of plants, looking for patterns of variation in crop plants and their wild relatives. He reasoned that areas that had the greatest diversity of a particular crop would most likely be the center of origin for that crop. On the basis of his research from 1916 to 1936, Vavilov proposed eight centers of origin for the major domesticated plants, six in the Old World and two in the New World. Examples of crops known to have originated in these centers are indicated in Figure 11.7. Vavilov's life ended tragically in a Soviet gulag. Biology in the Soviet Union under Stalin was dominated by Trofim Lysenko, who rejected established genetic theory in favor of his own outdated views. Vavilov's adherence to Mendelian genetics clashed with the established policies of the Soviet state, and he was sentenced to a Soviet prison camp, where he died in 1943, a martyr to the cause of scientific freedom.

Later work has expanded the number of centers and questioned Vavilov's conclusions. Evidence suggests that, although some crops have been domesticated more than once in different places, others did not originate where Vavilov indicated and still others were developed over vast regions. For example, certain New World crops, such as cotton and cassava, appear to have been independently domesticated in both Mesoamerica and South America. This search for the origin of certain crops is even more important today as plant geneticists strive to improve the gene pool of domesticated plants by tapping the genetic resources of wild strains (see Chapter 15).

CHAPTER SUMMARY

1. The first human societies were based on a foraging lifestyle, gathering wild plants and hunting animals for food. Archeological investigations have determined that the diet of Stone Age foragers was varied, especially in the variety of plants consumed. Studies on the diet and lifestyle of extant foragers, especially the !Kung San of the Kalahari Desert, reinforce the viewpoint that the foraging lifestyle more than satisfies the nutritional requirements yet allows time for activities not directed to food gathering and preparation.

2. The archeological record indicates that, at least in some parts of the world, certain groups began to shift from the nomadic, foraging lifestyle to the sedentary one of agriculture.

3. The earliest agricultural settlements, approximately 11,500 years old, have been found in the Far East along the Yellow and Yangtze River valleys. Sites in the Near East, in an area known as the Fertile Crescent, also document early agriculture. The New World dates of domestication from

the Tehuacan and Oaxaca valleys of Mexico show that agriculture started there at approximately the same time.

4. Domesticated plants and animals are genetically different from their wild relatives because they have been shaped by artificial selection. Many of the traits, such as nonshattering fruiting heads, do not enhance a plant's survival value but have been selected to suit humanity's needs.

5. Because domesticated plants have been modified greatly from their wild ancestors after thousands of years of artificial selection, it has often been difficult to pinpoint their area of origin. Nikolai I. Vavilov laid the foundation for detecting the centers of origin of domesticated plants when he proposed eight centers, six in the Old World and two in the New.

REVIEW QUESTIONS

1. What has been learned about foraging societies of the past by studying the !Kung?
2. Describe an early agricultural community in the Near East.
3. How do archeologists reconstruct diets of prehistoric peoples?
4. How have the theories about the origin of agriculture changed in recent decades?
5. What crops were domesticated in the New World? the Far East? the Near East?
6. You are an archeologist on a dig and discover in a cave the preserved seeds and fruits of a type of squash. These remains are dated to between 8,400 and 10,000 years old. What characteristics would you look for to determine whether this squash was a wild type, gathered by foraging, or a domesticated plant, cultivated in the fields?
7. A national seed company offers a reward of $10,000 to the first gardener who develops a pure white chrysanthemum. How would you go about breeding a chrysanthemum for its color? What is the underlying process and its mechanism of action?

©SafakOguz/iStock/Getty Images

CHAPTER 12

The Grasses

Grains of bread wheat, Triticum aestivum, *which is one of the most widely cultivated cereals in the world, supply a major percentage of the nutrient needs of the human population.*

KEY CONCEPTS

1. Grasses are members of the monocot family Poaceae, whose characteristic grains are a vital food source.
2. Whole grains with the bran and germ intact are nutritionally superior to their refined counterparts, which contain only endosperm.
3. Wheat, corn, and rice, the major cereals, outrank all other plants as food sources for human consumption.
4. Grasses are also indispensable components of forage crops and landscaping designs as well as major sources of biofuels.

CHAPTER OUTLINE

Characteristics of the Grass Family 188
    Vegetative Characteristics 188
    The Flower 188
    The Grain 188
Wheat: The Staff of Life 189
    Origin and Evolution of Wheat 190
    Modern Cultivars 191
    Wheat Genome 191

A CLOSER LOOK 12.1 The Rise of Bread 192

    Nutrition 193
Corn: Indian Maize 193
    An Unusual Cereal 194
    Types of Corn 194

A CLOSER LOOK 12.2 Barbara McClintock and Jumping Genes in Corn 196

    Hybrid Corn 197
    Ancestry of Corn 197
    Corn Genome 199
    Value of Corn 199
Rice: Food for Billions 200
    A Plant for Flooded Fields 200
    Varieties 201
    Rice Genome 202
    Flood-Tolerant Rice 202
Other Important Grains 202
    Rye and Triticale 202
    Oats 204
    Barley and Tritordeum 204
    Sorghum and Millets 204
Other Grasses 205
    Forage Grasses 205
    Lawn Grasses 205
Bioethanol: Grass to Gas 206
    Corn 206
    Sugarcane 207
    Cellulosic Ethanol 207
Chapter Summary 208
Review Questions 209

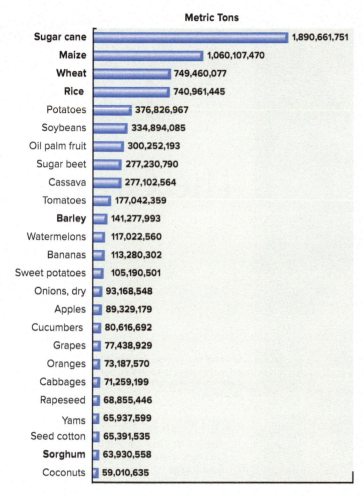

Figure 12.1 Annual world crop production figures (in metric tons) reveal that six of the top 25 crops are grasses (in boldface type).
Source: FAO Production Yearbook, 2016.

The grass family is of greater importance to humanity than any other family of flowering plants. The edible grains of cultivated grasses, or **cereals**, are the basic foods of civilization, with corn, rice, and wheat the most extensively grown of all food crops. Other important cereals are barley, sorghum, oats, millet, and rye; most are among the top 25 food crops (fig. 12.1).

CHARACTERISTICS OF THE GRASS FAMILY

There are approximately 780 genera and close to 12,000 species of grasses found throughout the world, making the grass family one of the largest and most widely distributed plant families. Grasses are the dominant plants in prairies and savannas but can also be found wherever plants can grow, under a wide variety of environmental conditions from arctic marshes to tropical swamps. In fact, 25% of the world's vegetation belongs to the grass family.

Vegetative Characteristics

Members of the grass family, the Poaceae, are usually herbaceous, having linear leaves with parallel venation typical of monocots (fig. 12.2). Leaves usually have an alternate arrangement, and the base of each leaf forms a sheath that wraps around the stem. The stems, or **culms**, are often hollow between the nodes and usually unbranched. Many species also have horizontal stems (either aboveground **stolons** or underground **rhizomes**) that can propagate the plant vegetatively by giving rise to new shoots (see Chapter 14). Both annual and perennial species of grasses occur, with most cereals being annuals and most pasture and lawn grasses being perennials. The primary root system is fibrous, and adventitious roots may also form either from the lower nodes of erect stems (as **prop roots**) or from rhizomes or stolons.

The Flower

The flowers of grasses are borne in inflorescences, typically spikes, racemes, or panicles. The tassels of corn and the heads of wheat are common examples of grass inflorescences. The individual flowers are small, inconspicuous, and incomplete, with sepals and petals lacking completely or replaced by small structures called lodicules (fig. 12.2). Each flower normally has three stamens, and the gynoecium has a single ovule in the ovary but two styles and stigmas. The stigmas are enlarged and feathery, and the mature stamens pendant; these features facilitate wind pollination. Surrounding each flower are two bracts, the outer **lemma** and inner **palea**; the flower and the two bracts together make up a **floret**. One to 12 florets are arranged on a **spikelet**, which also may be subtended by two bracts called **glumes**. Often, a slender bristle can be seen extending from either a glume or a lemma (occasionally the palea); this structure is known as an **awn**.

The Grain

The typical fruits for the grass family are **grains**, which are dry, single-seeded indehiscent fruits (fig. 12.3). The bracts that surrounded the flower (or even the spikelet) now surround the grain and are called **chaff**. The outer wall of the grain, consisting of the fruit wall fused to the seed coat, is known as the **bran**. Interior to the bran is a layer of enlarged cells known as the **aleurone layer**, which is normally high in protein. If the seed is allowed to germinate, this layer provides the enzymes that break down stored food for the growing embryo. The majority of the seed is occupied by endosperm, which contains stored food mainly in the form of starch. The cotyledon transfers food to the embryo, which is surrounded by the coleoptile and coleorhiza sheaths. The embryo, with its sheaths, is often referred to as the **germ**.

The large amount of stored food in the grain makes this family valuable as a food crop. In the economically important cereals, the endosperm is mostly starch, and in the refining process, the chaff, germ, and bran (usually with the aleurone layer attached) are removed, leaving only the starchy endosperm. Commercially, this refined product is available as white flour, corn starch, and white rice. In whole-grain products,

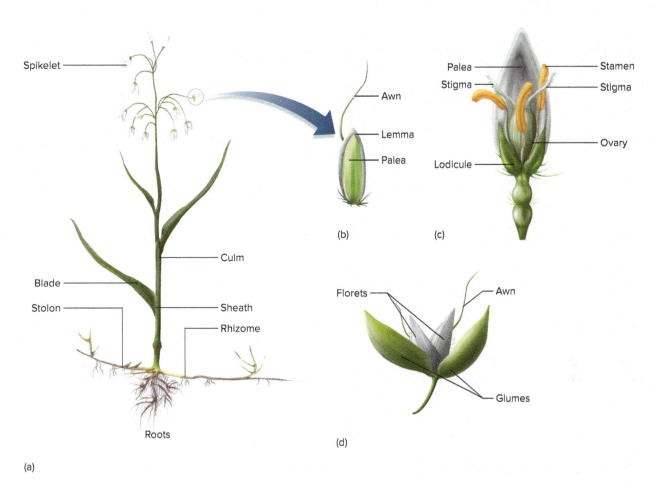

Figure 12.2 Typical grass plant. (a) Whole plant. Some grasses can reproduce vegetatively through stolons or rhizomes; however, the major cereal crops lack these structures. (b) Each flower or floret is surrounded by a lemma and palea. (c) The grass flower consists of three stamens and one carpel with two separate styles and stigmas. (d) Glumes subtend each spikelet that bears one or more florets.

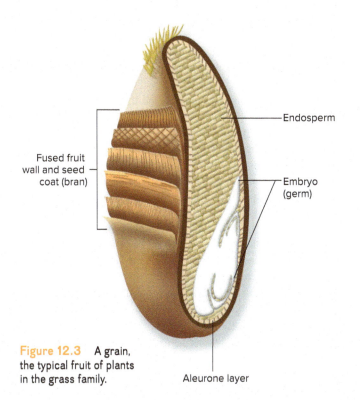

Figure 12.3 A grain, the typical fruit of plants in the grass family.

only the chaff is removed and the entire grain is used, providing certain nutritional advantages over the refined material. The bran provides fiber as well as some protein from the aleurone layer, and the germ is a source of vitamins, proteins, and some oils. Thus, brown rice, whole-wheat flour, and even popcorn are more nutritious than refined grains. Their nutrient status is important because these three grains, whether whole or refined, are the dietary staples for the world's population, providing more than 50% of the Calories consumed and the chief sources of both carbohydrates and proteins.

WHEAT: THE STAFF OF LIFE

Wheat is one of the most widely cultivated cereals in the world and supplies a major percentage of the nutrient needs of the human population. Bread is literally the "staff of life" in many cultures. The vegetative appearance of the wheat plant is typical of most grasses, but the spikes are tightly packed with grains, which usually have long awns, giving the fruiting head a bearded appearance (fig. 12.4). Wheat does best in temperate grassland biomes that receive about 30 to 90 cm (12 to 36 in.) of rain a year and have relatively cool temperatures. Some of

Figure 12.4 Wheat, one of the most widely cultivated cereals in the world.

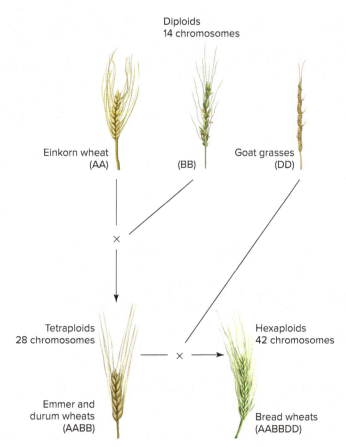

Figure 12.5 Evolution of domesticated wheat.

the top wheat-producing countries are China, India, Russia, the United States, Canada, France, Ukraine, and Pakistan. Wheat is one of the oldest domesticated plants, and it laid the foundation for Western civilization. Domesticated wheat had its origins in the Near East at least 9,000 years ago, and several wild species of wheat can still be found in northern Iraq, Iran, and Turkey.

Origin and Evolution of Wheat

The origin of domesticated wheat is a story of hybridization and polyploidy involving several species of wheat, *Triticum* spp. and species of closely related goat grasses in the genus *Aegilops*. (In some sources, goat grasses are also considered members of the genus *Triticum*; however, molecular studies indicate they are different genera.) In contemporary wheat species, three groups can be identified on the basis of chromosome number: one group has 14 chromosomes; the second group has 28; and the remaining group has 42 chromosomes. This forms a polyploid series with a base chromosome number of 7 ($n = 7$) including diploid species (14 chromosomes), tetraploid species (28 chromosomes), and hexaploid species (42 chromosomes).

Triticum monococcum is a diploid species of wheat known as einkorn wheat. It was one of the first cultivated species of wheat, and wild forms are still found. The domesticated form differs from the wild type in that the grains are nonshattering (they tend to stay on the stalk) in domesticated einkorn. Most of the other species of wheat are polyploid hybrids that arose naturally through crosses with species of goat grass, which also have a diploid chromosome number of 14. A related wild species of wheat, *T. urartu,* which is also diploid and also called einkorn wheat, is one of the ancestors of our modern bread wheat.

It is believed that *Triticum urartu,* hybridizing with one of the goat grasses, *Aegilops speltoides,* gave rise to the tetraploid emmer wheat, *Triticum turgidum* (fig. 12.5). To fully understand the development of this tetraploid species, consider the diploid chromosome status of einkorn to be represented by AA and that for the goat grass BB (fig. 12.5). A hybrid between them would be AB with 14 separate chromosomes, and if that hybrid doubled its chromosomes, it would be AABB, the chromosomal makeup of tetraploid wheat. Tetraploid wheat evolved naturally long before the origin of agriculture. Molecular studies suggest that this may have occurred more than 100,000 years ago, and wild species of emmer are still found in the Near East. Early societies in this area were cultivating emmer along with einkorn wheat. Domesticated emmer has nonshattering grains that are covered by clinging bracts (hulled), but other, later varieties of tetraploid wheats, such as durum, which is widely grown for pasta flour, have naked grains. Naked grains separate easily from the surrounding bracts, a trait referred to as free threshing.

The next stage in the evolution of wheat was the result of another hybrid cross, this time between an emmer wheat and another goat grass (*Aegilops tauschii*). Crossing the tetraploid AABB with the diploid goat grass DD produced the hybrid ABD. This hybrid doubled its chromosomes, thereby forming the hexaploid AABBDD, which appeared at least 8,000 years ago in the Near East (fig. 12.5) and today is known as bread wheat, *T. aestivum*. The genetic contributions of *A. tauschii* were a higher protein content in the endosperm and greater tolerance to environmental conditions. Among the proteins in the wheat grain is **gluten**, an important component of flour for the elasticity it provides. This elasticity in wheat flour, coupled with a leavening agent (yeast, baking soda, or baking powder), allows dough to rise, making it suitable for breads and cakes (see A Closer Look 12.1: The Rise of Bread). While the gluten content is important for baking, it is a problem for people with gluten sensitivity. Recall from Chapter 10 that individuals with celiac disease must follow a lifelong gluten-free diet. This means eliminating all wheat, rye, and barley from their food. Current wheat breeding programs are attempting to solve this problem by developing varieties of wheat with reduced gluten levels.

Modern Cultivars

The two types of wheat that are widely cultivated today are durum and bread wheat. Durum wheat (*T. durum,* also called *T. turgidum* subspecies *durum*) is grown in the northern United States (especially North Dakota), Canada, southern Europe, and parts of India. It has a high gluten content and yields semolina flour, which is used to make spaghetti, macaroni, and noodles. *Triticum aestivum,* bread wheat, is the dominant type of wheat grown, making up approximately 90% of the world production. The flour from bread wheat is used for bread, pastries, and breakfast cereals. There are thousands of cultivars of wheat, enabling the crop to be grown under a variety of environmental conditions.

Wheat cultivars can be categorized by their growing conditions and their protein content. Hard wheat has a higher protein content (higher gluten), and the flour is usually used in bread making. Soft wheat has lower protein, yielding a soft flour that is better for pastries. Hard wheats are generally grown in areas with limited rainfall; greater moisture is needed to grow soft wheat. Planting time is another criterion. Spring wheat is planted in the spring and is harvested in the fall; winter wheat is planted in the fall, it overwinters (requiring a cold period before flowering), and then it is harvested the following spring or summer. In areas where the winter is very severe, spring wheat is grown. In North America, these areas include North and South Dakota, Minnesota, Montana, and Canada as far north as the Arctic Circle. South of this area, winter wheat is planted, thereby dividing the wheat-growing regions of the continent into winter and spring wheat belts.

New cultivars are constantly being developed to improve characteristics such as disease resistance and yield. Wheat is highly susceptible to many diseases, particularly stem rust of wheat caused by the fungal pathogen *Puccinia graminis* (see Chapter 23). This fungal disease goes back millennia; rust epidemics were recognized in ancient Greece and Rome. The most recent concern is Ug99, a race of *P. graminis* that was first detected in Uganda in 1999 and has since spread to other countries in Africa and the Middle East. This is an especially virulent race of the pathogen, and 90% of the wheat grown around the world is vulnerable to it. Plant pathologists and wheat researchers from many countries are working to develop varieties of wheat resistant to original Ug99 and newer variants. Promising new lines of wheat have been developed and others are in progress. Unfortunately, it takes several years to develop sufficient seeds for widespread planting. Some resistant wheat varieties have been developed; however, they only represent 5% to 10% of the wheat grown in the areas of Africa and Asia with Ug99. In addition, these plants are not necessarily resistant to all variants of the pathogen. Plant pathologists agree that wheat breeders must develop varieties of wheat with several resistant genes. In 2013, researchers identified a resistant gene in *T. monococcum* and another in *A. tauschii.* Transferring these genes to bread wheat may help develop stable resistance to Ug99 and all variants. Furthermore, a new aggressive strain of a related fungus, *Puccinia striiformis,* which causes a similar disease, stripe rust of wheat, has also been causing problems over the past few years in wheat-growing areas around the world.

In addition to rust resistance, present-day crosses between wheat and goat grasses are also yielding new varieties that show resistance to other fungal pathogens. High-yielding varieties of wheat developed since the 1970s have made a major impact on world food supplies and have turned some wheat-importing countries into wheat-exporting ones (see Chapter 15). Researchers need to stay constantly vigilant to prevent fungal pathogens from destroying these gains.

Wheat Genome

Although *Triticum aestivum*, bread wheat, is one of the most important crops in the world, it is the last of the major cereal crops to have its **genome** (the entire genetic information, DNA, of an organism) completely sequenced. To resolve the genome of an organism, the DNA is extracted and then processed into small pieces for sequencing (see A Closer Look 7.2: Try These Genes on for Size). Once sequenced the fragments are then assembled into longer and longer segments until the complete list of nucleotides is determined. The difficulty in achieving the full sequence of the bread wheat genome is due to the large size of the genome and its polyploid origin. The genome contains 17 billion base pairs, which is more than five times larger than the human genome. As described earlier, bread wheat is a hexaploid species derived from hybridization of three diploid species, and the resulting genome is a composite of three similar subgenomes A, B, and D. The similarity complicates the sequencing efforts because it is difficult to distinguish genes from the subgenomes.

Researchers from many countries have being working on deciphering the wheat genome since 2004. Over this time, several groups of researchers have each taken different

A CLOSER LOOK 12.1

The Rise of Bread

Bread is the basic food for many cultures and often supplies more than half of the dietary Calories. Breads can be as different as the cultures that produce them: the corn tortillas of Mexico, chapatis of India, Scandinavian crisp breads, croissants of France, pumpernickels of Germany, and Jewish bagels. These are among the thousands of types of bread that are basically made from mixing flour with water to make a dough. Any type of starchy meal can be used to prepare a dough that can be baked into a bread; however, the cereals are the foremost source of bread flours, and wheat flour is most commonly used for leavened bread (box fig. 12.1).

Making a leavened bread requires flour that contains sufficient quantities of gluten. Gluten is a complex of proteins consisting largely of gliadin and glutenin. If flour containing gluten is mixed with water, the dough becomes elastic. When yeast is added and undergoes **fermentation** (anaerobic respiration), the resulting carbon dioxide is trapped as small gas bubbles, stretching and expanding the elastic dough. The dough rises and, when baked, results in leavened bread. Other leavening agents include baking powder and baking soda, which chemically produce the carbon dioxide bubbles. Of all the cereal grains, only wheat and rye have sufficient gluten to produce a leavened bread, and wheat is preferred for its higher gluten content. In preparing gluten-free breads or pastry, xathan gum or guar gum is usually added; both gums are polysaccharides that can replace the elastic properties of gluten and allow the bread or pastry to stick together and to rise. Xathan gum is produced by the bacterium *Xathomonas campestris* when growing on sugar and is widely used as a thickening and stabilizing agent in prepared foods. During growth, the bacterium produces a polysaccharide gel that can be extracted, dried, and pulverized into a powder. Guar gum is produced from the ground endosperm of guar beans, *Cyamopsis tetragonoloba*, a member of the Fabaceae (bean family). Like xathan gum, this polysaccharide is also widely used as a thickening and stabilizing agent.

The Egyptians are credited with discovering leavened bread using wheat flour almost 4,000 years ago. Of all the grains used by the Egyptians, only wheat flour had the potential to produce a leavened bread (rye was unknown to the ancient Egyptians). At the time, grains had to be parched

Box Figure 12.1 Bread is the staff of life.

or toasted in a fire before threshing. Applying heat to the grain made the glumes easy to remove but also changed the gluten, making it inelastic and unable to trap any carbon dioxide. The resulting meal produced only flat breads. A new free-threshing form of wheat that could be threshed without heat set the stage for the discovery of leavened bread. It is assumed that yeast (from the air or possibly even from some beer added for flavor) was accidentally introduced into a dough prepared from these unparched grains (see Chapter 23). If the dough was set aside for a time before baking, it would rise and so produce a lighter, tastier bread.

These early breads were prepared from whole grains that were coarsely ground or milled on grinding stones. Flour milled in this way was coarse, with chaff, bran, germ, and even small pieces of grinding stone. As milling and sifting techniques improved over the centuries, the flour became more and more refined, and by the nineteenth century, iron roller mills had replaced stone mills. The roller mills further refined the flour by removing the nutrient-rich germ and more of the bran. Thus, the refined flour was mainly starchy endosperm. The advantage of refined flour was a longer shelf life, since the oils from the germ in stone-ground flour became rancid within a few weeks. Unfortunately, the lower nutritional value of the resulting bread was detrimental for people relying on bread for a large portion of their diet.

approaches, and each group has contributed to the task. One group, the International Wheat Genome Sequencing Consortium (IWGSC), includes geneticists, wheat breeders, and growers from many countries. The IWGSC resolved to sequence individual chromosomes. In July 2014, the IWGSC published a draft survey of the gene content on each chromosome. By comparing the bread wheat genes with its closest relatives that represent the A, B, and D subgenomes, the researchers also found that none of the subgenomes dominated gene expression. In August 2018, the IWGSC published the complete annotated genome sequence of bread wheat. The newly finished genome identified over 107,000 genes.

Wheat breeding programs in many parts of the world are striving to improve wheat productivity. Information from the wheat genome will aid researchers identify genes for desired traits. This should speed up the development of new strains that are more productive than today's varieties and more resistant to drought, pathogens, and pests.

Nutrition

Compared with the other major cereals, wheat is a nutrient-rich food. In terms of the whole grain, wheat is missing only four of the known essential nutrients; those absent are vitamins A, B_{12}, and C and iodine. The protein content is good, depending on the variety, with an average of 12.9%, but recall that cereal grains have incomplete proteins because the lysine and tryptophan contents are low. The nutrients, however, are not evenly distributed in the grain; many of the nutrients are concentrated in the bran and germ. Although eating the whole grain provides the greatest nutritional benefits, most of the wheat now consumed in the United States is refined; with bran and germ gone, many of the nutrients are lost (tables 12.1 and 12.2). To compensate somewhat for this nutritional deficit, refined white flour is often enriched with iron and four B vitamins (thiamine, riboflavin, folic acid, and niacin). However, the consumption of whole-wheat breads and cereals is on the increase, a good sign that Americans are becoming more aware of the nutritional benefits of whole grains.

CORN: INDIAN MAIZE

When the Europeans came upon the New World late in the fifteenth century, they found that the dietary staple was a cereal unknown to them and called maize by the indigenous peoples. Later to be named *Zea mays* by Linnaeus, this crop was grown from southern Canada to southern South America and formed the basis of the New World civilizations. (Although commonly called corn in the United States, this name is ambiguous. In many countries, *corn* refers to the most commonly grown cereal: in England, *corn* refers to wheat, and in Scotland it refers to oats. In North America, the terms *corn* and *maize* have been used interchangeably; this practice will continue in this text.)

Table 12.1
Nutrients in Whole, Refined (Unenriched), and Enriched Wheat Flours per Cup

	Whole Wheat (120 g)	White Unenriched (125 g)	White Enriched (125 g)
Calories	400	455	455
Protein (g)	16.0	13.1	13.1
Fat (g)	2.4	1.3	1.3
Carbohydrates (g)	85.2	95.1	95.1
Calcium (mg)	49	20	20
Phosphorus (mg)	446	109	109
Iron (mg)	4.0	1.0	3.6
Potassium (mg)	444	119	119
Thiamine (mg)	0.66	0.08	0.55
Riboflavin (mg)	0.14	0.06	0.33
Niacin (mg)	5.2	1.1	4.4

Source: Data from *USDA Handbook No. 456.*

Table 12.2
Vitamins and Minerals Lost during Refining of Wheat

Nutrient	Lost (%)
Cobalt	88.5
Vitamin E	86.3
Manganese	85.8
Magnesium	84.7
Niacin	80.8
Riboflavin	80.0
Sodium	78.3
Zinc	77.7
Thiamine	77.1
Potassium	77.0
Iron	75.6
Vitamin B_6	71.8
Phosphorus	70.9
Copper	67.9
Calcium	60.0
Panthothenic acid	50.0
Molybdenum	48.0
Chromium	40.0
Selenium	15.9

Source: Data from Henry A. Schroeder, Losses of Vitamins and Trace Minerals Resulting from Processing and Preservation of Foods, *American Journal of Clinical Nutrition* 24: 562–573, 1971.

An Unusual Cereal

European herbalists of the sixteenth century were puzzled by the appearance of corn when they compared it with their more familiar cereals. Corn is much larger than other cereals and is unusual for having separate staminate and carpellate inflorescences (fig. 12.6). The tassel at the apex of the stalk is the staminate inflorescence arranged in a panicle, with each floret consisting of three stamens surrounded by bracts. The carpellate inflorescence, a thickened spike, is borne on a lateral stalk and gives rise to the familiar ear of corn. The grains (kernels), which can be of various colors, are naked (they lack bracts around each grain), but the entire ear is tightly covered with specialized bracts known as husks. The silks can be seen beneath the husk; each silk is actually the style and stigma of an individual carpellate flower. Corn is poorly adapted for survival under natural conditions because its ensheathing husks totally prevent seed dispersal. The closely packed kernels in a fallen ear (even if they germinate) would produce seedlings under such intense competition that few would survive. Modern corn literally could not survive without human intervention.

Types of Corn

Several types of corn are grown today and can be characterized mainly by the nature of the starch present in the endosperm. Starch has two components, **amylose** (an unbranched chain of glucose molecules) and **amylopectin** (a highly branched chain of glucose molecules). Hard starch has a higher percentage of amylose than does soft starch.

The types of corn are classified as popcorn, flint, flour, dent, sweet, waxy, and pod (fig. 12.7a). Most of these were actually cultivated by the Native Americans before the Europeans reached the New World. One of the oldest types, and possibly the most primitive, is popcorn, whose kernels swell and burst when heated. This trait was useful for primitive peoples because heating made the grains edible without the need for arduous grinding.

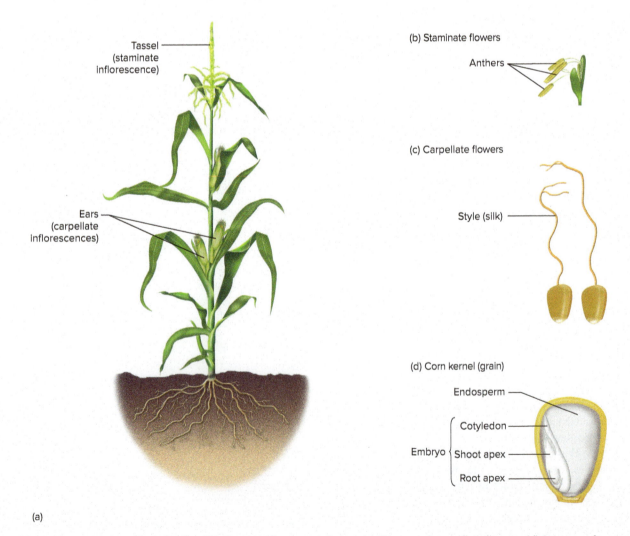

Figure 12.6 Corn. (a) View of whole plant. (b) Close-up of staminate flower. (c) Close-up of carpellate flowers. (d) Anatomy of a corn kernel.

Flint corn also has hard starch near the outer part of the kernel and was the predominant corn grown by the Native Americans in northern areas of North America when the European settlers arrived. Flour corn, also known to the Native Americans, is similar in appearance to flint corn but has a softer endosperm that makes it easier to grind and prepare a dough by mixing with water. However, the softer endosperm makes the kernels more susceptible to insect damage. Both flint and flour corn are no longer grown extensively, having been replaced by dent corn. Dent corn has both hard and soft starch, with the hard starch along the sides and the soft restricted to the top and center. Upon drying, the kernel shows a characteristic dent as the soft starch shrinks. Today, dent corn is the most widely grown type in the Corn Belt, which encompasses most of middle America, and is primarily used for animal feed, corn starch, and corn meal.

Sweet corn contains a high concentration of sugar instead of starch in the cells of the endosperm. The sweet-tasting kernels are a popular vegetable, either fresh on the cob, canned, or frozen. Waxy corn is a relatively new kind of corn that first appeared in China and was introduced into the United States early in the twentieth century. It is named for the glossy, waxlike appearance of the cut kernels that is due to the presence of only amylopectin in the endosperm. Pod corn is a rare type of corn that occasionally appears in a field of corn. Unlike the other forms of corn, each kernel in pod corn is covered with glumes (fig. 12.7a). It is a botanical curiosity and is not grown commercially. Some botanists believe that pod corn is a primitive type of corn that may be ancestral to the modern ones.

Corn is a summer annual that grows best under moderate conditions of temperature and moisture, but it is cultivated under more diverse environmental conditions than any other crop. An average temperature of 23°C (73°F) or above, with lots of sunshine, and 37 to 50 cm (15 to 20 in.) of rain spread over a 3- to 5-month growing season are ideal. Corn requires a nutrient-rich soil, and fields under cultivation for several years become depleted. The most frequently grown types of corn have thousands of cultivars, each with specific traits. For example, varieties of corn could be selected for early maturation, color of the kernels, size of the kernels, size of the ears, dwarf stalks, extra sweet taste (for sweet corn varieties), resistance to certain pests, and so on.

The color variety of the kernels was a trait that impressed the European settlers. Kernel color is a complex characteristic that depends on pigmentation of the endosperm, aleurone layer, and pericarp. The familiar yellow kernels result only from yellow pigment in the endosperm; this pigment is apparently lacking in the white varieties. The multitude of colors and patterns visible in Indian corn used for autumn decorations result from pigments in the aleurone layer and pericarp (see A Closer Look 12.2: Barbara McClintock and Jumping Genes in Corn). Although both of these layers can be colorless, the pericarp may be red, orange, brown, or variegated, and the aleurone layer can be various shades of red, blue, purple, bronze, and brown. The mixture of colors from the endosperm, aleurone, and pericarp results in the kernel's

(a)

(b)

Figure 12.7 Variations in corn. (a) Prinipical varieties of corn (from left to right): popcorn, sweet corn, flour corn, flint corn, dent corn, and pod corn. (b) The multitude of colors and patterns visible in Indian corn result from pigments in the aleurone layer, endosperm, and pericarp.

Popcorn has extremely hard kernels with hard starch surrounding a core of soft starch. The endosperm cells in the center contain a large percentage of water; upon heating, the water turns to steam, building up pressure in the kernel. At a certain point, the whole kernel explodes and turns inside out. The moisture level inside the kernel is critical for successful popping, with the optimum moisture content between 13% and 14.5%.

During the colonial period, New Englanders often ate popcorn served with milk and maple sugar for breakfast, and it has been suggested that popcorn was served at the first Thanksgiving. By the mid-nineteenth century, popcorn was considered more of a snack food and remains popular to this day. In fact, in recent decades "gourmet popcorn" has become fashionable, with varieties ranging from chocolate to taco flavored. With the high fiber of a whole grain, popcorn is healthier than other snack foods; however, butter and other toppings can increase the fat and Calorie content.

A CLOSER LOOK 12.2

Barbara McClintock and Jumping Genes in Corn

For many reasons, corn has been a major research tool for geneticists:

- Corn is easily grown and displays a great deal of variability.
- Each kernel represents a different genotype, so that a whole population exists on a single ear.
- Corn has large chromosomes.
- Chromosomal mutations can be easily studied with a light microscope.

Even in beginning biology classes, the inheritance of kernel color can illustrate basic genetic principles. One of the most intriguing aspects of corn genetics was the discovery of jumping genes by Dr. Barbara McClintock (box fig. 12.2).

Barbara McClintock (1902–1992) was born in Hartford, Connecticut, and received her doctorate in botany from Cornell University in 1927. Most of her professional life was spent at Cold Spring Harbor Laboratory on Long Island, New York. This research center is famous for its significant work in genetics, molecular biology, and biochemistry as well as virology and cancer studies. Most of Dr. McClintock's research involved corn genetics and the behavior of chromosomes, and during her long career she made many important discoveries. She was particularly interested in the inheritance of complex color patterns in Indian corn. McClintock concluded that the color patterns she observed were possible only if genes could move around from chromosome to chromosome. These jumping genes, called transposable elements (or transposons), are fragments of chromosomes that move at random from one chromosomal position to another. When a transposon inserts on a chromosome, it alters or controls the normal expression of the neighboring genes.

McClintock reported preliminary results of her work on these transposable genes in the late 1940s and published major articles in 1950, 1951, and 1953. Her work was viewed with a great deal of skepticism because it contradicted the prevailing theory that chromosomes consisted of genes in fixed positions. In addition, many scientists had difficulty accepting the controlling or regulatory nature of these genes. Vindication came with reports on regulatory genes in bacteria in the 1960s and transposable genes in bacteria in the 1970s. Today it is known that transposons occur in almost all organisms, both prokaryotic and eukaryotic, and they are often abundant constituents of the genome. In addition, several different types of transposons

Box Figure 12.2 Indian corn with red- and white-striped kernels. The color and pattern are evidence of transposons (jumping genes) in the genetic makeup of the corn.

occur; most contain genes for one or more proteins whose only function is to move or copy the transposon to a new location.

In addition to her discovery of jumping genes, Barbara McClintock is also recognized as the first geneticist to consider the existence of epigenetic inheritance. Recall from Chapter 7 that epigenetics is the study of inherited changes that are not caused by changes in the DNA sequence. McClintock showed an inherited but reversible change of gene expression in corn during mitosis in genetically identical cells, thereby recognizing that genes could be silenced. This observation was made before the structure of DNA was described and decades before the mechanisms of epigenetics were determined.

Widespread recognition of Dr. McClintock's work finally came in the 1980s with several awards, culminating in the Nobel Prize in Physiology or Medicine for 1983. Barbara McClintock was the first woman to receive the Nobel Prize in this category for work done on her own. The Swedish institute that names the Nobel laureates in medicine compared the significance of her research to that of Gregor Mendel's. They said her studies

> reveal a whole world of previously unknown genetic phenomenon [sic] . . . She was far ahead of the development in other fields of genetics. . . . Her most important results were published before the structure of the DNA double helix and the genetic code had been discovered.

appearance (fig. 12.7b). A black kernel results from a dark red pericarp over a blue or brown aleurone, and a greenish kernel results from a blue aleurone over yellowish endosperm.

Hybrid Corn

Corn has always been an important crop in the New World and is the most widely grown crop in the United States. The majority of corn grown today is hybrid corn. The major characteristics introduced into the hybrids include two to three ears per stalk (as opposed to one ear, which was typical of most varieties at the beginning of the twentieth century) for greater productivity and stronger stalks with standard positioning of the ears for easier mechanical harvesting. These improvements have been so valuable that the planting of hybrid corn increased from only 1% in 1930 to virtually 100% today.

Attempts to increase corn production early in the twentieth century led to the development of **hybrid** corn. Hybrid corn results from crossing inbred lines; these hybrids are hardier because of hybrid vigor, or heterosis. Inbred lines consist of genetically homozygous plants that are produced by self-fertilization for several generations. Each inbred line reliably produces certain desired traits, but inbred lines are, unfortunately, weaker and less productive with each generation. Crossing different inbred lines results in offspring that have the desirable traits of both parents as well as restored hybrid vigor (fig. 12.8). Because the original hybrid seeds were produced on inbred lines with small ears and, therefore, a small number of seeds, commercial production was impractical. Seed production was improved by making crosses between two single hybrids to produce a vigorous double hybrid that had a large number of kernels and still possessed the desired traits. Today, the inbred lines are improved, and it is possible to produce abundant seeds from a single hybrid cross. A disadvantage of planting hybrid corn is the need to purchase new hybrid seeds each year. On the other hand, a new enterprise was created, as some farmers have specialized in planting the inbred lines to produce the hybrid seeds.

In the production of hybrid seeds, a row of one inbred line (to serve as the female parent or seed parent) is planted next to a different inbred line (that serves as the male or pollen parent). The staminate inflorescence must be removed or detasseled to prevent self-pollination of the female line. This labor-intensive practice was the standard for many years; however, male-sterile lines were developed later that made detasseling unnecessary. A drawback to the male sterility method became apparent in 1970 when a new strain of a fungal pathogen, *Bipolaris maydis* (*Cochliobolus heterostrophus*), that causes Southern leaf blight appeared. Linked to male sterility was an increased susceptibility to this disease. By this time, 70% to 90% of the corn grown in the United States had been developed from seeds that had a male-sterile parent and carried the increased susceptibility. Disaster followed when the blight struck and destroyed approximately 15% of the U.S. corn crop. In some states the devastation was even greater, with 50% loss reported (see Chapter 15). Since that time, other male-sterile lines have been developed that are not as susceptible. Also, detasseling has made a comeback with some seed companies.

Ancestry of Corn

Unlike wheat, whose ancestors are fairly well identified, corn's origin has been a botanical mystery that has been the subject of speculation and controversy for a great many years. Wild wheat can still be found in the Near East, but no obvious equivalent exists for corn anywhere in the natural environment of central Mexico where it was domesticated between 6,000 and 10,000 years ago.

One school of thought, first proposed by George W. Beadle in 1928, holds that teosinte is the ancestor of modern corn. Teosintes are wild grasses native to Mexico and Central America that share several characteristics with corn. Both teosinte and maize have terminal staminate inflorescences and lateral carpellate spikes. Teosinte, a much smaller plant, has multiple stalks with a tassel at the apex of each stalk, in contrast to the single stalk of modern corn (see fig. 15.2). While a single corn plant normally forms only a few large ears, each with multiple rows of kernels, teosinte produces numerous small ears, each with 6 to 10 kernels (fig. 12.9). The kernels are triangular in outline and have a very hard outer fruit case. This fruit case surrounds the kernels in teosinte but is reduced to a cupule found at the base of each kernel in modern corn and remains attached to the corn cob. The spike of teosinte shatters easily at maturity to disseminate the kernels, very different from *Zea mays*.

In addition to sharing some vegetative similarities, corn and teosinte are closely related species that are able to hybridize and form fertile offspring. They have the same number of chromosomes ($2n = 20$), and chromosomes in the hybrids pair normally during meiosis. Beadle's hypothesis states that only five mutations would have been necessary to change teosinte into modern corn with the two major ones being

1. a mutation to a nonshattering spike and
2. a mutation to a soft or reduced fruit case.

In fact, an occasionally seen teosinte mutant (tunicate form) actually has a reduced fruit case with the kernels covered by soft glumes. This tunicate form also has less of a tendency to shatter. In breeding experiments carried out by Beadle and coworkers, crosses between teosinte and modern corn produced plants with small, primitive ears similar to ones found in 7,000-year-old archeological specimens.

The opposing school of thought had been led by Paul C. Mangelsdorf, who first suggested an alternative ancestry in 1939. Mangelsdorf believed that both modern corn and the annual teosinte were descended from a cross between an ancestral wild corn, a primitive pod popcorn now extinct, and *Z. diploperennis*, a diploid perennial teosinte discovered in 1979. (Although another perennial teosinte has been known much longer, it is a tetraploid, and crosses with corn result in sterile triploid hybrids.) In the F_2 generation of experimental crosses between *Z. diploperennis* and a primitive Mexican popcorn, there were a

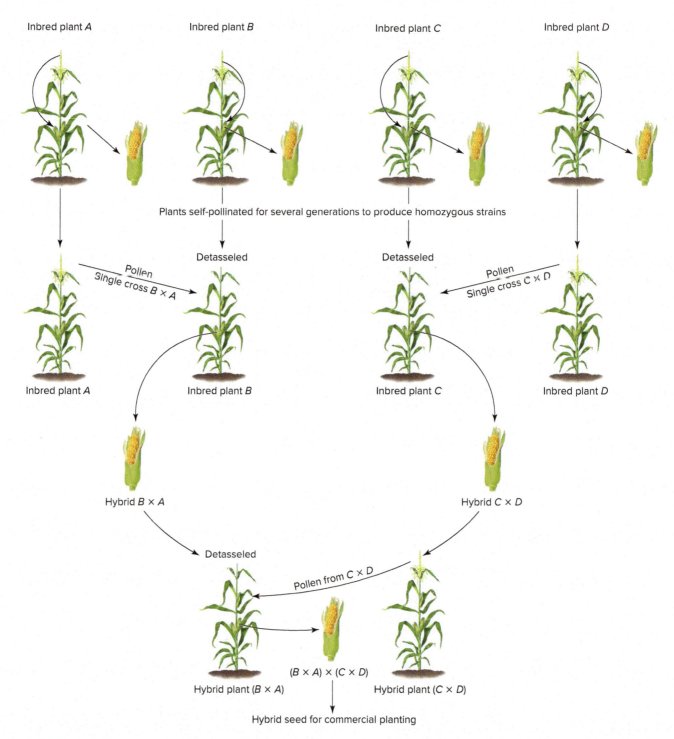

Figure 12.8 Hybrid corn is essential for commercial seed production. Often the double-cross method is used, incorporating genes from four inbred strains.

variety of fertile hybrids, some of which resembled annual teosinte and others, modern corn. Mangelsdorf saw this result as proof of his hypothesis. (In a previous version of this hypothesis, Mangelsdorf had suggested another grass, *Tripsacum*, as the genus that crossed with the wild pod popcorn; however, genetic studies and pollen analysis discredited this view.)

Molecular research with several species of teosinte has shed more light on the origin of maize. John Doebley of the University of Minnesota examined three perennial and three annual teosintes found in Mexico. Two of the annuals (*Zea mays* subspecies *mexicana* and *Zea mays* subspecies *parviglumis*) appeared most similar to maize, and consequently, their genetic profiles were examined for a comparison with that of maize (*Z. mays*). This analysis revealed that the subspecies *parviglumis* is essentially indistinguishable from maize and according to this researcher, confirms that teosinte,

Figure 12.9 Teosinte, *Zea diploperennis*. Immature ear on left; in center, ear is cut open; on right, mature fruit cases with grain visible on bottom.

specifically the subspecies *parviglumis,* is the undisputed ancestor of modern maize.

Scientists are trying to determine what specific gene changes occurred during the evolution of modern maize from teosinte. In 1983, Hugh Iltis suggested that the ear of corn evolved not from the slender female spike of teosinte but rather from the central spike of the male tassel. Iltis suggests that this morphological change came about by means of a "catastrophic sexual transmutation." Molecular evidence suggests that five major gene changes could differentiate maize from teosinte, with the major gene identified as teosinte branched 1 (*tb 1*), which controls plant architecture, especially lateral branch development. In maize, this gene causes an increase in apical dominance by repressing the growth of axillary organs and also enables the formation of female inflorescences.

Although the molecular evidence is compelling, the controversy continues. Another researcher has produced a hybrid between *Tripsacum* and *Zea diploperennis* that seems to show characteristics of primitive corn.

Corn Genome

In November 2009, collaborators from numerous U.S. institutions completed sequencing the *Zea mays* genome. The genome consists of 2.3 billion base pairs and includes more than 39,000 genes, about 18,000 to 20,000 more than the human genome. For this 4-year endeavor, the researchers used an inbred line of corn called B73, which is the most important parent line for hybrid corn; descendents of B73 are widely planted in the United States and the rest of the world. In a companion study, Mexican researchers reported on the sequencing of an ancient Mexican landrace of corn, known as *Palomero,* and compared it with B73.

Among the findings was confirmation that about 85% of the corn genome consists of transposons dispersed across the genome (see A Closer Look 12.2: Barbara McClintock and Jumping Genes in Corn). The research also found that the genome shows tremendous diversity, with some genes present in only one line. Many genes were also identified that are not found in other higher plants. Researchers believe that the corn genome can provide valuable information to improve corn productivity by scanning the genome for genes that allow corn to be successful under drought conditions or with reduced fertilizer use. This sequencing achievement will have major implications for advancing basic research as well, since corn has been a model organism for genetic and chromosome studies for the past 100 years.

Value of Corn

Much of the corn grown in the United States is used as animal feed. Only a small part of the corn harvest is eaten directly as a vegetable, with most processed commercially for food products or industrial applications, including biofuel production (fig. 12.10). As food for either humans or livestock, corn is a good nutrient source of carbohydrates, fats, and proteins; however, the protein content is lower than that of wheat (7%–10%), and like other cereals, corn is low in lysine and tryptophan. Corn geneticists have been trying to improve the protein balance in corn and have produced some varieties with higher lysine content. Corn is also deficient in niacin, and what little niacin is present is in an unavailable form. Human diets based largely on corn can lead to the deficiency condition pellagra (see Chapter 10).

Corn starch, corn meal, corn flour, corn oil, and corn syrup are all processed from corn kernels. These products make their way into thousands of prepared foods, so that the average American is consuming corn one way or another in almost every meal. Corn-based breakfast cereals and snacks are prevalent in the American marketplace, with the snack foods alone accounting for over a $2 billion share annually. Corn oil, largely polyunsaturated, is obtained from the germ and used as salad oils, cooking oils, salad dressing, and margarine. USDA researchers in Iowa have developed varieties of corn that are high in oleic acid, a monounsaturated fatty acid. It is hoped that the commercialization of these varieties will result in the production of heart-healthy corn oil similar to olive oil and

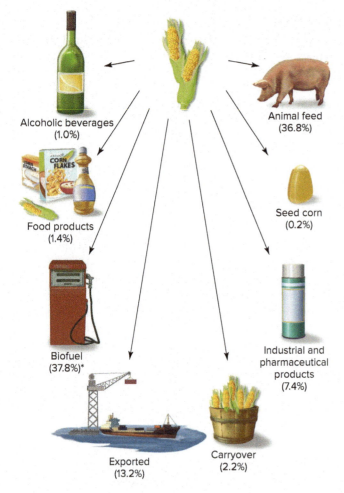

Figure 12.10 The multiple destinies of the U.S. corn crop, based on projected use of 2017 harvest.

*This percent includes a by-product of ethanol biofuel production known as distillers grain. This is the leftover mash from fermentation, which is used as a high-protein animal feed supplement.

Thinking Critically
Researchers spent decades trying to identify the ancestor of modern corn and believe that teosinte is the likely plant.

What is the value of knowing the ancestor of corn or other domesticated crops?

RICE: FOOD FOR BILLIONS

Rice feeds more people worldwide than any other crop, and it is estimated that more than 2 billion people, mainly in Asia, rely on rice as a dietary staple. The oldest evidence of rice cultivation dates back more than 8,000 years in eastern China with some researchers suggesting a separate domestication in northern India. *Oryza sativa* is the main cultivated rice, although over 20 species in the genus are known. The only other cultivated species of note is *O. glaberrima,* which is grown in West Africa. (The grasses known as wild rice in North America, *Zizania aquatica* and *Z. texana,* are unrelated to cultivated rice and are, in fact, still mostly harvested from the wild by Native Americans.)

Oryza sativa is believed to have originated from a wild ancestor, *O. rufipogon* in lowland tropical areas that were subject to periodic flooding, and it is under these conditions that rice is still most productive. By contrast, rice is now a worldwide crop, able to grow under many different environmental conditions, with 11% of the world's arable land devoted to rice cultivation. It was introduced into North America early in the colonial period and soon became an important crop in the Carolinas. For about 200 years, it was confined to the southeastern states, spreading to Louisiana, Texas, Arkansas, and California in the late nineteenth and early twentieth centuries. Although the United States produces less than 2% of the rice grown, it is one of the world's largest rice exporters.

A Plant for Flooded Fields

The rice plant is a large, multistalked annual growing to an average height of approximately 1 meter (3 feet), with the typical vegetative appearance of grasses (fig. 12.11a). Each stalk is terminated by a panicle bearing grains surrounded by bracts. A distinctive characteristic of rice is the presence of air chambers in the stem that permit the diffusion of air from stomata in the leaves through the stem and eventually down to the roots. This adaptation, seen in many aquatic plants, permits rice to survive in flooded or waterlogged soils.

Although rice can be grown like other cereals without flooding (upland rice), in most areas of the world it is cultivated in flooded fields or paddies with 5 to 10 cm (2 to 4 in.) of standing water (lowland rice). This method of growing rice is an ancient one, dating back several thousand years. In fact, rice farmers in some parts of Asia are known as "farmers

canola oil (see Chapter 10). Besides its use directly in food, corn starch is used as the base for producing corn syrup, which has become one of the most popular sweeteners in prepared foods (see Chapter 14). In addition, corn is the base for fermented beverages (see Chapter 24), including chicha (a South American beer) and bourbon (corn mash whiskey). Industrial uses of corn starch are almost limitless, with laundry starch, pharmaceutical fillers, glues, lubricants, ethanol for use in biofuel, and biodegradable plastics and packing materials among the many products. During the past two decades, the industrial uses of corn have increased dramatically, especially ethanol for fuel, creating many new markets for corn growers.

Corn breeding programs are aimed at creating high yields; varieties adapted to infertile soils, saline soils, or drought; and varieties resistant to insect pests. The insect-resistant varieties were developed through genetic engineering and are now widely planted in the United States. The use of these varieties has made headlines over concerns about their safety in our food supply and in the environment (see Chapter 15).

Figure 12.11 Rice, *Oryza sativa*, is the dietary staple for over 2 billion people. (a) Close-up of fruiting stalks. (b) Rice paddy in Indonesia.

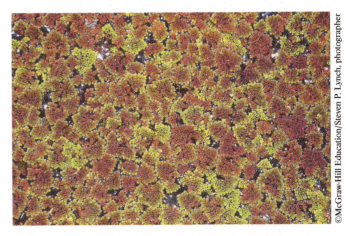

Figure 12.12 *Azolla* was originally noticed as a weed in rice paddies, but today it is deliberately introduced to cut down on the need for expensive fertilizer.

of 50 centuries." The paddies are diked with earthen dams and filled by rain or irrigation (fig. 12.11b). Young seedlings started in seedbeds are transplanted, usually by hand, into the paddies.

One weed of the rice paddies is known to be beneficial for cultivation. This weed is *Azolla*, a small, aquatic fern that is inhabited by a symbiotic nitrogen-fixing cyanobacterium, *Anabaena azollae* (fig. 12.12). Recall from Chapter 9 that symbiosis is the intimate association of two species living together in a relationship that can be beneficial to one or both organisms. Nitrogen-fixing species are able to convert atmospheric nitrogen into a form of nitrogen that can be utilized by plants (see Chapter 13). Since all plants require nitrogen compounds for growth, this natural source of nitrogen reduces the need for costly fertilizers. Chinese and Vietnamese rice farmers observed that the rice crop improved when this weed was present and used it as a "green manure" for centuries, even though they were unaware of the nitrogen fixation that was occurring.

When the grains are almost mature, the fields are drained to prepare for harvesting. Traditionally, harvesting has been done by hand using sickles. Threshing follows to free the grain from the outer bracts or chaff; **winnowing** then separates the grain from the fragments of chaff. In the United States and other industrialized countries, all stages of rice production—from seeding (by airplane in some farms) through processing of the grain—are highly mechanized, as opposed to the labor-intensive method still employed in developing nations.

Brown rice is the dehulled whole grain and is nutritionally superior to white rice, which has been milled and polished to remove the bran and germ. Brown rice contains more protein (8.5% to 9.5%) than white rice (5.2% to 7.6%) and more vitamins, especially thiamine and vitamin B_1. As pointed out in Chapter 10, a diet based on polished white rice can result in beriberi. Unfortunately, polished white rice is preferred for its taste, quicker cooking, and longer shelf life.

Varieties

There are thousands of varieties of *O. sativa*, which differ in growing conditions or in the color, shape, size, aroma, flavor, and cooking characteristics of the grain. These varieties can be grouped into two major subspecies, *indica* and *japonica*, with a third, *javanica*, (also called tropical japonica), that is not as widely cultivated. The *indica* varieties are primarily grown in the tropics, and they produce long grains that do not stick together when cooked. By contrast, *japonica* varieties, which are grown in cooler subtropical to temperate regions, have short grains that are sticky when cooked. The stickiness of the cooked grains is dependent on the composition of starch in the endosperm; drier grains have more amylose. *Javanica* varieties are cultivated in Indonesia and are characterized by tall, thick stems and large grains. In the United States, *japonica* varieties are grown in California, while southern states grow varieties that are intermediate between *japonica* and *indica* (medium-grain rice). New varieties have been developed that are high yielding, disease resistant, and early maturing. Research has recently led to the development of "golden rice," varieties that are able to synthesize beta-carotene in the endosperm (see Chapter 15). Recall that this pigment is the precursor to vitamin A. It is believed that the widespread use of golden rice will reduce vitamin A deficiency among the poor in developing countries and thereby reduce the incidence of blindness caused by this deficiency (see Chapter 10).

Rice Genome

In 2002, the sequencing of the rice genome was announced by two groups of researchers: one group, from the Beijing Genomics Institute, sequenced a strain of *indica* rice, and the second group, from the biotech company Syngenta, sequenced a strain of *japonica* rice. The sequencing of the rice genome was given priority because it has the smallest genome of any of the major cereal crops—430 million nucleotides. The corn genome is five times larger, and the wheat genome is 40 times larger; however, preliminary comparisons suggest a great deal of similarity with the rice genome. This similarity makes rice an excellent model for characterizing the genes of cereal crops and identifying those of interest to agriculture. It is hoped that knowledge of the rice genome will help scientists develop new varieties that are resistant to diseases, insect pests, salinity, and drought and thereby improve the productivity of all cereals, the most important foods for humanity.

Flood-Tolerant Rice

Although rice is often grown in paddies with a few inches of standing water, most varieties cannot tolerate complete submergence for more than 3 days. Unfortunately, about 20% of the rice-growing areas of the world are susceptible to flooding from monsoons, particularly in low-lying regions of South and Southeast Asia. Approximately 4 million tons of rice are lost each year due to flooding. In the mid-1990s, David Mackill and other researchers from the International Rice Research Institute (IRRI) in the Philippines identified a variety of rice from India that is resistant to flooding; it can tolerate complete submergence for up to 17 days and then renew growth when the water has subsided.

While the discovery was good news, this variety has low yields. For several years, IRRI geneticists used traditional plant breeding to cross this flood-tolerant rice with locally grown varieties that had higher yields. The results were disappointing. Although the hybrids were flood tolerant, the yields remained low, and the taste was not acceptable to the farmers. Pamela Ronald, a plant pathologist from the University of California at Davis working with the IRRI researchers, determined that the ability to tolerate submergence is due to the presence of genes named *Sub 1*. Using precision breeding (a modern form of breeding that requires knowledge of molecular markers—see Chapter 15), these researchers were able to transfer *Sub 1* into the high-yielding local varieties. Field tests in India and Bangladesh showed that these varieties perform very well and taste the same as traditional types. Research continues transferring the *sub 1* gene into other local varieties, so that submergence-tolerant rice will soon be widely available for planting in flood-prone regions.

> **Thinking Critically**
>
> The domestication of wheat, rice, and corn formed the basis for civilizations in the Near East, the Far East, and the New World.
>
> *What characteristics of these cereal crops are so valuable that they formed the basis for civilization?*

OTHER IMPORTANT GRAINS

Rye and Triticale

It is believed that rye first came to human attention as a weed of cultivated wheat and barley fields about 5,000 years ago in southwestern Asia. It has always been noted for its hardiness, especially to cold and drought. Its ability to thrive in marginal areas was even noted by the Greek botanist Theophrastus, who reported the commonly held belief that wheat grown on poor soil would turn to rye. As wheat cultivation spread through Europe, weedy rye thrived in the colder regions and eventually became cultivated instead of wheat. *Secale cereale* is the only cultivated species of rye, although several wild species of *Secale* are known. The plant resembles wheat, but the spikes are slender and more elongate (fig. 12.13a). The grains yield a flour that is suitable for making leavened bread, although the gluten content is lower than that of wheat and the bread is somewhat soggy and heavy. During the Middle Ages, cultivation of rye became widespread in northern Europe and gave rise to the familiar black bread of the peasants. Rye breads are still very popular in Sweden, Poland, Germany, and Russia. Because of the lower gluten content, rye flour is commonly mixed with wheat flour to prepare commercial "rye breads" sold in the United States today.

Crosses between wheat and rye have been made for over a 100 years to produce an intergeneric hybrid known as *Tritosecale,* or triticale (fig. 12.13b). Although the hybrids produced in the nineteenth century were sterile, techniques developed since the 1930s have permitted the breeding of triticale varieties that produce viable seeds. Much interest has centered on these hybrids, which combine desirable characteristics of each parent: the hardiness, disease resistance, and better protein quality of rye and the higher yield of wheat. Varieties of triticale have been developed that have a protein content equal to that of wheat (average 12.9%) but higher than that of rye (average 10.75%). Although rye has a lower protein content than wheat, it has a better biological value because the lysine content is higher, and the lysine content of triticale approaches that of rye. One disappointment of triticale has been its poor performance as a bread flour. Although gluten is present, it tends to break up when the dough is kneaded, producing a bread that is heavy and not springy. Consequently, triticale flour is usually mixed with wheat flour for breads.

Figure 12.13 Other commercially important grains include (a) rye, (b) triticale, (c) oats, (d) barley, (e) pearl millet, and (f) sorghum.

Oats

The inflorescence of the cultivated oat plant, *Avena sativa*, has a distinctive appearance when compared with other cereals. It is a branched panicle with an open, delicate appearance (fig. 12.13c); the spikelets are long, with up to seven florets (two is common). Because the oat grain is eaten unrefined, it is highly nutritious, with close to 15% protein and a good mix of vitamins, minerals, and oil. Although the exact place and time of its domestication are unknown, *A. sativa* was being cultivated in Europe around 4,500 years ago. Oats do well in moist, temperate climates and possess some degree of cold hardiness. In addition to *A. sativa*, there are several other cultivated and wild species of the genus.

Oats have always been considered a good food for horses but historically have had mixed acceptance as food for humans. While some societies, such as the ancient Romans, considered it only a food for animals, others developed many dishes based on this grain. This is especially true in Scotland, where every celebration calls for an oat-based food. In the United States, oats were primarily eaten as a breakfast food until the mid-1980s, when it was suggested that oat bran has cholesterol-lowering properties. The soluble fiber that makes oatmeal so gummy is the effective component in the bran (see Chapter 10). To cash in on this health claim, food companies began adding oat bran or oats to many processed foods and even creating novel oat bran products.

Barley and Tritordeum

Barley is one of the oldest domesticated crops and was brought into cultivation, along with wheat, approximately 9,000 years ago in the Near East. The cultivated barley plant, *Hordeum vulgare*, is similar in appearance to wheat, with long awns that give barley a bearded appearance (fig. 12.13d). Although other species of *Hordeum* exist, they are of minor importance for cultivation. Although the spikes of some barleys have only two rows of grains, most cultivars today are the six-row type, with spikelets in threes on alternate sides. Barley can grow under a wide range of environmental conditions, tolerating cold temperatures, high altitude, low humidity, and saline soils.

Although barley was an important food for the ancient peoples of the Mediterranean region, today most barley is used as animal feed, with about one-third of the crop used in the production of malt for brewing beer (see Chapter 24). A small percentage of barley is polished to make pearl barley, a common ingredient in vegetable soups. Whole barley grains contain 13% protein, but the protein level drops to 8% during the refining of pearl barley. Although direct human consumption of barley is almost insignificant, the importance of barley for livestock and in the brewing industry accounts for its fourth place among cereal crops and eleventh place among the world's crops. The International Barley Genome Sequencing Consortium published the barley genome sequence in 2012. The genome is about 5 billion bases and an estimated 32,000 genes. Researchers believe this information will help produce new varieties of barley that will be better able to adapt to higher temperatures and drought and be resistant to various barley pathogens and pests.

A new cereal crop was developed by Spanish plant breeders by crossing durum wheat (*Triticum durum*) and wild barley (*Hordeum chilense*). This intergeneric crop known as tritordeum (genus: *Tritordeum*) was produced through traditional plant hybridization and is hexaploid. Research began in the late 1970s, and in 2013 tritordeum reached the market place in a variety of breads, cakes, other pastries, cereals, pasta, and pizza. It is also being used in brewing. Tritordeum was first grown in Spain and Italy, but by 2016 it was also cultivated in Portugal, France, and Turkey. In addition, tritordeum flour is now milled and sold in a several other European countries. Tritordeum products have a slightly sweeter taste than wheat products, a nutty flavor, and a soft texture. The flour has high protein and high fiber, and it is rich in total carotenoids, especially lutein, giving the flour a yellow color. Tritordeum has seven to ten times more lutein than durum wheat and bread wheat, respectively; genes from the wild barley parent are the source of the higher lutein levels. Lutein is an antioxidant pigment believed to be involved in eye health although the exact role is still unknown. Tritordeum has lower gluten levels than wheat and may be a good option for those wishing to reduce dietary gluten; however, it is still not suitable for those with celiac disease.

In terms of growing qualities, early varieties of tritordeum had lower yields than durum wheat; however, 30 years of plant breeding have created high-yield cultivars. Today several hundred lines of tritordeum are available for planting. Tritordeum is drought resistant and heat tolerant, and relatively disease resistant. Because it requires less water, it has a low environmental impact.

Sorghum and Millets

Sorghum and millets are cereal grains that are seldom used as food for humans in North America, but they are important staples in other parts of the world (figs. 12.13e and f). There are many cultivated varieties of sorghum, with most belonging to the species *Sorghum bicolor*. The vegetative appearance of sorghum is similar to that of corn, but sorghum has perfect flowers that are borne in a terminal inflorescence. Varieties of sorghum can be grown for their grain, for their syrup, as forage, or as a biofuel crop. The grain, which is most often used as animal feed in the United States, is ground and eaten as mush or baked into flat cakes for human consumption in Africa and India. The sweet sorghums, or sorgos, are used either as forage or for syrup (sorghum molasses). A special type of sorghum known as broomcorn is grown for its stiff inflorescence branches that are used to make brooms.

In January 2009, a team of international scientists published the genome sequence of *S. bicolor*. This was only the second cereal grain to be sequenced and the first plant that uses the C_4 Photosynthesis Pathway, which is more efficient at fixing CO_2 than C_3 plants. Recall that the C_4 Pathway is often found in plants growing in arid environments with high light intensity and high temperature (see Chapter 4). The information on the genome will be of value to scientists hoping to improve sorghum productivity as well as to those studying other C_4 plants, such as sugarcane and *Miscanthus*, which is a source of bioethanol production. Sorghum, native to tropical Africa, is a drought-tolerant crop; the genome sequence may also shed light on the genetic basis of drought tolerance that could be transferred to other crops.

In the United States, millet is commonly used as a forage grass and as birdseed, but it is grown extensively in parts of India, Africa, and China as a staple cereal. The term *millet* actually refers to several genera of grasses that were originally domesticated in the Old World and are exceptionally tolerant of drought conditions. Pearl millet, *Pennisetum glaucum*, is widely cultivated in India and Egypt, where it is ground to make a flour for bread. Foxtail millet (*Setaria italica*) is the second most widely grown millet and the most important in East Asia. It has been cultivated in China for over 8,000 years. The small grains can be cooked like rice and also ground into flour. In 2012, researchers published the genome sequence of this millet. The genome is small with only about 500 million bases and is considered a model organism for closely related switchgrass and napier grass, which are important biofuel grasses. The nutritional value of the millets is comparable to that of the other cereal grains; however, studies indicate that the lysine content is higher.

OTHER GRASSES

This chapter has focused on cereals that feed people, but grasses play other important roles as well. Wild and domesticated herbivores rely on both the grains and the vegetative parts of grasses as food. The landscapes of lawns, parks, and playing fields are dominated by grasses. Other grasses provide society with both sugar from sugarcane (see Chapter 4) and building materials from bamboo (see Chapter 18).

Forage Grasses

Grasses provide not only food for humanity but also nourishment in the form of cereal grains (as discussed previously) or **forage** for livestock. There is actually more land dedicated to forage crops than to all other crops cultivated, but much of this land is marginal and either too hilly, rocky, wet, or dry for other crops. Forage, any vegetation consumed by domestic herbivorous animals, includes grasses and legumes (see Chapter 13). The majority of forage plants are herbaceous perennials and are grown for their vegetative structures (leaves and stems), not for their grains or seeds. The nutritive value of the forage grasses is not in the form of stored starch (as in the grains) but cellulose and hemicelluloses from the vegetative cell walls. Herbivores are able to digest these compounds through the action of symbiotic microorganisms within their digestive tract. Forage can be consumed directly during grazing; the crop can also be cut, dried, and baled as hay; or it can be harvested and fermented by bacteria to produce silage. Hay and silage have value as stored food reserves when climatic conditions prevent grazing.

In many parts of the world, natural grasslands provide pasture for grazing animals; however, in other areas large acreages are planted with forage crops. For example, Kentucky bluegrass, *Poa pratensis*, mixed with clover is grown as forage on many horse farms. Although native to Europe, this grass was introduced into North America during the colonial period and adapted well to cool, humid climates. It is considered one of the best forage grasses and is also one of the foremost lawn grasses. Many other introduced grasses native to Europe, such as timothy grass (*Phleum pratense*) and fescues (*Festuca* spp.), are also planted as forage in the eastern half of the United States. However, in the central United States many native grasses, such as big bluestem (*Andropogon gerardii*), little bluestem (*A. scoparius*), and blue grama (*Bouteloua gracilis*), are important forage species. Unfortunately, overgrazing in some native pasturelands has decimated the native species to such an extent that inferior species have supplanted them.

Thinking Critically

Foraging and lawn grasses often have different vegetative and reproductive characteristics than the cereal crops.

Are there any characteristics of forage or lawn grasses that would be of value in a food crop?

Lawn Grasses

Imagine the ideal golf course, an emerald green rolling turf manicured to perfection; this would be a lawn that any homeowner would envy. In addition to its beauty and recreational uses, a lawn cuts down on mud and dust and cools the surface. These lawn plants, of course, are grasses, but unlike other crops, the harvest (grass clippings) is discarded. The cultivation also differs in that the plants are not spaced out but closely crowded together. This crowding maximizes competition, and nutrients and water are often supplemented to ensure luxurious growth. To maintain these "green carpets," Americans spend billions of dollars each year on sprinkling systems, maintenance, fertilizers, and herbicides. Also, the overuse of chemical fertilizers and herbicides contributes to the ever-growing problems of pollutants in the watershed (see Chapters 13 and 26).

Lawn grasses differ in tolerance to drought, temperature extremes, shade, humidity, and salinity. These perennial grass species also vary in their ability to grow in certain soil types and to spread vegetatively, and in appearance there are differences in blade texture, color, and density (table 12.3). For the North, Kentucky bluegrass is the most common lawn grass, while Bermuda grass (*Cynodon dactylon*) is preferred in the southern states.

Table 12.3 Common Lawn Grasses and Their Characteristics

Grass	Desirable Characteristics	Region of United States Commonly Grown
Bahia grass	Heat and drought tolerant; coarse texture	South
Bermuda grass	Drought tolerant; coarse texture	South
Kentucky bluegrass	Fine texture; beautiful color	North
Perennial ryegrass	Winter grass; fine texture	South
Red fescue	Drought and shade tolerant; fine texture	North
St. Augustine	Drought tolerant; coarse texture	South
Tall fescue	Coarse texture; drought tolerant	North
Zoysia	Dense growth; low water need	South

Frequent mowing is essential to maintain any lawn. All too often, grass clippings end up in landfill areas, filling up the sites. One alternative to bagging clippings is the use of mulching mowers, which shred clippings so finely that they decompose readily and return nutrients to the soil. Composting of grass clippings, along with other plant refuse, is another alternative that produces a nutrient-rich humus through microbial decomposition. This organic compost can be used later to enrich the soil.

BIOETHANOL: GRASS TO GAS

The search for clean, renewable energy sources is a high priority for the United States and other countries around the world. The U.S. goal is to produce 36 billion gallons of renewable and alternative fuels by 2022 from all possible sources. Biofuels, such as bioethanol and biodiesel, from plants figure prominently in this search as replacements for petroleum-based fuels. Most of the biodiesel produced in the United States is from soy oil (see A Closer Look 13.2: Harvesting Oil); however, several other oil crops and even algae (see Chapter 22) are also sources of biodiesel. Ethanol can be produced from many different plants through fermentation; however, corn, sugarcane, and other members of the grass family are major sources of bioethanol. Currently, the United States is the leading producer of bioethanol, followed by Brazil. These two countries produce 85% of the world's ethanol fuel.

In the United States, most gasoline for cars contains about 10% ethanol and is often called gasohol or E10. Similar percentages are used in many other countries; however, flexible-fuel vehicles can run on various ethanol-gasoline blends. Internal sensors automatically detect the mixture in use and adjust the engine combustion to match. In the United States, flexible-fuel vehicles can run on E85, which is an 85% ethanol, 15% gasoline blend, while in Brazil many cars run on 100% ethanol, E100. Due to the lower energy content of ethanol compared with gasoline, any of these mixtures will result in a decrease in miles per gallon.

Corn

In 2017, the United States produced 58% of the world's supply of ethanol for fuel, and 98% of this production was from corn. This has been the fastest-growing segment of the biofuels industry. In 2006, ethanol distillers used 14.3% of the 2005 corn crop for ethanol production. This increased dramatically during the next 12 years, and 2018 estimates for the 2017 corn harvest indicated that 37.8% of the crop was used to produce over 15.8 billion gallons of fuel ethanol. Although ethanol production is generating a great deal of interest, this is not a new venture. The U.S. government has subsidized the ethanol industry since 1978. Under the United States Renewable Fuel Standards program, the original goal was the production of 15 billion gallons of bioethanol by 2022. This was first met in 2016, six years ahead of schedule. Corn-based ethanol products continue strong; however, many researchers have questioned the wisdom of using corn to relieve the U.S. dependence on fossil fuels. Even if all the corn grown in the United States were utilized for ethanol production, gasoline use in motor vehicles would decrease by only about 12%.

The energy efficiency of growing corn for ethanol production has also been examined. Fossil fuels are used for the production of the fertilizer needed to grow corn, for the diesel engines used in the combines that harvest the corn, for the trucks that carry the corn to the distillery, and for heat at various steps during distillation. Some scientists, such as David Pimentel at Cornell University, claim there is no energy gain and may even be energy loss in using corn for ethanol production. Others claim there is a small net energy gain; one study found that ethanol production from corn yielded about 10% more energy than was used in the production. A 2015 report from the USDA indicated there is a positive net energy balance for corn ethanol, especially when some of the corn residue is used to produce energy for processing. That report estimated ratios of 2.1 to 2.3 BTU of ethanol for 1 BTU of energy inputs.

Other scientists question the ethical issue of using corn for ethanol production when the crop would be better used to feed the millions of starving people in the world. Economists are also predicting that the demand for corn directed toward ethanol production will cause the price of corn to increase; this will result in higher prices for some food staples as well as meat, poultry, and dairy products.

Sugarcane

In 2017, Brazil produced 26% of the world's bioethanol using sugarcane as the source; this is considered the most successful biofuel industry to date. Scientists agree that bioethanol produced from Brazilian sugarcane shows a significant net gain in energy; 8 BTUs of ethanol are produced for each BTU input. As a result, this is considered the only sustainable form of bioethanol. The success of this bioethanol industry is due to multiple factors. The sugarcane varieties are high yielding, disease and pest resistant, and often drought tolerant. Hundreds of varieties are currently cultivated, with ongoing research focused on developing new ones. Also, bagasse, which is the sugarcane waste left after sucrose extraction, is burned to provide electricity to run the ethanol processing facility.

Sugarcane has been grown in Brazil since the sixteenth century, and ethanol from sugarcane has been in gasoline for cars since the 1930s. Beginning in 1976, the Brazilian government has mandated blending ethanol into gasoline, and the percentage has gradually increased. Since July 2007, fuel used in cars in Brazil must contain at least 25% ethanol; the 25% ethanol/75% gasoline blend is known as E25, and higher-percentage blends are also available. Brazilian car manufacturers have produced flexible-fuel vehicles capable of operating with fuels ranging from E25 to E100. In 2009, 94% of the cars sold in that country were flexible-fuel vehicles, with more than 30 million in use by May 2018. Drivers often select the blend to use based on fluctuating prices of ethanol.

The Brazilian government is planning to expand its biofuels industry by 2020. Ecologists are concerned about potential land use changes that could indirectly threaten the Amazonian rain forests and savannas (see Chapter 26). Although the sugar industry has no plan to expand into vulnerable areas, farmers and ranchers displaced by sugar plantations could encroach on these ecosystems. A 2017 study found that the expansion into areas, which are not protected or reserved for food cultivation, could replace more than 13% of the global crude oil production and reduce CO_2 emission by at least 5%.

Cellulosic Ethanol

A number of scientists believe that using plant material for ethanol production will not become efficient until the ethanol can be made from cellulose instead of sugars and starches. Recall that the polysaccharide cellulose is a polymer of glucose. Breaking down cellulose from corn stalks, wheat straw, wood chips, other waste materials, and native grasses into glucose will be a more energy-efficient process and will reduce greenhouse gas emissions. To meet the 2022 goal for renewable and alternative energy sources, 16 billion gallons are projected to come from cellulosic ethanol sources. Pilot projects for converting cellulose to ethanol at several facilities are utilizing fungal, enzymatic, or chemical degradation of cellulose into glucose followed by fermentation, but improvements are needed in the process to maximize glucose production. Research is currently focused on switchgrass and *Miscanthus*; however, other grasses as well as crop residue, such as corn stover and wheat straw, are other possible sources. Several cellulosic ethanol refineries are now operational, although production is well behind projected growth. The 2017 output was only 10 million gallons. Large-scale production of cellulosic ethanol is still in the future.

Switchgrass, *Panicum virgatum*, is a perennial grass native to North America. It naturally occurs in all states in the continental United States except California, Oregon, and Washington. The grass is widely planted for livestock, since it is useful as both pasture and hay. In addition, switchgrass is also valuable for erosion control in sand dunes, strip-mine sites, and other disturbed areas. Switchgrass typically grows 1.5 to 1.8 meters (5 to 6 feet) tall, although lowland varieties can be taller. It has deep roots and is drought resistant. The species is hardy even in poor soils and poor climatic conditions; it grows rapidly and requires little fertilizer or herbicides (fig. 12.14). Switchgrass is considered a good candidate for cellulosic ethanol production due to its rapid growth and

Figure 12.14 Researchers measure switchgrass density in a field in Mississippi.

the large amount of plant material, biomass, produced. It is believed that switchgrass has the potential to produce approximately 1,000 gallons of ethanol per acre, compared with 400 gallons for corn, although some studies have suggested the ethanol yield would not be that high. In terms of net energy yield, switchgrass is a big winner. Scientists from the USDA found that switchgrass yielded 540% more energy than was used to grow it. Researchers are developing higher-yielding varieties to maximize biomass production in different regions of the country. Researchers are also looking at other ways to improve the yield of ethanol by developing switchgrass varieties with reduced lignin. The presence of lignin in cells with secondary walls is a drawback to the breakdown of cellulose due to the cross-linkage of lignin with other wall components. Plants with reduced lignin should increase cellulose degradation and the yield of fermentable sugars. Further testing is needed to compare growth and productivity of these plants under field conditions.

Another potentially important bioethanol source is *Miscanthus giganteus*, a fast-growing perennial grass that is beginning to receive a great deal of research interest. The genus *Miscanthus* is native to parts of southern Asia and Africa, with about 15 species; *M. giganteus* is a naturally occurring sterile hybrid of *M. sinensis* and *M. sacchariflorus*. It is propagated by rhizome cuttings, and stems can grow up to 3.5 meters (12 feet) tall during the growing season. It has high rates of photosynthesis and can tolerate poor soils. *Miscanthus* is currently used as an energy crop in several European countries, where it is mixed with coal in coal-powered generating plants. The mixtures typically contain up to 15% *Miscanthus* biomass. Field experiments in Europe over the past 20 years have shown that it is capable of growing and producing high yields in a wide range of soil and climate types, although some varieties are limited by frost and drought. Experiments at the University of Illinois found that *Miscanthus* outperformed other bioethanol sources, producing three times more biomass than a switchgrass variety. Although the researchers acknowledged that higher switchgrass yields had been recorded in other studies, they felt *Miscanthus* could become the most productive bioethanol source available. Research needs to be done to develop high-yielding varieties suitable for various locations.

CHAPTER SUMMARY

1. Grasses are members of the family Poaceae. The characteristic grains, with large amounts of stored food, are a vital food source, providing more than 50% of the Calories consumed by the world's population. Wheat, rice, and corn outrank all other plants as food sources for human consumption. Whole grains, with the bran and germ intact, are nutritionally superior to their refined counterparts, which contain only endosperm.

2. Wheat is one of the most widely cultivated crops in the world and is one of the oldest domesticated plants. Domesticated wheat had its origins in the Near East at least 9,000 years ago, and wild species of wheat can still be found in northern Iraq, Iran, and Turkey. Contemporary wheat species include diploid, tetraploid, and hexaploid species. The polyploid species arose through hybridization between diploid wheat and goat grasses. Wheat is nutritionally superior to other grains but still lacks some essential nutrients.

3. Corn, a New World native, was the dietary staple of many Native American tribes before European explorers arrived in America. Corn is an unusual grass, having separate male and female spikes. Many types of corn are cultivated, but all varieties are members of a single species, *Zea mays*. Corn was first domesticated in an area of central Mexico at least 5,500 years ago, but the origin of corn has been a botanical mystery. Today most botanists believe that teosinte is the ancestor of corn, though other theories have been advanced.

4. Rice feeds more people worldwide than any other crop. *Oryza sativa* is the main cultivated rice species, but over 20 species in the genus are known. Rice cultivation dates back over 8,000 years and evidence suggesting separate domestication has been found in both eastern China and northern India. In most areas of the world, rice is cultivated in flooded fields or paddies, although it can be grown without flooding.

5. Other economically important grains include rye, triticale, oats, barley, sorghum, and millet. Triticale is a hybrid between wheat and rye that was developed to combine the hardiness and disease resistance of rye with the higher yield of wheat. Oats have always been considered a good food for horses but historically have had mixed acceptance as food for humans. In the United States, oats were primarily eaten as only a breakfast food until the mid-1980s, when it was suggested that oat bran has cholesterol-lowering properties. Direct human consumption of barley is almost insignificant, but the importance of barley for livestock and in the brewing industry accounts for its eleventh place among the world's crops.

6. Grasses also provide nourishment for livestock in the form of forage. There is actually more land dedicated to forage crops than to all other crops cultivated, but much of this land is marginal and either too hilly, rocky, wet, or dry for other crops. Lawn grasses are indispensable components of landscaping designs and differ from crops in cultivation because plants are closely crowded together, maximizing competition between plants. Various species differ in tolerance to drought, temperature extremes, shade, humidity, and salinity and are therefore adapted to different climatic regions.

7. Bioethanol figures prominently in the search for clean, renewable energy sources. Although many plants can serve as a source for bioethanol production, members of the grass family are some of the most important sources. Currently, almost all the bioethanol produced in the United States is from corn, while sugarcane is the source for Brazilian bioethanol. A major aim of the biofuels industry for the next few years is the development of cellulosic ethanol, with switchgrass and *Miscanthus* two likely sources.

REVIEW QUESTIONS

1. Describe the vegetative and reproductive characteristics of a member of the Poaceae.
2. List the structural components of a cereal grain and the nutritional value of each part.
3. Trace the evolution of modern-day bread wheat from its wild ancestors.
4. What are the different types of corn, what accounts for the differences, and which type is thought to be the most primitive?
5. How is the cultivation of rice different from the cultivation of other crops? What is the value of *Azolla* in the cultivation of rice?
6. What is triticale? What advantages does it have over wheat or rye?
7. Besides cereal crops, what are other important economic uses of grasses?

CHAPTER

13
Legumes

©Karen McMahon

Developing pods of the Kentucky coffeetree (Gymnocladus dioicus).

KEY CONCEPTS

1. The legumes are second only to the cereals in their importance in human nutrition and are an excellent source of high-quality protein.
2. Nitrogen fixation is important for generating nitrogen compounds that can be used by plants in both natural and agricultural ecosystems.
3. Due to the wonders of chemistry, the soybean has been transformed into a variety of food products and has an ever-increasing role in the Western diet.

CHAPTER OUTLINE

Characteristics of the Legume Family 211

A CLOSER LOOK 13.1
  The Nitrogen Cycle 212

Important Legume Food Crops 214
  Beans and Peas 214
  Peanuts 215
  Soybeans 217

A CLOSER LOOK 13.2
  Harvesting Oil 219

Other Legumes of Interest 221
  A Supertree for Forestry 221
  Forage Crops 221
  Beans of the Future 222

Chapter Summary 223
Review Questions 223

CHAPTER 13 Legumes

Legumes are members of the bean family, Fabaceae, which includes all types of **beans and peas as well as soybeans, peanuts, alfalfa, and clover.** This large, widely distributed family also includes various trees and ornamentals, such as black locust, wisteria, lupine, and the Texas bluebonnet.

CHARACTERISTICS OF THE LEGUME FAMILY

Most members of this dicot family share a very similar flower and fruit structure (fig. 13.1). The five-petaled flower is irregular, with bilateral symmetry, and has been described as either butterfly-shaped or boat-shaped. The fruit is a **pod**, or legume, with one row of seeds; the seeds contain two prominent food-storing **cotyledons**. The two halves of a peanut are clear examples of cotyledons. Although the leaves of some legumes are simple, most are pinnately or palmately compound.

The seeds of many legumes are an important food staple worldwide because they are rich in both oil and protein. They are higher in protein than any other food plant and are close to animal meat in quality. In fact, they are often called "poor man's meat" because they are an inexpensive source of high-quality protein. The high protein content of legumes is correlated with the presence of **root nodules** (fig. 13.2), which contain nitrogen-fixing bacteria. These bacteria, which are species of the genus *Rhizobium*, are able to convert free atmospheric nitrogen into a form that can be used by plants in the making of protein and other nitrogen-containing compounds (see A Closer Look 13.1: The Nitrogen Cycle).

Because of the presence of nitrogen-fixing bacteria, the cultivation of legumes enriches the soil. For this reason, farmers often rotate legumes with crops that deplete soil nitrogen. Soybeans are often rotated with corn in the Corn Belt region of the United States. Sometimes leguminous crops are even plowed under as a "green manure" instead of being harvested. Before the advent of commercial fertilizers, these practices were more common than they are today, but they may gain renewed importance as fertilizer costs continue to rise and environmental awareness increases. Without the need for the massive application of fertilizers, legumes can be cultivated worldwide, in even the poorest soils. Ecologists have even recommended planting fast-growing leguminous trees to reclaim eroded or barren areas.

(a)

(b)

Figure 13.1 (a) Flowers and pod of the hyacinth bean. Flowers of the hyacinth bean resemble sailboats, a shape common to many species in the legume family. (b) Peas (seeds) in an opened pod.

Figure 13.2 Nitrogen-fixing bacteria, *Rhizobium* spp., inhabit nodules on the roots of many legumes.

A CLOSER LOOK 13.1

The Nitrogen Cycle

Nitrogen, one of the essential elements for all living organisms, is a major component of amino acids, proteins, nucleic acids, and other organic compounds. Nitrogen gas (N_2) makes up approximately 78% of the air we breathe, but unfortunately most living organisms cannot use this form of nitrogen to make these cellular components. Certain bacteria and cyanobacteria have the enzymatic ability to reduce nitrogen gas to ammonium (NH_4^+), which cells can convert to other nitrogen-containing compounds. This process is called **nitrogen fixation**, and the organisms are called nitrogen-fixing. Some species of nitrogen-fixing organisms can live freely in the soil, while others are found in symbiotic associations with higher plants, as in the root nodules of legumes. The small water fern *Azolla* is known to have a symbiotic association with a nitrogen-fixing cyanobacterium (see Chapter 12).

Plants lacking a symbiotic nitrogen-fixing partner must rely on the nitrogen compounds present in the soil (box fig. 13.1). During the decomposition of dead plants and animals and their waste products, microorganisms break down the proteins and other complex nitrogen-containing organic molecules into ammonium. Although some plants can uptake ammonium directly, nitrifying bacteria present in the soil quickly convert ammonium to nitrite (NO_2^-) and then to nitrate (NO_3^-). Nitrate is the form of nitrogen usually absorbed by plants. Most commercial fertilizers contain a mixture of both ammonium and nitrate.

Other natural sources of nitrogen compounds originate from volcanic activity and from lightning. These atmospheric compounds are washed into the soil by rain and contribute to the cycling of nitrogen. Not all bacterial conversions make nitrogen available to plants; denitrifying bacteria found in wet soils actually break down ammonium and nitrates, returning nitrogen gas to the atmosphere. Nitrous oxide (N_2O) is an intermediate in many of these bacterial conversions, and this gas is also released into the atmosphere.

While nitrogen fixation by bacteria and cyanobacteria is a normal component of the nitrogen cycle, humans have influenced this cycle through the widespread use of legumes, which have increased natural fixation, and through the use of chemical fertilizers. In 1909, German chemist Fritz Haber developed a technique to synthesize ammonia in the laboratory using hydrogen and nitrogen gases with high heat and high pressure. Soon after, Carl Bosch designed a method for the industrial-level production of ammonia using Haber's process. Both scientists received Nobel Prizes for their work, and the industrial production of ammonia is usually called the Haber-Bosch process. This discovery was important to the war effort because ammonia was needed to produce gunpowder and explosives. The agricultural use of factory-fixed nitrogen as fertilizer expanded significantly after World War II, growing from a few million tons in the 1950s to approximately 120 million metric tons per year today. Each year over $100 billion is spent on fertilizer.

World agricultural systems have become increasingly dependent on nitrogen fertilizer. The use of chemical fertilizers has been one of the reasons for the success of modern agriculture, including the Green Revolution, and possibly one of the factors contributing to recent world population increases (see Chapter 15). The food supply that sustains one-half of the world's population is dependent on factory-fixed nitrogen. Although a significant amount of the chemically fixed nitrogen is absorbed by crops, much is washed away as runoff and contributes to the dangerous overfertilization of our waterways.

The fertilizing effect causes huge population increases of algae (algal blooms) living in the water. When the algae eventually die, bacteria break down the remains of the algae, using up oxygen in the water. The results are low-oxygen, low-life areas often called "dead zones," which are increasing in frequency around the world (see Chapter 22).

Unfortunately, there are no means available to grow crops without nitrogen. Scientists hope to find more efficient ways to fertilize crops: by monitoring the amount of nitrogen in the soil to minimize fertilizer use, by increasing crop rotation, by increasing the use of organic fertilizers, and by increasing legume cultivation. In addition, researchers are examining the effectiveness of using biofertilizers for various crops. Biofertilizers include nitrogen-fixing bacteria and cyanobacteria as well as mycorrhizal fungi and other useful microorganisms. When these organisms are applied as seed or soil inoculants, they lead to improved nutrient uptake and enhance plant growth. Meanwhile, in coming decades, world dependence on chemical fertilizers will increase even more to feed the 2 billion people who will be added to the world population over the next 30 years.

Humans have also impacted the nitrogen cycle through the burning of fossil fuels. During their combustion, nitrogen in the air is combined with oxygen to form nitric oxide

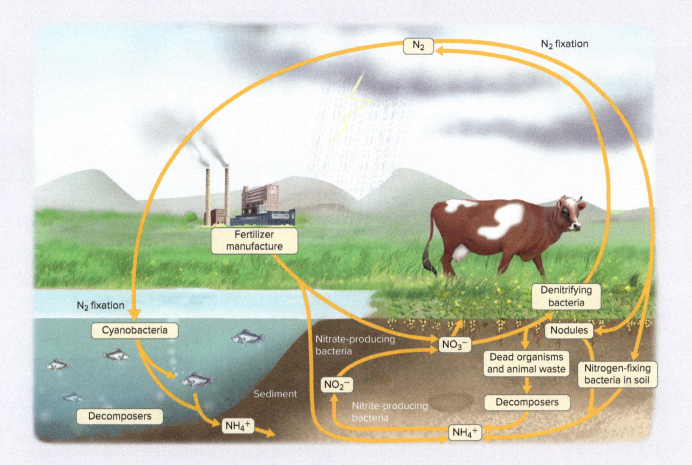

Box Figure 13.1 The nitrogen cycle. Nitrogen is cycled through the environment via the actions of microorganisms. Nitrogen-fixing cyanobacteria and bacteria, free-living or associated with legumes, convert nitrogen gas into ammonium (NH_4^+). Bacteria and other microorganisms decompose dead organisms and animal wastes, releasing ammonium. Other bacteria convert ammonium into nitrite (NO_2^-) and nitrate (NO_3^-), while denitrifying bacteria release N_2 back to the atmosphere.

(NO), which contributes to both photochemical smog and acid rain. Also, some fixed nitrogen from the fuel itself is released into the atmosphere. Nitrous oxide (N_2O) is also added to the atmosphere by the actions of soil bacteria on the nitrates and ammonia in fertilizer. Nitrous oxide has been identified as one of the greenhouse gases responsible for global warming. Although there are many natural and human sources of nitrous oxide, the greatest source is from agricultural soils. It has been estimated that agriculture accounts for 67% of the emissions, and this amounts to 4.5 million metric tons of nitrous oxide per year. More efficient application of nitrogen-based fertilizers can help reduce these emissions.

Through these processes, anthropogenic (caused by human activity) sources are adding significant amounts of fixed nitrogen to the global ecosystem. In fact, human actions are adding as much or more fixed nitrogen than all the natural sources combined. For the first time in history, humans are dominating one of the major biogeochemical cycles, and we need to have a clearer understanding of the consequences.

Thinking Critically

The human influence on the nitrogen cycle has serious environmental repercussions.

What would be the effects of reducing fertilizer use to solve the environmental problems that have been created?

> **Thinking Critically**
>
> One of the goals of genetic engineering is to develop crops that are able to provide their own supply of nitrogen through symbiosis with *Rhizobium*, as do the legumes.
>
> *What benefits to agriculture and the environment would be realized if cereals and other staples became nitrogen-fixing crops?*

Figure 13.3 Beans, peas, and lentils are the seeds of legumes. They come in a wide variety of shapes, sizes, and colors.

IMPORTANT LEGUME FOOD CROPS

As you recall from the discussion on the origin of agriculture, legumes have been cultivated for thousands of years in both the Old and New Worlds (see Chapter 11). A possible reason for their long history as food crops may be that their seeds, which are easily harvested, have a low water content and, when dry, are easily stored for long periods of time. These features, plus their high protein content and ease in growing, make legumes ideal crops.

Beans and Peas

To most people, the word *legume* brings to mind beans and peas; these are, in fact, some of the oldest and most common food crops (table 13.1). Beans come in all shapes, sizes, and colors: kidney beans, lima beans, pinto beans, navy beans, green beans, wax beans, and butter beans are just a few of the many types familiar to us in the United States (fig. 13.3).

Add to this list the hundreds of varieties found in other parts of the world and you can see the diversity implicit in the term *bean*.

Beans are a good source of protein, with values ranging from 17% to 31% and the average about 25%. Although the dry seeds were considered the only edible part for thousands of years, some of the most popular varieties today, such as green beans and wax beans, have edible pods. Beans are warm-season annuals requiring a modest amount of rainfall. Like all legumes, they can tolerate most types of soil and can be grown worldwide.

One bean of particular interest is *Vicia faba*, the broad bean or Windsor bean, which is an Old World species. It has been cultivated and eaten for several thousand years in the

Table 13.1
Some Common Edible Beans and Peas

Common Name	Scientific Name	Value
Adzuki beans	*Vigna angularis*	Popular in Japan and often used in desserts and confections
Anasazi beans	*Phaseolus vulgaris*	Dramatic red and white heirloom bean of the American Southwest
Black-eyed peas	*Vigna unguiculata*	Popular favorite in the South—a *must* on New Year's for good luck
Black turtle beans	*Phaseolus vulgaris*	Small, jet-black bean, the ingredient in black bean soups popular in Latin American cuisine
Chickpeas	*Cicer arietinum*	Also known as garbanzo and ceci; common in Middle Eastern and Mediterranean foods
Kidney beans	*Phaseolus vulgaris*	Best known in chili; most consumed legume in the United States
Lentils	*Lens culinaris*	Used in soups and stews; most important legume in India
Lima beans	*Phaseolus lunatus*	A New World crop native to South America and named after the Peruvian capital
Mung beans	*Vigna radiata*	Widely cultivated in India and China; best known as bean sprouts in Asian cooking
Navy beans	*Phaseolus vulgaris*	Smallest white bean; celebrated in Boston baked beans
Pigeon pea	*Cajanus cajun*	Ingredient in dal of India
Pinto beans	*Phaseolus vulgaris*	Mottled pink and brown beans; used in refried beans and other Tex-Mex dishes
Tepary beans	*Phaseolus acutifolius*	Small white or brown beans native to American Southwest desert; extremely tolerant of heat and drought

Mediterranean region; however, a disease called favism is associated with its consumption. In susceptible individuals, eating broad beans or even inhaling the pollen can produce favism—technically, hemolytic anemia (the lysis of red blood cells). The disease is actually caused by an inherited enzyme deficiency common in Mediterranean people but is aggravated by the type of alkaloids found in broad beans.

The tepary bean (*Phaseolus acutifolius*) is a small brown or white bean native to the American southwest desert region and the traditional food of the Tohono O'odham (formerly the Papago). In fact the name tepary comes from the Tohono O'odham word for the bean *t'pawi*. The bean with a sweet, nutty taste thrives in the severe heat and dryness of the desert. The tepary is high in protein (49%) and a good source of iron, niacin, and calcium. It is also an integral part of the southwest culture as exemplified by the legend of how the coyote was running across the desert with a bag of the beans when he tripped and the bag ripped releasing the beans to the night sky. The Milky Way was formed as a result of the thousands of glittering white tepary beans. Although many Tohono O'odham abandoned the traditional diet in the last 50 years, there are attempts to re-introduce this dietary staple to the Tohono O'odham and others by groups such as Native Seeds, RAFT (Renewing America's Food Traditions), and the Tohono O'odham Community Action Committee (TOCA).

The term *pea*, like *bean*, denotes dozens of different kinds of edible seeds that have been cultivated for millennia. Those most familiar to us include green peas, split peas, black-eyed peas, lentils, chickpeas (garbanzos), and snow peas (fig. 13.3). Again the varieties, like snow peas with edible pods and green peas with fresh (not dry) seeds, are of more recent origin. Nutritionally, peas are also a good source of protein, with averages about 21%. Unlike beans, peas are grown during the cooler seasons of the year in temperate zones. Biologically, the most famous pea is the garden pea, *Pisum sativum*, which Gregor Mendel used for his famous genetics experiments.

Native to India, the pigeon pea (*Cajanus cajan*) is an important crop in India, East Africa, and the Caribbean. It is a nitrogen-fixing legume that can tolerate drought conditions and has probably been cultivated for over 3,000 years. Large pods hold sweet-tasting seeds, which can be eaten fresh or dry. The pigeon pea is one of several legumes used to make dal, a thick paste made by boiling the split peas and flavored by a variety of spices. Nutritionally, the protein content, at 22%, compares favorably with that of other peas. In 1974, the International Crops Research Institute for the Semi-Arid Tropics (ICRISAT) in India selected the pigeon pea as one of five crops (along with sorghum, pearl millet, chickpea, and groundnut) for a breeding program to create improved varieties. Success was soon met in producing varieties of pigeon pea that matured quicker, 3 months rather than 9, and that were more disease resistant than traditional ones. But it has taken nearly 30 years to produce hybrid seed in the pigeon pea. Recall that the development of hybrid seed in maize led to improved productivity (see Chapter 12). In producing hybrid seeds, inbred lines are crossed to yield varieties that are hardier because of hybrid vigor. As with maize, it is desirable to create male sterile lines to control pollination. A breakthrough came when researchers found varieties of pigeon pea with a type of male sterility encoded by genes located in the chloroplasts. (This is an example of **cytoplasmic inheritance**, in which genes outside the nucleus code for traits. Since it is the egg, not the sperm, that passes on its cytoplasm and organelles to the zygote, genes are passed down through the maternal parent.) Breeding for the male sterile trait finally produced pigeon peas with flowers lacking all pollen. It was then possible to produce hybrid seed and by 2004 the new hybrid variety was ready for distribution. This was the first time hybrid seed had been created in a legume, and the information for the development of pigeon pea hybrids will be useful in developing other legume seed hybrids. Already, the productivity of hybrid pigeon peas in the field has nearly doubled. One disadvantage is that the hybrid seed must be purchased every time a field is planted. This is an added burden for small farmers, since the hybrid pigeon pea is double the cost of standard varieties. There is an ongoing project to teach the women in rural villages to produce the hybrid seed for themselves.

Peanuts

Peanuts, also known as goobers and groundnuts, are originally native to South America but are grown more extensively today in other parts of the world. Although the exact date of domestication is unknown, finely crafted gold and silver peanut-shaped jewelry was recently unearthed in Peru in the tomb of a Moche warrior priest (fig. 13.4). Carbon dating indicates that the tomb was from A.D. 290. This archeological discovery shows that the peanut played a prominent role in the ancient Moche civilization.

Figure 13.4 A necklace of solid gold and silver peanuts reveals the prominence of the peanut to the agriculture of the Moche, a people who inhabited northern Peru 1,800 to 1,000 years ago.

In the sixteenth century, Spanish explorers discovered peanuts growing in South America and took them back to Europe. From there, trading introduced the peanut to Africa, where it soon became widely cultivated. The slave trade returned the peanut to the New World, but this time to North America. In the United States, the peanut is a staple crop of the South, growing best in light, sandy soils and mainly cultivated in Georgia, North Carolina, Texas, Alabama, Virginia, and Oklahoma (fig. 13.5).

The peanut, *Arachis hypogaea*, is one of nature's more unusual plants. After pollination, the flower stalk elongates downward, pushing the developing fruit into the soil. It is here, underground, that the fruit matures into a pod, characteristically with two seeds (peanuts) in a shell (fig. 13.6). The whole growing cycle takes about 5 months. Two varieties commonly grown in the United States are the larger-seeded Virginia peanut and the smaller-seeded Spanish peanut, which has a slightly higher oil content.

With 45% to 50% oil and 25% to 30% protein, the peanut is a highly nutritious seed that is used in many ways. It is a favorite in the United States, with over 1 billion pounds consumed annually, mainly as a snack food, in candy, and in peanut butter. In fact, half of the U.S. peanut crop is used to make peanut butter. Peanut butter is a uniquely American food, first developed by a St. Louis physician in the 1890s as a nutritious and easily digested food for invalids who had difficulty chewing.

Today, peanut oil is found in margarine, shortening, salad dressing, cooking oil, certain soaps, and a variety of cosmetic and industrial products, such as shaving cream, plastics, and paints. (Unfortunately for some, the widespread use of the peanut and its oil in a variety of foods can be deadly. In Chapter 21, allergies to the peanut and other plant-based foods are discussed.) Even after the extraction of oil, the pressed cake that remains is used as a livestock feed that is rich in protein. Hogs, in particular, have such a fondness for peanuts that they will uproot them in fields if given the opportunity.

The versatility of the peanut is due in large part to the work of George Washington Carver (1864–1943), who developed more than 300 food and industrial uses and encouraged its cultivation in the South. Carver (fig. 13.7), the son of a slave, developed an interest in agriculture during his college years

Figure 13.6 An uprooted peanut plant shows how peanut pods develop underground.

Figure 13.5 Oklahoma is one of several southern states where the peanut is extensively grown, a fact celebrated by a statue of the peanut in Durant, Oklahoma.

Figure 13.7 George Washington Carver introduced and promoted the cultivation of the peanut to the Civil War–ravaged South.

and spent his whole teaching and research career at Tuskegee Institute in Alabama. He sought to revitalize agriculture in the South, whose soil had been exhausted during years of cotton cultivation, by introducing soil-enriching legumes, such as the peanut and the soybean. Carver developed a number of crops—the peanut, soybean, and sweet potato are just a few examples—to diversify agriculture in the South. Through his efforts, the scope of Southern agriculture was changed for the better. When Carver began his work, the peanut was not even recognized as a crop; today it is the second leading cash crop in the region.

The cultivated peanut is known to be a natural hybrid of two wild diploid species, *Arachis duranensis* and *Arachis ipaensis*, and occurred between 4,000–6,000 years ago in northern Argentina. At some point, chromosome duplication occurred in the hybrid resulting in a tetraploid, *Arachis hypogea*. The genome of the cultivated peanut was determined in 2014 by mapping the genomes of its two wild ancestors that account for 96% of all peanut genes.

Soybeans

Although peanuts will never lose their place in the American diet, roasted soy nuts are available in health food stores. The soybean, *Glycine max*, is relatively new to the West but has been esteemed in Asia for centuries (fig. 13.8a). It was considered one of the sacred crops of the ancient Chinese, and evidence suggests that it was domesticated in northern China at least 3,000 years ago. Soybeans were first taken to Europe in the seventeenth century by the German botanist Engelbert Kaempfer.

Although the soybean was introduced into North America in 1765, there was very little interest in growing the crop until the 1920s. During World War II, soybean oil substituted for imported fats and oils, and the nutrient meal was used to boost livestock production. Soybean production has continued to rise dramatically, making the United States a leading producer. Because of this spectacular rise from a second-rate crop, it has often been referred to as the Cinderella crop. Reasons for the success story are the versatility of the soybean and its suitability for growing in the Corn Belt region of the Midwest; it does best in warm, temperate climates with moderate amounts of rainfall.

The first draft of the soybean genome was released in January 2010. Soybean chromosomes have been duplicated at least twice in the history of this species. It has a relatively small genome, one-third the size of the maize genome. Another unexpected fact is that almost 60% of the soy genome is composed of transposable elements, repetitive stretches of DNA, nicknamed *jumping genes* because they can copy and insert themselves throughout the genome (see Chapter 12).

Since ancient times in Asia, soybeans have been consumed in hundreds of ways. Soybeans cannot be consumed raw because of the presence of a trypsin inhibitor. Trypsin is a digestive enzyme, and the presence of this inhibitor interferes with normal protein digestion in humans. When cooked, soybeans can be eaten whole because the inhibitor is inactivated by heat. Most often, however, soybeans are modified into a paste, curd, or "milk." One familiar soybean product is soy sauce; although many American brands are made synthetically, soy sauce traditionally is made by fermenting soybeans in brine.

(a)

(b)

Figure 13.8 (a) Soybean (*Glycine max*) pods with seeds. (b) The ingredient common to all these products is the versatile soybean.

In the preparation of soy milk, beans are soaked in water and puréed. The mixture is then heated, and the liquid soy milk is poured off. Soy milk provides a nondairy substitute for both milk and baby formula for lactose-intolerant individuals who are unable to digest the lactose that naturally occurs in cows' milk or for those allergic to the milk proteins. Tofu, or bean curd, is prepared from curds of soy milk and is extremely versatile, being used in main dishes in both Japanese and Chinese cuisine. Tofu on its own is bland tasting, but it acts as a sponge, soaking up flavors from other ingredients. It also serves as the basis for a large variety of soy cheeses, yogurt, sour cream, and other dairy spreads. Imitation cream cheese, cheddar cheese, Swiss cheese, and a host of other varieties are available in health food stores. An ice cream–like dessert called tofutti is another well-known tofu product. These are especially welcome products for lactose-intolerant individuals as well as for those wishing to avoid the saturated fat in dairy products (fig. 13.8b).

The use of tofu, soy milk, and other soy foods has received a great deal of attention because they contain isoflavones, which are **phytoestrogens**. Although these plant estrogens are much weaker than human estrogen, they may have effects on the body. Some research indicates that these compounds lower cholesterol levels, especially LDL levels, which could lower the risk of cardiovascular disease. Some studies have also found that isoflavones may reduce symptoms of menopause and reduce risks of osteoporosis by preventing bone loss in older women. Other studies have shown that isoflavones may inhibit the formation and growth of tumors. Some researchers believe that low rates of both breast cancer and prostate cancer among the Japanese are due to the consumption of a soy-rich diet. A 6-year study released in 2009 reported how diets high in soy foods affect the recurrence of breast cancer in women. The objective of the research was to investigate if the estrogen-like isoflavones in soybean might increase the return of breast cancer and the mortality rate of women previously diagnosed and treated, as it is well established that estrogen is linked to the development and progression of certain types of breast cancers. Unexpectedly, the consumption of soy in the diet was shown to dramatically decrease the reappearance of breast cancer and reduce the death rate from breast cancer.

Edamame is a soybean dish from Japan that has become popular in the West. It is made by boiling or steaming, then lightly salting unripe green soybeans still in the pod. Pods can be eaten whole, or if both ends of the pods are cut before cooking, the beans are popped out and eaten by hand.

Other common soybean foods in Asia are miso and tempeh. Miso is a fermented food of Japan prepared from soybeans, salt, and rice; the mixture is fermented by fungi for several months and then ground into a paste and used as a spread. Tempeh, fermented soybean cake that originated in Indonesia, is prepared by inoculating parboiled soybeans with mold and allowing them to ferment for a few days. The fungal mycelium binds the soybeans together into a cake, which can be sliced and cooked in various ways.

Soybeans are among the richest foods known, with 13% to 25% oil and 30% to 50% protein, depending on the variety. Overall, they have a higher protein content than lean beef. Although still used largely as animal feed, the soy protein is used more and more in the human diet. After the extraction of oil, the soy meal that remains is made into a flour and can be included with wheat flour in a variety of breads, pasta, baked goods, and breakfast foods. Replacing a small fraction of the wheat flour with soy flour significantly improves the protein content. The soybean has also played a vital role in relief efforts as a protein-rich food supplement to famine victims in many developing nations.

Another product is textured vegetable protein (TVP), produced by spinning the soy protein into long, slender fibers. TVP can pick up flavors from other substances and can therefore be used as a meat extender. With the addition of artificial flavorings and colors, TVP can be made into cholesterol-free imitation meats. Imitation bacon bits are made this way.

Soy oil is used extensively as cooking oil, as salad oil, and in the manufacture of margarine, shortening, and prepared salad dressings. It is such a widespread ingredient in so many foods that the average American consumes almost 23 liters (6 gallons) of soy oil a year. Many manufacturers are replacing unhealthy saturated fats with soybean oil in commercially prepared food.

Industrially, soy oil can be used in dozens of processes for the manufacturing of paints, inks, soaps, insecticides, and cosmetics (see A Closer Look 13.2: Harvesting Oil). Probably the most imaginative use of soy oil was the manufacture of a soybean-based "plastic" car by Henry Ford in 1940. Ford's commitment to the use of soybeans in car manufacturing was so great that at one point he stated that his goal was to "grow cars rather than mine them."

Lecithin, a common food additive, is a lipid extracted from soybeans. Added to many packaged foods, such as cake mixes, instant beverages, whipped toppings, and salad dressings, it stabilizes them and extends their shelf life. The use of soybeans should increase even more in the future. One possible market is developing countries. Attempts to improve the protein-deficient diets in these countries have included using soy products to enhance the nutritional value of the native foods.

Thinking Critically

Soybeans and soy products are becoming increasingly important components of the human diet.

Investigate the health benefits of using soy milk and soy-based cheeses to replace traditional dairy products. Are there any disadvantages to this practice?

A CLOSER LOOK 13.2

Harvesting Oil

Soybeans and peanuts are two legumes widely utilized for their oil. Soy oil is one of the leading vegetable oils in the world today in terms of production, industrial use, and human consumption. Soy oil has hundreds of uses, ranging from multiple food uses to the production of nonconsumables, such as soy-based inks, soy crayons, and biodiesel fuel. The top vegetable oils used for food and industrial purposes today are shown in Box Figure 13.2.

Utilization of plants for oil with multiple applications is not unique to modern society. Since earliest times, humans have utilized certain plants for their oil needs. Oils used by early cultures throughout the world included walnut, sesame, almond, linseed, soybean, coconut, and castor bean. Ancient civilizations in the Mediterranean area traditionally used olive oil for food, lighting, medicine, and other uses. Biblical texts state that clear oil from beaten olives was to be used for lighting the lamp within the Tabernacle. Olive oil was also used ritually for washing and anointing by the Hebrews, Egyptians, and Greeks. Olive trees (*Olea europea* in the Oleaceae, olive family) were domesticated over 6,000 years ago, and cultivation spread throughout the region in ancient times. Olive oil is unusual because it originates from olive fruits rather than seeds. After harvesting, the fruits are pressed to extract the oil. The first pressing done without heat produces "virgin" olive oil, while additional pressings use heat and produce oils of lower quality. Olive oil contains largely monounsaturated oil, which has been shown to lower LDL levels (see Chapter 10). This benefit has recently led to the increased consumption of olive oil. Additional studies have also linked the use of olive oil with lower risks of breast cancer.

Canola oil has also become increasingly important to the health-conscious consumer. When compared with other vegetable oils, canola oil has the lowest levels of saturated fats and relatively high levels of monounsaturated ones. Canola is a new name given to certain varieties of rapeseed plants. Rapeseed oil obtained from seeds of *Brassica napus* and *B. campestris* (in the Brassicaceae, the mustard family) has a long history of use in northern Europe and Asia but is relatively new to North America. Through the middle of the twentieth century, rapeseed oil was largely used as a lubricating oil. Varieties that were developed in the late 1960s and 1970s improved the nutritional qualities of the oil by reducing the levels of erucic acid and glucosinolates. The name *canola* (**Can**adian **o**il, **l**ow **a**cid) applies to these rapeseed varieties. Canola is a cool-season crop, with Canada being one of the leading producers in the world. Although the United States still imports some canola oil, the cultivation of rapeseed has increased dramatically, with the majority grown in North Dakota. Breeding programs are underway to develop varieties suited to the climate in various regions of the United States as well as to develop disease-resistant and herbicide-resistant varieties. Today canola is the third most consumed oil, and statistics indicate that the demand for canola oil will continue to be important in the coming decades.

Other vegetable oils that are important as edible oils include those from sunflower, cottonseed, sesame, safflower, corn, palm, and coconut, although not all of these are beneficial additions to the human diet.

Although oils are typically characterized by the nature of the component fatty acids, they can also be categorized as **drying oils**, **semidrying oils**, or **nondrying oils**. Drying oils react with oxygen in the air to form a thin, waterproof, elastic film; linseed oil, tung oil, soy oil, and safflower oil are commercially important drying oils. In general, the most unsaturated oils (many double bonds present in the triglyceride) will dry the quickest. Nondrying oils, which remain liquid for prolonged periods on exposure to the air, include peanut oil, olive oil, rapeseed (canola) oil, and castor bean oil. Semidrying oils are intermediate between the other types and dry slowly or at elevated temperatures. Cottonseed, sunflower, sesame, and corn oils are included in this category.

Linseed oil is a drying oil obtained from seeds of the flax plant (*Linum usitatissimum*), which is perhaps better known for linen fiber (see Chapter 18). It is one of the oldest cultivated oil crops; the oil was used by ancient Egyptians to form a waterproof coating for coffins. Today it is still widely used in paints, varnishes, and linoleum.

Physaria (lesquerella or bladderpod) is a crop being developed in the Southwest for its seed oil. *Physaria* (formerly classified as the genus *Lesquerella*) is a genus of shrubs in the Brassicaceae, with many species native to

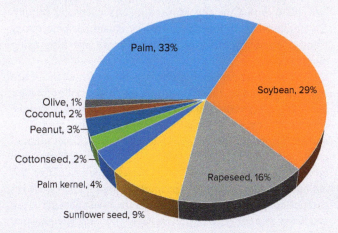

Box Figure 13.2 World consumption of major vegetable oils, 2017.

the Southwest and Mexico. Three species are being investigated for their oil production, but the interest in commercialization has focused on *Physaria fendleri* because of its growth and high yield. Lesquerella oil is a drying oil that is rich in hydroxy fatty acids, a feature that makes it ideal for use as a lubricant, protective coating, surfactant, and resin and in nylons and cosmetics. The molecular characteristics of the oil give rise to tougher and more durable coatings as well as water-resistant inks. The oil appears particularly promising for the cosmetic industry, where it substitutes for castor oil; lipstick produced with lesquerella oil is currently being manufactured. The future for this crop appears very promising.

The drying ability of soy oil has made it of value for many industrial processes, with recent successes in the production of inks. Soy oil is rapidly replacing the petroleum used as a carrier or vehicle for the pigments in traditional inks. The soy-based inks have lower emissions of volatile compounds and are brighter because the soy oil is clearer than petroleum. Although not suitable for all printing processes, soy inks have gained wide acceptance by the Newspaper Association of America for newsprint. The soy inks biodegrade more fully than traditional inks but, at present, are still slightly more expensive.

With the world's increasing demand for hydrocarbon products and with the ability to chemically modify a vegetable oil into a form that can replace petroleum hydrocarbons, vegetable oils are becoming even more important. Perhaps the best example of this trend is the development of **biodiesel** fuel. Biodiesel is a vegetable oil product that is being marketed as a cleaner alternative to conventional petroleum-based diesel fuel because it produces little soot or other pollutants. In 2017, the United States produced 1.8 billion gallons of biodiesel oil. This amount has already exceeded the original U.S. goal to produce 1 billion gallons of biodiesel by 2022 and is close to meeting the 2018–2019 goal of 2.1 billion gallons. Much of the biodiesel production in the United States uses soy oil as the base oil; however, other plant oils are also being used or developed. In temperate climates this includes canola, sunflower, and camelina oil, while in tropical areas *Jatropha* oil and palm oil are the principal sources for biodiesel production.

Camelina sativa, camelina or false flax, a member of the Brassicaceae has generated much interest for biodiesel production. The species is native to Europe and Central Asia and has been cultivated there for several thousand years. Historically it was important for use in oil lamps as well as a source of vegetable oil and animal feed. Today there is widespread interest in camelina as a biofuel crop and as an edible oil because it is high in omega-3 fatty acids. Camelina does best in cooler climates, and over the past several years the production in the northern United States and Canada has been increasing. *Jatropha* is a large genus of plants in the Euphorbiaceae (spurge family); they produce seeds that contain from 27% to 44% oil. There has been a great deal of interest in growing *Jatropha* in several African and South American countries because the plants can grow on marginal lands under semiarid conditions. However, existing species of *Jatropha* have produced disappointing results. There is considerable effort underway to develop high-yielding *Jatropha* hybrids. Two species of *Elaeis* in the Arecaceae (palm family) are important oil palms grown in tropical areas; Malaysia, Indonesia, and Nigeria are the principal producing countries. Palm oil is extracted from the fleshy pericarp of palm fruits, and palm kernel oil is extracted from the seeds. Palm oil is the leading vegetable oil (box fig. 13.2), with countless food and industrial uses. Recently, there has been expansion of palm plantations for biodiesel production. This expansion is a major environmental concern when tropical rainforests are cleared for palm plantations, leading to loss of diverse rainforest biomes as well as habitat destruction of endangered species (see Chapter 26). Biodiesel is also being produced from algae (see Chapter 22) and used cooking oils collected from fast-food restaurants and other establishments.

The use of vegetable oils in diesel engines is not new; in 1900, Rudolf Diesel, who first developed the diesel engine, showed that it could run on peanut oil. In operating performance, biodiesel functions like petroleum diesel and can be blended in any ratio with petroleum diesel fuel, although engine modifications may be needed. At this time, a blend of 80% diesel and 20% biodiesel, known as B-20, is being marketed and used in many areas. The use of biodiesel began in the late 1990s in Florida with the U.S. Postal Service fleet in Orlando and West Palm Beach. Since then, biodiesel usage has increased dramatically. This increase was encouraged when the federal Energy Policy Act was amended in 1998 to include B-20 biodiesel fuel use as a way for federal, state, and public utility fleets to meet requirements for using alternative fuels. Other countries around the world are also pursuing this cleaner fuel. In addition to the current use of biodiesel in cars, buses, and trucks, experiments have shown that biodiesel can be used in jet fuels, and several airlines have tested flights with biodiesel blends. Biodiesel home heating oil blends are being used in many areas; in fact, New York City and Massachusetts currently require 5% biodiesel in all homes heated by diesel oil. Emergency electric power generators in some areas are also using biodiesel.

Soybean biodiesel and corn ethanol (see Chapter 12) are now the two major alternative biofuels in the United States, but there are striking differences between the two in environmental costs. Growing corn calls for more pesticides and fertilizers than does soybean cultivation. Corn ethanol produces 10% more energy than is required to grow and convert the corn to fuel. (Other studies have calculated the net energy balance for corn-produced ethanol to be 25%.) In comparison, soybean biodiesel, with its lower energy costs, produces an energy gain of 93%. Greenhouse gas emissions are reduced by 12% when gasoline is replaced with corn ethanol, but burning soybean biodiesel in place of diesel lowers the same emissions by 41% or more.

OTHER LEGUMES OF INTEREST

Legumes have many and varied uses; in terms of sheer numbers, legumes are by far the most utilized plants. Above and beyond their worth as a food source, legumes are valued as timber, forage, spices, and ornamentals and as sources of medicines, insecticides, resins, and dyes (table 13.2). Following are a few noteworthy legumes.

A Supertree for Forestry

Leucaena, or lead tree, is a widely distributed genus of woody plants native to the New World. In the pre-Columbian era, natives of Mexico and Central America ate the protein-rich seeds in young pods of several *Leucaena* species either fresh or boiled. *Leucaena leucocephala* is a tropical tree that gained fame as one of the fastest-growing species of woody plants (fig. 13.9). Some varieties have been reported to grow as much as 9 meters (30 feet) per year, earning it the nickname "Jack's beanstalk." In many countries where firewood is the main source of fuel, *Leucaena*'s remarkable growth makes it a quickly renewable energy resource. Likewise, the tree can be grown for its wood, to be used in furniture or construction or converted into pulp to make paper and paper products.

Like many legumes, *Leucaena* has a symbiotic relationship with nitrogen-fixing bacteria. Archeological evidence suggests that the Mayans grew the trees alongside maize to restore soil fertility. As nitrogen compounds are produced by symbiotic bacteria, they are transported and accumulated in the leaves. Later, when the leaves fall, they quickly decay, releasing nitrogen compounds that enrich the soil.

Figure 13.9 Lead tree (*Leucaena leucophylla*) is valued for its versatility.

Undecayed leaves can even be packaged and sold as an inexpensive fertilizer. Six bags of dried leaves contain the same nitrogen as one bag of commercially produced ammonium sulfate. Because of the soil-enriching properties, *Leucaena* has been used as a "nurse" tree, providing nutrients and shade for coffee, pepper, cacao, vanilla, and other shade-loving crops.

Leucaena has also been utilized as a forage, since herbivores readily consume its nutritious foliage. Another potential market is developing this versatile tree for gum production. Gums are an ingredient of many prepared foods as well as many industrial products.

Forage Crops

Worldwide, many legumes are planted and grown exclusively as pasture or forage crops. Their high protein content, which makes them ideal as a source of food for humans, also makes them desirable as animal fodder. Combinations of carbohydrate-rich grasses and protein-rich legumes are grown in most pastures for direct consumption or for hay.

Table 13.2 Other Legumes of Interest and Their Uses

Plant	Scientific Name	Use
Carob	*Ceratonia siliqua*	Chocolate substitute
Copaifera	*Copaifera officinalis*	Resin for paints, lacquers
Fenugreek	*Trigonella foenum-graecum*	Spice
Indigo	*Indigofera tinctoria*	Dye
Licorice	*Glycyrrhiza glabra*	Extract
Mesquite	*Prosopis glandulosa*	Charcoal
Rosary pea	*Abrus precatorius*	Jewelry
Rosewood	*Dalbergia* spp.	Timber
Senna pods	*Cassia fistula*	Laxative
Tamarind	*Tamarindus indica*	Seasoning
Tuba root	*Derris elliptica*	Insecticide

Alfalfa (*Medicago sativa*) is probably the best known and most widely grown of these forage legumes (fig. 13.10). Commonly called lucerne in other countries, alfalfa has been cultivated as a forage crop since ancient times. It was introduced into Greece from Persia about 500 B.C. by the invading Medes. Linnaeus used this historical reference in assigning the scientific name *Medicago sativa*. The Romans recognized this crop as a superior feed for their horses; later the Spanish introduced alfalfa to the New World for the same purpose. One of the largest markets for alfalfa today is as a dehydrated feed for livestock (cattle, horses), pets, and laboratory animals (gerbils, rabbits, mice).

There has been recent interest in developing alfalfa products for human consumption. Alfalfa sprouts make their appearance at the salad bar and as a sandwich condiment. These tiny sprouts pack a powerful nutritional package of minerals, vitamins, and proteins. Alfalfa pills and extracts are a common commodity in health food stores. Alfalfa is also an important tool in biotechnology because it is easily grown in cell culture and readily incorporates novel genes through standard genetic engineering techniques (see Chapter 15). Other forage legumes include the clovers, sweet clovers, vetches, lespedezas, and bird's-foot trefoil.

Figure 13.10 Alfalfa (*Medicago sativa*) has been used as a forage since the days of the ancient Greeks and Romans.

Source: Keith Weller/USDA

Beans of the Future

Because of their nutritional value, there has been considerable interest and research in discovering "new" varieties of legumes. Ethnobotanists have spent years searching the globe—in particular, developing countries—for little-known crops that have the potential to become major food sources (see Chapter 15).

One of the interesting finds is the winged bean, *Psophocarpus tetragonolobus*. Native to Southeast Asia and New Guinea, this plant is valued because all parts of the plant are edible and highly nutritious. The pod, which has four extensions, or "wings," running along its length, is edible either raw or cooked. The flowers, when cooked, are said to taste like mushrooms; the tendrils that support the vine taste like asparagus; and the leaves are eaten like spinach. The root is tuberous and can be eaten like the potato, but its protein content is much higher. Of greatest value are the seeds, which are similar to soybeans, with a protein content of 37% and an oil content of 20%. Some scientists believe that the winged bean may be as important to the future of tropical agriculture as the soybean has proved to be in temperate agriculture.

The groundnut (*Apios americana*) is a nitrogen-fixing legume native throughout eastern North America and, historically, a major food source for the eastern tribes. Groundnuts were one of the foods introduced by the Native Americans to the starving Pilgrims of Plymouth Colony. Both its seeds and golfball-sized tubers were collected from wild stands and eaten. The plant itself is a twining vine with pinnately compound leaves and reddish flowers. Numerous tubers are produced along the plant's rhizome. Today there is interest in domesticating this plant for its nutritious (12% to 13% protein) and versatile tubers, which can be boiled, baked, or fried.

The yam bean, or Mexican turnip, is another underutilized legume that has been targeted for development. Also known as jicama (derived from its Aztec name), it is a tropical vine grown for its large, tuberous roots, which are good sources of proteins and carbohydrates (sugars and starch). Several cultivated species are native to South America, but *Pachyrhizus erosus* from Mexico and Central America is the primary agricultural crop. The vines, which can grow to a length of 6.5 meters (20 feet) or more, bear trifoliate leaves that are extremely variable in form, ranging from entire to deeply lobed. Flowers are either violet or white. Roots are generally harvested before seeds are produced and can reach a weight of 23 kilograms (50 pounds), but those harvested average about 1.8 kilograms (4 pounds).

Although perennial, jicama is grown as an annual in warm temperate regions. Currently, jicama is cultivated primarily in Mexico, Guatemala, El Salvador, and Honduras, where it was a crop of the Aztecs, Mayans, and other pre-Columbian cultures. Introduced to China in the seventeenth century, jicama cultivation spread to Vietnam, Thailand, and India. The traditional method of growing is to plant in flooded fields; it is often planted at the edge of rice fields. It can also be grown on dry land, and that is the preferred method for large-scale commercial production. Harvesting takes place when the roots have reached a desirable size. Depending on the local market, there is a demand for all sizes of roots: small, medium, and large. Harvesting occurs about 5 to 7 months after planting. Because the sugar content increases after harvesting, roots are stored for 2 or 3 days before consumption if the local market prefers a sweet taste.

Jicama is one of the most efficient nitrogen-fixing legumes, and in Central America it has been intercropped with corn. Seeds, pods, leaves, and stems contain rotenone, a natural insecticide (see Chapter 21) that affords the plant some protection from pests. (The tuberous roots also contain rotenone, but not in significant amounts.) There is interest in developing commercial enterprises for the extraction of rotenone from the seeds. The seeds are also good sources of high-quality polyunsaturated oil that is comparable to the oils of cottonseed and groundnut, but the high concentration of rotenone makes the oil inedible without processing.

The roots can be eaten raw or cooked in a variety of ways. In Mexico, slices of the fresh root are mixed in vegetable and fruit salads or sprinkled with lime juice and eaten alone. Slices of the roots are made into a popular sweet by soaking them in syrup. Jicama can also be used in stir-fry or pickled in vinegar. In India, the tubers are used in chutneys and to make a drink called *kheer.* In Asia, the pods, which are similar to those of soybean, are eaten. Jicama has also been used as a nutritious animal feed. There are several reports about jicama in herbal medicine. Preparations from the roots were used to cure fevers and to treat peeling or itching skin. Tinctures from the seeds have been used to treat mange, head lice, and cattle louse.

CHAPTER SUMMARY

1. Legumes are members of the Fabaceae or bean family and include several important food crops: peas, beans, soybeans, and peanuts. The highly nutritious seeds, and occasionally the pods, of legumes have become a food source second only to the cereals. In terms of nutritional quality, legumes are excellent sources of protein and, in many cases, oil.

2. The high protein content of many legumes is due to their association with nitrogen-fixing bacteria in their roots. Nitrogen fixation converts the unusable N_2 gas into ammonium (NH_4^+), which the plants can incorporate into the synthesizing of proteins.

3. Dozens of types of peas and beans are valued for their high protein content.

4. The peanut (*Arachis hypogaea*) has a South American origin and is unusual in that the plant itself "plants" its seed pods in the soil. The peanut is valued for both its oil and its protein content. George Washington Carver was single-handedly responsible for developing the peanut as a major crop in the post–Civil War South.

5. The protein content of soybeans (*Glycine max*) is one of the highest of all crops. The soybean has a long history of use in eastern Asia, where it has been modified by various treatments, making it an extremely versatile food.

6. Legumes are also valued as forage crops and as "green manure" to enhance soil fertility.

7. The search for new crops has uncovered several relatively unknown legumes, such as the winged bean, jicama, and groundnut, that may become the new Cinderella crops in the future.

REVIEW QUESTIONS

1. What are the distinguishing characteristics of the legume flower and fruit?
2. What is the role of nitrogen-fixing organisms in the nitrogen cycle? How do legumes fit in?
3. Why are soybeans called the Cinderella crop? List several products derived from soybeans.
4. Trace the history of the peanut from its South American origins to the present day.
5. What is the nutritional value of legumes?
6. What legumes are being developed as economically important world crops?
7. Why are seeds the usual source of most plant oils?

CHAPTER

14

Starchy Staples

©BananaStock/Alamy Stock Photo

Potatoes are the most important nongrain food crop. Originally native to the Andes Mountains in South America, potatoes are now grown around the world, with China the largest producer.

KEY CONCEPTS

1. Modified stems and storage roots can have several functions in plants: as food reserves, for asexual reproduction, and as the starting point for renewed growth after dormancy.
2. Starchy staples, with their high carbohydrate content, include some of the world's foremost crops and play major roles in the human diet.
3. Historically, the potato has been pivotal in the development of several societies, from the ancient Incan civilization in South America to the preindustrial countries of Europe, especially nineteenth-century Ireland.

CHAPTER OUTLINE

Storage Organs 225
   Modified Stems 225

A CLOSER LOOK 14.1 Banana Republics: The Story of the Starchy Fruit 226

   Storage Roots 226
White Potato 228
   South American Origins 228
   The Irish Famine 228
   Continental Europe 229
   The Potato in the United States 229
   Solanum tuberosum 230
   Pathogens and Pests 231
   Potato Genome 232
   Modern Cultivars 232

Sweet Potato 232
   Origin and Spread 233
   Cultivation 233
Cassava 234
   Origin and Spread 234
   Botany and Cultivation 234
   Processing 235

A CLOSER LOOK 14.2 Starch: In Our Collars and in Our Colas 237

Other Underground Crops 238
   Yams 238
   Taro 238
   Jerusalem Artichoke 238

Chapter Summary 240
Review Questions 240

Most plants store food reserves in the form of starch; often these reserves are stored in underground organs, in either some types of roots or modified stems. This is the case with most of the starchy staples considered in this chapter. Most of them are of tropical origin, even though today some are extensively cultivated throughout the temperate world. These crops, all propagated asexually and not planted from seed, tend to be highly productive, yielding many tons per acre. An added characteristic of underground crops is their ability to provide food insurance against accidental or natural disasters, such as fire, typhoons, or hail.

Since prehistoric days, humans have utilized these storage organs as valuable food sources, and today some of them are significant world crops. The potato and cassava are among the top 10 crops in terms of tonnage produced annually, and sweet potato is the fourteenth in tonnage. Nutritionally, the starchy staples are high in carbohydrates, mostly starch, but are low in protein and fat.

An important starchy fruit, the banana is considered in A Closer Look 14.1: Banana Republics.

STORAGE ORGANS
Modified Stems

Stems are not always vertical and aboveground; many plants have modified stems that can be horizontal and even underground. These modified stems have a variety of functions. Some are specialized for asexual reproduction; some are specialized for food storage; and some can function in both capacities. In many herbaceous perennial plants, these modified stems serve as a protected food reserve. When the aboveground vegetative structures die back, these food reserves are available for renewed growth on the return of favorable weather conditions. These modified stems, like erect stems, have recognizable nodes and internodes.

Stolons, or **runners,** are aboveground, horizontal stems that produce buds and roots at the nodes. These buds soon develop into plantlets; an entire area can be quickly invaded through this method of vegetative reproduction. One familiar example of plants that reproduce by stolons is the strawberry; in fact, the name is derived from *strayberry,* which refers to the stoloniferous habit (fig. 14.1a). The common lawn weed crabgrass and the familiar houseplant *Chlorophytum,* or spider plant, are two other stolon producers.

Horizontal stems that are underground, often just below the surface, are known as **rhizomes.** Reduced, scalelike leaves are present on the surface of rhizomes, and adventitious roots form all along the underside (fig. 14.1b). (Adventitious roots develop from organs other than roots.) Buds found at the nodes can give rise to new plants. In some, the rhizome is also a food-storage organ; ginger and iris are plants with this type of rhizome.

Tubers are the enlarged storage tips of a rhizome; the familiar white potato is a tuber. The "eyes" of the potato are actually buds located at the nodes, and each bud can give rise to a new plant (fig. 14.1c).

(a) Stolon (b) Rhizome (c) Tuber

(d) Bulb (e) Corm (f) Tuberous root (g) Taproot

Figure 14.1 Modified stems and roots.

A CLOSER LOOK 14.1

Banana Republics: The Story of the Starchy Fruit

Americans consider bananas a favorite dessert or snack food; however, for millions of people in tropical countries, bananas are an important dietary staple. The familiar sweet bananas are usually eaten raw, but starchy plantains, which are more important as food in the tropics, are traditionally cooked and eaten as a vegetable. Cultivation of the sweet banana is greatest in Central America, while Africa leads the world in plantain production. Unlike other starchy staples, bananas are true fruits, even though modern cultivars are seedless.

Bananas are native to Southeast Asia, the Indo-Malaysian region, and northern Australia and are believed to have been among the first cultivated plants in these areas more than 7,000 years ago. It is thought that Polynesians spread the banana throughout the Pacific Islands. Records from India indicate that bananas have been cultivated there for at least 2,500 years. About 2,000 years ago, Arabian traders introduced bananas into parts of Africa, where the crop flourished. In fact, the word *banana* comes from West Africa. Portuguese and Spanish colonizers spread bananas throughout tropical regions, and early in the sixteenth century they were introduced to the New World. Bananas became so well established that many early explorers thought they were native to the New World.

Four centuries after their introduction, bananas played a pivotal role in the history of Central America. Early in the twentieth century, large U.S. corporations, such as United Fruit Company, developed extensive banana plantations in Central America, along with corporate-run railroads and steamships. Over the next 50 years, United Fruit exerted tremendous control over the economies and governments of several countries that became known as Banana Republics. The rise of nationalism, starting in the 1950s, led to the decline in the influence of United Fruit.

Bananas are produced by various species in the genus *Musa* in the Musaceae, the banana family. *Musa acuminata* and *M. balbisiana* are the major species grown for consumption. Cultivars of *M. acuminata* generally produce sweet bananas, while cultivars of *M. balbisiana* produce plantains. Over 1,000 cultivars are recognized, and many are hybrids of the two species. Most cultivars today are sterile triploids that probably arose from the fusion of a haploid gamete with a diploid gamete that resulted from erratic meiosis. Bananas require a tropical climate with constant moisture, and members of the genus are cultivated throughout the tropics for the fruit, the fiber (see Chapter 18), or even the foliage, which is often used to wrap foods as a type of natural waxed paper.

The banana plant, often called a tree, is a large, herbaceous perennial that can reach 6 meters (20 feet) or more in height (box fig. 14.1). The "trunk" of this monocot is not woody but is actually a rosette of overlapping, tightly packed leaf bases that arise from an underground corm. The leaf blades themselves are also large, often about 2.5 meters (8 feet) or more in length. Although the leaf blade is complete when it first develops, it is soon torn into strips by prevailing winds.

After approximately 1 year, the apical meristem converts from vegetative growth to flowering, and the monoeocious inflorescence develops. The single flowering stalk contains 5 to 13 groups of flowers (often called hands or bunches), with each group covered by a large purple bract. Most groups contain carpellate flowers that develop parthenocarpic fruit, with staminate flowers confined to the end of the inflorescence. The fruit, which is technically a berry, requires about 80 to 120 days to mature. Bananas destined for market are picked green and shipped under controlled conditions to prevent ripening. Once yellow, bananas perish rather quickly.

Fruit production terminates the life of an individual plant, but new suckers or sideshoots develop from the corm. Since the fruits are seedless, these suckers are used in vegetative propagation. Suckers reach maturity in 9 to 12 months.

Bulbs and **corms** are modified stems found in monocots. Bulbs are erect, underground stems with both fleshy and papery leaves; food is stored in the fleshy leaves. Onions, tulips, daffodils, hyacinths, and lilies are familiar bulbs (fig. 14.1d). Not only will each bulb give rise to a new plant during the growing season, but the bulbs themselves can also multiply. New bulbs can develop in the axil of the leaf scales; these eventually will separate from the parent bulb.

Although commonly confused with bulbs, corms store food reserves in the stem, not in the leaves (fig. 14.1e). These erect underground stems lack the fleshy food-storage leaves and are covered only with dry, papery leaves. Like bulbs, corms can multiply in number by producing small corms, or cormels. Examples of plants that form corms are gladiolus, crocus, and taro.

Storage Roots

Tuberous roots are modified fibrous roots that have become fleshy and enlarged with food reserves. Like modified stems,

Box Figure 14.1 Inflorescence of a banana plant. Carpellate flowers are visible beneath the purple bract, and a hand of green fruit is visible above the bract.

Because of the asexual reproduction, all plants of one variety are genetically identical clones and share susceptibility to various pathogens. The worst threat to plantations is Panama disease, caused by a fungal pathogen, *Fusarium oxysporum* f. sp. *cubense*, that produces a vascular wilt. Growth of the fungus in the vessels of the xylem brings about the wilting; leaves turn yellow, then brown, and eventually die, even though water is available. Panama disease decimated Central American plantations from the early to mid-twentieth century. At the time, the banana variety grown throughout the region for export was Gros Michel, a variety that was susceptible to the pathogen. Most other banana varieties are also susceptible to the pathogen. Gros Michel plants were replaced with a resistant variety, known as Cavendish, that soon became widely grown throughout the tropics and today accounts for 40% of the cultivated bananas worldwide. Cavendish bananas also dominate the world export trade. A new strain of Panama disease fungus emerged in the 1990s, initially in South East Asia and slowly spread to Indonesia, Australia, China, the Philippines, and Mozambique. There have also been recent reports of this strain in the Middle East. Although the new strain of the Panama disease fungus has not been reported in Central America, the threat to bananas everywhere is real. There are no effective fungicidal treatments for Panama disease. Resistant varieties of banana are the only viable control measure. In 2017, researchers from Australia and the Netherlands developed transgenic Cavendish bananas carrying a resistant gene RGA2 from a wild relative of banana. Field trials identified varying levels of resistance in the transgenic lines with one line remaining disease-free for the 3-year trial. Additional field trials of other transgenic cultivars carrying the resistance gene are in progress. Researchers also hope to identify those transgenic lines that could be developed for commercial release. This research is a big step in saving bananas from Panama disease.

Another serious disease is black Sigatoka, caused by the fungus *Mycosphaerella fijiensis*. This leaf spot disease produces significant damage to the leaves, resulting in yield losses of 50% or more. It also causes premature ripening of the fruit, which is a serious problem for export trade. Although this pathogen can be controlled by fungicides, the cost is high. In many areas weekly applications of fungicides are necessary during the growing season. For many plantations, the total cost of chemical control often reaches $1000 per hectare (2.47 acres). Black Sigatoka threatens both dessert bananas and plantains; the goal of breeding programs is the development of resistant varieties. In August 2012, an international team of researchers published the results of their work sequencing the banana (*Musa acuminata*) genome. The genome contains over 500 million bases and approximately 36,000 genes. Among other benefits, this achievement may enable researchers to identify various genes responsible for disease resistance, thereby aiding in the development of banana varieties resistant to both fungal pathogens.

Nutritionally, the fruits are good sources of energy and potassium and are easily digestible. Bananas are rich in starch; however, in the sweet varieties, some of the starch is converted to sugar as the fruit ripens.

---

tuberous roots can function in asexual reproduction as well as in food storage. Tuberous begonias, dahlias, cassava, and sweet potatoes all produce tuberous roots (fig. 14.1f).

As described in Chapter 3, taproots are often food-storing organs for biennial plants such as carrots, rutabagas, and turnips (fig. 14.1g). They differ from tuberous roots in that taproots are enlarged primary roots. Because taproot crops are not dietary staples (although they certainly supplement the diet with added nutrients), they are not considered in this chapter.

Thinking Critically

Many variations of underground food-storage organs have evolved in plants: bulbs, corms, tubers, tuberous roots, and taproots.

What environmental factors would favor the development of underground storage organs? How would you investigate your hypothesis?

WHITE POTATO

South American Origins

Although the white potato, *Solanum tuberosum,* is often associated with Ireland or Idaho, its origins can be traced back to South America. About 10,000 years ago, native people in the Andes Mountains of southern Peru domesticated the potato from wild species. Cultivated potatoes were widely distributed throughout the Andes region, including the lowland areas of southern Chile. By the time the Spanish conquistadors, led by Francisco Pizarro, arrived in the 1530s, the potato was the staple of the great Inca civilization that extended thousands of miles along the west coast of South America. Although the Spanish were attracted by the Incan gold, silver, and jewels, the real Incan treasure was later realized to be the potato (fig. 14.2).

The potato was introduced to Spain sometime during the middle to late sixteenth century. The exact date of introduction is uncertain; however, detailed accounting records from a hospital in Seville indicate that potatoes were among the food items purchased in 1573, and by 1580 potatoes were widely available in the markets of that city. Once introduced, cultivation of the potato slowly spread throughout Europe. Grown first as a curiosity and later as a food for livestock, the potato was widely accepted as a food for humans only in the eighteenth century. The slow acceptance by Europeans can be traced to misinformation about its edibility. Because European relatives of the potato (nightshade, mandrake, and henbane) were known to be poisonous or hallucinogenic, many assumed that the tubers would also be deadly. (In fact, the tuber is the only part of the plant that is safe to eat; all the aboveground parts are, indeed, poisonous.) Others feared that the potato could cause a variety of diseases, such as leprosy, rickets, and consumption. Still others claimed that the potato was an aphrodisiac, stimulating lust and passion.

The Irish Famine

Nowhere else in Europe was the potato accepted as readily as in Ireland. It was an established crop as early as 1625 and was a dietary staple for the Irish peasant throughout the eighteenth century and the first half of the nineteenth century. The climate and soil in Ireland were ideal for the potato; even small plots of land could yield a large enough harvest to feed a family. The potato was so successful that historians link the subsequent population explosion (from 1.5 million to 8.5 million between 1760 and 1840) to the presence of this reliable food supply. The poor in Ireland subsisted on potatoes, some milk, and, occasionally, fish or meat; estimates are that the average adult consumed from 4 to 6 kilograms (8 to 12 pounds) of potatoes each day. This reliance on a single crop set the stage for disaster: the Irish potato famine of 1845–1849.

The most lethal pathogen of the potato is *Phytophthora infestans,* which causes the disease late blight of potato. The pathogen, a fungal-like oomycete (see Chapter 23), attacks and destroys the leaves and stem, causing them to blacken and decay in a short time, thereby stopping tuber growth. In addition, the tubers themselves are attacked and rot in the ground, or even later in storage. In cool, wet weather, the pathogen can kill a plant within a week. The disease first appeared in continental Europe in 1844, most likely inadvertently carried with new varieties of potato from Central or South America. It first appeared in Ireland in August of 1845 and caused devastation during the next few years.

The disease struck several times during the period of 1845 to 1849, resulting in widespread destruction of the potato crop and devastating famines among the Irish peasantry. It is estimated that over 1 million people died from starvation or from virulent diseases, such as typhus and cholera, that followed the famine. Another 1.5 million Irish emigrated to all parts of the world, but largely to the United States, resulting in a 25% to 30% decline of the population in Ireland in less than a decade.

Without some background in Irish history, it is difficult to understand how the famine was so devastating to the Irish population. The history between the Irish and their English rulers is one of conflict, dating back to a series of invasions and rebellions beginning in the Middle Ages. The lingering unrest in Northern Ireland today is the legacy of centuries-long turmoil. Ireland for most of this period was, in essence, a colony of the English; and as such, the land and people were shamelessly exploited. The Penal Laws enacted in 1695 are witness to this exploitation. Some of its provisions were

Catholics were not allowed to practice their religion.

Catholics could not serve in the armed forces.

Catholics could not practice law or hold office.

Figure 14.2 The potato has been cultivated for approximately 10,000 years in the Andean highlands of South America. This ancient pottery vessel from South America is in the form of a potato.

Catholics could no longer buy land.

Catholic estates were to be divided and subdivided among the male heirs (unlike the English system, in which the eldest son was the sole inheritor of large land holdings). If one of the heirs became Protestant, he could inherit the whole estate.

Catholics were barred from attending school and even from going abroad to study.

The Penal Laws deprived the Irish of religious freedom and economic opportunity. The Irish were turned out of their lands and forced to live as tenants on lands they once owned. Much of the country was held by absentee English landlords and populated by Irish tenant farmers. The best lands were reserved for the cultivation of cereals, such as barley, oats, and wheat, that were sold to England. Much of the Irish populace was reduced to dire circumstances even in the best of economic times. Potatoes were the only crop that could produce enough to feed a family on an extremely small acreage. Half of all the farms in Ireland in 1841 were less than 5 acres. A single acre of land could yield 12 tons of potatoes per year; a man, wife, and four children could live on 5 tons of potatoes, and the rest could be sold to pay taxes and rent. Even so, potatoes did not store all that well, and the poorest farmers were always on the edge of starvation by summertime before the fall harvest.

At the beginning of the famine years, Prime Minister Robert Peel secretly brought in maize from the United States and sold it at a low price to keep food prices down. Unfortunately, poor people had little money to buy food at any price. Also, many merchants complained that the government was underselling them and cutting into their profits. The policy of free trade took precedence over food distribution programs with the succeeding years of the famine. Public works projects were created to provide jobs, but the pay was so meager that a day's wage for backbreaking work was not sufficient to buy food. Workhouses accepted the destitute but forced wives to separate from husbands and children from parents. Anyone who entered a workhouse was forced to give up his land, which to a farmer meant relinquishing any hope of recovery in the future. For these reasons, many preferred to take their chances on the outside rather than enter a workhouse. The starving were forced to fend for themselves, eating so-called famine foods—roadside weeds, such as sorrel and nettle—or scouring the beaches for seaweed and mussels. Although the government in London refused to intervene, it tried to make landlords responsible for their starving tenants. With the passing of the Poor Law Extension Act, the Irish were taxed to pay for the costs of famine relief. Landlords were made liable for taxes owed by starving and impoverished tenants. The collected taxes would then be applied to famine relief. Some landlords went bankrupt in their efforts to take care of the famine-stricken tenants. Others offered paid passage to Canada for those unable to pay taxes. Still others followed a ruthless policy of eviction, known locally as extermination. The hovels of the poor were ripped apart, and the former inhabitants were forced off the land to certain starvation, disease, and death.

Many private organizations from the United States set up soup kitchens to feed the destitute. The Quakers especially were praiseworthy in their relief efforts. Also memorable is the response of Native Americans to the Irish famine. Although impoverished themselves, the Choctaw Nation collected $170 to assist the relief effort of 1847. In 1995, nearly 150 years after the famine, Irish President Mary Robinson visited the Choctaw capitol in Durant, Oklahoma, to thank tribal members for their past generosity.

Continental Europe

The potato was also widely grown in Europe, where cultivation was strongly encouraged by the aristocracy and leaders in several countries as a cheap, reliable food for the peasants. Frederick Wilhelm, the Elector in Prussia, promoted the cultivation of potatoes as early as the 1650s. In 1744, his great grandson, Frederick II (the Great) of Prussia, still attempting to gain acceptance for the potato, decreed that potatoes were to be cultivated in Pomerania and Silesia (now part of Poland) and distributed seed potatoes for planting. By 1770, the cultivation of the potato was extensive, partly due to crop failures of wheat and rye. During the War of Bavarian Succession (1778–1779), the opposing armies of Austria and Prussia faced off to a stalemate in Bohemia. Both armies consumed the local potato harvest until depletion of the food supply and cold weather forced them to retreat; today the war is also known as the *Kartoffelkrieg*, the Potato War.

In France, the champion of the potato was Antoine-Augustin Parmentier, who learned the value of the potato when he was a prisoner of war in Hanover during the Seven Years' War (1756–1763). Apparently, he survived only on potatoes during his time in prison. Upon his release, he returned to France and found widespread famine; he worked to popularize the potato as the solution to the famine. His efforts were rewarded when the French finally incorporated the potato into their daily diet after the French Revolution.

By the end of the eighteenth century, the potato had at last gained widespread acceptance as food throughout Europe. Production continued to increase during the first half of the nineteenth century, until the potato blight devastated the crops throughout continental Europe. Although famine resulted, the effects were not as severe as the Irish famine because the potato was not the sole dietary staple.

The Potato in the United States

Although native to the New World, *Solanum tuberosum* made its appearance in North America through the European colonists. There is some doubt as to the exact date of introduction; some claim that the potato was introduced to

Virginia as early as 1621, while others claim 1719 as the year of introduction to New England. Part of the uncertainty is related to confusion in historical records between the white potato and the sweet potato, which had been introduced to Europe around the same time. The word *potato* stems from the Arawak Indian word *batata*, which actually referred to the sweet potato, an edible root completely unrelated to the white potato.

In the United States today, potatoes are grown in virtually every state at some time of the year, but the top-producing states are Idaho, Washington, and Maine. About one-third of the U.S. harvest is consumed fresh, and one-half is processed to make frozen french fries, potato chips, dehydrated flakes, and other products, including potato starch. Almost half of all processed potatoes are made into frozen french fries. The deep-frying of potato strips presumably originated in France, and tradition holds that they were introduced to the United States by Thomas Jefferson during his presidency. Today, most french fries come pre-packaged and frozen, a convenience food whose popularity skyrocketed in the latter half of the twentieth century with the availability of home freezers and the prominence of fast-food chains.

Another favorite processed potato product, the potato chip, is said to have been developed by George Crum, a short-order cook in Saratoga Springs, New York, in the 1850s. The story goes that a customer complained about the thickness of his fried potatoes. In exasperation, Crum deep-fried superthin slices of potato, and the potato chip was born. These should not be confused with the British "chips," which are actually french fries; our potato chip is called a crisp in the United Kingdom.

Processed potatoes are nothing new in the history of the potato; Peruvian Indians from high in the Andes Mountains have prepared chuño, a kind of freeze-dried dehydrated potato, for the past 2,000 years. Potato tubers are spread on the ground when a heavy frost is expected. After freezing, the potatoes are allowed to thaw during the day and then trampled underfoot to express water and bitter compounds and to remove the skins. Processing methods vary after these initial steps. One method uses extensive washings to leach additional bitter compounds; this is followed by drying. Another method continues the freezing-thawing-trampling cycle for several days before final drying. The resulting chuño, from either method, is used primarily as a basis for soups, stews, and desserts. This dehydrated product can be stored for several years without spoiling and provides a stable food reserve.

Solanum tuberosum

The cultivated potato belongs to the genus *Solanum,* a large genus with more than 2,000 species in the Solanaceae, or nightshade family. Only 200 species in the genus are tuber bearing, and only ten of them have been cultivated, with most of these mainly confined to native regions in South America. In North

Figure 14.3 A few of the cultivars of *Solanum tuberosum,* including Yukon gold on the left and Peruvian purple in front and back.

America there is archaeological evidence that *Solanum jamesii* was consumed by native tribes in the southwest about 10,000 years ago. Although there is no evidence of early cultivation, *S. jamesii* continued to be collected from the wild into the twentieth century. *S. tuberosum* is the most widely cultivated potato; in fact, there are almost 6,000 cultivars (fig. 14.3) of this one species, but most commercial growers plant only a limited number of varieties. In the United States, of the 50 varieties that are grown commercially, 12 account for 85% of the potato harvest.

Today potatoes are grown in 160 countries around the world and rank right behind wheat, rice, and corn in importance to the human diet. Potatoes are a dietary staple for the poor in many parts of the world and are more productive on less land than other crops. China currently produces approximately 24% of the world potato crop, followed by India with 12%, and the Russian Federation at 8%. The Ukraine and United States follow with 6% and 5%, respectively, of the world total.

The potato plant is a bushy, herbaceous annual with an alternate arrangement of large, pinnately compound leaves. The white, pink, purple, blue, or yellow flowers are five-merous, a common dicot floral pattern. The fruit is a many-seeded berry, but normally the flower does not set seed because the precise temperature and day length requirements are not met under most northern agricultural conditions. Two types of stem are produced: the ordinary foliage-bearing stems and underground rhizomes that end in tubers (fig. 14.4). Anatomically, the tuber is a modified version of a dicot stem, with an enlarged pith, a ring of vascular tissue, a narrow cortex, and a thin periderm. In some potato chips, the vascular tissue often appears as a darkened ring, with the narrow cortex on the outside and the large pith within the ring. The potato is a cool-season crop, with maximum tuber production at temperatures ranging from 15° to 18°C (59° to 64°F). Tuber

CHAPTER 14 Starchy Staples   231

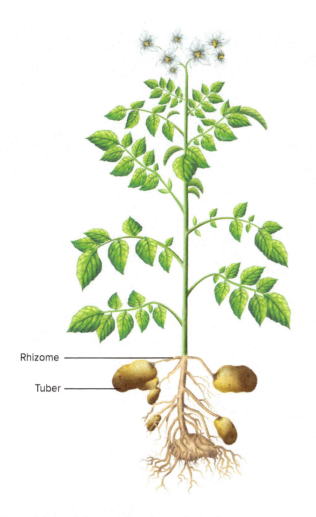

Figure 14.4 *Solanum tuberosum*, cultivated potato.

formation is reduced at higher temperatures and inhibited at 29°C (84°F); this has limited the usefulness of the potato as a food crop in lowland tropical countries.

Potato cultivation involves propagation by seed potatoes, which are small, whole tubers or cut tuber pieces containing at least one eye. (Recall that the eyes are actually buds.) Cultivation by seed potatoes is a method of asexual reproduction that produces plants genetically identical to the parent and maintains the desired traits within a cultivar. Seed potatoes are produced by certain growers who specialize in growing only seed potatoes; certified seed potatoes are ones that have been inspected and determined to be disease free. This is an important consideration because many potato diseases (in addition to late blight) can be transmitted through infected tubers. Although asexual reproduction is fast and produces plants with prescribed qualities, there are disadvantages as well. Genetically identical plants all share the same susceptibility to adverse environmental conditions and diseases. The devastation caused by *P. infestans* was compounded by the fact that most of the potatoes in Ireland were genetically identical because they were derived from one or two plants originally introduced into the country.

Pathogens and Pests

Thought to be long under control, *P. infestans* has again been threatening global potato production. It is estimated that current worldwide losses from this pathogen amount to $6.7 billion each year. One factor in the resurgence of late blight is the appearance of exotic strains that are resistant to the fungicides being used. The origin of these strains is Mexico. In 1976, drought in Europe caused a decline in the potato harvest. Mexican potatoes brought in to supplement the harvest were apparently the vehicle for the *Phytophthora* reintroduction. The late blight epidemic in the 1840s was caused by a single strain from Mexico with limited genetic variability. This strain was also incapable of reproducing sexually because it consisted of only a single mating type. With the latest introduction, the availability of the opposite mating type enables sexual reproduction to occur. Sexually reproduced spores can survive in the soil for months, perhaps years. Sexual reproduction also introduces genetically variable populations that may differ in their virulence and susceptibility to fungicides. These factors may account for the aggressiveness of these new strains. In 3 years, the initial infection spread from New York State in 1992 to 23 states by 1995, and since then over 20 new strains have been identified in the United States. During the 2009 growing season, a new strain spread through 29 states with losses to U.S. farmers estimated at $1.8 billion. Similar outbreaks of new strains have also occurred in other countries. Since 2008, severe late blight epidemics have occurred in southern India on both potato and tomato crops.

Potato crops around the world have been affected, and the gravity of this situation has not gone unnoticed by the International Potato Center in Lima, Peru, which in 1996 called for a global effort to control the resurgence of this devastating disease. One approach being undertaken is to look for resistance among wild species of potato. Scientists from the International Potato Center and the University of Wisconsin have determined the distribution of all wild potato species. Wild potatoes grow in the highlands from the southwestern United States to Argentina and Chile; the greatest species diversity occurs in Peru. Many potato species have been collected and studied intensively for ongoing breeding programs. A major aim of this research has been to identify potato genes that confer resistance to the late blight pathogen. Although several resistant genes have been identified, most of these provide resistance to only one strain of the pathogen. However, researchers from the University of Wisconsin found a wild potato species, *Solanum bulbocastanum*, that is resistant to all known strains of the pathogen; researchers also identified the gene that conferred blight resistance. They have recently cloned this gene and inserted it into a susceptible cultivated potato variety. The resultant transgenic potato plant was resistant to all strains of the pathogen tested. Also, in 2018 researchers from the United Kingdom identified a new late-blight resistant gene from a Mexican species of wild potato, *Solanum verrucosum*, which is known to be resistant to *P. infestans*. This research is promising because the new gene provided broad-spectrum resistance and is genetically different than any previously identified resistance genes. These exciting results may lead to the development of commercial varieties with late blight resistance.

In 2009, an international team of researchers reported sequencing the *P. infestans* genome and found that the genome is larger and more complex than other *Phytophthora* species. Scientists have been able to identify the virulence genes in the pathogen; it is thought that the products of these genes may be targets for resistance proteins produced by the potato plants. It is hoped that this research will lead to new methods of protecting crops and end the devastation caused by *P. infestans*.

Potatoes are also susceptible to a number of other diseases and are attacked by many insect pests. The Colorado potato beetle is the most serious insect pest of potatoes and other crops in the Solanaceae, including tomatoes, eggplants, and peppers. Both the adult beetle and the larvae feed on potato leaves; the loss of foliage retards the development of tubers and may even kill the plants. Unfortunately, these beetles have become resistant to many of the chemical insecticides used for control. As a result, farmers use various approaches to control the beetles, including natural insecticides, such as those produced by the bacterium *Bacillus thuringiensis* (Bt). Bt insecticides have been available commercially for more than 40 years, and scientists have used genetic engineering (see Chapter 15) to introduce the genes that control the production of the insecticidal proteins into several crop plants. In 1995, genetically engineered potato plants with Bt genes were approved for use in the United States, and the first such crops were grown in 1996. These plants, developed by Monsanto Company (a large agribusiness conglomerate), were called New Leaf potatoes. The genetically engineered potatoes were grown for five seasons; however, they amounted to only 2% to 3% of the U.S. potato crop. In addition, the controversy over genetically engineered foods resulted in a weaker market for the potatoes. As a result, Monsanto discontinued its line of New Leaf potatoes in 2001.

Potato Genome

In July 2011, the Potato Genome Sequencing Consortium, an international team of researchers, published the results of their multiyear efforts compiled from two different varieties of potato. They established that the *S. tuberosum* genome contains approximately 844 million bases and consists of more than 39,000 genes. As part of their efforts, the team also identified many of the genes responsible for tuber formation, as well as the location of many disease-resistant genes. Because potatoes are vulnerable to many pathogens and pests, the information from this research is providing new resources for potato breeders.

Modern Cultivars

Many of today's cultivars can be grouped into four familiar types: the round white, russet, round red, and long white. The round white is an all-purpose potato good for boiling, baking, or processing into chips, fries, or flakes; Superior and Katahdin are common round whites. Russets, also known as Idahos, are elongate, cylindrical tubers with a corky, russet-colored skin and mealy texture that make excellent baking potatoes. Their shape also makes them ideal for french fries. Russet Burbank is the top-selling russet potato. A few years ago, USDA researchers in Idaho released the Defender, a russet variety with resistance to late blight. This variety, developed by traditional breeding, is gaining popularity for processed potatoes. The last two types, round reds and long whites, are usually sold as new potatoes, which are harvested earlier in the growing season and have a very thin skin. By contrast, late potatoes are harvested later and have a thicker, corky skin. Popular round reds, Red LaSoda and Norland, good for boiling, steaming, and roasting, are distinguished by their red skin and firm flesh. White Rose, a common long white, is a good, all-around potato for home use; it has a tan, smooth skin and is often even more elongate than a russet. New varieties, such as Yukon gold and purple Peruvian, are good, all-purpose potatoes and add interesting color to the dinner plate (fig. 14.3).

Potatoes are rich in carbohydrates (about 25% of the fresh weight); parenchyma cells within the pith are filled with starch grains (see fig. 2.7). Although not rich in proteins (only 2.5% of the fresh weight), the potato is actually a good source of high-quality protein because of the balance of essential amino acids. Potatoes are virtually fat-free (if not fried or topped with butter and sour cream), have no cholesterol, and are good sources of potassium, iron, several B vitamins, and vitamin C. Much of the vitamins, minerals, and fiber of the potato occurs in the periderm and cortex; therefore, the nutritional value is enhanced if potatoes are eaten with skins.

From its South American origins, the potato has traveled an interesting route to its present status as the world's fifth leading crop.

SWEET POTATO

Although the common name implies some relationship to the white potato, the sweet potato (*Ipomoea batatas*) is a storage root in the morning glory family, the Convolvulaceae. In fact, the flowers of the sweet potato resemble those of the familiar morning glory in gardens. The sweet potato plant is an attractive vine that is easy to grow at home. (A root placed in water will soon produce sprouts, or slips, trailing shoots with adventitious roots.) Like the white potato, the sweet potato is propagated vegetatively. Slips, produced by the tuberous root in specially prepared beds, are transplanted into the field (fig. 14.5). In addition, sweet potatoes can also be propagated by vine cuttings. The crop requires a long, warm growing season because sweet potatoes are particularly susceptible to chilling injury.

Figure 14.5 *Ipomoea batatas*, sweet potato. (a) Vegetative propagation of sweet potato. Adventitious shoots, known as slips, develop from the tuberous root. (b) A single slip ready for planting.

Origin and Spread

Columbus discovered the sweet potato on his first voyage to the New World and introduced it to Spain on his return, at least 50 years earlier than the introduction of the white potato. There were several native names for the sweet potato in the New World, but in the Caribbean the Arawak Indians called it *batata*, which was eventually corrupted into the word *potato*. Originally, the name *potato* was exclusive to *I. batatas*, but *S. tuberosum* also became known as a potato because it, too, was an underground crop. After its introduction, the sweet potato became widely grown in Spain and other Mediterranean countries and was exported to northern Europe. The sweet potato was initially considered a delicacy in Europe; it was a favorite dish of England's King Henry VIII. It was also rumored to be an aphrodisiac, a claim that was later transferred to the white potato along with the name.

The sweet potato is native to tropical Central and South America; archeological evidence dates its cultivation back at least 8,000 years in Peru. It was widely grown as a staple crop in Central America and tropical South America prior to the European arrival in the New World. During the same period, it was also cultivated in several Pacific islands and New Zealand. Evidence of its history of cultivation in Polynesia suggests an earlier introduction, and archaeological research has suggested that this occurred approximately 700 to 800 years ago. Was the sweet potato introduced to Polynesia by early seafaring natives or by the natural dispersal of seeds by birds or ocean currents? Thor Heyerdahl's voyage from Peru to Polynesia in the reed raft *Kon-Tiki* in 1947 demonstrated that a voyage by primitive peoples was a possibility.

In 2013, Caroline Roullier and colleagues published molecular evidence supporting the early introduction by eleventh- to twelfth-century seafarers. Their research showed genetic differences between present-day sweet potatoes from South America and those from the Caribbean region. In the Polynesian islands, current sweet potato varieties showed mixed genetic lineage. However, when the researchers used molecular methods to examine herbarium specimens of sweet potato plants collected from the Polynesian islands by European explorers from the 17th to early 20th centuries, they found molecular markers indicating their origin only from South American sources. The researchers suggested that more recent introductions of sweet potatoes interbreeding with older varieties obscured the pre-Columbian South American origins. Further supporting evidence for this comes from the linguistic similarity between the names for sweet potato in Polynesian languages and the name used by the Quechua people of South America.

In 2018 new evidence on the evolutionary origin of the sweet potato renewed the decades-long discussion on the disjunct distribution of sweet potatoes. Research by Robert Scotland and colleagues from the University of Oxford in the United Kingdom and other scientists from the United States and Peru analyzed DNA from 199 specimens of sweet potato and their closest wild relatives. The research showed that sweet potatoes (*Ipomoea batata*) evolved from *I. trifida*, a wild weedy relative that does not produce any storage roots. The molecular data showed that the ancestor of sweet potatoes split from *I. trifida* about 800,000 years ago in Central America or northern South America. Scotland and colleagues indicated that other species in the genus *Ipomoea* also have disjunct populations between America and Polynesia. They also reported that *Ipomoea* species, closely related to sweet potato with similar seeds and fruit, are known to disperse by sea currents. Although their findings do not rule out the possibility of ancient seafarers introducing sweet potato to Polynesia, it does suggest that long distance dispersal by sea currents is a logical explanation.

Cultivation

Today the sweet potato is a significant crop throughout the tropics and has expanded to warm temperate regions; in some countries, it is used as livestock feed as well as an important food staple. China dominates the world's production of the sweet potato, with several other Asian countries distant seconds. In Asian countries such as Japan, the sweet potato has uses beyond its consumption as a starchy vegetable. It is often sliced and dried, then ground into a meal that can be used in cooking or fermented to make an alcoholic beverage. The sweet potato has also become an important component of the diet in several African countries. In the

United States, the sweet potato, originally promoted by George Washington Carver, is primarily grown in the South, with North Carolina the leading producer. Two distinct varieties of *I. batatas* are widely grown, a drier, starchier, and yellow-fleshed variety is favored in northern states, while a sweeter, moister, deep orange variety is preferred in the South. The latter variety is commonly called yams but should not be confused with the true yam, *Dioscorea* spp., an important tropical tuber. Varieties with purple or white flesh are also available although these tend to be more popular in other countries. Recently, USDA researchers in South Carolina have developed new varieties of sweet potatoes with less sweetness and a mild flavor, traits suitable for fries and chips. Although the market for sweet potato chips is still small today, the USDA researchers hope to change this situation with these new varieties, which will produce more nutritious snack foods.

Like the white potato, the sweet potato is a nutritious vegetable rich in carbohydrates and certain vitamins and minerals. The sweet potato is accurately named because some of the carbohydrates are present in the form of sugar; it contains about 50% more Calories than white potatoes but slightly less protein. In addition, the carotene-rich root is an excellent source of vitamin A and is a good source of vitamin C. Purple-fleshed varieties are also rich in anthocyanins, pigments that impart deep red, blue, and purple colors in plants. Anthocyanins are thought to function as antioxidants, which may provide some additional health benefits (see Chapter 10). Sweet potato consumption has been steadily increasing in recent years as American consumers are starting to realize that this nutrient-rich vegetable deserves a more prominent role in our diet than just at Thanksgiving dinner.

Thinking Critically

The sweet potato is said to have a disjunct distribution, found in two widely separated regions of the world. Species with disjunct populations have been of great interest to phytogeographers, scientists who study and attempt to explain the distribution patterns of plants. For example, plant species with disjunct distributions in Africa and South America have been used as evidence to support the theory that these continents were once part of a single landmass that later moved apart as a result of continental drift.

If you were to find a plant species with widely separated populations, what hypotheses could you come up with to account for the distribution? How would you investigate your hypotheses?

CASSAVA

Although cassava, *Manihot esculenta,* is unfamiliar to most North Americans except in the form of tapioca pudding, this tuberous root is an important starchy staple. This member of the spurge family (Euphorbiaceae) is a vital food to millions of people in the tropics. It is known by many different names, including manioc, tapioca, yuca, yucca root, and mandioca. Cassava ranks as the ninth leading world crop and is the third largest source of Calories for the human diet in tropical countries.

Origin and Spread

Manihot esculenta, a New World crop, had its origins in South America; wild populations occur in west central Brazil and eastern Peru. Cassava was a well-established crop in the New World tropics long before the arrival of the Europeans, and researchers have suggested that it was domesticated about 10,000 years ago. In South America today, Brazil is the leading producer of cassava. The Portuguese, who ruled Brazil, introduced cassava into West Africa in the sixteenth century, and it was introduced into East Africa in the middle of the eighteenth century. Extensive cultivation began in the twentieth century, and today Africa is a major producer of cassava. The Portuguese may also have introduced cassava into Asia via the port of Goa, on the Indian subcontinent, in the early eighteenth century. A later introduction from Mexico brought cassava to Indonesia and the Philippines. During the nineteenth century, cassava became a widely grown crop in areas of tropical Asia. Today Nigeria, Thailand, Brazil, Indonesia, and Ghana lead the world in cassava production. As with many other crops, cassava cultivation has spread far beyond the region of origin.

Botany and Cultivation

Manihot esculenta is a tall shrub with palmately compound leaves and numerous tuberous roots that are similar in appearance to sweet potatoes but usually much larger (fig. 14.6). Plants are propagated by stem cuttings that contain axillary buds; growth is fairly rapid, and little care is needed following planting. The propagation is distinctive because none of the root, the economically valuable part, is used, unlike the other starchy staples where part of the harvest must be saved for cultivation. Although usually propagated from stem cuttings, cassava can also be cultivated from seed, which can be a source of new genetic varieties.

Many varieties of this tropical crop are tolerant to a wide range of moisture and soil conditions. Although cassava is predominantly grown in the hot lowland tropics, varieties exist that can survive the colder temperatures of tropical highlands. Although generally grown in areas that have moderate to heavy rainfall, cassava requires well-drained soils to prevent root rot. It can tolerate dry periods that last as long as 6 months and

Figure 14.6 *Manihot esculenta,* cassava, and cassava products. The large tuberous roots can be used to prepare chips or processed into starch for tapioca pearls or flour.

Figure 14.7 In traditional preparation of cassava meal, farinha, the peeled roots are grated and the juice containing HCN is expressed through a "tapiti," a woven basket, shown here.

therefore does well in areas where rainfall is seasonal. This partly explains its popularity in the drier regions of Africa. Other advantages of cassava are its ability to grow well even in the nutrient-poor soils typical of the tropics and its tolerance of acid soils as well as soils high in aluminum. Cassava is also resistant to many insects and fungal pathogens. Especially noteworthy is its resistance to locusts, which have periodically plagued parts of Africa. Even when the locusts eat the leaves, the roots are protected and the plant rapidly regenerates. The most damaging pathogen for cassava cultivation is the cassava mosaic virus, which is found throughout the cassava growing regions in Africa. The disease is spread by white flies or by planting contaminated stem cuttings, and crop losses of 30% to 40% are common in infected areas. Another damaging virus is the cassava brown streak virus, common in East Africa. Some cassava varieties resistant to the mosaic virus are highly susceptible to brown streak virus.

Depending on the variety, roots may be harvested anywhere from 8 months to 2 years after planting; all the roots from a single plant may be harvested at one time or removed individually from the ground as needed. Flexible harvesting is possible because the tuberous roots remaining in the ground continue to grow and do not decay. Once harvested, however, the roots are subject to rapid decay and must be dried or processed within 24 to 72 hours.

Processing

Varieties are classified as either sweet or bitter, based on the concentration of poisonous hydrocyanic acid (HCN). If not removed, this toxin can cause death by cyanide poisoning. The HCN is liberated by the action of enzymes on **cyanogenic glycosides** present in cassava. There is no distinction between the sweet and bitter varieties other than the concentration of the toxins, and intermediate types exist. Environmental conditions are known to influence the production of cyanogenic glycosides, and what is a sweet variety in one locale can be a bitter variety when grown under different conditions. Sweet varieties with lower HCN levels can be eaten with little preparation; peeling followed by boiling, steaming, or frying will remove most of the toxins. Bitter varieties, which actually taste bitter because of the much higher levels of HCN, must undergo more extensive preparation to be detoxified before the roots are consumed. Traditional methods of treating the peeled bitter roots vary from country to country and include drying, soaking, boiling, grating, draining, and fermenting or combinations of these. Although the sweet varieties are often cooked and consumed fresh, they can be processed similarly to the bitter varieties to make a meal or flour. The processed products all keep better than the quickly deteriorating fresh roots. Because of its bulk and perishability, most of the fresh cassava harvested is used locally and not exported. Worldwide, about 65% of the cassava grown is used for human consumption, with roughly half of that used as processed products.

In South America, the traditional preparation produces a meal called farinha. The peeled roots are grated (sometimes following a lengthy period of presoaking) and squeezed through a long, cylindrical, woven basket known as a tipiti (fig. 14.7). One end of the tipiti is tied to a tree while the other end is tied to a pole that is used to stretch the tipiti, thereby expressing juice from the grated pulp. This method eliminates much of the HCN from the pulp. The resulting mash is roasted until dried; in this form, it can be stored for long periods and eaten many different ways. In the Caribbean Islands and the northeast coast of South America, the ground meal that results from processing is commonly made into a large circular flat bread. This bread is called *casabe,* from which the word *cassava* is derived.

In Africa, traditional methods of preparation produce gari; however, this involves one additional step—fermentation.

The peeled roots are grated, and the resulting pulp is placed in cloth sacks. Stones and logs are piled on top of the bags to squeeze out the juices containing HCN. This process takes several days, during which time the pulp naturally ferments, imparting a characteristically sour taste to gari. Following the fermentation, the pulp is roasted or fried until dry and then stored. Gari, like farinha, can be eaten dry, used in soups or stews, or made into a doughy porridge.

In Africa, there have been reported cases of cassava paralysis due to the effects of cyanide poisoning on the motor regions in the brain. One local name for the condition is *konzo,* meaning "tied legs." Konzo may arise if shortcuts are taken in cassava processing, with the result that substantial amounts of cyanide remain in the prepared cassava. Even with proper processing, small amounts of cyanide can be found in cassava, but these are normally rendered harmless by combining with the body's sulfur stores to form thiocyanate, an inert compound. The body's source of sulfur comes from protein, however, and if a diet is both protein deficient and based mainly on cassava, cyanide accumulates and paralysis results.

One of the simplest techniques of processing cassava is the traditional method used in Indonesia where peeled roots are sliced and dried in the sun, allowing the HCN to diffuse out. The resulting chips are called gaplek. They can be stored for long periods or ground into flour and used to make breads, cakes, cookies, noodles, and other products. These are just a few of the many traditional ways to prepare cassava. In many parts of the world today, mechanization is replacing these labor-intensive methods.

Starch is the main nutrient supplied by cassava (approximately 30% of the fresh weight); however, the root is very low in protein (1% or less) and can result in kwashiorkor, protein deficiency, among people who rely on cassava exclusively. In some African countries, cassava leaves are also consumed as a cooked vegetable and are a better source of protein, which can partly offset the low values in the root. Roots also contain moderate amounts of calcium and several B vitamins and are a reasonably good source of vitamin C; however, cassava does not provide sufficient quantities of other micronutrients; it has only 30% of the minimum daily requirements (MDR) for iron and zinc, and 10% of the MDR for vitamin A.

Cassava breeding programs are attempting to improve the nutritional quality of the roots through both conventional breeding and genetic engineering. Researchers in Brazil have crossed cassava with wild species of *Manihot,* and a hybrid with 5% protein has recently been developed. In addition, a variety of cassava has been bred that produces 50 times more beta-carotene than regular cassava. Farmers in Brazil are now testing this cultivar, which could help alleviate widespread vitamin A deficiency in tropical countries. Other hybrids between cassava and wild relatives have produced virus-resistant varieties that are now widely cultivated. Although cassava can tolerate dry periods, work is also underway to develop varieties with improved drought tolerance to increase productivity in dry climates. U.S. researchers are using genetic engineering to produce varieties of cassava that are rich in protein, beta-carotene, iron, zinc, and vitamin E. Field trials are underway to test these varieties. The researchers are hoping to combine all these traits into a single variety, but this may take many years to develop. Some of this research is being carried out by BioCassava Plus, a team of scientists from Africa and North America who are attempting to reduce malnutrition in regions of Africa where cassava is the dietary staple. The BioCassava Plus project is funded by the Bill and Melinda Gates Foundation.

In November 2009, a draft sequence of the cassava genome was released. The genome consists of 770 million nucleotide bases and contains an estimated 30,666 genes. Subsequently another cultivated variety and a wild subspecies of *Manihot esculenta* have also been sequenced, and ongoing efforts are focused on sequencing additional cultivars. It is anticipated that this information will speed up the development of cassava varieties with improved nutritional balance and with resistance to viral diseases.

Although in Africa almost all the cassava grown is used for human consumption, in Asia and the Americas a considerable quantity is used for animal feed and for commercial starch production. Processing cassava roots into dried chips or pellets for animal feed has been increasing in recent years, so that today 20% of world production is fed to animals. Southeast Asia is the leading producer of animal feed for export, with much of it going to European markets. Cassava starch has many applications in the food, textile, paper, and pharmaceutical industries (see A Closer Look 14.2: Starch).

As described in this chapter, cassava is an important dietary staple in many tropical countries. At one time, the only form of cassava most North Americans had eaten was tapioca pudding, made by cooking tapioca pearls with milk, eggs, sugar, and vanilla. The pearls are partly gelatinized cassava starch made by heating moist cassava flour in shallow pans; heating causes the wet granules of starch to stick together and form beads. Today cassava and cassava products are found in many grocery stores, health food stores, and ethnic markets. Cassava (tapioca) flour is widely available and is commonly used in gluten-free baking, and cassava chips, which are similar to potato chips, are becoming an increasingly common snack food (fig. 14.6).

Thinking Critically

All varieties of cassava contain cyanogenic glycosides, yet native peoples learned to process cassava to remove the toxins.

Speculate on how early cultures could have learned these techniques.

A CLOSER LOOK 14.2

Starch: In Our Collars and in Our Colas

Starch is the most common carbohydrate storage product found in plants. Although the primary focus in this chapter has been the importance of starch in the diet, it is a versatile compound that has many commercial uses in both food and nonfood industries.

Sugars are the direct products of photosynthesis but are generally not stored as such. Instead, accumulated sugars are stored in the form of starch. The starch forms as starch grains found in amyloplasts, an organelle abundant in parenchyma cells of storage organs. Chemically, starch is composed of thousands of glucose molecules bonded together in chains. There are two distinct components of starch, amylose and amylopectin, which differ in the structural configuration of their chains. Amylose is an unbranched chain of several hundred to a few thousand glucose units. On the other hand, amylopectin is a highly branched chain containing between 2,000 and 500,000 glucose units. Amylopectin is the major component (approximately 80%) of starch, with amylose making up the remaining 20%, but the ratio varies with the species and variety of plant. These two forms of starch also differ in their nutritional benefit. Amylopectin is broken down and absorbed into the bloodstream more rapidly than amylose. As a result, the ratio of amylopectin to amylose in foods influences the glycemic index. Foods with a higher level of amylose have a lower glycemic index (see Chapter 10). Amylose tends to be relatively soluble in water, while amylopectin is insoluble. When potatoes are boiled, the amylose is extracted by the hot water, turning it cloudy. By contrast, when sticky varieties of rice (with low amylose) are cooked, all the water is absorbed, and the grains are clumped together because of the film of amylopectin adhering to the grains. In 2010, the European Union approved the commercial cultivation of the Amflora potato in member countries. This genetically modified potato produces starch composed almost entirely of amylopectin and will be used exclusively for industrial extraction of starch, not for human food.

Starch can be extracted from almost any grain, tuber, or other storage organ, but commercially most starch is extracted from corn, potatoes, and cassava, with smaller contributions from wheat, rice, arrowroot, and sago. Although starch extraction varies from species to species, the following is a general description of the process. Basically, the plant material is ground or pulverized to liberate starch grains from the cells. The resulting pulp, a mixture of starch and plant debris, is sieved and washed to remove debris from the starch suspension. Finally, the starch is separated from the water (by filtration, settling, or centrifugation) and dried.

Commercial starch finds hundreds of industrial uses; among the foremost nonfood applications are the manufacture of adhesives and the production of sizings. Starch-based adhesives are used in the production of cardboard, paper bags, and remoistening gums for envelopes and stamps, to name a few. In the past, they were also used in the lamination of veneers to make plywood, but today resins have largely replaced starch adhesives. Sizings are substances used as fillers or coatings, and starch is one of the most commonly used types of sizing. Steps in the manufacture of paper, cloth, thread, and yarn involve sizing; these steps can strengthen the material, impart a smooth finish, or prepare the surface for dyes. Starch finds other uses in the pharmaceutical industry as a binding and coating agent for medicines and in the petroleum industry as drilling compounds. Last, laundry starch is a familiar starch-based product with uses in the home as well as in industry.

One of the increasingly important uses of starch is in the production of sugar-based sweeteners. Starch is hydrolyzed with weak acids or enzymes to produce a glucose (dextrose) syrup, which may be marketed as such or enzymatically converted to fructose. High-fructose syrups are in great demand in the food-processing industry as a replacement for sucrose (table sugar) because fructose tastes sweeter than equivalent amounts of sucrose and is less expensive. Some starch is used directly in food processing for the manufacture of puddings, gravies, and sauces; in other cases, starch is blended with flour in the production of bakery goods.

The breakdown and fermentation of starch by yeast produces alcohol that can be used for beverages, for industrial solvents, or as a petroleum fuel substitute. Since the 1980s, the replacement of gasoline by alcohol has received much media attention. As petroleum reserves dwindle and prices increase, the search for alternative fuels will continue to intensify and the use of plants for this process will increase (see Chapter 12). Starch, as one of the world's renewable resources, will continue to have an expanding market.

OTHER UNDERGROUND CROPS

Yams

Dioscorea spp. are mainly tropical tuber crops, true yams, that are important staples in many areas of the world, especially West Africa, Southeast Asia, the Pacific Islands, and the Caribbean. This large genus in the Dioscoreaceae (yam family) is composed of several hundred species, of which 10 are major food sources; evidence indicates that yams have been cultivated for over 5,000 years in tropical Africa. Tubers of these vines can vary from small ones the size of potatoes to massive ones often weighing over 40 kilograms (88 pounds) and 2 to 3 meters (6 to 9 feet) in length (fig. 14.8). Yams can be prepared in ways similar to the potato, with boiling and mashing the most common method. As much as 20% of the tuber is starch, with the protein content averaging about 2% and vitamins and minerals in negligible amounts.

Medically, the tubers were an important source of **saponins** (sapogenic glycosides), a type of steroid that had been used to make human sex hormones and cortisone. In fact, early research on the development of the birth control pill was based specifically on diosgenin extracted from yams and shown to inhibit ovulation. Today, synthetics have largely replaced the natural compounds.

Yams are planted on small hills, with their twining vines and cordate leaves supported by stakes. They grow best under humid and subhumid conditions, so their cultivation is mainly restricted to the wet lowland tropics, and as such, *Dioscorea* is a crop that does well in tropical rainforests. Propagated from eyes, the tubers of this monocot are harvested after a long growing season of 8 to 12 months. Harvesting tubers is a laborious task because the tubers are generally deeply buried. For this reason, yam production has declined in recent years, and yams have been largely replaced by cassava.

Figure 14.8 True yams, *Dioscorea* spp., are important staples in many tropical areas.

Taro

Poi, the traditional dish of the native Hawaiians, is prepared from the corm of taro (*Colocasia esculenta*), a member of the Araceae, or arum family (fig. 14.9). This plant is related to, and resembles, elephant's ear, a common landscaping plant in the southern United States. The large leaves of taro also play a role in Hawaiian traditions, since foods are wrapped and cooked in the leaves during a Hawaiian feast, or luau. In fact, the word *luau* is the Hawaiian name for these leaves.

In preparing poi, the corms are steamed, crushed, made into a dough, and then allowed to ferment naturally by microorganisms. This doughy paste is then eaten with the fingers or rolled into small balls; this food was one of the main staples in the traditional diet, with large amounts consumed daily. The corm may also be cooked similar to potatoes (baked, steamed, roasted, or boiled) or processed into flour, chips, and breakfast foods. Nutritionally, the corm contains approximately 25% carbohydrate (which includes about 3% sugar), 2% protein, and very little fat. The majority of the carbohydrate present is a fine-grained, easily digested starch. The corms are also a good source of calcium because of the presence of calcium oxalate crystals; if the corm is eaten raw, these crystals cause intense burning or stinging in the mouth and throat. Cooking destroys the crystals. The crystals are characteristic of the arum family and are found in familiar plants, such as *Philodendron, Dieffenbachia,* and jack-in-the-pulpit (see Chapter 21).

Taro is believed to have originated in Southeast Asia and spread both east and west during prehistoric times. The eastward spread took it to Japan, the Pacific Islands, and as far east as Hawaii. It spread westward to India and then to the Mediterranean region during the time of the ancient Greek civilization. Eventually, it spread to Africa, and from there it was taken to the West Indies and tropical America by African slaves. Today it is cultivated in the wet tropics, where it thrives under saturated soil conditions. It is propagated by planting either the tops of corms or cormels. Although not a major crop in the world market, taro remains locally important in some areas.

A similar member of the Araceae is yautia, or cocoyam (*Xanthosoma* spp.), native to tropical Central and South America and also referred to as taro in some areas. Like taro, the underground storage organs can be prepared in various ways and have nutritive value comparable to that of the other starchy staples.

Jerusalem Artichoke

Another carbohydrate-rich staple of note is the misnamed Jerusalem artichoke, which is the tuber of a North American sunflower (*Helianthus tuberosus*). Samuel Champlain, a seventeenth-century French explorer, found Native Americans cultivating the plant on Cape Cod in 1605 and brought it back with him to France. In the Algonquin language of the eastern Native Americans, it was called sun root. By the mid-1600s, the Jerusalem artichoke was a common vegetable and livestock feed in Europe and colonial America. The French particularly embraced the vegetable, and it peaked in popularity at the turn of the nineteenth century there.

There are several explanations for its name. One is that the Pilgrims of North America named this staple of the New World with regard to the "New Jerusalem" that they were carving out of the wilderness. It is also known that it was Champlain who said that the tuber tasted like an artichoke. The Italians added the name of *girasole* ("turning to the sun," sunflower) because its flowering heads follow the path of the sun during the course of the day. Later the English corrupted girasole artichoke (literally, "sunflower artichoke") to Jerusalem artichoke. Other names applied to the plant have been the French or Canada potato, topinambour, and lambchoke. In the 1960s, Frieda Caplan, a produce wholesaler from California, tried to revive the plant's appeal by renaming it the sunchoke, a name that is still used today.

Nutritionally, the Jerusalem artichoke contains about 10% protein, no oil, and surprisingly no starch. It is rich, however, in the carbohydrate inulin (76%), a polymer of the monosaccharide fructose. If the tubers are stored for any length of time, the inulin is digested to its component fructose. Since fructose is one and a half times sweeter than sucrose, Jerusalem artichokes have an underlying sweet taste. They have also been promoted as a healthy choice for diabetics because fructose is tolerated better by sufferers of diabetes than is sucrose. Interestingly, Jerusalem artichoke was a reported folk remedy for diabetes.

The plant itself is a weedy perennial that can reach heights of 1 to 4 meters (3 to 12 feet) by the end of summer. Its stems and leaves bear coarse hairs, and a single plant produces a profusion of sunflower heads that bloom in July through September. Reddish, knobby tubers are found at the end of elongate rhizomes. The tubers (approximately 7.5 to 10 cm, or 3 to 4 inches, lengthwise; 3 to 5 cm, or 1 to 2 inches, thick) store well in the ground and can be harvested as needed. Cultivated varieties such as Mammoth French White, Jerusalem White, Long White, and New White have tubers that are white and rounder than those of the wild type.

The tubers can be eaten fresh but are usually prepared by boiling, roasting, steaming, frying, or even pickling, a Pennsylvania Dutch specialty. They are as versatile as all the other starchy staples. The taste of the Jerusalem artichoke has been described as similar to that of a water chestnut.

The plant is cultivated by planting whole tubers or pieces of tubers. It is adaptable to a wide variety of soils and habitats and suffers from relatively few diseases, insect pests, and weeds. Either the tubers themselves or the aboveground growth can be used for forage. In France, the Jerusalem artichoke is also grown for wine and beer production. In North America, it has been promoted as an alternative crop to corn for the production of biofuels such as ethanol and butanol. Currently, the tuber is primarily sold as a gourmet vegetable in American markets, but there is much interest in further developing this distinctively North American staple.

This chapter described the major starchy staples cultivated today. With advances in plant breeding, improved varieties and new starchy staples are likely to be developed and promoted for tomorrow's table.

(a)

(b)

(c)

Figure 14.9 *Colocasia esculenta*, taro: (a) harvested plants, (b) corms, and (c) in cultivation.

CHAPTER SUMMARY

1. Many plants store large quantities of starch in underground structures that are modified stems or roots. Botanically, these structures have several functions: as food reserves, for asexual reproduction, and as the starting point for renewed growth after dormancy. Some of these storage organs are starchy staples that include some of the world's foremost crops. In addition to their direct consumption, potatoes and cassava are also sources of commercial starch, which has many applications in both food and nonfood industries.

2. Native to the highlands of South America, the potato was the dietary staple of the ancient Incan civilization and was introduced to Europe in the sixteenth century. Cultivation of these tubers spread slowly throughout Europe, becoming the food of peasants. The crop was so successful in Ireland that it led to a population explosion and, later, massive famines when late blight of potato destroyed the crops in the 1840s. Today, about one-half of the U.S. potato crop is processed to make french fries, potato chips, and dehydrated flakes. This use of potatoes continues a long tradition; Peruvian Indians have prepared a processed potato product, chuño, for over 2,000 years. The potato plant is a member of the nightshade family, the Solanaceae. Many members of this family are poisonous; in fact, the potato contains toxins in all parts except the tuber. Potatoes are cultivated by seed potatoes, which are small tubers or cut tuber pieces containing at least one "eye." Although this method of asexual reproduction is fast and produces plants with desired characteristics, all the offspring are genetically identical and share susceptibility to adverse environmental conditions and diseases.

3. Sweet potatoes are storage roots of a plant in the morning glory family and are unrelated to the white potato. The sweet potato is native to tropical Central and South America; archeological evidence dates its cultivation back at least 8,000 years in Peru. It was also cultivated in several Pacific islands and New Zealand. Evidence of its early cultivation in Polynesia may indicate an earlier introduction by seafaring natives or natural long distance dispersal. Although of relatively minor importance in the United States, the sweet potato today is a significant crop throughout the tropics and some warm, temperate regions.

4. Cassava is a tuberous root that is another important starchy staple vital to the food supply of millions in the tropics. Although native to South America, cassava is now cultivated throughout the tropics. Cassava varieties are classified as either sweet or bitter based on the concentration of hydrocyanic acid. Various methods of processing have been developed to detoxify cassava and make it safe to eat.

5. Other underground crops include yams and taro, which are important staples in tropical areas, especially the South Pacific islands. Jerusalem artichoke was a Native American staple that currently is of interest to nutritionists because it is rich not in starch but in inulin, a polymer of fructose. Unlike other starchy staples, bananas are fruits. Sweet bananas are often considered a dessert or snack food, but starchy plantains are important as food in the tropics, where they are traditionally cooked and eaten as a vegetable.

REVIEW QUESTIONS

1. Discuss the various plant organs modified for storage.
2. What has been the social impact of the white potato?
3. What is the difference between the bitter and sweet varieties of cassava? How does their processing differ?
4. How can the presence of the sweet potato in both South America and Polynesia be explained?
5. Contrast the nutritional value of the starchy staples to that of cereals and legumes.
6. Describe some of the commercial uses of starch.
7. What is the connection between yams and birth control pills?
8. Compare and contrast the cultivation of the potato, sweet potato, cassava, and taro.
9. In what ways is the banana different from other starchy staples? Consider the growth and storage organs.
10. In what ways is the Jerusalem artichoke different from the other underground crops considered in this chapter?

Source: Scott Bauer/USDA

CHAPTER 15

Feeding a Hungry World

USDA researcher examines transgenic peach and apple seedlings.

KEY CONCEPTS

1. The major challenge in agriculture today is producing enough food to feed the world's population.
2. Traditional breeding programs and biotechnology are being used to develop high-yielding and disease-resistant cultivars.
3. Germplasm, the genetic information encoded in plants, is a valuable asset to plant breeders and must be preserved.

CHAPTER OUTLINE

Breeding for Crop Improvement 242
The Green Revolution 243
    High-Yield Varieties 243
    Disease-Resistant Varieties 244
    History of the Green Revolution 245
    Problems with the Green Revolution: What Went Wrong? 245
    Solutions? 246
Genetic Diversity 247
    Monoculture 247
    Sustainable Agriculture 248
    Genetic Erosion 249
    Seed Banks 249
    Heirloom Varieties 250
    Germplasm Treaty 250
    Crops and Global Warming 251
Alternative Crops: The Search for New Foods 252
    Quinoa 252
    Amaranth and Chia 252
    Tarwi 253
    Tamarillo and Naranjilla 254
    Oca 254

Biotechnology 254
A CLOSER LOOK 15.1 Mutiny on the HMS *Bounty:* The Story of Breadfruit 255
    Cell and Tissue Culture 256
    Molecular Plant Breeding 257
Genetic Engineering and Transgenic Plants 257
    Herbicide Resistance 258
    Against the Grain 260
    Insect Resistance 260
    Bt Corn and Controversy 261
    The Promise of Golden Rice 262
    Other Genetically Engineered Foods 262
    Disease Resistance 263
    Farming Pharmaceuticals 264
    Nonfood Crops 265
    Genetically Modified Trees 265
    Regulatory Issues 266
    Environmental and Safety Considerations 266
Gene Editing: CRISPR/Cas9 268
Chapter Summary 270
Review Questions 270

In the past few years, stories in the news media have focused on the obesity epidemic in the United States and other developed countries; however, hunger and malnutrition still plague the populations in many developing countries. In 2017, the United Nations Food and Agriculture Organization estimated that 815 million people receive insufficient food to meet daily requirements, with the majority of hungry in developing countries. It is only when the situation reaches crisis proportions in a specific area that the media focus on the horrible conditions. It must also be remembered that the human population is still growing, and the need for food will continue to increase during the twenty-first century. By developing higher-yielding and disease-resistant crops, plant scientists will be at the forefront of the efforts to increase the global food supply.

BREEDING FOR CROP IMPROVEMENT

Since earliest times, humans have practiced plant breeding by selecting for certain traits. For most present-day crops, native plants were the starting point for selection. Even before the beginnings of agriculture, early peoples were probably selecting for certain sizes of fruits or seeds. An example of artificial selection can be seen with the trait for nonshattering heads of grain (see Chapter 11). While people were still foraging, it was easier to gather grain from plants with nonshattering heads, a recessive trait. These plants retained the grains while the wild type plants scattered seeds for optimum dispersal. (Wild type is the most commonly found phenotype for a particular trait in a population.) When human groups shifted from foraging to farming, they planted seeds that were on hand. Often, a large percentage of the seeds were from plants with the nonshattering trait. Soon the nonshattering trait dominated the population, thereby establishing this characteristic as standard for cultivation. At first the selection was not deliberate, but later farmers may have intentionally selected for traits such as large seeds or certain fruit or seed colors by carrying out various practices, such as cross-pollination (fig. 15.1).

Although crop improvement had been well established by the nineteenth century, the scientific basis for plant breeding was not understood until the widespread acceptance of Mendel's work. The development of hybrid corn in the early twentieth century is an example of the early use of selection on a wide-spread scale in modern agriculture (see Chapter 12).

Human selection often circumvents evolution through natural selection, but it may increase the rate of evolution for some plants. In some cases, the evolutionary change has been so great that the ancestral plants are not easily identified. The most notable example is, of course, corn (see Chapter 12). Although teosinte is the wild grass from which corn was domesticated, there are striking differences between the two species. Corn is totally dependent on humans for survival because it has lost the ability to disperse seeds in the natural environment, has much larger ears encased in husks, and has naked grains. These traits are lacking in teosinte (fig. 15.2).

Figure 15.1 Relief carving from Nimrud (ninth century B.C.) showing an Assyrian deity pollinating female date palm flowers. Artificial pollination and breeding of date palm flowers was apparently practiced at that time.

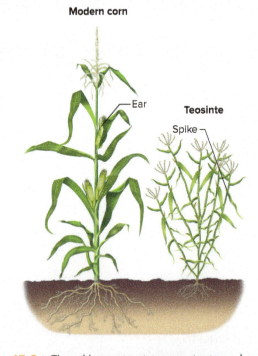

Figure 15.2 The wild grass teosinte gave rise to modern corn. Plants differ mainly in the spikes or ears. Although modern corn has just two or three large ears that remain attached to the one main stalk, teosinte has numerous small, brittle spikes (on many stalks) that shatter when the kernels are mature. Just a few genetic changes may have resulted in the evolution of modern corn.

Figure 15.3 Artificial selection has led to the development of many different crops from the wild type *Brassica oleracea*.

An interesting example of artificial selection is in the development of many different crops from a single species, *Brassica oleracea*. Cabbage, kale, broccoli, cauliflower, kohlrabi, and brussels sprouts were all developed from the wild type *Brassica* by selection techniques that emphasized different parts of the plant (fig. 15.3). Today, selection is still important in plant breeding programs; however, molecular techniques often reduce the long time periods required to establish new varieties.

THE GREEN REVOLUTION

Improvements in crop yield and quality achieved through plant breeding helped produce a modern and efficient agricultural system in the United States by the early part of the twentieth century. Yields increased throughout the twentieth century and continue to increase. In fact, over the past 60 years there has been a threefold increase in agricultural production. The use of high-yielding varieties is the center of this accomplishment; however, these varieties need excellent growing conditions to be productive, and the use of fertilizers, pesticides, mechanization, and irrigation is an integral part of the higher yields. These technologies have been transferred to developing nations, where tremendous increases have also occurred. This achievement is known as the **Green Revolution.**

High-Yield Varieties

The central thrust for solving the world food problems has focused on improvements of existing crops. During the past several decades major efforts have been made to increase the yields of wheat, rice, and corn as well as other crops, especially in tropical and subtropical areas. Much of the research has been carried out at international agricultural research centers located in developing nations (table 15.1). These centers not only select and develop new varieties but also are active in the collection and preservation of diverse species.

Wheat has been the principal crop behind the Green Revolution. Dr. Norman Borlaug (fig. 15.4), who is usually referred to as the Father of the Green Revolution, developed high-yielding dwarf strains of wheat at the CIMMYT (Centro

Table 15.1 International Agricultural Research Centers

CIAT	Centro Internacional de Agricultura Tropical; Cali, Colombia
CIFOR	Center for International Forestry Research; Jakarta, Indonesia
CIMMYT	Centro Internacional de Mejoramiento de Maiz y Trigo; Mexico City, Mexico
CIP	Centro Internacional de la Papa; Lima, Peru
ICARDA	International Center for Agricultural Research in Dry Areas; Aleppo, Syria
ICRAF	World Agriforestry Center; Nairobi, Kenya
ICRISAT	International Crops Research Institute for the Semi-Arid Tropics; Patancheru, India
IFPRI	International Food Policy Research Institute; Washington, DC
IITA	International Institute of Tropical Agriculture; Ibadan, Nigeria
ILRI	International Livestock Research Institute; Nairobi, Kenya
IPGRI	International Plant Genetic Resources Institute; Rome, Italy
IRRI	International Rice Research Institute; Manila, Philippines
WARDA	West Africa Rice Development Association; Monrovia, Liberia

Figure 15.4 Dr. Norman Borlaug (1914–2009), second from left, on a visit to Kenya in 2005, discusses the threat of a new strain of wheat rust with researchers from the Kenya Agricultural Research Institute.

Internacional de Mejoramiento de Maiz y Trigo—International Center for Maize and Wheat Improvement) sponsored by the Rockefeller Foundation in Mexico City. The high-yielding varieties were produced from crosses with a dwarf variety introduced from Japan after World War II. These varieties did well in Mexico and other areas, such as India and Pakistan. Within a 20-year period, the Mexican wheat harvest doubled, and similar accomplishments were realized in other countries. In India, the high-yielding varieties attracted so much attention that armed guards were needed to prevent people from stealing the seeds before they were ready for release. The rust-resistant cultivars that Borlaug developed have strong, stiff stems that can accept heavy applications of fertilizer without lodging (falling over during heavy winds or rain, especially near harvesting). The older, taller varieties tended to lodge from the weight of the grains when heavily fertilized, making harvesting very difficult. Similar success was achieved in the Philippines at the IRRI (International Rice Research Institute), where high-yielding dwarf rice cultivars were developed.

Starting in the late 1960s, food production increased dramatically in areas where these high-yielding crops were introduced, contributing to the chronicles of the Green Revolution. Breeding programs continue at these agricultural research centers, and scientists there are constantly developing and introducing new cultivars with improved resistance to pests and pathogens. Also, programs are now in place to improve crops, such as cassava and sweet potato, that are dietary staples in tropical countries. As long as the human population continues to grow, the need for increased food production will fuel the search for improved crops.

Although the Nobel Prize committee does not award a prize in agricultural sciences, Dr. Borlaug was awarded the Nobel Peace Prize in 1970 for his efforts to increase world food production. In making the award, the Nobel committee called the Green Revolution a "technological breakthrough which makes it possible to abolish hunger in the developing countries in the course of a few years."

The committee also stated that the Green Revolution contributed to the solution of the population explosion. Although the achievements were extraordinary, the Nobel committee may have been overly optimistic in assessing the early success of the Green Revolution, as we will discuss later.

Disease-Resistant Varieties

Crops are threatened by thousands of diseases that make the global food supply vulnerable. Some of the diseases are noninfectious and caused by such factors as mineral deficiencies or pollutants, but the majority are infectious diseases caused by a wide range of organisms, including nematodes, bacteria, viruses, and fungi. Even with the heavy use of various pesticides, approximately 50% of the world's food crops are destroyed in the field or in storage. As a group, fungal pathogens make up the most serious disease threat and rank immediately behind insects as the chief competitors for human food (see Chapter 23).

The field of science that deals with the study of these diseases is plant pathology. Although plant pathologists use many approaches to produce healthy plants, use of disease-resistant varieties of crop plants is the cheapest and most effective method. Breeding for disease resistance must be coupled to yield, quality, climate, consumer acceptability, and a host of other traits the farmer demands. The primary efforts of plant breeders are naturally focused on the world's major food crops, especially wheat, rice, and corn, and disease-resistant varieties have been commercialized for many areas of the world. Unfortunately, some of the achievements in this area have been undermined by the continued evolution of the fungal pathogens.

In many instances, a single dominant gene in the host plant has been found to be the source of the resistance. Some parasites produce toxins or enzymes that enhance their ability to attack the host plant and thus contribute to their success. A resistant plant may produce a molecule that blocks the parasite enzyme or alters the toxin-binding site so that the parasite cannot be successful. With time, a mutation may arise in the parasite that results in a modified enzyme that cannot be blocked or a modified toxin with improved binding ability. These genetic interactions between host and parasite constantly occur in the natural environment. Plant pathologists and breeders valiantly try to stay ahead of the newly evolving strains of pathogens that are capable of destroying various crops. This has been a problem especially for the development of rust-resistant varieties of wheat (see Chapters 12 and 23), and today researchers are trying to develop varieties resistant to Ug99, a virulent strain of the stem rust fungus first described in Uganda in 1999.

History of the Green Revolution

The earliest involvement of U.S. scientists with agriculture in developing nations came about in 1941, when Iowa State University set up an experimental station in Guatemala to study indigenous plants as sources of genetic material to improve U.S. crops. Also in 1941, the U.S. government sent four teams of agricultural scientists to travel throughout Latin America. The purpose was to determine how to strengthen the agriculture in the area, for the benefit of both Latin America and the United States, should the war in Europe continue to escalate. Rubber production was a major interest at this time because war with Japan would threaten the U.S. source of rubber. As a result of the inspection tour, experimental research stations were established in Peru, Ecuador, Nicaragua, and Guatemala. Agricultural programs received major emphasis, and U.S. foreign aid programs helped establish experimental stations within most non-Communist developing nations.

During the 1950s and 1960s, there was overwhelming enthusiasm for the agricultural programs and tremendous confidence that science could solve the food problems in the hungry nations. One of the most acclaimed programs was CIMMYT in Mexico, where Norman Borlaug carried out his wheat breeding experiments that led to the high-yielding dwarf cultivars.

William Gaud, the head of the U.S. foreign aid program, coined the term *Green Revolution* during a speech in 1968, and it quickly became the catchword for both the media and the scientific community. The Green Revolution was credited with miraculous accomplishments. Two years later, Norman Borlaug received the Nobel Peace Prize for his involvement in the Green Revolution. But the optimism about the Green Revolution during the 1960s, so eloquently expressed by the Nobel committee when it honored Dr. Borlaug, met the cold realities of the 1970s.

Problems with the Green Revolution: What Went Wrong?

The Green Revolution is not without its problems and critics, nor is it accessible to everyone. Today's high-yielding and disease-resistant crop varieties are high-impact crops; they are dependent on fertilizer and pesticide application, adequate water, and mechanized farming. Poor farmers cannot afford the seeds, fertilizers, pesticides, irrigation, farm equipment, and fuel needed to cultivate the high-yielding varieties.

Other problems are high energy costs, environmental damage, and loss of genetic diversity. The manufacture of inorganic fertilizers requires a great deal of energy (two barrels of oil per barrel of nitrogen fertilizer). In 1973, when the OPEC oil embargo resulted in major increases in the cost of petroleum products, the agricultural community suddenly realized the actual costs of high-impact agriculture. What had been profit soon became deficit. Because of their high cost, fertilizers are often reserved only for cash crops or export crops, not those grown for domestic consumption. The dramatic increase in fertilizer use over the past several decades has also caused serious environmental problems (see A Closer Look 13.1: The Nitrogen Cycle). Similarly, fossil fuels are required for the increasingly specialized and efficient farm machinery that has been developed to make planting and harvesting more efficient.

Application of pesticides for crop protection is also inherent in the Green Revolution. Insecticides, herbicides, and fungicides are all needed to maximize yield and were originally viewed as spectacular means to fight plant disease and increase yield. Today, the environmental and health effects of pesticide use and the problems with pest resistance to these chemicals are growing concerns (see Chapter 26).

For maximum yield, the miracle crops also require ample water, from either optimum rainfall or irrigation. In the United States, where high-yielding crops are standard, more than one-third of the water supply goes for irrigation. Some have questioned the wisdom of massive irrigation in desert areas such as central Arizona for growing lettuce and pecan trees. Is this the most rational use of water? Often, irrigation water is pumped from underground **aquifers,** such as the Ogallala Aquifer stretching from Texas to South Dakota (fig. 15.5), which had accumulated water for millions of years. The water from this and other aquifers is not being replaced as rapidly as it is being used. In some areas, water is removed so rapidly that sinkholes have developed as land collapses into the space left after the water has been pumped out.

Saline soils are another growing environmental problem around the world because irrigation water contains small amounts of salt, which remain when water evaporates. Over time the salt levels in the soil build up, making the cropland increasingly less productive. It is estimated that 25% of the irrigated farmlands in the United States have developed saline soils, and the estimate for saline soils around the world is 40%. Overall, almost 10 million hectares of land are lost annually because of salinity; this may be the most serious environmental factor limiting productivity.

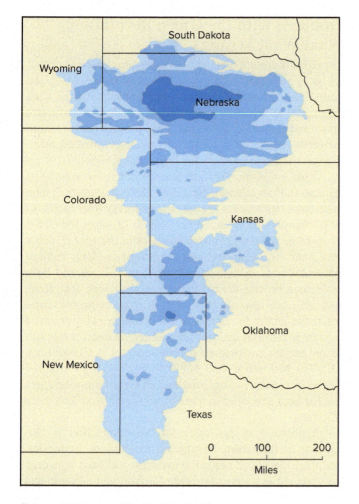

Saturated thickness of the Ogallala Aquifer:

▫ Less than 200 ft

▫ 200–600 ft

▫ More than 600 ft (as much as 1,200 ft in places)

Figure 15.5 Underground aquifers often supply much of the irrigation water used by farmers. Ogallala Aquifer is the largest known body of fresh water. Yearly use of this water, mainly for irrigation, equals the total annual flow of the Colorado River, resulting in significant drops in the water table: as much as 50 meters (162 feet) in some areas.
Source: U.S. Geological Survey.

earlier. Before the potato was introduced, the population of Ireland was relatively stable at about 1.5 million, although most people lived in total poverty. The potato was a cheap, nutritious food source that allowed the population to grow, resulting in a population explosion. The only check on this population growth was starvation, which followed the famine.

Increased food production in areas where the population growth rate is uncontrolled is a dangerous situation. It ultimately means that more people will starve and more environmental degradation will occur. Even in the late 1960s, when the Green Revolution was being heralded, some scientists realized that population growth was destroying all the efforts being made to feed the hungry of the world.

Solutions?

In early July 2018, the human population was approximately 7.6 billion with realistic projections of more than 8 billion by 2025 and 9.7 billion by 2050. The population will continue to increase into the twenty-second century. At the present time, approximately 97% of the world's population growth is occurring in developing nations. Over the past 40 years, many developing nations have made great progress in reducing fertility rates through active programs of population control. As a result, population increases predicted today are far lower than estimates even 20 years ago.

Although tremendous increases have taken place in worldwide agricultural output in the past 50 years, population increases have wiped out part of these gains. In order to feed the projected population, most scientists believe that food production will need to increase significantly in the next 25 years and continue to increase during this century. However, critics contend that the problem is food distribution, not capacity. World agricultural production is currently sufficient to feed the world population if it were well distributed. While some countries produce far more food than consumed, millions in other areas starve. Food availability varies from country to country and, even within a single country, from one area to another. Food aid is critical for relieving short-term shortages during emergencies, but it does not solve the problem. Improved farming techniques can yield more food than simple food aid can.

The answer to feeding the world's population must be found in the regions where most of the people live and where most of the future population growth is anticipated—the developing nations. Innovative techniques must be encouraged to reduce the high-impact aspect of high-yielding cultivars. The use of reduced-tillage and no-tillage (a new crop is planted without removing the debris from the previous crop) agricultural methods has greatly reduced soil erosion (fig. 15.6); however, the need for pesticides is often increased. These methods also increase the efficiency of water use, decrease the need for fertilizer, and reduce pollution from runoff and leaching. Biological controls, such as predators (ladybug, praying mantis) and pathogens (fungi, bacteria, and viruses), should be advanced to control agricultural pests and thereby reduce the use of pesticides. Organic fertilizers, such as manure and compost, can lower the need for expensive inorganic fertilizers,

Some scientists feel that the whole premise behind the Green Revolution needs to be reexamined. Increasing food supplies in developing nations during the 1960s and 1970s led to massive population increases in those countries. The situation has been described as being similar to the events that led to the Irish potato famine in the nineteenth century, when the destruction of the potato by the fungus *Phytophthora infestans* caused the population of Ireland (8.5 million in 1840) to decrease by approximately 2.5 million and left most of the survivors in total poverty. It has been suggested that the catastrophe was actually due to the introduction of the potato centuries

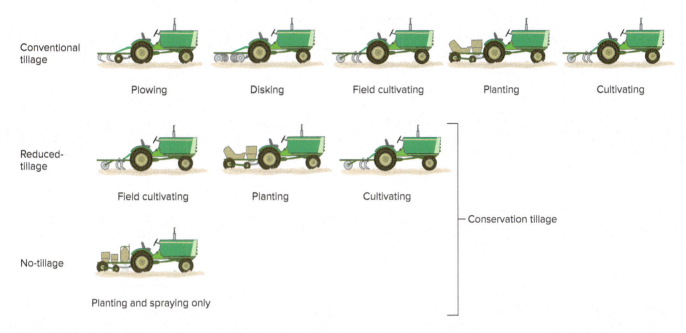

Figure 15.6 Reduced-tillage and no-tillage farming techniques have reduced soil erosion and energy use and also have decreased the need for fertilizers. However, the need for pesticides is often increased.

and nitrogen can be provided through biological fixation using improved strains of *Rhizobium* bacteria, free-living nitrogen-fixing bacteria, and by genetically engineering the nitrogen-fixing process so that nonlegumes can benefit. New techniques in dryland agriculture and drought-tolerant crops may increase the efficiency of water use and even permit sustained production in some areas when irrigation water becomes scarce.

Also, changes in current irrigation techniques could help conserve precious water resources. Current irrigation methods typically use a high-pressure system that sprays water above the crop. Approximately 35% of the water is lost to blowing wind and evaporation. Newer techniques conserve water by spraying water close to the ground; this practice increases irrigation efficiency to over 90%. Even greater water efficiency can be achieved through drip irrigation. This method carries water through horizontal pipes at ground level next to the crops or buried just beneath the surface. Water slowly drips through holes in the pipes right onto crop roots and stems; as a result, very little water is lost to evaporation.

GENETIC DIVERSITY

Although these methods will help improve productivity and reduce the environmental impact, there is still reason for concern. Some of the concerns center on the loss of genetic diversity in crop plants owing to the predominance of monocultural practices.

Monoculture

Monoculture, growing the same crop year after year in a large region, has become a mainstay of modern agriculture (fig. 15.7). Several forces increased the practice of monoculture

Figure 15.7 Field of genetically engineered golden rice.

in the twentieth century: mechanization, improvement of crop varieties, and chemicals to fertilize crops and control weeds and pests.

The mechanization of agriculture began with the steel plow and the steam engine. By the end of World War II, most commercial farms had switched to gasoline tractors. As machinery replaced draft animals, it was no longer necessary to produce as much livestock feed. Increasingly, more grain could be sold for profit, and the increased profit went into buying more farm machinery. With more sophisticated machinery, a single person could farm more land, and farms became larger. Agricultural machinery became increasingly specialized to perform a single job, often for a single crop. The financial investment in specialized machinery proved to be a strong incentive for the farmer to grow only the crop for which the machinery was designed. The introduction of high-yielding varieties and new crops has also contributed to monocultural practices. By concentrating on the growing of a single, high-yielding variety, a farmer could maximize productivity. The large-scale growing of high-demand crops, such as soybean and rapeseed (canola), further maximized profits and increased the risks associated with monoculture. The widespread use of chemical fertilizers and pesticides ensures that the crop will continue to produce year after year. Rotation of crops to encourage soil fertility or to reduce pest populations is no longer necessary.

The result of these practices is rampant monoculture. Often, the major crop grown over a region is dependent on only a few varieties. For example, half of the wheat acreage in the United States is planted with just nine varieties; four varieties account for 75% of the potato crop; three for 50% of the cotton; and six for 50% of the soybean. When genetic diversity is so limited, all individuals in the crop are essentially genetically identical and vulnerable to disaster. A single pest or disease can wipe out an entire crop because there is little genetic variation and, thus, no range in natural resistance. History has recorded time and again the dangers of monocultural practices: the Irish potato famine in the 1840s, the obliteration of Sri Lankan coffee plantations (see Chapter 16), the 1970 epidemic of corn leaf blight in the southern United States (see Chapter 12), and in 1984 the destruction of 18 million Florida citrus trees owing to a bacterial infection.

Sustainable Agriculture

Some researchers believe that the future in agriculture is to move away from the instability and unnaturalness of annual, monocultural cultivation and instead model agriculture after natural systems. According to this concept, agriculture becomes sustainable; that is, crops can be harvested without the degradation of the environment (see Chapter 26). Agriculture would switch from a monocultural field planted annually to a polyculture of perennial crops. Warm-season and cool-season grasses and nitrogen-fixing legumes would be grown in proximity. They would tend to mature at the same time to facilitate harvesting, but their life cycles would be different enough that they would not compete simultaneously for the same resources. Several perennials are being developed for polyculture planting. Illinois bundleflower (*Desmanthus illinoenis*) is a warm-season legume and a prodigious nitrogen fixer. Its seeds are high in protein and borne on erect stalks that facilitate harvesting. Research is ongoing to improve the taste of the seeds to make them suitable for human consumption. A hybrid between Johnson grass (*Sorghum halepense*) and the African annual *Sorghum bicolor* has produced a perennial grain sorghum. Perhaps it will be possible to develop a modern perennial corn from the perennial wild relative, *Zea diploperennis*. Perennial relatives for several other major crops are listed in Table 15.2

Table 15.2
Annual Crops and Perennial Relatives Used in Hybridization

	Annual Crop	
Common Name	Scientific Name	Perennial Relative
Barley	*Hordeum vulgare*	*Hordeum jubatum*
Chickpea	*Cicer arietinum*	*Cicer anatolicum*
		Cicer songaricum
Maize	*Zea mays*	*Zea mays* ssp. *diploperennis*
		Tripsacum dactyloides
Oat	*Avena sativa*	*Avena macrostachya*
Pearl millet	*Pennisetum glaucum*	*Pennisetum purpureum*
Rice	*Oryza sativa*	*Oryza rufipogon*
		Oryza longistaminata
Sorghum	*Sorghum bicolor*	*Sorghum propinquum*
		Sorghum halepense
Soybean	*Glycine max*	*Glycine tomentella*
Sunflower	*Helianthus annuus*	*Helianthus maximilliani*
		Helianthus rigidus
		Helianthus tuberosus
Wheat	*Triticum* spp.	*Thinopyrum* spp.
		Elymus spp.
		Leymus spp.
		Agropyron spp.

A field planted according to the suitable agricultural design would consist of mixtures of perennial crops that could be harvested for a number of years without replanting. This system has many advantages: soil erosion from yearly plowing is diminished; pest management is improved; and disease control is more effective. The model for this type of agriculture is nature itself. For example, the prairie is a sustainable ecosystem of grasses that lasted for millennia. The incredible fertility of prairie soil was the result of natural soil-building processes.

Genetic Erosion

Today's food crops have their origins in the wild ancestors first domesticated by early farmers thousands of years ago. As agriculture developed over the centuries, artificial selection resulted in local varieties of a crop, so-called **land races,** or traditional varieties. Land races possessed unique and valuable genetic characteristics, traits that allowed them to survive cold, drought, or disease. **Germplasm** is the encoded information of a plant, the genetic instructions that dictate not only the type of plant but also all of the traits unique to an individual plant. It is the raw material of the plant breeder, and without it, the creation of new, improved varieties is not possible. Unfortunately, as traditional varieties are abandoned and replaced with modern, high-yielding ones and as habitat destruction brings about the extinction of wild relatives, valuable genetic heritage is lost forever. This irreversible loss of genetic diversity is known as **genetic erosion** and is of monumental concern.

Genetic erosion has accelerated since the introduction of high-yielding varieties during the Green Revolution in the 1960s. In 1959, farmers in Sri Lanka grew more than 2,000 traditional varieties of rice. Today, that number has been reduced to five. India has lost more than 75% of its 30,000 land varieties of rice. It is estimated that the United States has lost 90% of the seed varieties brought here before 1950. In addition, human encroachment in natural habitats is expected to bring about the extinction of a quarter of the world's plants by the middle of the twenty-first century. What can be done to save this genetic diversity for future generations?

Seed Banks

One way nations and organizations have tried to offset genetic erosion is through the establishment of gene banks, or **seed banks.** In a type of botanical Fort Knox, seeds of both domesticated plants and their wild relatives are collected from around the world and deposited. In the United States, the main seed bank is in the National Center for Genetic Resources Preservation in Fort Collins, Colorado, established by the USDA in 1958. It is one of 20 such facilities located around the nation that make up the National Plant Germplasm System. More than 500,000 seed samples representing 8,000 species are carefully treated and housed here. To extend their viability, the seeds are dehydrated, sealed in aluminum foil bags, and placed in freezers at −18°C (0°F). Other seed samples are preserved cryogenically (fig. 15.8), placed in tanks of liquid nitrogen at temperatures as low as −150°C (−238°F). Although these

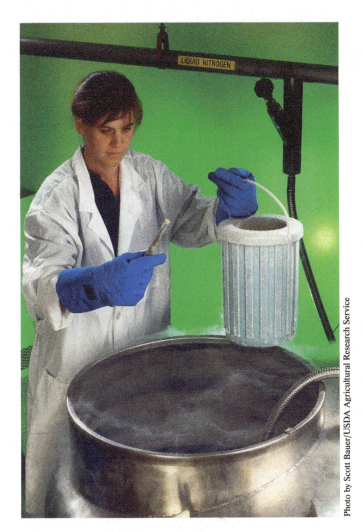

Figure 15.8 Seeds are cryogenically preserved in vats of liquid nitrogen.

procedures may extend seed longevity for decades, inevitably some seeds will not survive prolonged periods of cold storage. It is necessary at certain points to take the seeds out of storage, induce them to germinate, and then grow the plants to refurbish the seed supply. Some plants, such as the white potato, do not readily produce seeds; these plants are usually propagated by cuttings that do not store well. For these plants, arboretums and conservation gardens have become "botanical zoos," maintaining representatives of old varieties and endangered wild plants. The importance of this work is reflected by a growing network of seed banks throughout the world, in more than 100 countries, all dedicated to the cause of preserving the genetic diversity of crop plants and their wild relatives.

In 2008, Svalbard Global Seed Vault, nicknamed the Doomsday Vault, opened. Gene banks can be destroyed and their valuable contents lost due to war, natural catastrophes, equipment failure, lack of funding, or poor management. Funded by the Norwegian government and the Global Crop Diversity Trust, the Doomsday Vault has been designed to avoid these pitfalls. Located near the village of Longyearbyen on Svalbard, a group of Norwegian islands in the Arctic Circle, the site was chosen in part for its remoteness. Carved 120 m (394 ft) into the side of a mountain, the vault sits 120 m (426 ft) above current sea levels, high enough to avoid future flooding if ocean levels rise as ice sheets melt due to global warming. The vault is located in the permafrost with temperatures ranging from −4° to −6°C (24.8° to 21.2°F), so that even if there were a power failure, the seeds would remain frozen and safe. An entry tunnel opens into two large chambers, which are capable of holding 3 million seeds, the most complete collection in the world. The Global Crop Diversity Trust based in Rome is assisting developing nations in collecting, packing, and sending seeds to the vault as well as paying for operational costs. The Global Crop Diversity Trust is dedicated to securing and preserving free access to genetic diversity. Since opening in February 2008, nearly 500,000 unique seed samples from seed banks all over the world have been stored in the vault.

Seed banks have come to the rescue in preserving a variety of sorghum. Native to Ethiopia, the *zera zera* variety had been preserved in a gene bank there. Unfortunately, the seed bank was destroyed during a time of civil strife in Ethiopia. Scientists who returned to the original site to collect more seeds were disappointed to discover that the population had become extinct because of a combination of drought and human development in the area. Luckily, seeds had also been stored in a seed bank in India, preventing the loss of a variety critical to future sorghum breeding programs.

Similarly, samples of chickpeas, barley, lentils, fava beans, and seeds of other traditional varieties collected 30 years ago from the markets and farms of Afghanistan may restore this country's agricultural bounty. The International Plant Genetic Resources Institute of Rome collected seeds of land races in Afghanistan in the 1970s to be used in breeding crops for the arid regions of West Asia and North Africa. Later, most of the collection was transferred to the International Center for Agricultural Research in the Dry Areas (ICARDA) located in Syria. Duplicate samples were also deposited in Afghanistan's national gene bank, but this facility was destroyed in 1992. Much of the country's seed crop was depleted in 2001 after a third year of drought. Although irrigated agricultural fields have started to come back into production, many of the rain-fed fields are still devastated. It is hoped that these land races retrieved from ICARDA will thrive as they once did in the local environment when planted again in their homeland. By reseeding with the country's traditional varieties, the agricultural heritage of Afghanistan may be restored in the future.

Heirloom Varieties

Conservation organizations, such as Seed Savers Exchange and Native Seeds, are devoted to saving the seeds of traditional, or heirloom, varieties from extinction by encouraging their cultivation. Seed Savers stores more than 9,000 nonhybrid varieties of vegetable crops, and on their Heritage Farm in Iowa, they cultivate thousands of old-time varieties to supply seeds to persons interested in growing them. These varieties are not grown on most farms because they do not ripen uniformly and are not suitable for mechanized harvesting. Native Seeds has set out to preserve the traditional seeds grown by Native Americans. This collection has been used to reintroduce Sonoran panic grass, *Panicum sonorum*, from Mexican samples. Extinct in the United States since the 1930s, Sonoran panic grass was a food staple to the Cocopah Indians and now, thanks to Native Seeds, is being cultivated by them once again. The value of preserving these local varieties is exemplified by the Australian interest in the rust-resistant sunflowers grown by Havasupai Indians of the Grand Canyon region. Accompanying seed preservation is the preservation of the folklore and native knowledge about each plant, a legacy of knowledge to be passed on to future generations.

An heirloom variety of rice developed in South Carolina is once again being cultivated thanks to a U.S. gene bank. Carolina Gold is a rice variety that was widely grown in the Lowcountry of South Carolina from the 1770s to the 1860s. Named for the yellow color of its husks, this rice stands tall at 1.2 to 1.5 m (4 to 5 ft) and is easily lodged in high winds. After the Civil War, the cultivation of Carolina Gold rice declined due to competition from cheaper Asian imports and the loss of slave labor. The last crop grown was destroyed by a hurricane in 1911. In 1984, a Savannah eye surgeon who was an avid duck hunter contacted the USDA to find a source of Carolina Gold to grow as a food source to attract waterfowl. Fortunately, Carolina Gold had been donated to the gene bank in 1902. After growing the rice for 2 years, the USDA was able to send 6.4 kg (14 lb) of the Carolina Gold grains to seed 14 acres in 1986. Today, a foundation continues to promote the cultivation of this heirloom rice, which has found a market in the re-creation of historically accurate recipes. From DNA analysis, it has been determined that Carolina Gold most likely originated from Indonesia, and an expedition is set to identify the specific ancestor.

Germplasm Treaty

Germplasm is at the center of an international controversy, the question of ownership. Many developed nations on the cutting edge of agricultural biotechnology are home to relatively few major food crops. For example, the United States is the center of origin for only four: blueberries, cranberries, sunflowers, and pecans. Many developing nations, however, are rich in germplasm because they are the centers of origin to more than 96% of the world's food crops (see Chapter 11). Over the thousands of years that a plant population grows naturally in an area, a wealth of genetic variation develops. With accelerating genetic erosion, these genetic resources have become increasingly valuable to multinational seed companies.

Improving crops requires a constant infusion of fresh germplasm; the average life of a new crop variety in the United States ranges only from 5 to 9 years. Typically, a new variety with improved disease or insect resistance makes its debut, performs well for a few years, and then is replaced by another, more promising variety. Plant breeders must constantly search out traditional and wild relatives that possess desired traits. For example, plant breeders are interested in pig's weed (*Oryza nivara*), an endangered wild relative of rice in Sri Lanka. Some populations of this endangered weed are resistant to grassy stunt virus, a pathogen that is a constant threat to the rice crop in Asia. Given that most crops are grown outside their centers of origin, a potential conflict arises when access to this germplasm is needed.

Ethiopia is the only primary center of genetic diversity for the coffee tree. Colombia is the leading coffee-producing nation, yet it relies entirely on germplasm from Ethiopian trees. Is it proper for Ethiopia to forbid the export of its germplasm to Colombia? What is the price tag for the genetic resistance to yellow dwarf virus that was transferred from Ethiopian barley to protect California's crop? Should developing nations be compensated for their germplasm? Occidental Petroleum compensated China for its rice germplasm, and other multinational corporations have followed suit. Should seed companies share their profits with a developing nation if new varieties are created using germplasm from that nation? Does a nation of origin have access rights to advanced varieties developed with its germplasm? Some authorities have suggested that a nation's local crop varieties be identified through molecular tags to ensure that royalties can be collected. An alternative suggestion is that a certain percentage of the profits be turned back to the country of origin for use in preserving natural habitats and maintaining germplasm centers. Others argue that germplasm, whether obtained from wild or high-yield varieties, is a natural heritage to the world and should be shared freely among nations.

In June 2004, the International Treaty of Plant Genetic Resources for Food and Agriculture was enacted to protect valuable genetic resources. Ratified by 55 nations, the treaty defines plant genetic resources as "any genetic material of plant origin of actual or potential value for food and agriculture." Over 64 major crops and forages are currently protected. Under the administration of the United Nations Food and Agriculture Organization (FAO) in Rome, the treaty calls for global cataloging of plant genetic resources and assessment of the threats to these resources. The treaty encourages the collection of seeds and vegetative material for storage in seed banks to ensure a broad genetic base for crop research. The treaty also sets the conditions for access to protected plant genetic resources and the sharing of benefits developed from these resources. For example, if a commercial product is developed using protected plant genetic resources, the treaty provides for an equitable share of monetary benefits to the plant's country of origin, such as funding programs to help small farmers in developing countries.

Crops and Global Warming

Several troubling reports have been released recently concerning the effect of increasing CO_2 levels upon the nutritional values of some important grains and legume crops. Researchers looked at how elevated levels of CO_2, ranging from 363–386 ppm to 546–586 ppm (parts per million), affected the nutritional quality of important global food crops, grains (wheat, rice, maize, and sorghum), and legumes (soybeans and peas) as well as wild plants (goldenrod).

When wheat, rice, soybeans, and peas were grown at the higher CO_2 levels in line with projections for 2050, iron and zinc concentrations were reduced by 5%–10%. Additionally, protein levels in wheat and rice dropped by 5%–10%. Sorghum and maize, C_4 plants (Chapter 4), were less affected. The reasons for this drop in some crops and not others are unclear, but the consequences may be severe. Estimates report that 2 billion people currently suffer from iron and zinc deficiencies, and this number would be expected to rise as levels of these minerals decline in these global crops. Deficiencies of iron and zinc are associated with anemia, fatigue, and a weakened immune system. Protein is often the most limiting of the macronutrients, and children especially are susceptible to protein-deficiency diseases (Chapter 10).

Another study found that the soil concentration of selenium, a micronutrient (Chapter 10) essential to human health, is declining globally primarily as a result of rising CO_2 levels and global warming. Soils in arid regions tend to lose more selenium content, but soils high in clay or organic carbon (e.g., leaf litter) are better at retaining the element. It is estimated that two-thirds of the agricultural lands in the world will lose selenium by an average of nearly 9% as global temperature rise with increasing CO_2 levels. When reduced in soils, uptake of selenium by plants is typically lowered and consequently less selenium is passed up the food chain to humans. Currently, one in seven people are deficient in selenium and this number is expected to grow as global temperatures rise.

The protein quality of pollen is also decreasing with increasing CO_2 levels. A study analyzed the protein content of the common weed goldenrod (*Solidago canadensis*) over a span of 172 years using preserved specimens and recently collected flowers. The protein content dropped by one-third from 18% to 12%. During the same period, average atmospheric CO_2 concentrations rose from about 280 ppm to 398 ppm. Experimental studies in which goldenrod was grown in various concentration of CO_2 up to 550 ppm verified the accompanying protein loss in pollen documented in the historical record. Wild bees, as well as honey bees, obtain most of their protein content from pollen, important as food for their larvae, immune function, and other bodily functions. Protein loss in their food sources may be contributing to the worldwide decline in the population of these important pollinators of crops and wild plants.

Thinking Critically

Seed banks attempt to stockpile the germplasm of useful plants.

Often in times of famine or war, the seed supply for an agricultural region is lost, eaten by the starving. Explain how an international network of seed banks could be of vital use in the recovery of a famine-stricken or war-torn region.

ALTERNATIVE CROPS: THE SEARCH FOR NEW FOODS

Thomas Jefferson once said, "The greatest service which can be rendered any country is to add a useful plant to its culture." That sentiment is shared by persons who search the world for alternative food crops. It is estimated that there are approximately 50,000 edible plant species, but surprisingly, only 250 to 300 of those species are cultivated as food. The statistics are even more alarming when it is realized that only 22 domesticated plants are major crops and that just three crops—rice, maize, and wheat—provide over 50% of the Calories from plants in the human diet. Obviously, this reliance of the world's food supply on so few a number is a precarious practice. The United Nations Food and Agriculture Organization and other groups have initiated searches to discover local crops that have the potential to be developed into alternative food sources for the world market. These alternative crops usually possess characteristics—such as high nutritional quality, tolerance to harsh environmental conditions, or the ability to grow in poor soils—that make them attractive to plant breeders. The following is a discussion of several crops unknown to most of the world that may one day become staples in the world's larder.

Quinoa

Quinoa (*Chenopodium quinoa*, in the amaranth family, Amaranthaceae) has been a vital food crop in the high Andes of South America for centuries. In the languages of the Quechua and Aymara Indians, the descendants of the Incas, quinoa is referred to as the "mother grain" because quinoa and potatoes were the dietary staples of these highland communities. Quinoa has been described as resembling a cross between sorghum and spinach, an appropriate description because both the leaves and fruits are edible. It is a broad-leaved annual, 1 to 2 meters (3 to 6 feet) tall. Flowers are borne in terminal and axillary panicles. The fruits, called grains commercially, are actually achenes that have a hard, four-layered pericarp encasing each tiny seed. The seeds are high in protein, 12% to 18%, when compared with most cereal grains. The protein is also of an exceptional quality for a plant source. It is high in the essential amino acids lysine and methionine; lysine is deficient in most cereals, and methionine is typically lacking in legumes. The carbohydrate content of the seeds is also high, 58% to 68%, with approximately 5% sugar. Fat content ranges between 4% and 9%, half of which is the essential fatty acid, linoleic acid. The mineral content of calcium and phosphorus is higher in quinoa than in other grains. Bitter saponins in the pericarp must be washed (traditional method) or milled out (commercial preparation) in the processing of the seeds. After the saponins have been removed, the seeds can be cooked like rice and are often used as a high-protein substitute for rice and other grains. Quinoa grains have a delicate taste that compares with wild rice. Grinding the seeds yields a flour that can be mixed with wheat flour to make bread or used alone to make a gluten-free food product (see Chapter 10). Since the 1980s, quinoa has been cultivated in the Colorado Rockies and has made its entrance into the U.S. marketplace as a gourmet or specialty food, used in soups, pasta, breakfast cereals, and desserts (fig. 15.9a).

Amaranth and Chia

Another New World crop under development for the world market is grain amaranth. Amaranths were an important staple, along with corn and beans, for many indigenous populations in the pre-Columbian era, but their use declined after the arrival of the Spanish. It seems that the amaranth seeds, with human blood used as glue, were shaped into figurines used in Aztec religious ceremonies. Consequently, the cultivation and eating of amaranth were banned by the Spanish conquistadors to crush what they considered pagan and heretical practices.

(a)

(b)

Figure 15.9 (a) Quinoa, an ancient crop of the Incas, is becoming popular as a gluten-free substitute in pasta and breakfast cereals. (b) Oca tubers.

The genus *Amaranthus* includes at least 60 species, most of which are widely dispersed weeds. Amaranth can be brilliantly colored in shades of purple, orange, red, or gold; several varieties, such as love-lies-bleeding and Prince of Wales feather, are cultivated as ornamentals. The plants are tall, up to heights of approximately 2.5 meters (8 feet), and have broad, edible leaves that taste similar to spinach when cooked. The seeds of amaranth, however, are what attract the greatest attention as a potential nongrass cereal crop. Three especially promising species that plant breeders have targeted for development since the late 1970s are *A. cruentus*, *A. hypochondriacus*, and *A. caudatus*.

Unlike the major cereals (wheat, maize, and rice), amaranth, as a member of the Amaranthaceae, is only one of a handful of nongrass grains. Tiny and numerous (up to 50,000 per plant), the black, cream, tan, brown, or pink grainlike seeds are borne in elongate seedheads at the end of each stalk. The uncooked seed is indigestible because of a tough seed coat, but toasted, boiled, or popped like popcorn, it has a mild, nutty flavor. Amaranth seeds can also be ground into a flour that can be incorporated into baked goods.

Amaranth grains are easy to digest and have been traditionally given to those who are recovering from an illness or a fast. They have also been used by persons allergic to other grains. Nutritionally, the protein content of the seeds ranges from 12.5% to 17.6%, comparing favorably to the protein content of the three major cereals. In addition, amaranth protein is rich in the essential amino acid lysine. Amaranth flour from toasted seeds enriches wheat flour or corn meal by improving both the protein quality and quantity. From 6% to 10% of the seed is an oil that can be extracted from the germ. It is unsaturated and high in linoleic acid. Amaranth oil also contains significant quantities of squalene, a high-priced hydrocarbon usually extracted from shark livers. Squalene is the raw material for synthetic steroids and has uses in the cosmetic industry. In addition, amaranth starch grains are extremely small and might have applications in nonallergenic aerosols, in dusting compounds, or as a talcum powder substitute. Another desirable feature of the amaranth is its tolerance to unfavorable environmental conditions, such as heat, drought, and poor soil.

In 1977, research was begun in Pennsylvania by the Rodale Research Center to develop amaranth varieties suitable to the mechanized agriculture practiced in the United States. Varieties are being developed that are shorter and of uniform height, have seedheads that do not shatter, and have light-colored seeds that are preferred in cooking. Already, plant breeders have doubled amaranth productivity. It is used as forage for livestock and as a source of dyes. Popped amaranth grain can be eaten like popcorn. Breakfast cereals, breads, crackers, and pasta made from the flour can be found in health food stores throughout the United States. Amaranth flour is gluten-free, and this Aztec crop has found a new market in these food products.

Chia, another staple seed crop of the Aztecs, has been recently promoted to the global market as a high-energy super food. *Salvia hispanica* known as chia and *S. columbariae* known as golden chia are annual herbs in the Mint Family (Lamiaceae). *Salvia hispanica* is the species of commerce and produces small, oval seeds in variegated shades of brown, black, gray, and white. Native to southern Mexico and Guatemala, the Mexican state of Chiapas is actually named after chia. Major producers include Bolivia, Argentina, Ecuador, Nicaragua, Guatemala, and Australia.

The name *chia* is derived from the Nahuatl word *chian* meaning oily; the seeds are 25%–30% polyunsaturated oil, rich in the heart healthy omega 3 and 6 fatty acids. The seeds are also good sources of protein, fiber, and the B vitamins, niacin and thiamine. Documents from the sixteenth century describe chia as a crop widely cultivated by the Aztecs. Chia was one of their staple crops and so esteemed it was given as tribute and used in religious ceremonies. It has been recorded that the Aztecs would chew on chia before a battle and Hopi Indian runners would fuel up on chia as they traveled hundreds of miles.

Its current acclaim is due to its promotion as a super food; athletes of extreme sports, such as ultra-runners who traverse for epic distances, eat the seeds to sustain energy. Chia can be eaten raw or used as a topping or added to other foods such as bread, cereal, and nutrition bars.

Tarwi

Much interest has been expressed in developing tarwi (*Lupinus mutabilis*), a South American legume. The high protein (46%) and oil (20%) content of the seeds rivals that of the soybean. The quality of the protein is exceptionally high and it is rich in lysine; when tarwi is eaten with cereals, the ideal nutritional balance of essential amino acids is achieved. Despite this nutritional value, tarwi is largely unknown outside the highland areas of Peru, Bolivia, and Ecuador. Pre-Incan people domesticated this lupine more than 1,500 years ago. As a significant source of protein, tarwi, along with maize, white potato, and quinoa, made up the staple diet of the peoples of the Andean highlands.

The plant is a branching annual 1 to 2.5 meters (3 to 8 feet) tall with palmately compound leaves. The large, purplish blue flowers emit a honeylike fragrance. Each fruiting pod contains two to six round seeds, usually white. The seeds must first be processed (traditionally soaked in running water) to remove bitter alkaloids, but it may be possible to develop sweet varieties with reduced levels of alkaloids that require little or no processing. The polyunsaturated oil from the seeds has been described as being similar to peanut oil in properties and uses. In the Andean highlands, tarwi seeds are most often used in soups, stews, and salads or eaten by the handful as a snack. Tarwi can also be grown as a livestock feed.

Although tropical, tarwi grows well in temperate regions, since it is adapted to the cooler conditions of high altitudes. In addition, tarwi tolerates a wide range of environmental conditions, including frost and drought. As a nitrogen-fixing legume, it can be grown in soils of poor quality and, in fact, is an excellent green manure. Currently, tarwi is being grown experimentally in Europe, Australia, South Africa, and Mexico to develop its potential as a world crop.

Tamarillo and Naranjilla

Tamarillo and naranjilla are two relatives of the tomato from South America that possess the potential for global development. The tree tomato, or tamarillo, is another Andean crop making inroads into the world's marketplace. *Cyphomandra betacea* (Solanaceae) is a plant of the subtropics; it reaches heights of 1 to 5 meters (3 to 16 feet) and has large, shiny, heart-shaped leaves. Pinkish flowers hang on pendant branches. Fruits are egg-shaped and brilliantly colored; commercial varieties are typically bright red or golden yellow. The flesh may also be variously colored and contains numerous small, edible seeds. The taste of the tamarillo is described as tangy, although sweeter varieties are also available. Tamarillos are eaten peeled; they are often served as toppings on cakes or ice cream. They also can be eaten sliced and added to sandwiches and green salads. The fruits can be blended for juice and when mixed with milk, ice, and sugar make a delicious drink. Tamarillos are also good when cooked and are an ingredient in soups, stews, baked goods, and relishes. Tamarillo jams, jellies, preserves, and chutneys are another application for this fruit.

Nutritionally, tamarillos are high in several vitamins: A, B_6, C, and E. They are low in Calories, fewer than 40 per fruit. Unlike many Andean crops, tamarillos have begun to find a world market. Introduced into New Zealand, tamarillos have been popular there for more than 50 years. In fact, New Zealand is the origin of the name *tamarillo*. Commercial growers there have begun air shipping this fruit to North America, Japan, and Europe, hoping that tamarillos will duplicate the success of kiwifruit, an earlier New Zealand import.

Naranjilla (*Solanum quitoense*) is native to South America. The fruits were appreciated by the Incans, who gathered them from the wild and bestowed yet another name on this plant, *llullu*—or in Spanish, *lulo*, as it is still called in Colombia. It is a tall (nearly 2 meters, or 6 feet), white-flowered perennial with leaves and petioles attractively accented with purple coloration. The round, orange berries explain the name *naranjilla* ("little orange" in Spanish). The fruits are also fuzzy, a trait that accounts for another of its names—the apricot tomato. When the naranjilla is cut open, all comparisons fail because the flesh and juice of the fruit are a vivid green. Naranjilla is primarily grown for the juice, which is unique in its flavoring. Some people compare its tartness to that of citrus; others have said it is a combination of pineapple and strawberry. Nutritionally, the fruits are rich in vitamins A and C.

The plants grow at high altitudes (1,000–1,500 meters, or 3,000–4,500 feet) and so can tolerate cool temperatures. Naranjilla is grown primarily in Colombia, Ecuador, and Venezuela and has been introduced successfully to Panama, Costa Rica, and Guatemala.

Oca

The oca is a tuber crop from the Andes that one day might rival another Andean tuber, the white potato. A member of the Oxalidaceae, or wood sorrel family, oca (*Oxalis tuberosa*) is an herbaceous perennial with three-parted leaves similar to those of a shamrock. The tubers, actually tuberous rhizomes, resemble "stubby, wrinkled carrots" ranging from white to red (fig. 15.9b). The flesh of the tuber is white and, in some varieties, has a slightly acid taste that has been described as "potatoes that don't need sour cream." The acidity is due to the presence of oxalic acid, a substance also present in spinach. Other varieties are so sweet they are sold as "fruits." The tubers can be boiled, baked, or fried. In addition, they can be eaten fresh in salads or pickled. Nutritionally, the oca is similar to the white potato but with higher protein levels. Also, a large percentage of the carbohydrate content is sugar. As a highland crop, it is tolerant of temperatures and high altitudes that are prohibitive to most other crops. Ocas have potential as a livestock feed, especially for pigs. The oca has also become popular in Mexico and New Zealand. It is sold in New Zealand as "New Zealand yam," a name bestowed by that country.

BIOTECHNOLOGY

Biotechnology can be broadly defined as the use of living organisms to provide products for humanity. Applications as simple as using yeast to make bread and culturing other fungi to produce antibiotics can be included in biotechnology. The definition is often focused to particular ends, and for a plant breeder, a working definition of biotechnology might be the use of **cell culture** or molecular techniques, including **genetic engineering,** to create plants with new, useful characteristics. For the most part, the intent behind biotechnology is no different from that of traditional plant breeding: to identify a promising trait and breed or engineer it into a valuable crop plant. For example, disease resistance or herbicide resistance might be introduced into an established variety or large grain size bred into a variety that is already disease resistant. Current uses of biotechnology differ from traditional plant breeding in the speed of introducing new traits and the types of hybridization that are possible.

> ### Thinking Critically
> Alternative crops can be brought to the attention of global agriculture and can diversify the world's food sources.
>
> *People of all cultures are usually conservative about food choices, often rejecting offhand crops that have good nutritional and agricultural value but are unfamiliar to them. How would you ensure that a plant crop new to an area is accepted? Before answering, contrast the successful introduction of the kiwifruit (see Chapter 6) to the utter failure of the breadfruit (see A Closer Look 15.1: Mutiny on the HMS Bounty).*

A CLOSER LOOK 15.1

Mutiny on the HMS *Bounty:* The Story of Breadfruit

The introduction of new plant crops is always a risky proposition. People are generally conservative in their diet choices, choosing foods that are similar to the ones they know and rejecting those that differ. The attempted introduction of the breadfruit tree to the Caribbean is an example of food crop introduction that went horribly awry, precipitating one of the most famous mutinies in naval history. Native to Malaysia, the breadfruit tree (*Artocarpus altilis*) is widespread throughout the South Pacific islands, and its huge, multiple fruits are a traditional food staple of Polynesians. A member of the Moraceae (mulberry family), a single tree can bear up to 500 spherical, seedless green fruits, each weighing about 3 kilograms (6 pounds) (box fig. 15.1a). The unripe fruits have a high starch content and can be fried, boiled, or baked. Ripened breadfruits are sweet and mushy and can be eaten raw or as an ingredient in pies, puddings, and other desserts. Breadfruit can also be fermented to produce a sour condiment used as a spice.

The British naturalist Joseph Banks, who accompanied Captain James Cook on his first expedition to Polynesia and Australia, was one of the first Europeans to observe groves of breadfruit trees in Tahiti. Impressed with the quantity and nutritional value of breadfruit (a 100-gram serving contains 160 kilocalories, 2 grams of protein, and 37 grams of carbohydrates, plus calcium and other minerals), he suggested to Great Britain's King George III that it would be an ideal, inexpensive food for the slave population in the British West Indies. Plantations in these British island colonies depended on slaves to grow cash crops of sugar, coffee, cacao, and indigo. Banks's arguments were apparently persuasive; in 1789, Lieutenant William Bligh set sail to Tahiti on the HMS *Bounty* for the sole purpose of obtaining breadfruit trees for transport to the West Indies. During a 6-month stay in Tahiti, Bligh secured nearly a thousand potted breadfruit saplings, and when he set sail for the return voyage, the *Bounty* was said to look like a floating garden. The saplings were treated with special care; they were kept below deck at night to keep them warm and then shifted back to the deck in the morning for sunshine. The growing saplings required so much water that the crew's supply of drinking water was diminished. For a variety of reasons that have been the source of much speculation in the book *Mutiny on the Bounty* and three Hollywood movies, the crew, led by Master's Mate Fletcher Christian, rebelled. Bligh and the seamen loyal to him were set in an open boat 4,000 miles from the Dutch island of Timor (box fig. 15.1b). Miraculously, Bligh and his companions reached Timor unscathed. The breadfruit trees were not as lucky; they were unceremoniously dumped into the ocean by the mutinous crew.

Fletcher Christian and the mutineers sailed back to Tahiti. They persuaded several native women and men to join them and once again set sail, to find a refuge where they would be safe from the wrath of the British Navy. Their odyssey ended with the discovery of Pitcairn Island, where the descendants of the world's most famous mutineers still live.

Bligh was tenacious; after his triumphant return to England and the hanging of some of the captured mutineers, he went again to Tahiti and this time took 678 trees safely to the West Indies in 1793 aboard the HMS *Providence*. Sixty-six breadfruit saplings were delivered to the Bath Botanical Gardens in Jamaica, where their descendants and a few original specimens survive today. Despite Bligh's travails, the breadfruit did not win immediate acceptance by the slaves. They preferred the plantain, or cooking banana, over the exotic breadfruit. Or perhaps they resented, naturally enough, any food forced on them by their slave masters.

Box Figure 15.1 (a) The large, multiple fruits of breadfruit are a starchy staple in Tahiti. (b) Bligh and his loyal crew were set adrift in a longboat by the mutineers of the HMS *Bounty*.

Cell and Tissue Culture

The ability to develop whole plants from single cells has triggered major interest in cell culture. Although this technique has been available to scientists since the mid-1950s, the applications have grown tremendously since the 1980s. For cell or tissue cultures, small pieces of plant tissue are grown on a nutrient medium supplemented with plant hormones. After a few days, the cells begin dividing and produce a small, undifferentiated mass of tissue known as a **callus.** The cells in the callus continue to grow and divide for several weeks and eventually produce tiny plantlets with stems and roots. When large enough, these plantlets can be replanted in soil and grown to maturity (fig. 15.10). Most of the time, plantlets produced from a single callus are genetically identical, or clones; however, sometimes a mutation occurs, giving rise to a plantlet with different characteristics. These mutations are known as **somaclonal variants,** variants that develop from a single somatic cell (the one that originally produced the callus). Although the appearance of mutations is relatively rare among offspring produced through normal sexual reproduction (about one in a million), the rate of mutation in cell culture seems to be much greater.

Variants that develop by this technique may have useful traits. In one study at a New Jersey biotechnology company, 230 tomato plantlets were generated from callus that developed from tomato leaf tissue. When mature, 13 plants showed unique traits, including larger fruit, tangerine-colored fruit, and denser (fleshier) fruit. Mutant traits developed this way are passed on to the offspring of these plants and can lead to commercially successful new varieties.

Callus can be generated from various types of plant tissue. Parenchyma cells from the stem pith are frequently used, as are pieces of leaf tissue and root cortex parenchyma. The shoot meristem is also utilized for tissue culture applications, especially for producing healthy plants. Because viruses (and other pathogens) are often less common in the shoot tip, meristem culture has become an important method for generating virus-free plants.

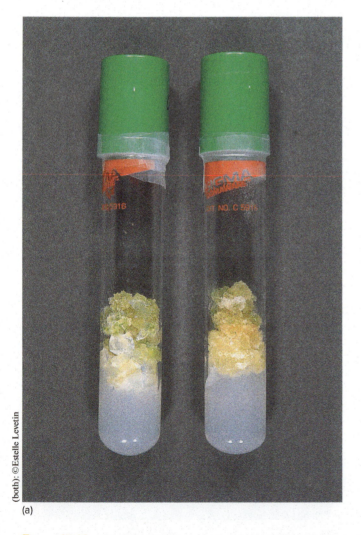
(a)

(b)

Figure 15.10 Plant tissue culture can be used to regenerate plants. (a) Callus growth from small pieces of plant tissue. (b) When hormones are added to the culture medium, the callus differentiates into a plantlet.

Gametophytes, especially pollen, can also be cultured to produce callus; however, the cells are haploid because they are products of meiosis. Callus produced this way is often exposed to the drug colchicine, which blocks microtubule assembly and, thus, prevents the spindle from forming during mitosis. Although the chromosomes are duplicated, the chromatids are not pulled to opposite poles of the cell, thereby doubling the number of chromosomes. The resultant callus tissue becomes diploid and homozygous for every characteristic.

Another method that gives rise to plantlets from cell culture involves the use of **protoplasts,** plant cells whose walls have been removed by enzyme treatment. These protoplasts can also be grown in culture in a nutrient medium. Protoplasts from different species have been fused together using various chemical treatments, and the nuclei have even fused, giving rise to new plants. This technique may be useful for crossing plants that cannot be crossed through normal sexual reproduction.

In addition to searching for new traits, scientists can select for specific mutant types by exposing developing callus cells or protoplasts to certain poisonous substances. The cells that survive have genes that provide resistance to the poison. The resulting plants and their offspring should also show resistance to this chemical. A useful application of this technique is in the development of varieties that are resistant to herbicides. This technique will permit widespread herbicide application to a cultivated crop without damage to the crop plant (see Herbicide Resistance later in this chapter). The same technique can be used to screen for cells resistant to a parasite toxin. Plants generated from these cells may be resistant to the parasite that produces the toxin.

Molecular Plant Breeding

Artificial selection using traditional plant breeding may take many generations to produce desired traits in the offspring. Modern breeding techniques use tools of molecular biology to more quickly select for, modify, or even insert the desired traits into crops. Many different techniques are used, including marker-assisted selection, precision breeding, gene editing, and genetic engineering. These techniques, along with the increasing number of fully sequenced plant genomes, help plant breeders who are trying to feed a hungry world.

The presence of some traits may not be immediately recognized, especially if the trait does not have an obvious phenotype. For example, disease resistance may not be directly visible in the offspring; it requires the presence of the pathogen and the right environmental conditions to foster disease. Also, the trait for large fruit is not visible until the plant completes its life cycle, and it requires optimum growing conditions. Sometimes the inheritance of a trait can be deduced indirectly through the presence of morphological markers, easily seen characteristics that are linked to the desired trait. (Recall from Chapter 7 that linkage is when genes for two traits occur on the same chromosome and are inherited together.) Plant breeders have used linkage based on visible phenotypes since the early twentieth century. Today the marker may be a specific DNA sequence that is linked to the gene of interest and is referred to as **marker-assisted selection.** This allows researchers to screen large numbers of offspring to locate those possessing the specific DNA sequence without waiting for the plant to mature and express the desired trait.

Precision breeding, or **cisgenesis,** transfers specific genes from one variety to another within the same species or between very closely related species. Both donor and recipient plants must be capable of naturally cross-breeding, but with natural crossing it would take many generations of backcrossing to produce the desired results. This technique can achieve the results in a fraction of the time. It differs from some types of genetic engineering in that no foreign genes are introduced from a different species. As a result, in some countries this technique is not subject to the regulations of genetically engineered crops; also, the plants are usually accepted by individuals opposed to genetic engineering. Plants created by precision breeding are called cisgenic plants. The development of flood-tolerant rice (see Chapter 12) was facilitated through the use of precision breeding.

Since the mid-1990s, the most widely used molecular techniques have involved genetic engineering to create transgenic plants. The next section will focus on the development of genetically modified plants and the controversies surrounding these techniques.

GENETIC ENGINEERING AND TRANSGENIC PLANTS

The field of molecular biology has provided science with an understanding of how genetic information is stored, replicated, and translated into specific proteins for cellular development, function, and control. The tools of genetic engineering, especially recombinant DNA technology (see Chapter 7), have made major changes to the field of plant breeding. While traditional methods of plant breeding permit the transfer of genes within a species or to closely related species, genetic engineering allows the transfer of useful genes from one organism to a totally unrelated plant species. Organisms that contain a "foreign" gene in each of their cells are called **transgenic**. Transgenic plants express the transferred gene because all organisms use the same genetic code. Early recombinant DNA research used bacteria to express genes from higher organisms. For example, the gene for human insulin was transferred into the bacterium *Escherichia coli,* which could then produce this valuable protein. Today, it is possible to transfer foreign genes into eukaryotic plant or animal cells, and the same gene for human insulin has been successfully transferred into tobacco plants, which are then able to transcribe and translate the human protein. Other beneficial products from transgenic tobacco are currently being tested. In a similar fashion, corn, soybeans, and potatoes have been genetically engineered to produce human antibodies, which are also in the testing phase of production. Although the development process of transferring genes to plants takes longer, start-up costs are lower and full-scale production can be achieved at a fraction of the cost.

The development of transgenic plants begins with the identification and isolation of the gene that controls a useful trait and the selection of an appropriate vector that will transport the gene into the plant cell. In many plants, foreign genes are introduced using a **Ti (tumor-inducing) plasmid**. A **plasmid** is a small, circular strand of DNA found in bacterial cells. The Ti plasmid normally occurs in *Agrobacterium tumefaciens*, a bacterium responsible for crown gall disease in plants. It invades plant cells and causes tumorlike growths. The tumor-causing genes of the plasmid can be removed without interfering with its ability to enter the nucleus of a plant cell. The plasmid then can be used to introduce desired foreign genes into the host plant cell (fig. 15.11).

One problem with the use of the Ti plasmid is the limited host range. Although there have been some reports of success with monocots, the method is efficiently useful only for certain dicot families. Another technique that is widely used today is the DNA gene gun, which shoots microscopic particles into intact cells. These tiny gold or tungsten particles are coated with plasmid DNA and shot directly into cells using a burst of helium. The cells repair the small puncture wounds, and the DNA becomes integrated into the DNA of the host cell. The cell can then be cultured to produce a plantlet. Although hundreds of cells may be shot, only a small percentage may be successfully transformed and express the foreign DNA.

Genetic engineering has yielded plants with useful traits (table 15.3). It is estimated that close to 75% of the processed foods in U.S. markets contain at least one ingredient from a genetically engineered (also called **genetically modified, or GM**) plant. By 2017, GM crops were grown on 75 million hectares of U.S. farmland and 189 million hectares in 24 countries around the world. This is an amazing accomplishment when it is realized that the first genetically engineered crop was developed at Washington University in St. Louis, Missouri, in 1982. In the spring of 1994, the first of these genetically engineered crops, the Flavr-Savr tomato, was approved for commercial marketing. The Flavr-Savr tomato was altered to slow the ripening process and prolong shelf life; this tomato was never commercially successful and was discontinued in 1997. Since that time, dozens of crops have been approved for sale in the United States and other countries; however, GM crops have been the focus of much controversy and have resulted in strongly polarized debate.

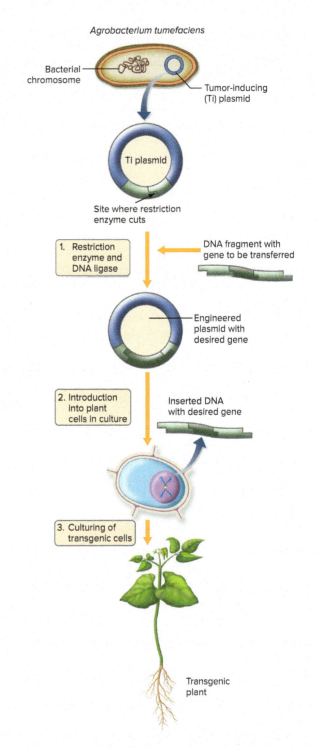

Figure 15.11 Transgenic plants are often developed using the Ti plasmid from *Agrobacterium tumefaciens* as a vector.

Herbicide Resistance

Control of weeds is a significant method of increasing yields because weeds compete with crops for soil nutrients, water, and sunlight. They also harbor pests and clog irrigation systems. A variety of chemical herbicides are used by farmers throughout the world, and their use is increasing. As the value of no-tillage and reduced-tillage agriculture for reducing soil erosion becomes more widely recognized, the need for herbicides will continue to increase. Broad-spectrum preemergent sprays are generally used before planting to kill all the vegetation in the field. Once crop plants are growing, farmers often use a combination of postemergent herbicides throughout the growing season; these kill a narrow range of plants and are carefully chosen to prevent damage to the crop under cultivation.

Table 15.3 Genetically Engineered Crops Currently Grown in the United States

Crop	Genetically Engineered Trait
Alfalfa	• Herbicide resistance • Reduced lignin
Apple	• Nonbrowning
Canola	• Herbicide resistance • Lauric acid production
Corn	• Bt gene for insect resistance • Herbicide resistance • Drought resistance
Cotton	• Herbicide resistance • Bt gene for insect resistance
Papaya	• Virus resistance
Potato	• Nonbruising and low acrylamide levels
Soybean	• Herbicide resistance • High oleic acid production
Squash	• Virus resistance • Herbicide resistance
Sugarbeet	• Herbicide resistance

A bacterial gene that confers resistance to the herbicide glyphosate has been introduced into various crop plants. Several other resistant genes have also been identified for other types of herbicides, including glufosinate, and these have been transferred to crop plants as well. This practice simplifies the application of one broad-spectrum herbicide throughout the growing season because the resistant crop plants are not threatened by the chemical. Today, herbicide resistance is found in 88% of transgenic crops grown around the world; 47% of these crops have only the trait for herbicide resistance and 41% have both herbicide and insect resistance (fig. 15.12). Herbicide-resistant (also known as herbicide-tolerant) soybeans, corn, cotton, and canola are the most widely grown transgenic crops, and herbicide-resistant soybeans account for about 50% of all transgenic crops. Other herbicide-resistant plants have been approved, and others are still in the testing stage.

In June 2005, the USDA approved herbicide-resistant alfalfa for commercial sale; this was the first genetically engineered perennial crop to be approved. Concern has been expressed that a GM perennial crop poses special environmental risks. There is also concern about the spread of the herbicide resistance to non-GM alfalfa. Because alfalfa is pollinated by honey bees, there would be no way to prevent bees from carrying GM pollen to nearby fields with non-GM crops. This would present major problems for the organic dairy industry

Global Area Planted with the Most Common Transgenic Crops in 2017

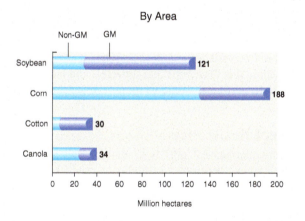
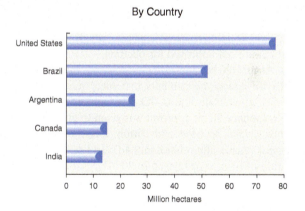

Percent of Total Transgenic Crop Area in 2017

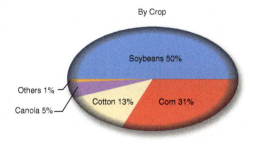
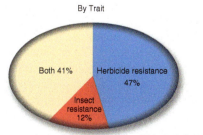

Figure 15.12 Transgenic crops in 2017: Most prevalent crops, traits, and countries where grown.
Source: ISAA Brief 53-2017. Global Status of Commercialized Biotech/GM Crops in 2017.

by possibly contaminating the forage for organic dairy cows. Approximately 5% of the U.S. alfalfa crop was planted with the transgenic variety in 2006. The Center for Food Safety and several other groups brought a lawsuit against the USDA in 2006, citing that the approval process was flawed because environmental issues were not addressed. In February 2007, a U.S. District Court judge ruled that the USDA improperly approved the herbicide-resistant alfalfa because it failed to adequately assess possible environmental impacts, and in May 2007 he issued a permanent injunction on the sale of GM alfalfa seed until the federal government prepared an environmental impact statement. In June 2010, the U.S. Supreme Court, in a 7 to 1 vote, lifted the nationwide ban on planting the herbicide-tolerant alfalfa seeds, thereby reversing the District Court injunction. The Supreme Court ruling was in response to a draft of the USDA environmental impact statement, which indicated that there was no significant impact in using the herbicide-tolerant seeds. However, the court ruled that the USDA needs to complete the impact statement before it can authorize widespread planting. This was completed, and in January 2011 the USDA approved the unregulated planting of herbicide-tolerant alfalfa.

Today, in the United States, 94% of the soybean crop, 80% of the corn crop, and 82% of the cotton crop are genetically engineered for herbicide resistance. These are the most widely adopted transgenic plants on the market (fig. 15.12). Recall that soy- and corn-based ingredients are widely used in processed foods (see Chapters 12 and 13). As a result, it is likely that most Americans have been consuming genetically engineered soy and corn products for many years. In Europe and Japan, herbicide-resistant soybeans are approved for use in food but are not approved for planting.

In August 2014, the USDA approved genetically engineered corn and soybean plants that carry resistance to two herbicides, glyphosate and 2,4-D (2,4-dichlorophenoxyacetic acid). In January 2015, approval was given to glyphosate- and dicamba-resistant soybean and cotton. Dicamba is an herbicide that is chemically related to 2,4-D. In the United States, cultivation of these crops began in 2016.

Against the Grain

In 1997, Monsanto began developing GM wheat with herbicide resistance, and in 2002 the company asked for USDA approval. Tests conducted by Monsanto as well as by other researchers suggested that the glyphosate-resistant wheat could increase yields by 15%. However, farmers as well as food manufacturers and overseas importers began raising objections to the biotech wheat. In fact, some countries, such as Japan and South Korea, objected so strongly that they indicated they would not purchase any wheat (conventional or biotech) from any country that grows biotech wheat because of the chance that grain mixing might occur. The United States is the world's largest wheat exporter, with about half its annual crop sold to foreign countries; the inability to sell the grain to other countries was the main objection raised by farmers. Food manufacturers protested for fear they would have to create separate production facilities for goods, one for biotech wheat in order to manufacture products such as cookies for U.S. consumption and a second one for conventional wheat sold to countries that banned biotech products. Some scientists were also concerned about the effects on future crops. When wheat fields are planted with a different crop the following season, volunteer wheat is regarded as a weed and inexpensive glyphosate-based herbicides are commonly used to kill it. Eradication of the glyphosate-resistant wheat would require the application of more expensive herbicides.

In May 2004, Monsanto postponed its plans to market herbicide-resistant wheat and indicated that the company would reconsider the project when there was consumer acceptance of the product.

In May 2013, an Oregon farmer found wheat plants growing in his field although they had not been planted. Surprisingly, the plants did not die when the he sprayed the field with glyphosate herbicide. Investigations by scientists at Oregon State University and the USDA found that these wheat plants were genetically modified to be herbicide resistant. The last field trials in Oregon conducted by Monsanto had been in 2001, and no trials had ever been conducted at that farm. Despite intensive investigation, the source of these plants remains a mystery. The discovery of these transgenic plants caused South Korea and Japan to temporarily ban the purchase of wheat from the northwest United Sates. Similar events occurred in Montana in 2014, in Washington state in 2016, and in Alberta, Canada in 2018. Again the source of these plants has remained a mystery. Meanwhile, researchers are developing other varieties of biotech wheat, including allergen-free varieties and drought-tolerant wheat. Many people believe that GM crops that show real and immediate benefits to the consumer would be more readily accepted.

Insect Resistance

Bacillus thuringiensis (Bt) is a common soil bacterium that is well known because of its ability to produce proteins with insecticidal properties. For over 50 years, Bt has been available as a safe, naturally occurring pesticide (in dried form) for field applications. The many subspecies of Bt differ in the type of insecticidal proteins that are produced. More than 1,000 strains of Bt have been isolated, and over 200 insecticidal proteins have been identified. Different strains of Bt have different insecticidal activities toward specific insect pests. Some subspecies produce toxins specific for moth larvae; others control beetle larvae; and others control mosquitoes and black flies. There are over 400 Bt preparations registered in the United States for control of various insect pests.

During the 1990s, Bt genes were transferred into more than 50 crop plants, with the resulting transgenic plants expressing the genes for various insecticidal proteins. In 1996, the first Bt transgenic crops were commercially planted, with approximately 1 million hectares of transgenic corn, cotton, and potatoes grown in the United States. The use of Bt transgenic corn and cotton has increased dramatically since 1996. Over 63% of the U.S. corn crop now expresses Bt toxins, and this is grown on more than 22 million hectares. For cotton, over 4 million hectares and 93% of the U.S. crop express Bt toxins. On the other hand, Bt potatoes never amounted to more than 2% to 3% of the U.S. potato crop. Bt potatoes were grown for five seasons but were discontinued by their developer, Monsanto Company, in 2001 (see Chapter 14). Other Bt crops are being tested but have not yet been approved in the United States. In 2009, China approved two varieties of Bt rice. At the present time, environmental safety studies are on-going prior to wide-spread commercial release. Bt eggplant was approved by the Bangladesh government in 2013, and the first commercial cultivation was in 2014.

The use of Bt crops offers several benefits: the toxins are continuously produced and persist for some time; fewer chemical insecticides are needed; and a greater range of insect pests is controlled. In the mid-1990s, 1.6 billion pounds of insecticide were applied worldwide at an estimated cost of nearly $8 billion. It is believed that this technology can save 50% to 70% of the cost of pesticides to protect these crops and reduce environmental damage from pesticide use (see Chapter 26). In addition, the Bt toxins are not harmful to mammals, birds, amphibians, or reptiles, and it is generally believed that there is little effect on nontarget organisms.

Bt Corn and Controversy

The widespread adoption of Bt corn is due to its ability to control a number of insect pests that attack corn, especially the European corn borer. The larvae of this moth affect all varieties of corn and some other crops as well. Yield losses and control costs for this insect pest amount to $1 billion annually for U.S. farmers. In addition to the damage done directly by the corn borer larvae, the open wounds on the plant are sites of infection by various fungal pathogens. In fact, the corn borer larvae themselves are known vectors of fungal spores pathogenic to corn. Most of the Bt corn varieties are grown for grain; fewer GM sweet corn and popcorn varieties are being grown in the United States.

Because corn-based ingredients are so widely used in processed foods, most U.S. consumers have eaten products containing Bt corn. Initially, there was little noticeable opposition to the use of transgenic corn; however, issues later arose regarding the safety of Bt corn. One issue is related to an environmental concern, and the other is a human health concern.

In 1999, a study by researchers from Cornell University charged that the use of Bt corn could be harmful to the monarch butterfly populations in North America. Monarch butterfly larvae feed exclusively on milkweed leaves. Because milkweeds can be found in and around cornfields, these scientists charged that the pollen from Bt corn plants could contaminate nearby milkweed plants. To show the dangers of this situation, they fed monarch larvae milkweed leaves covered with Bt corn pollen in a laboratory experiment. The larvae grew more slowly and suffered higher mortality rates when they ate milkweed leaves dusted with Bt corn pollen. Many people feared that this laboratory finding could have serious implications for monarch populations in the U.S. Corn Belt. Others criticized the Cornell study for several reasons: the amount of corn pollen on the leaves was not carefully measured; there was no indication that the corn pollination season would overlap with the presence of the monarch larvae in the field; and there was no mention of how much harm is done to the monarch larvae through the current use of conventional chemical pesticides that drift from cornfields.

In 2001, several follow-up studies were published in the *Proceedings of the National Academy of Sciences.* These studies showed that the expression of Bt toxins in pollen varied among Bt corn hybrids; only one type (called event 176) had high levels of Bt toxins in the pollen and showed significant harm to monarch larvae. In fact, event 176 was the variety used in the 1999 Cornell study. It should also be noted that event 176 is no longer commercially available because it provided protection from the European corn borer only early in the season. The effects of other Bt corn pollen on monarch larvae were found to be negligible. The survival of monarch larvae feeding on milkweed leaves within a Bt cornfield was significantly higher than the survival of monarch larvae feeding on milkweed leaves within a traditional cornfield sprayed with insecticide. Also, only a portion of the monarch population can be found in or near cornfields, although there is some overlap between the time of pollination and larval activity in the northern part of the monarch breeding range. Overall, these studies suggested that Bt corn has little impact on monarch butterflies in North America.

During the fall of 2000, headlines around the United States reported finding DNA from StarLink corn in taco shells and other foods. StarLink corn was a variety of Bt corn that had been approved for use in animal feed but not human food. This transgenic corn carried a Bt gene for one of the many insecticidal proteins produced by this bacterium. Although other varieties of Bt corn were approved for human food, routine testing of the potential allergenicity of the protein in StarLink before approval by the Environmental Protection Agency left some questions. The Bt toxin in StarLink was shown to be more resistant to digestion and heat than other Bt toxins. Because many food allergens are also resistant to digestion, the approval of StarLink corn for human food was delayed pending further allergy testing. Although the media stories described this toxin as a food allergen, the toxin in StarLink was never actually shown to be allergenic. In fact, tests conducted by the Centers for Disease Control and Prevention found no evidence of allergic reactions among 51 individuals who had self-reported allergic responses to StarLink corn.

The discovery of the StarLink corn in food products resulted in massive recalls of various corn products, and the final costs to farmers, the food industry, grain handlers, and the developers of StarLink may have exceeded $1 billion in losses. The incident raised consumer fears about the safety of genetically engineered crops and led to a negative perception of biotechnology in the minds of many people. It also brought an awareness about how GM crops are handled. Although farmers were required to grow StarLink corn separately from other corn, grain handlers apparently did not observe this segregation. StarLink was mixed with corn intended for human food. As a result, StarLink ended up in dozens of food products throughout the country, even though it represented only a small percentage of the U.S. corn crop. This experience with StarLink corn may help improve policies and practices related to the handling of GM crops.

The Promise of Golden Rice

One of the major challenges facing humanity is providing sufficient food to feed the growing population; however, more than just kilocalories are needed. It is estimated that 2 billion people are suffering from malnutrition, with deficiencies of vitamin A and iron the most prevalent. The development of major crops enriched with one or both of these nutrients has been a major goal for Ingo Potrykus, Peter Beyer, and other researchers at the Institute of Plant Sciences in Zurich, Switzerland. Researchers there have developed genetically engineered "golden rice" (fig 15.7), which are able to synthesize beta-carotene in the endosperm. It is believed that the presence of this pigment, which is the precursor to vitamin A, will reduce the incidence of vitamin A deficiency in developing nations. This deficiency is the leading cause of preventable blindness. It is estimated that 124 million children in developing nations are deficient in vitamin A and up to 500,000 children become blind each year because of this deficiency (see Chapter 10). At the present time, approximately 6,000 people die each day as a result of vitamin A deficiency.

In 1999, Potrykus and Beyer developed the first beta-carotene-producing rice by transferring a gene from daffodil and one from a bacterium. Although the yield of beta-carotene was low, they proved it was possible to transfer beta-carotene synthesis genes into rice endosperm. Golden rice 2, which was developed in 2005 by Syngenta, contains a gene from corn plus the bacterial gene and produces much greater amounts of beta-carotene. This enhanced rice can supply 30% to 50% of the vitamin A requirement for women and children. Also, a study published in 2009 found that beta-carotene from golden rice is effectively converted into vitamin A in humans. Researchers have been transferring this trait by conventional breeding into local rice varieties in the Philippines, India, Bangladesh, Vietnam, and Indonesia. Vitamin A deficiency is common in all these countries. Field trials have shown that while beta-carotene levels were good, grain yields were lower than desired. Research continues to improve yield on these golden rice varieties.

In 2018, golden rice was approved by Food Standards Australia New Zealand, Health Canada, and the United States Food and Drug Administration. Although there are currently no plans to grow or market golden rice in these countries, it is assumed that golden rice or food products derived from golden rice may enter the food supply in these countries through imports. Two target countries for growing golden rice are Bangladesh and the Philippines, and applications for regulatory approval have been filed in both of these countries.

Golden rice was developed to reduce vitamin A deficiency in developing nations; therefore, it has to reach poor farmers without cost. Although the technology to develop golden rice began in Switzerland, it built on prior work and continued with subsequent work by other researchers in various countries. Consequently, many individuals as well as industrial and governmental agencies in many parts of the world have agreed to make this technology available free of charge to the poor in developing nations. Although golden rice holds great promise for relieving malnutrition, many critics feel that malnutrition can be relieved more easily and sooner by building roads and delivering vitamin pills to at-risk people in developing nations.

While the development of golden rice was proceeding, the Bill and Melinda Gates Foundation provided funding to Peter Beyer and colleagues to continue research on golden rice in order to add vitamin E, more protein, and enhanced iron and zinc availability. Iron-deficiency anemia is the most common dietary deficiency disease in the world, with estimates of 2 billion people suffering from iron deficiency (see Chapter 10).

Other Genetically Engineered Foods

Soybean oil, which represents 30% of the oil consumed worldwide (see A Closer Look 13.2: Harvesting Oil), contains mainly polyunsaturated fatty acids, which decompose when heated. Manufacturers typically compensate for this decomposition by partially hydrogenating the oil; hydrogenation provides greater heat stability. Unfortunately, this process adds to the cost and creates saturated fats and trans fatty acids, which have been linked to increases in cholesterol levels.

A soybean variety has been genetically modified to suppress the production of polyunsaturated fatty acids. The result is an increase in the production of oleic acid, which is a monounsaturated fatty acid. The oil from this GM soybean has over 80% oleic acid, in contrast to the 20% in regular soybean oil. Recall that the consumption of monounsaturated fatty acids has been shown to lower LDL (bad cholesterol) levels in the blood, while the consumption of polyunsaturated fatty acids lowers both LDL and HDL (good cholesterol) levels (see Chapter 10). This high-oleic acid soybean oil is being promoted as heart-healthy. In fact, oleic acid is the main fatty acid in olive oil and may be responsible for the health benefits of olive oil use. In addition to the health benefits for the consumer, the high-oleic acid soybean oil offers savings for the manufacturer because hydrogenation is not needed. Also, the high-oleic acid soybean oil will be superior to the current oils for biodiesel fuel. This modified soybean has been grown in some areas of the United States for the past few years. The soybean industry hopes to expand the cultivation to 7 million hectares by 2023.

Researchers in Australia have been developing GM bananas, which contain enhanced levels of pro-vitamin A and iron. The goal of this project, funded by the Bill and Melinda Gates Foundation, is to improve vitamin and mineral deficiencies among the population of East Africa. It is hoped that these enriched bananas will be available in East Africa in a few years. Other scientists around the world are also making progress at improving the nutritional qualities of other crops with the same goal. The crops being modified include corn, potatoes, sweet potatoes, soybean, sorghum, tomatoes, and strawberries and the traits include protein levels, essential amino acid levels, and vitamin levels. For example, researchers in India are developing a GM potato that has more protein and high levels of the essential amino acids lysine and methionine by inserting a gene from amaranth.

Several GM crops recently have been approved in the United States and became commercially available for the 2015 growing season. Two apple varieties, Granny Smith and Golden Delicious, were modified to reduce the expression of the enzyme polyphenol oxidase; this enzyme naturally causes browning in apples as well as other fruits and vegetables when they are cut or injured. The molecule technique, called gene silencing or RNA interference (see Chapter 7), was used to reduce the level of the enzyme and thereby reduce browning in cut apples. This was accomplished by transferring genes from other apple varieties.

Three potato varieties were also approved; these varieties were modified to resist bruising and produce lower levels of acrylamide during frying. Acrylamide has been shown to be carcinogenic in rodents. Specifically, these varieties have reduced levels of the amino acid asparagine and reducing sugars, which produce acrylamide at high temperatures. The modified potatoes were produced using genes from other potato varieties through gene silencing techniques similar to those used with the apples.

An alfalfa variety with reduced lignin was approved in November 2014. Recall that alfalfa is a major forage crop around the world (see Chapter 13). Lignin found in the secondary walls of fibers, vessel elements, and tracheids is a complex molecule that provides rigidity to the walls of these cells and helps support the plant. However, lignin is not digestible by animals. In addition, alfalfa has high levels of lignin compared to other forage crops and the lignin content increases as the plant ages. These plants with reduced lignin will have greater digestibility and energy yield for the animals. The trait was accomplished by gene silencing to reduce the expression of one enzyme in the lignin synthesis pathway.

Scientists at several biotech companies and universities have been developing GM crops that are drought resistant. Drought-resistant corn was the first of these to be approved for use in the United States in 2011. The potential for drought-resistant crops is enormous. They will be a tremendous benefit to poor farmers in developing nations, where drought typically leads to famine. Many scientists consider drought tolerance the most important trait that could be engineered into crops because water availability is the greatest limiting factor for agricultural production worldwide. In addition to drought tolerance, research is also ongoing to develop crops that are tolerant of higher temperatures and high CO_2 levels, which are both predicted in current climate change models.

Disease Resistance

Development of disease-resistant crops is another aim of genetic engineering research. Any advance in disease control will substantially influence the world's food supply. Many early efforts focused on the development of virus-resistant plants. Viral genes have been introduced into various plants. These genes express viral coat proteins but turn off when the cell is infected. Although the mechanism is not completely understood, it prevents the reproduction of the virus. Virus-resistant tobacco has been grown in China since 1992, and virus-resistant squash and papaya are commercially available in the United States. This technique has also been successfully tested with alfalfa, pepper, rice, and other crops.

Plum pox is a viral disease that affects stone fruits, including plums, peaches, and apricots. The disease has a worldwide distribution and significantly limits stone fruit production wherever it becomes well established. Fruit yields are reduced, and the fruit that is produced may be unmarketable. It is one of the most devastating diseases of stone fruits; however, the severity of the disease is dependent on the strain of the virus and the susceptibility of the plant cultivar. The disease was first identified in Bulgaria in 1915 and has since spread to many parts of the world. It is especially problematic in Europe, where it is estimated that 100 million trees are infected with the virus. The disease was identified in Pennsylvania in 1999, but the outbreak was contained with stringent eradication measures. Over 100 hectares of infected trees were destroyed to prevent disease spread. Continued testing the following year in the surrounding areas turned up more infected trees, which were also destroyed. Subsequent testing has turned up fewer infected trees, but the possibility remains that ornamentals, such as flowering almond, harbor the virus in the surrounding areas. USDA scientists began research on a long-term solution by developing virus-resistant trees using both traditional plant breeding and genetic engineering. A plum pox–resistant transgenic plum tree is the first of these resistant trees (fig. 15.13), and it received final regulatory approval in 2010.

Figure 15.13 Transgenic plums contain a gene that makes them resistant to plum pox virus.

Late blight of potato caused by the pathogen *Phytophthora Infestans* is the most important disease of potato worldwide (see Chapters 14 and 23). Researchers in England have recently developed a GM potato that is resistant to the late blight pathogen. They transferred a gene from a wild South American potato that confers natural resistance to the disease into a common variety of red-skinned potato. The British researchers licensed the technology to an American company, which subsequently transferred the resistance gene into three widely grown potato varieties. These received USDA approval in 2015, and EPA and FDA approval in 2017. These disease-resistant potatoes should be available for marketing soon.

Genetic engineering is being used to produce rice plants resistant to a devastating disease known as leaf blight, caused by the bacterium *Xanthomonas oryzae*. The leaves of infected plants develop lesions and wilt in a few days. This disease is common throughout rice-growing regions of Asia and Africa and can destroy half the crop in heavily infected areas. Traditional breeding has been used for years to develop blight-resistant varieties of rice, but this time-consuming process may take more than 10 years to produce the desired traits in the offspring. In 1994, a gene for blight resistance was identified in a wild species of rice. The resistance gene, known as *Xa21*, was cloned and transferred into a susceptible variety using a gene gun. The transgenic rice was able to pass *Xa21* on to its offspring. The gene has also been introduced into three commercially important varieties of rice, and research is continuing to introduce the gene into locally important rice cultivars. It is believed that disease-resistant transgenic plants hold great promise for increasing rice production.

Fusarium head blight is a disease of wheat, barley, and other grains that is caused by the fungus *Fusarium graminearum*, although other *Fusarium* species can also cause the disease. The fungus affects the developing heads of grain, resulting in significantly reduced crop yields. In addition, mycotoxins can form in the grain, making it unusable for humans or animal feed (see Chapter 25). *Fusarium* head blight is responsible for limiting wheat production in many parts of the world. Resistant varieties of wheat are known, and wheat breeding programs are using both conventional plant breeding and genetic engineering to transfer resistance to commercially important wheat varieties (fig. 15.14).

In the future, identifying useful genes in plants may become easier as research expands on the knowledge of the DNA sequences for economically important species of plants. The National Plant Genome Initiative (NPGI) was developed by representatives from several federal agencies, including the National Science Foundation, the U.S. Department of Agriculture, and the National Institutes of Health. The long-term goal of the NPGI is "to understand the structure and function of genes in plants important to agriculture, environmental management, energy, and health." The genome research could lead to significantly increased food production in the coming decades; it has been suggested that this Gene Revolution will be even more productive than the Green Revolution.

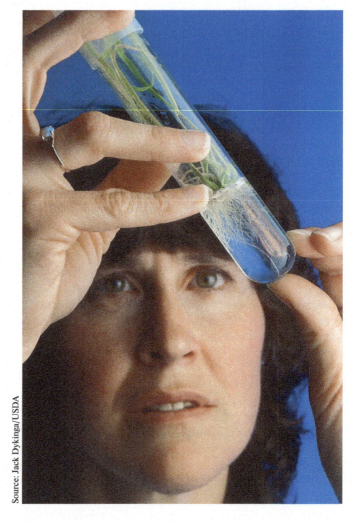

Figure 15.14 USDA researcher examines root growth on a genetically engineered wheat seedling possibly showing resistance to *Fusarium*, a fungal pathogen of many crops.

Farming Pharmaceuticals

The promise of biotechnology extends beyond food crops; researchers are also using genetically engineered crops to produce products that will keep people healthy. The term **molecular farming** is used to describe the growing and harvesting of genetically engineered crops that are producing pharmaceuticals. At the present time, hundreds of pharmaceutical products are in various stages of development, with much of this research focused on the development of edible vaccines.

Vaccines are major weapons in the fight against infectious disease; they have eliminated smallpox and, it is hoped, will eliminate polio in the near future. Also, most childhood diseases, such as measles, rubella, and whooping cough, are a thing of the past. However, 20% of the children in the world are not immunized. It is estimated that 2 million children, primarily in developing nations, die each year of diseases that could be prevented if the vaccines could reach everyone. High costs and the need for refrigeration are two factors standing in the way. Research is underway to solve these problems by developing genetically engineered plants that can be used as edible vaccines.

Vaccines work by preparing a person's immune system to destroy an invading pathogen. Generally, vaccines contain viruses or bacteria that have been killed or weakened and are injected into the body (or given orally) to sensitize the immune system. The immune system responds by producing specific antibodies, which are then ready to mount a defense if the pathogen is encountered in the future. Even a protein from the virus or bacterium can be enough to sensitize the immune system. When the gene for the protein is transferred into plants, the plant will express the protein and can be used to vaccinate people or animals. The first human clinical trials on an edible vaccine were performed in 1997. The results showed that volunteers who ate the GM potatoes carrying a gene from a bacterium showed immune responses against a bacterial toxin that causes diarrhea. Similar trials on other edible vaccines against Norwalk virus (which also causes diarrhea) and hepatitis B virus also showed promising results. Animal trials for a number of other vaccines have also been conducted. Potatoes, tomatoes, and bananas have all been used in experiments to produce these vaccines, and other foods under consideration include lettuce, carrots, wheat, rice, and corn. Technology still has a long way to go before these edible vaccines become widely available; researchers estimate that it probably will be another 5 to 10 years before commercial production will be possible. Scientists are very optimistic that these vaccines will substantially benefit health care in developing nations. In a similar move, many tobacco plants have been genetically modified to produce individualized vaccines to treat patients with non-Hodgkin's lymphoma. Clinical trials are in progress to evaluate these vaccines.

In March 2007, the USDA granted preliminary approval for field-testing transgenic rice that produces human immune system proteins in its grains. The pharmaceutical rice developed by Ventria Bioscience would be harvested and milled to flour. The transgenic proteins would be extracted from the flour for use in medicines to fight diarrhea, dehydration, and other conditions that kill millions of infants and toddlers each year. Some of these products are currently being tested in clinical trials and if final approval is granted, the rice will be grown in Kansas; the state was selected because no commercial rice production occurs in the state. This would provide a safeguard to the accidental introduction of the GM rice into the food supply.

In July 2018, researchers from Spain, England, and the United States published the results of their work in developing transgenic rice that produces three proteins with anti-HIV (human immunodeficiency virus) activity. The proteins produced by the GM rice include a human antibody and two lectins, which are carbohydrate-binding proteins; these proteins target different components of HIV to avoid the development of resistance. The authors suggest that crude extracts of the rice grains could be used directly in a topical application to prevent HIV infection. For developing countries where HIV infection is a major problem, this could provide relatively low-cost HIV prophylaxis.

Nonfood Crops

Genetic engineering has also produced transgenic plants for nonfood uses. One major success in this area has been with rapeseed plants. Conventional plant breeding produced canola varieties of rapeseed that are low in erucic acid and more suitable for human consumption (see A Closer Look 13.2: Harvesting Oil). Today, new canola varieties, developed by genetic engineering, are high in lauric acid, a short-chain fatty acid that is valuable for soaps, detergents, cosmetics, lubricants, and paints. These canola varieties were among the first genetically engineered nonfood crops that were available commercially.

In 2010, the European Union approved the commercial cultivation of the Amflora potato in member countries. While most potatoes contain a mixture of amylose and amylopectin, this GM potato produces starch that is almost pure amylopectin. The gene for one of the enzymes needed for amylose synthesis has been turned off. This GM potato will be used only for the industrial extraction of starch and animal feed, not for human food.

Researchers in several countries have developed genetically engineered plants for enhanced phytoremediation (see A Closer Look 4.1: Mineral Nutrition and the Green Clean). Some of the engineered plants include tobacco, thale cress, birch, poplar, Indian mustard, and rice. These GM plants can accumulate heavy metals or break down various toxic organic compounds. Although none of these have been approved yet for widespread planting, the potential environmental benefits of these transgenic plants are apparent because they can process pollutants as much as 100 times faster than currently used plants.

Genetically Modified Trees

Since the late 1990s, there have been ongoing greenhouse and field trials of GM trees in many countries. A number of different traits have been transferred into various species of trees; at least 24 tree species have been genetically engineered. In the United States, hundreds of field tests have been conducted or are underway on GM apple, cherry, citrus, papaya, pear, persimmon, pine, plum, poplar, spruce, and walnut trees.

Much work has been done on the development of fast-growing poplar trees with wood that has a reduced lignin content and increased cellulose content. This would be a major boon to the paper industry, since the removal of lignin is expensive and environmentally hazardous (see Chapter 18). In May 2010, the USDA approved field trials of GM eucalyptus trees at 28 field sites in seven states. The trees were modified to withstand colder temperatures than conventional eucalyptus trees, which are frost sensitive. Eucalyptus trees are naturally fast growing, and these GM varieties are proposed as raw materials for paper, pulp, and biofuels.

Other trees have been modified to grow faster or to be insect resistant, herbicide tolerant, or disease resistant. Scientists from the United Kingdom have developed a transgenic elm tree that is resistant to Dutch elm disease. This fungal disease has devastated elm populations in Europe and North America since early in the twentieth century (see Chapter 23). However, decades of conventional breeding programs have already developed elm trees resistant to Dutch elm disease.

Chestnut blight is another fungal disease that changed the landscape, destroying as many as 4 billion American chestnut trees in the early twentieth century (see Chapter 6). Scientists from the State University of New York–Syracuse have developed transgenic chestnut trees that are resistant to chestnut blight. The resistant trees contain genes from the Chinese chestnut tree as well as genes from other plants. The researchers have begun field tests and hope to one day restore the chestnut tree in American forests.

There are many environmental concerns relating to the widespread use of GM trees. Many contend that the potential long-distance dispersal of airborne pollen from GM trees and the long-term environmental impact are not known and urge delays in this technology. Although none of these trees have been approved for commercial planting in the United States or in the European Union, over 1 million transgenic trees have been planted in China. In 2002, China's State Forestry Administration approved the commercial planting of 1 million poplar trees modified with Bt genes. The trees were planted in an effort to reverse environmental damage caused by deforestation. Prior plantings of nonmodified poplar trees were plagued by insect pests, so researchers developed and planted the GM trees. Unfortunately, there are no plans to monitor the environmental impact of these trees.

Regulatory Issues

Today the U.S. Department of Agriculture (USDA), Food and Drug Administration (FDA), and Environmental Protection Agency (EPA) are all involved in the approval of genetically engineered crops. The USDA is responsible for monitoring field trials of GM crops and for evaluating the potential impact of crops in the natural environment. For example, it would try to determine if the crop could become a weed, transfer genes to weeds, or affect other organisms. If the GM crop is to be eaten by humans or animals, the FDA evaluates its food safety. In 1992, the FDA developed a policy on genetically engineered crops, recommending that "developers consult with the FDA about bioengineered foods under development." Additional procedures were added in 1996, instructing developers to submit relevant safety and nutritional information to the FDA for evaluation. Recently, the FDA proposed changing the policy of voluntary consultations to a requirement for developers to submit a "scientific and regulatory assessment of the bioengineered food 120 days before the bioengineered food is marketed." Basically, the FDA wants assurances that there is no substantial difference between the GM variety and the traditional variety. If the FDA feels that the GM variety is not equivalent, it must be labeled; however, if the health concerns are serious enough, the GM crop will not be marketed. Transgenic plants that produce pesticides (such as Bt crops) must be approved by the EPA. That agency oversees environmental impacts of the pesticides as well as the human health aspects. As a result, the EPA was the agency responsible for limiting StarLink corn to animal feed. In reviewing the environmental effects, the EPA evaluates any possible effects on nontarget organisms and the environmental fate of the pesticidal product. The EPA may also require implementation of a "resistant management plan."

Environmental and Safety Considerations

There are many issues and questions involved in the development and use of genetic engineering and transgenic plants. Some scientists hold that developing transgenic organisms is only an extension of traditional methods of hybridization and the products should be treated no differently. Critics argue that transferring genes from one species to another can seriously affect the environment, can harm human health, and benefits only large agribusiness conglomerates. Some anti-technology activists are opposed to all genetically modified crops. This feeling seems to be especially prevalent in parts of Europe where GM crops have been referred to as "Frankenfoods." Critics are concerned about private-sector monopolies of GM technology and contend that transgenic crops offer few advantages for consumers. In 2017, GM corn was grown in both Spain and Portugal. In addition, several countries in the European Union import GM soybeans, corn, and canola; however, all GM products in European countries are only used for animal feed.

Fear of GM crops led to a decision by Zambian officials to refuse emergency food aid from the United States in August 2002. Drought in southern Africa had led to wide-spread famine that affected about 13 million people, including 2 million in Zambia. The refusal of the food aid was due to concerns that the GM maize could contaminate local agriculture. Other countries in the area (Zimbabwe, Mozambique, and Malawi) would accept the grain only after it was milled into flour. By contrast, South Africa produces GM corn, soybeans, and cotton, and Sudan grows GM cotton.

Among the concerns expressed by the scientific community about GM crops are the length of time the altered organism will survive in the environment, how quickly it multiplies, how far it can travel, and what environmental effects could occur if the organism did become established in the wild. Could the altered organism pass on its new traits to other organisms in the wild? Could the gene for herbicide resistance be passed on to weedy relatives of cultivated crops through natural hybridization? Danish researchers showed that genetically engineered canola with herbicide resistance could cross with a weedy relative in the same genus. The hybrid offspring were fertile, and the herbicide resistance trait was passed on to additional generations of the weed. This transference can happen only if there are closely related weeds in the area, and in the United States it might be a problem for crops such as canola and squash. If weeds acquire resistance to a particular herbicide, the marketing advantage of that herbicide-resistant plant will quickly disappear and the herbicide will be rendered useless. One possible solution to the problem is to genetically engineer the genes into the chloroplast genome instead of into the nuclear genome. For some species, this approach would permit herbicide resistance (or other traits) to be expressed in all tissues with plastids but would prevent gene transfer with the pollen. However, this technique would only be useful for species with maternal inheritance of plastids. Another proposed solution would link herbicide resistance with another gene that would be harmful to weeds, should the transfer occur.

In addition to concerns about the transfer of herbicide-resistant genes from GM crops to related species, there is also concern about the natural development of resistance among weeds. The repeated use of a single herbicide greatly increases the development of herbicide resistance. In fact, herbicide resistance was first described in 1957, and 304 weed varieties have developed resistance to one or more herbicides. Recently, interest has focused on glyphosate, the active ingredient in a widely used herbicide. Recall that many GM crops carry genes for resistance to glyphosate. By the end of 2014, 38 species of weeds had evolved resistance to glyphosate in countries around the world. In the United States there are at least 48 million hectares afflicted with glyphosate-resistant weeds. The most widespread resistance has been seen in horseweed, *Conyza canadensis.* Horseweed, a member of the sunflower family (Asteraceae), is a common weed throughout North America and other countries. It can significantly reduce yields of agricultural crops and dominate natural vegetation. A single horseweed plant can produce about 200,000 wind-dispersed seeds. Although the evolution of resistance among weeds is not directly related to the use of GM crops, the widespread use of herbicide-tolerant GM crops, especially soybean, allows resistant weeds to flourish if other controls are not used. When glyphosate-tolerant soybean fields are sprayed with glyphosate, many weeds are killed. Resistant weeds are not killed by the application of this herbicide, and the weeds can grow and reproduce. Although glyphosate-resistant weeds may be killed by the use of other herbicides, the advantage of using GM herbicide-tolerant crops is lost. Researchers are just beginning to investigate possible ways to control the spread of glyphosate-resistant weeds, including the use of GM crops that carry resistance to two herbicides. As previously indicated, GM crops resistant to both 2,4-D and glyphosate or resistant to both dicamba and glyphosate have recently been developed. However, additional problems appeared once these new herbicide-tolerant crops were put into use. Spray drift can occur when 2,4-D and dicamba herbicides are applied. This means that nontarget crops can be seriously affected. When the dicamba-tolerant crops were sprayed with dicamba, crop damage occurred on nearby farms not growing dicamba-tolerant crops. By the end of 2017, there were over 2,000 complaints and dozens of lawsuits.

For most crops, farmers can save a portion of their harvest as seeds for the next year's planting. In developed countries, many farmers have abandoned this practice and instead buy new, improved varieties of seeds each year, but in developing nations, seed saving is the standard. Plants engineered with a terminator gene appear normal and produce seeds that appear normal. However, the seeds are not viable. The development of terminator genes has caused tremendous controversy. Although some people view the terminator gene as a wonderful safeguard to prevent unintentional spread of engineered genes into weeds, others view the gene as a major disadvantage and hardship because it forces poor farmers to buy new seeds each year. Terminator genes have been banned in India and denounced by major research institutes. Monsanto has publicly promised not to market any crops with terminator genes; however, other seed companies may be developing terminator technology.

In addition to the concerns about the monarch butterfly described earlier in this chapter, other issues about the extensive use of transgenic crops expressing Bt genes focus on the development of resistance among target insects. Recently, problems have surfaced with corn rootworm becoming resistant to Bt corn plants in the U.S. Corn Belt. Similar problems are occurring in Brazil, where the corn leafworm has developed resistance. In India, problems have surfaced with Bt cotton. The Bt genes in transgenic cotton are principally aimed at the control of various bollworm species, the major pests of cotton. In 2009, it was reported that pink bollworm had become resistant to the toxin in Bt cotton in Gujarat, a state in western India. The consequences of widespread use of Bt genes for major crops are largely unknown. Scientists contend that many alternate forms of the Bt gene exist if resistance develops. In tests, it has been shown that development of resistance can be delayed when a mixture of toxins is used. The EPA requires that farmers in the United States plant a refuge of non-Bt crops among the transgenic ones to delay the development of resistance. These refuges will allow a population of susceptible insects to survive and make it less likely that rare Bt-resistant insects will quickly increase. Researchers urge caution with the Bt technology because misuse of Bt toxins could rapidly make them ineffective for major plant pests. Resistance may be a major problem for farmers planting Bt crops and for those who rely on environmentally safe Bt insecticidal sprays for organic crops. In China, a problem has developed with secondary pests. The use of Bt cotton in China has been very successful, and farmers have been able to reduce pesticide use dramatically. However, the populations of other insects, especially mirid bugs, began increasing and they are now major pests. Pesticide use is again increasing to control these secondary insects.

A related issue involves the contamination of non-GM crops by pollen carrying Bt genes. Analyses of organically grown corn in several states have shown traces of Bt genes. Although a buffer zone is required around genetically engineered crops, corn pollen was apparently carried by wind from nearby Bt corn plants. In a similar manner, a report suggested that Bt genes had shown up in native corn varieties in Oaxaca, Mexico. GM corn can be imported to Mexico for food use but not for planting. However, investigators believe that pollen from fields of illegally grown Bt corn was carried to native varieties. The region of southern Mexico around Oaxaca is the center of origin for corn and, therefore, an important source of genetic variability. Follow-up studies in 2003 and 2004 failed to find a single transgenic construct from tests on over

150,000 seeds from more than 870 plants in 125 cornfields in the area. Although this finding was good news, some scientists are still concerned that Bt genes may have escaped and threaten the genetic diversity of corn in the region. Similar questions arise about the ability to control gene flow for other GM crops. The airborne dispersal of pollen from GM creeping bent grass, *Agrostis stolonifera,* was monitored during field trials in Oregon through the use of sentinel plants at various downwind locations and the analysis of wind patterns. Offspring of the sentinel plants displayed herbicide tolerance and carried the transgene inserted in test plants, indicating that the sentinel plants had been fertilized by pollen from the GM plants. The USDA subsequently prevented the commercial release of GM creeping bent grass. This shows the difficulty of containing gene flow in wind-pollinated species of genetically engineered plants.

An additional worry about genetically engineered crops is related to harmful traits that might be inadvertently transferred. One such instance came to light during tests of soybeans that had been genetically engineered with a Brazil nut gene to provide extra methionine for animal feed. When the GM soybeans were tested during development, the transferred protein caused allergic reactions in individuals known to be allergic to Brazil nuts. The project was canceled before any beans were sold.

The protocol for allergy testing of GM foods calls for answering the following series of questions. Does the transferred gene come from a known food allergen? Does the DNA sequence of the gene match that of any known allergen? Does the protein exhibit heat stability or stability to digestive enzymes (which are traits commonly found in food allergens)? If the answer to any of those questions is yes, the crop will not be approved for human food without extensive further testing. StarLink corn had one yes answer. The gene was not from a known allergen, and the gene sequence did not match that of any known allergen; however, the protein showed stability during testing. Although an allergenic protein could be transferred through genetic engineering, the testing protocol should be able to ensure that such a product will never get to market.

In April 2010, the National Research Council released a report on the impact of genetically engineered crops on American farmers. The study found that there have been substantial economic and environmental benefits since the crops were introduced in 1996. Economic benefits to the farmers include reduced costs for weed control, reduced crop losses from insect pests, reduced cost for pesticides and fuel, and improved worker safety.

Herbicide-resistant crops reduce tillage and, therefore, reduce runoff of agricultural chemicals, resulting in water quality improvements. Unfortunately, there has been no infrastructure in place to study this aspect. The study indicates that the reduced water pollution could be the "largest single environmental benefit" of transgenic crops. However, the report warned that the widespread use of the herbicide glyphosate would lead to the further development of resistant weeds and may eliminate the economic and environmental benefits of the herbicide-resistant crops. It recommends improved monitoring and management strategies, including the use of herbicide mixes.

The use of engineered crops with Bt toxins has resulted in a reduction in insecticide use, which can have positive effects on beneficial insects. Although repeated use of Bt crops could bring about resistance in target organisms, the use of a refuge of non-Bt plants, which is required by the EPA, has slowed the development of resistance. The report recommended a greater use of multiple toxins in Bt crops, which would further delay the development of resistant insects.

The study also identified needs and possible future applications of genetic engineering, suggesting that the technologies should be used to improve world hunger and malnutrition by increasing the nutritional qualities of crops. It also suggested the need to develop crops able to tolerate higher temperatures, drought, and other possible conditions related to climate change.

GENE EDITING: CRISPR/Cas9

As described above, genetically engineering crops express new traits that have been transferred from unrelated organisms. By contrast, gene editing may modify the DNA of an organism by deleting a segment of the DNA, by changing a few nucleotides, or even by introducing a gene from a naturally interbreeding relative of the crop plant. Although there have been previous methods of gene editing, a new gene editing tool was developed in 2012, the CRISPR/Cas9 system. This molecular tool was first identified in bacterial cells, where it is believed to function as a primitive immune system recognizing and cutting up the DNA of bacterial viruses. CRISPR (clustered regularly interspaced short palindromic repeats) refers to the segment of DNA in the cells where this tool was first described; Cas9 is a CRISPR-associated protein, an endonuclease enzyme that can create double-stranded breaks in DNA. The development of CRISPR/Cas9 from its discovery in bacteria to its initial use in gene editing spans a 20-year period from 1993 to 2012 and involves many scientists from nine countries.

The first reports of plant genome editing with CRISPR appeared in August 2013, and by August 2018 thousands of reports have been published regarding the use of CRISPR to modify plants. For gene editing applications the CRISPR/Cas9 system consist of the Cas9 enzyme and a guide RNA (gRNA) molecule. Part of the gRNA can recognize, through base pairing, a specific site on the DNA molecule in the plant cell of interest. Another part of the gRNA interacts with the Cas9 enzyme, which then cuts the DNA at this site. In eukaryotic cells when a double-stranded break occurs in the DNA, the cell's repair mechanisms are triggered. This can result in the insertion or deletion of nucleotides at the break site, often causing a frame shift mutation (see Chapter 7) and an altered trait. These precise double-stranded breaks have been used by plant geneticists for gene editing (modifications, deletion, or insertions). Protoplast cultures are frequently utilized in gene editing experiments, and the resulting cells can then be cultured and grown into mature plants displaying the altered traits.

Although no gene edited crops have reached the market place yet, much research is on-going. The following are just a few examples of the many gene editing projects that may become useful to farmers and consumers. Citrus canker is a bacterial disease caused *Xanthomonas citri* that affects orange, lime, and grapefruit trees. The pathogen causes lesions on the leaves, stems, and fruit. Researchers have shown that CRISPR/Cas9 gene editing of a grapefruit variety and an orange variety produced enhanced disease resistance against the pathogen. In a similar manner, researchers have shown that gene editing of rice plants with CRISPR/Cas9 improved disease resistance to rice blast, a disease caused by the fungal pathogen *Magnaporthe oryzae*. The fungus can produce lesions on all parts of the plant, but infection of the inflorescence can result in complete loss of the grain. Improved resistance is an especially noteworthy achievement because this is the most destructive disease of rice plants worldwide. In laboratories around the world, scientists are using CRISPR technology to develop other crops resistant to a variety of viral, bacterial, and fungal pathogens.

Researchers at Cold Spring Harbor Laboratory in New York are using gene editing to improve the growth and productivity of tomatoes. When most tomatoes are picked or naturally abscise, the fruit separates from the main stem at a node (or joint) on the pedicel below the fruit. This leaves the calyx and a short segment of the pedicel attached to the fruit. These tomatoes are referred to as jointed, and most tomato varieties are jointed. In commercial harvesting, the calyx and pedicel must be removed by hand to prevent the pedicel from injuring other tomatoes in picking or shipping containers. In jointless tomatoes, the pedicel and calyx remain attached to the plant not the fruit. For packaging and shipping, this is a desirable trait, and it also permits faster harvesting. In crossing the jointless varieties with existing jointed varieties using conventional breeding, the offspring often had undesirable features. The branches bearing flowers developed many extra branches with multiple flowers; however, this resulted in less fruit development due to a drain on the plant's resources. With CRISPR/Cas9 researchers have been able to edit the genes responsible for the multiple branching. The researchers are working to develop plants with optimum branches and flowers for each tomato variety.

Polyphenol oxidase is an enzyme that causes the rapid brown of many fruits and vegetables, such as apples, potatoes, lettuce, and mushrooms. This browning results in a decline of food quality, changing the nutritional properties and shortening the storage life of the product. Researchers at Pennsylvania State University have edited the genome of the common button mushroom, *Agaricus bisporus,* to reduce browning. In the button mushroom, there are six genes that code for polyphenol oxidase, and the researchers using CRISPR/Cas9 have disabled one of these genes. This edit reduces the browning by 30%. It has been suggested that this CRISPR edited mushroom may be the first product to reach the marketplace.

Before the development of CRISPR/Cas9 as a gene editing tool, crops modified with older gene editing technologies had been considered exempt from federal regulations governing genetically engineered crops because no foreign genes had been introduced. In April 2016, the United States Department of Agriculture (USDA) indicated that it will not regulate the CRISPR edited mushrooms; they can be grown and marketed without USDA review. This was the first CRISPR edited crop to be given an okay by the USDA. In March 2018, this ruling was reinforced when Sonny Perdue, U.S. Secretary of Agriculture, announced the USDA would not regulate plant varieties developed by gene editing techniques. Many plant geneticists and biotechnology companies were pleased with this ruling since they feel that varieties developed with CRISPR technology would not bear the stigma associated with GM crops. These plants may still undergo a voluntary review by the Food and Drug Administration as well as be subject to oversight by the Environmental Protection Agency. A very different outcome occurred in Europe. In July 2018, the Court of Justice of the European Union ruled that plants created with CRISPR technology must go through the same approval process and labeling as transgenic plants.

CRISPR/Cas9 technology is not without concern. A July 2018 publication found DNA damage including large deletions at sites far from the edited gene. This damage was detected in three types of CRISPR edited cells including mouse embryonic stem cells, mouse bone marrow cells, and human retinal epithelial cells. This issue is a serious concern if CRISPR technology is going to be used in gene therapy research.

Even though some of the questions are still largely unanswered, biotechnology will continue to play an important role in increasing agricultural productivity. It must also be remembered that current agricultural practices have considerable environmental impact due to pesticide use, fertilizer use, and irrigation; the benefits and risks of genetically engineered crops and other modified crops must be weighed against these. Biotechnology may not be the magic bullet that will feed the hungry in the world, but GM crops used with other methods described in this chapter can substantially enhance food production in developing nations. In fact, some scientists argue that biotechnology is the only answer to feeding the 2 billion people who will be added to the world population in the next 30 years. Former president Jimmy Carter stated it well: "Responsible biotechnology is not the enemy; starvation is."

Thinking Critically

The development of a transgenic plant begins with the selection of a useful trait and the isolation of the gene responsible for the trait. With the use of an appropriate vector, that gene is then transferred to a valuable crop plant.

What trait might you try to transfer to wheat if you were developing a transgenic wheat plant that could grow on soils poor in nitrogen? Explain why you have selected this trait.

CHAPTER SUMMARY

1. Approximately 11% of the world's population receives insufficient food to meet daily nutritional requirements. The major challenge in agriculture is producing enough food to feed the world's population. Dramatic improvements in crop yield have been achieved through breeding of high-yielding and disease-resistant varieties.

2. Since the 1970s, efforts have been made to increase the yields of major crops in developing nations. The central focus of the Green Revolution has been the use of high-yielding and disease-resistant varieties; however, these crops are high-impact crops. Yields depend on fertilizer and pesticide application, adequate water, and mechanized farming, making it economically impossible for poor farmers. Other problems include high energy cost, environmental damage, and loss of genetic diversity. Some scientists feel that the whole premise behind the Green Revolution needs to be reexamined. Increased food production is dangerous in areas where population growth is unchecked, because it ultimately means more people will starve.

3. The loss of genetic diversity in crop plants and their wild ancestors because of monocultural practices is a serious concern to plant breeders. The cultivation of just one or two high-yielding varieties for most major crops has resulted in the loss of thousands of traditional varieties, and with them their genetic heritage. Monoculture is, in itself, an unstable system in that a single pest or disease can wipe out the crop of an entire region of essentially genetically identical and similarly susceptible individuals.

4. Sustainable agriculture is a movement away from the instability of current monocultural practices to a more environmentally sound approach in which crops are harvested without degradation to the environment. The monoculture of annuals would be replaced with a mixture of perennial crops more closely resembling a natural system.

5. Germplasm is the genetic information encoded within a plant; it is the raw material of the plant breeder, and without it, the creation of new varieties is impossible. Genetic erosion results when germplasm is lost to the plant breeder because traditional varieties are no longer cultivated or when wild ancestors become extinct because of habitat destruction. Seed banks preserve valuable germplasms by providing a storehouse for the seeds of domesticated plants and their wild relatives. An approach taken by other groups is to encourage the continued cultivation of traditional varieties. As the incalculable value of germplasm is realized, the question of ownership and a country's proprietary rights to this natural resource are being challenged and defined.

6. Although there are approximately 50,000 species of edible plants, only three domesticated crops provide over 50% of the kilocalories from plants in the human diet. The search for locally used crops that have the potential to be developed for expansion on the world market is ongoing. Some alternative crops that might play a bigger role in feeding a hungry world are quinoa, amaranth, tarwi, tamarillo, naranjilla, and oca.

7. Biotechnology is being used along with traditional breeding to develop high-yielding and disease-resistant cultivars. Methods of biotechnology, including cell culture, marker-assisted selection, precision breeding, gene silencing, genetic engineering and gene editing, can be used to introduce desired traits (such as insect resistance) into an established crop plant. Transgenic plants contain one or more genes transferred from another species and express that gene by producing a foreign protein, while gene edited plants have deletions, substitutions, or other changes to the DNA in the plant.

8. In 2017, genetically modified crops were grown on 189 million hectares in 24 countries around the world, with the United States growing approximately 40% of the total. The main crops that have been modified are corn, soybean, cotton, and canola, and the main traits are herbicide resistance and insect resistance. GM soybean and canola display herbicide resistance, while corn and cotton are engineered with either or both traits. Other GM crops are currently a small percentage of the total.

9. The first genetically engineered plant was approved for commercial marketing in 1994, and today approximately 75% of the processed foods in U.S. markets contain at least one ingredient from a GM plant. Although there are many environmental and safety concerns about the use of GM crops, this technology along with gene edited plants have the potential to enhance food production in coming decades.

REVIEW QUESTIONS

1. Can increasing crop yields solve the world's food problems?
2. Discuss the positive and negative effects of introducing a gene for herbicide resistance into a crop plant.
3. Describe the procedure to produce a transgenic plant.
4. Why have some scientists considered the Green Revolution flawed?
5. Many varieties of common crop plants may be disease resistant. What other characteristics must the varieties have to be useful for widespread cultivation?
6. What is monoculture, and how has modern agriculture encouraged its spread? What are the dangers of monoculture? Cite examples of agricultural disasters that were the result of monocultural practices.
7. What is the significance of germplasm to agriculture? Outline the controversies concerning the control of germplasm.
8. How might the practice of sustainable agriculture improve crop productivity?
9. How will rising carbon dioxide levels and global warming affect the nutritional quality of some of the world's most important food crops?
10. What is the value of alternative crops to the world's food supply? What characteristics are important to consider in the search for alternative crops?
11. What are the distinctions between transgenic plants and gene edited plants?

UNIT IV

CHAPTER 16

Stimulating Beverages

Source: Stephen Ausmus, USDA-ARS

Camellia sinensis leaves can be processed by different methods to produce black teas, green teas, and oolong (semifermented) teas.

KEY CONCEPTS

1. Caffeine and caffeinelike alkaloids have a stimulating effect on the central nervous system.
2. *Coffea arabica*, *Thea sinensis*, and *Theobroma cacao*, plants naturally rich in caffeine, have long been used as sources of stimulating beverages and historically have played an important role in human affairs.
3. Today coffee, tea, chocolate, and cola are consumed globally and are a mixed blessing to the world's population.

CHAPTER OUTLINE

Physiological Effects of Caffeine 272
    Medical Benefits of Caffeine 273
Coffee 273
    An Arabian Drink 273
    Plantations, Cultivation, and Processing 274

A CLOSER LOOK 16.1 Climate Change and the Future of Coffee 276

    From Bean to Brew 277
    Varieties 277
    Decaffeination 278
    Variations on a Bean 279
    Shade Coffee vs. Sun Coffee 279
    Fair Trade Coffee 279
Tea 279
    Oriental Origins 279

A CLOSER LOOK 16.2 Tea Time: Ceremonies and Customs around the World 280

    Cultivation and Processing 280
    The Flavor and Health Effects of Tea 281
    History 282
Chocolate 283
    Food of the Gods 283
    Quakers and Cocoa 283

A CLOSER LOOK 16.3 Candy Bars: For the Love of Chocolate 284

    Cultivation and Processing 285
    The High Price of Chocolate 286
Coca-Cola: An "All-American" Drink 287
Other Caffeine Beverages 287
Chapter Summary 287
Review Questions 288

The human need for water is even more pressing than the need for food. From earliest times, we have desired to quench our thirst with drinks more flavorful than water alone. This desire, coupled with the fortuitous discoveries of caffeine-rich plants, led to the creation of stimulating beverages in many cultures. Our most familiar examples of stimulating beverages are coffee, tea, and cocoa, all rich in caffeine.

PHYSIOLOGICAL EFFECTS OF CAFFEINE

Caffeine and related stimulants belong to a class of chemicals called alkaloids, which are substances mainly produced by plants (see A Closer Look 1.2: Perfumes to Poisons and Chapter 19). Although the roles of some alkaloids are still uncertain, it is believed that in plants caffeine functions as an insecticide that discourages insect pests. In general, alkaloids have diverse physiological and psychological effects on animals; specifically, caffeine and caffeinelike alkaloids are stimulants to the central nervous system. Caffeine is known to speed the heartbeat, increase blood pressure, stimulate respiration, and constrict blood vessels. It has long been used to alleviate fatigue and drowsiness, thereby promoting alertness and endurance. Caffeine also improves athletic performance by drawing on fat reserves for energy and by increasing motor skills of conditioned reflexes.

On the other hand, not all effects of caffeine are positive. In some individuals, caffeine can cause insomnia, nervousness, irritability, and indigestion. There is no set dosage that produces these effects; in some individuals even one cup of coffee induces this "caffeinism." Research has shown a link between caffeine and birth defects in mice; therefore, pregnant women are usually advised to limit their intake of caffeine. The animals in the studies were given very high levels of caffeine; there is no evidence to suggest that the usual range of human caffeine intake would cause birth defects. Conversely, evidence does suggest that moderate to high caffeine intake, especially early in pregnancy, does affect fetal growth, resulting in babies with slightly lower birth weight. Some studies also suggest that women trying to become pregnant should limit caffeine consumption. However, this research is inconclusive because other studies have shown no connection between caffeine intake and infertility. Heart patients should also limit caffeine intake because it affects the cardiovascular system.

Like many drugs, caffeine is addictive and can cause withdrawal symptoms. Without caffeine, users become irritable, nervous, restless, and unable to work and often develop severe headaches. Even if you are not a coffee or tea drinker, you may be using caffeine; caffeine is a common additive in many soft drinks, energy drinks, and medications (table 16.1). In fact, there is concern about the caffeine consumption of children who drink large quantities of soft drinks. Two cans of many soft drinks contain the equivalent amount of caffeine found in a cup of coffee. This concern has resulted in the greater availability of caffeine-free beverages. At the same time, there has also been the introduction of new beverages with even higher caffeine content. Energy drinks, such as Red Bull, Monster Energy, and Rockstar, are soft drinks that are advertised to boost energy and are primarily marketed to young adults between the ages of 18 and 30. Caffeine is the major active ingredient in these beverages, which have increased in availability and popularity in recent years. In 2017, sales of energy drinks in the United States were $13.4 billion. The energy drinks come in dozens of flavors, and the caffeine is often augmented with other active ingredients, such as vitamins and herbal supplements. Caffeine levels range from 80 to 260 mg per container, which is two to six times the level of traditional soft drinks. Energy shots are another segment of

Table 16.1
Caffeine Content of Common Products

Product	Caffeine Content
Coffee	
Drip (5-ounce cup)	115 mg (60–180 mg)*
Instant (5-ounce cup)	65 mg (30–120 mg)*
Decaf (5-ounce cup)	3 mg (1–5 mg)*
Espresso (1.5- to 2-ounce cup)	100 mg
Tea	
Brewed (5-ounce cup)	40 mg (20–90 mg)*
Cocoa	13 mg
Soft Drinks, Energy Drinks, and Caffeinated Water (12-ounce serving)	
Coca-Cola/Coke Zero	34.5 mg
Diet Coke	46.5 mg
Dr. Pepper	39.6 mg
Mr. Pibb	40.8 mg
Mountain Dew	55.5 mg
Mello Yello	52.5 mg
Pepsi-Cola	37.5 mg
RC Cola	36.0 mg
Rockstar	120.0 mg
Monster Energy	120.0 mg
Jolt Energy	140.0 mg
Red Bull	120.0 mg
Water Joe	45.0 mg
Krank$_2$O	70.0 mg
Over-the-Counter Medications (per tablet or caplet)	
Excedrin, Extra Strength	65 mg
Anacin	32 mg
NoDoz	100/200 mg
Vivarin	200 mg
Midol	60 mg

*A range in caffeine content occurs because of variation in brewing time and strength.

the caffeinated beverage market. These are small, usually 50-ml (2-ounce) containers packed with as much or more caffeine as the full-size energy drinks, with some containing over 300 mg of caffeine per serving. Energy shots are marketed to those wanting a jolt of caffeine without having to drink a cup of coffee or a larger energy drink. Another new trend is the addition of caffeine to snack foods and chewing gum. It is easy to find trail mix, potato chips, mints, brownies, jelly beans, and other candy with added caffeine.

With the widespread availability of energy drinks, energy shots, energy foods, and caffeine pills, there is concern about caffeine toxicity, and many people may be unaware of the total amount of caffeine they are consuming. Even in healthy individuals, an overdose of caffeine can occur, resulting in increased heart rate, heart palpitations, anxiety, and jitteriness. A published study found that more than 20,000 emergency room visits in the United States during 2011 involved energy drinks. There have also been a few fatalities linked to consumption of high levels of caffeine from energy drinks, although most of the individuals involved had underlying medical problems. Energy drinks mixed with alcohol are considered especially unsafe because the effects of the caffeine will mask an individual's awareness of his or her intoxication. However, the alcohol will still be causing impaired functioning and judgment.

Medical Benefits of Caffeine

Although high doses of caffeine can result in toxicity, there are some positive health benefits related to moderate caffeine consumption. Moderate consumption is considered to be less than or equal to 400 mg per day for a healthy adult. Caffeine enhances the pain-relieving effects of aspirin and acetaminophen and is found in many over-the-counter and prescription drugs. Many headaches are caused by dilated blood vessels; caffeine is known to constrict these blood vessels and thereby alleviate the pain. Caffeine is also included in many diet pills for several reasons, including its action as an appetite suppressant and its properties as a weak diuretic.

Since the late 1990s, several large epidemiological studies have looked at the relationship between coffee, caffeine intake, and the incidence of Parkinson's disease. Parkinson's disease is a neurological disorder in which the cells of the brain that produce the neurotransmitter dopamine are destroyed over time. Characteristic symptoms include jerky movements, a shuffling walk, uncontrolled hand tremors, and a frozen facial expression. Although some cases have a genetic component, evidence points to an unidentified environmental factor. These studies indicate that coffee drinkers have a significantly lower incidence of Parkinson's disease than non-coffee drinkers. In one of these studies, nondrinkers had a two to three times greater incidence of the disease and a five times greater risk of Parkinson's than did men who consumed 840 milliliters (28 ounces) or more of coffee per day. Although other nutrients were investigated, the only one that seemed to account for the difference was caffeine. It has been suggested that caffeine may provide protection from unknown harmful environmental agents. Further research is needed to understand this reported health benefit of coffee and caffeine. Studies have also investigated the effect of coffee consumption on the occurrence of type 2 diabetes. In one investigation, people who drank more than five cups of coffee per day significantly reduced their risk of developing type 2 diabetes. It is not yet determined if it is the caffeine or some other component of coffee that accounts for this antidiabetic effect. Researchers have shown that coffee, tea, and hot chocolate all have antioxidant properties that have been shown to fight cancer, heart disease, and aging. The antioxidant effects are due to the polyphenols, especially flavonoids, present in these beverages. Research has shown that coffee is the most important source of antioxidants in the American diet, based on the antioxidant content and the consumption level.

COFFEE

Coffee is made primarily from the seeds of *Coffea arabica*, a tree native to the mountains of Ethiopia and a member of the coffee family (Rubiaceae). Consumption of coffee has a long history, but its origins are lost in legends. One popular myth states that goats actually discovered the stimulating properties of the plant. Let out to graze, they came back one day friskier than normal. Investigating, the goat herder discovered the animals had been eating the berries of a nearby tree. Trying some himself, he enjoyed the same stimulating effect and introduced the fruit to others. The fame of this berry soon spread.

At first, coffee berries were eaten whole; later they were crushed and mixed with fat and eaten as a stimulating food. The practice of roasting the seeds within the berries and producing what we would recognize as coffee began in the thirteenth century in Yemen.

Thinking Critically

Caffeine belongs to a group of compounds called alkaloids, which act as stimulants to the central nervous system of animals.

Caffeine is produced only by plants, yet it has profound effects upon the nervous system of animals. Speculate as to why plants, which do not possess a nervous system, have evolved to produce neurologically active compounds, such as caffeine.

An Arabian Drink

By A.D. 1500, coffee trees had been widely cultivated in Yemen, and coffee drinking had spread rapidly throughout the Arabian world (fig. 16.1). Although coffee was first used to keep worshippers awake through long vigils, coffee drinking

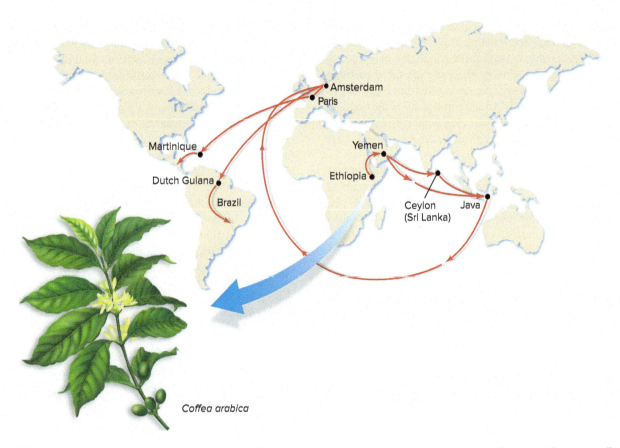

Figure 16.1 *Coffea arabica*, native to the mountains of Ethiopia, was spread through many tropical areas. Plantations became well established in the New World early in the eighteenth century.

soon acquired social aspects. Coffeehouses were established to accommodate this rapidly spreading habit. From the beginning, coffeehouses were controversial. Religious leaders felt that the time spent in the coffeehouses should have been spent in the mosques. Political leaders also felt threatened by the political discussions common in coffeehouses. Although at times efforts were made to outlaw coffeehouses, they remained an integral part of the Arab culture. In fact, some terms we associate with coffee, such as *mocha, kava,* and even *coffee* itself, are derived from Arabic terms.

Venetian traders introduced coffee to Europe in 1615, and by 1700 coffeehouses could be found throughout Europe. They were so popular in London that one could be found on every corner. The coffeehouses were not mere restaurants but centers for commerce, the arts, intellectual discourse, and political debate; they were often nicknamed penny universities. Tuition at these universities was a penny for a cup of coffee. Lloyd's of London, the famous insurance underwriters, began in a coffeehouse in 1688. Edward Lloyd opened a coffeehouse near the wharves on the Thames River; it soon became a convenient meeting place for seafaring men. Overhearing the gossip about ships and their cargoes, underwriters who also frequented the place began insuring the shipping trade. Edward Lloyd undertook no insurance business himself; he provided the facilities and shipping information for his customers to conduct their insurance business. After 300 years, Lloyd's of London, no longer a coffeehouse, is still operating. Although coffeehouses and cafes were still popular in most European countries, by the end of the eighteenth century coffeehouses had virtually disappeared from England, as tea became the caffeine beverage of choice.

The first of the North American coffeehouses opened in Boston in 1669. They did not achieve the intellectual status of their European counterparts but were meeting places for businessmen and merchants (fig. 16.2). Three hundred years later, during the 1960s, coffeehouses in the United States had a resurgence and became focal points of political thought and socially conscious folk music. Today, coffeehouses are popular again, specializing in gourmet and exotic blends.

Coffee drinking is still making inroads today; thousands of coffeehouses have opened in Japan in the past 30 years. Coffee may soon become the most popular stimulating beverage in the world.

Plantations, Cultivation, and Processing

Through the seventeenth century, coffee cultivation was limited to North Africa and the Arabian Peninsula; the city of Mocha was the center of this lucrative trade. The Dutch introduced coffee trees into their colonies in the East Indies and Ceylon (now called Sri Lanka). Java plantations soon became the major supplier of coffee beans to western Europe,

Figure 16.2 The London Coffee House, opened in the 1750s, was a major center of mercantile and political life in Philadelphia during the colonial period.

breaking the Arabian monopoly. From Java, a tree was taken to the Amsterdam Botanical Garden in 1706 (see fig. 16.1). An offspring from this tree went to the Jardin des Plantes in Paris. A cutting from this Parisian tree was taken to the French colony of Martinique, eventually giving rise to the first Caribbean coffee plantation. The Dutch had already introduced coffee into the New World in their colony of Dutch Guiana (now the independent country Suriname) in 1718. Coffee was introduced into Brazil about 10 years later when a Portuguese envoy brought seeds from the Guianas. In Brazil and other areas of tropical Latin America, coffee plantations flourished. Brazil is the world's leading coffee producer, followed by Vietnam. Coffee has been grown in Vietnam since the mid-nineteenth century. Production was disrupted during the Vietnam War but began again in the 1980s and has steadily increased, surpassing Colombia as the no. 2 producer.

Today the fate of coffee is directed by the International Coffee Organization, which sets and regulates the production and price of coffee for its member nations. Coffee is second to oil as the world's most widely traded commodity.

Coffee rust, caused by the fungal pathogen *Hemileia vastatrix*, has had a major impact on the history of coffee cultivation. Introduced in the 1850s, coffee rust had decimated the coffee plantations in Ceylon (now called Sri Lanka) by 1892. The coffee industry of Ceylon never recovered; tea plantations replaced those of coffee. Other coffee-growing areas, such as Indonesia and Africa, were also ravaged by this disease. Even today coffee rust continues to limit the production of *C. arabica*. From 2008 to 2013, outbreaks of coffee rust severely affected coffee trees in 10 Latin American and Caribbean countries, reducing harvests over 40% in some areas. These outbreaks were the worst since coffee rust first appeared in the region in the 1970s. Estimated losses were between $1 and $2 billion annually. The rust fungus damages the leaves, decreasing the photosynthetic capacity of the tree. This results in fewer coffee berries being produced. In addition to the direct loss of the coffee crop, the disease has a trickling effect on the whole economy. With reduced productivity, farmers hire fewer workers to pick the coffee berries. The livelihood of millions of people in Latin America are dependent on coffee.

Coffee rust is traditionally controlled by fungicide application. Funding is becoming available for more fungicides; however, replacing existing trees with rust-resistant varieties will be the enduring solution. Unfortunately, it will take the new trees at least 3 years before they begin bearing coffee cherries. After an outbreak of coffee rust in Colombia in 2008, many farmers began replacing their traditional coffee tree varieties with a resistant variety known as Castillo. This was successful, and Colombian coffee production has rebounded. In other Latin American countries, resistant varieties adapted to local conditions are also being used. However, in 2017 an outbreak of coffee rust occurred on coffee varieties that were thought to be resistant to the pathogen. Researchers have suggested that a new strain of the fungus had evolved to infect the plants. New coffee varieties need to be developed to resist this new pathogen strain. Researchers are also concerned about the impact of climate change on coffee rust severity (see A Closer Look 16.1: Climate Change and the Future of Coffee).

The coffee tree is a small evergreen tree or shrub with shiny, simple leaves. The plant bears clusters of small, white, fragrant flowers in the axils. After the flower is fertilized, it develops a berry, often called a coffee "cherry" because it is red at maturity (fig. 16.3a). Within the cherry is a fleshy, edible pulp surrounding two grayish seeds. Only the seeds are used in the production of coffee; the seeds are the coffee "beans" of commerce. Most coffee fruits produce two elongate seeds, or beans, but about 5% to 10% of the time a single pea-shaped bean, or peaberry, is found in the fruit. A peaberry forms when one of the beans fails to grow. Coffee made from peaberries has a more concentrated taste than the double-beaned type and is sold as a specialty coffee.

Coffee trees are cultivated in tropical and subtropical climates. They need about 150 to 250 cm (60 to 100 in.) of rainfall per year and grow best in the cool highlands (1,000 to 2,000 meters, or 3,000 to 6,000 feet) where the temperature is a stable 20°C (68°F). The plants cannot tolerate frost at all; a killing frost can devastate the coffee industry.

Trees start bearing fruit at 3 to 5 years of age and will continue bearing for about 35 years. A mature tree will yield 2.5 to 3.0 kilograms (5.5 to 6.6 pounds) of berries per year. The berries are usually picked by hand, especially in the steep, mountainous areas (fig. 16.3b). Mechanical harvesters can be used in plantations that are located on gently rolling hills. Each coffee-picking machine shakes off the cherries and performs the work of 100 laborers.

Once the cherries have been picked, depulping follows to free the beans from the fruit. There are two methods of depulping: the dry method used predominantly in Africa and the wet method favored in Latin America. In the dry method, the cherries are allowed to dry in the open for several days while the pulp ferments. After drying, the beans are freed mechanically and can then be roasted. In the wet method, the

A CLOSER LOOK 16.1
Climate Change and the Future of Coffee

The Intergovernmental Panel on Climate Change (IPCC) has strongly confirmed that the scientific evidence of climate change is unequivocal (see Chapter 26). Climate change has produced warmer temperatures in many areas of the world as well as more extreme weather events including extended droughts, stronger storms, and more intense rainfall. Research has shown that climate change is affecting crops around the world, including the major beverage plants described in this chapter.

Coffea arabica is cultivated in tropical and subtropical countries and does best in high elevations where temperatures are mild year-round; however, trees are highly vulnerable to environmental damage as well as pathogens and pests. In the tropical highlands, *C. arabica* trees will flourish when mean annual temperatures are about 20°C (68°F), although they can tolerate annual temperatures up to 23°C (73°F). Warmer annual temperatures speed up ripening of the coffee berries; however, this can lower the quality of the coffee beans. In addition, extended exposure to temperatures at or above 30°C (86°F) can harm the plants, causing leaf yellowing, stunted growth, and other abnormalities. Appropriate rainfall is also critical. Yearly precipitation of about 100 cm (approximately 40 in) is considered the bare minimum needed, with 150 to 250 cm (60 to 100 in) optimal. A dry season lasting about 3 months with about 4 cm (1.5 in) of rain per month promotes uniform flowering of trees and increases the yield of berries. A longer dry period reduces yield.

Evidence indicates that coffee production has already been impacted by climate change in many countries, including areas of Ethiopia, the native home for *C. arabica*. Ethiopia has been experiencing increasing temperatures and decreasing rainfall since the 1950s. Droughts have become more frequent, and during a drought in 2015, hundreds of hectares (one hectare equals 2.47 acres) of coffee trees died in the Arsi region of Ethiopia. Similar conditions have been reported in Brazil, the world's leading coffee producer. Over the past several years severe droughts have stricken the country, and coffee growing regions of Brazil were among the hardest hit resulting in decreased harvest and poorer quality beans.

Based on computer climate models, the IPCC indicated that by 2050 there may be substantial decreases in the area suitable for coffee cultivation if global temperatures continue to warm. One study estimated that worldwide the suitable areas may decrease by 50%, while other studies found the impact will vary by region. Across Central America areas suitable for coffee could be reduced between 38% and 89%. As lower-altitude plantations become unsuitable for arabica coffee, it may be possible to grow the trees at increasingly higher altitudes. In some countries cultivation has already been moved to higher elevations.

Warmer temperatures will also increase the threat of diseases, especially coffee rust. The uredospores are the main dispersal stage of the coffee rust pathogen, *Hemileia vastatrix*, spreading the pathogen from infected trees to healthy ones. Experiments conducted under controlled conditions in growth chambers found that the fungus produced 10,000 times more uredospores per infected leaf area under warmer temperatures. Colonization of leaf tissue also occurred more rapidly at warmer temperatures. In Central America increasing temperatures and intense rainfall events were believed to be responsible for a severe outbreak of coffee rust that began in 2011.

Insect pests can also cause increasing problems under warmer temperatures. The coffee berry borer, *Hypothenemus hampei*, is the most important pest affecting coffee. Females of the species bore a hole in coffee berries and deposit their eggs there. When the eggs hatch, the larvae feed on the coffee seeds severely reducing the quality and marketability. Warmer temperatures have increased the number of generations of coffee borer during the growing season and have also increased the distribution range of these insects to higher elevations.

The livelihood of millions of small farmers in tropical countries is dependent on coffee. Adapting to climate change, coffee farmers may need to establish new plantations at higher altitudes or replace existing trees with new varieties. These are long-term investments that are not easy to accomplish without abundant resources. To accommodate the anticipated climate changes, action is needed now. The development of new varieties is necessary, and research institutions in many countries have started working on this. In 2017 the genome of *C. arabica* was sequenced; this achievement should be beneficial for the development of new varieties with greater thermal tolerance or disease resistance.

This material has focused on the effects of climate change on arabica coffee; however, similar problems are also occurring with the cultivation of robusta coffee. In addition, climate change is also affecting both tea and cacao plantations, and research is on-going to develop new varieties of these crops as well.

Figure 16.3 The fruit of the coffee tree is a berry usually called a coffee "cherry." (a) A cluster of coffee cherries ready for picking. (b) Cherries are often picked by hand, especially in mountainous regions. This picker in Colombia carries a basket of cherries.

> **Thinking Critically**
>
> Coffee cultivation has followed an interesting path, from Ethiopia to the East Indies to Central and South America.
>
> *What major crops have gone in the opposite direction, from a South American origin to dominance in other parts of the world?*

cherries are floated in large tanks to remove debris and then are mechanically depulped. The residual pulp clinging to the beans is allowed to ferment for up to 24 hours. This fermentation is not alcoholic but refers to natural enzymatic changes brought about during processing and should not be confused with anaerobic fermentation involved in the production of beer and wine. Once fermentation is completed, the beans are washed and dried, and the seed coats are mechanically removed. The beans, green without the seed coat, are ready for roasting. During the mechanical depulping process, the leftover fruit is normally discarded. It is often piled up and left to decompose or dumped into nearby rivers. Recently, a company developed a method to process the fruit pulp into flour that is rich in protein, iron, and fiber and gluten-free. The flour can be used in baking and has a slightly fruity flavor with a small amount of caffeine. The coffee flour is currently being produced in factories in several coffee-growing countries and should become available for purchase very soon.

From Bean to Brew

The flavor and aroma of coffee depend on the fine art of roasting. The temperature and timing of roasting are critical for producing the different varieties of coffee. Light roasts, which are generally preferred in North America, are accomplished at temperatures of 212°–218°C (414°–424°F). Dark roasts are produced at higher temperatures; the higher the temperature, the darker the beans. Dark brown Vienna roasts are produced at 240°C (464°F) and black French roasts at an even higher temperature of 250°C (482°F). As a general rule, the lighter the bean, the milder and sweeter the coffee; the darker the bean, the stronger and less sweet the flavor. The strength of the brew is related only to the flavors and aromas and not to the caffeine concentration.

These changes result from the chemical reactions that take place when the bean is roasted. One reaction that occurs is the conversion of starches to sugars; this begins when the temperature reaches 207°C (405°F). At slightly higher temperatures, 212°–218°C (414°–424°F), the sugars begin to caramelize and the bean turns brown. At 238°C (460°F), carbonization begins as the sugar burns, leaving only the carbon that darkens the bean. Other reactions occurring during roasting involve the release of substances such as the essential oil caffeol, which gives coffee its characteristic aroma. At roasting temperatures above 240°C (464°F), some oils are driven to the surface. Roasting also helps break down the cell walls, which aids in grinding.

Unroasted beans can be stored for long periods, but once roasted, the beans rapidly deteriorate in flavor; some can detect the deterioration in even a month's time. Ground beans have an even shorter shelf life, but vacuum packaging slows down the deterioration. Once the package has been opened, ground coffee should be refrigerated to preserve the flavor. However, coffee connoisseurs insist that using freshly roasted and freshly ground beans is the only way to ensure a good cup of coffee.

Varieties

The majority of coffee beans are produced by *C. arabica* trees, but it is not the only species in the genus. There are at least 120 other species of *Coffea,* but only two others are cultivated: *C. canephora* (*robusta*) and *C. liberica. C. arabica* is known for producing the choicest and mildest coffees. Approximately

60% of the coffee grown is arabica, although there are over 100 varieties being cultivated. Latin American countries predominantly cultivate arabica varieties.

Coffea canephora, mainly grown in Africa and Asia, has a stronger and more robust taste and is used primarily in instant and decaffeinated coffee. Even though it produces an inferior coffee, canephora is resistant to coffee rust and is grown in regions where the fungus is endemic. Also, it is easier to harvest, and it produces much higher yields per acre than arabica. These factors account for its increased production in recent years. *Coffea liberica* produces a bitter coffee, and its use is limited. Like canephora, liberica grows well at lower altitudes and is cultivated in West Africa and Malaysia. Liberica is also resistant to the coffee rust fungus. Because of the resistance to the pathogen, both canephora and liberica are being used in breeding programs with arabica varieties. The goal is to produce hybrids carrying resistance from either canephora or liberica and flavor attributes of arabica.

Most brand-name coffees are a blend of several varieties of beans. Select coffee beans are available at gourmet coffee shops, so that connoisseurs can create their own blends. Varieties are so distinctive that professional coffee tasters can identify the country of origin and even the region where the bean was grown. Most coffee experts agree that high-mountain-grown coffees are among the best, and Jamaica Blue Mountain is one of the finest.

One of the most expensive coffee with the most unusual method of production is *kopi luwak,* or Indonesian civet coffee. The Indonesian civet is a catlike animal with a particular penchant for eating ripe coffee cherries. The pulp is digested, while the seeds (coffee beans) pass through the digestive tract of the civet and are excreted, relatively intact, in the dung. The seeds are collected from the dung, thoroughly cleaned, and roasted. During the passage through the civet, digestive enzymes partially break down some of the seed proteins; this removes some of the bitterness and improves the flavor of the final beverage. Coffee connoisseurs claim this is the world's finest coffee. Civet coffee is produced in Indonesia and the Philippines, and in 2018 the price for *kopi luwak* beans ranged from $180 to $600 per pound. However, buyers need to be cautious because fake *kopi luwak* beans are often sold on the Internet and in markets. For several years, *kopi luwak,* ranked as the world's most expensive coffee; however, by 2018, it had fallen to fourth place. Currently, the most expensive coffee is Black Ivory Coffee, which is made from arabica beans grown in northern Thailand. The arabica coffee cherries are fed to elephants that process them during digestion. Once deposited, the cherries are washed and dried in the sun. Black Ivory Coffee is extremely rare with only around 300 pounds (150 kg) produced each year. The coffee is mainly sold to select 5-star hotels, although some is available for individual consumers at approximately $800 to $1000 per pound.

Instant coffee was created by a Japanese chemist in 1901 but did not gain popularity until years later. During World War I, instant coffee was shipped to U.S. troops overseas; this was the first widespread use of instant coffee. Instant coffee is manufactured by dehydrating brewed coffee by various means. The most common method is by spray drying; the hot coffee is sprayed through nozzles into a tall room. As the coffee falls, the water evaporates, leaving a dry powder. The powder can be made to appear granular, looking more like ground coffee, by mixing it with steam or water. Freeze-drying is the latest innovation in the dehydrating process. Freezing under a vacuum dehydrates the coffee and produces the coffee crystals.

Decaffeination

With today's more health-conscious public, decaffeinated coffee has increased in popularity. No longer restricted to a few bland-tasting instant brands, decaffeinated coffee can also be found as roasted beans and ground coffee blends.

The decaffeination process is surrounded by controversy. Convinced that an addiction to coffee had killed his father, the German chemist Ludwig Roselius began a search in 1900 to find compounds that could decaffeinate coffee beans without ruining the taste. (Arabica beans typically are about 1% caffeine by weight, whereas robusta [canephora] beans have twice that amount.) From his work, Roselius found various solvents that could extract the caffeine from unroasted green beans that have been softened by steam. Methylene chloride became the solvent of choice for many years because it selectively removes the caffeine but does not affect the sugars and other flavor ingredients in coffee. In 1989, however, the FDA banned its use in hair sprays because laboratory animals that had inhaled methylene chloride developed cancer. The FDA still allows its use for some specialty coffees if the solvent concentration in the beans after processing is below 10 parts per million.

Despite this allowance, many coffee manufacturers switched to ethyl acetate. It is often touted as a natural solvent because this chemical is found naturally in coffee fruits. Another process is the Swiss-Water decaffeination method, in which green coffee beans are set to soak in an aqueous solution that contains all the chemical components found naturally in beans except for caffeine. Caffeine present in the beans then naturally diffuses out of the beans into the water.

The newest method of decaffeination is the supercritical carbon dioxide process. In the supercritical state, carbon dioxide acts both as a liquid and as a gas. First, the beans are soaked in water, which doubles their size and dissolves some caffeine present. Next, the beans and water are placed at the top of a large extraction vessel that is filled with supercritical carbon dioxide. As the beans move down to the bottom of the vessel, a process that takes about 5 hours, the carbon dioxide in its supercritical state can penetrate the beans, dissolve the caffeine, and diffuse out. This method removes about 97% to 99% of the caffeine present. None of the flavor molecules are removed from the beans, only the caffeine. As the process ends, the decaffeinated beans are removed and the caffeine is extracted from the carbon dioxide for use in soft drinks and pharmaceutical preparations.

Although the benefits of decreasing caffeine intake are clear, there is some concern that drinking decaffeinated coffee might elevate blood cholesterol levels. This claim has not been proved.

Variations on a Bean

Specialty coffees are popular throughout North America. These include espresso, latte, mocha, and cappuccino. A variation of coffee preparation that originated in Italy, espresso is produced by brewing a single cup of coffee at a time, as hot water under pressure is quickly forced through finely ground and densely packed, dark-roasted arabica beans. This method produces an intensely flavored and concentrated coffee that is drunk in single-ounce (30-milliliter) servings. A caffe latte is essentially espresso with steamed milk—approximately 7 ounces (210 milliliters) of steamed milk for every ounce (30 milliliters) of espresso. Caffe mocha is a latte with less steamed milk but with chocolate syrup added. Cappuccino is also made with espresso, with less steamed milk than a latte but with a large cap of foamed milk. The name came about because the large cap of frothed milk resembled the hood of the Capuchin friars.

Shade Coffee vs. Sun Coffee

Coffee trees have traditionally been grown in the shade of other trees; however, many growers have been switching to "sun coffee" trees, which have higher yields. Ecologists are concerned about this switch because shade coffee trees in northern Latin America are the usual winter nesting sites for many U.S. birds. The sun trees, by contrast, shelter as few as 3% of the number of bird species. The diversity of other organisms is similarly reduced in sun coffee plantations, and the need for expensive fertilizers and pesticides is significantly increased. The environmental cost associated with these chemicals is a related issue that cannot be ignored. In addition, soil erosion and agricultural runoff are commonplace. Unless steps are taken soon, many of the coffee-growing areas are likely to undergo serious environmental degradation.

Currently, two organizations are encouraging the production of shade-grown coffee through certification programs. One program is sponsored by the Smithsonian Migratory Bird Center and offers a Bird Friendly seal of approval for coffee that is 100% organic, shade-grown coffee. The second certification is from the Rainforest Alliance, which is an international nonprofit organization devoted to the preservation of tropical forests. Both programs include various criteria for forest composition, including species diversity, tree height, canopy cover, and canopy levels. Consumers can help promote shade-grown coffee by looking for the Bird Friendly or Rainforest Alliance seal when purchasing coffee.

Fair Trade Coffee

The coffee trade is a major international enterprise worth over $19 billion per year. Coffee is produced in over 60 countries, mainly in the developing world, and it is estimated that 25 million farmers are growing coffee. Unfortunately, many small farmers are barely able to survive financially because the prices they get for their coffee are often less than their production costs. Many farmers are forced to sell their coffee to middlemen for only 25 to 50 cents per pound. Fair Trade certified coffee is an attempt to solve this problem. Coffee that is Fair Trade certified must meet certain standards. The importer must pay the farmer a minimum price of $1.40 per pound; however, this amount increases if regular coffee is trading above this level. The Fair Trade price then becomes 10 cents higher than the regular trading price and 30 cents higher for organic Fair Trade coffee. In addition, the importer must also provide technical assistance and credit to the farmer. Other Fair Trade practices include helping with community development, health, education, and training for environmentally friendly and sustainable farming practices.

The Fair Trade movement originally began in Europe in the 1960s and achieved significant growth in the late 1980s with the establishment of fair trade labeling by organizations in various countries. In 1997, the Fairtrade Labeling Organization (FLO) was created as an umbrella organization to set international fair trade standards. TransFair USA, a separate organization, oversees and certifies coffee and other products in the United States and Canada. Due to growing consumer demand, Fair Trade certified coffee can now be found in an increasing number of stores and coffee shops, although it still represents only a small percentage of the global coffee trade. In addition to coffee, Fair Trade certified labels can be found on a variety of products, including tea, cocoa, rice, sugar, vanilla, bananas, pineapples, and mangoes. While the goals are admirable, the Fair Trade movement is not without its critics. Many have questioned the effectiveness of the endeavor. Recent studies found that the Fair Trade certification increases income for skilled growers and larger farm owners; however, the study found little evidence that workers, especially unskilled seasonal coffee pickers, received any benefit.

From the accidental discovery made by a goat herder long ago, coffee has made a global impact on history, social customs, economics, international trade, and the human diet.

TEA

Oriental Origins

According to the Chinese, tea was discovered by the Emperor Shen Nung in 2737 B.C. when a tea leaf accidentally fell into water that was being boiled for drinking. A different legend comes from India. A saintly Buddhist priest vowed to spend 7 years without sleep in contemplation of Buddha. After 5 years, he found it increasingly difficult to stay awake. In desperation, he grabbed the leaves of a nearby bush and began to chew them. The leaves, from a tea plant, refreshed him and allowed him to fulfill his vow. Whatever the origin, today tea is the world's most popular beverage, next to water. Every day, more than 3 billion cups or glasses of tea are consumed around the world, sometimes with great fanfare (see A Closer Look 16.2: Tea Time).

A CLOSER LOOK 16.2

Tea Time: Ceremonies and Customs around the World

While the midmorning coffee break is a familiar tradition in the United States, other cultures have developed customs and ceremonies involving tea. The British high tea is a late afternoon light meal that may include such delicacies as scones, crumpets, cakes, and finger sandwiches, accompanied by piping hot brewed tea, often served with milk.

For centuries, Russians brewed tea using a metal urn called a samovar to boil the water. The tea was traditionally served in tall glasses, often with a slice of lemon; in fact, the Russians were the first to flavor tea with lemon. When lemons were not available, a spoonful of jam was added to the hot tea.

In Japan, the tea ceremony is a highly stylized ritual that symbolizes the Zen Buddhist concept that universal truths lie in simple tasks (box fig. 16.1). In some houses, the ceremony is held in a special room that is austere to focus attention on the details of the ceremony. Green tea is poured, whipped to a frothy consistency, and served from lacquered bowls. The tea ceremony has not changed in centuries and is considered a link to the traditions of the past.

Chai is a sweet and spicy tea drink from India that is becoming trendy across the United States. In India, it is the common drink of the working classes, served to field laborers and commuters at India's train stations. Chai is prepared by boiling black tea, water, milk, and spices together. Favorite spices include cinnamon, clove, cardamom, black pepper, orange peel, ginger, nutmeg, star anise, bay leaves, and vanilla. The origin of this tea tradition was probably a preventive measure to inhibit the transfer of disease by thoroughly boiling all ingredients together. The name *chai* itself is derived from another Chinese word for tea, *ch'a*. Chai is the name for tea in several languages: Hindi, Russian, Greek, Turkish, Arabic, and Swahili.

Box Figure 16.1 Server prepares tea in the tea ceremony.

Cultivation and Processing

Tea* is made from the dried tip leaves of the species *Camellia sinensis,* a small tree or shrub native to the area adjoining Tibet, India, China, and Myanmar (Burma); it is a member of the tea family (Theaceae). The plant, if left unpruned, uncut, and unpicked, would grow to a height of 7.5–10.5 meters (24–34 feet). It can be grown from sea level to over 1,800 meters (5,905 feet), but the finest teas come from plantations or estates at the higher elevations. The plants flourish in tropical or subtropical climates that have abundant rainfall and no danger of frost (fig. 16.4).

Each tea plant is pruned to encourage shrubby growth. The plants are usually kept at approximately 1 meter (3–4 feet), with a flat top that facilitates easy plucking or hand harvesting of the leaves. For best-quality teas, only the terminal bud and top two leaves of each branch are harvested. Plucking stimulates the plants to produce new shoots or flushes with tender young leaves and buds. New flushes will appear every 1 to 2 weeks (fig. 16.4).

Harvested tea leaves may be treated in one of four ways to produce black tea, green tea, white tea, or oolong tea. In the United States, 90% of the tea is black tea, or fermented tea. The processing of black tea begins with withering. The freshly picked leaves are taken to racks, where hot, dry air is passed over them for 12–24 hours. During this time, the leaf loses much of its water content. After withering, the leaves are rolled, usually by machine; rolling breaks up the cells, releasing enzymes that start the next process, fermentation.

---

*A tea is simply an infusion—the liquid resulting when the leaves, roots, or flowers of a plant are steeped in boiling water and then strained. However, the strict use of the word *tea* is often restricted to an infusion made from the leaves of the tea plant itself, while those made from plants other than *Camellia sinensis* are usually referred to as a *tisane* or *herbal tea*. Also, herbal teas are usually caffeine-free.

CHAPTER 16 Stimulating Beverages

Figure 16.4 Tea plantation in China with tea pickers plucking leaves.

The leaves are then spread out in cool, humid fermentation rooms for up to several hours. Fermentation brings about chemical changes that turn the leaf a copper color. The last stage is the firing or drying of the leaves. The leaves are passed through hot air chambers that stop fermentation, reduce the moisture content, and turn the leaf black. After drying, the leaves are sorted into sizes or grades. The smallest tip leaves are called orange pekoe; larger leaves are pekoe and souchong. Broken leaf pieces are sieved out and sorted according to size; these are used predominantly in tea bags.

Green teas are not fermented. The leaves are not withered; after plucking, the leaves are steamed, rolled, and dried. These leaves remain green.

Oolong teas are semifermented; they are lightly withered to permit a partial ferment, resulting in leaves that are a greenish brown.

White teas are the latest members of the tea family to become popular. Although white teas have been consumed for centuries in China, only recently have these varieties become known worldwide. White teas are produced from select varieties of tea plants on which the leaf buds and young leaves are covered with tiny white hairs (trichomes). These buds and young leaves are plucked early in the spring on dry, sunny days. White teas are not steamed. The leaves are withered and then dried. The leaves remain a silvery white color, and after brewing, the beverage is pale yellow.

The Flavor and Health Effects of Tea

The essential oils, theanine, and tannins in the leaf determine the flavor of the tea. Essential oils are volatile substances that contribute to the essence or aroma of certain species. The flavorful aroma of a steaming cup of tea is due to its essential oil, theol. Certain types of tea have additives that modify the flavor and aroma. Jasmine tea, a semifermented tea, has a delightful taste and fragrance due to the addition of jasmine blossoms during the withering process. Earl Grey tea, a traditional British favorite, has bergamot oil added; other teas may have mint leaves or spices added. Theanine is an unusual nonprotein amino acid that occurs naturally in tea and a few other *Camellia* species. This amino acid is synthesized in the roots of the tea plant; it is transported through the plant and accumulates in the developing tea leaves. Theanine contributes to the flavor of tea, and it is thought to be responsible for the relaxation and improved concentration associated with tea drinking.

Tannins are polyphenolic compounds found in a great many plants in addition to tea. They are believed to discourage herbivores from eating the leaves and have been widely utilized as stains, dyes, inks, and tanning agents for leather. The tannins in tea are responsible for staining the teapots and teeth of habitual tea drinkers. Black teas, in particular, are very rich in tannins, which are responsible for the color changes that occur during processing. During fermentation, the colorless tannins turn copper colored, then black when heated.

In addition to tannins, tea also contains several other types of polyphenols, of which flavonoids are the most abundant. The antioxidant properties of flavonoids are well known. (Recall from Chapter 10 that antioxidants can eliminate free radicals that damage cells, including those that damage DNA.) One cup of brewed tea will generally provide 150 to 200 mg of flavonoids, and the health benefits of tea are mainly attributed to the presence of flavonoids. Some of the main flavonoids present in tea are catechins, and one cup of green tea contains 90 to 100 mg of catechins. The most abundant catechin in green tea is epigallocatechin-3-gallate (EGCG). When tea leaves are processed to produce black tea, most of the catechins are converted to other flavonoids, such as theaflavin, which contributes to the color and flavor of black tea. The polyphenolic compounds in white tea are similar to green tea and would produce similar health benefits.

Although there are many beneficial health effects associated with drinking tea, the two most extensively studied have been the reduced risks for cardiovascular disease and cancer. A large number of epidemiological studies in many countries have examined the correlation between tea consumption and various forms of cardiovascular disease. Although many of these studies have produced contradictory results, with some showing a reduced risk and other studies showing no effect, the overall evidence does suggest that greater consumption of green or black tea results in a reduced risk of cardiovascular disease. Researchers have proposed possible mechanisms of action such as reducing cholesterol levels, reducing blood pressure, and preventing platelet aggregation; however, no definite conclusions can be drawn at the present time.

Dozens of epidemiological studies on the relation between tea consumption and cancer have also been published in recent years with similar inconclusive results. However, many researchers believe that drinking tea, especially green tea, reduces the risk of colon, breast, ovarian, prostate, lung, and other cancers (see A Closer Look 10.2: Eat Broccoli for Cancer Prevention). Animal studies and tissue culture studies of cancer cell lines have demonstrated the effectiveness of green tea extracts and tea polyphenolics, especially EGCG, for inhibiting tumor proliferation. Although various molecular mechanisms have been proposed as the mode of action, much more research needs to be done to determine if green tea extracts can become effective cancer treatments.

The stimulating effects of tea are due to the caffeine and theophylline present in the leaf. In fact, when caffeine is metabolized in the human body, a small amount of it is converted to theophylline. Although theophylline is structurally similar to caffeine, the medicinal properties of theophylline are better known than its stimulating properties. Theophylline has been used to treat asthma for many years by directly relaxing the smooth muscles of the bronchial airways and thus opening constricted air pathways. This bronchodilating action helps relieve wheezing, coughing, and other respiratory symptoms. In an emergency, asthmatics can get some relief by drinking a few cups of tea or coffee.

History

Scattered reports of this Chinese drink first reached Europe in the mid-fifteenth century, but tea did not actually arrive in western Europe until 1610 when Dutch traders brought it to Holland. It was in the 1650s that it finally reached England, the European country most associated with tea. The London coffeehouses were the first to bring the beverage to the public following their success in introducing coffee just a few years earlier. The British fondness for tea developed relatively quickly; in 1680, the British imported a mere 100 pounds of tea. By 1700, the imports had jumped to 1 million pounds, and by 1780, 14 million pounds, making tea the British national beverage.

Tea first made its appearance in North America about 1650, brought to New Amsterdam by the Dutch, and was soon introduced into the British colonies as well. At a cost of $30 to $50 per pound for dried tea leaves, the beverage first became popular only in the homes of the well-to-do. Despite the cost, tea drinking had spread throughout the colonies by the late 1600s, and by the mid-1700s, the colonists were avid consumers. Once they developed the tea-drinking habit, the colonists were dependent on imports for their tea leaves.

Although the British East India Company was the official import company, the colonists usually preferred the lower and duty-free prices offered by smugglers. In 1773, the East India Company complained to Parliament that it had a large surplus of tea and was anxious to improve the colonial tea trade. Parliament responded with the Tea Act of 1773, allowing the

Figure 16.5 Currier and Ives lithograph from 1846 depicts the destruction of tea during the Boston Tea Party.

East India Company to sell the tea directly to the colonies without paying any of the taxes imposed on the colonial merchants. Thus, the East India Company hoped to monopolize the colonial tea trade by underselling the colonial tea merchants. At the same time, this measure would make selling black market tea less lucrative and would drive the smugglers out of business.

The Tea Act aroused indignation in the colonies. The colonial merchants felt their livelihood was threatened, and it renewed resentment toward British taxation. At this time, no tax, no matter how small, was acceptable to the colonists without representation. The most famous response was the Boston Tea Party on December 16, 1773. Colonists disguised as Mohawk Indians boarded the British ships and dumped the cargo of tea into Boston Harbor (fig. 16.5). The aftermath of this incident contributed to the events that led to the Declaration of Independence and the Revolutionary War.

One lasting influence of the Boston Tea Party and the Revolutionary War was to make the United States a nation of coffee drinkers, but iced tea has become one of America's favorite cold beverages. In 1904 at the St. Louis Exposition, a young Englishman named Richard Blechynden was introducing teas from the Far East. A sweltering heat wave dampened the crowd's enthusiasm for hot tea. Blechyden experimented by pouring the hot tea into tall glasses filled with ice; it was an instant hit. Since then, iced tea has become the national summertime drink.

The year 1904 was significant in the history of tea for another reason as well—the invention of the tea bag. Thomas Sullivan, a New York tea importer, sent out samples of tea in small silk bags instead of the more expensive tins that were usually sent. To Sullivan's surprise, the orders came in not for the tea itself but for the tea packaged in silk bags. His customers had discovered that tea could be easily made by pouring boiling water over the silk bags. Silk is no longer used, but the tea bag industry has grown so much that tea made this way makes up more than half of all the tea consumed in the United States today.

CHOCOLATE

Although coffee and tea are known only as beverages, the third major caffeine plant gives us a confection as well as a beverage. In fact, it is the confection that makes the cacao tree so widely cultivated today. The source of all the chocolate and cocoa in the world is the seed of *Theobroma cacao;* this member of the Malvaceae or mallow family (previously the Sterculiaceae) is native to tropical Central and South America. (Note: The cacao tree should not be confused with the coca bush, which is the source of cocaine.)

Food of the Gods

The cacao tree had been cultivated by the native peoples from southern Mexico through Central America and into northern South America for centuries before the Spanish Conquest. According to Aztec mythology, the god Quetzalcoatl gave cacao beans to the Aztec people (fig. 16.6). The cacao beans were offered as gifts to the gods and were used to make a beverage consumed by noblemen and priests on ceremonial occasions. The scientific name *Theobroma* reflects this ancient tradition; it literally means "food of the gods."

In 1502, Christopher Columbus was the first European to be introduced to the cacao bean. Although he learned that the natives used these beans as money and to prepare a spicy beverage, he did not appreciate their significance. In 1519, when the Spanish conquistador Hernán Cortés invaded Mexico, he found Montezuma, the Aztec emperor, drinking a liquid called *chocolatl* from a golden goblet. In the mistaken belief that Cortés was the reincarnation of the god Quetzalcoatl, the Aztecs showered him with riches and offered some of this esteemed beverage. Chocolatl was made from roasted and coarsely ground beans from the cacao tree, combined with various spices, including chili peppers and vanilla beans. Boiling water was added, and the mixture was whipped to a foamy consistency. The resulting spicy, bitter beverage had no resemblance to today's cocoa, although it may have been similar to mole sauce, used in Mexican cooking.

Cortés introduced this beverage to Spain when he returned in 1528. The Spanish court modified the recipe and added sugar, making it more palatable to European tastes. The recipe was highly guarded, and the Spanish had a monopoly on the cacao beans; these factors kept cocoa a Spanish beverage for many years. By 1650, cocoa was served throughout Europe, and soon it was competing with coffee and tea. Although popular, cocoa never rivaled coffee or tea as a beverage. The high fat content in the cocoa bean made processing difficult and produced a greasy beverage that many people found unappetizing. Some of these problems were solved in 1828 when a Dutch chemist, Conrad van Houten, developed a process to remove some of the fat, or cocoa butter. In 1847, an English company, J. S. Fry & Sons, added cocoa butter and sugar to the ground beans to make chocolate. This was the creation of the first chocolate bar (see A Closer Look 16.3: Candy Bars).

Quakers and Cocoa

In the nineteenth century, Quaker businessmen entered the cocoa business and used their business skills to improve social conditions for factory workers while making cocoa and chocolate inexpensive and available to the masses. George Fox founded the Society of Friends, better known as the Quakers, in England in the seventeenth century. The Quakers made a commitment to improve society by promoting justice and equality, forgiveness and understanding. They also criticized the established Church of England for what they perceived as its many abuses of power. Because of their controversial beliefs, they suffered religious persecution in England and were barred from many professions (e.g., medicine, law, military, and politics). With so many professions closed to them, many Quakers entered business, and several entrepreneurial families built cocoa factories. Their names are familiar to chocolate lovers: the Frys, Cadburys, and Rowntrees.

Joseph Fry was a young apothecary who came to Bristol, England, in 1748. He later became a chocolate manufacturer, in part because he believed that cocoa was of great medicinal and nutritional value. His wife, Anna, and son, Joseph, continued the business, which evolved into the famous J. S. Fry & Sons chocolate manufacturers. By the latter half of the 1800s, they were the largest chocolate manufacturers in the world. They had even won the right to be the sole suppliers of cocoa and chocolate (a replacement for the daily grog ration) to the British Royal Navy.

The Cadbury chocolatiers were both Quakers and rivals of J. S. Fry & Sons. Their founder was John Cadbury, who in 1824 set up a coffee and tea shop in Birmingham. He also sold the cocoa drink, and by 1847, the family had moved into cocoa and chocolate manufacturing. In 1853, Cadbury's was granted the privilege of being the chocolate supplier to the royal family.

Figure 16.6 The Aztec god Quetzalcoatl, envisioned as a plumed serpent, was believed to have given cacao beans to the Aztec people.

A CLOSER LOOK 16.3

Candy Bars: For the Love of Chocolate

There have been many changes in chocolate candy since the first was created in 1847. In 1875 in Switzerland, Daniel Peter, in collaboration with Henri Nestlé, created milk chocolate by adding condensed milk. Milton Hershey, the foremost American chocolatier, modified the Peter process by adding whole milk. He not only produced chocolate bars but also is credited with creating the first candy bar, in the late 1890s. The candy bar is a conglomeration of nuts, fruit, caramels, or other ingredients, usually with chocolate. Soon the marketplace was filled with competitor candy bars. One early competitor that is still sold today is the Goo Goo Cluster, concocted in 1912 in Nashville, Tennessee. World War I gave a further boost to the candy bar habit when chocolate bars were handed out to the doughboys for quick energy. Other early candy bars capitalized on the health craze that swept the nation in the 1920s, with claims of aiding digestion and names such as Vegetable Sandwich.

At 500 Calories per 100 grams (3.5 ounces), chocolate is a high-energy food. Milk chocolate is about 57% carbohydrate, 32% fat, and 8% protein, with traces of vitamins and minerals (box fig. 16.2). The nutritional value of candy bars improves when nuts and fruits are added.

Eating chocolate is an undeniable pleasure for people, but the indulgence has often been linked with such undesirable side effects as obesity, tooth decay, acne, migraines, and food allergies. Some of these connections can be dismissed for lack of substantiating evidence. What cannot be dismissed are the many desirable health effects from cocoa and chocolate.

The so-called addiction to chocolate has recently been ascribed to the presence of cannabinoids, substances similar to the active principle in marijuana, which are associated with feelings of euphoria. In addition, cocoa butter recently has been discovered to contain several healthful ingredients. It consists mainly of the saturated fatty acid stearic acid as well as oleic and palmitic fatty acids. Unlike most saturated fatty acids, stearic acid does not appear to elevate blood cholesterol levels (see Chapter 10). Cocoa also is rich in flavonoids, powerful antioxidants that reduce LDL levels (the cholesterol associated with cardiovascular disease); it lowers blood pressure; and it reduces the risk of cardiovascular disease, with the highest antioxidant levels in dark chocolate. A 2007 study showed that young women who were given cocoa that was high in flavonoids had a significant increase in blood flow to the brain compared to a control group not given cocoa. The researchers hope this can lead to treatment for dementia patients who show decreased blood flow to the brain. Chocolate is also rich in vitamin E, another antioxidant.

Despite television commercials and the ubiquitous candy machines, Americans are not the greatest consumers of chocolate. We each consume a mere 5 kilograms (11 pounds) per year, not much compared with the 11.4 kilograms (25 pounds) consumed in Germany. What is the mystique of chocolate? Perhaps it is the stimulating effects of the alkaloids. Chocolate contains both caffeine and theobromine, an alkaloid closely related to caffeine but milder in its effects. Perhaps it is the feeling of a love affair. Some people suggest that phenylethylamine, a component of chocolate that is also found in the human brain, appears to increase in the brain when someone falls in love. Some chocoholics may be addicted to the feeling of "falling in love" rather than to the candy itself. (If you love your dog, however, don't feed it chocolate. Theobromine is highly toxic to dogs, although fatal levels require quite a large dose.)

Box Figure 16.2 Slabs of chocolate are mechanically sliced into bars in this modern chocolate factory.

The Cadburys also created Bournville, a model village for their employees, in the suburbs of Birmingham, where they had built their factory. Today, the original village still stands, along with a chocolate theme park called Cadbury World. Joseph Rowntree, the founder of another Quaker chocolate dynasty, also built a model factory town: New Eastwick in the suburb of York. In both villages, the workers had access to free education and health care for themselves and their families. There were, however, no pubs in these factory towns because the Quakers were strongly opposed to alcohol. The living conditions in these villages were much better than what a typical factory worker of the time could expect. For example, in 1919 only 7.2 babies out of every 1,000 born in Bournville died, whereas the infant mortality statistic for England and Wales was 12.1 per 1,000.

Fry & Sons remained in the center of Bristol but exercised their social conscience by boycotting the cacao plantations of Portuguese West Africa, where workers were subjected to conditions that approached slavery. They continued the boycott until conditions improved for the workers.

Cultivation and Processing

Theobroma cacao is a small tree in the understory of tropical forests. Optimum growing conditions require a wet climate and warm temperatures, restricting cultivation to a zone 20 degrees north and south of the equator. Three commonly grown varieties are criollo, forastero, and trinitario. Although criollo is grown in Central America, Venezuela, and Colombia, its susceptibility to disease and its modest productivity have limited its cultivation. Forastero is by far the most cultivated variety in Brazil and West Africa, and trinitario is a hybrid between the criollos of Trinidad and the Amazon forasteros. Today the Ivory Coast and Indonesia lead the world in cacao bean production; other tropical countries in West Africa and South and Central America are also major contributors.

The tree is characterized by football-shaped pods that form directly on the main trunk from small white or pinkish flowers (fig. 16.7a). The pods take 4 to 6 months to mature; depending on the variety, the mature pods turn yellow, orange, or red. Inside these fruits are 30 to 40 ivory-colored seeds, or beans, surrounded by a white, sweet, sticky pulp (fig. 16.7b). It is the delicious pulp that first attracted human interest, and it is the pulp that still attracts animals such as bats, rats, and squirrels. They gnaw through the fruits to find the pulp; in the process, the bitter, alkaloid-containing seeds are discarded. Cacao seed dispersal is entirely dependent upon animals because the fruits never fall from the tree.

Cacao plantations in many of the prime growing areas are plagued by a low rate of fruit production. As few as 1% to 3% of the hundreds of flowers in bloom on a single tree bear fruit. The problem lies in a lack of natural pollinators. In the Amazon basin, where cacao is thought to have originated, the pollinator is a biting midge that thrives in the leaf litter and humid conditions of the rain forest. However, most cacao production now takes place in drier climates and on

(a)

(b)

Figure 16.7 *Theobroma cacao*, the source of cocoa and chocolate. (a) Podlike fruits developing on a cacao tree. (b) A cacao pod split open to reveal the pulpy seeds, or beans.

well-manicured plantations. In Hawaii, where midge populations are encouraged with rotting banana litter between the rows of cacao trees, pollination success has risen to 15%. There is even an attempt in Costa Rica to grow cacao more naturally in a managed rain forest.

Several types of fungi are capable of destroying cacao pods, and some of them are attacking much of the world's cacao crop in South America, Africa, and Asia: black pod rot is caused by a species of *Phytophthora;* frosty pod rot is an infestation of *Moniliophthora roreri;* and witches' broom is caused by *Crinipellis perniciosa.* Because chemical fungicides are expensive and have had limited success, researchers are employing biological control strategies to combat the fungal diseases of cacao. In this approach, scientists have identified beneficial fungi that suppress some types of pathogenic fungi. Many of these belong to the genus *Trichoderma. Trichoderma* and other biocontrol fungi release biofungicides that have been shown to control the problem fungi in field and laboratory tests.

When the cacao pods are ripe, they are harvested by machete and split open. In some plantations, a mechanized pod opener has replaced the machete. The pulpy seeds are removed and allowed to ferment for up to 1 week, depending on the variety. The chocolate taste and aroma are not found in the fresh beans; they develop as the beans ferment. The beans, now chocolate brown, are dried either mechanically or in the sun and shipped to processing centers in Europe and North America.

The beans are classified according to quality and origin; the best come from Chuao Valley in Venezuela and from Ecuador, where a special type of criollo bean is grown. These are often reserved for the finest-quality chocolates.

The processing begins with the roasting of the beans at temperatures of 120°–140°C (248°–284°F) for 20 to 50 minutes, which develops the rich color and full flavor of chocolate. After roasting, the seeds are cracked open, freeing the large cotyledons, or nibs, from the seed coat and embryo. The nibs are crushed to produce a dark brown, oily paste, the chocolate liquor. This liquor can be solidified into squares of baking chocolate, or the cocoa butter can be removed from the liquor with heat and pressure to produce a brown cake, which is pulverized into cocoa powder. During this processing, alkali is added to neutralize the acidity of the cocoa. This step, called dutching, also increases the solubility and darkens the color of the cocoa powder.

Cocoa butter has many uses. In addition to it being added to the liquor to produce confectionery chocolate, it is the main ingredient for white chocolate. Interestingly, the FDA does not consider white chocolate to be "chocolate" because it contains no chocolate liquor. Cocoa butter is also used in a variety of suntan lotions, soaps, and cosmetics.

Today, the majority of cacao beans are processed to make chocolate. The recipe for chocolate starts with the chocolate liquor; sugar, cocoa butter, vanilla, and often milk are added during conching. The conching process, which may last for several days, involves a mechanical kneading and stirring that gives chocolate its velvety smoothness (fig. 16.8). After conching, the liquid chocolate is poured into molds and cooled (fig. 16.9).

Figure 16.8 The conching process imparts smoothness to chocolate.

Figure 16.9 Far removed from the tropical cacao tree, these delectable chocolate confections gladden the heart of many "chocoholics."

The High Price of Chocolate

In the spring of 2001, news media around the world reported a shameful tale that lies hidden behind the delicious taste of chocolate. Slave trading was revealed when a slave ship loaded with children was discovered off the coast of Benin in West Africa headed for cacao plantations in the Ivory Coast. Almost half the world's cacao is grown in West Africa, especially in the Ivory Coast. On some Ivory Coast plantations, cacao beans were harvested by young boys, typically 12 to 16 years old, but often younger. Many of them were lured from their homes and sold to plantation owners. The children were passed along by intermediaries and eventually sold to cacao (and coffee) plantations, where they worked as field hands, domestic workers, and even prostitutes. Children who escaped from some of the plantations described the deplorable living conditions, lack of food, hard work, and beatings.

Although the slave ship brought worldwide attention to these practices, human rights activists and news media in some parts of Africa had been suggesting for years that child slavery was escalating in the cacao plantations. According to antislavery activists, as much as 40% of the chocolate consumed at that time was produced with the help of child slaves. Companies such as Nestlé, Cadbury, and Hershey buy beans from traders and processors, not from individual farmers. By the time the cacao beans arrive at large chocolate companies in the United States and Europe, those harvested by slaves have been mixed in with those harvested by paid workers. As a result, consumers have no way of knowing whether the chocolate came from slave plantations.

The current status of child slavery in West Africa is uncertain; some sources indicate that child trafficking is still occurring. However, the focus has shifted to the overall status of child labor in the Ivory Coast and Ghana, with estimates of up to 1 million children working on farms harvesting cacao beans. Many are young children working on family farms. Some of these children are as young as 10 years old, and most have never attended school. Government organizations, international NGOs (nongovernment organizations), and the leading chocolate companies are working to stop child labor in this region. They are training farmers, building schools, and even building housing for teachers in small communities. Until farmers get fair prices for their cacao beans, this practice is likely to continue. Fair Trade, UTZ Certified, and Certified Organic cacao beans are grown without the use of child labor. Look for these certifications the next time you purchase chocolate.

COCA-COLA: AN "ALL-AMERICAN" DRINK

Coca-Cola, the drink that has become synonymous with American culture, begins with the seeds of the kola tree (*Cola nitida*), found in West Africa. The tree, a relative of the cacao tree, bears pods that usually contain eight seeds. The fleshy seed coats are removed, and the seeds are allowed to ferment before they are dried and pulverized. In Africa, the seeds, with their high caffeine content, are used as a stimulant and an appetite depressant. The seeds can be pulverized to prepare a tea, or they can be chewed whole. In addition to the caffeine, small quantities of kolanin, which acts as a heart stimulant, are also present in the seed.

Coca-Cola first made its appearance on May 8, 1886, in Atlanta, Georgia. Dr. John Stith Pemberton, a pharmacist, concocted a beverage using carbonated water, caramel for coloring, an extract from coca leaves, and an extract from the powdered kola seeds. Although coca extracts are still used, since 1903 the cocaine has been removed before the extracts are added. Other ingredients, including sugar, vanilla, cinnamon, and lime juice, are also used, but the exact formula is a highly guarded secret. The original recipe, in Dr. Pemberton's handwriting, was locked in an Atlanta bank vault until December 2011 when it was moved to a vault in the World of Coca-Cola museum, also in Atlanta. Visitors to the museum can see the locked vault, but not the recipe.

OTHER CAFFEINE BEVERAGES

Throughout the world, other caffeine beverages are consumed by local populations. Maté and guarana are two well-known beverages in South America. Maté, or Paraguay tea, is made from the leaves of a holly, *Ilex paraguariensis,* and is popular throughout Central and South America. Guarana, made from the pulverized seeds of *Paullinia cupana,* is primarily consumed in Brazil. Among the Jivaro Indians in Peru, a ritual morning hot drink is prepared from the caffeine-rich leaves (up to 7.5%) of *Ilex guayusa.* The leaves of other hollies are also known to contain small quantities of caffeine and have been used to make beverages. Most notable is *Ilex vomitoria,* the yaupon holly, whose leaves were sold for tea in the southern states during the Civil War blockade.

CHAPTER SUMMARY

1. Caffeine and caffeinelike alkaloids have a stimulating effect on the central nervous system. Plants naturally rich in caffeine have a long history of use for alleviating fatigue and drowsiness. Historically, many of these plants have played an important role in human affairs.

2. Coffee is obtained from the seeds of *C. arabica,* a tree native to Ethiopia. Coffee drinking spread to Europe from the Middle East early in the seventeenth century, and by 1700, coffeehouses could be found throughout Europe. Coffee plantations were established in South America before 1720. The trees flourished in the hospitable climate and soil, and today Brazil and Vietnam are the world's leading coffee producers. Coffee rust is a threat to *C. arabica* wherever it is grown. Depulping, fermentation, and roasting are required to produce the characteristic flavor of coffee, with the roasting process the most critical and most variable. Instant coffee and decaffeinated coffee are two twentieth-century innovations in the long history of coffee.

3. Tea, made from dried tip leaves of *C. sinensis,* is the world's most popular beverage. The tea plant is a shrub native to the area around Tibet, India, China, and Myanmar (Burma). It flourishes in tropical or subtropical areas where there is abundant rainfall. The essential oils, theanine, and tannins in tea are responsible for its distinctive flavors and aromas, but the stimulating effects are due to the caffeine and theophylline present. Tea was introduced to western Europe in the seventeenth century. Its popularity was most evident

in England, where it became the national beverage. By the mid-1700s, the British colonists in North America were also avid tea drinkers. The Tea Act imposed by Parliament laid the foundation for the Boston Tea Party and other events leading to the Revolutionary War.

4. The seeds of *T. cacao* are the source of both a beverage and a confection, cocoa and chocolate. The cacao tree is native to tropical Central and South America and was used to prepare a spicy beverage consumed by Aztec noblemen and priests. In the sixteenth century, Spanish conquistadors introduced the beverage to the Spanish court, where it was modified by the addition of sugar. In Europe, cocoa never rivaled coffee or tea, but the popularity increased after the first chocolate bar was developed in the mid-nineteenth century.

5. The kola tree, native to West Africa, is the source of a caffeine beverage that has been consumed increasingly since the early twentieth century. Although the widespread popularity of cola drinks is apparent in contemporary society, maté, guarana, and other caffeine beverages are consumed by local populations throughout the world.

REVIEW QUESTIONS

1. Describe the physiological action and effects of caffeine in the human body.
2. What is the importance of caffeine in contemporary society? Do you think decaffeinated products will diminish this impact?
3. Contrast the geographical spread of coffee plantations with the history of coffee consumption.
4. Describe the processing of green, black, and oolong teas.
5. Trace the steps from freshly picked coffee cherries to a freshly brewed cup of coffee.
6. How have the products of the cacao tree changed from the time of Montezuma to the present?
7. What is the impact of energy drinks in the beverage market?
8. Describe the health benefits of coffee, tea, and chocolate.
9. What is the importance of the Fair Trade movement to these beverage plants?

©Karen McMahon

CHAPTER

17

Herbs and Spices

Lavender (Lavandula) *flowers are the source of an essential oil that is used in hair products, skin lotions, and perfumes.*

KEY CONCEPTS

1. Essential oils are volatile substances that contribute to the essence, the aroma, or the flavor of herbs and spices.
2. The desire for spices had a significant impact on the history of world exploration, colonization, and trade.
3. In temperate regions, the use of herbs goes back to prehistoric time, and four plant families provide the majority of herbs in use today.

CHAPTER OUTLINE

Essential Oils 290
History of Spices 290
    Ancient Trade 290

A CLOSER LOOK 17.1 Aromatherapy: The Healing Power of Scents 291

    Marco Polo 291
    Age of Exploration 292
    Imperialism 293
    New World Discoveries 293
Spices 293
    Cinnamon: The Fragrant Bark 293
    Black and White Pepper 294
    Cloves 295
    Nutmeg and Mace 295
    Ginger and Turmeric 296
    Saffron 297
    Hot Chilies and Other Capsicum Peppers 297
    Vanilla 299
    Allspice 300
Herbs 300
    The Aromatic Mint 300
    The Parsley Family 303
    The Mustard Family 303
    The Pungent Alliums 304

A CLOSER LOOK 17.2 Sweet Talk 305

Chapter Summary 306
Review Questions 307

289

The aroma of a freshly baked cinnamon bun, the zip of a peppery stew, the tang of oregano in pizza sauce, and the distinctive flavor of real vanilla ice cream are all familiar **sensory inputs that embellish life.** Today the herbs and spices responsible for these tastes and others are taken for granted, but in the past the desire for spices was a driving force that actually shaped human history. Although there is no clearcut distinction between herbs and spices, **herbs** are generally aromatic leaves, or sometimes seeds, from plants of temperate origin; **spices** are aromatic fruits, flowers, bark, or other plant parts of tropical origin. Herbs and spices are mainly associated with cooking, but they also are used in herbal medicine, as natural dyes, and in the perfume and cosmetic industries.

ESSENTIAL OILS

The characteristic scents of aromatic plants are due to the presence of **essential oils,** volatile substances that contribute to the essence or aroma of certain species. These compounds give flavor to foods and are the components of value for perfumes and medicines (see A Closer Look 17.1: Aromatherapy). A simple distillation of plant parts can extract these volatile oils and thus capture the essence of a plant. Essential oils are widely distributed in plant organs but are most commonly found in leaves, flowers, and fruits, where they occur in specialized cells or glands. An essential oil is considered a type of **secondary plant product,** a compound that occurs in plants but is not critical for the plant's basic metabolic function. This sets secondary products apart from primary products, such as sugars, amino acids, proteins, and nucleic acids, without which plants could not exist. (See A Closer Look 1.2: Perfumes to Poisons.)

Chemically, most essential oils are classified as **terpenes,** a large group of unsaturated hydrocarbons with a common building block of C_5H_8 (isoprene molecule). Essential oils are end products of metabolic pathways, but the exact role of these substances in plants varies. Clearly, the essential oils in flowers attract pollinators by their alluring scents, but the function of other essential oils in different plant parts has been debated. At one time, it was believed that many essential oils were merely waste products of metabolism that accumulated. Now it is thought that many essential oils play a significant role in discouraging herbivores, particularly insects, and inhibiting bacterial and fungal pathogens.

Although the flavor of herbs and spices may have been the initial reason for their use by people, their antimicrobial properties may have been an added benefit for their continued use. One study, which summarized many laboratory tests, has shown compelling evidence that herbs and spices inhibit or kill many foodborne bacteria. Onions, garlic, allspice, and oregano were found to be the most effective for killing bacteria. In addition, this study by Cornell University researchers looked at the historical use of spices in different countries and found a correlation with climate. They determined that more spices were traditionally used in hotter countries where food spoilage occurs quickly without refrigeration. These researchers suggest that food preservation may have been the main reason spices were incorporated into the human diet.

HISTORY OF SPICES

There is no evidence of how primitive people first discovered herbs and spices, but it is reasonable to assume that they were attracted to the pleasant aromas of these plants and found many uses for them. The ancient Egyptians used herbs and spices extensively in medicine, in cooking, for embalming, and as perfumes and incense. The Ebers Papyrus, dated about 3,500 years ago, is a scroll that lists the medical uses of many plants and includes anise, caraway, mustard, saffron, and many other familiar herbs and spices. Cinnamon and cassia are also mentioned in Egyptian records. These two spices, native to Southeast Asia and China, are evidence that an active spice trade was already in existence.

Ancient Trade

During the time of the Ancient Greek civilization, the spice trade was flourishing between the Mediterranean region and the Far East. Spices such as cinnamon and cassia, as well as black pepper and ginger from India, were sought by the Greeks. The intermediaries for this trade were Arab merchants who brought the spices mainly by caravan from India, China, and Southeast Asia. To protect their monopoly, the Arabs invented misleading and fanciful stories about the source of the spices. For example, the Arabs claimed that pepper grew only under waterfalls protected by dragons.

When Alexander the Great conquered Egypt, he established the port city of Alexandria, which became the leading trading center for spices from the East and a meeting place for traders from Europe, Asia, and Africa. Alexandria remained one of the most important spice trading centers for centuries.

Spices were even more prominent in the Roman Empire as their culinary use expanded, along with their application in medicine and luxury items (perfumes, bath oils, and lotions). The Romans went to great expense to procure these precious spices; large amounts of gold and silver flowed to the East in payment. When Nero's wife, Poppaea, died in Rome during the first century, a year's supply of cinnamon was burned at her funeral as a tribute; this spice was considered the most precious commodity in the imperial storehouse. After the first century, Rome began trading directly with India by ship, via the Red Sea, to the Indian Ocean, thus breaking the centuries-old Arab monopoly on spice trading. It is worth noting that both the Greeks and Romans made extensive use of native herbs as well as the more exotic spices; in fact, almost all the herbs and spices that are in use today were known to these ancient civilizations.

As the Roman civilization spread its influence through Europe, they introduced exotic spices to the local tribes. Already familiar with many temperate herbs, these tribal people quickly developed a taste for these luxuries. When Alaric

A CLOSER LOOK 17.1

Aromatherapy: The Healing Power of Scents

Aromatherapy is a holistic approach to healing using essential oils extracted from plants. Holistic medicine looks at the whole individual, and treatments emphasize the connection of mind, body, and spirit to overcome sickness. The concept behind aromatherapy is that essential oils can be absorbed through the nose, lungs, or skin and thereby trigger certain physiological responses. Proponents believe that the oils can affect both the immune system and the endocrine system and influence mood, memory, emotions, performance, and general well-being. Many oils are also antiseptic, analgesic, or anti-inflammatory and can be used to treat cuts, burns, and insect bites (table 17.A).

Although the term *aromatherapy* may be unfamiliar, many people are acquainted with products such as Vicks VapoRub for treating congestion and coughs due to colds. This product is applied externally to the throat and chest or used in a steam vaporizer to allow the vapors to be inhaled. The active ingredients are camphor, menthol, and eucalyptus oil—all essential oils from plants. The use of VapoRub and similar products is a familiar form of aromatherapy.

The use of herbs and spices for medicines, as well as for cooking and dyes, dates back to antiquity. Essential oils have also been used in perfumes for thousands of years; in fact, the earliest applications of perfumes were as incense (see A Closer Look 5.3: Alluring Scents). The current revival of aromatherapy can be traced to French researchers in the early to mid-twentieth century. Rene Gattefosse, a French perfume chemist, discovered the healing powers of lavender oil after a lab accident during which he burned his hand. Gattefosse began to investigate the properties of lavender oil and other essential oils. He published a book on essential oils and coined the term *aromatherapy*. Several years later, other French scientists continued the research on essential oils, including their applications in massage.

The antibacterial and antifungal properties of many essential oils are well established and are related to their function as secondary metabolites in the plant. However, rigorous scientific testing to substantiate other healing claims has not been done. If the claims of aromatherapy are valid, then individuals must use pure essential oils that meet exact botanical, biochemical, and pharmacological specifications. For example, there are three varieties of rosemary, *Rosmarinus officinalis*. Although the varieties are all the same species, they have notable differences in their essential oils, producing very different effects. Unfortunately, the industry is not regulated, and pure, unadulterated essential oils are not always available. In addition, not all essential oils are suitable for aromatherapy. Certain oils are potentially toxic because of their high ketone content; others contain phenols and can be caustic. The aromatherapist must be knowledgeable in all these matters for the oils to be safe and effective.

In the United States, producers and distributors of essential oils are not permitted to advertise any health claims regarding their products. By contrast, in France essential oils are often prescribed by physicians for various ailments and are prepared by pharmacists. As the use of these essential oils becomes more familiar in the United States and undergoes scientific testing, the incorporation of aromatherapy into mainstream medicine may occur here as well.

Table 17.A Commonly Used Essential Oils for Aromatherapy

Essential Oil	Common Uses
Bulgarian rose	Antiseptic, insomnia, relaxation
Cypress	Antiseptic, asthma, coughing, relaxation
Eucalyptus	Anti-inflammatory, arthritis, relaxation
Frankincense	Coughing, bronchitis
Geranium	Dermatitis, relaxation, depression
Ginger	Bronchitis, arthritis, stimulant
Juniper	Antiseptic, aches, pains, relaxation
Lavender	Antiseptic, respiratory infections, relaxation
Marjoram	Respiratory infections, relaxation
Pine	Asthma, arthritis, depression
Roman chamomile	Toothaches, arthritis, tension
Rosemary	Bronchitis, depression, mental alertness
Sandalwood	Acne, bronchitis, depression
Tea tree	Respiratory infections, acne, depression

the Visigoth threatened Rome, he demanded and received 3,000 pounds of pepper in addition to gold, silver, silks, and other valuable items. His demand highlights the value of the spice and illustrates the extent and volume of the spice trade. When Rome fell in A.D. 476, the trade between Europe and the East virtually disappeared; several centuries elapsed before the spice trade actively resumed.

Marco Polo

During the Dark Ages, exotic spices from the East were rare, and the people of Europe had to rely, for the most part, on native temperate herbs. Many of these herbs were valued for medicinal as well as aromatic purposes and were grown in monastic and royal gardens throughout Europe. Merchant travelers during this period kept a limited supply of spices flowing from the Arab trading centers to Europe. Later, the Crusades, beginning in 1095, increased the importation of spices and other goods from the Near East. Venice and Genoa, the great merchant cities of Italy, rose to prominence during this time. One Venetian trader in particular, who spurred on the European desire for spices and valuables from the East, was Marco Polo.

In 1271, Marco Polo began his travels at 17 years of age, in the company of his father and uncle, who were making a return journey to the court of Kublai Khan in China. They spent 25 years in the Orient and witnessed firsthand many of the riches. A few years after his return to Venice, Marco Polo was taken prisoner during a war between Venice and Genoa. During this year of captivity in Genoa, he dictated the memoirs of his adventures, later published as *The Travels of Marco Polo*. In the book, he described in striking detail the spice plantations of Java, the immense pepper stores in China, and the abundance of cinnamon, pepper, and ginger on the Malabar Coast of India. His accounts whetted the European appetite for the riches of the East and lured more and more travelers eastward in search of these spices. New overland routes were established, and soon explorers were searching for sea routes to the East.

Age of Exploration

No one was more determined to find a sea route than Prince Henry of Portugal, better known as Henry the Navigator, who sought to break the Venetian-Muslim trade monopoly. He established a school of navigation in 1418, where he gathered the leading astronomers, cartographers, geographers, and navigators of the day. Although he died before a sea route was discovered, his efforts laid the groundwork for the Age of Exploration that followed. In 1486, Bartolomeu Dias discovered the Cape of Good Hope at the southern tip of Africa, proving that a sea route to India was possible; Vasco da Gama made the possibility a reality when he reached the west coast of India in 1497 (fig. 17.1).

While the Portuguese were exploring southern routes to reach Asia, the Genoan Christopher Columbus, under the flag of Spain, sailed west in search of the spices of the East. From his first voyage in 1492, he was convinced that he had discovered the route to China and Japan. Although he did not return with valuable oriental spices, he still was able to persuade the Spanish monarchs, Queen Isabella and King Ferdinand, to finance three more voyages. Christopher Columbus never did find the black pepper and cinnamon for which he so ardently searched, but he firmly established Spain's claim to the New World and introduced a great many plants, including yams, sweet potatoes, cassava, kidney beans, maize, capsicum peppers, and tobacco, to Europe. Columbus died in obscurity in 1506, still believing that he had reached the East and never realizing that he had stumbled upon the New World, a discovery worth far more than the spices he was seeking. Several years later, the Portuguese Ferdinand Magellan, also sailing for Spain, led the expedition that circumnavigated the globe (1519–1522) and discovered a western route to the Spice Islands (now known as the Moluccas). Many of the important oriental spices, including cloves, nutmeg, mace, and pepper, are native to these Spice Islands (fig. 17.1).

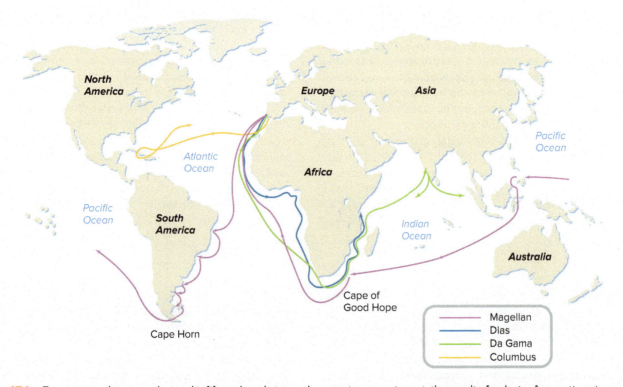

Figure 17.1 European explorations during the fifteenth and sixteenth centuries were, in part, the result of a desire for exotic spices.

Imperialism

During the sixteenth century, Portugal monopolized the spice trade to Europe through its outposts in India, China, Japan, and the Spice Islands. The explorations of da Gama, Pedro Alvares Cabral (1460–1526), and Alfonso de Albuquerque (1453–1515) and the subsequent military victories established Portuguese control over the major spice trading centers throughout the East. The Portuguese were ruthless in their control, often enslaving native populations to labor in the plantations. Without competition, the price of spices soared throughout Europe, and the revenues brought such tremendous wealth and power to Portugal that other European nations sought to break the Portuguese stranglehold and share the riches.

The Dutch and English eventually broke this control early in the seventeenth century. By 1621, the Dutch had forced the Portuguese from the Spice Islands, securing control over nutmeg and cloves for the Netherlands. By the end of the seventeenth century, the Portuguese were left with holdings only in Goa (India) and Macao (a small island off the coast of China), and the British were only minor players in the spice trade. The Dutch were the dominant force in the East Indies, Ceylon (Sri Lanka), and the Persian Gulf spice markets. Their control over the spices was even harsher than that of the Portuguese. For example, to artificially create a scarcity and inflate the price of nutmeg and cloves, they uprooted 75% of those trees on the Spice Islands. Cinnamon from Ceylon, more than any other spice, brought huge profits to the Dutch East India Company, the officially sanctioned trading conglomerate. As in the case of the Portuguese, the country that controlled the spices controlled great wealth and power.

In the latter half of the eighteenth century, the Dutch monopoly began to break down for many reasons: the British and French began spice plantations in their own colonies; the British East India Company gained a strong foothold on the Malabar Coast of southwest India that produced pepper, ginger, and cinnamon; and a British blockade during the war between England and the Netherlands resulted in near bankruptcy for the Dutch East India Company. By the end of the eighteenth century, the nearly 200 years of Dutch control had ended.

By the early nineteenth century, the English East India Company had control over most of the spice-rich Orient, but the decentralization of the spice trade had begun and a spice monopoly would never occur again. In the late eighteenth and early nineteenth centuries, even the United States became involved through clipper ships that sailed from New England to the island of Sumatra in Indonesia in search of spices, principally pepper. Many fortunes were made from this lucrative spice trade centered in Salem, Massachusetts.

New World Discoveries

During the sixteenth century, Spain greatly expanded its influence in the New World, conquering and subjugating native populations in Central and South America. The New World spices, introduced first to Spain, included allspice, vanilla, and several varieties of capsicum peppers, such as chili peppers and paprika. The capsicum peppers could be grown in many parts of the temperate and tropical world and soon became important in the diets of many people in Europe, Africa, and Asia. Allspice remains an exclusively New World spice because it has not been successfully grown elsewhere. However, since the mid-nineteenth century, vanilla has been cultivated in many tropical areas. Trade in these New World spices never had the allure, importance, or adventure associated with the spices from the Far East.

> **Thinking Critically**
>
> Most ancient societies used herbs and spices for medicinal and culinary purposes.
>
> *Speculate on how early humans would learn about these properties in plants.*

SPICES

The spices of the Old World that dominated trade and were the impetus to exploration still remain prominent commodities throughout the world, as are the New World spices. The following discussion focuses on those spices that have the most widespread use (table 17.1).

Cinnamon: The Fragrant Bark

Cinnamon is one of the oldest and most valuable spices known; its use is documented in ancient Egyptian, biblical, Greek, Roman, and Chinese accounts. It was one of the main spices sought in the early explorations. This spice comes from the bark of an evergreen tree, *Cinnamomum zeylanicum,* in the laurel family (Lauraceae) native to India and Sri Lanka where it grows best under wet tropical conditions. The similar spice cassia, which may also be called cinnamon, comes from several related species but primarily *Cinnamomum cassia,* native to Southeast Asia. Trees under cultivation are kept small and bushy, but in the wild they can reach a height of 12 meters (39 feet). Two-year-old stems and twigs are cut, and the bark is removed with a uniquely curved knife. The outer layer of the bark is scraped away, and the inner bark curls into quills (cinnamon sticks) as it dries (fig. 17.2). Imperfect quills and trimmings are ground into powdered cinnamon. Cassia differs in that the entire bark is used to make the quills. In the United States, much of what is called cinnamon may actually be cassia. Today these spices are usually associated with baking, but they have also been used in medicines, perfumes, and scents (fig. 17.3).

Table 17.1
Common Spices, Their Scientific Names and Families, and the Plant Part Used

Spice	Scientific Name	Family	Part Used
Allspice	*Pimenta dioica*	Myrtle	Fruit
Black pepper	*Piper nigrum*	Pepper	Fruit
Capsicum peppers	*Capsicum annuum*	Tomato	Fruit
	Capsicum baccatum		
	Capsicum chinense		
	Capsicum frutescens		
	Capsicum pubescens		
Cassia	*Cinnamomum cassia*	Laurel	Bark
Cinnamon	*Cinnamomum zeylanicum*	Laurel	Inner bark
Cloves	*Eugenia caryophyllata*	Myrtle	Flower
Ginger	*Zingiber officinale*	Ginger	Rhizome
Mace	*Myristica fragrans*	Nutmeg	Aril
Nutmeg	*Myristica fragrans*	Nutmeg	Seed
Saffron	*Crocus sativus*	Iris	Stigma
Turmeric	*Curcuma longa*	Ginger	Rhizome
Vanilla	*Vanilla planifolia*	Orchid	Fruit

(a)

(b)

Figure 17.2 (a) Illustration from a sixteenth-century French text depicts a cinnamon harvest. (b) Cinnamon comes from the scraped-off bark of trees in the laurel family.

Black and White Pepper

Black pepper, the most widely used spice today, was once a precious and most desired commodity. Both black and white pepper are obtained from the dried berries of *Piper nigrum*, a climbing vine in the Piperaceae (pepper family) native to India and the East Indies where it thrives in a hot, wet climate.

For black pepper, berries are picked green just before ripening; the berries are allowed to dry for a few days, and during this process they turn black and shrivel (fig. 17.4). They may be sold in this form, as peppercorns, or ground. Because the biting flavor is due to volatile oils, peppercorns begin to lose their flavor after grinding; for this reason, the taste

CHAPTER 17 Herbs and Spices

Figure 17.3 Ceremonial spice box, or *b'samim*, used in the Jewish Havdalah ceremony marking the end of the Sabbath. Spices (notably, cloves and cinnamon) are placed inside the box, which is passed around during the ceremony, so that participants can enjoy the aromas. Symbolically, this ceremony promises that the fragrant memories of the Sabbath will be with participants throughout the coming week.

Figure 17.4 Peppercorns of *Piper nigrum* can be green, white, black, or red depending on how they are processed.

of freshly ground pepper is often preferred. White pepper is obtained by allowing the berries to ripen on the vine; after harvesting, the outer hull is removed, leaving a grayish white kernel that is ground. White pepper is slightly milder, lacking the pungency of the black. Red peppercorns are ripe pepper fruits; the red color of the berry is retained during processing.

Figure 17.5 The dried flower buds of *Eugenia caryophyllata* become the familiar clove "nail."

After harvesting, they are either dried or pickled. The flavor of red pepper is considered to be more pungent and aromatic than green pepper. Red peppercorns are new to the marketplace and difficult to find.

Cloves

Cloves were valued in ancient China, where they were used to sweeten the breath of court officials before they addressed the emperor. They are native to the Spice Islands. Centuries later, when the Dutch controlled these islands, they confined the production of cloves to a single island in order to drive up prices. It has been estimated that 60,000 native islanders were killed in the process of destroying the clove plantations. A Frenchman managed to steal some clove seeds in 1770 and established plantations on French colonies. Cloves are the unopened flower buds (fig. 17.5) of *Eugenia caryophyllata*, an evergreen tree in the myrtle family (Myrtaceae). The buds have to be picked with care because, once opened, they are useless as a spice. After picking, the buds are dried and marketed as whole cloves or ground and used in desserts, beverages, meats, pickling, sauces, and gravies. In Indonesia, cloves are mixed with tobacco for cigars and cigarettes. In addition to its use as a spice, extracted clove oil has been used in medicines, disinfectants, mouthwashes, toothpastes, soaps, and perfumes; however, synthetics are now replacing the natural oil.

Nutmeg and Mace

Nutmeg and mace are two spices obtained from a single plant, the nutmeg tree, *Myristica fragrans* of the nutmeg family (Myristicaceae), native to the Spice Islands. Nutmeg trees are usually dioecious but monoecious individuals can be found. The fertilized pistillate flowers produce yellow fruits that split into halves when ripe, exposing a large, glossy, dark brown

Figure 17.6 (a) The fruit of *Myristica fragrans* is the source of two spices, nutmeg and mace. Mace is derived from the netlike red aril that is wrapped around the single seed. The shelled seed is the source of nutmeg. (b) Nutmeg seeds.

seed covered by a scarlet net (fig. 17.6). The net is technically an **aril,** a lacy outgrowth from the base of the seed. The leathery aril is dried and ground to produce the aromatic spice called mace (not to be confused with the aerosol chemical used in self-defense). After the removal of the aril, the seed is cured by drying, then shelled to remove the seed coat. The resulting nutmegs are sold either whole or ground (fig. 17.6b). The two spices have similar properties, with a strong and spicy, but slightly bitter, aromatic flavor and are used in baking sweets as well as meat and vegetable dishes.

Both nutmeg and mace have received some notoriety as potential hallucinogens. Achieving a hallucinogenic state requires consumption of very large quantities of either spice. The essential oils are believed to be the hallucinogenic agents, but because of the toxicity of these compounds, the hallucinations are accompanied by many unpleasant side effects, including nausea, vomiting, dizziness, and headaches.

Although nutmeg and mace were not known to the ancient Western civilizations, they had reached Europe by the twelfth century and were two of the precious spices of the Middle Ages. Later, the Portuguese, and then the Dutch, controlled production of these spices in the Spice Islands until the French smuggled seedlings to their island colonies. Yankee traders in the nineteenth century developed a profitable scam by producing fake wooden nutmegs, which they sold as the real thing. The nickname of Connecticut, the "Nutmeg State," reflects this historical anecdote.

Ginger and Turmeric

Two very different spices that come from the same family are ginger and turmeric. Ginger is obtained from the rhizomes of *Zingiber officinale* of the ginger family (Zingiberaceae), a small, herbaceous perennial native to tropical Asia but cultivated throughout the tropics. Not only do the rhizomes provide the spice, but small portions of the rhizome also are

Figure 17.7 Ginger rhizome, source of the familiar spice.

used for vegetative reproduction and give rise to aboveground shoots (fig. 17.7). Nine months to 1 year after planting, the greatly enlarged rhizomes are harvested and cleaned and sold fresh or peeled, dried, and ground. The aroma and taste of ginger are characteristically spicy, hot, and pungent, and the best ginger today is said to be from Jamaica. The Spanish introduced this spice into the New World, where it grew so successfully that by 1547 Jamaica was exporting ginger to Spain. Ginger is also a versatile spice used in baked goods, especially gingerbread, Asian dishes, pickles, vegetables, meats, poultry, and ginger ale. Ginger has been used medicinally to relieve nausea; ginger ale is a common folk remedy to ease morning sickness and upset stomachs.

Also native to tropical Asia, *Curcuma longa* is the source of turmeric, another spice obtained from a dried rhizome. Turmeric is also used as a brilliant yellow dye to color both food and fabric. The propagation, harvesting, and processing of turmeric are similar to the methods described for ginger. Although turmeric is not as familiar as ginger, it is a common ingredient in prepared yellow mustard, the main spice in curry powder, and often a substitute for the more costly saffron. Turmeric is also used in Middle Eastern and East Indian cooking and even is used medicinally and cosmetically in parts of Asia.

Turmeric has a long history in the Indian Ayurvedic system as a remedy for a variety of complaints from wound healing to stomach ailments. The biologically active ingredients in turmeric have been identified as curcumin and related compounds known as curcuminoids. Curcumin has been shown to have antioxidant, anti-inflammatory, antiviral, antibacterial, and antifungal properties that may be effective in fighting certain cancers and other diseases. In fact, the University of Texas MD Anderson Cancer Center has begun using curcumin as an adjunct therapy to treat pancreatic cancer and multiple myeloma (cancer of plasma cells in the bone marrow). Cancer patients are advised to ingest 8 grams of curcumin per day, about 40 times what is consumed in the average Indian diet. Other clinical trials are testing the effectiveness of curcumin in preventing Alzheimer's disease and colon cancer. There is also preliminary evidence that curcumin may be successful in treating many inflammatory diseases, including arthritis, colitis, and allergies.

Saffron

The modern world's most expensive spice is obtained from the delicate stigmas of an autumn-blooming crocus, *Crocus sativus*, in the iris family (Iridaceae) (fig. 17.8). The species is native to eastern Mediterranean countries and Asia Minor. Although saffron is not one of the exotic tropical spices from the Far East, it is a spice that was much desired by the ancient civilizations of Egypt, Assyria, Phoenicia, Persia, Crete, Greece, and Rome and may be the oldest spice used in the region. It was an important commodity in their spice trading and was valued as a yellow dye. In the East, the yellow dye symbolized the epitome of beauty and its aroma was considered sublime. The word *saffron* is from the Arabic word *zafaran*, which means yellow. Arabs introduced saffron to Spain in the tenth century, and today Spain dominates world production, with 70% of the market.

Crocus sativus is propagated by corms; small cormlets are planted, giving rise to the familiar crocuses. Every autumn the appearance of the purple flowers signals the beginning of the saffron harvest. The blooming period is short, about 2 weeks, and the flowers must be picked in full bloom before wilting; often the critical time period for harvesting is limited to a few hours. Once picked, the flowers must be carefully stripped of their orange-red three-parted stigmas; again, haste is important to remove the stigmas before the petals wilt. Most of this backbreaking and delicate work of harvesting and stripping traditionally has been done by hand, but mechanization is making inroads in Spain. After the stigmas are removed, they are dried by slow roasting and sold either as saffron threads (whole stigmas) or powdered. The stigmas from 150,000–200,000 flowers yield 1 kilogram (75,000–100,000 flowers, 1 pound) of the spice. In 2018, an on-line major retailer was selling saffron to the United States market for approximately $5.95 per gram (or $149.94 per ounce), validating its claim as the world's most costly spice. Hoping to cash in on the lucrative saffron market, unscrupulous merchants have been known to adulterate saffron with turmeric, marigold, safflower petals, or other substances. Researchers in Spain are studying ways to boost production of saffron. Although much of the research has focused on traditional plant breeding, some scientists are using molecular techniques to identify genes that might increase the number of stigmas.

The flavor and aroma of saffron are pungent, slightly bitter, and musky. A small amount of saffron imparts a delicate and enticing taste and color to many different foods. It is widely used in French, Spanish, Middle Eastern, and Indian cooking, being an essential ingredient in bouillabaisse, paella, arroz con pollo, saffron cakes and buns, and challah.

Hot Chilies and Other Capsicum Peppers

Not to be confused with their namesake black pepper, capsicum peppers from the New World have become equally important in international cuisine since their discovery by Columbus. When Columbus found the capsicum fruits to be as pungent as the black pepper from the Far East, he believed that his voyage west in search of spices had been justified, and it was with high hopes that he took these New World fruits back to Spain. After their introduction to Spain, their cultivation and use spread throughout Europe, Asia, and Africa as their versatility became realized, and eventually they became as prominent in these regions as they were in the New World. Evidence indicates that capsicum peppers had been widely cultivated for thousands of years by the people of tropical America. Fragments of a 9,000-year-old chili pepper were discovered in a Mexican cave. Fossil starch grains of

Figure 17.8 The delicate orange-red stigmas of *Crocus sativus*, the source of saffron, must be separated carefully by hand.

Figure 17.9 Capsicum peppers come in a wide variety of colors, shapes, sizes, and degrees of hotness.

Peppers	Scoville Heat Units
Aji	30,000–50,000
Banana (sweet)	<1
Bell	<1
Capsaicin	16,000,000
Cayenne	30,000–50,000
Cherry	100–500
Chiltepin	50,000–100,000
Habanero	100,000–300,000
Jalapeño	2,500–5,000
Jamaican hot	100,000–200,000
Naga jolokia	855,000–1,041,427
Pimento	<1
Scotch bonnet	100,000–250,000
Tabasco	30,000–50,000
Thai	70,000–80,000

Table 17.2 Scoville Ratings for a Variety of Capsicum Peppers

domesticated chili peppers have been found in Peru, Ecuador, Central America, Mexico, and the Caribbean, with the oldest dating back to more than 6,000 years ago. When Cortés invaded Mexico, the native peoples considered capsicum peppers indispensable elements in their diet, along with maize, tomatoes, and beans, and were growing numerous varieties that were used with every meal.

Capsicum peppers are the fruits of plants belonging to a single genus in the tomato family (Solanaceae), *Capsicum*, which includes five cultivated species and hundreds of varieties (fig. 17.9). *Capsicum annuum* is the most widely cultivated of these species throughout the world and includes the mild sweet bell peppers as well as many varieties of hot peppers, such as cayenne. This species is a small, bushy, herbaceous annual that produces small, white flowers similar to those of tomato or potato. The fruits are berries that vary considerably in shape, size, and color among the hundreds of varieties; the immature fruits are green, and the mature fruits vary in color from yellow to orange to bright red and in shape from long and narrow to almost spherical. *Capsicum frutescens*, cultivated mainly in the tropics and warm temperate areas, generally has a more fiery taste and includes the varieties of peppers used in making Tabasco sauce. *Capsicum chinense*, despite its scientific name, has South American origins. This species includes the habanero, one of the hottest chili peppers known. *Capsicum baccatum* is the most widely grown pepper in South America, where it is called *aji*. Grown in the highlands of Central and South America, *C. pubescens* is the least known of the domesticated chilies.

The biting taste of capsicum peppers is due to the mixture of seven related alkaloids, of which capsaicin is the most prevalent. Capsaicinoids are found mainly in the seeds and placental area (where seeds attach to the ovary wall).

The capsaicin content is negligible in sweet bell peppers but found in such high concentrations in hot chili and jalapeño peppers that even handling or cutting the peppers can irritate the skin. Capsaicin is so potent that it can be tasted in concentrations as low as 1 part per million. In 1912, Wilbur Scoville developed a method to record the hotness of various capsicum peppers. A panel of subjects tasted extracts from a pepper. The sample was continuously diluted and tested until pungency could no longer be detected. Scoville Heat Units were then assigned (table 17.2). The hottest is naga jolokia or the ghost chili, a hybrid created in India between *C. chinense* and *C. frutescens*, with a Scoville rating ranging from 855,000 to over 1,0000,000. This means that an extract of this pepper is still detectable even in a 1,000,000 to 1 dilution! Today, HPLC (high-pressure liquid chromatography) can directly measure the concentration and type of capsaicinoid present. The potency of capsaicin has been utilized in two completely different applications: police use it as a pepper spray to subdue unruly persons, and many people use it in creams that are applied to relieve the pain of arthritis, shingles, cluster headaches, and other ailments.

Capsaicinoid levels vary in wild chilies even within the same species. New work has discovered that in Bolivian populations of *Capsicum chacoense*, higher capsaicinoid concentrations protect seeds from a pervasive fungus (*Fusarium*), which destroys about a third of the seeds produced. However, the higher the capsaicinoid level, the more susceptible the plant is to drought and its seeds are more prone to ant attack. Hotter chilies apparently have more stomata in their leaves and consequently transpire more water, which lowers their ability to survive drought. At the same time, hotter chilies produce seeds with thinner seed coats, making it easier for ants to bite into the seed and devour the embryonic plant.

How capsaicin works as an analgesic has not been entirely worked out, but there are intriguing associations between how the body perceives capsaicin and extreme heat. Many cooks have experienced a burning sensation if the juice of hot chilies touches the skin. It appears that capsaicin and hot temperatures of 43°C (109°F) trigger the same heat-sensitive pain nerve fibers; thus, we perceive the taste of a chili pepper as hot. Application of capsaicin as an analgesic cream either eventually desensitizes or may actually destroy the nerve fibers. Either way, an arthritis sufferer applying capsaicin cream or the habitual chili eater feels less heat or pain over time.

Pepper fruits are also excellent sources of vitamin C; even one pepper is more than enough to satisfy the daily requirement. The amount of vitamin C is actually higher in peppers than in citrus fruits; in fact, vitamin C was first chemically isolated from paprika in 1932 by Albert Szent-Gyorgyi, who won a Nobel Prize in 1937 for other work.

Many varieties of capsicum peppers are sold whole, either fresh or dried, while powders are prepared by grinding the dried fruits of several varieties. Included in this last group are paprika, red pepper (both ground and crushed are sold), cayenne, and chili powder, which is actually a blend of spices in addition to the ground chili peppers. The versatility of the capsicum peppers is reflected by the number of different cuisines that are associated with them. Many Hungarian, Italian, Mexican, Cajun, Indonesian, Indian, and Asian dishes utilize some type of capsicum pepper or spice. Chilies can also be grown as ornamentals for their colorful fruits. It is a tradition in New Mexico to string red chilies into ristras, which are hung near the entranceway of homes as a symbol of hospitality.

(a)

(b)

Figure 17.10 (a) Fruits of the vanilla vine, the source of vanilla flavoring. (b) Cured vanilla beans.

Thinking Critically

The presence of capsaicin in the seeds and placental region of the fruit is responsible for the sharp, biting flavor of chili peppers.

What would be the value for the plant of synthesizing capsaicin or a similar compound?

Vanilla

Another important spice associated with the New World tropics is vanilla, the only spice obtained from a member of the orchid family (Orchidaceae). Vanilla orchids are perennial vines native to the humid tropical rain forests of Central America and Mexico. They produce elongate fruits or capsules, commonly called pods, that are processed into the vanilla "beans" of commerce (fig. 17.10a). *Vanilla planifolia* is the main vanilla orchid of commerce and is cultivated in Madagascar, the islands of the Indian Ocean (Réunion and Seychelles), and Mexico.

Vanilla tahitensis is the source of Tahitian vanilla, much esteemed in the luxury gourmet market. The Tahitian vanilla orchid has been shown by genetic analysis to be a natural cross between *V. planifolia* and another vanilla species, *V. odorata*. In pre-Columbian times, the Mayans cultivated this variety in the tropical rain forest of what is present-day Guatemala. Spanish explorers later introduced *V. tahitensis* along with other domesticated plants from the New World to the Philippines via Manila galleons, a seafaring trade route between the Pacific coast of Mexico and the Philippines from 1565 to 1815. Later, *V. tahitensis* was most likely introduced by French sailors into Tahiti, where the orchid has flourished.

To ensure good pod production, growers hand-pollinate the flowers, particularly in areas outside the native range where the natural pollinators (certain bees and hummingbirds) are not present. The pods are picked while still green and then undergo a traditional curing for several months. The first stage of this process involves sweating, alternately heating the pods in the sun and then wrapping them in blankets throughout the night. A slow drying process follows, during which the final aroma is developed. Uncured or unfermented vanilla beans lack the characteristic vanilla flavor, which is due to vanillin, a crystalline compound synthesized during curing. Interestingly, vanillin is one of the precursors in the chemical pathway that synthesizes capsaicin, the compound found in chili peppers. When properly cured, the pods are black (fig. 17.10b) with crystals of vanillin on the surface. Alternative methods of rapid curing are used in some areas, but these do not always produce a satisfactory product. Although whole beans are available, most are processed into vanilla extract by percolating cut beans in a 35% alcohol-water solution. Pure vanilla extract contains more than just vanillin; other flavoring compounds are leached out of the bean and contribute to the final extract. These compounds may vary according to the growing conditions of a region and account for taste and quality differences; many feel that Mexican vanilla is top quality.

Less than 1% of the production of vanillin on the world market is derived from the orchid. For imitation vanilla extract, vanillin is synthesized chemically from a variety of compounds, such as clove oil, lignin from wood pulp, or coal tar. The chief advantage of imitation vanilla is the lower cost; however, it lacks the subtle flavor of pure vanilla extract. In addition, extracts of tonka beans, *Dipteryx odorata,* have been passed off as vanilla extracts. This is a dangerous substitute, because the tonka beans contain coumarin, a blood thinner that could cause internal hemorrhaging.

Recently, a research group genetically engineered two common species of yeast, fission (*Schizosaccharomyces pombe*) and baker's (*Saccharomyces cerevisiae*), with the ability to convert glucose into vanillin. Researchers introduced a combination of bacterial, human, and fungal genes into the yeast in order to create a microbe that can produce vanillin cheaply and without undesirable by-products. As stated previously, microorganisms have been used to make vanillin, but this process is costly and produces environmentally harmful waste. The researchers also inactivated a gene naturally present in the yeast that converts vanillin into an undesirable form. As vanillin itself becomes toxic to the yeast when it accumulates in high concentrations, the researchers added another gene that adds a sugar group to the vanilla. The added sugar group makes vanillin nontoxic and does not affect the taste of vanilla. The result is that engineered yeast can produce more vanilla without dying off, boosting production rates.

Long before the Europeans arrived in America, vanilla was an important commodity among the Aztecs. The beans were used as flavoring, perfume, medicine, and even a means of tribute. It is believed that Bernal Diaz, a Spanish explorer with Cortés, was the first European to report on the use of vanilla in the preparation of *chocolatl* (see Chapter 16). Vanilla beans were taken back to Spain, and soon their use spread throughout Europe. Mexico continued to be the leading producer of vanilla beans for several centuries until it was discovered that the flowers could be hand-pollinated. Today Madagascar leads the world in cultivation of vanilla beans.

In late 2018, a Bronze Age tomb in Israel dating 3500 years ago, was found to contain jugs with residues of vanillin, the principal ingredient in natural vanilla extract. The presences of vanilla extract in the tomb suggests that vanilla was developed as a food flavoring, ingredient in medicine, and/or an embalming agent by the ancient peoples of Mesopotamia. There are approximately 110 species of vanilla orchids found in tropical areas around the world and analysis of the chemical profile of the vanillin found in the tomb matches that of present-day vanilla orchid species growing in East Africa, India, and Indonesia. It is believed that ancient trade routes most likely brought vanillin or vanilla beans to the Middle East from India or perhaps from East Africa. This is the first time that evidence supports that the vanilla spice was developed in areas other than Mexico and Central America.

Allspice

The third New World spice discovered by Spanish explorers was allspice, the dried berries of *Pimenta dioica,* an evergreen tree in the myrtle family (Myrtaceae). Although there is some evidence that Columbus may have come across the fragrant leaves of a *Pimenta* tree on his second voyage, the spice long used by the Mayan civilization was not discovered by Europeans until the 1570s. Allspice is named for its multifaceted flavor, which is similar to a combination of cinnamon, nutmeg, and cloves. The berries, which are picked nearly ripe and then dried, resemble peppercorns; the other common names for the plant—Jamaica pepper, clove pepper, pimento, and Jamaica pimento—reflect this resemblance (*pimenta* means pepper in Spanish). Unlike capsicum peppers and vanilla, this spice has never been successfully cultivated outside the Western Hemisphere; Jamaica controls the world production of allspice today. This versatile spice is used whole or ground. Ground, it is an ingredient in baked goods, cooked fruits, sauces, and relishes; whole, it is best known for its use in pickling vegetables and meats.

HERBS

In recent years, there has been a renaissance in the use of herbs in everyday life, going beyond the kitchen to find applications in shampoos, cosmetics, soaps, potpourris, and medicines. Herbs are usually the aromatic leaves or sometimes seeds of temperate plants; however, other plant organs are sometimes considered herbs as well (fig. 17.11a). Throughout the centuries, thousands of plants have been used as herbs, but the following discussion will focus only on four well-known families.

The Aromatic Mint

The mint family (Lamiaceae) is the source of many important and familiar herbs: spearmint, peppermint, marjoram, oregano, rosemary, sage (fig. 17.11b), sweet basil (fig. 17.11c), thyme, savory, and others (table 17.3). Members of this

Figure 17.11 Anne Hathaway Herb Garden, Woodward Park, Tulsa, OK. (a) Many of the most flavorful and useful culinary herbs can be attractively grown in an herb garden: (b) sage, (c) sweet basil, (d) dill.

Table 17.3
Common Herbs, Their Scientific Names and Families, and the Plant Part Used

Herb	Scientific Name	Family	Part Used
Anise	*Pimpinella anisum*	Parsley	Fruit
Basil	*Ocimum basilicum*	Mint	Leaves
Bay leaves	*Laurus nobilis*	Laurel	Leaves
Caraway	*Carum carvi*	Parsley	Fruit
Cardamom	*Elettaria cardamomum*	Ginger	Seed
Celery	*Apium graveolens*	Parsley	Fruit
Chervil	*Anthriscus cerefolium*	Parsley	Leaves
Chives	*Allium schoenoprasum*	Amaryllis	Leaves
Cilantro	*Coriandrum sativum*	Parsley	Leaves
Coriander	*Coriandrum sativum*	Parsley	Fruit
Cumin	*Cuminum cyminum*	Parsley	Fruit
Dill	*Anethum graveolens*	Parsley	Fruit, leaves
Fennel	*Foeniculum vulgare*	Parsley	Fruit
Fenugreek	*Trigonella foenum-graecum*	Bean	Seed
Garlic	*Allium sativum*	Amaryllis	Bulblets
Horseradish	*Armoracia rusticana*	Mustard	Root
Lavender	*Lavandula angustifolia*	Mint	Flowers
Leek	*Allium porrum*	Amaryllis	Leaves
Marjoram	*Origanum majorana*	Mint	Leaves
Miracle Fruit	*Synsepalum dulcificum*	Sapodilla	Fruit
Mustard	*Brassica alba; Brassica nigra*	Mustard	Seed
Onion	*Allium cepa*	Amaryllis	Bulb
Oregano	*Origanum vulgare*	Mint	Leaves
Parsley	*Petroselinum crispum*	Parsley	Leaves
Peppermint	*Mentha piperita*	Mint	Leaves
Rosemary	*Rosmarinus officinalis*	Mint	Leaves
Sage	*Salvia officinalis*	Mint	Leaves
Savory	*Satureja hortensis*	Mint	Leaves
Shallot	*Allium ascalonicum*	Amaryllis	Bulb
Spearmint	*Mentha spicata*	Mint	Leaves
Star anise	*Illicium verum*	Star anise	Seed
Sweetleaf	*Stevia rebaudiana*	Aster	Leaves
Tarragon	*Artemesia dracunculus*	Aster	Leaves
Thyme	*Thymus vulgaris*	Mint	Leaves

cosmopolitan family include mainly herbaceous plants and small shrubs, characterized by square stems and aromatic, simple leaves with numerous oil glands. The Mediterranean region is an important center of origin for the mint family, and a variety of these herbs have been used for thousands of years by the civilizations that developed in this area. With the long history of use, each herb has evolved a distinctive biography, folklore, and usage, but space limits the discussion to a brief overview of this family.

Typifying mint flavors are spearmint and peppermint, *Mentha spicata* and *M. piperita,* respectively. The dried leaves and the distilled oils of these herbaceous perennials find widespread use as flavorings and perfumes in gums, candies, cookies, cakes, cigarettes, toothpastes, mouthwashes, antacids, soaps, jellies, ice creams, teas, and other drinks. Mint flavorings have become indispensable in everyday life; it would be illuminating to count the number of times mint is encountered in just one day. Menthol is the most abundant constituent of peppermint oil; for large-scale commercial extraction of menthol, the related Japanese mint, *Mentha arvensis,* is frequently used because of its higher menthol content.

Both marjoram and oregano were used by the ancient Egyptians, Greeks, and Romans, who employed the herbs for both cooking and medicine. These perennial herbs are members of the genus *Origanum,* but each has a different taste; marjoram, *O. majorana,* is mild, and oregano, *O. vulgare,* has a biting flavor. The dried leaves of both herbs are widely used with meats and vegetables; oregano has become especially popular through its use in pizza and spaghetti sauce.

Sweet basil, *Ocimum basilicum,* is one of the oldest herbs known (fig. 17.11c). This native of India has been considered a sacred plant in the Hindu religion. The Greeks referred to basil as the herb of kings; indeed, the word *basil,* perhaps, is derived from the Greek *basileus,* meaning king. The dried leaves of this annual, noted for its sweet, aromatic flavor, are widely used in a variety of foods but are most commonly associated with tomato dishes.

Lavender (*Lavandula*) is another important member of the mint family and was used by the Romans to scent their bathwater, giving rise to its name from the Latin *lavare,* to wash. The lavender is a small shrub with gray-green leaves and pink to purple flowers. It is native to the Mediterranean region and thrives under sunny conditions and in well-drained soils. There are two commercially important species—common, or English, lavender (*L. angustifolia*) and French, or Spanish, lavender (*L. stoechas*). Many of the cultivated varieties are hybrids of these and other species. Flowers, either fresh or dried, are the parts of the plant used for potpourri and sachets or to flavor vinegars, jams, or sugar for a variety of confections. The essential oil is also obtained from the flowers and has long been valued for perfumes and its antiseptic properties, which hasten healing (see A Closer Look 17.1: Aromatherapy).

Recently, the essential oil of lavender has been linked to gynecomastia, abnormal breast development in boys. All of the patients were under the age of 10 and had used a personal care product, such as a shampoo, hair gel, or soap, which contained lavender oil as a main ingredient. Once the boys were advised to stop using products with lavender oil, breast development stopped and the condition reverted to normal in a few months. In laboratory studies, lavender oil mimicked the action of the female hormone estrogen in the body, which stimulates breast tissue development. At the same time, lavender oil inhibited the male hormone androgen, which inhibits breast development. These results confirm that lavender oil is an endocrine disrupter. Tea oil is another botanical oil that is reported to affect sex hormones in the same way. Other studies have noted an increase in early breast development in young girls. Perhaps the use of lavender oil or other plant oils is the cause.

The Parsley Family

The temperate cosmopolitan parsley family, the Apiaceae, which gives us carrots and parsnips, also provides many familiar herbs: parsley, caraway, dill (fig. 17.11d), fennel, celery, anise, coriander, cilantro, cumin, and chervil (table 17.3). Annual, biennial, or perennial, the members of this family are easily recognized by their umbels (flat-topped inflorescences) and alternate compound leaves. The characteristic fruit for the family is a dry indehiscent **schizocarp,** which splits into two one-seeded identical halves commercially referred to as seeds. For many plants in this family, these fruits are used as the herb; however, for others the fresh or dried leaves are the desired parts. Like the herbs in the mint family, the herbs in the Apiaceae have a long history of usage, but only a few will be considered.

Parsley, *Petroselinum crispum,* is probably the best known and most widely used member of this family, valued for its use as a garnish as well as its flavor. Native to the Mediterranean region, parsley was revered by the early Greeks as a symbol of both victory and death and, as such, was used in crowns for champions and wreaths for tombs. It was the Romans who first valued parsley as a culinary herb, and today it is an almost indispensable ingredient in many dishes. On the other hand, the attractive, dissected, and usually curly green leaves are too often used just to enhance the visual presentation of food and then discarded uneaten. This is unfortunate because the leaves are high in vitamins A and C and several minerals and are said to sweeten the breath after a garlic-laden meal.

Dill, *Anethum graveolens,* provides both leaves (dill weed) and fruits (dill seeds) that are widely used as seasonings (fig. 17.11d). Individual plants are generally not grown for both because when dill weed is desired, the plant is harvested before flowering. Dill oil is another important commodity from this plant that finds its greatest use in the pickling industry. This annual, native to the Mediterranean area and Europe, has finely dissected, feathery leaves and umbels with bright yellow flowers, making it an attractive garden plant.

The distinctive flavor of caraway seeds in rye bread is familiar to most people. Native to most parts of Europe, Asia, North Africa, and India, the biennial caraway, *Carum carvi,* is one of the oldest herbs known. The use of these seeds is thought to have originated with the ancient Arabs, who called the seeds *karawya,* which is the source of the English word. In addition to flavoring bread, caraway seeds are used in cheeses, soups, sausages, and a variety of meat or vegetable dishes. Anise, cumin, celery, coriander, and fennel are additional members of the Apiaceae also valued for their schizocarps.

The Mustard Family

Members of Brassicaceae (mustard family) are important vegetable crops, such as cabbage, broccoli, cauliflower, brussels sprouts, turnips, and radishes as well as two flavorful herbs or condiments, mustard and horseradish (table 17.3). This temperate cosmopolitan family is especially abundant in the Mediterranean area. Members of the family are easily recognized by their characteristic flowers, each with four petals arranged in a cross, which accounts for the old family name, the Cruciferae.

Seeds of *Brassica nigra, B. alba,* and related species are the source of one of the most familiar seasonings, mustard.

Mustard produced from the seeds of *B. alba,* white mustard, is somewhat milder tasting than the more pungent product of *B. nigra,* black mustard. Both plants are annuals native to Europe and western Asia that have a long history of use as both condiments and medicinals. Forerunners of today's prepared mustard can be traced back to the late Middle Ages, when crushed mustard seed was mixed with vinegar to prepare a sauce. In many countries today, prepared mustard is the primary product of these species (recall that the bright yellow color of many brands of prepared mustard is due to turmeric). Whole seeds and ground seeds are two other ways that mustard is marketed; the whole seeds are primarily for pickling, but the ground mustard finds its way into hundreds of recipes. The sharp, tangy taste of mustard is the result of reactions between sinigrin (in black mustard) or sinalbin (in white mustard) and myrosin, an enzyme. In the presence of water, these sulfur-containing components react to produce volatile oils that give mustard its characteristic taste. Unless prepared mustard is acidified (as with vinegar), its flavor quickly deteriorates.

Recently, the use of mustard has been increasing. Among herbs and spices, mustard is currently the world's most heavily traded commodity, and specialty mustards are some of the trendiest new items in gourmet shops and upscale markets. These condiments are made by blending mustard with other herbs and spices or even with fruits. In addition, the popularity of Dijon-style mustard (made from ground black mustard seeds and white wine) is beginning to rival that of the familiar yellow varieties.

Horseradish, *Armoracia rusticana,* is an herbaceous perennial native to Europe that has been cultivated for centuries (fig. 17.12). Although horseradish has a long history of use as a medicinal plant, its use as a condiment dates only from the Middle Ages in Denmark and Germany. Horseradish sauce is prepared from the taproots, which are white and faintly resemble a large, misshapen carrot. The pungent aroma and hot, biting taste are due to the interaction of two components, sinigrin and myrosin (identical to those found in black mustard), which combine to produce a volatile oil. In intact roots, the volatile oil is not produced because the two components occur in separate cells; however, when the roots are scraped or grated, the components are liberated and free to interact. The volatile oil that is produced diffuses easily on exposure to air, and the grated condiment quickly loses its pungency.

Sinigrin, which contributes to the potency of both mustard and horseradish, is similar to the chemical responsible for the caustic effects of mustard gas, a poisonous gas used in chemical warfare during World War I and during the Iran-Iraq war of the 1980s.

Although its common name implies that it is another type of horseradish, Japanese horseradish (*Wasabia japonica* = *Eutrema japonica*) is indeed a member of the mustard family but is only distantly related to the common horseradish. It is native to the mountain stream banks of Japan and has become more available in the West as Japanese cuisine becomes widespread. Real wasabi is one of the most uncommon and difficult vegetables in the world to grow. Although a perennial, it is most often cultivated in water for 1 to 2 years before it is harvested. Only the rhizome is used. In Japan, wasabi is highly desired because of its fiery pungency and is commonly eaten with seafood. It may also be used as an ingredient in dressings, dips, sauces, and marinades. Prepared fresh, it is grated and served with raw fish. It can also be used in cooking as a dried powder because, unlike the common horseradish, it retains its taste even when dried.

Some medicinal properties have been attributed to wasabi. Isothiocyanates, the compounds that confer the pungent taste to wasabi (also horseradish and mustard), have been shown to be antimicrobial, and eating wasabi regularly may prevent tooth decay. Other benefits ascribed to isothiocyanates are cancer prevention and alleviation of asthma symptoms.

Thinking Critically

Many essential oils have antibacterial and antifungal properties.

How would you investigate whether or not a plant extract can be used as an antiseptic?

The Pungent Alliums

Members of the Amaryllidaceae (Amaryllis family) are found worldwide and are largely herbaceous perennials that arise from rhizomes, bulbs, or corms. A single genus, *Allium* from central Asia, is the source of many familiar zesty herbs (table 17.3): onions (*A. cepa*), garlic (*A. sativum*), leeks (*A. porrum*), shallots (*A. ascalonicum*), and chives (*A. schoenoprasum*) (fig. 17.13). The following discussion will focus on a brief consideration of the history, botany, chemistry, and medicinal and culinary use of onion and garlic. Onion, a

Figure 17.12 The large taproots of horseradish are grated to make horseradish sauce.

Figure 17.13 Inflorescences of chives (*Allium schoenoprasum*).

biennial, produces a single large bulb, whereas the perennial garlic produces a composite bulb, with each clove of garlic called a **bulblet.** Both species produce an inflorescence consisting of small, three-parted flowers, which also make them attractive ornamentals.

The pungent flavor and scent of onion and garlic are due to the presence of various volatile sulfur compounds that are released when the tissues are cut. In intact onions and garlic, these compounds are inactive and are released only through the action of the enzyme allinase. The main active ingredient released from garlic is allicin; that from onion is known as the lacrimatory factor. These molecules are highly reactive and easily change into numerous other sulfur-containing compounds that have a wide range of biological effects. The most familiar effect is the tearing caused by the lacrimatory factor of onions. (Cutting onions under running water will wash away some of the lacrimatory factor, and cutting them cold will retard the enzyme activity.) Another undesirable effect of these sulfur compounds is the lasting aroma of onion and garlic on the breath after ingestion. After digestion, the sulfur compounds are transported by the bloodstream into the lungs, where they may diffuse into the exhaled air. The antibacterial and antifungal effects of garlic and onion extracts are also examples of the biological action of these sulfur compounds.

Onion and garlic are two of the oldest cultivated plants used for both culinary and medicinal purposes, with a history even predating the ancient Egyptian civilization. The Ebers Papyrus, an Egyptian medical reference from 3,500 years ago, listed 22 uses of garlic as a treatment for various ailments. In the herbal medicine traditions of both India and China, onions and garlic have been prescribed for centuries as remedies for numerous conditions. Modern research has shown that these folk remedies have a sound scientific basis. The organic sulfur compounds from garlic do inhibit the growth of many disease-causing bacteria and fungi and inhibit the formation of blood clots. (Blood clots have been linked to certain forms of cardiovascular disease as the cause of heart attacks and strokes.) Researchers have found that allicin blocks certain enzymes that infectious organisms use to damage or invade tissues. Allicin has also been found to interfere with enzymes that synthesize cholesterol in the body. This may be why several earlier studies found that garlic consumption lowers cholesterol levels, especially LDL levels, in the test tube and in laboratory animals. However, a recent study questioned these results in the human body. Approximately 200 adults with moderately high blood cholesterol levels were divided into four treatment groups. Each group was given raw garlic (one clove), powdered garlic, aged garlic extract, or a placebo for 6 days per week over a period of 6 months. Blood cholesterol levels showed no significant decrease in either LDL cholesterol or triglyceride level in any of the treatments. Critics contend that eating garlic will maintain good cholesterol levels and prevent the development of high cholesterol, but there were never any claims that consumption of garlic could reverse already elevated cholesterol levels.

Americans have gone wild over garlic. Whether the reason is the awareness of the health benefits or the increased interest in ethnic and gourmet foods, the result is that in the United States garlic consumption tripled from the 1970s to the 1990s. Garlic-laden dishes are now staples at many restaurants. The Stinking Rose restaurants in San Francisco and Los Angeles specialize in garlic dishes from appetizers to desserts; you can begin your meal with garlic roasted in olive oil and spread on sourdough bread and polish it off with garlic ice cream. Garlic lovers can also attend the many garlic festivals that are springing up across the United States each year.

Sweetleaf and miracle fruit, two herbs valued for imparting a sweet taste to foods, are discussed in A Closer Look 17.2: Sweet Talk.

This chapter focused on the history and culinary uses of herbs and spices; however, these plants also had widespread use as dyes (see A Closer Look 18.2: Herbs to Dye For), perfumes (see A Closer Look 5.3: Alluring Scents), cosmetics, and medicinals (see A Closer Look 17.1: Aromatherapy). In fact, the foundation of modern medicine is the herbal medicine practiced by many societies, and this aspect of herbs will be examined in detail in Chapter 19.

A CLOSER LOOK 17.2

Sweet Talk

Two herbs have been in the spotlight of late because they impart a sugary taste to foods: stevia and miracle fruit. Stevia is a natural, noncaloric sweetener that has become popular recently in the United States but has a long history of use in South America. *Stevia* is also the name of a group of small shrubs and herbs, members of the Asteraceae (sunflower family), which are found in tropical to subtropical regions in western North America and South America. The most important species in cultivation is *Stevia rebaudiana*, commonly called sweetleaf, a small perennial, woody shrub native to eastern Paraguay and characterized by opposite, sessile leaves with only four flowers in the head inflorescence. *S. rebaudiana* is not common in the wild but has been extensively cultivated in China, Malaysia, Singapore, South Korea, India, Taiwan, Thailand, Paraguay, Brazil, the United States, Canada, and Europe.

Also known as *yerba dulce*, the leaves of *S. rebaudiana* have long been chewed as a treat or dried and added to sweeten tea and medicines by indigenous peoples in South America. French chemists first isolated stevioside in 1931 as the active ingredient. Stevioside has been identified as a glycoside (see Chapters 1 and 19) composed of approximately 60% glucose and 40% steviol, a chemical unique to this group of plants. Steviol has no taste at all but, combined with glucose, it produces an extremely sweet sensation that is said to be up to 300 times sweeter than sucrose. Five other glycosides have since been identified in the leaves of *S. rebaudiana* as contributing to the sugary taste: rebaudiosides A, D, and E and dulcosides A and B. Rebaudioside A has been reported as the sweetest, least bitter, and most stable of the six compounds.

Stevioside has been used as an alternative sweetener in Japan since the 1970s when artificial sweeteners, such as cyclamate and saccharin, were banned because some studies linked them to cancer. It is also popular in China and South Korea as well as in Brazil, Argentina, and Paraguay as a sweetener for soft drinks, ice creams, cookies, candies, pickles, and chewing gum. Stevioside has zero calories and is a popular weight-loss aid. Because stevia has been shown to reduce blood glucose levels and enhance insulin secretion in patients with type 2 diabetes, it has been used extensively in diabetic products. Stevioside, unlike table sugar, does not contribute to dental plaque or cavities because it is not fermentable by bacteria. In 2008, stevioside and rebaudioside A were approved for use as food additives in the United States.

Miracle fruit is a red berry (box fig. 17.2) of the shrub *Synsepalum dulcificum* (Sapotaceae or Sapodilla family), native to West Africa, that can modify sour tastes such as those found in citrus fruits or vinegar and make them taste sweet. It has a long history of use in West Africa to improve the palatability of sour maize dishes and to sweeten acidic beverages. The active ingredient has been identified as miraculin, which in itself is not sweet. Miraculin is a protein that can bind to taste receptor cells found in the taste buds of the tongue. At low pH, miraculin binds the hydrogen ions released by acidic foods. This action blocks the hydrogen ion receptors on tastant cells and prevents detection of sourness. Instead miraculin activates the sweet receptors in the taste bud; the effect of intense sweetness can last from 1 to 2 hours until the miraculin is washed away from the tongue by saliva. Recently, the gene that codes for miraculin has been isolated from *S. dulcificum* and used to create transgenic tomato plants that make miraculin, improving tomato flavor by reducing the acidity of the fruit.

Box Figure 17.2 The berry of miracle fruit masks sour flavors in foods with the sensation of sweetness.

CHAPTER SUMMARY

1. The characteristic scents of aromatic plants, such as herbs and spices, are due to volatile substances called essential oils, which can be extracted to obtain the essence of the plant. Spices generally have a tropical origin and have been used for a variety of purposes since ancient times.

2. The ancient Egyptian, Greek, and Roman civilizations used spices extensively for cooking, embalming, and medicine and as perfumes. Trading routes for the purpose of obtaining these spices from India and China have a long history. Exotic spices became rare in Europe during the Dark Ages, but trade routes to the East were reopened during the Crusades. Marco Polo's account of his travels to the court of Kublai Khan renewed the European desire for spices. The Portuguese, under the direction of Prince Henry the Navigator, were the first European power to travel by sea to India via the Cape of Good Hope. Christopher Columbus sailed west in an attempt to reach the spice-rich lands of the East.

3. The history of spices is the history of domination of the spice-rich lands by various European powers. Spice plantations were established in European colonies to cash in on the lucrative spice trade. Some of the most sought-after spices were cinnamon, black pepper, and cloves. Capsicum peppers, vanilla, and allspice from the New World were added to the list of spice treasures.

4. Herbs are usually the aromatic leaves of plants from temperate regions. The mint family is the source of many familiar herbs: peppermint, spearmint, marjoram, and oregano, to name a few. The parsley family is another important herb family; herbs in this family include parsley, dill, and caraway. Mustard and horseradish are members of the mustard family. Allicin is the sulfurous compound in onion, garlic, and related plants in the onion family that is the source of tearing and unpleasant aromas but also medicinal effects.

5. The desire for sweetness in foods has stimulated interest in the herbs sweetleaf and miracle fruit.

REVIEW QUESTIONS

1. What are secondary plant products? How have they been a benefit to society?
2. How did spices influence world history?
3. Distinguish between herbs and spices.
4. What plant parts are processed to prepare the following: cinnamon, nutmeg, cloves, black pepper, vanilla, parsley, mustard?
5. Discuss the medicinal value of herbs.
6. What familiar herbs are obtained from the mint and parsley families?
7. Although Columbus was searching for a source of black pepper, he introduced Europeans to the capsicum peppers. How have these measured up as a source of valued seasonings?
8. Fruit-eating birds, such as cardinals, finches, and mockingbirds, disperse the seeds of the chiltepin, wild chilies, and some of the hottest peppers known. In contrast, small mammals assiduously avoid these hot fruits. What does this observation tell you about the sense of taste in birds? In mammals?
9. How does miracle fruit which does not taste sweet impart a sweetness to sour foods?

CHAPTER

18

Materials: Cloth, Wood, and Paper

At the restored Shaker village at Pleasant Hill in Kentucky, a craftsman crafts brooms from broom corn and jute.

KEY CONCEPTS

1. Many textiles are woven from plant fibers; one of the most important discoveries made by ancient peoples was to weave cloth from plant fibers.
2. Plants have also been the source material for woven baskets, tools such as brooms, and natural dyes.
3. Wood and wood products, which supply construction materials and fuel to much of the world population, rank right behind food plants in terms of their importance to humanity.
4. Paper, the chief medium of communication in contemporary society, is produced from the pulp of trees and other plants.

©Karen McMahon

CHAPTER OUTLINE

Fibers 309
   Types 309
A CLOSER LOOK 18.1 A Tisket, a Tasket: There Are Many Types of Baskets 310
   King Cotton 312
   Linen: An Ancient Fabric 315
   Other Bast Fibers 316
   Miscellaneous Fibers 317
   Rayon: "Artificial Silk" 319
   Bark Cloth 319
A CLOSER LOOK 18.2 Herbs to Dye For 321

Wood and Wood Products 323
A CLOSER LOOK 18.3 Good Vibrations 324
   Hardwoods and Softwoods 326
   Lumber, Veneer, and Plywood 327
   Fuel 329
   Other Products from Trees 329
   Wood Pulp 330
Paper 331
   Early Writing Surfaces 331
   The Art of Papermaking 333
   Alternatives to Wood Pulp 333
Bamboo 334
Chapter Summary 335
Review Questions 336

In chapters 10 through 17, we examined the use of plants to provide sustenance, but plants clearly furnish society with more than food and drink. Since earliest times, they have also been used to provide shelter, clothing, and fuel. Later, as civilizations developed, plants also supplied various media for written communications. This chapter will present a brief examination of some of the materials that plants contribute to society.

FIBERS

Plant fibers are one of the most useful plant materials; they have been used for millennia to make cloth, rope, paper, and numerous other articles. Baskets are also made from weaving plant materials and are discussed in A Closer Look 18.1: A Tisket, a Tasket. The term *fiber* actually has several meanings. Recall from Chapter 3 that, anatomically, fiber cells are long and tapering, with extremely thick secondary cell walls. The composition of the fiber cell wall is chiefly cellulose, although lignin, as well as tannins, gums, pectins, and other polysaccharides, may also be present. The plant fibers of commerce often are not individual fiber cells per se but stringy, elongated masses of plant material that are actually collections of fiber cells or entire vascular bundles.

The most valuable fibers are those that are nearly pure cellulose and white. Cellulose is an extremely strong material, with properties of tensile strength that rival that of steel. (Tensile strength measures the resistance of a fiber to tearing apart when subjected to tension.) Fibers that are high in lignin are generally of poorer quality, browner, and of lower mechanical strength.

A breakthrough discovery in the field of plant fiber occurred when the gene that makes cellulose in plants was identified. The *RSW1* gene codes for the enzyme responsible for cellulose production in plants. It was isolated in *Arabidopsis*, a member of the mustard family that is used extensively in genetic investigations. This gene had been identified in the cotton plant previously but its function had not been definitively proved. Isolation of the gene is the first step in manipulating the cellulose content of plants in ways that may benefit humanity. It may be possible in the future for scientists to add more cellulose to fiber crops to make a stronger cloth.

Types

Fibers can be classified according to their use. Fibers that have been used to weave cloth are known as textile fibers; cordage fibers are used in making rope; and filling fibers are used as stuffing in upholstery and mattresses. Fibers come from many sources. Natural fibers are mainly of plant or animal origin, although mineral fibers, such as asbestos, can also be used. Animal fibers, such as wool and silk, have a protein makeup, while plant fibers are composed primarily of cellulose. Some synthetic fibers are created using natural sources as a base. For example, cellulose from wood pulp is processed chemically into rayon.

Plant fibers can be classified according to where they are found on a plant. **Surface fibers** are found on the covering of seeds, leaves, or fruits; cotton cloth is made from seed hairs covering the surface of cotton seeds. Linen and ramie are made from **bast** or **soft fibers**, clusters of phloem fibers found in the inner bark of dicotyledonous stems. **Hard fibers** or **leaf fibers** are obtained from the vascular bundles or veins in leaves; these consist of both xylem and phloem as well as the ensheathing fibers. Monocotyledonous leaves are the usual source of hard fibers; sisal and Manila hemp are examples of this type of fiber. Hard fibers usually have a higher lignin content than soft fibers (table 18.1).

Table 18.1
Important Plant Fibers and Their Uses

Plant	Scientific Name	Family	Type of Fiber	Use
Coir	*Cocos nucifera*	Arecaceae	Surface	Filling, cordage
Cotton	*Gossypium hirsutum* *Gossypium barbadense*	Malvaceae	Surface	Cotton cloth
Flax	*Linum usitatissimum*	Linaceae	Bast	Linen
Hemp	*Cannabis sativa*	Cannabaceae	Bast	Hemp cloth, canvas, cordage
Jute	*Corchorus* spp.	Malvaceae	Bast	Burlap
Kapok	*Ceiba pentandra*	Malvaceae	Surface	Filling
Manila hemp	*Musa textilis*	Musaceae	Leaf	Cordage
Piña	*Ananas comosus*	Bromeliaceae	Leaf	Cloth
Ramie	*Boehmeria nivea*	Urticaceae	Bast	Ramie cloth
Sisal	*Agave sisalana* *Agave fourcroydes*	Asparagaceae	Leaf	Cordage, matting

A CLOSER LOOK 18.1

A Tisket, a Tasket: There Are Many Types of Baskets

Perhaps the inspiration for basket making came from observing the natural world. Early humans may have first utilized discarded birds' nests until they devised a method to construct their own containers. It is known that the techniques of basketry were developed across the world because baskets have been found in every culture. Although few ancient baskets exist today because the materials used in basketry are perishable, some remnants have been found in arid regions. Ancient Egyptian baskets have been discovered that date back 8,000 to 10,000 years, and baskets from the Great Basin region of Utah and Nevada in the American Southwest have been dated to be as old. These findings indicate that basketry is one of the oldest of human crafts and predates pottery. Although homemade natural baskets have largely been supplanted in the industrialized world by synthetic containers, there is a renewed interest in the artistry of basket making in part due to the popularity of country decor, the revival of Native American culture, and the renewed appreciation of handcraftsmanship.

Baskets come in all shapes and sizes to serve the practical needs of a culture (box fig. 18.1). The Shaker-designed catheads, and the smaller kittenheads, are baskets that rest on four outside points and thus resemble a cat's profile. The skep, or beehive, is a large, conical basket woven from rye straw and placed in the garden to keep bees for their honey and pollinating activities. The rattle-lid basket is a lidded basket ingeniously designed by Native Americans of the Pacific Northwest. The hollow lid of the basket was filled with pebbles, seeds, or other small objects that rattled whenever the basket was moved or opened. Baskets have been constructed to gather wild fruits, grind corn, winnow rice, collect eggs, trap fish, cradle babies, sow wheat, transport goods, and even boil water. Baskets have been made into hats that shield the wearer from the rain. The Pomos of California were renowned for their jewel baskets decorated with feathers and shells and given as gifts to special people or as a memento for an important event. Baskets have been made to commemorate special events in the life of a person. Among the Hopi of the American Southwest, the bride's family and friends make flat, circular plaques of wicker to present to the groom and his family.

In many cultures, basket making was a womanly task. There were exceptions, however. In the Pomo tribe of what is now northern California, both men and women made some of the finest North American baskets, but the type of basket made was divided along gender lines. Men made the utilitarian types of baskets, and women made decorative and ceremonial baskets.

Basketmakers relied on the plants available locally for the materials to construct their baskets (table 18.A). In the tropics, baskets were constructed from palm and coconut leaves; in temperate forests, the materials of basketry were the bark, twigs, or roots of trees. In prairies, grasses were utilized. A basket can be constructed of a single plant material, but often several plants are utilized.

There are three basic techniques to basketry: plaiting, twining, and coiling. Although most societies specialized in one technique, more than one method can be found in a single basket. One of the most familiar methods is plaiting. Plaiting requires wide, flat strips. Warps are the strips of material that either are placed vertically or radiate out from the center. The warps make up the framework of the basket. Wefts are the more flexible, horizontal strips that are woven over and under the warps. Plaiting was the dominant basket-making technique in much of the eastern United States. Thin, flat strips of wood, called splints, were the primary material. Splints of black ash (*Fraxinus nigra*) were preferred because newly cut trunks of young trees, when pounded with a mallet, split naturally along growth rings to form long strips. Other woods used for splints were white oak, hickory, and maple. Splint basketry was also common in Europe, and basket-making techniques were shared between the European settlers and the Native Americans in the eastern region.

Extracting the Fiber

Various methods are employed to extract fibers from the source material. Surface fibers are usually separated mechanically from plant material by **ginning**, in which machines tear the fibers loose. Many soft fibers are extracted from the stems through **retting**, a process that uses microbial action to degrade away the soft tissues, leaving the tough fiber strands intact and freed. The process of **decortication**, in which the unwanted tissues are scraped away by hand or machine, is the way hard fibers are commonly extracted.

Spinning into Yarn

Once fibers have been freed from the source material, they are cleaned of plant materials and dirt. Then the fibers are combed and laid parallel to each other to form a strand. Next, the strand is stretched or pulled with the fingers or from the

Box Figure 18.1 Examples of basketry. (a) Coiled basket of yucca leaves from the American Southwest. (b) Birch bark basket from Siberia.

Table 18.A Some Popular Plant Materials Used in Basketry

Plant	Part Used
Bamboo, sweetgrass, esparto grass, reeds, sedges, rushes, cattails, rattan	Stem
Black ash, black gum, white oak, chestnut, hickory, maple	Wood splints
Birch, cedar, willow, cherry	Bark
Palm, coconut, yucca, pandanus, sisal, raffia	Leaves
Rye, rice, oat	Straw
Pine, spruce, sedge, yucca	Roots
Willow, sumac, cottonwood	Shoots

In twining, two or more strands of thin, flexible weft are woven around a warp, which is also thin and flexible. The native inhabitants of the Aleutian Islands, a chain of small treeless islands that stretches from the southwest coast of Alaska toward Russia, were some of the finest basketmakers in the world. They used wild rye grass (*Elymus mollis*) as both the warp and the weft. Their baskets were so finely and tightly woven that they resembled linen cloth.

Making a coiled basket requires plant material that is naturally long or that can easily be lengthened by constantly adding material. The plant material is manipulated to form a continuous spiral, and each succeeding row is secured to the preceding one with stitches. The Hopi of the North American Southwest produce baskets with very wide coils, 2.54 cm (1 in.) or wider, by bundling the plant material. Coils can also be made by braiding the plant material together, a technique that makes a stronger basket than one that is made from loose coils.

Often, baskets were decorated. In the southwestern region of the United States, the Hopi, Navajo, and Apache weave the naturally black devil's-claw (*Proboscidea parviflora*) with light-colored willow or grass to make geometric or abstract designs. The Cherokee of the southeastern United States usually dyed splints and alternated them with undyed splints for a contrasting pattern. Baskets have been adorned with shells, feathers, and in the northeastern region of North America, porcupine quills. The famous lightship baskets of New England had lids decorated with carved ivory or ebony.

weight of a spindle, and the individual strands are twisted together to form the yarn or thread. The simplest method of spinning fibers into yarn is to roll the fibers together between the palms of the hands or against the thigh. The invention of the spindle made spinning easier. It is the rotation of the spindle that twists and holds the fibers together, forming the yarn. The finished yarn is then woven into cloth or used for other purposes.

Sometime around A.D. 750 in India, the first spinning wheel was invented by mounting a spindle on a frame; the spindle rotates when the wheel is turned. The spinning wheel was faster than the hand spindle and made the yarn more uniform. Chinese inventors later added a treadle, a pedal to operate the spindle by foot rather than by hand, that further speeded up the process. One drawback to these early spinning wheels was that the spinner had to stop periodically to wind yarn on the

Figure 18.1 A woman spins fibers into yarn with the flyer spinning wheel at the restored Shaker Village at Pleasant Hill, Kentucky.

Figure 18.2 Cotton bolls mature 50 to 80 days after fertilization. In an opened cotton boll, the fluffy, white mass consists of seed fibers.

spindle. The invention of the flyer in sixteenth-century Europe overcame this problem. A U-shaped device affixed to the end of the hollow rod that holds the spindle, the flyer turns at a different rate than the spindle (fig. 18.1). In this manner, yarn can be wound and twisted on the spindle at the same time. During the time of the Industrial Revolution, spinning machines using multiple spindles mechanized the process of spinning.

King Cotton

The process of rendering cotton fibers into cloth was discovered independently during prehistoric times in both the Old and New Worlds. Archeological evidence indicates that cotton was harvested from wild populations in coastal Peru as early as 10,000 years ago and domesticated by 4,500 years ago. It was also grown and used by native peoples of the American Southwest. In the Old World, cotton cloth dates back 5,000 years on the Indian subcontinent. Cotton spread westward from India to the nations of Assyria and Babylonia. The Romans and ancient Greeks knew of cotton, but its use was limited because they preferred linen. Cotton became a Muslim industry in the Near East in the ninth and tenth centuries, and *muslin,* the term for a fine-weaved cotton, reflects this early association. It was the Arabs who introduced cotton cultivation to Muslim Spain. The Arabic influence also is seen by the fact that the word *cotton* is derived from the Arabic *qutn.* Cotton was introduced into Florida in 1556, and by 1607 it was grown as a commercial venture in Virginia. Less than 100 years later, cotton was a significant crop in the southern colonies. Cotton, for the most part, was a minor cloth in Europe until the introduction in the eighteenth century of New World varieties that, with their longer seed hairs, were more suitable for spinning. Currently, the top cotton-producing country in the world is the People's Republic of China. Other major cotton-producing countries are the United States, India, the republics of the former Soviet Union, Pakistan, Egypt, Brazil, and Turkey.

The Cotton Plant

Today cotton is the most popular natural fiber, accounting for half the world's textiles, and is one of the most economically important nonfood plants in the world. The cloth itself is woven from seed fibers obtained from species of *Gossypium.* The cotton plant is a member of the Malvaceae, or mallow family, of tropical and temperate distribution. There are 20–30 species in the genus *Gossypium* native to parts of Asia, Africa, Central and South America, and Australia.

Cotton is a shrubby plant with palmately lobed leaves. In tropical climates, most species are perennials, but in temperate zones they are grown as annuals. Cotton grows best in a warm climate, with sufficient water during the growing season. Depending on the species, flower color ranges from white to purple. The fruit of cotton is a capsule—in commercial circles, a boll (fig. 18.2)—and when it splits open along five seams, it reveals a white mass of fibers. The fibers are hairs that extend from the seed coats of each of the 10 or so seeds in every fruit. As many as 20,000 seed hairs may grow from a single seed. Under natural conditions, the fibers enable seeds to become airborne and thus dispersed by wind currents. Each cottonseed hair, actually a single seed coat cell, is a twisted, hollow strand of cellulose up to 7.62 cm (3 in.) in length that flattens at maturity (fig. 18.3). The hairs begin to develop from the integument of the ovule after the fertilization of the flower. As the hair grows, the wall continuously thickens to the point that the lumen in

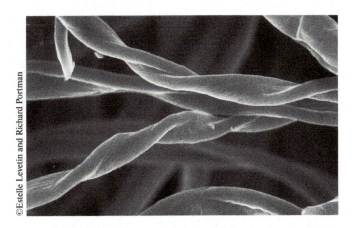

Figure 18.3 Scanning electron micrograph of cottonseed fibers reveals their natural twist, which makes them suitable for spinning.

> **Thinking Critically**
>
> Cultivation of naturally multicolored cotton in the Andean region of Peru dates back to the pre-Columbian era. Unfortunately, many of the farmers have abandoned the *algodon nativo* cultivar and instead plant coca bush for the lucrative cocaine trade.
>
> *Devise an economic plan to encourage Andean farmers to cultivate colorful cotton rather than the coca bush.*

a mature fiber is practically nonexistent. Two types of seed hairs cover the seed surface: the lint or staples (long, slender fibers) and linters (shorter, fuzzy hairs). High-quality cotton is made from the longest of lints. The purity of the cellulose (90%) and its natural twist make cotton an excellent fiber for spinning into yarn.

Old and New World Varieties

Commercially important species of cotton are *G. hirsutum* and *G. barbadense* from the New World and *G. arboreum* and *G. herbaceum* from the Old. From these, numerous cultivars have been developed. The Old World cottons are diploid species ($2N = 26$) and produce a short lint, about 1 to 2 cm (0.4 to 0.8 in.) in length. *Gossypium herbaceum* is a species that originated in southern Africa, eventually spreading to India and giving rise to *G. arboreum.*

Both New World species are tetraploids ($2N = 52$), and there is some evidence to indicate that they may be the result of a hybrid cross involving *G. herbaceum* and *G. raimondii* from northern Peru. *Gossypium hirsutum* is the predominant type grown in the world today. Also known as upland cotton, this species arose in Central America and Mexico and spread to South America. Lint from upland cotton varies from short to long, 2.2 to 3.0 cm (0.9 to 1.1 in.). *Gossypium barbadense* is believed to have originated from the Andean region of South America, where its use dates back to as early as 10,000 years ago. The species includes the cultivars known as Sea Island, pima, and Egyptian cotton. The lint of *G. barbadense* is especially long and silky, and its fibers are used to produce high-quality, luxury cotton cloth.

Commercial varieties of cotton that have naturally brown lint suitable for machine spinning have become available from native varieties developed by the natives of Mexico before the Spanish conquest. Fibers from these varieties produce a naturally colored cotton cloth of greens and browns, eliminating the need for chemical dyes.

The Cotton Gin

Cotton bolls mature 50 to 80 days after fertilization; defoliants are sprayed on the plants, leaving only the bolls on the plant for picking. Picking cotton was formerly a labor-intensive process, but today it is performed primarily by machine. The seed hairs may be picked from the open bolls, or the entire boll can be harvested. The harvested bolls or fibers are next sent to a gin. During ginning, the lint is removed from the seeds. At first, ginning was painstakingly done by hand. The first type of gin invented, probably in India, was a roller gin. Lint was drawn between two closely set rollers; the lint passed through the rollers easily but the seeds did not. Unfortunately, this gin did not work with all species of cotton. Today modern gins are larger versions of the saw gin, invented by Eli Whitney in 1793. The saw gin (fig. 18.4a) is essentially a roller studded with spikes and covered by a metal mesh. The spikes draw in the lint, but the seeds do not pass through the mesh and are left behind. Whitney's gin made it possible to separate cotton fiber from seed much more quickly. Before this invention, it took a whole day of ginning by hand to yield a single pound of cotton (fig. 18.4b). In contrast, a Whitney gin could produce 22.5 kilograms (50 pounds) of cotton fiber per day!

Eli Whitney's invention had its greatest effect in the cotton-growing states of the South. Before 1790, the growing of long-staple cotton varieties, those that could be ginned most easily by hand, was limited to coastal regions in the South. The soil and climate of the southern interior were suitable only for the cultivation of short-staple varieties. The cotton gin made the production of short-staple cotton economically feasible and expanded the Cotton Belt in the South. Mechanized cotton production also came of age at the same time that Europe's demand for raw cotton grew. Consequently, more fields in the South were planted with cotton, and with each new acre of cotton planted, the demand for slave labor to grow and process the cotton also grew. Slavery, which had been dying out, underwent a resurgence as a direct result of the cotton boom, as seen in Table 18.2. Immediately before the Civil War, cotton production was 4 million bales, making "King Cotton" the number one crop in the United States, a crop supported by enslaved labor.

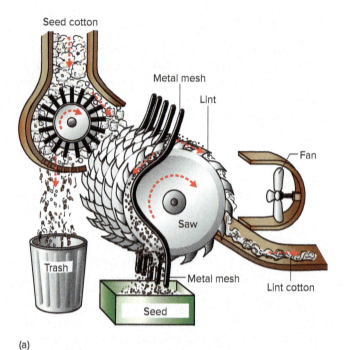

Table 18.2
Slavery and Cotton Production in the South from 1790 to 1860

Year	Number of Slaves	Cotton Produced (millions of pounds)
1790	697,124	1.5
1800	893,602	35
1810	1,191,362	85
1820	1,538,022	160
1830	2,009,043	331
1840	2,487,355	834
1850	3,204,313	1,001
1860	3,953,760	2,275

Source: Data from Bailey, Ronald, "The Other Side of Slavery: Black Labor, Cotton and Textile Industrialization in Great Britain and the United States," in *Agricultural History*, vol. 68, no. 2, 1994, 35–50.

Cotton seeds can be used to make livestock feed; also, the seeds are the source of cottonseed oil used as a cooking oil. The ginned fibers are packed into large bales (fig. 18.4c) and graded for quality. The graded bales are shipped to the appropriate yarn or cloth manufacturers, where the lint is straightened (carded) and then sorted into parallel bundles of similar size (combed) in preparation for spinning into yarn.

Finishing and Sizing

The finishing process may be applied to either the yarn or the woven cloth; finishes may alter the appearance or modify the function of the textile. Most plant fibers are bleached to remove the natural color, and cotton is no exception. Different methods of bleaching have been used throughout history, many of which sound bizarre by today's standards. The standard method in eighteenth-century England was to soak the cloth or yarn in sour milk and cow's dung, and then steep it in lye. A bath in buttermilk followed, and after washing, the cloth was spread on grass until exposure to the sun bleached it white. A breakthrough in the process came about in 1774 when the bleaching effect of chlorine was discovered.

Mercerization is a finishing process for cotton that improves its strength, luster, and affinity for dyes. Named after John Mercer, who patented the process in the mid-1800s, mercerization consists of passing cotton cloth through a bath of caustic soda (sodium hydroxide). The flat cotton fibers swell into a round shape, and the fiber becomes stronger and more lustrous. Mercerized cotton also has a greater affinity for dyes.

Permanent press is an example of a shape-retentive finish. Cellulose fibers, such as cotton, wrinkle and crush easily. A permanent press finish involves the application of chemicals that cross-link the cellulose fibers. This gives the fabric a

Figure 18.4 (a) Operation of the cotton gin. The saw draws in the lint, but the seeds are held back. (b) Unginned (left) and ginned (right) cotton fiber. (c) Ginned cotton fiber baled and ready for shipment.

built-in memory, so that the shape of the garment is retained even after laundering, with little or no ironing required.

Sizing is the application of materials to the yarn or fabric that produce stiffness or firmness. It makes the fabric smoother and gives increased sheen; it also protects the fabric from soiling. Cotton is usually sized with starch, dextrin, or resin.

Bioengineered Cotton

Research has advanced rapidly in the development of transgenic cotton hybrids (see Chapter 15). Bollgard is the trade name for the first Bt cotton hybrid created by Monsanto, a U.S. company based in St. Louis. Bt hybrids incorporate the toxin-producing gene *Cry1Ac* obtained from the bacterium *Bacillus thuringiensis*. Bt toxin is an effective insecticide against several particularly noxious pests of cotton: cotton bollworm, pink bollworm, spotted bollworm, and tobacco budworm. Since their introduction in 1996, Bt cotton hybrids have become one of the most widely planted transgenic crops.

In the United States and China, where pesticide use is common and crop loss due to insect pests ranges from 12% to 15%, Bt cotton yields are only 10% higher compared to conventional varieties, although Bt cotton does reduce the use of costly chemical pesticides, which results in both economic and environmental benefits (see Chapter 26). Bt cotton in India, however, presents a much different outcome. Bollworm pressure in India is high; researchers estimate that 50% to 60% of the cotton crop is lost to pests, yet the use of expensive chemical pesticides in the small farms of India is limited. Averaged over 4 years, Bt cotton harvests had a 60% greater yield than did non–Bt cotton hybrids. The non–Bt cotton had to be sprayed with pesticides three times as often as the Bt hybrids. (Bt cotton still must be sprayed with pesticide because the Bt toxin does not cause 100% mortality in the American bollworm and does not provide resistance against sucking insect pests, such as aphids and whiteflies.) Bt cotton is expected to have comparable yields over conventional varieties in other developing regions in South and Southeast Asia and sub-Saharan Africa.

Agracetus, another biotechnology company, is working on transgenic cotton plants that combine the breathability and feel of cotton cloth with the low-maintenance and heat-retaining properties of polyester. They have created transgenic cotton plants that fill the hollow middle of the cellulose seed hairs with a small amount of polyester material. A cotton-polyester blend fiber harvested directly from the field may not be too far off in the future.

Linen: An Ancient Fabric

The oldest plant fiber used to make cloth may be flax (*Linum usitatissimum*) of the Linaceae, or flax family. The stem fibers of this annual have been woven to make linen for thousands of years. Recently 30,000-year-old wild flax fibers were discovered in a cave in the western part of the Republic of Georgia. The discovery indicates that Stone Age hunter-gatherers were using flax fibers to make cords for binding stone tools, weaving baskets, and spinning thread. Some of the fibers appeared to have been dyed in various colors including yellow, green, black, red, and blue. Apparently, the hunter-gatherers possessed extensive knowledge about natural plant dyes, which they used to make colorful textiles.

Linen was also known to many ancient civilizations in Mesopotamia, Assyria, and Babylonia. But it was ancient Egypt that was known as the land of linen. Fragments of Egyptian linen have been found that are 6,500 years old. Egyptian linen was worn by priests and royalty, used as mummy cloths, and exported to other countries as the material of choice for sails. The ancient Greeks and Romans grew some flax for linen. Prehistoric textiles from the American Southwest also indicated the use of flax. In medieval Europe, Flanders became the first important linen center because Belgium has an ideal climate for growing flax. Ireland is another country noted for the quality of its linens since the seventeenth century. Today linen accounts for only 2% of the world's textiles; the leading areas of flax cultivation are the republics of the former Soviet Union, the People's Republic of China, and western Europe.

The Flax Plant

Flax is delicate in appearance; its straight, slender stems support sessile gray leaves and bell-shaped flowers of white or blue (fig. 18.5a). The fruit, or boll, is a capsule containing 10 brown seeds. Flax is harvested after 100 days when the stalks are a golden color. Two types of flax are grown commercially: one for its seed oil (linseed oil) and one for its fibers. Linseed oil is used in the manufacture of paints, stains, varnishes, and linoleum. The oil cakes are used as livestock feed. The oil has been used medicinally as a laxative and poultice. Oilcloths, which predated plastic tablecloths, were made by applying linseed oil to canvas.

Unbranched varieties of flax are grown for linen production. The stems in these varieties reach a height of 1 meter (3 feet), with fibers as long. Flax is a bast fiber, composed of the phloem fiber cells found in the stem. Each flax stem contains between 15 and 40 fiber bundles, with each bundle containing 10 to 40 fiber cells, each 1.3 to 3.8 cm (0.5 to 1.5 in.) long. The length of the fibers is preserved by pulling the flax stems up by the roots from the ground, either by hand or by machine (fig. 18.5b). Gathered into bundles, the harvested stems are stood upright in the field to dry. Next, seed bolls are removed from the now-dried flax straw by rippling, pulling the heads of the plants through a comb or a special type of threshing machine.

Processing Flax

Retting extracts the fibers from the stem by employing microbial decomposition to break down the outer part of the stem. The word *ret* is derived from the Dutch *roten*,

Figure 18.5 (a) A field of flax (*Linum usitatissimum*) in bloom. (b) Flax stems are harvested by pulling machines to avoid damaging the fibers. (c) Scutched flax fibers are inspected and sorted.

which means to rot. According to the traditional method of dew retting, the flax straw is spread over the fields to allow the interaction of dew or rain, sunlight, and microbes to rot away the soft, nonfibrous tissues of the stem. Dew retting usually takes 4 to 6 weeks, depending on the weather conditions. Immersing the flax straw in ponds or slow-moving rivers accomplishes the same purpose in 1 to 2 weeks. The retting process can be accelerated to a matter of days by soaking the straw in tanks of warm water. Chemicals can be added to the tanks to speed the retting process to a few hours. Most flax fibers are retted using a combination of dew and chemicals.

After the retted fibers have been dried by the sun or in a kiln (oven), they undergo processing with a flax brake, a type of pounding used to free the fiber from the stem. Most flax brakes have been replaced by machines that crush the dried and retted flax as it passes through rollers. Any shives, or stem particles, that remain are scraped off during scutching (fig. 18.5c). The now-freed fibers are hackled, or combed, to separate the short fibers (tow) from the long fibers (line). The flax fibers, which are naturally yellow (hence, the expressions *flaxen-haired* and *tow-headed*), are bleached in the sun or by chemicals. The flax fibers are then spun, either wet or dry, into yarn; wet spinning is said to produce the best-quality yarn.

Flax Fibers and Linen

Flax fibers consist of bundles of fiber cells held together by gums and pectins. Each fiber consists of about 70% cellulose, with the balance made up of waxes and pectins. The fibers are smooth and straight, not twisting as are cotton fibers. Line fibers are quite long, averaging 46 to 56 cm (18 to 22 in.) in length. Tow fibers are always 30 cm (12 in.) or shorter. The best flax fibers are a pale yellow and require little or no bleaching. Flax fibers are naturally lustrous (due to the waxes) and extremely strong. Linen is durable and is used for both clothing and home furnishings, such as drapes and upholstery. The flax fiber tends to draw moisture, a characteristic that has made linen a choice fabric for towels. On the downside, the fibers lack elasticity and resilience, making linen fabrics prone to wrinkling. In the process of cottonization, the individual fiber cells are separated out by dissolving the pectins between the cells. Flax fiber processed in this way is softer and can be blended with cotton, rayon, wool, or silk.

Other Bast Fibers

Ramie, hemp, and jute are, like flax, bast fibers, and they have been used for a variety of purposes. Ramie (*Boehmeria nivea*) is a perennial shrub and member of the nettle family. Also called China grass, it has been cultivated for centuries in several Asian countries. After the ramie stalks have been cut, the bark is peeled or beaten away. Sun drying bleaches the fibers white; a bath of caustic soda removes pectins and waxes. Ramie fibers are some of the longest (45.7 cm, or 18 in.), strongest (eight times stronger than cotton), and most lustrous fibers. However, the fibers are somewhat brittle, and removing the large amounts of pectins and gums in the stems

Figure 18.6 Women retting jute fibers in Bangladesh.

is difficult. Ramie has recently made inroads into the Western world; its cultivation has been introduced into parts of Europe and North and South America. Ramie fibers are often blended with cotton or rayon in the making of sweaters and other knitted fabrics.

Jute to Burlap

Obtained from the stems of *Corchorus* species in the Malvaceae, or mallow family, jute fiber is used to make burlap, ropes, wall coverings, carpet backing, upholstery lining, and inexpensive clothing. The coarse "sackcloth" worn by penitents in medieval Europe was made from jute. Jute is an annual plant, native to Asia, that thrives under the monsoon conditions of the wet tropics. The stems must be retted to free the fibers. The fibers vary in color from yellow to brown and are difficult to bleach; for that reason, many jute fabrics are in natural colors. The fibers are not as strong as other bast fibers; they range from 1.7 to 6.7 meters (5 to 20 feet) in length but are usually broken into smaller fragments during processing. About 90% of the world's supply of jute comes from Bangladesh and India. Jute is of particular importance to Bangladesh (fig. 18.6) because it is the main export crop of one of the world's poorest nations.

Another Use for Cannabis

The source of hemp is a plant more often associated with drugs than fabrics: *Cannabis sativa* (see Chapter 20). The processing of hemp fiber is similar to that of flax. The natural color of the fiber is dark brown, and the length of the fiber varies widely. Staminate plants produce the longest fibers, up to 2 meters (6 feet) long; however, some fibers are as short as 2 cm (3/4 in.). Hemp fibers are used primarily to make industrial fabrics, such as canvas (the word is derived from *Cannabis*), ropes, and twines. Although jeans today are made from 100% cotton, their inventor, Levi Strauss, originally produced his working clothes from hemp. The hemp cloth was imported from Nimes, France, and was known as "serge de Nimes," which was later corrupted into *denim*. The globally recognized term *jeans* is a derivative of the French pronunciation of the Italian city of Genoa, another city known for its manufacturing of hemp cloth.

Industrial hemp was an agricultural staple for hundreds of years; its cloth was used in sailing ships and covered wagons. Its cultivation has been illegal in the United States for decades because of drug enforcement policies; however, finished hemp products may be imported. Recently, there has been a revitalization of the hemp fiber industry. Fashion trend starters, such as Calvin Klein and Giorgio Armani, are selling hemp clothing and bed linens at premium prices. Adidas even manufactures athletic shoes made with hemp cloth. There has been a grassroots movement to grow hemp fiber once more in the United States. Supporters for the legalization of industrial hemp point out that the fiber crop is extremely low in THC, the principal hallucinogenic in marijuana. Most states define industrial hemp as cannabis with levels of THC of 0.3% or less; West Virginia sets the limit at 1% THC or less. The cultivation of industrial hemp has also been championed by the eco-friendly consumer because it requires neither the irrigation nor the pesticide application that cotton does. Canada made hemp cultivation legal in 1998. At the end of 2018, the Agricultural Improvement Act, also known as the Farm Bill, became law; one of its provisions was to legalize the production of industrial hemp in the United States. The Navajo Nation passed a motion that permits hemp to be grown on tribal lands. The DEA (Drug Enforcement Agency) has not yet issued licenses permitting U.S. farmers to grow what the federal government considers a controlled substance.

Thinking Critically

Drug enforcement agencies long argued that growing hemp for fiber should be prohibited because they feared that it would be easy for pot growers to camouflage drug plants in hemp fields.

Do you believe that drug enforcement agencies are correct in arguing that lifting the ban on hemp cultivation will make it more difficult to curtail marijuana use?

Miscellaneous Fibers

Manila Hemp

Hemp should not be confused with a leaf fiber, Manila hemp, much valued for making seaworthy ropes. Manila hemp, or abaca, is obtained from *Musa textilis,* a member of the

Figure 18.7 Traditional apparel in the Philippines is woven from leaf fibers of the piña, a close relative of the commercial pineapple.

banana family. The fibers are extracted from the huge petioles; the fibers themselves can be up to 5 meters (15 feet) in length. Their natural color varies from off-white to dark brown; however, they can be bleached and dyed a wide variety of colors. In the Philippines, where most Manila hemp is grown and produced, the fibers have been used to make lightweight clothing, cigarette filters, and teabags, but today most Manila hemp is made into marine rope.

Pineapple Cloth

The shirts, shawls, scarves, and other traditional apparel of the Philippines are made of a gossamer-like material known as piña fiber, which comes from the leaf fibers of the pineapple plant (fig. 18.7). Special varieties of pineapple (*Ananas comosus*) are grown for their leaves; in fact, fruit development is discouraged by pinching off young pineapples as they appear. Traditionally, the leaves are scraped with knives to uncover the fiber bundles. More scraping and beating cleans and separates the fibers. Machinery and chemical retting are modernizing the labor-intensive process. Piña fibers are 5 to 10 cm (2 to 4 in.) long, white, soft, and lustrous.

Sisal

Sisal, another of the leaf fibers, comes from the desert *Agave*. Two species are commonly used: *Agave sisalana* (sisal) and *A. fourcroydes* (henequen). The fibers of these plants, native to Central America, were originally used to make rope and coarse garments. Today the fiber is used primarily for making rope, string, and floor mats. Sisal fibers are approximately 1 meter (3 feet) long (fig. 18.8).

Figure 18.8 (a) Leaves of *Agave sisalana* (shown with flowering inflorescences) are the source of fibers to make sisal used in rope and floor mats. (b) Sisal fibers are laid out to dry in the sun in Mexico.

Kapok

Kapok is a surface fiber that has saved lives, for it was once the stuffing in life preservers. The kapok tree (*Ceiba pentandra*) is a majestic giant that is native to the tropical rain forests of South and Central America and is grown on plantations in several tropical countries. The fibers develop within the large pods and grow inward from the ovary wall after fertilization, eventually completely covering the seeds. When the pods split open, the hairs enable the seeds to become airborne and thus dispersed. Workers gather the pods before they open; ginning is not necessary because the fibers separate easily from the seeds. A coating of cutin makes the fibers waterproof, and a large lumen makes them lightweight; a life vest filled with kapok fibers does not waterlog and will support up to 30 times its weight. Kapok fibers are too fine and slippery to be spun but are used for padding and stuffing in upholstered furniture, mattresses, and pillows because they are nonallergenic.

Figure 18.9 Coir is the fibrous material from a coconut that is used to make floor mats.

Coir from Coconuts

The familiar coconut is the source of coir, a seed fiber. This fibrous material makes up the mesocarp of the coconut fruit. The natural color of the coconut is a dark brown, and the fiber is often left natural because it is difficult to bleach to lighter colors. The fibers are too coarse for garments but are valued for their durability in items such as ropes and doormats (fig. 18.9).

Rayon: "Artificial Silk"

The idea of making synthetic fibers is an old one; Robert Hooke, the English scientist who coined the term *cell* (see Chapter 2), suggested the possibility in his *Micrographia*, published in 1664. Two hundred years passed before technology made the suggestion of producing cloth from artificial fiber a reality. The first synthetic fiber was rayon, originally called "artificial silk." Rayon is a human-produced fiber of pure cellulose. The viscose process for making rayon was invented in 1891 by British scientists, and their method, improved and modified over time, is still employed today.

The original source of cellulose in the production of viscose fibers was cotton linters; wood fibers are commonly used today. Wood is processed into cellulose pulp and pressed into thin sheets about 6 meters (20 feet) square. The sheets are soaked in an alkali solution, converting the pulp to alkali cellulose. The alkali cellulose is shredded into fine crumbs and aged; treatment with carbon disulfide produces a material called sodium cellulose xanthate (viscose). The viscose is dissolved in an alkali bath, forming a thick, honey-colored solution. This spinning solution is passed through the fine holes of a spinerette into a coagulating bath of salts and sulfuric acid that solidifies the liquid into solid fibers (fig. 18.10). The fibers have a high luster; sometimes delustering agents are added to reduce their shiny appearance. Viscose may be dyed in solution before the fibers have been spun out.

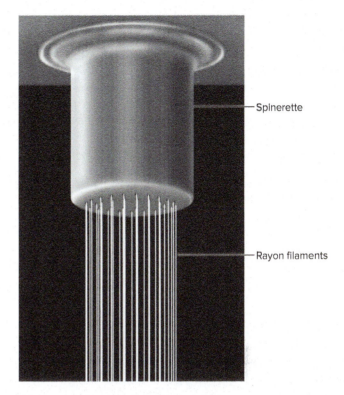

Figure 18.10 Rayon filaments form when viscose is forced through a spinerette into a coagulating bath.

Manufacturers can also put a crimp in viscose fibers to create fibers with better spinning quality.

Because rayon is a synthetic fiber, manufacturers can modify the size and shape of the fibers, producing rayon fibers that vary considerably. The strength and elasticity of rayon fibers are low; rayon tends to wrinkle and stretch easily. Like cotton, rayon has good moisture absorbency, which makes it easy to dye and comfortable to wear. Rayon blends well with cotton, giving the cloth a lustrous appearance. Adding rayon to ramie produces a linenlike fabric at a cheaper cost.

For the first time, researchers have identified a process to extract cellulose fibers from rice straw and spin them into yarn. The method removes some of the lignin and hemicellulose in the rice straw but leaves enough of these substances to bind the cellulose and build longer fibers. The fibers are 2.5 to 8 centimeters (1 to 3 inches) in length and can be spun to make yarn. The rice straw fabric has the feel of linen. The cultivation of rice discards more than 560 million tons of rice straw annually. It is estimated that about 80 million tons of fiber could be produced from this waste product, reducing the need for petroleum-based synthetic fibers, such as polyester and nylon.

Bark Cloth

Rendering bark into cloth is an age-old craft that developed independently in several regions. Polynesia, Africa, and Central and South America are areas particularly well known for the production and use of bark cloth.

Western explorers, such as Captain James Cook and botanist Joseph Banks, admired the beauty of bark cloth worn by the indigenous peoples they encountered in their travels. In Polynesia, the general term for such cloth is *tapa,* and typically it is made from the inner bark of the paper mulberry. Paper mulberry, or *Broussonetia papyrifera,* is a member of the Moraceae, or fig family, and the source of some of the finest bark cloth in the world. Native to East Asia, the plant was introduced to the Pacific region by early colonizers. In Hawaii, the paper mulberry is one of the so-called *canoe plants* taken to the islands by the Polynesians who settled there. The tree grows best in moist, rich soil and can grow as tall as 15 meters (50 feet), but when cultivated for the purpose of making bark cloth, it is severely pruned. Trunks are cut to ground level, and the young, slender shoots that sprout from the stump are trimmed of side branches to ensure that there are no knots in the bark. The trees, propagated from stem cuttings, are planted close together in sheltered areas to promote straight trunks with little branching. After 6 to 10 months, the shoots are harvested when they are about 3 meters (10 feet) high.

First, the bark is stripped from the tree in one piece. Next, the outer bark is scraped off. The inner bark of phloem, or bast fibers, is soaked to soften the fibers and remove impurities. For the finest cloth, fermentations alternate with soakings to further soften the fibers. In the Hawaiian Islands, bark cloth is known as *kapa,* which means beaten thing—an apt description, for the inner bark is beaten into a clothlike material. The women place strips of sodden inner bark on a hollowed log or anvil. Strips are overlapped so that, when beaten, they felt together to form a single large piece of cloth. Beaters of wood are marked with grooves similar in design to a kitchen mallet. A "watermark" is applied as a finishing touch to the kapa with a special mallet that is carved with an intricate geometric design (fig. 18.11a).

The finished kapa is then laid out in the sun to be dried and bleached. Decorations are painted freehand or stenciled on with a bamboo stamp bearing an elaborate pattern. Dyes are obtained from nature. In Samoa, the cloth might be decorated with natural plant dyes of brown from *Bischofia javanica,* red from seeds of the annatto (*Bixa orellana*), yellow from the rhizome of turmeric (*Curcuma longa*), or black from the candlenut tree (*Aleurites moluccana*). The soft, flannel-like cloth is used to make bedsheets, draperies, wall hangings, and a variety of clothing for men and women—loincloths, sarongs, skirts, and capes.

Bark cloth was also used as paper by the Mayans and Aztecs. After the Spanish conquest, the entire written storehouse of poetry, literature, history, economics—virtually all the recordings of the Aztec culture—was burned in 1529 by order of Fray Juan de Zumarraga, the first bishop of Mexico. Visitors to Mexico today can still find bark cloth in the brightly colored bark paintings called *amate* (fig. 18.11b). *Amate* is derived from the word *amatl,* which means paper in the Nahuatl language and was the name applied to the trees from which the inner bark was taken. The bast was originally obtained from fig trees (*Ficus cotinifolia, F. padifolia,* and others) native to the region, but today other species are often used. After the inner bark has been extracted, it is soaked in a mixture of water and ash, which both softens the fibers and eliminates impurities. The bark is then boiled for several hours. Next, the fibers are rinsed and placed in a grid pattern on a wooden board. A rectangular stone is used to beat the fibers to mesh them together and to create paper of the proper thickness. The sheet of fibers is then allowed to dry in the sun. The color of the finished bark paper varies from off-white to dark tan, depending on the species of bark used. Village scenes or figures of animals and plants are painted on in vibrant colors to complete the decorative artwork.

Dyes to color fabrics and other products were originally obtained from natural materials, including plants and fungi, as presented in A Closer Look 18.2: Herbs to Dye For.

(a)

(b)

Figure 18.11 Bark cloth. (a) Hawaiian kapa cloth. (b) Amate, brightly colored painting on bark cloth, is a traditional folk art in Mexico.

A CLOSER LOOK 18.2

Herbs to Dye For

For millennia, natural materials were the only source of dyes until the availability of synthetic aniline dyes from coal tar in the latter half of the nineteenth century. Even today, natural dyes are still important in many traditional societies. Dyes can be classified according to how easily a dye is affixed to the fiber. Direct dyes are soluble in water and readily picked up by the fiber. Turmeric and safflower are direct dyes that yield a yellow hue. In contrast, mordant dyes do not impart their color directly; the fiber must be treated with a chemical agent, the mordant. A mordant fixes the dye to the fabric; often, the color of the dye can be changed by using different mordants. Many lichens and club mosses contain alum (potassium aluminum sulfate), a mordant used since antiquity. Another natural source of mordant is the oak gall, rich in tannins. An oak gall is an abnormal growth on oak leaves and branches. It is caused by a female insect laying her eggs on the plant. The plant responds by producing a growth that encases the developing insects. The feeding activity of the insect stimulates the increased production of tannins in the gall.

A vat dye is insoluble and must be rendered soluble by the action of chemical agents or microorganisms. When dye-saturated fabric is exposed to air, the dye is oxidized. The color is permanent and does not fade in sunlight or after washing. Indigo blue is a vat dye discovered long ago in Asia.

Natural dyes are obtained from both animal and plant sources; two of the best known animal dyes are Tyrian purple and cochineal red. Tyrian purple derives its name from the ancient Phoenician city of Tyre on the Mediterranean coast. The source of the dye is the mucous gland of several species of whelk, a type of shellfish. The dye was in much demand but extremely costly; the color of "royal purple" became synonymous with position and wealth in the ancient Greek and Roman societies.

The Aztecs obtained cochineal red by boiling the females of a type of scale insect that is commonly found on the *Opuntia* cactus.

Herbal dyes have also been used extensively throughout history. Blue is the rarest of all hues, and as the source of a deep, rich blue, indigo (*Indigofera tinctoria*) was one of the most valuable herbal dyes. The leaves of this legume from the Old World tropics are the source of the dyestuff. The Aztecs used different species of this plant as a source of blue dye. Indigo was one of the last of the herbal dyes to be replaced by synthetics. Woad (*Isatis tinctoria*, mustard family) is the temperate counterpart of indigo; in fact, the dye principle (indigotin) in the leaves of the two plants is the same. Native to Europe, woad was the primary source of blue until the introduction of indigo in the eighteenth century. Woad, as were many dye plants, was used as a cosmetic. Julius Caesar's *The Gallic War* described the practice of the inhabitants of the British Isles of staining their bodies blue with woad before battle.

Red herbal dyes are as scarce as blue; the madder plant (*Rubia tinctorum*, coffee family) is an excellent and ancient source of this much-sought-after color. A mordant-type dye is obtained from the slender roots of this herb. The famous "red coats," the uniforms that became synonymous with the British army, are a historical example of the importance of madder dye. Logwood, *Haematoxylum* spp., is one of many dye plants known to the Aztecs. It is a small tree in the bean family (Fabaceae). Chips of the wood yield colors of blue, black, gray, brown, and purple. Logwood for dyeing purposes was one of the early exports to Europe, beginning in the sixteenth century. The dye hematoxylin is one of the few natural dyes that has not been replaced by a synthetic for commercial use. Hematoxylin is indispensable as a cell stain in the preparation of tissues for microscopic study.

Many common household, yard, and garden plants are sources of hidden colors (box fig. 18.2); a few are listed in Table 18.B.

Fungi are also a valuable source of pigments, especially the colorful pattern of pigments left by various species of wood-rotting fungi. Wood colored and patterned by fungal growth is known as *spalted* wood and has been used in woodcrafts since at least the 1500s in Europe. Two types of wood-decaying fungi are important in creating spalted wood: brown and white rots. The brown-rot fungi grow primarily on the softwoods (gymnosperm trees such as pines, firs, and spruces), whereas the white-rot fungi usually decay hardwoods (angiosperm trees such as oaks, maples, and ash). Both brown rot and white rot are classified as basidiomycetes in the division Basidiomycota based on the type of sexual spore produced by this group (see Chapter 23). Both types of wood-rotting fungi grow on dead or severely stressed trees. Wood decayed by brown rot becomes crumbly and brown; white rot makes the wood mushy and lightened.

White-rot fungi impart colored markings to the wood, called zone lines. Zone lines form when pigments, most commonly melanin, are deposited at the boundary between two different fungal colonies, usually between two different white-rot fungi, growing on the decaying wood. The colors of zone lines vary from black, brown, yellow, orange, red, or green and can be thick or thin. Beyond zone lines, some fungi secrete substantial quantities of pigments into the wood, changing its color. It has been hypothesized that the fungi release these pigments as chemical warfare to prevent other fungi from invading the wood; as a chemical reaction to components of the wood; or as a response to environmental stress. These colored woods, often in shades of blues or greens, had long been prized in the making of wood-inlay dioramas, a technique that fits together small pieces of wood to make a larger picture. Interest in handcrafted wood inlay and fungal-stained wood declined in the early 1900s when machine-manufactured products became widely available. Not until the 1970s, when wood turning became a craft, did spalted wood (now expanded to include naturally colored wood with zone line markings as well as wood colored by fungal pigments) become much desired.

For the last 10 years, researchers have experimented with inducing fungi to produce zone lines and pigmented wood under laboratory conditions. Zone lines produced in the lab can replicate in a few hours what would take two years in the field. For pigmented wood, the fungi that secrete the most desirable colors (pinks or greens) can be induced to release pigments when cultivated on standard malt agar with the addition of sterilized white-rotted wood chips. The pigment is then extracted and can be applied to wood or used to dye textiles.

Box Figure 18.2 (a) Natural dyes extracted from a variety of plants have produced these stunning colors of yarn (Shaker Village at Pleasant Hill, Kentucky). (b) Wood-decaying fungi release pigments at zone lines, creating spalted wood as seen in this ash bowl.

Table 18.B   Some Common Household, Yard, and Garden Dye Plants

Plant	Scientific Name	Part Used	Color
Black walnut	*Juglans nigra*	Hulls	Dark brown, black
Coreopsis	*Coreopsis* spp.	Flower heads	Orange
Lilac	*Syringa* spp.	Purple flowers	Green
Red cabbage	*Brassica oleracea*	Outer leaves	Blue, lavender
Turmeric	*Curcuma longa*	Rhizome	Yellow
Yellow onion	*Allium cepa*	Papery brown outer layers	Burnt orange

WOOD AND WOOD PRODUCTS

Wood and wood products rank right behind food plants in over-all importance to society. Construction materials, furniture, musical instruments, paper, fuel, charcoal, and synthetic materials (e.g., rayon, cellophane, and cellulose acetates) are just a few of the products that come directly or indirectly from trees (see A Closer Look 18.3: Good Vibrations). Although metals, concrete, and plastics have replaced wood for many of the traditional uses, the strength of wood, its insulating properties, its versatility for construction, and its natural beauty ensure the continued need for lumber well into the future, yet forestland is currently being lost at a rate that jeopardizes the future availability of wood.

Around 30% of Earth's land surface is covered by forests that supply the wood and wood products used by humanity. The world's forests are being cut down at an alarming rate; in 2018 the United Nations Food and Agriculture Organization estimated that 7.5 million hectares (18.7 million acres) are lost each year. Forest destruction is not new. It has been going on since humans learned to harness fire. However, it continues to get worse. Some forestry scientists estimate that 80% of the world's forests have already been destroyed, cleared for agriculture, firewood, or lumber. In some areas, wood is being used at a faster rate than forests can regenerate it, a concern that has been expressed for many years. The Civilian Conservation Corps Edition of *The Forestry Primer*, published by the American Tree Association in 1933, stated the following:

> We find that our forests are going faster than they are being replenished. This is due to cutting for our needs and to destruction by forest fires, insect pests and diseases. Owing to our constant increase in population we have ever new demands for what the forest yields us. Some forest uses change, such as when people build houses of concrete and steel instead of wood, but new uses are found so that the trees of the forest still remain tremendously important to the progress of our nation. We simply could not do without them.

Although some reforestation programs are in place, many ecologists believe that new forestry management practices are needed to maintain forests as complex ecosystems. In many areas, thousands of acres of diverse old-growth forest have been clear-cut, removing all vegetation. Although reforestation may be practiced, typically only a single species of commercially valuable timber trees is planted in the cleared areas (see Chapter 26). Only about 13% of the world's forestland is being managed, and only about 2% of the world's forests are protected in forest reserves. In Latin America, only one tree is planted for every 10 cut, and in Africa the percentage is even lower.

The preservation of tropical rain forests is a major concern among many biologists. Some environmental groups have called for boycotting lumber from tropical trees; however, harvesting for timber is responsible for only a small fraction of the rain forest destruction. Most tropical forests are cleared for agriculture or ranching. The majority of trees that are cut for wood tend to be used locally for fuel, with less than 15% used for lumber. Most of the lumber is also used locally, and less than one-third enters international trade. Still, the preservation of desirable species for timber must be kept in mind when harvesting trees, and many ecologists feel that sustainable harvesting can be achieved. This problem is often more complex than simply replanting a seedling when a mature tree is cut down, and mahogany trees serve as a good example of the dilemma. Mahogany, *Swietenia* spp., is native to tropical forests in Latin America and is desirable for fine furniture because of the color and quality of the wood. In the Plan Piloto Forestal in the Yucatan Peninsula of Mexico, attempts are being made to use sustainable harvesting of mahogany. Because tropical forests are very diverse, with many different species, few mahogany trees occur in any one area. If one mahogany tree is harvested, it should be replaced with a seedling; however, these seedlings cannot survive in the shade of a mature forest. Sustainable harvesting of mahogany requires the removal of many other trees to provide sufficient sunlight for the mahogany seedling. This situation has created controversy among ecologists, who feel that more environmental damage is done to the forest by clearing an area than is done by simply removing a single tree. If maintenance of biodiversity is the major objective, then unsustainable harvesting may be the better option. Clearly, finding the solutions to the preservation of tropical forests is a complex issue. With the world population continuing to grow, the demands for agricultural land and forest products also continue to grow, especially in developing nations. Better forest administration is essential to protect and sustain this valuable resource (see Chapter 26).

Although forest destruction continues, there is growing awareness of the problem. In November 2006, the United Nations Environment Programme began a worldwide tree planting initiative called *Plant for the Planet: Billion Tree Campaign*. This program encourages individuals, communities, businesses, organizations, and governments to plant native trees that are adapted to local environments. The original goal of the campaign was to plant 1 billion trees during 2007. That goal was easily met; and by mid-2018 participants from 225 countries had planted over 15 billion trees. This project was inspired by Dr. Wangari Maathai (1940–2011), the

Thinking Critically

The rapid destruction of tropical rain forests is a serious biological concern. Although some of the forests are harvested for lumber and firewood, other areas are being cleared for much-needed cropland to feed growing populations.

What alternatives are there for agriculture that do not necessitate destruction of the rain forest?

A CLOSER LOOK 18.3

Good Vibrations

Our dependence on plants for food, fuel, and fiber is easy to understand; what may be less obvious is the widespread use of plants by people around the world to fashion musical instruments, ranging from primitive folk instruments to those found in a symphony orchestra. Although the origins of music are lost in prehistory, almost all cultures have music, and plant-based materials have been used in musical instruments since earliest times. Music, as is all sound, is produced by some type of vibrating material. A musical instrument will magnify the vibrations and launch them into the air. Many musical instruments actually have two or more vibrating components. For example, the strings and sounding board (or resonator) in a guitar or violin combine to produce sounds loud enough to be heard. One or more components are frequently of botanical origin (box fig. 18.3).

Some of the first instruments produced were percussion instruments, such as rattles or wooden sticks that are struck together. Advances over simple sticks are notched sticks that are scraped together and hollowed-out logs, which form simple drums. Primitive wind instruments include hollow stems of bamboo, reeds, and other plants, and gourds have been used as the resonator for primitive stringed instruments. It is not possible here to describe all the plants that have been used in musical instruments, so we will focus on three types of plant material that have a variety of uses in many instruments: gourds, reeds, and wood.

Gourds are hard-shelled fruits produced by several members of the cucumber family (Cucurbitaceae). Other plants in the same family are squashes, pumpkins, and melons. Gourds come in various shapes and sizes and are widely distributed in tropical to temperate areas. In the United States, some small gourds are familiar as decorative objects, but among cultures in other parts of the world, gourds have had more practical applications as bottles, bowls, cups, ladles, and even hats. The use of gourds in musical instruments may be as extensive as other uses. Today, they are still used as folk instruments in many parts of the world. The simplest and most obvious instrument made from a gourd is a rattle, which may be one of the oldest musical instruments. When a gourd is dried, the seeds will produce a sound when the gourd is shaken. Often, pebbles, shells, or additional seeds are added to increase the sound, and a wooden handle may be added to round gourds. Maracas, which are associated with Latin American music, are traditionally decorated gourds. In some African countries, gourd rattles are often covered with a loose net that has beads or small shells attached. These add to the sound when the rattle is shaken. Other percussion instruments made from gourds include a variety of drums. Often, a membrane, such as an animal skin, is stretched across an opening in a large gourd. A variety of

Box Figure 18.3 Violins made by Antonio Stradivari from the late seventeenth to the early eighteenth century are among the most valuable musical instruments in the world today.

wind instruments, including whistles, flutes, horns, and trumpets, have been fashioned from gourds. The greatest diversity of gourd-based instruments are stringed instruments that use gourds as the resonators. These range from one-stringed musical bows to banjos, fiddles, and even harps. In India, the sitar is the most commonly used instrument of this type; it has seven main strings and 11 or 12 sympathetic strings. The most elaborate gourd instruments are marimbas and xylophones, which are composed of a series of wooden bars that are struck with mallets. Resonators are mounted beneath the wooden bars. Originally, gourds formed the resonators; however, in modern xylophones and marimbas tubular resonators are used.

Reeds are widely used in musical instruments and also have a long history. *Reed* is the common name for certain plants in the grass family (Poaceae) that grow in wet areas. *Arundo donax*, commonly called the giant reed, was originally native to the Mediterranean region but is grown around the world today. The hollow stems of this reed have a long history of use as primitive flutelike instruments. In other parts of the world, bamboo stems or other large grasses were utilized in the same manner. The Pan pipe was a primitive musical instrument also made from *A. donax* stems. Several small pieces of stem were held together by crosspieces, with their open ends in a single row. Each piece of stem was a different length and each, therefore, produced a different note. Pan pipes can still be found as folk instruments in southeastern Europe and have been made popular in the United States by a burgeoning interest in South American music. It is also believed that the Pan pipe formed the basis for the first pipe organ over 2,000 years ago.

Today, *Arundo donax* is best known as the source of the reeds used in woodwind instruments. In wind instruments, the air vibrates in a tubular resonator; however, different methods are used to cause the air to vibrate. In woodwind instruments such as the clarinet and saxophone, a reed is used to make the column of air vibrate. The reed is attached to the top of the instrument over a hole in the mouthpiece. When the musician blows into the mouthpiece, the reed beats between the hole and the player's lips, causing the column of air in the resonator to vibrate. The reed used in these instruments is a small piece of *A. donax* stem, gradually thinned at one end to a broad delicate tip. Oboes, bassoons, and English horns use two reeds that are connected and vibrate jointly. The small space between the reeds allows for passage of a thin air stream, and the two reeds vibrate against each other.

The characteristics of *Arundo donax* stems are the key to the properties of the reeds. The stems appear to be remarkably resilient and elastic as well as resistant to moisture. The quality of the reed and therefore the ultimate quality of the sound depend on the stage of development when the stem was harvested. Two-year-old stems are often preferred. After harvesting, the stems are cured or dried, a process that traditionally takes several months to years. The stems are then cut into small sections and shipped to manufacturers, who trim the sections into basic reed shapes. Musicians carefully trim purchased reeds to obtain the precise sound wanted for a given piece of music.

In addition to the qualities of the reed, many musicians feel that the material used for the instrument also has a significant effect on its sound quality. Before the twentieth century, the body (resonator) of all woodwind instruments was made of wood. Today oboes, clarinets, and bassoons are still made of wood, but various metals are often used for other instruments in this group. African blackwood, or grenadilla, is the wood of choice for concert-quality clarinets, oboes, and flutes. Grenadilla is often called the tree of music, but there is concern about the continued availability of this wood. At one time, grenadilla trees covered vast areas of eastern Africa, from Ethiopia to South Africa. Today these trees are threatened by overharvesting, the spread of deserts, and increasing population pressures. The African Blackwood Conservation Project is an international agency that seeks to replenish supplies of grenadilla wood by reintroducing seedlings in Tanzania and other parts of Africa.

The type of wood is also of primary importance for the sound quality of stringed instruments. A great variety of instruments are included in this category: violin, viola, cello, bass, guitar, banjo, ukulele, harp, and piano, along with a number of folk instruments. In each of these, the sound is produced when a string (made from wire, silk, or gut) vibrates after being plucked, stroked with a bow, or struck with a hammer. The sound is amplified by a resonator, which is usually made of wood.

The ancestor of all stringed instruments is the musical bow, made from a curved branch with a string attached to both ends and similar to a hunter's bow. It was played by striking or plucking the taut string in front of the player's open mouth, which functioned as a natural resonator. Musical bows have been used by cultures in many parts of the world, and the oldest record of a musical bow is a 15,000-year-old rock painting found in a cave in France. As time went on, improvements were made by using resonators such as gourds, coconuts, tortoise shells, and wooden boxes, which were placed between the taut string and arched bow. Although various stringed instruments, such as harps and lyres, can be traced back to antiquity, the modern violin family (violins, violas, cellos, and basses) appeared in Europe only early in the sixteenth century. Since that time, they have been integral elements of Western music and are considered the heart of the symphony orchestra.

Perhaps the most valuable stringed instruments in the world today are those made by Nicola Amati, Antonio Stradivari, and Giuseppe Guarneri from the late seventeenth century to the early eighteenth century in Cremona, Italy (box fig. 18.3). The violins, violas, and cellos produced by these Italian luthiers (makers of stringed instruments) often will sell for millions of dollars whenever one becomes available. In 2012, a violin by Giuseppe Guarneri sold at auction for over $16 million to an anonymous buyer. The new owner generously gave the instrument to violinist Anne Akiko Meyers on lifetime loan. For centuries, musicians and physicists have been debating the reasons why the sound produced by these instruments is so unparalleled. Some believe it is the varnish; others think it is the way that the wood is cured; but many are convinced it is the wood itself.

Aged spruce (*Picea abies*) wood with a straight grain is usually used for the top plate (front) of a violin; maple (*Acer platanoides*) wood with a curl or flame across the grain is used for the bottom plate (back) of the violin. Spruce and maple have been the woods of choice for fine instruments for centuries, so what is different in a Stradivarius? When a violin (or other stringed instrument) is played, there are several vibrating elements. The strings vibrate when stroked with a bow, and then the top and bottom plates of the violin vibrate along with the air in between. Thinning a plate by even a fraction of a millimeter from just a small area of a plate can dramatically change the vibration, the harmonics, and the sound of the instrument. In addition, no two pieces of wood are exactly the same, so even boards cut from the same tree will have different vibrational properties because of minor differences in the stiffness and density of the wood. Selecting the proper pieces of wood and tuning them (trimming to the appropriate thickness) take many years of training. Stradivari and Guarneri were able to tell by ear when the plates were perfected. Modern luthiers, such as Carleen Hutchins (1911–2009), used scientific instruments in an effort to accomplish the same thing. Hutchins made stringed instruments and studied the acoustics of violin plates for more than 50 years. She believed that finding the specific areas on the plates where the vibrations occur permits more precise tuning. Her work also showed that the relationship between the vibrations of the front and back plates is important. Using this knowledge, she constructed and tested hundreds of instruments and even developed new members of the violin family.

In the 1980s, Tillman Steckner, a Canadian violin maker, suggested that the soundboards have irregular variations in thickness. Steckner believes that Stradivari and other masters detected that the grain was denser in certain areas of the wood and may have thinned the wood in these areas to produce a more uniform resonance. He suggests that the masters may have held the violin plates up to the light to locate the denser areas; Steckner has used this technique to improve the quality of harpsichord soundboards.

Henri Grissino-Mayer from the University of Tennessee has suggested that the climate was responsible for creating unique wood. From the mid-1400s to the mid-1800s, Europe was under the influence of the Little Ice Age, with the coldest period from 1645 to 1715. The trees during this period produced wood with compact, narrow tree rings. As a result, the spruce wood used by Stradivari and other seventeenth- and eighteenth-century luthiers was unusually dense and produced exceptional instruments.

Joseph Nagyvary, a biochemist by training and luthier by avocation, believes that the difference in the wood may be due to microbial degradation that causes more holes in the cell walls making up the wood. Nagyvary suggests that the wood used by the seventeenth-century instrument makers was floated in rivers from the Alps down to Venice, where the logs sat in saltwater lagoons, possibly for extended periods of time. He has tried various treatments to simulate the immersion. In addition, Nagyvary feels that the varnish may also influence the sound of the instruments from Cremona. Nagyvary found finely ground minerals in the varnish of a cello and viola from this period and has added these minerals to the varnish he uses.

Francis Schwarze, a professor of wood research in Switzerland, has developed a technique of wood treatment that has made some exceptional modern violins. Schwarze treats the wood with two species of wood-rotting fungi for 6 to 9 months. This treatment alters the acoustic properties of the wood. He believes that violins made from the treated wood produce a sound equal to a Stradivarius violin. Professor Schwarze hopes that his method will soon be able to provide excellent-quality instruments at an affordable price for young musicians. These instrument makers have all found ways to improve the tone quality of modern instruments; however, only time will tell if any one of these modern luthiers has truly discovered the secret of a Stradivarius.

For the finest sound quality, a truly exceptional violin, viola, cello, or bass must be matched with a truly exceptional bow. Just like the sounding board on the instrument, the bow resonates, contributing to the total sound produced. The best bows are those made from the wood of *Caesalpinia echinata*, the pernambuco tree, which is native to forests along the coastal plain of Brazil. However, the destruction of tropical forests is threatening the existence of the pernambuco in a situation similar to the grenadilla trees in Africa. Bow makers from around the world formed the International Pernambuco Conservation Initiative to save the species from extinction.

It is clear that plant materials are important to music, be it classical, jazz, rock, or folk. Next time you listen to a symphony orchestra or hear someone play an acoustic guitar, remember how plants contribute to the sound of music.

---

founder of Kenya's Green Belt Movement, which has planted over 30 million trees in 12 African countries since 1977. In recognition of her work for reforestation and women's rights, Dr. Maathai received the Nobel Peace Prize for 2004.

Hardwoods and Softwoods

Wood is secondary xylem consisting largely of dead cells involved in the transport of water and minerals as well as support. The overall value of wood resides in the strength and durability of the cell wall materials, principally cellulose and lignin.

Two categories of forest trees and lumber are recognized, **hardwoods** and **softwoods**. *Hardwood* refers to angiosperm trees, and *softwood* refers to gymnosperm trees (conifers). It is estimated that around 35% of the world's forest area is dominated by softwoods, and in North America, they cover more area than hardwood forests. In a literal sense, the terms are meaningless because some types of gymnosperm wood are actually harder (denser) than some hardwoods. However, as a group, angiosperm wood is generally denser than gymnosperm wood. The hardness indicates the sturdiness of the cell walls, which is largely a reflection of the amount of lignin. The harder the wood, the more resistant to wear.

Some hardwoods are so dense that it is actually difficult to drive a nail into the wood, thus making it less desirable for construction. The hardest and densest wood is lignum vitae, *Guaiacum officinale*, with a specific gravity of 1.37. The specific gravity of water is 1.0, so this wood will sink. The lightest and least dense wood is balsa wood, *Ochroma pyramidale*, with a specific gravity of 0.17. Pine and other woods used for construction generally have a specific gravity of 0.35 to 0.45; oak and other woods for furniture have a specific gravity of 0.60 to 0.70.

In softwood, tracheids and some parenchyma cells are the only cell types present. In the tree tracheids combine the functions of both support and conduction, and parenchyma cells carry out various metabolic activities. In hardwood trees, tracheids and vessels are principally involved in the conduction of water and minerals; fibers are purely support elements. Xylem parenchyma occurs in longitudinal strands or in rays.

Recall from Chapter 3 that in woody plants the vascular cambium produces a ring of secondary xylem during each growing season. These annual rings result from seasonal changes in cambial activity and are often used for determining the age of a tree or past climatic events (see A Closer Look 3.1: Studying Ancient Tree Rings). The centermost region of secondary xylem in a tree is known as **heartwood** (fig. 18.12). It is usually darker than the surrounding area because the cells often contain tannins, gums, and resins, which accumulate in this older area and help prevent decay. The function of the heartwood is support; this region is no longer involved in the transport of water and minerals. The region of secondary xylem outside the heartwood is **sapwood**, which functions in both support and conduction. The sapwood also stores carbohydrates and performs other vital functions. Because this region is actively involved in conduction, sapwood cells are normally wet. For lumber use, heartwood is usually preferred. Not only is it more resistant to decay but it also is drier and less likely to shrink and warp.

(a) Radial cut   (b) Tangential cut

Figure 18.13 The direction in which the logs are sawn determines the patterns on the finished wood. The radial cut (a), also known as quarter-sawn, produces parallel lines on the board, and the tangential cut (b), called plain-sawn, produces wavy bands.

Many characteristics of wood are determined by the thickness of the cell walls and the proportion of vessels, tracheids, and fibers. The frequency and distribution of these components are distinctive for various species and contribute to the characteristic grain. The prominence of the annual rings and the direction of cutting also contribute to the grain. Tangential and radial cuts of the same species show variations in the pattern (fig. 18.13). In tangential cuts, known as plain-sawn, the wood is cut tangential to the annual rings and therefore perpendicular to the rays. The slices are parallel and the grain is straight. This method produces the most common and least expensive type of lumber. The annual rings appear wavy and irregular and produce a pleasing grain pattern. In radial cuts, the wood is cut along the rays or the radius of the log. The annual rings appear as closely spaced parallel strips and the rays as scattered patches running at right angles across the straight grain. This method is also known as quarter-sawn, because the log is first cut into quarters and then along the radius. Quarter-sawn lumber is more expensive because it produces fewer boards per log, but it wears more evenly and is more stable. Another important feature is the presence of knots, which are the bases of branches that have been covered over by subsequent lateral growth of the tree trunk.

Figure 18.12 Cross section of a *Juglans cinerea* (butternut tree) tree trunk clearly shows the darker heartwood in the center surrounded by the lighter sapwood.

Lumber, Veneer, and Plywood

The United States and Canada are the leading lumber-producing countries in the world. About half the wood harvested in the United States is used as lumber, primarily for

(a) **RED OAK** — A heavy, strong hardwood that is mostly straight grained with a course texture. The heartwood is light brown with a reddish cast, and conspicuous rays appear as flecks in the wood. Red oak is used in flooring, furniture, doors, kitchen cabinets, molding, paneling, and railroad ties.

(b) **WHITE PINE** — Soft, lightweight wood with a straight, even grain and coarse texture. Color ranges from cream to reddish-brown and tends to darken with age. This softwood is used in boxes, crates, construction, furniture, boats, doors, windows, paneling, molding, plywood, particleboard, and paper pulp.

(c) **PECAN** — Hard, strong, and shock resistant. Sapwood is pale, and heartwood may range from pale to reddish brown, often showing wide variations in color. Generally straight grained but can be wavy and flowing in places. This hardwood is used for furniture, cabinets, flooring, paneling, ladders, tool handles, and sports equipment such as baseball bats, skis, and bows.

(d) **BLACK WALNUT** — Medium-density hardwood that is strong and durable but not excessively heavy. The heartwood is medium to dark chocolate brown with some darker streaks; the wood has a straight grain with occasional wavy or curly areas and is used in cabinets, furniture, flooring, paneling, doors, gunstocks, and veneer.

(e) **MAPLE** — Fine-textured hardwood with a straight, even grain, although wavy or curly areas occur. Wood is valued for being very hard and resistant to shock. The cream-to golden-colored sapwood is used as lumber for gym floors, bowling alleys, countertops, kitchen cabinets, butcher blocks, furniture, musical instruments, baseball bats, bowling pins, stairs, and doors.

(f) **MAHOGANY** — Tropical hardwood valued for its beauty, strength, and reddish-brown color. Wood has a natural luster with a fine grain that can be straight or wavy. Mahogany is widely used in furniture, especially antique reproductions. Also used in cabinets, veneers, musical instruments, turned objects, and carvings.

(g) **DOUGLAS FIR** — Softwood with a medium density, straight grain, and light orange-pink color. The wood is known for its strength and durability. It is widely used for home construction, railroad crossties, boxes, crates, pallets, cabinets, doors, boats, furniture, fences, paneling, flooring, and plywood.

(h) **REDWOOD** — Known for its distinctive reddish-brown heartwood with a remarkably straight grain. This softwood is moderately strong and highly resistant to insects and decay. Redwood finds its greatest use in outdoor furniture, construction, bridge timbers, decking, siding, doors, paneling, veneer, plywood, shingles, posts, and novelty items.

(i) **SITKA SPRUCE** — Lightweight softwood known for very high strength and consistent straight grain. The wood has a fine, even texture and ranges from creamy white to light golden brown in color. This wood is widely used in aircraft, ships, furniture, crates, boxes, and stringed instruments. Sitka spruce is an excellent conductor of sound in musical instruments.

(all): ©Estelle Levetin

Figure 18.14 The characteristics of each type of wood make it ideal for specific purposes.

construction, with a considerable amount used to make furniture (fig. 18.14). When the tree is living, about 50% of the weight of wood comes from water. Before the wood can be used, the water content must be reduced to 10% or less. The wood can be dried (seasoned) in the open air or in specially built kilns.

The greatest use of softwood lumber is for home construction. Pines generally figure high on the list of important gymnosperms and are valued for their light but strong wood. Worldwide there are 95 species in the genus *Pinus,* with 35 species native to North America. Historically, white pine, *Pinus strobus,* had been the most important timber tree in the United States. During the colonial period, the tree was valued as a source of masts for British sailing ships. This species remained the most important pine species until the twentieth century, when overcutting and the introduction of a fungal disease, white pine blister rust, led to its decline. Today, long leaf, loblolly, and slash pines (often called yellow pines in reference to the wood color) are leading pines in the southern United States and are a major source of lumber; ponderosa pine fills this role in the West.

Douglas fir (*Pseudotsuga menziesii*) is another of the most desired timber trees in the world. This strong wood is used in the production of plywood and is an important source of large beams. Douglas fir grows throughout the Rocky Mountain belt but reaches its greatest development in the Pacific Northwest, where the species rivals redwoods in size and grandeur. The diameter and height of these trees enable the harvesting of many large, knot-free boards. The species is in heavy demand by the lumber industry, and although large stands still occur, the species may be facing overharvesting. Other important softwoods include spruce, hemlock, bald cypress, and red cedar.

The oaks are the most economically important hardwoods in the United States. The genus *Quercus* is a large one, with species placed into two general groups; the white oak group, with rounded lobes on the leaves, and the black or red oak group, with pointed lobes. The white oak group is more important commercially and includes post, bur, and white oaks. White oak, *Quercus alba,* is the most prized of all species. The wood is very heavy, durable, and attractive. It is widely used in furniture, cabinets, flooring, trim, and whiskey barrels. Blackjack, scarlet, red, and pin oaks are among the more familiar trees in the black oak group, with *Quercus rubra,* northern red oak, the most commercially valuable. Although it is not as strong as white oak, it is used for general construction, flooring, furniture, posts, and railroad ties. Other hardwood trees of value are black walnuts, hickories, maples, sweetgums, tulip trees, and birches. Before the introduction of the chestnut blight fungus into North America, the American chestnut was also a widely used hardwood (see Chapter 6). Examples of how wood from different tree species can be used are listed in Figure 18.14.

Furniture may be made from solid wood or constructed using a **veneer**. A veneer is a very thin sheet of a desired wood that is glued to a base of less expensive lumber. The veneer provides the look of fine wood at lower cost. Veneers are also used decoratively to produce exceptionally beautiful matched designs in fine furniture. Some of the most popular woods for veneers are black walnut, black cherry, bird's-eye maple, mahogany, and teak.

Plywood consists of three or more layers of thick veneer glued together. The grains of alternate layers are at right angles to each other. Because wood is strongest along the grain, the layering produces a sheet or board that is more uniformly strong than a comparable piece of solid wood. The result is a lightweight but strong building material for roof and wall sheeting, subflooring, shelves, cabinets, boxes, and signs. About 20% of harvested lumber is used in the production of plywood, with Douglas fir and various species of pine the most common sources.

Fuel

Throughout the history of human civilization, wood has been the chief source of fuel until relatively recent times. In many developing nations, the vast majority of harvested wood is still used as fuel. Over 2 billion people depend on wood or charcoal for 90% of their energy needs for heating and cooking. Because of this dependence, it is estimated that each year approximately 50% of the world's harvested wood goes to fuel.

Firewood gathering is hastening the demise of many tropical forests because over 1 billion people in the tropics depend on wood for fuel. Local supplies are being cut faster than trees can regenerate. For example, in India, it has been estimated that the forests can sustain an annual harvest of approximately 39 million metric tons of wood; however, the demand in India is for 133 million metric tons annually. In addition, it is often necessary to travel long distances with heavy loads to meet fuel requirements. As wood becomes scarcer or the price of fuel rises, the poor will lose their ability to thoroughly cook their food. This may facilitate the spread of disease-causing organisms. This situation will only get worse as populations in developing countries continue to increase.

In addition to being burned directly, wood can also be converted into charcoal by partial combustion in an oven or other enclosure that restricts airflow. Charcoal is almost pure carbon and burns at much higher temperatures than wood and can even be used for smelting ores into metals. Charcoal production was first developed over 7,000 years ago and ushered in the Age of Metals. During the Middle Ages, the forests of southern Europe were decimated partly for the production of charcoal, which was used for smelting iron ore to make cannons.

Harvested wood is also converted into wood pellets which can be burned in home stoves or boilers and can also be burned in power plants. In both the United States and European countries, efforts are underway to reduce emissions from power plants that contribute to global warming. This is being accomplished by reducing the amount of coal burned and increasing the use of renewable resources such as plant biomass. Wood pellets are often the biomass being burned in the power plants, either combined with coal (cofired) in older power plants or alone in new ones with much of the pellets supplied by the forestry industry in the southeast United States. In 2016, the United States exported about 5.2 million tons of wood pellets to Europe. The United Kingdom is by far the largest imported of wood pellets at 80% of the total U.S. exports, followed by Denmark and Belgium; these three countries account for 98% of the wood pellets shipped to Europe. In the United States, the use of wood pellets in power plants is currently only a small component of the U.S. energy industry.

The use of wood pellets is not without critics. Some researchers have shown that cofiring wood pellets with coal may not be economically feasible. The European Union is actively promoting the switch to renewable energy sources and are subsidizing the cost of wood pellets. In the United State, no subsidies are available. Either subsidies for wood pallets or taxes on coal would be needed to compensate for the higher cost of wood pellets. Some people question the contribution of burning wood to global warming. Proponents contend that the emissions from burning wood are balanced by the planting of new trees, others raise doubts about this assumption. Ecologists are also concerned about the increased use of wood for fuel when deforestation is a major environmental concern. These are clearly complex issues as countries try to switch to renewable fuels and also protect our forests.

Other Products from Trees

Resins include a broad collection of compounds that are composed of polymerized terpenes mixed with volatile oils (see Chapter 17). Resins are insoluble in water and apparently function naturally in furnishing protection to the tree. They discourage herbivores and make the wood resistant to some decay-causing fungi. Fossilized resins produce amber, often considered a botanical jewel (see A Closer Look 9.2).

Although hardwood trees, especially members of the Fabaceae, produce resins, the best known commercial resins are extracted from conifers. Resin is produced in ducts or canals that occur throughout the tree. Resin oozes out when the tree is cut; it is easy to see and smell the sticky resin in a pine tree by just breaking off a cluster of needles. Commercial extraction is performed by slashing the bark, and the crude exudate collected is also known as pitch. When pitch is heated, some volatile components evaporate easily. These condense into turpentine, which is used as a thinner for oil-based paints and as an organic solvent. After the volatile compounds are removed from the resin, the remainder is known as rosin, a material frequently used by musicians and baseball players. The bow of a string instrument is drawn across a block of rosin to make it slightly sticky. This stickiness increases the friction between the bow and the strings, resulting in more vibration and improved tone quality. The baseball player's bag of powdered rosin improves the pitcher's grip on the ball. Also, the stickiness of Band-Aids is due in part to rosin.

Pine pitch has been used for waterproofing since ancient times. It was commonly used to waterproof wooden ships; as a result, pitch, resin, and turpentine have been called naval stores. During the colonial period in U.S. history, the British obtained large quantities of naval stores from North America to support England's large fleet. The extensive pine forests of North America were one of the reasons the British opposed independence for the American colonies.

Another nonwood product from trees is **cork**. As a tree increases in diameter, the epidermis is replaced by periderm (see Chapter 3). The major component of this tissue is cork, or phellem, produced by the cork cambium. Cork cell walls contain suberin. At maturity the protoplasm dies and the cells become air-filled. Cork is good insulation and possibly provides protection against fire damage in intact trees. The characteristics of cork were known for several thousand years. During early Greek and Roman times, cork was used as a seal for jars and casks and for flotation devices when crossing rivers. Cork has some remarkable properties. Among other things, it is lightweight, good insulation, chemically inert, and long lasting. Although plastics and other synthetics have replaced some of the uses of cork, no synthetic substitutes exist with all its properties.

Commercial sources of cork are from the bark of the evergreen oak, *Quercus suber*, a tree native to the western Mediterranean region, with the greatest production from Portugal and Spain. The layers of cork may be several inches thick in *Q. suber*, which is one of the few species whose bark can be stripped repeatedly without significant damage. The outer bark can first be removed when a cork oak tree is about 20 to 25 years old, and subsequent strippings can then take place every 10 years for several hundred years (fig. 18.15). A large tree can yield up to one ton of cork at a single

Figure 18.15 *Quercus suber*, the commercial source of cork. Cork from the bottom of this tree has been recently harvested.

stripping. The initial cork is coarse, but subsequent strippings result in a closer-grained product. The stripped bark is seasoned briefly, then boiled to remove tannic acid and make it pliable. After drying, the cork can be trimmed, graded, and sent to market.

Wood Pulp

Wood pulp is a watery suspension of pulverized wood containing tracheids, vessels, and fibers in hardwood pulp or just tracheids in softwood pulp. In industrialized nations, approximately 50% of the harvested wood goes into wood pulp, with the vast majority of pulp used in the manufacture of paper. About one-fourth of the wood pulp is produced mechanically by grinding the wood with water, making a slurry. The mechanical process produces the greatest yield, but paper produced by such pulp is weak and yellows quickly. Newsprint, catalogs, and paper towels are manufactured by this process.

A principal goal of processing wood for pulp production is to remove as much lignin as possible from the wood. Lignin, which constitutes 25% to 35% of the wood, is brown and continues to darken with age, making it unsuitable for quality writing paper. Approximately three-fourths of the wood pulp is produced chemically by various methods that dissolve lignin and other noncellulose components. In one method, wood chips are dissolved in sodium hydroxide; sulfites or sulfates are employed in other processing methods. Some of the dissolved lignin may be used in the production of other compounds, such as vanillin, a vanilla substitute, and dimethyl sulfoxide. However, most of the separated lignin causes disposal problems, as do some of the chemicals used in the process. In addition, some pulp is also bleached to produce a white paper; unfortunately, the discharged chemicals from this step also add to pollution problems. Chlorine, traditionally used in the bleaching process, reacts with the pulp to form organochlorides, including dioxin, which is highly toxic. Dioxin was identified in the runoff from paper mills in the mid-1980s. Even small amounts may be harmful because dioxin can magnify in the food chain and accumulate in fatty tissues of fish, birds, and other animals (see Chapter 26). Recently, the wood pulp industry has shifted away from the use of chlorine to a variety of other chemicals, especially chlorine dioxide, to solve this problem. Chlorine-free pulping is now the prevailing method used in most parts of the world and accounts for 75% of the pulp bleaching. The use of wood-rotting fungi in pulp production may eliminate the pollution problem in the future (see A Closer Look 23.2: Dry Rot and Other Wood Decay Fungi). In addition, scientists have been developing fast-growing poplar and eucalyptus trees that have been genetically engineered to produce wood with reduced lignin content and increased cellulose content (see Chapter 15). This may also decrease the environmental hazards of pulp production.

Wood pulp is also used in the manufacture of cardboard and fiberboard as well as rayon, cellophane, and cellulose acetate. Eucalyptus trees are often planted specifically for rayon production because these fast-growing trees can be harvested for pulp in 7 to 10 years. Cellophane and cellulose acetate are made by treating dissolved cellulose with acetic acid or acetic anhydride. The resulting material can be spun into fibers or rolled into sheets. Cellulose acetate is often used in the manufacturing of molded "plastics," such as eyeglass frames, toothbrush handles, combs, car steering wheels, and pens.

The southern Atlantic states and the Northwest are the leading pulp-producing regions of the United States, with the southern pines the most important pulp species. Western softwoods, including spruce, fir, western hemlock, and Douglas fir, supply a significant proportion of pulp as well.

PAPER

Although the "information superhighway" provides worldwide computer linkup and instantaneous electronic exchanges, paper is still the major medium of written communication in contemporary society. The United States accounts for over one-third of the world's production and use of paper and cardboard. Each year about 1 billion trees are cut down to satisfy the demand for paper and paper products, with each American directly or indirectly using approximately 317 kilograms (700 pounds) of paper. That amounts to 868 grams (1.9 pounds) of paper each day. In the United States 76 million metric tons of paper are produced each year. Among other uses of paper, there are 2 billion books, 350 million magazines, and 24 billion newspapers published. Written communication still links millions of people. What a contrast to its humble start as symbols on clay tablets!

Early Writing Surfaces

Contemporary historians generally assume that the first written language was developed by the Sumerians around 5,000 years ago. Possibly because of economic or administrative needs, Sumerians came up with the idea of writing on clay. Although early examples were crude and totally pictographic, their system of writing became a conventional phonetic system over time. Clay tablets, prisms, and cylinders were the writing surfaces used by Sumerians. (An example of a Sumerian cylinder is seen in fig. 24.9.) Angular sticks were used to make impressions in the soft clay, which was then baked, making the writing permanent.

Egyptian hieroglyphic writing is dated to about 100 years after the earliest Sumerian records. It has been suggested that the development of Egyptian writing was stimulated by the Sumerian example. The writing surface immortalized by the ancient Egyptians was papyrus.

Papyrus, *Cyperus papyrus,* is a sedge (fig. 18.16a) that grows naturally in Egypt, Ethiopia, the Jordan River valley, and Sicily. The papyrus writing surface was originally developed about 4,500 years ago; it was made of thin slices of the plant's cellular pith that was beaten and then laid lengthwise with other layers crosswise on it. The mat was moistened with water, then pressed and dried. In the final step, the papyrus sheet was rubbed smooth with ivory or a smooth shell (fig. 18.16b). The finished sheets were made into rolls, often up to 9 meters (30 feet) in length.

The Greeks appeared to have learned about papyrus around 2,500 years ago. The use of papyrus by the Greeks, and later the Romans, continued until the fourth century, when it was superseded by parchment. After that, it was still used for official and private documents until the eighth or ninth century. Today papyrus is used only for decorative pieces (fig. 18.16c) and lives on semantically as the origin of the word *paper.*

By the beginning of the first century, Roman writing implements were varied and included wax-coated wooden tablets for temporary writing and school materials. Letters were scratched on the waxed surface with a metal or bone stylus. For permanent records, writing was done on papyrus with a reed cut to a fine point and dipped in ink and also on parchment or vellum with flat brushes and reeds.

Parchment is prepared from the skins of sheep, calves, or goats; vellum is a finer-quality parchment from kids, lambs, and young calves. In the preparation of parchment, the animal skin is cleaned and the hair removed. Both sides of the skin are scraped and smoothed and finally rubbed with powdered pumice. Parchment has been used for about 2,200 years. It gradually replaced papyrus and was itself replaced by paper after the development of printing. Today parchment is still used for formal honorary documents and some diplomas.

Papyrus reeds.	Peel the stems down to the pith.	Cut the inside pith into strips.	Put a second layer across the top, at right angles to the first.
Smooth the sheet with a scraper.		Beat with a mallet until the layers From a smooth, fat sheet.	Join the sheets by overlapping the ends And beating them.

Figure 18.16 Papyrus sheets were made from strips of pith from the stem of *Cyperus papyrus* (a) and used as a writing surface by ancient cultures. (b) Steps in making papyrus sheets. Today papyrus is used only for decorative pieces (c) like this modern painting from Egypt.

The Art of Papermaking

Paper is prepared from pulp, a slurry of plant cells that are separated and dispersed in a watery suspension. Although many plant materials can supply pulp, including straw, leaves, or rags, today most paper is prepared from wood pulp. The cells are matted into a thin layer that may be filled with clay or talc for added body, coated with sizing (such as starch) for smoothness, and then compressed. The cells must be long enough to mat well once the water is drained off. Tracheids, vessels, and fibers are the usual cells involved in the process; however, in the papermaking vernacular, these are all referred to as fibers. True papermaking can be traced back to China, early in the second century, where paper was made using a process similar to that used in contemporary production. In the New World, the Mayans and Aztecs independently invented paper using fibers from native plants (see Bark Cloth in this chapter).

Early Chinese scholars had traditionally written on strips of wood with a stylus. Later, woven silk and other cloths were used as writing surfaces. According to tradition, paper was first made in the year A.D. 105 by Ts'ai Lun (A.D. 50–118), a eunuch attached to the Eastern Han court of the Chinese Emperor Ho Ti (A.D. 89–105). The material used by Ts'ai Lun was the inner bark of the paper mulberry tree, *Broussonetia papyrifera*. The bark was ground to a pulp with water and sieved through a mold made of bamboo strips with a cloth mesh. The art of papermaking expanded to include other materials, such as bamboo, hemp, and rice straw, in addition to the paper mulberry bark. Use of various materials produced papers of different quality and different textures.

For about 500 years, papermaking remained a Chinese property. It was introduced to Japan in 610, to Central Asia in 750, and to the Near East and Egypt around 800. The Moors introduced the use of paper to Europe, and the first European paper was made in Spain around 1150. Linen and cotton rags were the principal source of material for paper. The craft spread slowly through Europe over the next few centuries. The introduction of movable type by Johann Gutenberg in the mid-fifteenth century greatly accelerated the spread of literacy, thought, and education, which were components of the Renaissance. This dissemination of knowledge provided a major stimulus for the papermaking trade.

The increased use of paper in the seventeenth and eighteenth centuries resulted in shortages of linen rags and stimulated the search for cheaper substitutes. The first machinery to replace the hand-molding process was developed by the French inventor Nicholas Robert in 1798. Henry and Sealy Fourdrinier, British papermakers, improved on Robert's design in 1803, and papermaking machinery still bears the Fourdrinier name (fig. 18.17). (The Fourdrinier screen is a continuous belt of wire cloth onto which pulp is deposited.) The use of wood pulp was introduced around 1840, ending the search for an inexpensive and abundant raw material. The search for alternative sources of pulp is once again active because of the threats of deforestation.

Figure 18.17 The Fourdrinier paper machine. Pulp is deposited onto a moving wire screen, shown here. The water drains through the screen, leaving a mat of fibers that make up a sheet of paper.

Alternatives to Wood Pulp

Today wood pulp is the major source of the world's paper supply, even though many other sources of pulp exist. For hundreds of years, cotton and linen have been used to produce writing paper. Fine-quality stationery and paper for permanent records still contain a large percentage of rags. Since the early 1990s Levi Strauss and Company has been recycling denim scraps to produce paper for company stationery and other items. Rice straw, a by-product of rice cultivation, is another possible source of pulp, as are bamboo, bagasse from sugarcane (see Chapter 4), and even hemp fiber.

Hemp, *Cannabis sativa*, was one of the earliest plants used by the ancient Chinese papermakers. It continued to be used until the early twentieth century. Hemp paper was even used in the first drafts of the Declaration of Independence and the U.S. Constitution. There is a renewed interest in using hemp for paper production. The plant is fast growing even in poor soils and requires few or no pesticides or herbicides. Bleaching is unnecessary, thereby reducing environmental problems. In addition, the archival potential of hemp paper is excellent. Whereas paper produced from wood pulp has a shelf life of 25 to 100 years, hemp paper will hold up for 1,500 years. Varieties of hemp that are used for industrial purposes have only trace amounts of tetrahydrocannabinol (THC), the psychoactive compound (see Chapter 20). Nevertheless, until recently it was illegal to grow any variety of *Cannabis sativa* in this country, and U.S. companies imported hemp fiber from China and Europe. However, that situation is changing. In 2014, President Obama signed a bill making hemp farming legal for pilot programs in states that have already legalized the crop. Since then 40 states have passed legislation allowing for either development of a hemp industry or to promote research into the development of a hemp industry. On the Federal level, the Senate version of the 2018 Farm Bill contained provisions for legalizing hemp farming. Considering its potential for paper as well as for fabric, hemp farming is likely to expand expand in the near future.

Thinking Critically

The use of wood pulp for paper is contributing to deforestation and environmental pollution from pulp production. From the environmental viewpoint, there is a major emphasis on using alternative sources of pulp.

What are some potential problems that may occur if farmers begin using cropland for the production of a herbaceous source of pulp?

One of the most promising alternatives is kenaf, *Hibiscus cannabinus*, a herbaceous plant in the Malvaceae, the mallow family. It grows from seed to mature size, which is approximately 4 meters (12 feet) tall, in about 4 to 5 months. By comparison, southern pine, a main source of wood pulp, takes 7 to 15 years to reach harvesting size. Moreover, fiber yield from an acre of kenaf is three to five times the yield for an acre of pine. Kenaf is also relatively disease resistant and drought tolerant. The U.S. Department of Agriculture has suggested that kenaf is the most viable fiber plant for U.S. paper production (see Chapter 26).

Another source of pulp is recycled paper. Many types of paper can be soaked in water to release the component pulp fibers, which are reused to make new paper. Paper recycling can reduce deforestation and pollution; in 2016, about 67% of the paper and paperboard used in the United States was recycled. This includes approximately 73% of the newspaper, 91% of the cardboard, and 66% of the office paper. At the present time, approximately 37% of the paper produced in the United States uses recycled materials. The continued development of recycled paper and other wood pulp alternatives is essential to reduce deforestation and ensure the availability of wood and wood products for future generations.

In addition to pulp production, agricultural wastes have recently been used for new applications. Cereal straw (from wheat, rice, and corn), sugarcane bagasse, sunflower hulls, and other materials have been utilized to produce particleboard, fiberboard, and additional types of composite panels.

It has been estimated that over 150 million tons of waste fibers, called agrifibers, are available yearly. Previously, this waste was plowed under, thrown away, or burned after harvest. Some states in the United States have banned the burning of agricultural wastes in an effort to reduce air pollution, leaving farmers with fewer alternatives. Agrifiber seems to be a possible solution.

This environmentally friendly, tree-free alternative to wood has been gaining in popularity, although the market is still small. For production of particleboard and other products, the agrifiber is mixed with a resin, such as polyurethane or soy-based resins. The resulting panels can be used for many applications, such as cabinets, furniture, door cores, and partitions. The addition of a hardwood veneer to an agrifiber core can also provide the look of wood for numerous products.

Dakota burl is a unique composite produced from sunflower hulls, left after processing oil seeds or confectionary seeds, plus urethane. The colors of the hulls result in a composite that is similar in appearance to traditional burled wood. Like the other types of agrifibers, Dakota burl panels can be used for a wide variety of applications and worked with standard woodworking tools. The panels are not intended for outdoor applications or for kitchen and bathroom countertops. Several companies now specialize in building office furniture from renewable resources, including these composite sunflower hull panels.

BAMBOO

It has been said that "bamboo is all things to some men, and some things to all men." Over 1,000 applications for this treelike grass have been described; no other plant has so many uses. Large stems are used as posts and rafters in houses, and split sections form the side walls. Long sections of stem are ideal for water or drain pipes, and short sections are useful for containers or for musical instruments. Split bamboo is woven into hundreds of products, such as baskets, screens, mats, fans, and carpets. Young shoots of small bamboo varieties are cooked, pickled, or even preserved in sugar. In some Asian countries, the farmer lives with bamboo from cradle to grave.

There are over 1,000 species of bamboo found in about 90 genera of the Poaceae, the grass family (fig. 18.18). Species occur in North and South America, Africa, Australia, and southern Asia. The greatest diversity occurs in China, with about 300 species in 26 genera; however, India has the largest reserves of bamboo, approximately 10 million hectares. Although the species vary greatly in size, shape, and color, they all have a woody stalk, or culm, produced from an underground rhizome. Most species produce new stalks every year. In general, the culms have a hollow pith that is solid at the nodes. The strong, lightweight stalks are the reason that bamboo is so useful. Unlike trees, there is no secondary growth. Bamboo does not increase in diameter after its initial growth. A bamboo culm will reach full size in 6 to 12 weeks

Figure 18.18 Bamboo.

but cannot be cut until it is 3 to 5 years old. New culms contain high percentages of water and will shrink and crack if cut and dried.

Species range in size from small plants (just a few inches high) to giant grasses, 40 meters (120 feet) in height and 30 cm (12 in.) in diameter. One striking feature of bamboo is the incredible rate of growth of new stalks; however, the most astounding feature is related to its flowering. Most species of bamboo flower only at long intervals of 30, 60, or even 120 years. All the plants of one species, wherever they are located, will come into flower around the same time. Specimens from a single culture have been taken to different parts of the world, such as the Royal Botanical Gardens at Kew (in England) and rain forests in Jamaica. The specimens continue to flower in synchrony with each other, suggesting an internal control mechanism. A Japanese scientist traced the flowering of black bamboo in historical records for more than 1,000 years. The first record was in the year A.D. 813; other documents indicated flowering every 120 years. After flowering, the aboveground parts of bamboo die back; it often takes 5 years or more for groves to become reestablished from either rhizomes or newly shed seeds.

Within China, the most useful bamboo is a large cane called *mao chu*, or hairy bamboo (*Phyllostachys pubescens*), which accounts for two-thirds of the country's bamboo production. It is used to make furniture and even reinforcement rods in heavy construction. Tonkin, or tea stick, bamboo (*Arundinaria amabilis*) is highly prized, and China exports 5,000 tons of it each year. It is widely used, especially in Europe, for garden stakes for fruits and vegetables, ski poles, fishing rods, and furniture.

Bamboo has been intricately linked with the giant panda. High in the mountains of the Sichuan Province of southwestern China, a native species of bamboo furnishes food for pandas; they consume both culms and leaves. After mass flowering and the resulting death of the culms, pandas often migrate to lower elevations, seeking other species of bamboo. Because of human pressures and widespread cultivation, alternative sources of food may be scarce. At these lower elevations, poaching becomes an additional threat to the survival of the panda.

Chinese bridges have traditionally been made from bamboo. Cables made of split bamboo strips woven or twisted together have incredible strength and have been used on hanging bridges for centuries. In fact, the bridge over the Min River in China hangs from bamboo cables that are about 7 inches in diameter. This bridge, considered one of the engineering marvels of the world, has been in use for over 1,000 years.

Bamboo has a long history of use in papermaking and today is the source of pulp for approximately two-thirds of all paper made in India. The products range from kraft paper (heavy brown) to fine writing stock. Although the yield from an acre of bamboo is not as great as the yield from pine, it is harvestable in 3 or 4 years, whereas a pine tree might take 15 years before it can be harvested. Expansion of this application of bamboo may be another viable alternative to wood pulp for a paper-hungry society.

CHAPTER SUMMARY

1. In addition to food and drink, plants furnish society with the raw materials to provide shelter, clothing, and paper. Plant fibers have been extracted and treated to weave cloth and manufacture rope for millennia. Cotton fiber is actually elongated seed hairs, and the techniques to render cotton fibers into cloth were discovered independently in both the Old and New Worlds. *Gossypium hirsutum* and *G. barbadense*, cotton species native to the New World, make the finest cotton cloth. These species are widely cultivated and have largely replaced Old World cultivars. The development of Eli Whitney's cotton gin brought about the mechanization of cotton production and with it an increased demand for slave labor, setting the stage for the American Civil War in the 1860s.

2. Linen, made from bast fibers of flax (*Linum usitatissimum*), was the cloth of choice for many ancient civilizations. Retting uses microbial action to free the flax fibers from the stem.

3. Other bast fibers that have been used to make cloth are ramie, jute (burlap), and hemp. Manila hemp, piña, and sisal are leaf fibers that have been used to make cloth as well as a variety of other products, such as rope and mats. Kapok is a surface fiber too fine to be spun but useful as padding and stuffing; coir is a seed fiber from the coconut valued for ropes and mats. Rayon was the first of many artificial fibers; it is made from extracting cellulose from wood fibers and molding the cellulose into yarn.

4. Bark is another material that has been used to make cloth and paper. Examples are the kapa cloth of the native Hawaiians and the amate folk paintings of Mexico.

5. Herbal dyes were one of the main sources of color in the ancient world, replaced by synthetic dyes only in the latter half of the nineteenth century. Wood decay fungi release pigments to wood, resulting in spalted wood that is much desired in woodcrafts and a source of fungal pigments that can be used to dye textiles.

6. Wood and wood products supply construction materials and fuel to much of the world population. The strength of wood, its insulating properties, its versatility, and its natural beauty sustain the demand for lumber, yet forestland is being lost at a rate that jeopardizes the future availability of wood. Pine and Douglas fir are the leading softwoods produced in the United States. The greatest use of softwood lumber is for home construction. Oaks are the most economically important hardwoods in the United States, with white oak the most prized oak species. The wood is heavy, durable, and attractive. It is widely used in furniture, cabinets, and flooring. Black walnut, hickories, maples, sweetgum, and birches are other important hardwoods. Each year a large segment of the world's wood harvest is used as fuel, especially in the developing nations. Over 2.0 billion people depend on wood or charcoal for the majority of their energy needs for heating and cooking. Other products from trees include resins and cork.

7. In industrialized nations, approximately 50% of the harvested wood goes into wood pulp, with the majority of pulp used in the manufacture of paper. Processing pulp removes the lignin. Although removing lignin is necessary to produce a quality writing paper, it has resulted in some environmental problems caused by disposing of the residue and by processing chemicals. It has been suggested that wood-rotting fungi may be able to substitute for some of the pollution-causing chemicals. In the United States each year, about 1 billion trees are cut down to satisfy the demand for paper and paper products. Although most paper today is prepared from wood pulp, this source of pulp has been used only since the mid-nineteenth century. Many plant materials can supply pulp, including straw, leaves, and rags. Overuse of forest resources is renewing the search for alternative sources of pulp.

8. Bamboo is a treelike member of the grass family with a long history of use in construction, furniture, baskets, screens, carpets, paper, and food. The versatility of bamboo is due to the strong, lightweight stalks that have an astounding rate of growth. Bamboo species range from small plants to giant grasses and are found throughout the world. The greatest diversity occurs in China, which has about 300 species.

REVIEW QUESTIONS

1. What are the various types of fibers in industry?
2. Describe the processing of cotton. How was Eli Whitney's invention of the cotton gin connected to the American Civil War?
3. List and briefly describe the steps in the processing of flax.
4. In what products would the fibers of sisal, jute, hemp, piña, coir, or Manila hemp most likely be found?
5. State a few prominent examples of herbal dyes. What is a mordant? What is spalted wood?
6. What is wood pulp? How is it used? What are other possible sources of pulp?
7. Why is bamboo considered a multipurpose grass?
8. Distinguish between the terms *hardwood* and *softwood* and the terms *heartwood* and *sapwood.*
9. What are the major uses of wood in the world today? Why is there concern about the future?
10. Distinguish between papyrus sheets and true paper. How is each prepared?
11. What is sustainable harvesting? What are some of the problems associated with using sustainable harvesting of tropical trees?

UNIT V

CHAPTER 19

Medicinal Plants

Artemisia annua leaves are the source of artemisinin, which has become a component of the standard treatment for malaria in most parts of the world.

KEY CONCEPTS

1. The connection between botany and medicine is an old established one; even today, 25% of prescription drugs are of plant origin.
2. Alkaloids and glycosides have been identified as the therapeutically important components of most medicinal plants.
3. Investigating folk remedies has led to important medical discoveries that have saved millions of lives.

Source: Scott Bauer/USDA

CHAPTER OUTLINE

History of Plants in Medicine 338
    Early Greeks and Romans 338
    Age of Herbals 338

A CLOSER LOOK 19.1 Native American Medicine 340

    Modern Prescription Drugs 340
    Herbal Medicine Today 342
Active Principles in Plants 343
    Alkaloids 343
    Glycosides 343
Medicinal Plants 343
    Foxglove and the Control of Heart Disease 343

Aspirin: From Willow Bark to Bayer 345
Malaria, the Fever Bark Tree, and Sweet Wormwood 346
Diabetes, French Lilac, and Metformin 349
Snakeroot, Schizophrenia, and Hypertension 350
The Burn Plant 351
Ephedrine 351
Cancer Therapy 352
Herbal Remedies: Promise and Problems 354
Chapter Summary 356
Review Questions 357

No one knows where or when plants first began to be used in the treatment of disease, but the connection between plants and health has existed for thousands of years. Evidence of this early association has been found in the grave of a Neanderthal man buried 60,000 years ago. Pollen analysis indicated that the numerous plants buried with the corpse were all of medicinal value. An accidental discovery of some new plant food or juice that eased pain or relieved fever might have been the beginning of folk knowledge, which was passed down for generations and eventually became the foundation of medicine.

HISTORY OF PLANTS IN MEDICINE

The earliest known medical document is a 4,000-year-old Sumerian clay tablet that recorded plant remedies for various illnesses. By the time of the ancient Egyptian civilization, a great wealth of information already existed on medicinal plants. Among the many remedies prescribed were mandrake for pain relief and garlic for the treatment of heart and circulatory disorders (see Chapter 17). This information, along with hundreds of other remedies, was preserved in the Ebers Papyrus about 3,500 years ago.

Ancient China is also a source of information about the early medicinal uses of plants. The Pun-tsao, a pharmacopoeia published around 1600, contained thousands of herbal cures that are attributed to the works of Shen-nung, China's legendary emperor who lived over 4,500 years ago. Another legendary emperor, Huangdi, is the source of other herbal remedies from about the same time period. In India, herbal medicine dates back several thousand years to the Rig-Veda, the collection of Hindu sacred verses. This has led to a system of health care known as Ayurvedic medicine. One useful plant from this body of knowledge is snakeroot, *Rauwolfia serpentina*, used for centuries for its sedative effects. The active components in snakeroot have been used in Western medicine to treat high blood pressure and schizophrenia.

The Badianus Manuscript, another important record of herbal medicine, is an illustrated document that reports the traditional medical knowledge of the Aztecs. It was created in 1592 to be offered as a gift to Charles I, king of Spain. The herbal is the work of an Aztec healer whose Christian name was Martinus de la Cruz and who, as an instructor of medical arts, rose to the rank of Physician of the College of Santa Cruz, a center of higher learning for indigenous peoples established by the Spanish. The manuscript is named after Juannes Badianus, an Indian instructor of Latin at the college, who translated the herbal from the native Nahuatl into Latin. The 13 chapters of the herbal cover nearly 100 afflictions and their treatments, using preparations from plants, animals, and minerals. The plant illustrations are often recognizable as familiar species that have verifiable physiological effects. For instance, *Dioscorea*, or tropical yam, which is known to have sapogenic glycosides, is noted. The document is especially valuable because much of the Aztec heritage was destroyed by the Spanish conquistadors during the period of conquest.

In all parts of the world, indigenous peoples discovered and developed medicinal uses of native plants, but it is from the herbal medicine of ancient Greece that the foundations of Western medicine were established.

Early Greeks and Romans

Western medicine can be traced back to the Greek physician Hippocrates (460–377 B.C.), known as the Father of Medicine. Hippocrates believed that disease had natural causes and used various herbal remedies in his treatments. Early Roman writings also influenced the development of Western medicine, especially the works of Dioscorides (first century A.D.). Although Greek by birth, Dioscorides was a Roman military physician whose travels with the army brought him in contact with many useful plants. He compiled information on these plants in *De Materia Medica,* which contained an account of over 600 species of plants with medicinal value. It included descriptions and illustrations of the plants along with directions on the preparation, uses, and side effects of the drugs.

Many of the herbal remedies used by the Greeks and Romans were effective treatments that have become incorporated into modern medicine. For example, willow bark tea, the precursor to aspirin, was used to treat gout and other ailments. Unfortunately, some treatments from ancient times have been lost. Historians have reported that silphium was used as an effective contraceptive by Greek and Roman women for over 1,000 years. This plant, a member of the genus *Ferula* and known as giant fennel, was the contraceptive of choice until the third or fourth century, when the plant was collected to extinction (fig. 19.1).

Dioscorides's work remained the standard medical reference in most of Europe for the next 1,500 years because little new knowledge was added during the Middle Ages. Although medical botany was nearly at a standstill in Europe, progress was being made in the Islamic world. Most notable is the early eleventh-century Persian, Avicenna, who wrote *The Canon of Medicine*, which included new information on herbal medicine.

Age of Herbals

The beginning of the Renaissance in the early fifteenth century saw a renewal of all types of intellectual activity. Botanically, this change was expressed in the revival of herbalism, the identification of medicinally useful plants. This revival, coupled with the invention of the printing press in 1450, ushered in the Age of Herbals (see Chapter 8). Some of the most richly illustrated herbals (fig. 19.2a) were those written by the four German "fathers of botany" in the sixteenth century: Otto Brunfels, Jerome Bock, Leonhart Fuchs, and Valerius Cordus. During the same period, the Englishman John Gerard published his famous work *The Herball or Generall Historie of Plantes* in 1597. Other English herbals published in the mid–seventeenth century were John Parkinson's *Theatrum Botanicum*, considered one of the best, and Nicholas Culpeper's *The Complete Herbal,* one of the most popular herbals of the day. All the herbals focused on the medicinal uses of plants but also

CHAPTER 19 Medicinal Plants   339

Figure 19.1 A wealth of knowledge about medicinal plants existed among ancient civilizations. These ancient Cyrenian coins illustrate silphium plants. Silphium was widely used by women as an effective contraceptive but had been collected to extinction by the third or fourth century.

included much misinformation and superstition (fig. 19.2b). For example, Culpeper's herbal had a strong astrological influence and revived the **Doctrine of Signatures.**

The most famous advocate of the Doctrine of Signatures was the early sixteenth-century Swiss herbalist Paracelsus. He believed that the medicinal use of plants could easily be ascertained by recognizing distinct "signatures" visible on the plant that corresponded to human anatomy. For example, the red juice of bloodwort should be used to treat blood disorders, and the lobed appearance of liverworts suggests their use in treating liver complaints. The belief behind the Doctrine of Signatures has been developed independently among many different cultures; however, there is no scientific basis to this concept.

Figure 19.2 (a) Herbals from the Middle Ages contained richly illustrated descriptions of plants. This illustration of *Juniperus* is from *De Historia Stirpium Commentarii Insignes* (or *Notable Commentaries on the History of Plants*) published in 1542 by Leonhart Fuchs. (b) The "humanoid" form of this mandrake root suggested its use in promoting male virility and ensuring conception.

A CLOSER LOOK 19.1

Native American Medicine

Long before Europeans colonized the New World, Native American tribes had developed an impressive array of medicinal plants. They had learned by trial and error which plants could heal or treat the diseases and injuries that afflicted them. Many tribes shared their knowledge with the settlers, with the result that some of these plants became well-established herbal remedies (box fig. 19.1a). At the present time, much interest has been focused on identifying and saving medicinal plants from tropical rain forests; however, the wealth of medicinal plants from Native American tribes has not been fully explored. The potential of this resource is typified in the following plants.

Goldenseal (*Hydrastis canadensis*) was widely used among various woodland tribes. The ground-up yellow rhizome of this plant has been valued as a dye, an insect repellent, and an antiseptic or antibiotic wash for treating wounds, mouth sores, and eye inflammations. It was also used internally for stomach and liver ailments. Goldenseal's therapeutic action is due to two alkaloids, hydrastine and berberine. Studies have shown that hydrastine can reduce blood pressure; berberine has antibacterial properties.

Witch hazel (*Hamamelis virginiana*) is a fall-flowering shrub or small tree native to the forests of eastern North America. Leaves, twigs, and bark have been used to prepare infusions to treat various aches and pains from sore muscles to sore throats. Today, a water or alcoholic extract known as aqua hamamelis is used as a topical astringent and antiseptic. In this form, it is probably more extensively used than any other herbal remedy from North America.

Native to North America, nine species have been identified in the genus *Echinacea* (hedgehog, so named for the spiny bracts in the central cone of the flower head). Three species of this composite are commonly used in herbal preparations: *E. angustifolia*, *E. pallida*, and *E. purpurea*. Only the purple coneflower (*E. purpurea*) is cultivated; the others are collected from the wild, and many populations have become endangered by overzealous and unscrupulous harvesters. Purple coneflower was used medicinally by the Plains Indians as an antidote for snakebites as well as for bites and stings from other venomous animals (box fig. 19.1b). They also used echinacea as a smoke remedy for headaches. An extract of the rootstock was also applied topically for other skin ailments, and a tea was drunk for treatment of various infectious diseases. Echinacea is currently one of the top-selling herbal products in U.S. health food stores. Germany has been the leader in echinacea research since seeds collected from North America were taken there in the 1930s. Echinacea is widely used for the treatment and prevention of colds and flu. Clinical studies have suggested that dosing with echinacea may lessen the symptoms of flu and colds, decrease their number, and speed up recovery time.

In a study released in 2003, *Echinacea purpurea* was not shown to be more effective than a placebo in reducing either the severity or duration of colds in children ages 2–11. Those children treated with echinacea did, however, have fewer subsequent colds than the control group. Skin rashes developed in a small number of the children receiving echinacea treatment.

Recent studies of cultured human epithelial cells showed that respiratory viruses resulted in the release of cytokines, chemicals that promote inflammation; many cold and flu symptoms are due to these pro-inflammatory cytokines. Echinacea extracts inhibited the release of cytokines from these epithelial cells and also showed antiviral activity. A scientific review published in 2014 indicated that echinacea provided a slight benefit for preventing colds.

Bloodroot (*Sanguinaria canadensis*) is a spring-flowering perennial herb in the poppy family found in the eastern woodlands of North America (box fig. 19.1c). The blood-red sap from the rhizome contains several alkaloids, which account for its medicinal properties. Although the sap is toxic in large doses, a tea brewed from the rhizome was used to clear the breathing passages of persons suffering from coughs, laryngitis,

---

During the eighteenth century, as scientific knowledge progressed, a dichotomy in medicine developed between practitioners of herbal medicine and regular physicians. At about the same time, a similar split occurred between herbalism and scientific botany, the study of plants above and beyond their medical applications.

Modern Prescription Drugs

Although herbalism waned in the eighteenth and nineteenth centuries, many of the remedies employed by the herbalists provided effective treatment. Some of these remedies became useful prescriptions as physicians began experimenting with therapeutic agents. William Withering was the first in the medical field to scientifically investigate a folk remedy. His studies (1775–1785) of foxglove as a treatment for dropsy (congestive heart failure) set the standard for pharmaceutical chemistry.

In the nineteenth century, scientists began purifying the active extracts from medicinal plants. One breakthrough in pharmaceutical chemistry came when Friedrich Serturner isolated morphine from the opium poppy in 1806. Continuing this progress, Justus von Liebeg, a German scientist, became

Box Figure 19.1 Native American tribes have employed an impressive variety of plants as traditional medicines. (a) Native American healer sharing herbal remedies. (b) Purple coneflower, *Echinacea purpurea*. (c) Bloodroot, *Sanguinaria canadensis*.

and bronchitis. It was also used to treat rheumatism and is a powerful emetic. Perhaps the most acclaimed use has been as an anticancer agent. In the 1850s, a study by J. W. Fell showed that breast cancer could be successfully treated with a procedure that included bloodroot sap. Later, extracts found applications in the treatment of skin cancer. More recent uses of bloodroot have involved one of its alkaloids, sanguinarine, which is used as an antiplaque agent in mouthwashes.

Many Native American tribes use forms of aromatherapy in healing and ritual ceremonies (see A Closer Look 17.1: Aromatherapy). Cedars or junipers, *Juniperus* spp., have been especially important for many tribes that regard it as a sacred "tree of life" and use it as incense for both ritual purification and medicinal benefits. Among the Yuchi tribe, dried cedar leaves and twigs with small female cones are mixed with dried flowers of cudweed (also called everlasting) and burned as incense for the treatment of allergies. Other plants, such as sweetgrass and white sage have also been used as sacred, ceremonial incense and as medicines by various tribes.

a leader in pioneering the field of pharmacology. This increased knowledge of the active chemical ingredients led to the formulation of the first purely synthetic drugs based on natural products in the middle of the nineteenth century. In 1839, salicylic acid was identified as the active ingredient in a number of plants known for their pain-relieving qualities and was first synthesized in 1853. This achievement led to the development of aspirin, which is the most widely used synthetic drug today.

Although the direct use of plant extracts continued to decrease in the late nineteenth and the twentieth centuries, medicinal plants still contribute significantly to prescription drugs. It is estimated that 25% of prescriptions written in the United States contain plant-derived active ingredients (close to 50% if fungal products are included). An even greater percentage is based on semisynthetic or wholly synthetic ingredients originally isolated from plants. Today there is a renewed interest in investigating plants for medically useful compounds, with some of the leading pharmaceutical and research institutions involved in this search. A Closer Look 19.1: Native American Medicine examines several medicinal plants that are under pharmaceutical investigation. In addition, many people in the United States and other countries are currently using herbal remedies that are sold as dietary supplements (see Herbal Remedies: Promise and Problems later in this chapter).

Herbal Medicine Today

While Western medicine strayed away from herbalism, 75% to 90% of the rural population in the rest of the world still relies on herbal medicine as their only health care. The long tradition of herbal medicine continues to the present day in China, India, and many countries in Africa and South America. In many village marketplaces, medicinal herbs are sold alongside vegetables and other wares. Practitioners of herbal medicine often undergo rigorous and extended training to learn the names, uses, and preparation of native plants.

The People's Republic of China is the leading country for incorporating traditional herbal medicine into a modern health care system. The resulting blend of herbal medicine, acupuncture, and Western medicine is China's unique answer to the health care needs of more than a billion people. Plantations exist for the cultivation of medicinal plants and the training of doctors; active research programs also investigate potentially useful specimens. Thousands of species of medicinal herbs, some collected from all over the world, are thus available for the Chinese herbalist. Chinese apothecaries contain a dazzling array of dried plant specimens, and prescriptions are filled not with prepackaged pills or ointments but with measured amounts of specific herbs. These colorful apothecaries can also be found in U.S. cities such as San Francisco, New York, and Seattle with their large Chinese populations (fig. 19.3).

While China melded traditional practices with Western medicine, in India traditional systems have remained quite separate from Western medicine. At Indian universities, medical students are trained in Western medicine; however, much of the populace puts its belief in the traditional systems. In addition to Ayurvedic medicine, which has a Hindu origin, Unani medicine, with its Muslim and Greek roots, is another widely practiced herbal tradition. Economics is also a factor in the reliance on indigenous cures because the cost of manufactured pharmaceuticals is beyond reach for most of the population.

The renewed interest in medicinal plants has focused on herbal cures among indigenous populations around the world (fig. 19.4), especially among indigenous peoples in the tropical rain forests, which have been studied by some of the world's leading ethnobotanists, including Mark Plotkin, Walter Lewis, Michael Balick, and Paul Cox. They have spent time with local tribes to learn their medical lore. It is hoped that these investigations will add new medicinal plants to the world's pharmacopoeia before they are lost forever.

Tropical rain forests are of special concern because the widespread destruction of these ecosystems threatens to eliminate thousands of species that have never been scientifically investigated for medical potential. As forests are cut down, native healers must travel farther and spend more time searching for the medicinal plants they have traditionally used. This gradual loss of the primary health care system among indigenous peoples is a serious concern. In addition to the destruction of the forests, the erosion of tribal cultures is also a threat to herbal practices. As younger members are drawn away from tribal lifestyles, oral traditions are not passed on. Mark Plotkin has compared this loss of knowledge to the burning down of a library containing books that are one of a kind and irreplaceable. One step to preserving this knowledge was taken by the government of Belize, which established the Terra Nova Forest Reserve. This Central American reserve is a 6,000-acre sanctuary dedicated to the survival of medicinal plants and the traditional healers who use them.

Figure 19.3 A Chinese apothecary shop in Brooklyn, New York, selling Korean ginseng along with dozens of other herbal remedies.

Figure 19.4 Statue representing the contribution of ethnobotany to medicine at the Missouri Botanical Garden in St. Louis, Missouri.

ACTIVE PRINCIPLES IN PLANTS

The medicinal value of plants is directly connected to the vast array of chemical compounds manufactured by their various biochemical pathways. These compounds are considered **secondary plant products** and include **alkaloids, glycosides, terpenes,** and **phenolics** (see A Closer Look 1.2: Perfumes to Poisons). Although scientists formerly believed that many of these compounds were simply waste products of metabolism, it is now known that they do, in fact, have important functions. Some secondary products discourage herbivores; others inhibit bacterial or fungal pathogens. One way of classifying medicinal plants is by the chemical makeup of the active principle. Several of the plants described in this chapter contain alkaloids or glycosides.

Alkaloids

Alkaloids are a diverse group of compounds of which more than 3,000 have been identified in 4,000 species. Although widely distributed throughout the plant kingdom, most alkaloids occur in herbaceous dicots and also in the fungi. Three higher plant families particularly known for the occurrence of alkaloids are the Fabaceae (legume family), the Solanaceae (nightshade family), and the Rubiaceae (coffee family).

Although they vary greatly in chemical structure, alkaloids share several characteristics: they contain nitrogen; they are usually alkaline (basic); and they have a bitter taste. They affect the physiology of animals in several ways, but their most pronounced actions are on the nervous system, where they can produce physiological or psychological results or both. Although some alkaloids are medically important, others are hallucinogenic or poisonous. It should also be noted that the difference between a medicinal and a toxic effect of many alkaloids (or any drug) is often the dosage. Common alkaloids include caffeine, nicotine, cocaine, morphine, quinine, and ephedrine. (Note that the names of most alkaloids end in *-ine.*) Although there are legitimate medical applications for morphine and cocaine, these alkaloids are discussed in Chapter 20.

Glycosides

Glycosides are widespread in the plant kingdom and are second in importance as medicines or toxins. They are so named because a sugar molecule (*glyco-*) is attached to the active component. The active portion of the molecule varies greatly, but the sugar is generally glucose. Glycosides are categorized by the nature of the nonsugar or active component, with the following types most common: **cyanogenic glycosides, cardioactive glycosides,** and **saponins.**

Cyanogenic glycosides release cyanide (HCN) upon breakdown. Recall that cassava contains cyanogenic glycosides, which must be removed before consumption. Also, the seeds, pits, and bark of many members of the rose family (apples, pears, almonds, apricots, cherries, peaches, and plums) contain amygdalin, the most abundant cyanogenic glycoside. The pits of apricots are a particularly rich source of amygdalin and are ground up in the preparation of laetrile, a controversial cancer treatment. Theoretically, laetrile releases HCN only in the presence of tumor cells and thus selectively destroys them. This claim has not been substantiated; therefore, laetrile has not been approved for cancer therapy in the United States.

Both cardioactive glycosides and saponins contain a **steroid** molecule as the active component. Cardioactive glycosides have their effect on the contraction of heart muscle, and in proper doses, some can be used to treat various forms of heart failure. The best-known cardioactive glycoside used medicinally is digitalis, which will be discussed next. On the other hand, some of the deadliest plants, such as milkweed and oleander, contain toxic levels of cardioactive glycosides. Saponins are less useful medically and can be highly toxic, causing severe gastric irritation and hemolytic anemia. One useful saponin is diosgenin from yams (*Dioscorea* spp.), which can be used as a precursor for the synthesis of various hormones, such as progesterone (one of the female sex hormones and an ingredient in birth control pills) and cortisone (see Chapter 14).

MEDICINAL PLANTS

Although hundreds of plant extracts are still an important part of the pharmaceutical industry, the descriptions that follow are a limited selection from the many medicinal products that have had an impact on society in either the past or the present. Many of the compounds derived from plants are listed in Table 19.1.

Foxglove and the Control of Heart Disease

Today in the United States, several million heart patients rely on digitalis as the primary treatment for their condition. Most are unaware of the plant source or the efforts of an eighteenth century English physician who was the first to scientifically investigate a folk remedy. William Withering was an English country doctor who had an extensive knowledge of the local flora and, in fact, wrote a book on the plants of the British Isles. Although he was aware of the flowering plant known as foxglove (*Digitalis purpurea*), his interest in it as a medicinal herb came about when he learned of a folk remedy for dropsy. As Withering later stated,

> In the year 1775 my opinion was asked concerning a family receipt for the cure of dropsy. I was told that it had long been kept a secret by an old woman in Shropshire who had sometimes made cures after the more regular practitioners had failed.

Dropsy is a condition characterized by severe bloating due to fluid accumulation in the lungs, abdomen, and extremities. Today, it is known that the fluid retention is caused by congestive heart failure, a failure of the heart to pump sufficiently. Although the folk remedy had 20 herbal ingredients,

Table 19.1
Plants of Known Medicinal Value

Scientific Name	Common Name	Family	Active Principle	Medicinal Use
Aloe vera	Burn plant	Asphodelaceae	Aloin	Skin injuries
Artemisia annua	Wormwood	Asteraceae	Artemisinin	Malaria
Atropa belladonna	Belladonna	Solanaceae	Atropine	Diverse uses, including relaxing muscles, dilating pupils, relieving asthma
Camptotheca acuminata	Chinese happy tree	Nyssaceae	Camptothecin	Cancer therapy
Cannabis sativa	Marijuana	Cannabaceae	Tetrahydrocannabinol	Glaucoma; antinausea during chemotherapy
Catharanthus roseus	Madagascar periwinkle	Apocynaceae	Vinblastine; vincristine	Leukemia, lymphoma, and other cancers
Chondrodendron tomentosum	Pareira	Menispermaceae	Tubocurarine	Muscle relaxant during surgery; cerebral palsy; tetanus
Cinchona spp.	Fever bark tree	Rubiaceae	Quinine	Malaria
Colchicum autumnale	Autumn crocus	Colchicaceae	Colchicine	Gout
Digitalis purpurea	Purple foxglove	Plantaginaceae	Digitoxin	Congestive heart failure
Dioscorea spp.	Yam	Dioscoreaceae	Diosgenin	Steroid drugs for contraception and inflammation
Ephedra spp.	Ephedra	Ephedraceae	Ephedrine	Bronchial asthma; bronchitis
Erythroxylum coca	Coca	Erythroxylaceae	Cocaine	Local anesthetic for eye surgery
Eucalyptus globulus	Eucalyptus	Myrtaceae	Eucalyptol	Cough suppressant
Galega officinalis	French lilac	Fabaceae	Guanidines	Diabetes
Ginkgo biloba	Ginkgo	Ginkgoaceae	Ginkgolides, bilobalide	Alzheimer's disease
Hydnocarpus spp.	Chaulmoogra	Flacourtiaceae	Hydnocarpic acid	Leprosy
Papaver somniferum	Opium poppy	Papaveraceae	Morphine; codeine	Severe pain; cough suppression
Pilocarpus pennatifolius	Jaborandi	Rutaceae	Pilocarpine	Glaucoma
Podophyllum peltatum	Mayapple	Berberidaceae	Podophyllin; podophyllotoxin	Warts; cancer
Rauwolfia serpentina	Snakeroot	Apocynaceae	Reserpine	Hypertension
Salix alba	White willow	Salicaceae	Salicin	Pain; fever; inflammation
Sanguinaria canadensis	Bloodroot	Papaveraceae	Sanguinarine	Antiplaque mouthwash
Taxus brevifolia	Pacific yew	Taxaceae	Taxol	Ovarian, breast, and other cancers

Withering was aware that foxglove leaves were the active component. He began treating dropsy patients with digitalis tea prepared from ground leaves and described one such patient:

> nearly in a state of suffocation, her pulse extremely weak and irregular, her breath very short and laborious, her countenance sunk, her arms of a leaden colour, clammy and cold. Her stomach, legs, and thighs were greatly swollen; her urine very small in quantity, not more than a spoonful at a time, and that very seldom.

Within a week, she no longer showed any symptoms. For the next 10 years Withering conducted careful experiments with powdered digitalis leaf to determine dosage, preparation, and effectiveness in treating dropsy. This research culminated in the publication of *An Account of the Foxglove and Some of Its Medical Uses: With Practical Remarks on Dropsy and Other Diseases* in 1785. Withering believed that foxglove primarily acted as a diuretic in relieving dropsy. Although he did note that digitalis "has a power over the motion of the heart," he never positively made the connection between dropsy and heart failure.

Digitalis purpurea, the purple foxglove, is an attractive biennial in the snapdragon family (Plantaginaceae) with a spike of large, purple, bell-shaped flowers (fig. 19.5). As do most biennials, it overwinters as a basal rosette of leaves, which gives rise during the next growing season to the flowering stalk, which may be a meter (3 feet) in height. Because of the attractive flowers, foxglove is often used as a garden ornamental. The leaves contain over 30 glycosides, with digoxin and digitoxin the most medically significant. In addition to *D. purpurea*, other *Digitalis* species, particularly *D. lanata*, produce these

Figure 19.5 Foxglove, *Digitalis purpurea*.

Figure 19.6 For hundreds of years, plants containing salicylates, such as black willow, *Salix nigra*, were used in herbal remedies for the treatment of pain and fever.

cardioactive glycosides. The concentration of glycosides in the leaves is highest before flowering; after the leaves have been dried and powdered, the glycosides can be extracted.

The cardioactive glycosides in digitalis slow the heart rate while increasing the strength of each heartbeat, so that more blood is pumped with each contraction. The resulting improvement in circulation decreases edema in the lungs and extremities and increases kidney output. Although the digitalis glycosides are an effective treatment for congestive heart failure, they do not cure the underlying heart condition. It should also be noted that there is a fine line between a therapeutic and a toxic dose of digitalis. Even patients under a doctor's supervision can experience toxic side effects and must be carefully monitored for the correct dosage. The most dangerous toxic effect is a life-threatening arrhythmia (a rapid and irregular beating of the heart).

Aspirin: From Willow Bark to Bayer

Although aspirin is presently the most widely used synthetic drug, with Americans alone consuming 80 million pills a day, its origin is most definitely botanical. The bark of willow trees (*Salix* spp.) had long been known among many cultures as an effective treatment for reducing fever and relieving pain. The ancient Greeks used an infusion of bark from white willow (*Salix alba*) to treat gout, rheumatism, pain, and fever; this later became a well-known folk remedy throughout Europe. Many Native American tribes had independently discovered the healing powers of willow bark.

The evolution of the drug from willow bark tea into synthetic aspirin began in England in the eighteenth century. Reverend Edmund Stone experimented with willow bark tea for 6 years and found it beneficial for the treatment of fever and chills. In the early nineteenth century, French and German chemists sought to isolate the active compound from willow bark. In 1828, salicin was first isolated, and over the next decade, the extraction method was refined. Salicin is a glycoside of salicylic acid; salicylates occur widely in species of *Salix* (fig. 19.6) as well as meadowsweet (*Spirea ulmaria*, or *Filipendula ulmaria*), poplars (*Populus* spp.), and wintergreen (*Gaultheria procumbens*).

The next step was the laboratory synthesis of salicylic acid in the mid-nineteenth century by several German scientists. Now that salicylic acid was widely available, an inexpensive treatment existed for many ailments. It was used for rheumatic fever, gout, rheumatoid arthritis, and osteoarthritis. Today, it is still used but is primarily applied topically to the skin for the removal of warts, corns, and various skin ailments. In 1898, while searching for a similar compound that caused less gastric distress to help his father who was afflicted with arthritis, Felix Hoffmann, a chemist at Bayer Company, came across acetylsalicylic acid in the chemical literature. The new compound was more palatable and was soon given the name *aspirin,* a word with an interesting derivation. The *a* is from the acetylsalicylic acid, and the *spirin* is from *Spirea,* the plant from which salicylic acid was first isolated. (The names *salicin* and *salicylic acid,* of course, reflect the *Salix* origin.)

Aspirin is valued for its three classic properties as an **anti-inflammatory, antipyretic** (fever reducing), and **analgesic** (pain relieving), although the mode of action in the body is not completely understood. It has also found new uses in the prevention of heart attacks, strokes, and colon, ovarian, and esophageal cancer. Aspirin may also delay the development of cataracts in the elderly and enhance the immune system in protecting the body against bacteria and viruses.

Of these uses, probably the greatest attention has been given to the beneficial effects of an aspirin a day in the prevention of heart attacks. Various studies have shown that the administration of aspirin after a heart attack or stroke reduces the risk of a second heart attack or stroke. The subsequent death rate among the group treated with aspirin was significantly lower than for the control group. Additional studies have shown that aspirin given to healthy middle-aged men reduces the incidence of a first heart attack by 44%. Although the initial studies were conducted with men, more recent studies have confirmed similar beneficial effects for women.

In 1970, John R. Vane and others of the Royal College of Surgeons in London found that low doses of aspirin suppress the aggregation of blood platelets. This aggregation is a necessary step in the formation of blood clots, which can block blood vessels and lead to heart attacks and strokes. (Vane later received a Nobel Prize for his work.) The mechanism behind this and other actions of aspirin is the suppression of prostaglandins, a group of hormonelike substances that are widely produced throughout the body and have a number of regulatory activities. Prostaglandins are not stored in the body but are released from normal or injured cells or cells that have been stimulated by other hormones. Prostaglandins mediate physiological responses affecting functions such as digestion, reproduction, circulation, and the immune system. An overproduction of prostaglandins can lead to headaches, fever, menstrual cramps, blood clots, inflammation, and other complaints. Specifically, aspirin inhibits the enzyme cyclooxygenase (COX), which promotes the production of prostaglandins.

Although aspirin is incredibly versatile, it is not without drawbacks. It has been known for quite some time that aspirin irritates the stomach. Scientists have found ways to coat, or buffer, it; however, this gastric distress is still one of its drawbacks. Again, it is the suppression of prostaglandins that is responsible for the irritation. In the stomach, prostaglandins prevent the overproduction of acid and promote the secretion of mucus that blocks self-digestion of the stomach lining.

A second drawback, discovered in the 1960s, resulted in limiting the use of aspirin for children and teenagers. Children who had taken aspirin while recovering from chicken pox or influenza occasionally developed unusual symptoms, including vomiting and change in mental alertness. This condition is known as Reye's syndrome and, although rare, can be fatal. It is unclear how aspirin is involved in this syndrome.

Since its initial discovery in willow bark, salicylic acid has been identified in a great many plants. It normally occurs in low concentrations in plants and is capable of being translocated in the phloem. In fact, salicylic acid is a naturally occurring plant hormone involved in several reactions, including plant protection. Plants respond to pathogen attack by activating a network of defenses. At the site of infection, this response might include strengthening the cell walls, synthesizing enzymes that attack the pathogen, or synthesizing other antimicrobial compounds. Plants also have more broadly based defense mechanisms known as **systemic acquired resistance,** which helps protect against secondary infection. Salicylic acid is the signal that turns on this systemic response, which results in the synthesis of specific proteins that increase resistance. In infected plants, salicylic acid levels rise significantly throughout the plant, activating the resistance proteins. The external application of salicylic acid or even aspirin (acetylsalicylic acid) to plants will also stimulate this immune response. Recently, researchers discovered that stimulation of salicylic acid in an infected plant can also turn on responses in neighboring plants. Some of the salicylic acid was converted to methyl salicylate, a volatile compound that readily evaporated from the diseased areas of the plant. Healthy plants nearby absorbed the airborne methyl salicylate molecules and converted them back into salicylic acid. This stimulated defenses, making the healthy plants more resistant to the pathogen. Researchers hope that this knowledge may one day be utilized to reduce the use of toxic fungicides and other pesticides on agricultural crops. An aspirin spray may keep the pathogen away!

Malaria, the Fever Bark Tree, and Sweet Wormwood

Malaria, known since antiquity, is still one of the world's most prevalent diseases. One to three million people die each year of malaria, and the majority of these deaths are of young children in sub-Saharan Africa. About 3 billion people live in parts of the world where malaria is endemic, and over 200 million infections are diagnosed each year. Malaria is largely confined to tropical and subtropical countries in Asia, Africa, and Central and South America (fig. 19.7a). In the past, however, malaria reached epidemic proportions in temperate areas of North America, Europe, and Asia and may again as a result of global warming (see Chapter 26).

Records of malaria can be found in ancient Egyptian papyri, which date back over 3,500 years, and in the work

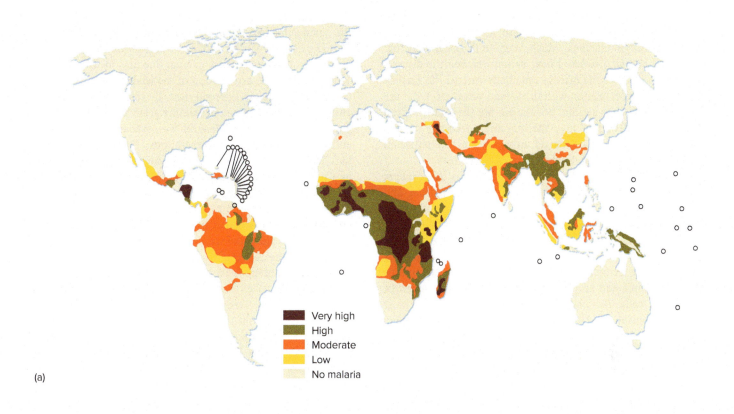

(a)

- Very high
- High
- Moderate
- Low
- No malaria

1. When a female *Anopheles* mosquito, harboring the malaria parasite in her salivary glands, bites an individual, the sporozoite stage of the parasite is injected. The sporozoites are carried to the liver in the bloodstream.

2. In the liver, merozoites are produced and released into the bloodstream.

3. The merozoites invade red blood cells and multiply.

4. After a period of time the red blood cells rupture, releasing the new generation of merozoites, which then infect other red blood cells. The cycle is synchronous, with simultaneous rupture of red blood cells and release of merozoites and toxins, causing the periodic fever and chills.

5. Male and female gametocytes are formed in some red blood cells. If the gametocytes are ingested when an *Anopheles* mosquito bites an infected individual, they can complete sexual reproduction and begin the infection cycle anew.

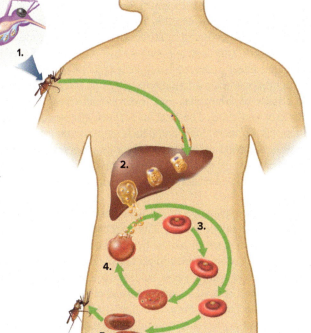

(b)

Figure 19.7 Malaria is usually considered the world's most prevalent disease. (a) Currently, malaria is endemic in tropical and subtropical countries. (b) Life cycle of *Plasmodium vivax*, a protozoan responsible for one form of malaria.

Source: Data from World Health Organization, 2005.

of the Greek physician Hippocrates. The Greeks also noted the higher incidence of the disease among people living near swamps or marshes. Much later, Italians believed that breathing bad air (*malaria*) near swamps caused the disease, and their name for the disease eventually entered the English language. The correlation with the swamp was correct, but the mosquitos near the swamp, not the air, are the vectors of the disease. The details of how the mosquitos transmit malaria were discovered by Sir Ronald Ross, an English physician working in India during the late nineteenth century.

Four species of protozoans in the genus *Plasmodium* cause malaria in humans: *P. vivax, P. ovale, P. falciparum,* and *P. malariae*. The French physician Alphonse Laveran first observed the unicellular parasite in human blood in 1881. The four species each cause a slightly different form of malaria, varying in the recurring onset of fever and chills, the most characteristic symptoms of the disease. Other symptoms include anemia and an enlarged spleen due to destruction of red blood cells. *Plasmodium falciparum* is the species most often responsible for fatalities. Untreated, *P. falciparum* infection can result in cerebral malaria, which is characterized by convulsions, seizures, and coma and can lead to death.

The disease is initiated through the bite of a female *Anopheles* mosquito that has previously fed on the blood of an infected individual. The mosquito carries a stage of the parasite in its salivary glands and injects the parasite into the bloodstream with its bite. The parasite multiplies in the liver and eventually releases merozoites (the asexual stage in the life cycle of the parasite) into the bloodstream. The merozoites invade red blood cells and multiply; after a period of time, the red blood cells rupture, releasing a new generation of merozoites that infect other red blood cells (fig. 19.7b). The cycle is synchronous, with simultaneous rupture of red blood cells and release of merozoites and toxins, which cause the periodic fever and chills. The interval between symptoms is most commonly 48 hours. Male and female gametocytes are formed in some red blood cells. These do not develop any further in the victim, but if they are ingested by an *Anopheles* mosquito, they can complete sexual reproduction and begin the infection cycle anew.

Until the application of the fever bark tree, there was no effective treatment for malaria. The fever bark tree, called *quina-quina* (meaning bark of barks) by the Incas, is native to the eastern slopes of the Andes Mountains. This small evergreen tree with broad, shiny leaves belongs to the Rubiaceae, the coffee family. The fever-reducing powers of the tree were well known to the Incas of Peru, and they shared this knowledge with Jesuit missionaries. The Jesuits later used infusions from the bark to treat people infected with malaria. Legend has it that in 1638 one of the treated individuals was the Countess of Chinchon, wife of the viceroy of Peru. Her miraculous recovery spread the reputation of the bark, which was soon introduced to Europe. (Later, Linnaeus, believing the legend, named the genus *Cinchona* in honor of the countess.) A new trade developed, briefly under the monopoly of the Jesuits, in harvesting and shipping the "Jesuits' bark" to the Old World. By the end of the seventeenth century, the powdered bark of the quina-quina tree was the standard treatment for malaria.

The demand for the bark was tremendous because it took 2 pounds of powdered bark to treat one person with malaria. This resulted in an exploitation of the wild stands of *Cinchona* trees that threatened their survival. In 1820, two French scientists isolated the alkaloid quinine from the bark; within a few years, the purified alkaloid was available commercially and replaced whole bark preparations. The demand for the bark increased even more. Four out of a total of 36 alkaloids in *Cinchona* bark actually have antimalarial properties, but quinine is the most effective of these compounds. The concentration of quinine in the bark varies considerably among the 40 species within the genus, and even among different populations of the same species.

During the nineteenth century, the British and Dutch established plantations in India and Java, respectively. High-yielding strains of *Cinchona ledgeriana,* whose bark contains up to 13% quinine, were first developed in Bolivia. The Dutch purchased seeds from these strains to start their plantations. The Dutch dominated the world quinine trade until sources were cut off during World War II, and synthetics filled the void. One of these synthetics, chloroquine, became the most widely used drug for the treatment of malaria; it was less toxic and more effective than quinine. Chloroquine also lacks some of the side effects often associated with quinine treatment: tinnitus (ringing in the ears), blurred vision, depression, and miscarriages. Unfortunately, the widespread use of this synthetic has fostered the development of chloroquine-resistant strains of *P. falciparum* in large areas of Central and South America, Africa, and Asia. Because these strains are not resistant to quinine, it may be used, in combination with other drugs, for these resistant infections.

Quinine acts on the merozoite stage, killing the parasite in the bloodstream. It is also effective as a prophylactic, preventing the initial infection of red blood cells in travelers visiting malaria-infested areas. The prophylactic value of quinine brought about the development of tonic water used by the British in India. The British colonists made the quinine water more palatable by adding gin; this practice popularized gin and tonic as a favorite drink in the tropics. The greater reliance on synthetics, both for prophylaxis and treatment, has funneled most of the quinine production today into flavoring for beverages.

Beginning in the late 1960s, scientists in China began investigating the antimalarial properties of plants used in traditional Chinese medicine. The lead scientist in this project was Youyou Tu, a pharmaceutical chemist. Although hundreds of plant extracts were evaluated, progress was slow until extracts from *Artemisia annua* (sweet wormwood) were identified. The species, a member of the Asteraceae (sunflower family) is native to China, where it is known as *qinghao*; however, the species has become naturalized and can be found in most temperate areas of the world. *Qinghao* was long known in China to reduce fever, and in 1971 Youyou Tu and her colleagues found that the extracts were 100% effective in killing malaria parasites in infected mice and monkeys. The chemists discovered that this plant contains artemisinin, the compound, which is toxic to the malaria parasites. Artemisinin is a terpene (see A Closer Look 1.2: Perfumes to Poisons), which occurs in glandular trichomes found on the leaves and stem. This was the first natural antimalarial since quinine; it shows fewer side effects than quinine and is effective against chloroquine-resistant strains of *P. falciparum*. In 2015 Youyou Tu was awarded the Nobel Prize in Physiology or Medicine, jointly with two other scientists for an unrelated discovery.

In addition to artemisinin, a number of semisynthetic derivatives have also been developed and are being used to treat chloroquine-resistant strains of *Plasmodium* in endemic areas of the world. Artemisinin rapidly clears the malaria parasites from the blood; however, it has a short half-life, so high rates of reinfection are possible. In addition, the liver stages of malaria are not affected, so artemisinin is not an effective prophylactic. Over the past few years, there have been reports of artemisinin resistance developing in several countries. As a result, the World Health Organization now recommends combination therapy with artemisinin and a quinine derivative or other class of compound as the first-line treatment of uncomplicated falciparum malaria. Within the past decade, control efforts with insecticide-treated mosquito nets have significantly reduced the number of new cases in sub-Saharan Africa during the last 10 years. This has raised the hope that malaria can be eradicated; artemisinin and its derivatives will likely play a prominent role in these efforts.

Diabetes, French Lilac, and Metformin

Diabetes mellitus (commonly called diabetes) is a metabolic disorder that is characterized by elevated blood glucose (blood sugar) levels. Recall from Chapter 4, glucose is the main source of energy for living cells; the food we eat is the source of the glucose in our blood. The hormone insulin, which is produced in the pancreas, helps glucose get into our cells. If the body does not make enough insulin or if the insulin is not being used well, then glucose levels build up in the blood. Excess blood sugar can eventually damage eyes, kidneys, and nerves. It can also lead to heart disease, strokes, and blindness. The symptoms of diabetes include frequent urination, increased thirst, blurred vision, and unexplained weight loss. However, sometimes symptoms are so mild, they are overlooked.

The three major types of diabetes are type 1 diabetes, type 2 diabetes, and gestational diabetes. Type 1 diabetes is also called insulin-dependent diabetes because the pancreas does not make enough insulin, or any insulin. This occurs when the immune system attacks and destroys the beta cells in the pancreas that make insulin. This condition is usually diagnosed in children although it can occur at any age. Individuals with type 1 diabetes must take insulin, either by injection or through an insulin pump.

The most common form of diabetes is type 2 diabetes. In type 2 diabetes, the body does not make enough insulin or the insulin does not function properly, which is known as insulin resistance. In insulin resistance, cells do not respond well to insulin and, therefore, cannot easily take up glucose from the blood; this principally affects the cells in muscles, fat, and liver. Type 2 diabetes is usually diagnosed in middle-aged or older people although it can occur at any age. Individuals who are overweight, physically inactive, or have a family history of diabetes are more likely to develop type 2 diabetes. An estimated 27 million adults in the United States have type 2 diabetes; this estimate represents 90% of all adult diabetes cases. The World Health Organization estimates that 422 million, worldwide, have type 2 diabetes; many in the medical community consider this a global epidemic.

Gestational diabetes occurs in some women who are pregnant, and it usually goes away after the baby is born. However, women who have had gestational diabetes have a greater risk of developing type 2 diabetes later in life.

Some medical historians suggest that the first written reference to diabetes is in the Ebers papyrus about 3500 years ago; however, the text there is vague, mentioning only a condition producing "plentiful urine" and suggesting various remedies for treating this condition. In ancient India an Ayurvedic medical text from 2500 years ago references two types of diabetes, and ancient texts from China have similar references. In these early societies, various plants were used to treat this condition. Later, the writings of Greek physician Aretaeos of Cappadokia in 100 A.D. provide an accurate description of diabetes and its symptoms. In fact, Aretaeos is the physician who named the condition "diabetes."

Over the centuries, hundreds of plants have figured prominently in herbal remedies for treating diabetes. Among the plants used by various societies are *Aloe vera* (burn plant), *Catharanthus roseus* (Madagascar periwinkle), *Coriandrum sativum* (coriander), *Juniperus communis* (juniper), *Eucalyptus globulus* (eucalyptus), *Allium sativum* (garlic), and *Glycyrrhiza glabra* (liquorice).

Possibly the most important herbal remedy is *Galega officinalis*, commonly known as French lilac, goat's rue, Italian fitch, or professor-weed. This herbaceous member of the Fabaceae (bean family) was originally native to the Middle East but spread and became naturalized in much of Europe. It was introduced to the United States in the 1890s and now is found in 10 states where it is considered an invasive weed. In medieval Europe, French lilac was used as a traditional medicine for several ailments and can be found in various herbals during this period. In 1772 British botanist John Hill included an account and illustration of *Galega* in *The Vegetable System* and described its use for treating symptoms of diabetes. This species continued to be used as a traditional remedy for diabetes into the twentieth century.

In the late nineteenth century, it was discovered that *G. officinalis* contained guanidine and related compounds. In 1918 guanidine was shown to reduce blood sugar in animals but was too toxic for medical use. In the 1920s several compounds related to guanidine were synthesized including metformin. Testing in 1929 showed that metformin was effective in lowering glucose levels. However, the potential use of this compound was not fully appreciated at this time because insulin was becoming widely available to treat diabetics and overshadowed other therapies.

Little additional research was done on the use of metformin for diabetes until 1956 when Jean Sterne, a physician at the Aron Laboratories in France, began investigating the mode of action of several guanidine derivatives in normal and diabetic animals. Sterne and colleagues chose to focus on metformin, based on its effectiveness and fewer side effects than other guanidine derivatives. Clinical trials of the drug with diabetes patients showed that metformin could replace the need for insulin in some patients with type 2 diabetes and lower the dose needed for others. In those with type 1 diabetes, it did not eliminate the need for insulin. In 1957, Sterne published the results of his research. Aron Laboratories soon began marketing metformin under the brand name Glucophage (meaning glucose eater), and within a couple of years it was available in a few European countries. During this same time, similar compounds, phenformin and buformin, were being developed and heavily marketed by larger pharmaceutical companies; although effective in lowering blood glucose levels, these drugs were later withdrawn from most countries due to serious side effects. Metformin was approved for use in the United States in December 1994 and was widely available early in 1995. Today, metformin is the drug of choice worldwide for managing type 2 diabetes and is the most prescribed oral medication for treating this condition. Metformin is also prescribed to those individuals considered to be at high risk for developing type 2 diabetes and to treat gestational diabetes.

Metformin reduces blood glucose levels by at least three modes of action. The main one is believed to be through decreasing the production of glucose in the liver by inhibiting gluconeogenesis. Gluconeogenesis is a process that results in the formation of glucose from non-carbohydrate substrates, including glycerol, certain amino acids, and other molecules. This is one of the mechanisms that helps maintain blood glucose levels in the body. The inhibition of gluconeogenesis occurs through complex effects on mitochondrial enzymes. Secondly, metformin improves insulin sensitivity by increasing the activity of insulin receptors and, thereby, increasing the uptake of glucose into cells, especially in liver and skeletal muscle cells. This action also lowers blood glucose levels. Research shows that a third mode of action is a reduction in the amount of glucose absorbed by the gastrointestinal tract.

Metformin has additional benefits for patients with type 2 diabetes. Coronary artery disease is the leading cause of death in patients with type 2 diabetes. In a large epidemiological study, it was shown that metformin reduces heart attacks by 39% and coronary deaths overall by 50%. At this time, it is not known if this is a result of reduced blood glucose levels or if there are direct benefits involved. Metformin also appears to decrease the risk of kidney damage, and in some overweight or obese individuals, metformin also results in modest weight loss.

In addition to type 2 diabetes, metformin may be used to treat polycystic ovary syndrome (PCOS), a condition that results from an imbalance of hormone levels in a woman's body. This causes cysts to form on the ovaries, irregular menstrual cycles, and other symptoms. PCOS is the most common hormonal disorder among women in their reproductive years and is one of the major causes of infertility. Insulin resistance often occurs in PCOS, therefore women with PCOS have an increased risk of developing type 2 diabetes. Studies have shown that metformin can improve the hormonal imbalance as well as restore regular menstrual cycles and ovulation in up to 50% of the women with PCOS. Metformin improves the insulin resistance and has also been shown to prevent gestational diabetes in patients with PCOS.

From its humble origins as an herbal remedy in medieval Europe to metformin, which is on the list of essential medicines from the World Health Organization, *Galega officinalis* has truly had a remarkable history.

Snakeroot, Schizophrenia, and Hypertension

Snakeroot, *Rauwolfia serpentina* (fig. 19.8), exemplifies the belief of the Doctrine of Signatures. Because the long, coiled roots resemble a snake, healers believed that the root could be used for treating snakebites. For more than 4,000 years, Hindu healers in India used the root of this rain forest shrub for the treatment of snakebites, insect stings, and even mental illness. Tea brewed from the leaves was known to have a soothing

Figure 19.8 *Rauwolfia serpentina* is the source of the drug reserpine and other alkaloids that have been used in the treatment of hypertension and schizophrenia.

effect and was often used to induce a meditative state. Despite all these claims, *R. serpentina* came to the attention of Western medicine only about 63 years ago. In 1952, the alkaloid reserpine was the first active principle isolated from the roots, but today dozens of alkaloids are known, with rescinnamine and deserpidine also important pharmaceuticals.

The sedative effects of reserpine (it depresses the central nervous system) made it valuable as one of the first tranquilizers prescribed for schizophrenia. A side effect observed during the administration of reserpine to mentally ill patients was a reduction in their blood pressure. This side effect became the principal application of reserpine—a treatment for hypertension. In fact, at one time many drugs prescribed for controlling high blood pressure contained *Rauwolfia* alkaloids. These alkaloids act on the nervous system by blocking neurotransmitters; this results in the dilation, or relaxation, of the blood vessels, causing the blood pressure to drop. The sedation and depression associated with their use as hypotensive agents are now considered negative side effects and must be monitored carefully. Because newer drugs have fewer side effects, reserpine is only used today when other options are not available.

Originally native to forests in India, *R. serpentina* is now cultivated in India as well as in Thailand, Bangladesh, and Sri Lanka. Other species, including *R. vomitoria* from Africa, are also grown for their alkaloids and, in fact, have a long history of use by native healers. Although the *Rauwolfia* alkaloids can be synthesized in the laboratory, it is much cheaper to extract them from natural sources.

The Burn Plant

A folk remedy familiar to most people in contemporary society is the application of *Aloe vera* sap for minor burns and cuts. Members of the genus *Aloe* have been used for thousands of years as treatments for various skin ailments, including rashes, sunburns, direct burns, scalds, and minor wounds. Members of the genus are succulent perennials originally native to Africa, and their medicinal use dates back to the ancient cultures in Africa and the Mediterranean area. After its introduction to the Western Hemisphere, it also became widely adopted by indigenous groups for its medicinal value. Aloes usually have basal rosettes of large, succulent, sword-shaped leaves that may give rise to a vertical inflorescence of brightly colored, tubular flowers. *Aloe vera* (*A. barbadensis*), the burn plant or medicine plant (fig. 19.9), is probably the best-known member of the genus, but many other *Aloe* species are also used.

When cut, the succulent leaves yield a thick, mucilaginous sap (also referred to as a gel) that can be soothing when applied to injured skin. The sap contains numerous compounds, including several anthraquinone glycosides, collectively referred to as aloin, and glucomannans, which are polysaccharides composed of glucose and mannose. One glucommanan is acemannan which is currently believed to be the compound with possibly the greatest healing effect on skin. Studies have shown that aloe sap promotes faster healing with less scarring by stimulating cell growth and inhibiting bacterial

Figure 19.9 Compounds in the sap of *Aloe vera* are known to enhance healing of burns and other skin ailments.

and fungal infection in injuries ranging from deep dermal burns to radiation burns. Other studies have shown that compounds in the sap inhibit pain, itching, and inflammation. The sap is also useful in treating skin and mouth ulcers, eczema, psoriasis, ringworm, athlete's foot, and poison ivy rashes.

Aloe sap is also used as a powerful purgative for the relief of constipation. The anthraquinones in the sap apparently irritate the gastrointestinal tract, resulting in its purgative action. The action is drastic, however, and is therefore often used in combination with other drugs to mollify the effect. Aloe sap has also been used internally as a traditional treatment for diabetes. In a recent investigation, dried aloe sap showed some promise in lowering blood glucose levels among a small group of noninsulin-dependent diabetics.

For many years, the cosmetic industry has capitalized on the moisturizing effects of the sap, and it can be found in a variety of skin creams, shampoos, sunscreen lotions, and bath oils. Aloe has been cultivated since the fourth century B.C., when Alexander the Great reportedly conquered the island of Socotra in the Indian Ocean to obtain a supply of aloe. Today, there are large plantations in Texas and Florida to supply this time-proven folk remedy.

Ephedrine

Ephedrine is an alkaloid produced by members of the gymnosperm genus *Ephedra* (see Chapter 9). Various species of *Ephedra* have a long history of use in herbal medicine as a decongestant for asthma, bronchitis, and other respiratory ailments. In China, *Ephedra sinica,* known as Ma Huang, has been used for several thousand years as a decongestant as well as for its properties as a stimulant. *Ephedra* species were also used by native peoples in Russia and India and by several Native American tribes.

The alkaloid was first identified in 1887, and ephedrine use was incorporated into modern prescription drugs during the 1920s. The decongestant properties are well known, and this alkaloid still is found in both over-the-counter and prescription decongestants and asthma medications. The decongestant effects result from the fact that ephedrine relaxes bronchial muscles. In addition, ephedrine is a central nervous system stimulant that is similar in action to adrenaline; it increases heart rate and blood pressure and also increases blood flow to the heart and brain. Today pseudoephedrine, another *Ephedra* alkaloid, is often preferred as the active ingredient in decongestants because the stimulating properties are not as great. Both ephedrine and pseudoephedrine have been chemically synthesized, and the synthetic versions are used in both prescription drugs and over-the-counter medications.

The stimulating effects of ephedrine are similar to effects produced by amphetamines, and this similarity has led to abuse of herbal *Ephedra* products. In fact, some *Ephedra* products are described as "herbal ecstasy" because they can be used as a substitute for the illegal street drug ecstasy. Ephedrine was also found in over-the-counter weight-control supplements and herbal energy boosters because it activates thermogenesis, the process by which the body burns calories to generate heat.

For several years, the U.S. Food and Drug Administration (FDA) issued warnings that ephedrine can be a dangerous drug. Herbal teas and weight-loss pills that contain ephedrine cause irregular heartbeats, dizziness, headaches, heart attacks, strokes, seizures, and psychotic episodes. More than 150 people have died and thousands have suffered harmful effects from dietary supplements containing ephedrine. In April 2004, the FDA banned the use of ephedrine in dietary supplements and herbals. The ban also includes pseudoephedrine and other, similar alkaloids found in species of *Ephedra* and other plants. Although FDA warnings had been previously issued about ephedrine and other herbal compounds, this action was the first FDA ban of a dietary supplement. (See Herbal Remedies: Promise and Problems later in this chapter.) Prior to the FDA action, several states had banned the sale of *Ephedra* products, and other states regulated sales by listing the products as controlled substances. The ban was challenged in U.S. District Court by Nutraceutical Corporation, a company that makes an *Ephedra* product with less than 10 mg of ephedrine. In a 2005 judgment, the FDA was prevented from taking any action against the company for the sale of dietary supplements containing 10 mg or less of ephedrine or similar alkaloids per daily dose.

The FDA ban on *Ephedra* products does not apply to prescription drugs or over-the-counter medications. However, in April 2004, Oklahoma passed a law regulating the availability of all over-the-counter cold and allergy remedies containing the decongestant pseudoephedrine. The law restricts sales of these medications to pharmacies and requires purchasers to present photo identification and to sign for the purchase. This legislation was part of a crackdown on illegal methamphetamine production because pseudoephedrine can be used to prepare methamphetamine. Other states soon followed with similar legislation, and in 2006 Congress passed a federal law restricting sales of over-the-counter medication containing either ephedrine or pseudoephedrine.

> **Thinking Critically**
>
> The abuse of products containing *Ephedra* extracts led to increased concerns about the unregulated marketing of herbal remedies.
>
> *Investigate the stores in your area that sell herbal remedies. What plant extracts are available that have known physiological or psychoactive effects?*

Cancer Therapy

Cancer is a diverse group of diseases characterized by uncontrolled growth of abnormal cells. Humans have suffered from cancer for thousands of years, and treatments have existed for just as long; even the Ebers Papyrus gives instructions for the treatment of cancer. Today in the United States, cancer is the second leading cause of death, and the search for cancer cures is relentless. Over the years, plants have figured prominently in folk remedies, and as reported in the late 1960s, more than 3,000 plant species have been used. In the United States, the search for plant sources of anticancer drugs began in the late 1950s under the National Cancer Institute, later joined in its efforts by the U.S. Department of Agriculture. Thousands of plants have been scientifically screened, and several have become part of standard chemotherapy for different forms of cancer (see table 19.1). But the search is not over because only a small percentage of known plant species has been investigated. It is especially crucial for this work to continue in the tropics, where rain forest destruction is proceeding at an alarming rate and most species have not been described scientifically.

Vinca Alkaloids

One remarkable story in the fight against cancer is the development of vinblastine and vincristine as treatments for various forms of leukemia and lymphoma. These alkaloids were isolated from the Madagascar periwinkle, *Catharanthus roseus* (*Vinca rosea*), a perennial herb native to the tropics (fig. 19.10a). This species had been used by traditional healers as a treatment for diabetes. Investigating this claim in the 1950s, scientists at the University of Western Ontario in Canada and Eli Lilly Pharmaceuticals in Indianapolis found no evidence of usefulness in treating diabetes. Extracts from the leaves, however, were found to be effective against leukemia cells, and soon the alkaloids responsible were identified. The concentration of alkaloids in the leaves is very low; it takes 53 tons of leaves to produce 100 grams (3.5 ounces) of vincristine.

Today, these alkaloids are widely used and are major chemotherapeutic agents. The alkaloids act on the microtubular level, where they interfere with spindle formation and thus prevent mitosis. Vincristine is especially effective for treating acute childhood leukemia, and vinblastine is useful in treating

Figure 19.10 Two important sources of anticancer drugs are (a) periwinkle, *Catharanthus roseus,* and (b) Pacific yew, *Taxus brevifolia.*

Hodgkin's disease. Both alkaloids are also used in chemotherapy for other types of cancer.

Taxol

Another anticancer drug that is the focus of extensive research is taxol, a terpene obtained from the bark of the Pacific yew, *Taxus brevifolia* (fig. 19.10b), a known poisonous plant (see Chapter 21). Like the vinca alkaloids, taxol acts on the microtubules, although the mode of action is different. The antitumor properties were first discovered in the 1960s during a screening program of the National Cancer Institute, and taxol was first approved for the treatment of ovarian cancer in 1992 and breast cancer in 1994. Today, taxol is a major chemotherapy agent used to treat patients with breast, ovarian, lung, head, and neck cancer as well as advanced forms of Kaposi's sarcoma. Clinical trials are still ongoing to determine the effectiveness of taxol for other forms of cancer.

The supply of taxol was initially a problem because it had not been synthesized in the laboratory. The original source of the alkaloid was the bark from mature Pacific yew trees, slow-growing conifers of old-growth forests in the Pacific Northwest. Concern about the destruction of these ancient forests prompted alternative approaches to obtaining a continual supply of taxol. Other, more common species of *Taxus,* including the common yew and Japanese yew, also contain taxol, with some occurring in the needles as well as the bark. Another promising direction is the establishment of taxol-producing tissue cultures of bark cells. One report indicated that cultures yielded seven times more taxol than is naturally produced in the bark of the tree. This method may eventually become an important and inexpensive source of taxol and other plant-derived medicines. In 1994, taxol was synthesized, putting to rest the environmental concerns. Today semisynthetic taxol can be prepared in the laboratory and is available in virtually unlimited amounts for the treatment of various types of cancer. Clinical results from taxol use have been so successful that researchers have been actively involved in developing a family of taxol-like compounds, referred to as taxanes. It is believed that taxanes may be ultimately less expensive, easier to produce, and more effective than taxol. One taxane, docetaxel, has been approved and is now in widespread use for treating breast, lung, prostate, stomach, head, and neck cancers.

Camptothecins

Camptotheca acuminata is a deciduous tree that is native to southern China and Tibet, where it is known as *xi shu,* or the Chinese happy tree. In China, preparations from *Camptotheca* have been used for centuries to treat various forms of cancer. In the late 1950s, the screening program conducted by the National Cancer Institute showed that compounds from *Camptotheca* had anticancer properties. Camptothecin, an alkaloid, was isolated from the bark and wood in 1966, and animal studies confirmed the anticancer properties of this compound.

Although camptothecin was promising in animal studies, its early clinical trials were suspended because of high toxicity and severe side effects. However, in 1985, it was discovered that camptothecin had a unique mode of action as a topoisomerase I inhibitor. Topoisomerase enzymes are required for the replicaton and repair of DNA and are therefore required for cell division in the rapidly proliferating cancer cells. Inhibition of topoisomerase triggers a cascade of events that results in cell death. This discovery led to renewed interest in camptothecin. In the late 1980s, researchers developed several semisynthetic derivatives that had fewer side effects. Two of these derivates, topotecan and irinotecan, were approved by the FDA in 1996. Topotecan was initially approved for the treatment of advanced ovarian cancer, and irinotecan was approved for metastatic colon and rectal cancers. These compounds are

> **Thinking Critically**
>
> The success of taxol in treating ovarian and breast cancer has renewed interest in searching for medicinal plants. At the same time, forest destruction is eroding the biodiversity of the planet.
>
> *What steps can be taken to ensure that useful plant species will be preserved?*

Figure 19.11 Herbal remedies for sale on a street in Grenada, Spain.

now also used in the treatment of other cancers. In addition, several other semisynthetic derivatives are in the final stages of clinical trials. Today camptothecins are among the most promising new anticancer drugs available.

Camptothecin has been synthesized in the laboratory; however, the chemical synthesis is not cost-effective. As a result, natural sources are needed for the production of camptothecin derivatives. Although camptothecin was initially isolated from bark and wood, it was later found in all parts of the tree, with high levels occurring in young leaves. Harvesting of *Camptotheca acuminata* trees resulted in a serious decline of this species in China, where it is now considered a protected species. *Camptotheca* cultivation has been started in several southern states in the United States. In addition, a new variety of a related species, *Camptotheca lowreyana*, has been developed by researchers in Texas. This new variety is cultivated as a shrub and has high yields of camptothecin in the leaves, which can be harvested 6 months after planting.

Other plant-based compounds are also in use as anticancer drugs or are in various stages of testing through clinical trials. Meanwhile, ethnobotanists continue to search for other promising folk remedies that will become the "taxol" of tomorrow.

Herbal Remedies: Promise and Problems

Since the 1990s, one significant change in the attitude of health-conscious individuals has been the dramatic increase in the use of alternative medical treatments (fig. 19.11). *Alternative medicine* refers to a wide range of therapies outside the mainstream of traditional Western medicine and includes such treatments as aromatherapy, acupuncture, biofeedback, chiropractic manipulation, herbal medicine, hypnosis, and massage therapy. Plants and plant extracts (often called botanicals) figure prominently in alternative treatments in both herbal medicine and aromatherapy (see A Closer Look 17.1: Aromatherapy.)

Herbal remedies are considered dietary supplements by the FDA. Traditionally, *dietary supplements* referred to vitamins and minerals or other essential nutrients; however, the Dietary Supplement Health Education Act (DSHEA) of 1994 expanded the category to include other products, such as herbs, other botanicals, amino acids, and metabolites. Sales of herbal remedies amount to over $7 billion per year in the United States and constitute over 20% of the total sales for dietary supplements.

The FDA requires extensive testing and clinical studies of prescription drugs and over-the-counter drugs to determine their safety, proper dosage, effectiveness, possible side effects, and interactions with other substances. Dietary supplements are *not* subject to these tests. The DSHEA holds the manufacturer of a dietary supplement responsible for ensuring that the product is safe but does not require FDA approval before sale. The FDA will, however, take action if a dietary supplement is later shown to be unsafe. For example, in 2001, the FDA advised manufacturers of dietary supplements to remove comfrey (*Symphytum officinale*) from products and to alert consumers of the possibility of liver disorders from use of this herb. Other warnings were issued for *Aristolochia*, kava (see Chapter 20), and St. John's wort because of interaction with prescription drugs. In addition, a ban of *Ephedra* products went into effect in April 2004 (see Ephedrine earlier in this chapter).

Like medicinal plants, herbal remedies contain secondary compounds that can have powerful physiological effects on people. Although not considered drugs by the FDA, they may, in fact, offer health benefits or cause adverse reactions. Herbal remedies also have limitations and are not right for everyone. Recent studies indicate that over 40% of the U.S. population uses some form of alternative medicine, especially to treat chronic conditions, such as back pain, headaches, depression, allergies, or asthma. Of those who use alternative medicine, a large percentage do not report the use to their primary care physician. This failure to report may result in problems because some herbal remedies interfere with or intensify the action of prescription and over-the-counter drugs. For example, the use of *Ginkgo* is not recommended for individuals taking aspirin or anticoagulants because the combination may cause internal bleeding. It is essential that physicians be aware of what herbal remedies their patients are taking, so that they can inform their patients of possible interactions with

Table 19.2 Top-Selling Herbal Remedies in 2016

Common Name	Scientific Name	Use
Horehound	Marrubium vulgare	Cough
Cranberry	Vaccinium macrocarpon	Urinary tract infections
Echinacea	Echinacea purpurea	Immune system stimulant
Green tea	Camellia sinensis	Antioxidant
Black cohosh	Actaea racemosa	Menopause symptoms
Garcinia cambogia	Garcinia gummi-gutta	Weight loss
Flax or flaxseed oil	Linum usitatissimum	Constipation, high cholesterol
Ginger	Zingiber officinale	Nausea
Ivy leaf	Hedera helix	Congestion
Turmeric	Curcuma longa	Inflammation

medications. The lack of rigorous scientific study of herbal remedies is starting to be addressed; independent researchers not connected to the manufacturers are beginning to examine the efficacy of these compounds as well as their possible side effects. So far, the studies have shown that most herbal remedies are safe but that few have positive health benefits. The top-selling herbal remedies in the United States in 2016 are shown in Table 19.2, however, it should be noted that the list of best-selling supplements changes often. Three widely used supplements are described in this section, although these are not currently on the top 10 list.

St. John's Wort

St. John's wort (*Hypericum perforatum*) is the latest treatment for sufferers of depression, a mental illness that affects millions of Americans each year. A perennial native to Europe, it is so named because it blooms with bright yellow flowers around June 24, the birthday of St. John the Baptist. St. John's wort has a long history of use in herbal medicine and is cited in the classical works of Hippocrates, Pliny, and Dioscorides. The herb has been credited for its effectiveness in healing wounds and relieving digestive upsets, but what has attracted the interest of physicians and medical researchers today is its effectiveness as an antidepressant. Although relatively new to most North American practitioners, St. John's wort has been prescribed by European physicians for about 25 years. In Germany, St. John's wort is the leading treatment for mild and moderate depression. Clinical trials in Great Britain showed that patients treated with the herb showed significant improvement. One possible drawback to St. John's wort use is the concern that it may induce photosensitivity in certain individuals (as it does in sheep; see Chapter 21). Like many antidepressant medications, St. John's wort apparently lifts spirits by raising levels of certain mood-enhancing neurotransmitters in the brain, particularly serotonin. It has been known for some time that patients suffering from depression have reduced levels of serotonin. For the same reason, there is also interest in developing St. John's wort as a weight-loss aid. Patients on medications that raise serotonin levels have been observed to lose weight.

The results of a recent examination into the effectiveness of St. John's wort in alleviating major depression was published in *JAMA*, the *Journal of the American Medical Association*. In contrast to previously reported studies, this study found that there was no difference between the group treated with St. John's wort and the placebo group; however, there have been several criticisms of this study. St. John's wort has been recommended for patients who are mildly to moderately depressed, not for the treatment of major depression. Patients in the *JAMA* study were severely and chronically depressed for more than 2 years. Also, the study did not compare the effectiveness of St. John's wort with that of pharmaceutical drugs; pharmaceutical drugs might not have helped the participants, either. The effectiveness of St. John's wort for the treatment of mild depression continues to be questioned.

It has also been recently reported that patients who take St. John's wort supplements experience a drop in blood concentration of certain prescribed medications. Apparently, hyperforin, one of the active ingredients in St. John's wort, activates the enzyme CYP3A in the liver. CYP3A is known as the garbage disposal enzyme because its function is to break down toxins and other foreign substances that enter the body. It is also involved in the metabolism of a large number of prescription drugs. Physicians have noted that certain drugs—such as theophylline for asthma, the immunosuppressive cyclosporin, the anticlotting drug warfarin, and birth control pills—become ineffective if the patients are also taking St. John's wort. This could potentially be a dangerous situation.

Ginkgo

A tree that has often been described as a living fossil, *Ginkgo biloba* (see fig. 9.9b in Chapter 9), is offering hope for Alzheimer's patients. Native to Asia, ginkgo has been a staple of Chinese herbalists for 4,000 years. In China, a tea made from the leaves has been used to treat asthma, bronchitis, and a wide variety of other conditions. Western medicine is currently evaluating ginkgo as a treatment for dementia, age-related cognitive declines, and cognitive performance in healthy adults. EGb 761 is a standardized extract of Ginkgo leaves, which contains 24% ginkgo-flavonol glycosides and 6% terpene lactones, and it is a widely used variety of Ginkgo supplement in both the United States and Europe. Ginkgo may be one of the most well researched herbal supplements available today. Unfortunately, the results have been inconsistent, with some studies showing positive benefits from ginkgo supplements and others showing no benefits. The largest study to date was an 8-year, multicentered study with over 3,000 subjects who were aged 75 and older. Ginkgo extract taken twice a day was not effective in reducing the rate of dementia or the incidence of Alzheimer's disease when compared to a placebo. By contrast, a recent meta-analysis of data from seven clinical trials, which included 2,625 patients, found that Ginkgo extract EGB 761 showed evidence of effectiveness in treating patients with dementia. Another study also found that EGB 761 was effective in alleviating neurosensory symptoms in patients with dementia. Animal studies have also shown enhanced learning and memory in mice treated with ginkgo extracts. Clearly, more work needs to be done to resolve these contradictory findings.

Ginkgo leaf extracts have several actions on the body; ginkgo's primary effects are on the circulatory system. The aggregation of platelets (blood-clotting factors) is inhibited. Increased blood flow to the brain delivers more oxygen and nutrients and arrests the death of nerve cells due to vascular impairment. Data also indicate that ginkgo extracts can counteract destructive free radicals. But there is no strong evidence that ginkgo extracts can permanently improve cognitive function in young, healthy adults.

On the downside, extracts of *G. biloba* can magnify the effects of anticoagulants, such as aspirin and warfarin (Coumadin) and thus induce bleeding. Patients who are planning to undergo surgery should cease taking ginkgo supplements at least 36 hours before the operation to avoid serious, perhaps life-threatening, complications.

Saw Palmetto

Saw palmetto (*Serenoa repens*) is a species within the Arecaceae, or palm family. It is indigenous to the southern coastal region of the United States and the West Indies and is familiar as the state tree of South Carolina. The palm may reach a height of over 3 meters (10 feet) and has large, palmately lobed, fan-shaped leaves (fig. 19.12). Flowers are white and fragrant. Medical interest centers on the reddish brown berries, each with a single seed.

Benign prostate enlargement (BPE) is a noncancerous condition in which the prostate gland increases in size. This condition affects 50% of all 50-year-old men and 90% of males aged 70 and older. Because the prostate gland encircles the urethra, once the gland enlarges, urination becomes difficult, and because the bladder cannot be completely emptied, sleep is disturbed by frequent trips to the bathroom. The culprit for BPE is dihydrotestosterone (DHT), a variant form of the male hormone testosterone. Dihydrotestosterone is also involved in male pattern baldness. In BPE, dihydrotestosterone induces the cells of the prostate gland to replicate, causing enlargement.

In some clinical trials, saw palmetto alleviated these symptoms by increasing urinary flow without shrinking the prostate or decreasing sex drive. It has been suggested that free fatty acids in the saw palmetto fruit inhibits the enzyme alpha reductase, which converts testosterone into DHT. It also prevents DHT from binding to receptor sites on the membrane of prostate cells. The results may be an overall reduction of DHT in the patient and the relief of symptoms. The saw palmetto extract

Figure 19.12 Berries of saw palmetto (*Serenoa repens*) have proven useful in treating symptoms of benign prostate enlargement.

is suggested to be effective for more sufferers, works faster in relieving symptoms, and has fewer side effects than the commonly prescribed drug finasteride, which is often associated with decreased sex drive and impairment of erectile function.

Thinking Critically

Herbal remedies have moved from the realm of health food stores to drugstores and supermarket pharmacies. In contrast to its regulation of the pharmaceutical industry, the FDA (Food and Drug Administration) does not set standards for the manufacturing and sale of herbal remedies, nor does it require exhaustive research to substantiate purported health benefits advertised on the labels of herbal supplements.

What advice would you give a person who is choosing to self-medicate using herbs? How can the educated consumer be assured of the efficacy of herbal preparations? What changes in the law would you suggest?

However, one study that looked into the use of saw palmetto supplements to alleviate BPE brings into question whether or not this herb is an effective treatment. Over 200 men who suffer from BPE were given either saw palmetto extract or a placebo daily for 1 year. There was no significant difference in symptoms experienced by the two groups after a year. Critics counter that, since the active ingredient in saw palmetto preparations may vary in content, it may be difficult to know if these results apply to all saw palmetto supplements. Subsequently, several additional studies have also shown that there were no significant differences in symptoms between subjects taking the saw palmetto supplements and those taking a placebo.

CHAPTER SUMMARY

1. From the earliest civilizations, native plants have supplied medicinal compounds. Today, 25% of prescription drugs are of plant origin, and an even greater percentage are based on synthetic or semisynthetic ingredients originally isolated from plants. While Western medicine strayed away from herbalism, 75% to 90% of the rural population in the rest of the world still relies on herbal medicine as their only health care. Today there is renewed interest in studying the medicinal plants used by indigenous populations in remote areas of the world, especially the rain forests. Investigating these folk remedies may lead to important medical discoveries that could save millions of lives.

2. Alkaloids and glycosides have been identified as the therapeutically important components of most medicinal plants. Alkaloids are a diverse group of compounds that contain nitrogen, are alkaline, and have a bitter taste. Glycosides are another diverse group of compounds; the feature they share is the presence of a sugar molecule attached to the active component.

3. Digitalis, the primary treatment for several heart ailments, is derived from foxglove, a biennial in the snapdragon family. The leaves contain more than 30 glycosides, with digoxin and digitoxin the most medically significant compounds.

4. Today aspirin is purely synthetic, but it is based on traditional remedies from willow bark. Salicylic acid, the active principle in willow bark, was identified early in the nineteenth century, and laboratory synthesis was achieved about 30 years later. In 1898, the Bayer Company developed acetylsalicylic acid, a modification they named aspirin.

5. Two plant compounds figure prominently in the treatment of malaria, quinine from the bark of the *Cinchona* tree and artemisinin from trichomes on the leaves and stem of *Artemisia annua*. Currently, artemisinin combination therapy, consisting of artemisinin and a quinine derivative, is the recommended treatment for uncomplicated malaria cases.

6. Historically, many plants, including *Galega officinalis*, have been used to treat diabetes. Today metformin is the treatment of choice for patients with type 2 diabetes. Although metformin is a synthetic compound, it is based on guanidine, a natural compound found in *Galega* extracts.

7. In 1952, the alkaloid reserpine was isolated from the roots of *Rauwolfia serpentina*, a plant used by Hindu healers for thousands of years. Although first used as a tranquilizer, reserpine and other *Rauwolfia* alkaloids had greater use in the treatment of hypertension.

8. The sap of *Aloe vera* contains several glycosides that are used to treat minor burns and cuts. This herbal remedy, which has been used since ancient times, promotes healing and reduces scarring. The moisturizing effect of the sap has resulted in its inclusion in skin creams, shampoos, lotions, and other cosmetics.

9. Ephedrine and pseudoephedrine, alkaloids from various species of *Ephedra*, are found in prescription and over-the-counter decongestants. Ephedrine is also a potent stimulant, and abuse of this compound in herbal remedies led to its ban by the U.S. Food and Drug Administration.

10. Active compounds from several plants have become standard treatment for various forms of cancer, and researchers continue to look for and develop new anticancer compounds. The alkaloids vinblastine and vincristine, isolated from the Madagascar periwinkle, are remarkably effective in the treatment of certain forms of leukemia. Taxol, from the bark of the Pacific yew, has proved useful in the treatment of ovarian, breast, and other forms of cancer. Camptothecin derivatives from *Camptotheca acuminata* are becoming standard chemotherapy drugs.

11. Herbal remedies are a major component of alternative medical treatments in the United States. These widely available dietary supplements often contain secondary compounds that can have powerful effects on the human body but are generally marketed without rigorous scientific studies. Although many herbal remedies show great promise, some of them may interact with prescription and over-the-counter drugs.

REVIEW QUESTIONS

1. Trace the historical connection between medicine and botany. Is this still a valid connection today?

2. What parts of the world still rely primarily on herbal medicine? Why? Do you think it is effective? Do you think more funds should be allocated to investigating herbal remedies?

3. What active principles in plants account for their medicinal properties?

4. Discuss the history and use of the following medicines: digitalis, quinine, and reserpine.

5. Trace the development of aspirin from an herbal remedy to the world's most widely used synthetic drug.

6. How has cancer therapy improved through the use of plant-derived medicines?

7. Trace the development of metformin from an herbal remedy to the leading treatment for type 2 diabetes.

CHAPTER 20

Psychoactive Plants

Opium poppy capsules are slashed to release a milky latex; when dried it turns brown in color, is scraped off the fruits, and is identified as crude opium.

KEY CONCEPTS

1. Psychoactive compounds affect the central nervous system by acting as stimulants, hallucinogens, or depressants.
2. Most psychoactive drugs are alkaloids that have been employed by humanity for millennia.
3. Although many of these compounds have legitimate medicinal value in the proper dosage, the abuse of psychoactive drugs has led to devastating physiological, emotional, economic, and societal consequences.

©mafoto/iStock/Getty Images

CHAPTER OUTLINE

Psychoactive Drugs 359
The Opium Poppy 360
    An Ancient Curse 360
    The Opium Wars 360
    Opium Alkaloids 361
    Heroin 361
    Withdrawal 362
Marijuana 362
    Early History in China and India 362
    Spread to the West 363
    THC and Its Psychoactive Effects 364
    Medical Marijuana 365
    Legal Issues 365
Cocaine 366
    South American Origins 366
    Freud, Holmes, and Coca-Cola 366

A CLOSER LOOK 20.1 The Tropane Alkaloids and Witchcraft 367

    Coke and Crack 369
    Medical Uses 369
    A Deadly Drug 369
Tobacco 370
    A New World Habit 370
    Cultivation Practices 371
    Health Risks 371
Peyote 374
    Mescal Buttons 374
    Native American Church 375
Kava—the Drug of Choice in the Pacific 375
    Preparation of the Beverage 375
    Active Components in Kava 375
Lesser Known Psychoactive Plants 376
Chapter Summary 377
Review Questions 377

The use of psychoactive plants dates back to the earliest societies when humans first stumbled onto the mind-altering properties of plants. These plants were immediately valued for their ability to numb pain, relieve fatigue and hunger, or free the spirit from earthly concerns. Clearly, these plants were used for millennia before written records appeared. Six-thousand-year-old Sumerian clay tablets refer to opium as the joy plant.

PSYCHOACTIVE DRUGS

Psychoactive drugs affect the central nervous system in various ways by influencing the release of **neurotransmitters** (chemical messengers within the nervous system) or mimicking their actions. Neurotransmitters are released from neurons, or nerve cells, into synapses, the gaps between neurons. The neurons sending the neurotransmitters are called presynaptic neurons. Once in the synapse, neurotransmitters attach to specific receptors on the surface of the receiving, or postsynaptic, neuron. The action that takes place in the postsynaptic neuron after attachment depends on the type of neurotransmitter. Some neurotransmitters are inhibitory; others are excitatory. After the action is completed, the neurotransmitter is either degraded by enzymes or taken back to the presynaptic neuron via a transporter molecule and stored for future use.

On the basis of their effects, psychoactive drugs can be classified as **stimulants**, **hallucinogens**, or **depressants**. An important point to remember is that there may often be only subtle distinctions between medicinal, psychoactive, and toxic doses. For example, morphine and cocaine have legitimate medical uses for pain relief and local anesthesia, respectively. They also can be abused for their psychoactive effects and, when taken in overdose, can result in death.

Stimulants excite and enhance mental alertness and physical activity; they reduce fatigue and suppress hunger. Cocaine, with its intensely potent effects, and the much milder caffeine are well-known plant-derived stimulants. Hallucinogens such as peyote, marijuana, and LSD produce changes in perception, thought, and mood, often inducing a dreamlike state. Depressants dull mental awareness, reduce physical performance, and often induce sleep or a trancelike state. Opium and its derivatives, morphine and heroin, are classic depressants.

The term **narcotic** must also be considered when discussing psychoactive drugs. By strict definition, a narcotic drug is one that induces central nervous system depression, resulting in numbness, lethargy, and sleep. Under this definition, narcotics include opiates as well as alcoholic beverages and kava. In contemporary use, the term *narcotic* is applied to any psychoactive compound that is dangerously **addictive**; therefore, the stimulant cocaine is classified as a narcotic. Addictive compounds generally elicit one or more of the following: **psychological dependence**, **physiological dependence**, and **tolerance**. Psychological dependence is a desire to reexperience the pleasure induced by the drug. In physiological dependence, the body has a need for the drug to avoid withdrawal symptoms. Finally, tolerance is the need for ever-increasing amounts of the drug to produce the same effect.

Researchers have discovered that all addictive drugs affect an area in the brain called the mesolimbic dopamine system, or the reward circuit. Dopamine is one of about 100 neurotransmitters produced by the brain and is the primary neurotransmitter of the reward circuit. This pathway begins near the base of the brain at the ventral tegmental area (VTA) and extends to a region near the front of the brain called the nucleus accumbens (NA). The VTA sends axons—long, cellular extensions of neurons—into the nucleus accumbens. Neurons of the VTA produce the neurotransmitter dopamine and send it via the axons to the NA. Dopamine is released into the synapses and attaches to the receptors on dopamine-sensitive cells of the NA.

Despite their various physiological effects on the body, all addictive drugs increase dopamine levels in the NA, although the mode of action differs with the type of drug. Normally, neurotransmitters are accumulated at some point after their release and returned to their site of origin. Cocaine and other stimulants prevent the return of dopamine to the VTA by disrupting the transport system. In contrast, heroin and other opiates inhibit the neurons that normally shut down the production of dopamine in the VTA. Either way, dopamine levels in the NA stay elevated. Excess dopamine produces the intense euphoria that addicts experience and desire when abusing drugs.

Tolerance associated with addictive drugs can also be explained by the reward circuit. Chronically elevated levels of dopamine in the NA caused by chronic drug abuse induce dopamine-responsive cells in the NA to release a signal molecule called cyclic AMP. Cyclic AMP in turn activates a transcription factor, called CREB, that triggers certain genes to dull the reward circuit by inhibiting the dopamine-producing neurons in the VTA. Because less dopamine is being released, the addict finds that taking the same dose of the drug does not produce the same level of euphoria. In other words, the drug abuser experiences tolerance. Research is uncovering the roles of other transcription factors and neurotransmitters to understand better the physiological basis for dependence and other aspects of drug abuse.

In the majority of psychoactive plants, the active compounds are alkaloids (although marijuana is a notable exception). As discussed in Chapter 19, alkaloids are distributed throughout the plant and fungal kingdoms but predominate in certain families. This chapter will focus on psychoactive compounds in angiosperms, and the fungal compounds will be discussed in Chapter 25 and the alcoholic beverages produced by fungi in Chapter 24. Angiosperm families that are well known for containing psychoactive alkaloids include the Solanaceae (nightshade family), Papaveraceae (poppy family), Rubiaceae (coffee family), Erythroxylaceae (coca family), Myristicaceae (nutmeg family), and Convolvulaceae (morning glory family).

> **Thinking Critically**
>
> Psychoactive drugs may be physiologically or psychologically addicting, or both.
>
> *Many people believe that, the more dramatic the physical withdrawal symptoms, the more dangerously addictive a drug must be. Do you agree or disagree?*

THE OPIUM POPPY

The use of *Papaver somniferum*, opium poppy, for relieving pain was well known by early societies, as evidenced by their written documents. Babylonian and Egyptian writings, artifacts, and statuary attest to the fact that opium was revered for its analgesic and sleep-inducing properties. Hippocrates, Dioscorides, and Galen, leading contributors to early Western medicine, endorsed its use. Early peoples also valued its ability to allay the worries and sorrows of life. Surely, these ancient peoples were aware of the strongly addictive nature of compounds from this plant, and this curse has followed the opium poppy to contemporary society.

An Ancient Curse

The opium poppy is a large annual herb in the poppy family (Papaveraceae), with each stem topped by a showy flower of white, pink, red, or purple petals. After pollination, the ovary matures into a capsule. If the capsule is sliced when still green, it exudes a milky latex rich in potent alkaloids (fig. 20.1). The dried latex, which turns brown, can be peeled from the capsule and is known as crude opium. Originally native to the Middle East, it is now grown both legally and illegally throughout much of the world, with India the leading legal grower. In addition to the legal and illegal cultivation of drugs, poppy is also grown for its seeds and seed oil. Poppy seeds are common toppings for breads and rolls as well as ingredients for many muffins, cakes, and cookies. Poppy seed oil is a common cooking oil in some parts of the world. Although the quantity of alkaloids in the seed is negligible and therefore not at all addictive, it may be sufficient to register positive on some drug tests.

Opium has been eaten, drunk, and smoked for centuries. The usual method of preparation was to dissolve the crude opium in wine. Ancient pottery vessels for opium wine, pipes for smoking opium, and other artifacts decorated with poppy capsules have been dated back to ancient Cyprus and Knossos. Later, Hippocrates and other early physicians advocated poppy wine as a medicine for many complaints. During the Middle Ages, a common method of preparing opium was to dissolve it in alcohol, and this tincture, later known as laudanum, became a popular medication for centuries.

Laudanum usage reached its peak in the nineteenth century when it was consumed not only as a medicine but

Figure 20.1 The flower and capsule of the opium poppy, *Papaver somniferum*. Crude opium is the dried, milky latex from the incised capsule.

also as a mind-expanding drug by many in the so-called privileged class, including writers, artists, and intellectuals. The famous British poets Elizabeth Barrett Browning and Samuel Coleridge were both known laudanum addicts, as was Thomas De Quincey, who analyzed his addiction in *Confessions of an English Opium-Eater*. The French composer Hector Berlioz glorified the opium dream in his *Symphonie Fantastique*. The use of laudanum was so widespread during this time that little shame was attached to the addiction.

The Opium Wars

Opium use is directly tied to the history of China. The Chinese knew of opium for centuries but used it primarily for medicinal purposes. After A.D. 1600, opium use began to increase when tobacco smoking was introduced to China. Originally, the tobacco was mixed with opium, but gradually the amount of tobacco was decreased until the opium was smoked alone. Opium smoking spread slowly, and eventually many Chinese became addicted, even though it was banned by the emperor.

The addiction was fueled by the British desire for a favorable trade balance with China. The British wanted silk, tea, and porcelain from China; however, the Chinese demanded payment in silver because they had no interest in European goods. The drain of silver from Great Britain was intolerable to the British, and their solution was to meet the escalating Chinese need for opium. Opium was being widely cultivated in India under the auspices of the British East India Company. Merchant clipper ships from various countries, including the United States and Britain, brazenly smuggled the opium into China through the port of Canton. The trade was lucrative to everyone involved in the growing, shipping, and marketing of opium, and soon the flow of silver was reversed.

Alarmed by the rising addiction rate, the Chinese government sought to stop the opium trade. In 1839, the Chinese authorities confiscated and destroyed all the opium in Canton harbor. The British retaliated by sending warships, beginning the first Opium War, which lasted from 1839 to 1842. Chinese opposition was crushed, and the British received major concessions, including the right to trade in opium, payment for the destroyed opium, the opening of more ports for foreign trade, and the establishment of Hong Kong as a British colony.

Ten years later, the second Opium War took a further toll on China when Britain, other European countries, and the United States were granted additional concessions. The British opium trade continued until 1913 when moral pressure inside both China and Britain brought an end to the trading that had cost China so much. By this time, however, China was cultivating its own poppies, and about 25% of the population was addicted. The opium problems continued and were amplified during the Japanese occupation, to end only after the establishment of the communist People's Republic of China.

Opium Alkaloids

More than 20 alkaloids have been identified in opium, with morphine and codeine probably the most significant. Morphine was first isolated in 1806 by Friedrich Serturner, a German scientist who named the compound after Morpheus, the Greek god of dreams. This, in fact, was the first active principle isolated from a plant. Morphine was soon recognized for its analgesic value and is still unsurpassed in its ability to deaden pain. Initially, morphine was administered orally; however, its full medical potential was not realized until after the development of the hypodermic syringe in the middle of the nineteenth century. It is known today that morphine is rapidly inactivated and excreted when taken in oral preparations. Although the exact mode of action is not known with certainty, opiate binding sites that normally bind endorphins are found in the brain, spinal cord, and intestines. Endorphins, chemicals produced in the brain, modify mood by reducing sensations of pain and enhancing pleasurable feelings. Morphine depresses the areas of the brain involved in the perception of pain and reduces the anxiety that accompanies pain because it is able to bind to the endorphin receptor sites and mimics endorphin's actions. Morphine is a general central nervous system depressant and, in overdose, can lead to death by completely suppressing the respiratory center in the brain.

Like opium, morphine is strongly addictive. This became evident when thousands of injured Civil War soldiers became dependent after injections of morphine. In fact, this dependency was so widespread that morphine addiction became known as soldiers' disease. Physician-induced addiction became common because morphine was used to treat so many ailments during the last half of the nineteenth century. This problem was forcefully depicted in Eugene O'Neill's classic play *A Long Day's Journey into Night,* which showed the ravages of his mother's morphine addiction. Among the many applications beyond the control of pain, morphine was frequently prescribed for diarrhea (it slows peristalsis in the digestive tract) and coughing (it suppresses the cough reflex in the central nervous system).

With the growing awareness of the potential for addiction, morphine use declined around the turn of the century. Currently, morphine is still the drug of choice for the control of intense pain from severe burns or visceral pain during the postoperative period, cases of terminal cancer, and kidney stones. Morphine is also administered in some cases to patients suffering from heart failure.

Medically, codeine is one of the most commonly used opiates. It has value as an oral analgesic but is only one-fifth as strong as morphine. It works well in combination with nonopiate analgesics, such as aspirin and acetaminophen, and is especially useful in cough syrups because it suppresses the cough reflex. Because it is also addictive (but much less so than morphine), products containing codeine are available only by prescription in the United States.

Heroin

In 1898, the Bayer company introduced heroin, which they believed was a nonaddictive opiate with analgesic properties superior to morphine and cough-suppressant properties superior to codeine. This semisynthetic derivative of morphine was widely dispensed in many over-the-counter medicines for about two decades. Cough syrups with heroin were promoted for their effectiveness in adults and children alike (fig. 20.2). Eventually, physicians questioned the alleged nonaddictive nature of heroin, and by 1917 it was no longer available in cough syrups.

Contrary to the early promotion, heroin is actually six times more addictive than morphine, and its misuse has resulted in long-standing drug and crime problems. Heroin is not used medicinally in the United States, nor is it manufactured here legally. However, heroin is used medicinally in other countries to control severe pain. India is the largest legal producer of both opium alkaloids and heroin for medical purposes. Myanmar (Burma), Laos, and Thailand (the Golden Triangle) as well as Pakistan, Afghanistan, and Iran (the Golden Crescent) traditionally have been the source of most of the illegal opium. Currently, Afghanistan is the largest illicit grower of the opium poppy, supplying 70% of the world's opium.

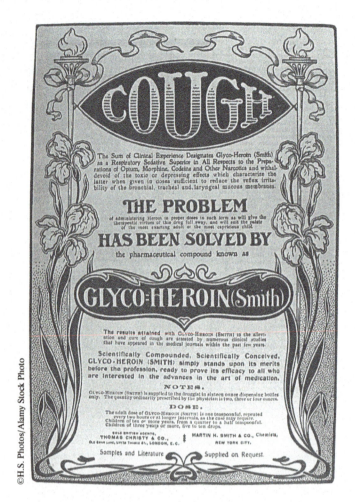

Figure 20.2 As this advertisement from the turn of the nineteenth century attests, heroin was an ingredient in many medications before its addictive properties were recognized.

Most of the heroin in the United States comes from two sources: Mexico, the main supplier to the western states, and Colombia, which is the primary source for the eastern United States. Evidence indicates that the organizations in Colombia that were previously involved in trafficking cocaine to the United States are now involved in the heroin trade. The heroin from Colombia is much cheaper and purer than the heroin from prior suppliers. In the 1970s, a bag of heroin cost just under $30 and was about 30% pure. In 2004, the price of the same quantity of heroin dropped to $4, and its purity ranged from 80% to 90%. Also, highly purified heroin can be snorted or smoked rather than injected, avoiding telltale needle tracks. Partly because of the low cost, heroin addiction is on the rise in the United States, especially among suburban teenagers.

Withdrawal

As with any addictive substance, opiates will trigger withdrawal symptoms when the drug is no longer taken. The intensity of these symptoms depends on the type of opiate, dosage, and length of the habit. Withdrawal from heroin addiction produces the most severe symptoms, which begin to occur within 4 hours and reach a peak at 36 to 72 hours. Common symptoms include increased respiration, perspiration, runny nose, goose bumps, muscle twitches, insomnia, vomiting, and diarrhea. In fact, the phrase "going cold turkey" used to describe the abrupt cessation of a drug habit refers to the goose bumps prominent in a person undergoing opiate withdrawal. The physiological trauma that the body suffers during withdrawal can be alleviated in some patients by gradually weaning them from heroin addiction by substituting methadone, a synthetic opiate that is less addictive and has milder withdrawal symptoms.

MARIJUANA

Cannabis sativa is one of the oldest cultivated plants and has been used for various purposes. In addition to its use as a hallucinogen, the plant has been used medicinally and for its fiber (see Chapter 18), its oil, and its seed. Marijuana is highly variable, and some taxonomists believe it should be divided into at least three separate species based on size and growth habit, from short, branched forms to plants over 8 meters (25 feet) tall. Marijuana plants are dioecious annuals with inconspicuous staminate and carpellate flowers on separate individuals. Inflorescences occur in the axils of upper leaves; the palmately compound leaves are distinctive, with three to seven (usually five) toothed leaflets (fig. 20.3). The plants are known for their **resin** production by glandular trichomes, with the maximum amount of resin coating the unfertilized carpellate flowers and adjacent leaves. The golden resin is the source of the hallucinogenically active compounds, and the potency of resin varies greatly depending on genetic strains and growing conditions.

The species is believed to have originated in central Asia and today has a cosmopolitan distribution that is the greatest among all the psychoactive plants. Its wide cultivation has led to many common names for the species, including marijuana, hemp, grass, pot, hash, hashish, bhang, charas, ganja, ma, kif, and dagga.

Early History in China and India

The use of *Cannabis* can be traced back to ancient China, where the hemp plant was valued for its fiber (for cloth, paper, and rope) and medicinal properties. About 5,000 years ago, the legendary Emperor Shen Nung recommended marijuana for the treatment of rheumatism, gout, malaria, beriberi, absentmindedness, female disorders, and constipation. There is also some evidence that the Chinese knew of and used the plant for its hallucinogenic properties, but this practice was not as prevalent there as it was in India.

Earliest documented records of marijuana's use as a hallucinogen can be traced to the Scythians (ancient nomadic Slav horsemen from central Asia), who burned *Cannabis* and inhaled the hallucinogenic smoke. In a ritualistic ceremony, the Scythians piled the plants on burning coals in a small tent made of sheepskin pelts. When the vapors accumulated, the Scythians lifted the sheepskin and inhaled. Pots of charcoal

Figure 20.3 *Cannabis sativa* is dioecious; the carpellate plants and flowers (right) produce a higher concentration of the THC hallucinogen than the staminate plants (left).

with remains of *Cannabis* have been found in Scythian excavations and dated to around 500 B.C.

Marijuana use spread from its central Asian homeland to Asia Minor, northern Africa, India, and elsewhere. The first written mention in India is found in the Sanskrit *Zend-Avesta* around 600 B.C. India has valued *Cannabis* for its medicinal properties, and folk medicine uses it extensively to treat a variety of ailments. It was the hallucinogenic properties, however, that gave *Cannabis* an integral role in religious ceremonies for achieving a contemplative state and placed it as one of the five sacred plants of ancient India. Three grades of *Cannabis* have been recognized in India: bhang, ganja, and charas. Bhang, the least potent of these, consists of the dried, cut tops that are ground with spices to prepare a drink or candy. Even today, bhang is sold outside Hindu temples in India. Ganja is prepared from the resin-rich pistillate flowers and tops of specially bred high-yielding strains of marijuana and is usually smoked. Charas, the most potent, consists of the pure resin itself (also known as hashish) from these special strains and is also smoked.

The use of *Cannabis* spread throughout the Muslim world, into the Middle East and Africa. Hookahs (water pipes) for smoking hashish were commonplace in bazaars and marketplaces, scenting the air with marijuana's sickly sweet aroma (fig. 20.4). Arabic literature is replete with references to *Cannabis*, as can be seen in the famous *Thousand and One Nights*, in which the potency of the drug was indicated by the expression "had an elephant smelt it, he would have slept from year to year."

Undoubtedly the most notorious legend about marijuana concerns the twelfth-century Persian Al-Hasan ibn-al-Sabbah, a leader of a Muslim sect. This fanatical religious sect, whose followers became known as Hashishins (after their leader), swore to kill all enemies of their faith. Legend has it that they were worked into a murderous frenzy by smoking *Cannabis* (the accuracy of this account is questionable). Nevertheless, the name of this sect, *hashishins,* left a legacy in the word *assassin,* as well as the name for the resin, *hashish.*

Spread to the West

Although northern Europe had a long history of using hemp fiber, the psychoactive use of *Cannabis* was not well known before the nineteenth century. It is rumored that Napoleon's troops, returning from a campaign in North Africa, popularized hashish smoking. By mid-century, Paris hashish clubs were frequent meeting places of intellectuals, writers, poets, and artists. Members believed that the psychoactive properties of hashish enhanced their creative abilities.

Figure 20.4 Alice, in the Lewis Carroll classic *Alice in Wonderland*, "stretched herself up on tiptoe, and peeped over the edge of the mushroom, and her eyes immediately met those of a large blue caterpillar that was sitting on the top with its arms folded, quietly smoking a long hookah, and taking not the slightest notice of her or anything else."

Marijuana was probably introduced to the United States around the turn of the twentieth century, possibly from Mexico or the Caribbean Islands. Mexican laborers brought the marijuana-smoking habit with them when they immigrated. In the 1920s, it spread among the urban poor in the South but was not in vogue until jazz musicians popularized smoking marijuana cigarettes. Not only was marijuana used recreationally, but it was also a popular ingredient in many over-the-counter medications.

During the 1930s, concerns about the dangers inherent in marijuana use led to the establishment of laws prohibiting its use. The Federal Bureau of Narcotics launched an "educational campaign" to make the public aware of the dangers of marijuana use. This campaign greatly distorted the problem and grossly exaggerated the dangers. These concerns culminated in the Federal Marijuana Tax Act of 1937, which controlled the legal sale of the plant and resulted in the virtual elimination of *Cannabis* from the nation's pharmacopoeia.

The dramatic increase in marijuana use came about during the social revolution of the 1960s. The youthful counterculture of this period embraced marijuana as the recreational drug of choice. In reaction to the pervasive usage of marijuana during the 1970s, many communities relaxed their laws concerning personal use and possession. In 2001, Great Britain relaxed its marijuana laws, downgrading it from a class B to a class C drug. This classification means that marijuana still remains illegal, but the maximum penalties for possession are greatly reduced. Also under the new law, the police will no longer be able to arrest someone on the street for possession. Prosecutions will be carried out by court summons. In 2014 in the United States, 54% of Americans were in favor of supporting the legalization of marijuana. As of June 2015, the District of Columbia and the states of Colorado, Washington, Oregon, and Alaska have legalized the recreational use of marijuana.

THC and Its Psychoactive Effects

The chemistry of *Cannabis* is complex and includes a unique class of phenolic compounds called cannabinoids. These compounds, over 60 in number, are pure hydrocarbons, with 21 carbon atoms in an elaborate two- to three-ring structure. Although cannabinoids have been known since the beginning of the twentieth century, it was not until 1964 that the psychoactive component of marijuana was identified as delta-9-tetrahydrocannabinol (THC), present in the resin. The concentration of THC in plants varies considerably, depending on genetic strain, sex, climate, and growing conditions. Some of the most potent varieties of *Cannabis* are usually grown under hot, dry conditions that seem to optimize THC production. Marijuana grown in Mexico, Colombia, and India generally has higher "street value" because of its higher THC concentrations. Sinsemillas, seedless strains developed illicitly in California during the 1970s, however, are reported to have some of the highest levels of THC. Newer, highly potent dwarf hybrids have also been illicitly developed for clandestine indoor cultivation. These sinsemilla plants, which are hybrids resulting from crossing *C. sativa* and *C. indica* (a hardy species of *Cannabis* native to Afghanistan), are cultivated by cloning carpellate plants in high-tech indoor gardens. Controlling the light and carbon dioxide levels makes it possible to bring about abundant flowering in as little as 2 months. This type of densely planted, environmentally controlled indoor garden is known as a Sea of Green.

The long-sought-after mind-altering effects of *Cannabis* include a sense of euphoria and calmness. Some of the most descriptive accounts of the intoxication were written by Walter Bromberg, relating both personal experience and detailed observations.

> Within a few minutes he begins to feel more calm and soon develops definite euphoria; he becomes talkative... is elated, exhilarated... begins to have ... an astounding feeling of lightness of the limbs and body... laughs uncontrollably and explosively ... without at times the slightest provocation... has the impression that his conversation is witty, brilliant.

Perception of time and space may also be distorted, and minutes may seem like hours.

Adverse effects of marijuana use have been studied since the 1930s, with research efforts intensifying after the identification of THC as the active component. Even moderate use of marijuana impairs learning, short-term memory, and reaction time. Because THC is fat soluble, it accumulates in body tissues, and measurable amounts may remain in the body for days after inhalation. The effects of marijuana on the male reproductive system are decreased sperm production and decreased testosterone levels. In pregnant women, THC is known to cross the placenta and might be damaging to the fetus; however, mutagenicity studies thus far have been inconclusive. Marijuana or hashish is usually inhaled in smoke and, as such, can damage lung tissue, much as cigarette smoking does. More research is needed to determine if long-term harm can be traced to marijuana use.

Many of the effects of marijuana can be understood with the discovery, in the late 1980s, that many regions of the human brain contain cannabinoid receptors, sites on brain cells that receive cannabinoids. (Receptors are found on the surface of a nerve cell. If the shape of the receptor is complementary to the shape of a specific neurotransmitter, a response is elicited in the receptor cell.) One identified receptor is CB_1, the receptor specific for delta-9 THC, the primary psychoactive ingredient in marijuana. CB_1 receptors are located in high densities in many areas of the brain, such as the cerebellum, the substantia nigra, and the globus pallidus. These regions are all involved in motor coordination and may account for the relief of tremors and spasticity (muscle rigidity) that treatment with THC provides to persons afflicted with Parkinson's disease or multiple sclerosis. CB_1 receptors are also found in the hippocampus, an area of the brain that governs short-term memory, and the amygdala, an area involved in emotional memory. The presence of CB_1 receptors in the hippocampus may explain why THC-treated rats behave on some memory tests as if they had no hippocampus.

CB_2 is the second THC receptor to be identified and is found primarily on the surface of cells in the immune system and blood stem cells in the bone marrow. It has been discovered that the body makes its own compounds that bind to the CB_1 and CB_2 receptors: anandamide, which binds to CB_1, and 2AG (2-arachidonyl glycerol), which binds primarily to CB_2. Both anandamide and 2AG are endocannabinoids, cannabinoid-like substances produced by the body. THC essentially mimics these compounds; however, unlike the endocannabinoids, which are short-acting compounds, THC affects thousands of receptors throughout the body, producing stronger and longer-lasting effects.

Medical Marijuana

Over the centuries in various cultures, marijuana has been employed to treat numerous ailments. In contemporary medicine, marijuana is used to treat glaucoma and as an aid to chemotherapy. Glaucoma is a group of eye diseases characterized by increased ocular pressure (pressure within the eye) and resulting damage to the optic nerve that may lead to blindness. In fact, it is the leading cause of blindness in the United States. Both smoking marijuana and ingesting preparations with THC have been shown to significantly reduce ocular pressure in patients with glaucoma. Patients undergoing chemotherapy for cancer often experience side effects of nausea, vomiting, and loss of appetite. These side effects have been reduced by the use of marijuana cigarettes as well as marinol, a synthetic form of THC. It is also used to counteract the weight loss associated with AIDS wasting syndrome.

Many multiple sclerosis (MS) patients suffer severe, burning pain as the fatty acid sheaths covering the nerves in their bodies deteriorate. Also, they frequently have painful muscle spasms and stiffness, which interrupts sleep. Studies of MS patients have shown that taking THC alone or in combination with another marijuana derivative, cannabidiol (CBD), relieves symptoms. CBD holds promise for other medical conditions as well because in experimental trials it has been shown to be an effective anti-inflammatory and a potent antioxidant, and it promotes wound healing but with fewer psychoactive effects. Sativex is a new drug that contains both THC and CBD dissolved in alcohol. It is sprayed under the tongue and absorbed through the mucosa of the mouth and throat. Marijuana grown in the United States for illicit purposes is usually low in cannabidiol because growers have selected plants for high levels of THC at the expense of cannabidiol.

Testing has also shown THC to be effective in killing off cancerous cells of the brain, skin, blood, and prostate in tissue culture by inducing the tumor cells to undergo programmed cell death, or cell suicide. THC has also been shown to inhibit the growth of blood vessels to a tumor. Without blood vessels, tumor cells are deprived of oxygen and nutrients and soon die. Another promising use for THC is in the treatment of post-traumatic stress disorder (PTSD). THC binds to receptors on the memory-producing regions of the brain, which allows sufferers of PTSD to forget unpleasant experiences and end nightmares.

These and other medical applications have resulted in a storm of controversy over the medical dispensing of marijuana cigarettes. Activists are pressuring the federal government to loosen controls over the use of marijuana to ease the suffering of patients with AIDS, cancer, multiple sclerosis, and glaucoma.

Legal Issues

As of January 2019, 33 states and Washington D.C. (Alaska, Arizona, Arkansas, California, Colorado, Connecticut, Delaware, Florida, Hawaii, Illinois, Louisiana, Maine, Maryland, Massachusetts, Michigan, Minnesota, Missouri, Montana, Nevada, New Hampshire, New Jersey, New Mexico, New York, North Dakota, Ohio, Oklahoma, Oregon, Pennsylvania, Rhode Island, Utah, Vermont, Washington, and West Virginia) have enacted laws or passed referenda that effectively allow patients to use medical marijuana. Additionally, other states, for example Georgia, allow residents to possess medical marijuana if they suffer from certain medical conditions.

Despite the possible medical benefits of marijuana use, however, in May 2001 the U.S. Supreme Court unanimously ruled that medical use of marijuana is not a valid exception to the federal law that classifies marijuana as an illegal substance. Writing for the court, Justice Clarence Thomas wrote that the controlled substance statute "includes no exception at all for any medical use of marijuana." Although voting with the majority, three justices (John Paul Stevens, David Souter, and Ruth Bader Ginsburg) wrote a concurring opinion stating that the decision went too far. They believe that there should have been a medical necessity option for a patient "for whom there is no alternative means of avoiding starvation or extraordinary suffering." In a split decision, the Supreme Court in 2005 once again ruled that the federal government can still ban the use of medical marijuana. This ruling means that those who take marijuana for personal medicinal treatment in states that allow its use still risk prosecution by the U.S. Drug Enforcement Agency. In 2006, the FDA advised against the use of marijuana for medical purposes, stating that there is no evidence from either animal or human studies to support its safety or efficacy. The controversy over medical uses of marijuana will continue for the foreseeable future.

COCAINE

Cocaine is the major alkaloid (1 of 14) of *Erythroxylum* (*Erythroxylon*) *coca* and *E. novogranatense,* the coca plant, a small tree or shrub with shiny evergreen leaves. The alkaloid occurs in the leaves, which can be harvested two or three times a year. (Alkaloids similar in structure to cocaine are also found in some members of the nightshade family, as discussed in A Closer Look 20.1: The Tropane Alkaloids and Witchcraft.) A member of the Erythroxylaceae, the coca plant is native to the Andes Mountains of South America, where the leaves have been chewed for centuries.

South American Origins

Archeological evidence, in the form of figurines depicting coca chewing, indicates that the plant has been used for at least 3,500 years in the Andes (fig. 20.5). To the Incas, the coca plant was important both socially and economically. According to myth, a god created the coca plant to alleviate hunger and thirst among the people. Because the great Incan civilization considered it sacred, coca chewing was mostly restricted to the ruling classes. Soldiers, workers, and runners were also permitted to chew the leaves for endurance. By the end of the fifteenth century, the use of coca was widespread among the Incas; however, casual chewing was considered a sacrilege. Economically, the coca leaves were used as a form of payment and could be given in exchange for potatoes, grains, furs, fruits, and other essential goods.

The early Spanish explorers noted the practice of chewing leaves with added ashes. (One component of ash is calcium carbonate, or lime, which releases the cocaine from the leaves.

Figure 20.5 Incan artifact depicts a coca user.

This practice of adding lime is still used by the Indians of the Andes region today.) After the Spanish Conquest, the native populations were enslaved and forced to work in mines under incredibly harsh conditions and with little food. The Spanish soon learned the value of coca leaves; the productivity and endurance of the enslaved Indians increased dramatically when they were given coca leaves to chew. King Philip II of Spain declared coca leaves necessary to the well-being of the Indians and set aside land in the Andes for coca cultivation. Although specimens of coca leaves had been sent back to Spain, the leaves did not retain their potency in storage. As a result, coca chewing never gained popularity in Europe.

Freud, Holmes, and Coca-Cola

The contemporary social history of *E. coca* actually begins in the late 1850s, when cocaine was isolated from coca leaves in a German laboratory. Although the anesthetic properties were quickly realized, it was the stimulating properties that brought it into the limelight. In 1884, Sigmund Freud became

A CLOSER LOOK 20.1

The Tropane Alkaloids and Witchcraft

A group of alkaloids with similar structure and similar physiological action can be found predominantly in members of the Solanaceae (nightshade family). These are known as the *tropane alkaloids* and include atropine, hyoscyamine, and scopolamine. These alkaloids have a variety of physiological effects on the body:

they relax smooth muscles;
they dilate the pupils of the eye;
they dilate blood vessels;
they increase heart rate and body temperature;
they induce sleep and lessen pain;
they stimulate and then depress the central nervous system; and
some can even induce hallucinations.

As a group, the tropane alkaloids are extremely toxic, capable of inducing coma and death due to respiratory arrest. Because of the toxicity of these solanaceous plants, Europeans feared the whole plant family and were reluctant to accept the potato and tomato (see Chapters 6 and 14), the New World relatives. One unique property of the tropane alkaloids is their ability to be absorbed through the skin, and they have been administered this way. All three of these tropane alkaloids occur in varying concentrations in deadly nightshade, or belladonna (*Atropa belladonna*), henbane (*Hyoscyamus* spp.), mandrake (*Mandragora officinarum*), and Jimsonweed (*Datura* spp.).

Atropa belladonna, a branching herbaceous perennial native to Europe and Asia, has a long history of use as a medicinal, psychoactive, and poisonous plant. One use of the plant, reflected in its specific epithet *belladonna*, was the practice by Mediterranean women of applying the plant's juice to the eyes. The result was dilation of the pupils to produce an alluring effect, hence the name "bella donna" or beautiful lady. This physiological response is due to the atropine in the plant and is used medicinally today by ophthalmologists. Atropine has multiple medicinal applications:

as an antispasmodic for treating Parkinson's disease, epilepsy, and stomach cramps;
as a bronchodilator for treating asthma;
as a heart stimulant following cardiac arrest; and
as an antidote for various poisons or overdoses.

The last use is ironic, in light of its toxic properties and its historical use as a poison itself. The application of botanical poisons probably reached its height during the Italian Renaissance, when special "schools" gave instructions in their preparation from deadly nightshade, henbane, mandrake, and other plants. These three plants were also used by witches and sorcerers of the Middle Ages to prepare magic potions. These decoctions would induce hallucinations and frenzies during witches' convocations. Images of witches flying through the air on broomsticks and transforming themselves into animals originated as delusions of their drug-induced state (box fig. 20.1a).

Along with its witchcraft applications, henbane was used medicinally in Europe for centuries as a sedative and pain reliever, especially effective for toothaches (box fig. 20.1b). Mandrake is an herbaceous perennial with a distinctive branching taproot that resembles the human form (see fig. 19.2b). Because of the Doctrine of Signatures, this plant was assumed to have magical powers to treat male and female sexual complaints. Malelike roots were thought to ensure sexual prowess, and femalelike roots supposedly cured infertility.

Unlike other solanaceous plants, *Datura* species have a cosmopolitan distribution and have been extensively used by many indigenous peoples for both medicinal and hallucinogenic purposes. In the New World, several species of *Datura* have an extensive history as sacred hallucinogens. The generic name is based on a Sanskrit word, *dhatura*, meaning poison and reflecting its toxic properties. *Datura stramonium* is probably the most widely distributed species and is cultivated for its scopolamine content, which is used today for motion sickness and for its sedative effects. The common name for this species, Jimsonweed or Jamestown

an enthusiastic advocate of the drug as a stimulant and as a means of combating morphine addiction. Freud published several papers on cocaine, which he described as a "magical drug." He used cocaine himself, recommended it to friends and family, and prescribed it for his patients, although years later he became disenchanted with the drug.

At the same time, cocaine was gaining wide popularity in the United States and could be found in over-the-counter medicines, tonics, elixirs, and even beverages. One reason for cocaine's popularity in over-the-counter preparations for colds, asthma, and catarrh (hay fever) was its effectiveness in shrinking mucous membranes and draining sinuses. It was also promoted as a panacea for ailments from headaches to hysteria. Coca-wine, notably Vin Mariani, became one of the most popular beverages of the late nineteenth century. In both the United States and Europe, luminaries such as Jules Verne, Thomas Edison, and John Philip Sousa praised its exhilarating properties. Another beverage that incorporated

(a)

(b)

(c)

Box Figure 20.1 (a) A witch is being prepared for the sabbat in this engraving from the eighteenth century. The broomstick was used to administer the salve of tropane alkaloids to the body. (b) Flowers of black henbane (*Hyoscyamus niger*). (c) Fruit of *Datura stramonium*, also known as Jimsonweed, thornapple, or Jamestown weed.

weed (box fig. 20.1c), refers to an incident of accidental poisoning of British sailors in colonial Virginia in 1676. They mistook *Datura* for an edible plant and suffered the consequences of consuming this potent hallucinogen,

> for they turn'd Fools upon it for several Days. . . . In this frantick Condition they were confined, lest they should in their Folly destroy themselves . . . and after Eleven Days, return'd to themselves again, not remembering anything that had pass'd.

cocaine was Coca-Cola, created in 1886 by an Atlanta pharmacist, John Stith Pemberton, who marketed his beverage as a "brain tonic." By the turn of the century, the negative effects of cocaine use were evident, and cocaine was eliminated from the Coca-Cola recipe in 1903 (see Chapter 16).

The widespread use of cocaine in the late nineteenth century is also evident from its appearance in the literature of the day. One of the most familiar examples can be found in Arthur Conan Doyle's Sherlock Holmes stories, in which the legendary detective was a cocaine user.

By the beginning of the twentieth century, cocaine's image was tarnished, and it was recognized as a dangerous drug. It was included under the Harrison Act of 1914, the first federal antinarcotic law that regulated the use of cocaine, opium, morphine, and heroin. Cocaine use gradually declined and was not a problem in the United States until the late 1960s.

Coke and Crack

By the mid-1970s, cocaine use had increased dramatically, becoming the favorite drug of the middle and upper classes by 1980. Under the misconception that consumption posed no long-term danger, cocaine use grew alarmingly during the 1980s. Cocaine-related deaths of sports figures and other celebrities gradually began to change the public's perception. Today there is a realization that cocaine is one of the most dangerous drugs, but cocaine abuse is still a serious national problem.

Cocaine makes its way into the United States from plantations in South America, with Colombia, Ecuador, Peru, and Bolivia the major producers. Although cocaine production is illegal (except for legitimate medical applications), hidden coca plantations supply 400 tons of the drug annually. Coca plants grow best in the hot, humid highlands of these countries, where leaves can be harvested every 35 days. The leaves are put into large vats containing a dilute solution of sulfuric acid to extract the alkaloids, and the mixture is stirred several times a day for about 4 days. The liquid is removed and mixed with a variety of chemicals to produce cocaine base. Although the yield from this extraction is low (only about 400 grams per acre of leaves), the purity is high. The extract is 75% pure cocaine. Clandestine labs refine the base into cocaine hydrochloride, a white powder that is shipped north to Mexico and the Bahamas and then on to the United States.

The cocaine hydrochloride is cut with various adulterants (additives to reduce the concentration); the street drug generally averages 12% cocaine hydrochloride. In this form, the powder can be snorted and the alkaloid absorbed through the mucous membranes of the nose. Freebasing and crack were modifications in the 1980s designed to produce quicker and stronger highs. Freebasing purifies the powder, accomplished in part by boiling it in an ether solution to produce pure cocaine, the freebase. (Ether is highly flammable and explosive. The use of any open flame or spark near ether is foolhardy.) The freebase is then smoked in a water pipe to produce an intense high, which can reach the brain in 15 seconds. Crack is a form of freebase prepared by heating a cocaine hydrochloride solution with baking soda. The resulting compound forms solid chunks, which can be broken into tiny "rocks," each costing a mere fraction of the cocaine powder. Like freebase, crack is smoked and produces a high within seconds. Crack is an especially addictive form of cocaine. Since the late 1980s, crack houses for the distribution and smoking of the drug have plagued cities in the United States.

Cocaine use rose steadily throughout the 1970s, peaking in the next decade. For most of the 1990s, cocaine use steadily declined; however, during the first years of the twenty-first century, use of cocaine has exceeded the peak years last seen in the 1980s. The Department of Health and Human Services' *2002 National Survey on Drug Use and Health* estimated that 2 million Americans use cocaine, of whom more than one-fourth use crack.

Medical Uses

Albert Niemann, who first isolated cocaine in the 1850s, is also credited with realizing its anesthetic properties. He noted that the purified crystal tasted bitter and soon numbed the tongue. Cocaine acts as a local anesthetic, temporarily blocking the transmission of nerve impulses at the site of application. Many of the widely used synthetic local anesthetics, such as Novocain (chemically known as procaine) and Xylocaine (lidocaine), are structurally similar to cocaine. Another valuable property of cocaine is its ability to constrict blood vessels and therefore reduce blood flow when applied locally. This effect has made cocaine the anesthetic of choice for ear, nose, and throat surgery, and it was formerly used for eye surgery as well.

A Deadly Drug

Cocaine is a powerful stimulant to the central nervous system that produces a short-lived euphoria. The high is accompanied by a burst of energy and alertness similar to that produced by an intense adrenaline rush. The resulting omnipotent feeling is the appeal that cocaine has for many users. The duration of the high varies with the method of administration: snorting—up to a few hours; injection—15 to 30 minutes; smoking—up to 1 hour. After the high, users experience depression and lethargy.

Cocaine has multiple physiological effects on the body: increased heart rate, respiration, blood pressure, and body temperature and dilation of the pupils. Cocaine's effects are all related to its action on neurotransmitters. Cocaine blocks the reuptake of certain neurotransmitters; therefore, they continue to act, stimulating the nervous system and producing the hyperactive effects associated with cocaine use.

These effects can be deadly. Cocaine abuse has resulted in sudden death due to heart attacks, cerebral hemorrhage, respiratory failure, and convulsions. It is impossible to predict a safe dose for a given individual, regardless of whether the person is a first-time user or a chronic abuser. A safe dose one time may cause a fatality with the next use.

A less deadly but debilitating effect of chronic cocaine use is a psychosis that is often indistinguishable from schizophrenia, with accompanying paranoia and hallucinations. Heavy users develop insomnia and appetite loss. The destruction of nasal membranes is another side effect of chronically snorting cocaine. A constantly runny nose, inflamed sinuses, and perforation of the nasal cartilage are symptoms of this destruction.

Cocaine is now considered one of the most addictive drugs. During the 1980s, medical opinion changed from the belief that cocaine is only psychologically addictive to the knowledge that it is physiologically addictive as well. Chronic users often develop tolerance, one classic criterion for addiction. Withdrawal symptoms may not be as obvious as in other drug addictions but include depression, irritability, and drug craving.

High in the Andes, coca chewing is still legal, and fresh coca leaves can be purchased at local markets for about $1 per pound. Traditional ways of chewing coca with lime have not changed for centuries among the descendants of the Incas. Leaves are also brewed as herbal teas for a mild stimulating beverage and as a folk remedy for altitude sickness. The growing of coca for the illegal drug trade, however, has become a multibillion-dollar business for the drug cartels. Some authorities believe that the only way to stop the drug traffic is to eradicate the source—the coca plant.

Cutting, burning, and herbicides have been used without much impact in the vast, mountainous areas where an untold number of farmers raise coca. Biological controls have been considered, and researchers have studied the use of caterpillars that feed on coca leaves. Perhaps the control will come about naturally; a fungal disease that attacks the roots of the coca plant has reportedly been destroying many plants in Peru more effectively than eradication efforts.

A new coca movement has been started by Bolivian President Evo Morales to introduce legal coca products to the world marketplace. In Bolivia, where chewing coca leaves and drinking coca tea are still popular practices, Morales plans to petition the United Nations to lift the ban on the international trade of coca leaf. An Aymara Indian, Morales began his political career as a union leader representing Bolivia's coca growers. After his election in December 2005 as Bolivia's first indigenous president, Morales began promoting new legal uses of the coca leaf in products such as toothpaste, tea, herbal treatments, medicines, soaps, and flour. His slogan is "Yes to coca, no to cocaine."

Thinking Critically

Caffeine, ephedrine, and cocaine are central nervous system stimulants, and all can be considered psychoactive, yet only cocaine was described in depth in this chapter.

How do these alkaloids act on the body? How do they differ in potency, and why is cocaine the only one of the three stimulants considered a major drug problem?

TOBACCO

Nicotine, the major alkaloid in tobacco, is the most addictive drug in widespread use, and cigarette smoking is a notoriously difficult habit to break. The detrimental effects of tobacco smoking on health have been known for some time, yet the addictive powers of the drug compel many people to continue this unhealthy practice. The rights of smokers versus the rights of nonsmokers is a hotly debated issue as more and more places are opting for a smoke-free environment. None of this controversy was foreseen when Columbus and his crew encountered the inhabitants of Cuba smoking *tabacos*.

A New World Habit

Nicotiana is a genus of approximately 65 species, with the majority native to the New World and the remaining species found in Australia. The two species that have been widely cultivated are *N. tabacum*, which is 1 to 3 meters (3 to 9 feet) in height, and *N. rustica*, a shrubby plant that is just over a meter (3 feet) in height. Both are tetraploid hybrids that originated in South America from natural hybridization of closely related *Nicotiana* species, followed by polyploidy. Members of the Solanaceae (nightshade family), these *Nicotiana* species are herbaceous annuals in temperate areas and have extremely large leaves subtending a terminal inflorescence of white, yellow, or pink tubular flowers. Nicotine is synthesized in the roots and translocated to the leaves; it is the leaves that become tobacco after curing.

Natives in the Andes region of South America first cultivated tobacco around 5,000 to 3,000 B.C. Archeological evidence indicates that tobacco was the first narcotic used in South America. Tribes throughout the continent utilized the leaves by chewing, by smoking, by using it as snuff (through the nose), or by drinking infusions in various rituals and medicines.

In South America, cigars were made from cured strips of tobacco leaves wrapped in banana leaves or corn husks with some cigars more than a meter (39 in.) in length with the tobacco wrapped around a stick or the midrib of a banana leaf. Licking was another method for imbibing tobacco. Tobacco tea was prepared, then the water was boiled off to make a syrup or a jelly thickened with manioc flour. The resultant product, known as ambil, was applied to a dipping stick and licked off or rubbed directly onto the gums or the teeth. Fields and people would be fumigated with tobacco smoke to kill agricultural pests and lice, an effective treatment because nicotine is now known to be a powerful insecticide.

In Central America, the Mayans farmed tobacco and smoked both for pleasure and for ritual. Two Mayan gods were depicted as habitual smokers. The Aztecs believed that tobacco could cure a number of diseases. They also used tobacco smoke in ritual human sacrifice and engulfed warriors in tobacco smoke to incite their fury in battle. The tobacco habit had reached all through the northern parts of North America by 2,500 B.C. Its importance is indicated by the fact that several tribes, such as the Blackfoot and Crow, who were hunter-gatherers and did not farm, did cultivate tobacco. Tobacco was also used medicinally for toothache and snakebites, but smoking a tobacco pipe was primarily a part of every major decision of the family or the tribe, or between tribes. The Omaha had both war pipes and peace pipes.

When Columbus set ashore on the island of Cuba in November 1492, his men encountered natives "drinking smoke" by inserting burning, rolled-up leaves of *tabacos* into their nostrils and inhaling. Soon the plant was introduced to Spain, and within a short time thousands were addicted to the smoking habit.

Introduction to France followed soon after, with credit generally given to Jean Nicot, French ambassador to Portugal. Nicot sent ground leaves and seeds of *N. rustica* to the French court for medicinal use as snuff. Nicot popularized the use of tobacco, and Linnaeus honored him in naming the genus 200 years later. John Hawkins, Francis Drake, and Walter Raleigh have all been credited with introducing and popularizing tobacco smoking in England. While in Spain cigar smoking was preferred, pipe smoking became popular in England. Soon, plantations in the colony of Virginia were established for growing tobacco, and in 1629, over 1½ million pounds of tobacco leaves were shipped to England. Tobacco became a major commodity in the colonies. Indentured servants from England and African slaves were used in the cultivation and processing of tobacco leaves on the large plantations in Maryland, Virginia, and the Carolinas. Many fortunes were made by the tobacco aristocracy, and many of the founding fathers (George Washington, Thomas Jefferson, and Patrick Henry) were tobacco growers.

Cultivation Practices

Tobacco growing demands much care and attention. The plants are susceptible to many fungal, viral, and nematode pathogens, and conditions must be controlled to minimize the spread of disease-causing agents. Seeds are planted in specially prepared seedbeds, and seedlings are later transplanted to the fields (fig. 20.6). When the plants begin to flower, the inflorescence and any side branches are removed to promote maximum leaf expansion. The leaves are harvested and taken for curing as they mature.

The method of curing varies with the intended use of the tobacco, but curing is basically a fermentation-and-drying process during which the characteristic aroma and flavors of tobacco develop. The leaves are hung to dry at elevated temperatures; during the curing period, the chlorophylls disappear, resulting in a yellow-brown color. The leaves are sorted and graded, then aged for several months to several years to fully develop the aromas and flavors. The final step is cutting and blending different grades as well as hundreds of flavoring additives to produce the desired mixtures for cigarettes, cigars, pipe tobacco, and chewing tobacco.

Although there were prototypes during the Mayan and Aztec eras, the modern cigarette made its appearance in Europe after the Crimean War. The development of a cigarette manufacturing machine in 1881 produced a ready-made, inexpensive tobacco product for the masses; however, it was not until after World War I that cigarette smoking dominated all other uses of tobacco in the United States. Per capita cigarette consumption in the United States reached a peak in 1963 at 42%. Although the number of smokers has declined to 25%, there is evidence that the number of cigarettes smoked each day has actually increased despite massive evidence on the negative health effects. Worldwide, approximately 1.1 billion smokers (18% of the world population) consume 5 trillion cigarettes a year.

(a)

(b)

Figure 20.6 (a) Tobacco fields in North Carolina. (b) The flavor and aroma of tobacco develop during the curing process.

Health Risks

When tobacco was first introduced to Europe, the Indian claim that tobacco had magical and curative powers was accepted by many. However, one vehement critic of tobacco use was King

James I of England, who wrote in *A Counterblaste to Tobacco* in 1604:

> A custome lothsome to the eye, hateful to the Nose, harmful to the braine, dangerous to the Lungs, and the blacke stinking fume thereof, neerst resembling the horrible Stigian smoke of the pit that is bottomlesse.

Despite this royal condemnation, tobacco use increased exponentially in the subsequent decades. For the next 350 years, claims about the healing powers (a panacea for most ailments) of tobacco persisted, and even in 1948 the *Journal of the American Medical Association* was stating that

> more can be said in behalf of smoking as a form of escape from tension than against it ... there does not seem to be any preponderance of evidence that would indicate the abolition of the use of tobacco as a substance contrary to the public health.

By the 1950s, evidence was starting to accumulate that cigarette smoking was indeed harmful to health. But this view was not widely accepted until the Surgeon General's Report of 1964 that stated cigarette smoking is linked to cancers of the lung and larynx and to chronic bronchitis. It is now recognized that 90% of the lung cancers in men and almost 80% in women are due to cigarette smoking. Later reports concluded that smoking causes a number of other diseases, such as cancers of the bladder, esophagus, mouth, and throat; cardiovascular diseases; and negative reproductive effects.

Forty years after the 1964 report, the surgeon general released *The Health Consequences of Smoking*, which stated that tobacco smoking causes disease in nearly every organ of the body because the toxins from smoking travel throughout the circulatory system. The report cited research studies that indicated for the first time a number of diseases that were not previously associated with smoking: cancers of the stomach, cervix, pancreas, and kidney; acute myeloid leukemia; pneumonia; abdominal aortic aneurysm (ballooning of the wall of an artery); cataracts; and periodontitis (gum disease). Smoking is the leading preventable cause of death and disease in the United States.

Since the American public was first warned about the health consequences of smoking in 1964, 12 million Americans have died of smoking and smoking-related diseases. Compared to nonsmokers, male smokers die an average of 13 years earlier, and women who smoke shorten their lives by approximately 14.5 years. Quitting smoking has immediate and long-term benefits, such as lowering the heart rate and improving circulation. Smokers who quit at age 65 or later reduce the risk of dying of a smoking-related disease by nearly 50%.

Both coronary heart disease and stroke, the first and third leading causes of death in the United States, are caused by smoking. Toxins from smoking are released into the bloodstream and initiate changes in the vessel walls that promote atherosclerosis. Atherosclerosis can lead to blood clots that block the flow of blood to the heart or brain, resulting in heart attacks or strokes. Chronic obstructive pulmonary disease, including chronic bronchitis and emphysema, is the fourth leading cause of death in the United States. Ninety percent of the deaths from these diseases of the lungs are attributable to smoking.

Clinical studies have demonstrated that women who smoke often experience difficulty in becoming pregnant. Smoking during pregnancy increases the incidence of pregnancy complications, low-birth-weight babies, and premature and stillborn births. Nursing mothers may pass nicotine to their infants through breast milk. Babies exposed to secondhand smoke are at a higher risk for sudden infant death syndrome. Postmenopausal women who smoke have a lower bone density and increased risk of hip fractures than women who do not smoke.

Tobacco is still dangerous even if it is not smoked. Chewing tobacco is strongly linked to oral cancer. Long-term users have a 50-fold increase in the risk of developing oral cancer.

Smokers are not the only ones affected by tobacco smoke. Much research has focused on the health effects in passive smokers, nonsmokers who are inhaling environmental tobacco smoke (secondary smoke). Children of smokers have an increased risk of respiratory disorders and reduced lung function, and in adults, passive smoking is a cause of lung cancer. In fact, it was the concern over the hazards of passive smoking in airplanes that prompted the ban on all smoking on domestic flights in the United States.

Tobacco smoke is a complex mixture of over 2,000 chemicals, plus extremely fine particulates, all of which can enter the lungs. Included in the mixture are nicotine, tars, carbon monoxide, hydrogen cyanide, aromatic hydrocarbons, and a host of other hazardous substances (table 20.1).

The alkaloid nicotine is addictive in any form of tobacco, and its addictiveness is as tenacious as that of other narcotics.

Table 20.1 Some Hazardous Components of Tobacco Smoke

Carbon monoxide	Acrolein
Nicotine	Acetaldehyde
Carcinogenic tars	Acetone
Sulfur dioxide	Acetonitrile
Ammonia	2,3-butanedione
Nitrogen oxides	Crotononitrile
Vinyl chloride	Ethylamine
Hydrogen cyanide	Hydrogen sulfide
Formaldehyde	Methacrolein
Radionuclides	Butylamine
Benzene	Dimethylamine
Arsenic	Endrin

Nicotine is volatilized during burning and absorbed upon inhalation. It exerts a stimulant action on the central nervous system, promoting arousal in certain areas of the brain. Nicotine also acts on the peripheral nervous system, first stimulating then blocking transmission of nerve signals. Sensations of heat, pain, and hunger are inhibited while an increase in adrenaline causes increased heart rate and blood pressure and constriction of blood vessels. Nicotine was isolated early in the nineteenth century and is known to be one of the most toxic plant poisons. In its pure form, ingestion of just a few drops is lethal, causing death by respiratory paralysis. The poisonous properties have been put to use as an especially effective insecticide. Nicotine has also been implicated in various forms of cancer, with evidence indicating that nicotine is converted into carcinogenic products in the body.

Besides nicotine, other dangerous components of tobacco smoke include tars, organic substances produced during the burning of tobacco leaves. Tars are known carcinogens, and cigarette manufacturers have attempted to reduce the tar, as well as the nicotine, content through filters. Smoking low-tar or reduced-nicotine "light" cigarettes does not substantially decrease the incidence of smoking-related disease. Poisonous gases are also found in cigarette smoke; carbon monoxide, hydrogen cyanide, and formaldehyde are just a few of the gaseous components. Carbon monoxide, in particular, has been shown to be associated with coronary heart disease and the retardation of fetal growth.

Like withdrawal from any addictive substance, withdrawal from nicotine is difficult, and many approaches have been used, with limited success. One innovation is the use of transdermal nicotine patches. In this form, the nicotine can be absorbed through the skin; the patches provide a steady supply of the alkaloid, eliminating the other dangers associated with smoking. The concentration of nicotine in the patches is gradually reduced, weaning the patient from the addiction over several months. The obvious drawback to the patches is the continued use of a dangerously addictive substance, but many lives may be saved if it does, in fact, facilitate quitting. Fifteen to 20 years after a smoker quits, the risks of developing lung cancer are 80% less than for active smokers.

In the past decade the tobacco industry has been put on the defensive. Internal company documents have been released that indicate companies have known for years that nicotine is addictive, and tobacco executives have also admitted knowing that smoking causes lung cancer, heart disease, and emphysema. By the end of 1998, all 50 states and the District of Columbia had reached settlements over lawsuits filed against the five largest tobacco companies. States sued to recoup Medicaid money paid out for tobacco-related diseases. Although Florida, Minnesota, Mississippi, and Texas had previously settled with the tobacco companies, the other states participated in the Master Tobacco Settlement of 1998. Collectively, the settlements cost the tobacco companies over $200 billion. Federal legislation has been proposed to reduce teen smoking, permit the FDA to control nicotine, and provide penalties for the tobacco companies if teen smoking is not reduced. Four of the major tobacco companies pledged to "work constructively" with the federal government in enacting tobacco-related laws.

In early 2005, the World Health Organization Framework Convention on Tobacco Control (WHO FCTC) was ratified by 168 countries. Ratifying nations are required to ban tobacco advertising and promotion. The treaty also requires large warning labels on cigarette packages and urges governments to increase taxes on tobacco products. It also commits nations to protect nonsmokers from secondary smoke in indoor workplaces. A fund will also be established for poorer nations to use in educating the public about the dangers of tobacco smoking. Tobacco use is the second leading cause of death globally and is responsible for the deaths of approximately five million people worldwide every year. Although smoking rates have fallen in North America and Europe in recent years, they have continued to rise in Asia, central and eastern Europe, and Latin America. According to WHO figures, deaths due to smoking will rise to 10 million per year within two decades and 70 percent of those deaths will occur in developing nations. The United States is one of 12 countries that signed but did not ratify the treaty; another 12 countries have not signed the FCTC treaty.

New evidence indicates that addiction to cigarettes may occur more rapidly than at first thought. Previous research suggested that nicotine addiction took hold when a five-cigarette-per-day habit developed. In a study of new smokers among adolescents, 10% experienced symptoms of addiction within 2 days of their first cigarette. Another study reported that 25% of first-time smokers reported symptoms of addiction after smoking just one to four cigarettes. It has been demonstrated that smoking increases the number of receptors in brain tissue that have an affinity for nicotine. Autopsies of smokers have a 50% to 100% increase in the number of nicotine receptors in certain areas of the brain. Laboratory rats exposed to the nicotine equivalent of one to two cigarettes per day exhibited an increase in nicotinic receptors by the second day of exposure. Adolescents can experience nicotine withdrawal symptoms due to nicotine's immediate effect on brain chemistry in just 2 days after their first cigarette, despite the fact that the five-per-day habit usually takes 2 or more years to develop.

Thinking Critically

On January 1, 1998, a California law went into effect banning smoking in bars. The intent of this law was to protect the health of workers in the bars. This follows similar legislation banning smoking in restaurants and other public places.

What is the rationale for these laws? Does the scientific literature support the reasoning for proposing this legislation?

Electronic, or e-cigarettes, first appeared in the U.S. market in 2007. A battery operated device heats up and volatilizes a liquid, containing nicotine and flavoring agents plus solvents, into a fine mist, or *vape*, that is inhaled. The e-cigarette was first promoted as a healthier tobacco product because no smoke is produced. They were also advertised as an aid to help wean cigarette users from their smoking addiction. Both claims have not withstood the scrutiny of recent studies; there is no evidence that vaping helps smokers quit the nicotine habit.

In fact, research has shown that vaping delivers a mixture of toxic chemicals into the lungs which increases inflammation, impairs the immune system, and makes germs more resistant to the body's defenses.

Traces of the solvents used to dissolve the nicotine and flavorings are known lung irritants; the high temperature used to produce the vapors breakdowns some solvents into carbonyls, known cancer-causing agents. Also the vapors deliver fine particles into the smallest and deepest respiratory passageways at a rate of 3 milligrams per cubic meter, a level which is approximately 100 times higher than the Environmental Protection Agency's 24 hour exposure limit for particulates. Particulates trigger inflammation in the lung and chronic exposure can cause asthma and other diseases. Studies have shown that vaping of flavorings and solvents by themselves produce free-radicals (see Chapter 10) damaging lung cells and impairing the immune system. Most troubling is that the e-cigarette habit has become extremely popular with teens and often leads teens to try cigarette smoking.

PEYOTE

As do many of the psychoactive plants discussed in this chapter, peyote has a long history of use, dating back at least 3,000 years in Mexico. Early accounts from sixteenth-century Spanish missionaries described the plant and its use as a hallucinogen in religious rituals among many Indian tribes. Bernardino de Sahagun indicated

> Those who eat or drink it see visions either frightful or mirthful; the intoxication lasts two or three days and then ceases. . . . It sustains them and gives them courage to fight and not to feel hunger nor thirst; and they say that it protects them from all dangers.

Although the Spanish initially attempted to suppress peyote rituals, the practices have continued to this day.

Mescal Buttons

Peyote (from the Aztec word *peyotl*) is the cactus *Lophophora williamsii*, native to Mexico and southwestern Texas. The crown of the spineless gray-green cactus is sliced and dried into "mescal buttons" (fig. 20.7). Dried, the buttons can

Figure 20.7 Peyote cactus (*Lophophora williamsii*) held by a native American healer.

keep indefinitely, retaining their hallucinogenic properties, because the active ingredients are not volatile. The buttons are softened by saliva in the mouth and then swallowed; alternatively, the buttons may be soaked in water and the water consumed.

Peyote contains up to 30 alkaloids, with mescaline the most active hallucinogen in the group. Again, the nervous system seems to be affected because mescaline is similar in structure to a neurotransmitter. Investigations have shown that the effects of mescaline alone differ from the action of consuming the whole mescal button, where the alkaloids may interact synergistically. Intoxication induced by peyote may consist of strikingly brilliant visions, as described so vividly by Havelock Ellis in 1898:

> At first there was merely a vague play of light and shade which suggested pictures. . . . Then, in the course of the evening, they became distinct, but still indescribable—mostly a vast field of golden jewels, studded with red and green stones, ever changing. . . . I would see thick, glorious fields of jewels, solitary or clustered, sometimes brilliant and sparkling, sometimes with a dull rich glow. Then they would spring up into flowerlike shapes beneath my gaze, and then seem to turn into gorgeous butterfly forms or endless folds of glistening iridescent, fibrous wings of wonderful insects. . . . Most usually there was a combination of rich sober color, with jewel-like points of brilliant hue.

Not all peyote experiences consist of pleasant visions. Other writers describe anxiety and horribly frightening visual hallucinations of snakes and demons. The intoxication is often accompanied by physical effects, including nausea, tremors, chills, and vomiting. Because of the interplay of alkaloids, peyote hallucinations are extremely complex, inducing a range of sensory experiences.

Native American Church

The use of peyote by members of the Native American Church has been a topic of controversy for over a hundred years. The origins of the Native American Church can be traced to the 1870s, when the Kiowa and Comanche Indians learned of peyote from tribes in Mexico and took the plant back with them to Oklahoma (then Indian Territory). The tribes established a whole new cult, with a combination of Christianity and peyote rituals. The cult spread rapidly through the Southwest to the Plains States, and eventually even to Canada, incorporating elements from the cultures of various tribes. As the cult grew and spread, it met with much opposition, especially from missionaries and government officials. In an attempt to protect their religious rights, Native Americans formally incorporated their movement into the Native American Church in 1918. Even though the Native American Church was officially recognized as a church and was protected by the Bill of Rights, the opposition to peyote use continued. A long legal battle culminated in a Supreme Court ruling that gave the members of the Native American Church the right to use peyote as a sacrament. However, a more recent Supreme Court decision restricted that right. In 1990, the Court upheld Oregon's right to outlaw peyote even for religious use, stating that the First Amendment does not permit people to break the law for the sake of freedom of religion. The controversy continues.

KAVA—THE DRUG OF CHOICE IN THE PACIFIC

Kava (*Piper methysticum*) is a small shrub in the Piperaceae, or pepper family, and in the same genus as black pepper. Kava roots are used to prepare an intoxicating beverage that has been consumed for thousands of years by peoples throughout the islands of the South Pacific. In Hawaii, kava is known as *awa*, and it is one of the plants brought by early Polynesian settlers when they arrived in Hawaii. In Hawaiian, *awa* means bitter and refers to the pungent taste of the beverage.

The beverage is a depressant, and about 10 minutes after consumption the effects of kava can be felt. A small quantity makes people relaxed and friendly, but unlike alcohol, it does not impair alertness. However, kava is a powerful soporific, and too large a quantity produces a deep, dreamless sleep.

Preparation of the Beverage

Kava has had social, ceremonial, and sacred uses throughout the islands. Kava was used at meetings to settle disputes between neighbors as well as in sacred healing ceremonies. Customs differed throughout the South Pacific islands, and in some areas only men were allowed to drink kava.

Traditional preparation of the beverage was steeped in ritual. Young, unmarried women chewed small pieces of the root until the fibers were completely broken down, then spit the wad into a ceremonial bowl. After several wads were accumulated, water was added to the bowl, and the mixture was stirred. When the desired potency was reached, the pulp was strained and the beverage was consumed in a coconut shell. Today the root is finely ground and then mixed with water or put in a blender with water. The mixture is then coarsely strained; however, kava powder is also available that requires no straining. Although the preparation has changed, many of the customs still remain throughout the islands.

Active Components in Kava

When fresh, the root is thick and the wood soft, but the root hardens as it dries. Plants must be allowed to grow and spread for 2 to 3 years before the roots can be harvested. This time is needed for increases in size as well as potency. The best roots are those that have been left in the ground for up to 20 years. The active ingredients are kava lactones; 10 have been identified, but six seem to be important. Different strains of kava appear to have different proportions of these compounds, and the ratio of these lactones is what produces the effects of the beverage. Eighty-two varieties of kava have been described in the South Pacific.

Kava also has a long history of use as a medicinal plant in the South Pacific. It has been used primarily as a tranquilizing elixir that produces relaxation and sleep; however, the leaves, stem, and bark have applications for treating a variety of ailments. It has also been employed as a muscle relaxant to relieve stiffness and muscle fatigue. Outside the South Pacific, kava extracts were utilized in Germany for several decades as an antianxiety drug. Recent clinical tests support the efficacy of kava extracts for treating anxiety of nonpsychotic origin.

During the 1990s, herbal supplements containing kava were widely used for its calming effects as antianxiety drugs and sleep aids. It was a top-selling herbal remedy in both the United States and Europe. This popularity caused a significant economic boom throughout the South Pacific, especially in Vanuatu and Fiji. Starting in 2001, medical authorities in many countries became concerned about the safety of kava. This issue first arose in Europe, where 25 to 30 reports of liver toxicity were linked to kava use. The adverse effects included acute liver failure, hepatitis, and liver cirrhosis. Several patients required liver transplants, and one fatality occurred. Germany, Switzerland, France, the United Kingdom, Canada, and other countries have banned the use of kava in herbal remedies, but some reports suggest that the ban is being ignored. In March 2002, the U.S. FDA issued a consumer advisory about the dangers associated with kava consumption.

Some researchers have suggested that the ban on kava was based on flawed evidence and have recommended lifting the ban. In the Pacific Islands, there is no indication of liver toxicity associated with traditional uses of kava, which is prepared from water extracts of powdered roots. By contrast, kava remedies available in Western countries are prepared from ethanol or acetone extracts of the roots. It has been suggested that the different methods of preparation may explain the difference in toxicity, although some researchers maintain that liver toxicity has not been specifically studied on the South Pacific Islands. Meanwhile, Pacific Island countries are suffering economically from the current ban on kava.

LESSER KNOWN PSYCHOACTIVE PLANTS

The discussion thus far in this chapter has focused on psychoactive plants that have become major problems in modern society. These discussions have by no means exhausted the list of known psychoactive plants (table 20.2). This section describes some lesser-known plants that have potent effects on the central nervous system.

Certain members of the morning glory family, the Convolvulaceae, also figured prominently as hallucinogens and were revered as powerful medicines among the Aztecs; these morning glories continue to be used by tribes in Mexico. The seeds of *Ipomoea violacea,* as well as other *Ipomoea* and *Rivea* species, contain amides of D-lysergic acid similar to, but far milder than, the potent hallucinogen LSD (see Chapter 25). Known as *ololiuqui* among the Aztecs, morning glory seeds were used in divination as well as religious and magical rituals. The shaman would consume a drink prepared from the seeds and, in the hallucinogenic state that followed, would divine the cause of a person's illness.

Nutmeg, *Myristica fragrans,* has a long history as a culinary spice but has also been valued in some circles for its hallucinogenic properties. It was known to practitioners of Ayurvedic medicine in India as the "narcotic fruit." It was also mixed with betel nut or tobacco snuff for its intoxicating effects and was occasionally used in other areas as a substitute hallucinogen when the drug of choice was unavailable. Because the active principle is volatile, the reaction to nutmeg is unpredictable, and thus potency varies greatly. Side effects, however, are predictable and extremely unpleasant (headache, nausea, dizziness, vomiting, irregular heartbeat, and others), making nutmeg an unpopular choice as a hallucinogen.

In addition to the potent psychoactive substances described in this chapter, caffeine and caffeinelike alkaloids should not be overlooked; they, too, affect the central nervous system. In fact, caffeine is the most widely used psychoactive substance in the world. The impact of these substances on society as well as physiological actions were fully discussed in Chapter 16. Also, many other psychoactive plants, which are only now being investigated, have been used by cultures throughout the world, particularly in the tropical rain forests. These studies should provide further evidence that will reinforce the pivotal role of psychoactive drugs in human society.

Table 20.2
Plants of Known Psychoactive Value

Scientific Name	Common Name	Family	Active Principle	Physiological Action
Atropa belladonna	Belladonna	Solanaceae	Atropine; scopolamine; hyoscyamine	Hallucinogen
Cannabis sativa	Marijuana	Cannabaceae	Tetrahydrocannabinol	Hallucinogen
Datura spp.	Jimsonweed	Solanaceae	Atropine; scopolamine; hyoscyamine	Hallucinogen
Erythoxylum coca	Coca	Erythroxylaceae	Cocaine	Stimulant
Hyoscyamus spp.	Henbane	Solanaceae	Atropine; scopolamine; hyoscyamine	Hallucinogen
Ipomoea spp.	Morning glory	Convolvulaceae	Lysergic acid	Hallucinogen
Lophophora williamsii	Peyote	Cactaceae	Mescaline	Hallucinogen
Mandragora officinarum	Mandrake	Solanaceae	Atropine; scopolamine; hyoscyamine	Hallucinogen
Myristica fragrans	Nutmeg	Myristicaceae	Volatile oils	Hallucinogen
Nicotiana spp.	Tobacco	Solanaceae	Nicotine	Stimulant/depressant
Papaver somniferum	Opium poppy	Papaveraceae	Morphine; codeine; heroin	Depressant
Piper methysticum	Kava	Piperaceae	Lactones	Depressant

CHAPTER SUMMARY

1. The use of psychoactive plants in human society dates back millennia. The active principle in psychoactive plants works by affecting the release of neurotransmitters in the central nervous system. Depending on their effect, psychoactive drugs can be classified as stimulants, depressants, or hallucinogens. A psychoactive drug is classified as addictive if it causes one or more of the following actions: psychological dependence, physiological dependence, or tolerance. The active principles in most psychoactive drugs are classified as alkaloids.

2. The opium poppy (*Papaver somniferum*) is the source of a milky latex from the green capsules that, when dried, becomes crude opium. Opium has been eaten, drunk, and smoked for centuries. More than 20 alkaloids have been identified in opium, but the chief ones are morphine and codeine.

3. Heroin is a semisynthetic derivative of morphine and the most strongly addictive of all the opiates. Heroin addiction is a major drug problem in most Western cities.

4. Marijuana (*Cannabis sativa*) has an equally long history as a hallucinogen in India. The resin found in the upper leaves and flowers of the carpellate plants contains the psychoactive and nonalkaloid principle THC.

5. Cocaine is derived from the leaves of the coca plant (*Erythroxylum coca*), a shrub native to South America. Chewing a wad of the leaves to experience mild stimulating effects has been practiced for at least 3,500 years in the Andes and is still practiced among native groups today. The extraction of cocaine from the leaves in the 1850s encouraged the spread of cocaine use in Europe and the United States around the turn of the twentieth century. The enchantment with cocaine use faded as the negative effects became apparent. Unfortunately, cocaine use in the more dangerously addictive forms of freebase and crack became popular in the 1980s. The lessons of the dangers of cocaine had to be relearned.

6. The most addictive drug in widespread use is nicotine, the major alkaloid in tobacco (*Nicotiana* spp.), native to the New World. The smoking of tobacco has long been associated with the rituals of many native tribes. It is now recognized that the habit of cigarette smoking increases the risk of many diseases in nearly every organ of the body. Passive smokers, especially the children of smokers, have been shown to have higher incidences of respiratory disorders and even lung cancer. Adolescents become addicted to nicotine more rapidly than those who begin smoking as adults.

7. The use of peyote (*Lophophora williamsii*) has long been associated with sacred rituals of many Indian tribes in Mexico. In the United States, the eating of mescal buttons has been a practice and right of members of the Native American Church, protected under the auspices of religious freedom.

8. Certain herbs that contain the tropane alkaloids have long been associated with the practice of witchcraft in medieval Europe; many of these drugs have medicinal qualities, and they are used in modern medicine today.

REVIEW QUESTIONS

1. Distinguish between stimulants, hallucinogens, and depressants.
2. What is meant by the term *addictive?* Which psychoactive plants are known to be addictive?
3. What active principle accounts for the psychoactivity in most plants? Explain.
4. Give the history and use of the following psychoactive plants: opium poppy, marijuana, coca bush.
5. Give examples of how psychoactive plants have been used in religious, magical, and healing rituals.
6. Trace the history of tobacco smoking, and describe the health effects of this habit.

CHAPTER 21

Poisonous and Allergy Plants

The crown of thorns (Euphorbia milii) *is protected from animals by both needlelike spines and a milky sap, which can irritate the skin and eyes.*

KEY CONCEPTS

1. Poisonous plants are everywhere in the environment, including some of the most common wild plants, yard plants, and houseplants; many accidental poisonings, especially of young children, occur each year.

2. Many plants have evolved chemicals that protect them from insects and other pests; some of these chemicals are being used as natural insecticides.

3. Certain plants are capable of triggering allergic reactions ranging from respiratory allergies to contact dermatitis.

©Karen McMahon

CHAPTER OUTLINE

Notable Poisonous Plants 379
    Poisonous Plants in the Wild 379
    Poisonous Plants in the Backyard 382
A CLOSER LOOK 21.1 Allelopathy—Chemical Warfare in Plants 383
    Poisonous Plants in the Home 386
    Plants Poisonous to Livestock 386
Plants That Cause Mechanical Injury 387
Insecticides from Plants 388
Allergy Plants 389
    Allergy and the Immune System 390
    Respiratory Allergies 390
    Hay Fever Plants 391
    Climate Change and Allergy Plants 394
    Allergy Control 394
    Contact Dermatitis 395
    Food Allergies 397
    Latex Allergy 398
Chapter Summary 399
Review Questions 399

The focus of this chapter is on plants that adversely affect the health of humans and other animals. Plants are the source of toxins so deadly that fatalities can result from miniscule amounts, yet poisonous compounds have also been utilized in medicine and as natural insecticides. Other harmful effects include allergic reactions that can be debilitating or even fatal.

NOTABLE POISONOUS PLANTS

As discussed in earlier chapters, the ability of ancient peoples to identify edible plants was the first step in the development of foraging societies. Poisonous plants would be recognized and the information passed on to avoid future calamities. There is also evidence that, from the earliest times, poisonous plants were utilized as a method of capturing prey. In fact, the word *toxic* is derived from the ancient Greek word *toxikon,* meaning arrow poison.

At various times in history, plant poisons were also employed for the more nefarious purpose of disposing of one's enemies. The Ebers Papyrus lists the many toxins known to the Egyptians of 3,500 years ago, and in ancient Athens a cup of poison hemlock was the standard method of capital punishment. Knowledge about toxic plants reached a peak in the Renaissance in Europe, when so-called succession powders often ensured an untimely death and a calculated ascent to power. Members of the De Medici family of Italy, especially Catherine De Medici, were renowned poisoners of this time.

As modern society has distanced itself from its natural surroundings, knowledge of toxic plants has shrunk to a small circle of trained professionals. Most people would be surprised to learn that many common weeds, landscape plants, and houseplants contain deadly poisons. It is estimated that there are thousands of plants and fungi that produce toxic substances, with hundreds of these found in the Americas alone. Why are so many plants poisonous? The obvious answer is that making at least some part of the plant toxic affords protection against grazing animals and herbivorous insects. Alkaloids and glycosides play a prominent role as poisonous compounds; remember that many of the medicinal and psychoactive plants can also be fatal at certain doses (see Chapters 19 and 20).

Poisonous Plants in the Wild

Strychnine, an alkaloid obtained commercially from the seeds of the Asian-Indian tree *Strychnos nux-vomica* (Loganiaceae), is a stimulant of the central nervous system, especially the spinal cord. This powerful nerve toxin induces muscle spasms and convulsions and is unique in the way it magnifies sensations of sight, smell, touch, and hearing. In fact, any sudden stimulus, such as a loud noise, can trigger a seizure in an affected individual. Historically, strychnine has been used to treat a variety of ailments (constipation, impotence, barbiturate poisoning), but its medical use today is confined to neurological research. It also has been used illegally to enhance athletic performance. Strychnine is still employed, however, as a rodent poison, particularly for moles, and is the main ingredient in the common poison "peanuts" for the garden.

Curare is the arrow poison employed by many South American tribes in hunting game. It is applied to the tip of the arrow, and if the arrow finds its target, the Indians say "he to whom it comes falls" (fig. 21.1). Curare paralyzes quickly. Various methods of preparation, and over 70 different species, have been used by South American tribes to make curare. However, the bark of two woody vines of the lowland tropical rain forest, notably *Strychnos toxifera* and *Chondrodendron tomentosum* (Menispermaceae), figure most prominently. The French physiologist Claude Bernard was the first to study the action of curare, in the nineteenth century, correctly concluding that it blocks nerve impulses at the junction of nerve and muscle. In 1942, H. R. Griffith was able to demonstrate the anesthetic value of an extract of curare. It was the prior

Figure 21.1   In many tribes of South America, hunters use darts poisoned with curare.

isolation of the chemical tubocurarine from curare by British researchers that made this application possible because tubocurarine, by itself, produces only a partial, reversible paralysis. An injection of curare, or the alkaloid tubocurarine, produces immediate muscle relaxation by blocking nerve impulses. Movement is impossible because of skeletal muscle paralysis; swallowing and speaking become difficult; and if the person goes untreated, death by asphyxiation follows because of impairment of the diaphragm. The medicinal value of curare as a muscle relaxant is unsurpassed and has found its greatest application in surgery, where adequate muscle relaxation can be achieved without excessive general anesthesia. This muscle toxin also has value in the treatment of spastic cerebral palsy, myasthenia gravis, polio, and tetanus.

Two genera of the Apiaceae—poison hemlock (*Conium maculatum*) and water hemlock (*Cicuta* spp.)—found in many areas share the distinction of being among the most poisonous wild plants in North America (fig. 21.2). (These plants should not be confused with hemlock, the common name of the conifer *Tsuga canadensis* of northern forests, which is not known to be poisonous.) Both are large, perennial herbs with big, white, compound umbels and pinnately compound leaves. They are often confused with each other and with such edible relatives as wild carrot and parsnip. Poison hemlock may reach heights up to 2 meters (6 feet). It has highly dissected, fernlike leaves and hollow, grooved stems with purple spots. Poison hemlock is notorious as an agent of death in ancient Athens. Socrates was condemned to death nearly 2,400 years ago for allegedly corrupting the young with his teaching philosophy and was forced to drink the juice of poison hemlock as punishment. The alkaloid coniine is a central nervous system stimulant that affects the body in a manner similar to a nicotine overdose; paralysis creeps from the lower limbs upward. Death is due to paralysis of the diaphragm and subsequent respiratory failure.

The water hemlock, as its name implies, grows in wet or swampy areas. It differs in appearance from the poison hemlock in that the leaves of the water hemlock are usually twice pinnately compound with coarsely toothed leaflets. The toxin is an alcohol, cicutoxin, which is found in highest concentration in the yellow sap that exudes from cut roots. Cicutoxin produces violent convulsions, and unless the person is treated promptly, death invariably follows.

Giant Hogweed (*Heracleum mantegazzianum*) is another dangerously toxic member of the carrot family (Apiaceae). Native to the Caucasus Mountains and southwestern Asia, this perennial herb was brought to New York in 1917 as an ornamental for the garden. Everything about giant hogweed is huge with heights up to 5 meters (16.5 feet) and hollow, ridged stems 5–10 cm (2–4 inches) in diameter. At the nodes, coarse white hairs and reddish-purple blotches distinguish the giant hogweed from related species. Its leaves are alternately arranged and pinnately compound with three lobed, toothed leaflets and can be up to 3 meters (10 feet) long and 1.5 meters (5 feet) wide. Compound umbels up to 0.8 meters

(a) Poison hemlock (b) Water hemlock

Figure 21.2 Two poisonous members of the carrot family are (a) poison hemlock and (b) water hemlock.

(2.5 feet) across have numerous tiny white flowers that bloom in summer. A single giant hogweed can produce over 80,000 two-seeded schizocarps (see Chapter 17). The flat oval fruits are marked with prominent brown stripes.

Giant hogweed is an aggressively invasive plant whose gigantic size shades out endangered native species. It is predominantly found in eastern and northwestern North America but has been recently reported spreading into the central region of the United States and Canada. The giant hogweed prefers moist soils and is often found near riverbanks where the dispersed fruits float and are transported to new locations.

It is the sap of giant hogweed that is toxic, containing furocoumarins, chemicals that can blister and cause severe burns on the skin and if it gets into the eyes, blindness. It may take up to 2 hours after contact for the sap to be absorbed into the skin. Exposure to Ultraviolet A (UV-A) radiation is necessary to trigger the furocoumarins in the cells. Activated furocoumarins bind to thymine and cytosine, nucleotide bases in DNA. Vital cellular processes such as transcription and translation (see Chapter 7) are inhibited and the skin cells soon die, resulting in burns and blisters. This condition is known as photophytodermatitis in which light activates plant-derived chemicals causing skin inflammation. Scars and brown stains may permanently mark the skin. Furocoumarins are found in other plants, notably citrus fruits such as lemons and limes, and members of the carrot family such as cow parsnip and wild parsnip, but the concentration is much higher, and consequently more damaging, in the sap of the giant hogweed.

The only treatment after contact with the sap of giant hogweed is to quickly and thoroughly rinse the skin with soap and water and to wash away the furocoumarins before they have penetrated the skin cells. If that is not possible, exposure of the affected site to sunlight should be avoided for several days to prevent initiating the furanocoumarin reaction.

There are some medicinal applications for furocoumarins in treating psoriasis and other skin ailments. The furocoumarins can be applied directly to the affected skin and then activated by a focused blast of UV-A to kill the targeted cells. Until the furocoumarins are broken down, the patient would have to limit exposure to sunlight to avoid damaging healthy skin cells.

Milkweeds (*Asclepias* spp.) are common weeds and easily recognized by opposite or whorled leaves, milky juice when the plant is cut, dense umbels of flowers, and follicles with silk-tufted seeds (fig. 21.3). The poisonous compounds in milkweeds include the resinous galitoxin as well as cardioactive glycosides similar to the drug digitalis (see Chapter 19). Livestock and humans, especially children, have been poisoned by eating milkweed. The larvae of the monarch butterfly can eat milkweed without ill effects; in fact, these caterpillars feed exclusively on certain milkweeds. Not only are these cardioactive glycosides not harmful to the caterpillars, but they are also passed on to the adult, or butterfly, stage intact. The presence of these cardioactive glycosides causes the unfortunate bird who eats a monarch to become ill and vomit for several hours. Birds quickly learn to avoid eating the distinctively colored

Figure 21.3 The larval stage of monarch butterflies feeds exclusively on milkweeds (*Asclepias*), passing on cardioactive glycosides to the adult butterfly.

butterflies, going so far as to also avoid viceroy butterflies, which have similar orange and black markings.

An even more dramatic example of an animal usurping a plant's toxins for chemical defense can be seen in the story of the rattlebox moth. This moth of the southeastern United States is so named because the black and yellow rattlebox caterpillar can feed on the poisonous rattlebox plant, *Crotalaria,* a species within the Fabaceae, or bean family. Unlike other animals, the rattlebox caterpillar can ingest monocrotaline, the primary toxic alkaloid of the rattlebox plant, without suffering the usual toxic results of liver impairment or respiratory distress. Instead, the alkaloid is sequestered within the tissues of the caterpillar, and a preliminary taste wards off spiders and other predators. The story does not end there. After the male caterpillar pupates and becomes an adult moth, it has in its tissues enough monocrotaline to be protected from predators. Female moths do not have this inherent chemical protection, but monocrotaline is transferred from the male during the 9-hour copulation. The avoidance of this chemical is so pronounced that wolf spiders have been observed to cut free monocrotaline-protected moths entrapped in their webs.

The consumption of a poisonous lichen, tumble-weed shield lichen (*Parmelia*), was implicated in the deaths of more than 300 elk in Wyoming. Lichens are composite organisms created when an alga and a fungus develop an intimate association (see A Closer Look 23.1: Lichens). They are often found growing on bare soil, the surface of rocks, and the trunks and limbs of trees. Often, they are mistaken for mosses.

The afflicted elk began to stagger, eventually collapsing. Unable to rise, the diseased elk died shortly after the appearance of symptoms despite all efforts to save them. The elk apparently foraged on large quantities of this common ground lichen, which, when analyzed, had higher than normal concentrations of usnic acid, a secondary product. Captive elk that were fed the lichen displayed the same toxicity symptoms, but those fed the lichen with the usnic acid removed did not become ill. Preliminary data indicate that usnic acid in some way destroyed the elks' muscle tissue.

Secondary products are produced by the fungal part of the lichen and are found deposited on the surface of hyphae, thread-like structures that make up the body of the fungus. Six hundred compounds have been identified so far, and most of these are unique to lichens. Usnic acid is produced by a number of yellow-green lichens and appears to be biologically active. It is both antiviral and antibacterial. It is also phytotoxic, a property that may be useful in the development of a natural herbicide to control the growth of weeds in agricultural fields. In laboratory studies, there was a bleaching of the cotyledons of seedlings treated with usnic acid. Usnic acid has been found to inhibit the synthesis of carotenoids, accessory pigments necessary for photosynthesis. Without carotenoids, the energy transfer in Photosystem II ceases and other changes occur, resulting in the degradation of chlorophyll (see Chapter 4). The inhibition of photosynthesis eventually causes the death of the seedling. Usnic acid has also been an ingredient in some weight-loss supplements because it unlinks electron transport and ATP synthesis during cellular respiration. Oxygen intake, body temperature, and metabolic rate all rise, and fat is "burned off," but no ATP is made. The FDA is considering banning the usage of usnic acid as a diet aid because of several reported cases of liver failure. The natural role of usnic acid in lichens may be to deter herbivores as well as inhibit the growth of plant competitors.

Thinking Critically

Many cultures incorporate poisonous plants into their diets by devising methods to detoxify the poisonous components. Several cultures have evolved novel methods to process hydrogen cyanide from bitter varieties of cassava (see Chapter 14). Milkweed pods are a favorite food of the Omaha Indians; the cardioactive glycosides are rendered harmless during cooking.

Speculate as to how poisonous plants were first identified by indigenous peoples, why cultures chose to eat poisonous plants, and how methods of detoxification were devised for these plants to become a mainstay of the daily diet.

Plants may also release compounds that are inhibitory to the growth of neighboring plants, a phenomenon discussed in A Closer Look 21.1: Allelopathy—Chemical Warfare in Plants.

Poisonous Plants in the Backyard

Most homeowners are unaware that many of the commonplace landscape plants are poisonous and sometimes fatal, especially to young children. One of the deadliest is the ever-green shrub oleander, widely planted in the southern states and California for its showy red, pink, white, or yellow flowers. All parts of *Nerium oleander* (Apocynaceae) contain dangerous levels of more than 50 toxic compounds, including two cardioactive glycosides, oleandroside and nerioside, which are similar in action to digitalis. The toxins are found even in the smoke from burning leaves and branches, and breathing it can be injurious. Unwary picnickers have died just from the toxin absorbed by their hot dogs roasted on green oleander sticks.

Other common poisonous shrubs are the yews, *Taxus* spp., most often grown as hedges, shrubs, or small trees. These conifers present an attractive picture in the landscape, having dark green, needlelike leaves that contrast with the red cup-shaped arils surrounding each black seed (fig. 21.4). The poisonous taxine alkaloids are found in all parts of the plant except the aril. Taxine acts on the central nervous system, and after the initial symptoms of dizziness and dry mouth, the heart begins beating erratically and breathing becomes labored. Death is due to cardiac or respiratory failure. Young children are attracted to the bright red arils, and it has been reported that eating only one or two seeds may be fatal to a small child.

Thinking Critically

The Pacific yew (*Taxus brevifolia*) is one source of taxol, a compound that is effective in treating breast and ovarian cancers (see Chapter 19).

Why is the investigation of poisonous plants often a promising avenue for discovering drugs that are effective in the treatment of cancer?

Figure 21.4 Poisonous taxine alkaloids are found in all parts of the yew shrub except the red, cup-shaped arils surrounding each black seed.

A CLOSER LOOK 21.1

Allelopathy—Chemical Warfare in Plants

Certain plant species have been shown to produce and release chemical compounds into the environment that have deleterious effects on the growth and development of competing plants. This biochemical interaction between plants is known as **allelopathy**. The chemicals, typically volatile terpenes or phenolic compounds, are classified as secondary compounds and may be found in roots, stems, leaves, fruits, or seeds. The route of release is varied: the allelopathic compounds may be

leached from leaves and stems by rain;
released through microbial decomposition of litter;
volatilized into the air from leaves; or
exuded into the soil from roots.

Long before allelopathy was identified and confirmed, its consequences were noted in the natural environment. In *Natural History*, Pliny (A.D. 23–79) observed that no plants grew beneath walnut trees, although plants thrived under the equally dense shade of alder trees. Much anecdotal information has also been gathered over the years about crops whose growth was inhibited by certain species. For example, A. P. de Candolle, a plant taxonomist of the nineteenth century, observed that thistles (*Cirsium*) growing in a field inhibited the growth of oats and that the presence of ryegrass (*Lolium*) had a similarly detrimental effect on wheat.

Research in the twentieth century provided the explanation for Pliny's observations. A chemical compound present in the leaves and green stems of black walnut trees (*Juglans nigra*) is leached by rainfall into the soil, where it is then hydrolyzed and oxidized to the allelopathic compound juglone. Juglone has been shown to be highly toxic to many plants and to be a significant inhibitor of seed germination. These allelopathic effects result in a zone of inhibition around each walnut tree that is inhospitable to many plant species.

Allelopathic interactions play an important role in desert plant communities. Some desert shrubs, such as brittlebush (*Encelia farinosa*), release allelopathic chemicals that prohibit the growth of annuals to such an extent that a bare zone is noticeable beneath the leaf canopy and in the immediate surrounding area. Another desert shrub, guayule (*Parthenium argentatum*), releases a root exudate that inhibits guayule seedlings as well as those of many other species. The subsequent spacing of desert vegetation resulting from these types of allelopathic interactions reduces competition for scarce water resources through the elimination of potential competitors.

Nowhere have the effects of allelopathy on vegetational patterns been more thoroughly investigated than in the California chaparral. A bare zone extending 1 to 2 meters (3 to 6 feet) clearly encircles and delineates the dominant shrubs, sagebrush (*Salvia leucophylla*) and artemesia

Rhododendrons and azaleas are broad-leaved, flowering shrubs of the heath family (Ericaceae) that contain poisonous compounds so toxic that leaves, tea made from the leaves, flowers, pollen and nectar, and even the honey made from it, are deadly. Grayanotoxins (andromedotoxins) are responsible for the toxicity symptoms by first stimulating and then blocking the nervous regulation of the heart. Writings from ancient Greece record the poisoning of soldiers from wild rhododendron honey. Humans have not been the only casualties; bees have also been poisoned from rhododendron nectar. Wild relatives of rhododendron are the American laurels (*Kalmia* spp.), which also contain grayanotoxins and have a history as poisonous plants among certain Native American groups. The Delaware Indians made a tea from the mountain laurel for use as a suicide potion. Recently, the story of poisoned honey involving a native shrub and a food chain, including two introduced insects, was finally deciphered in New Zealand. The first report of honey poisoning in New Zealand was recorded in 1857 not long after the introduction of the European honey bee (*Apis mellifera*) to the islands in 1839. Victims suffered from deliriums and seizures, symptoms similar to those who had been poisoned by ingesting the berries of tutu (*Coriaria* spp.), a group of shrubs native to New Zealand. Tutin was later identified as the culpable potent neurotoxin in the shrubs but the link to the honeybee took more investigation.

Another introduced insect, the passionvine hopper (*Scolypopa australis*) native to Australia, turned out to be the critical link in producing the poisoned honey. The passionvine hopper feeds on the sweet sap of the tutu shrub. Phloem sap is under pressure and when the mouthpart of the passionvine hopper penetrates the phloem sieve tube some of this high pressure liquid is forced out as droppings from the digestive tract. The sweet droppings, known as honeydew, are a desirable food source for other insects such as the honeybee. Unfortunately, the passionvine hopper's honeydew is laced with tutin and the honeybees carry the neurotoxin back to the honeycomb. Unlike other cases of poisoned honey, the honeybees do not feed directly on the nectar or pollen of tutu flowers.

A strange twist to this story is that only some who eat tutin-laced honey come down with poisoning symptoms almost immediately after consumption but others do not experience

(*Artemesia californica*). For 3 to 8 meters (10 to 25 feet) beyond the bare zone, the growth of herbs is inhibited, and they are sparsely distributed and stunted (box fig. 21.1). Scientists have demonstrated that the leaves release volatile terpenes, predominantly camphor and cineole, producing a vapor cloud around the shrubs. The terpenes are then adsorbed from the atmosphere by soil particles during the dry, hot summer. With the onset of winter rains, the terpenes are absorbed into the cuticle of seeds or roots, where they produce toxic effects inhibiting germination and growth. The fires that periodically sweep through the California chaparral destroy both the shrubs and their allelopathic terpenes present in the soil. Consequently, the soil is no longer contaminated and there is uninhibited growth of annual herbs and grasses for several years until the shrub stands reestablish themselves and the allelopathic pattern again develops.

Increasingly, researchers are looking at allelochemicals to eliminate weeds without the use of commercial herbicides. *Intrigue*, a variety of fescue (*Festuca rubra*) turfgrass, produces a dense lawn of fine-textured, bright green leaves. Additionally *Intrigue* is slow-growing, requiring little mowing, and can be grown in full sun but is tolerant of shade. Most importantly, *Intrigue* has kept field plantings 95% weed-free. The roots of this fescue release quantities of m-tyrosine, a variant of the amino acid tyrosine, which although readily absorbed by the roots of common weeds, completely stunts their growth.

Allelopathic chemicals from the red bottlebrush have been the model for a new herbicide against broadleaf weeds in corn fields. In 1977, a California scientist noticed that no plants grew beneath the red bottlebrush shrub (*Callistemon citrinus*) in his garden. Further investigation revealed that the roots of the bottlebrush secrete an allelopathic chemical identified as leptospermone. The synthetic version of leptospermone is mesotrione, sold under the trade name Callisto™. Leptospermone and mesotrione act by interfering in the synthesis of plastoquinone, one of the electron carriers in the light reaction of photosynthesis (see Chapter 4). Corn plants are not affected by these herbicides because they naturally possess enzymes that break down the herbicide into harmless by-products. Bottlebrush is a genus of 34 species in the Myrtaceae, the myrtle family. The majority of *Callistemon* species are native to Australia but grown in gardens worldwide because of their striking flowers. Petals are inconspicuous, but each flower in an elongate cluster produces masses of brightly colored stamens, giving the entire inflorescence the appearance of a bottlebrush.

Box Figure 21.1 The allelopathic effects in the chaparral are shown as clear zones encircling the shrubs.

symptoms until nearly a day later. Researchers discovered two forms of tutin: tutin glycoside with one sugar unit attached and tutin diglycoside with two sugar units attached. The sugars shield the consumer from the toxic effects of tutin. Once the sugars are broken off, the neurotoxin is released and symptoms appear soon. It is unclear if the sugars are hydrolyzed by the action of the acidic environment of the stomach, by microbe inhabitants of the digestive tract, or by the action of human enzymes.

The Fabaceae, or bean family, is known as a major source of alkaloids. Therefore, it is not surprising that a number of poisonous plants are legumes. The rosary pea (*Abrus precatorius*), lupines (*Lupinus* spp.), black locust (*Robinia pseudoacacia*), *Wisteria* spp., vetch (*Vicia* spp.), and golden chain (*Laburnum anagyroides*) are just a few legumes implicated in human poisoning. *Sophora secundiflora,* a shrub or small tree with pinnately compound leaves, stands out from the other poisonous legumes because of its long history of use as a hallucinogen by Indians of the North American Southwest. The bright red seeds (mescal beans) were the focal point of the Red Bean Dance practiced by the Arapaho and Iowa tribes as well as by Indians in northern Mexico. Ingesting the seeds produces visions that were believed to reveal the future. Eating mescal beans is a dangerous practice because the beans contain the highly toxic alkaloid cytisine. Cytisine is similar to nicotine, and high doses cause death due to paralysis of the respiratory muscles.

Surprisingly, many of the herbaceous ornamentals planted around homes, parks, and office buildings and cherished for their beauty contain deadly poisons. Spring-flowering members of the Liliaceae and Amaryllidaceae, such as tulip, hyacinth, star-of-Bethlehem, daffodil, and narcissus, fall into this category. Few would suspect that the delicate, bell-shaped flowers of lily-of-the-valley (*Convallaria majalis*), as well as the leaves, bulbs, and red berries, conceal toxic concentrations of a host of glycosides and saponins. The saponins are irritants to the digestive tract that cause vomiting, diarrhea, and a burning sensation in the mouth and throat. But it is the cardiac glycosides that cause fatalities by slowing the heartbeat to the point of cessation. One of them, convallotoxin, is reportedly one of the most toxic of all the known cardioactive glycosides. In the past, lily-of-the-valley preparations have been used medicinally in the treatment of cardiac insufficiency, but this application has been abandoned in favor of the more reliable digitalis from foxglove.

The spurge family (Euphorbiaceae) is another family that is potentially harmful because of its characteristic milky latex or sap, which it exudes when the plant is cut or bruised. Poinsettia (*Euphorbia pulcherrima*), crown-of-thorns (*E. milii*), pencil tree cactus (*E. tirucalli*), and snow-on-the-mountain (*E. marginata*) are all known to cause irritation to the skin and mucous membranes. Another member of this family, the castor bean (*Ricinis communis*; fig. 21.5a), is deadly because of its large and colorful, yet highly toxic, seeds. The mottled red, white, or brown seeds are extremely poisonous. Ingestion of just one seed may kill a child; eating three seeds would be fatal for most adults. The toxic protein ricin, most concentrated in the seeds, inhibits protein synthesis in the intestinal wall and causes the clumping of red blood cells. It is said to be the most deadly natural poison known. Symptoms do not develop until several hours or even days after the seeds are eaten. (Note: The seed coat must be broken to release the toxins; swallowing seeds whole provides some protection because the extremely tough seed coat prevents release of the poisons.) Nausea, vomiting, diarrhea, and burning of the mouth and throat are the first symptoms to appear. These are followed by electrolyte imbalances, retinal hemorrhaging, internal hemorrhaging in the digestive system and lungs, and extensive damage to the liver and kidneys. Death results from kidney failure. Interestingly, castor bean oil has been a staple of folk medicine from ancient times and is still used today as a laxative. It is not poisonous because ricin is not present in the oil.

An international case of ricin poisoning took place in London in 1978. Georgi Markov, an exile in Great Britain from Bulgaria, was a respected writer and dissident who frequently criticized the Bulgarian communist regime in broadcasts from the BBC (British Broadcasting Corporation) and Radio Free Europe. On September 7, Markov was waiting for a bus in London when he felt a sharp stab to his right thigh. He turned and saw a man with an umbrella, who apologized to Markov as he left in a taxi. Later that night, Markov experienced high fever, abdominal pains, vomiting, and diarrhea. He was rushed to a hospital, but physicians were unable to determine the cause of his symptoms or relieve his suffering. The next day, Markov went into shock, and he died 3 days later. An autopsy found, lodged in Markov's thigh, a small pellet with two holes. It was later determined that the pellet contained less than 0.5 ml of ricin solution. Apparently, the pellet had been injected into this thigh by a specially modified umbrella (fig. 21.5b) by an agent of the Bulgarian secret police. Ricin is a subject of renewed interest because of its potential as an agent in bioterrorism. It is classified by the CDC (Centers for Disease Control and Prevention) as a category B biological agent. Purified ricin is easy to disseminate in food and drink or via an aerosol and readily available from the estimated 50,000 tons of castor bean meal that is produced around the world in the production of castor oil.

Ricin is composed of two polypeptide chains, A and B, joined by a disulfide (S–S) bond. The B chain is the part that binds to a receptor found on the plasma membrane which allows the ricin to enter cells. The ricin receptor is found in

(a)

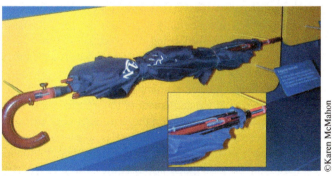

(b)

Figure 21.5 Many common yard plants and houseplants are poisonous: (a) castor bean (*Ricinus communis*) plant, (b) replica of the modified umbrella used to assassinate Georgi Markov with a deadly ricin pellet (International Spy Museum, Washington DC).

the plasma membrane of most cells of the human body, which explains why ricin is so toxic in a relatively short period of time. The A chain is the actual toxin. Ricin is taken into cells in a membrane-bound vesicle. Once inside the cell, the vesicle containing ricin is transported to the endoplasmic reticulum, and it is here that the A chain is cleaved from the B chain. The A chain then is released into the cytoplasm and goes directly to the ribosome. The effect of ricin is to lyse ribosomes, rendering them unable to synthesize proteins. Most cells die in 48 hours after ricin exposure. The Golgi apparatus reacts to the presence of ricin by sequestering the toxin in vesicles and exporting it from the cell, but the effort is futile. Purified ricin can kill an adult in very small quantities, approximately 25 milligrams per kilogram of body weight. Research is proceeding on developing ricin vaccines that prevent the A chain from entering cells.

Poisonous Plants in the Home

Poisonous plants are everywhere in the environment, including the home. Most calls to poison control centers concerning children ingesting houseplants involve the aroids (Araceae), some of the most popular houseplants and almost all poisonous. The two aroids most frequently cited are philodendron (*Philodendron* spp. and *Monstera* spp.) and dumbcane (*Dieffenbachia* spp.). The philodendrons are either vines or erect plants admired for their heart-shaped or dissected leaves. The dumbcanes are distinguished by speckled leaves on a moderately stout, erect stem. All parts of these plants contain crystal needles (raphides) of calcium oxalate. If swallowed, these crystals cause painful burning and swelling of the lips, tongue, mouth, and throat, which may persist for several days. Talking, swallowing, and even breathing may become labored. In fact, the difficulty in speaking after ingesting *Dieffenbachia* accounts for its common name, *dumb*cane. There are also indications that these plants contain toxic proteins that may intensify the pain and edema. Calcium oxalate crystals are also the culprits in the poisonous wild aroids jack-in-the-pulpit (*Arisaema triphyllum*) and skunk cabbage (*Symplocarpus foetidus*).

Plants Poisonous to Livestock

Poisonous plants are not only a concern in human health but can also affect the well-being and economics of livestock. Numerous plants have been identified as poisonous to sheep and cattle and, subsequently, to the humans who eat the flesh or drink the milk of poisoned animals. White snakeroot played a devastating role in the settlement of the frontier lands west of the Appalachian Mountains in the early nineteenth century. White snakeroot (*Ageratina altissima*) is a white-flowered, perennial herb of the Asteraceae (sunflower family) found abundantly in the rich, moist woods in the eastern half of the United States (fig. 21.6). All parts of the plant contain the toxic alcohol tremetol, which is highly soluble in oils or fats.

Cattle that ingested white snakeroot developed a condition known as trembles, characterized by a foul breath smelling like acetone, sluggishness, and muscular weakness.

Figure 21.6 White snakeroot (*Ageratina altissima*) is the cause of milk sickness.

In some areas, entire herds were wiped out by trembles. It was eventually observed that humans who ate the meat or drank the milk of afflicted cattle often developed a related condition, known as milk sickness, with symptoms similar to trembles. Milk sickness was epidemic in certain areas of the frontier. In one county of Ohio, 25% of the population died from this malady. In fact, Nancy Hanks Lincoln, the mother of Abraham Lincoln, was a casualty of this disease. Tremetol was isolated from snakeroot in the early twentieth century, and the connection between trembles and milk sickness was fully explained.

Tremetol affects the metabolism of the liver by blocking the breakdown of lactic acid, a temporary end product of cellular respiration, and thereby preventing the Krebs Cycle. The subsequent accumulation of lactic acid lowers the pH of the body, causing acidosis, which accounts for the acetone-like breath. This blocking of the exergonic metabolic pathways also explains the lethargy. Tremetol's solubility in products high in lipids, such as milk, also explains two additional observations: that lactating cattle were less affected by trembles than their calves and that young children were often the primary victims of milk sickness. This disease is now, fortunately, only of historical importance because the practice of milk pooling from many cows and dairies dilutes any tremetol present to nontoxic doses and because of the successful eradication of white snakeroot from pastures.

The leguminous locoweeds, or milk vetches (*Astragalus* spp. and *Oxytropis* spp.), have been a bane to ranchers in the western half of the United States and are considered to be some of the most toxic plants for horses, sheep, goats, and cattle.

Loco is the Spanish word for crazy and refers to the staggering and trembling behavior of poisoned animals, which walk into things and react to unseen objects. They are finally overcome by paralysis. The poisonous compounds are an unusual group of alkaloids that affect certain cells of the central nervous system, explaining the behavioral changes observed.

Indian hellebore (*Veratrum* spp.) has been widely used externally as a local anesthetic, and it was used internally as a purgative in certain religious ceremonies by many Native American tribes of the Pacific Northwest. In the past, hellebore extracts have been employed in conventional medicine to treat hypertension. Despite these valuable applications, Indian hellebore is known to be a highly toxic plant for both humans and livestock. It is a known teratogen (agent that causes deformities in the fetus) in sheep. Ewes that feed on the western *Veratrum californicum* early in their pregnancy commonly give birth to so-called monkey-faced lambs, which have the facial deformity of a single eye. The active compounds in *Veratrum* are a number of alkaloids found throughout the plant but most highly concentrated in the rhizome.

PLANTS THAT CAUSE MECHANICAL INJURY

Plants can also cause mechanical injury, usually by puncturing the skin via spines, thorns, prickles, burrs, or hairs. The foremost examples of plants in this category are the cacti. In the family Cactaceae, spines are modified leaves (fig. 21.7a) and seem to be a mechanism to ward off animal predation. Especially dangerous are the hardly noticeable glochids—minute, barbed hairs—which, once embedded in the skin or mouth, are difficult to remove and may lead to infections.

A recent study analyzed barbed and non-barbed spines within the cactus family (Cactaceae). It discovered that barbed spines were more effective in piercing a target and also required more effort to be removed from the target than non-barbed spines. Barbed spines were also more prone to detach from the plant and implant in the target. Non-barbed spines apparently have a purely defensive purpose in warding off herbivores from consuming edible plant parts but barbed spines

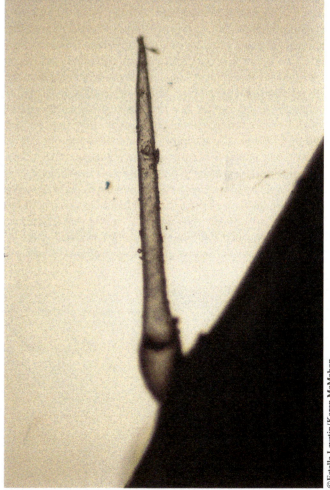

(a)   (b)

Figure 21.7 Plants can inflict injury through spines, thorns, or stinging hairs. (a) Spines of a cactus. (b) The hairs of the stinging nettle inject irritants that cause a painful burning sensation.

have not only a defensive function but may also be used to anchor vegetative propagules on animal targets for the purpose of asexual reproduction and dispersal.

Other plants possess stinging hairs that not only puncture the skin's surface but also inject chemical irritants, causing a painful burning sensation. The burning sensation is followed by itchiness, which may last for several hours. Leaves and stems in members of the stinging nettle family (Urticaceae) are covered with hollow hairs similar in operation to a hypodermic syringe (fig. 21.7b). When the tip of a hair penetrates the skin, chemicals, such as a histamine-like compound, are squeezed from the basal reservoir and injected into the skin, immediately producing a red swelling similar to a mosquito bite or hive (see Allergy and the Immune System later in this chapter).

Other plants produce compounds that can cause damage simply by contact when the plant is bruised or by being transported to the skin after the plant is digested. The end result is to make the skin extremely photosensitive; on subsequent exposure to ultraviolet light, symptoms vary from a mild sunburn to a blistered secondary burn. The weedy St. John's wort (*Hypericum perforatum*), which contains the specific phototoxin hypericin, was a serious problem for livestock in California during the first half of the twentieth century (fig. 21.8a). Sheep that foraged on St. John's wort became severely affected, with the skin in the face swelling to such an extent that the condition was commonly referred to as bighead. If the lips become affected, the animal is unable to feed and starves to death. Fortunately, a biological control program using beetles that feed exclusively on St. John's wort has successfully curtailed its spread. Preparations from St. John's wort show promising results in treating depression (see Chapter 19), although there is concern that some patients might experience photosensitivity reactions.

The milky latex of many members in the Euphorbiaceae has long been known to irritate the skin. In fact, unscrupulous beggars during the Middle Ages burned their skin with spurge juice to fake leprosy and evoke more sympathy and charity from the public. The manchineel tree (*Hippomane mancinella*), native to Florida and the Keys, produces a latex extremely caustic to skin that can even cause blindness if it gets in the eyes. Juice from snow-on-the-mountain (*Euphorbia marginata*) can burn skin on contact; in fact, it has been used to brand cattle (fig. 21.8b).

INSECTICIDES FROM PLANTS

It is well known that plants produce a great diversity of secondary compounds; in fact, it has been estimated that the total approaches 400,000. Many of these compounds reduce the palatability of the plant to insects; for example, the accumulation of alkaloids or saponins produces a repellent, bitter taste. The high concentration of tannins in oak leaves dissuades insect predation by both their bitter taste and their tendency to bind protein, reducing the digestibility of the leaves. In other cases, the secondary compounds are outright toxic, such as cyanogens, or affect the hormonal balance and life cycle of the insect, such as neem oil. Whatever the specific mechanism, secondary compounds afford the plant a means of chemical defense against predation. Research in this area has focused on the possibility of using these natural chemical defenses on a larger scale to protect agricultural crops from losses due to insect predation and to destroy those insects that are vectors of both human and live-stock diseases. Botanical insecticides have the advantage of being nontoxic to humans and most livestock, and they are biodegradable.

Two of the oldest and best-known groups of botanical insecticides still in use today are pyrethrum and rotenone. Pyrethrum is a powder made from grinding the dried flower heads of *Chrysanthemum cinerariaefolium* commercially grown in Africa and South America. It was introduced into the United States in the 1800s. Later, the active ingredients were identified and named pyrethrin I, pyrethrin II, cinerin I, and cinerin II; these are now used mainly in purified form as well as forming the basis for the synthetic pyrethroids. Pyrethrins act as a nerve poison, quickly paralyzing many common household insects, and they are common ingredients of flea

(a)

(b)

Figure 21.8 Both (a) St. John's wort and (b) snow-on-the-mountain are plants that can injure through irritant juices.

collars and quick-acting aerosol sprays. They are also available as a garden dust, which can be used safely on vegetables and fruits, because they are not known to pose any threat to human health. (Some individuals, however, are strongly allergic to these compounds.) Pyrethrum, unlike some of the pyrethroids, breaks down readily in sunlight.

Pyrethroids also differ in having greater toxicity to insects. Found in runoff from farms and lawns, pyrethroids do not dissolve readily in water but they do accumulate in stream sediments. This buildup has been shown to reach concentrations that are toxic for beneficial aquatic insects and crustaceans. The EPA has scheduled a reevaluation of its approval for several pyrethroids that will be completed within the next few years.

Rotenone and related compounds called rotenoids are derived from the roots of several tropical legumes; for example, tuba root (*Derris*), grown in Malaysia, and cubé root (*Lonchocarpus*), from South America. Originally, the aqueous extracts of these roots were used by indigenous peoples to catch fish. Rotenone causes nervous system paralysis in fish, making them surface in a helpless condition (fig. 21.9b). The fact that rotenone is highly toxic to fish and might contaminate water resources has somewhat limited its use as a crop insecticide. Nonetheless, it is a widely used garden insecticide favored for its effectiveness against caterpillars and a host of common pests and for its nonpersistence in the environment.

Much research has focused on the potential of the neem tree, or margosa (*Azadirachta indica*; Meliaceae, or mahogany family), as a source of natural insecticides. This tree is native to India and Myanmar (Burma) and is now widespread throughout the tropics. The medicinal applications of neem oil date back to Sanskrit records, and neem leaves and fruits have been used to repel insects in the home and in crop storage bins. As early as 1932, neem cake (the residue after the oil has been extracted from the seeds) was utilized as an insect repellent in Indian rice paddies. Chemical investigation of the neem tree has identified a number of compounds acting as antifeedants or growth regulators against a variety of pests. One of the most potent feeding deterrents isolated is azadirachtin from neem seeds.

Investigations have shown neem oil to be effective against more than 200 species of insect pests and a number of viruses, bacteria, and fungi. It has been approved to fight tooth decay, and a neem-based toothpaste is now on the market. There are even indications that neem oil can be used as a spermicide in birth control.

In addition to their pesticide applications and medicinal uses, neem trees are being used to reforest semiarid tropical areas. The neem tree is valued for its lumber, which is strong and resistant to termites, and its ability to reduce erosion. Also, recent research has shown that the use of neem cake as a soil additive will inhibit nitrification. In some tropical areas, ammonium fertilizer is rapidly converted to nitrate; the nitrate is lost from the field and enters the watershed. Neem cake inhibits this nitrification and prevents nitrogen loss, thereby increasing the efficiency of fertilizer application. Research is underway in many laboratories to identify further uses for this tree.

ALLERGY PLANTS

Certain plants or plant parts are capable of causing allergic conditions in sensitive individuals. Some plants trigger respiratory allergies due to the inhalation of airborne pollen; other plants cause contact dermatitis (skin rash) due to direct

(a)

(b)

Figure 21.9 Several insecticides have been developed from plants. (a) These products contain either the natural pyrethrins obtained from *Chrysanthemum cinerariaefolium* or the synthetic pyrethroids. (b) Rotenone obtained from a number of plants is applied to a stream to remove invasive fish. The rotenone has been dyed green to track its spread.

association with plant oils or resins; and still other plants bring about food allergies that cause various types of symptoms.

Allergic reactions are considered hypersensitivity reactions. They are just one type of response produced by the immune system, the body system that differentiates "self" (the body's own cells or molecules) from "nonself" (foreign cells or molecules).

Allergy and the Immune System

The immune system is a complex system of specialized cells and organs that defend the body against attack by foreign invaders, such as bacteria, viruses, fungi, and parasites. The basis of the immune system is the ability to distinguish between "self" and "nonself." Almost every cell within the body bears characteristic surface molecules that identify it as "self." When foreign substances are detected, the immune system responds with a proliferation of cells that either attack the foreign substance or produce proteins, called **antibodies,** that bind to the substance. Any substance capable of triggering an immune response is called an **antigen.** An antigen can be an invading pathogen or even a small portion or product of that organism.

In some people, however, harmless substances, such as ragweed pollen, cat dander, or fungal spores, are perceived as a threat and stimulate the immune system to produce an allergic response; in these cases, the antigens are known as **allergens.** Allergies are inherited disorders. Although sensitivity to a particular substance is not inherited, the tendency to develop allergies is genetically controlled.

White blood cells are important components of the immune system; one class of these is **lymphocytes,** which are produced by stem cells in the bone marrow that give rise to all blood cells. One type of lymphocyte, **B-lymphocyte,** functions by producing antibodies, also known as **immunoglobulins (Ig).** Antibodies are a major constituent of the blood, making up about 20% of the protein in the blood plasma. A human can produce 10 million types of antibody molecules; however, an individual B-lymphocyte synthesizes only a single kind of antibody. Each type of antibody is capable of binding to only a single antigen.

All antibody molecules share the same structural features; each is composed of four polypeptide chains. Two chains are called light chains, which are identical to each other. The other two chains are called heavy chains; they also are identical to each other. In the antibody molecule, the two light chains and the two heavy chains are joined by bonds to make one Y-shaped antibody molecule (fig. 21.10). Both the light and heavy chains have a variable region that joins to form two antigen binding sites. The variable regions differentiate one type of antibody from another and provide the antibody with the capability of binding to a certain antigen.

There are five classes of antibodies—IgA, IgD, IgE, IgG, and IgM—each with a different function in the immune system. The **IgE** class of antibodies is involved in allergic reactions, and people who suffer from allergies have elevated levels of

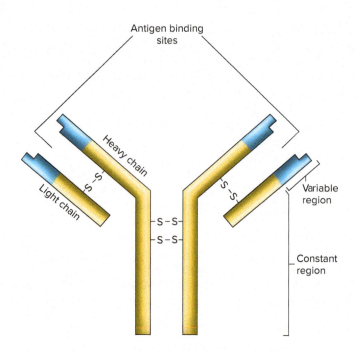

Figure 21.10 An antibody, or immunoglobulin, molecule consists of two identical light chains and two identical heavy chains joined to make a Y-shaped molecule.

these antibodies. When an allergic individual is first exposed to a particular allergen, lymphocytes produce specific IgE antibodies that are capable of binding to this substance. Unlike the other classes of antibodies, IgE antibodies do not circulate in the blood but are attached to the surface of basophils and mast cells (fig. 21.11). Basophils circulate in the bloodstream, while mast cells line the surface of the respiratory tract, intestines, and skin. When a particular allergen is encountered again, it will bind to the IgE antibodies on the surface of these cells, triggering a series of reactions inside the cells. Chemicals, such as histamine, which are initially contained in vesicles (granules), are released from these cells and are responsible for the symptoms associated with allergies.

Respiratory Allergies

Approximately 20% to 25% of the human population suffers from one or more significant allergies, with hay fever and allergic asthma the most common. These allergic diseases are major causes of illness, and although no complete data exist, it is believed that more than 55 million people are affected in the United States today. Treatment of these conditions is a multibillion-dollar-a-year business.

The name *hay fever* was coined in the early nineteenth century, when the condition was first recognized to coincide with the time of the year when hay was cut and baled. Because the condition has little to do with hay and seldom produces a fever, the term *allergic rhinitis* is usually considered more appropriate. The typical symptoms of this condition are sneezing, runny nose, nasal congestion, and postnasal drainage, along with red, itchy, puffy, and teary eyes. Although these

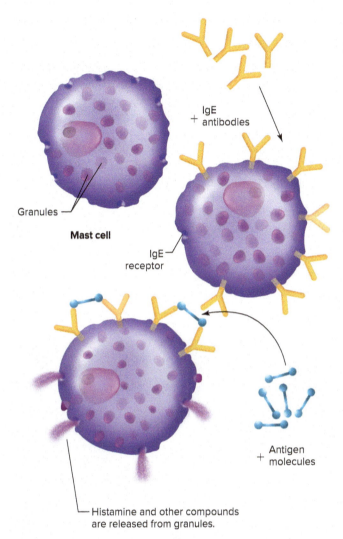

Figure 21.11 Mechanism of an allergic reaction. After the first encounter with an allergen, IgE antibodies are produced and attach to the surface of mast cells lining the respiratory tract, skin, and intestines. When the allergen is encountered again, it will bind to the IgE molecules and cause the release of histamine and other chemicals that are responsible for allergy symptoms.

75% increase in asthma prevalence. The exact reason for the increase is not known; in part, it is due to better diagnosis, but scientists believe that is not the full answer. Greater exposure to indoor allergens is thought to be one factor in this increase, especially in the inner city, where poverty may prevent proper medical treatment. In the United States alone, over 24 million people suffer from asthma, with a yearly economic cost of $80 billion including direct medical costs, missed work, and missed school.

Respiratory allergies can be triggered by exposure to pollen, fungal spores, dust, animal dander, insect, and dust mite allergens; however, the discussion in the next sections will focus on plants that produce allergenic pollen. Fungal spore allergies will be considered in Chapter 25.

Hay Fever Plants

Pollen from plants that are considered important triggers of hay fever and asthma have several characteristics in common; the pollen is airborne, lightweight, abundantly produced, widely distributed, and allergenic.

With very few exceptions, plants that cause hay fever are wind-pollinated. They produce abundant quantities of lightweight pollen that is passively dispersed into the air. (Recall that anemophilous, or wind-pollinated, plants generally invest metabolic energy into producing large quantities of pollen, while entomophilous, or insect-pollinated, plants produce showy petals and nectar to attract pollinators.)

Most allergenic pollen is 10 to 50 μm in diameter, a size range that helps keep the pollen aloft for extended periods. Once airborne, some pollen can be carried hundreds of kilometers. Such long-distance transport has been documented by aerobiologists, scientists who study airborne particles of biological origin. Eventually, all pollen settles out, the distance transported dependent on pollen size, prevailing winds, and other atmospheric conditions. Large pollen grains tend to settle out very quickly; smaller ones are carried greater distances.

Although long-distance transport does occur, most pollen grains settle out relatively close to their source, and allergy symptoms in susceptible individuals are closely correlated with proximity to the source. The most important hay fever plants are those that grow or are planted in large numbers close to human populations.

To be allergenic, the pollen must contain actual allergens. These compounds are pollen proteins, or glycoproteins (proteins with attached sugars). It is generally accepted that the role of some of these proteins in the pollen grain is to act as recognition factors for stimulating growth of the pollen tube when pollen lands on a receptive stigma. When a pollen grain is inhaled, it is usually deposited in the nasal passages or upper respiratory tract. The pollen wall proteins diffuse out of the pollen grain and into the respiratory tract, where they can trigger IgE-mediated responses in allergic individuals. The same molecules cause no reactions in people without allergies.

symptoms do not seem serious, allergic rhinitis contributes to innumerable hours of suffering and loss of productivity. In addition, chronic or untreated hay fever can lead to asthma and other, more serious, long-term conditions. Approximately 40 million Americans suffer from hay fever, and in the United States, hay fever is responsible for more than 17 million visits to physicians and annual costs of $18 billion.

Asthma is a disease characterized by inflammation of the airways, bronchial constriction, and excessive mucus secretion, resulting in wheezing, coughing, and choking. This chronic breathing disorder is responsible for up to 5,000 deaths per year in the United States. Asthma is an immunological-based disease; an allergen, a viral infection, stress, or exercise can trigger an asthmatic episode. One disturbing trend is that the incidence and severity of asthma have been increasing throughout the world. Since 1980, there has been a

Ragweed

Undoubtedly, the most notorious hay fever plant is ragweed, *Ambrosia* spp., a widespread genus in the Asteraceae, the aster or composite family. Interestingly, the generic name means "food of the gods" in Greek, surely an unlikely designation for the most important cause of allergic rhinitis in North America. Ragweeds are annual or perennial herbaceous plants with lobed or divided leaves. They range from minute plants 15 to 20 cm (6 to 8 in.) in height to giant ragweed, which can be 4 meters (12 feet) tall. Although 25 species of ragweed occur in North America, most allergy problems are caused by *Ambrosia artemisiifolia* (short ragweed) and *A. trifida* (giant ragweed), two species that account for more hay fever than all other plants together (fig. 21.12). Both are pioneer plants that are well adapted to invading disturbed soils. In natural environments, these species are restricted; however, in areas where human intervention has cleared existing vegetation for farming, housing developments, shopping centers, or roadways, ragweed quickly becomes established.

Ragweed is monoecious, with staminate and carpellate flowers produced on the same plant. The small staminate flowers (each with five stamens) are grouped into inflorescences with 10 to 20 florets borne together in a cupule or involucre of bracts (fig. 21.13a). Approximately 50 to 100 involucres occur on each of the many terminal flowering spikes. The thousands of staminate flowers on each plant release a total of approximately 1 billion pollen grains (fig. 21.13b).

In northern areas of the United States and in Canada, ragweed pollen release begins in early August, peaks by early September, and is completed early in October. An early, killing frost in late September often shortens the ragweed season in northern areas. In southern areas of the United States, the ragweed season begins later in August and continues until later in the fall. In Florida, as well as other areas of the Gulf Coast and the Southwest, the ragweed season may be much earlier or later, with some species of ragweed actually flowering year-round. The prevalence of this genus in the United States

Figure 21.12 (a) Short ragweed, *Ambrosia artemisiifolia*, and (b) giant ragweed, *Ambrosia trifida*, cause more hay fever than all other plants together.

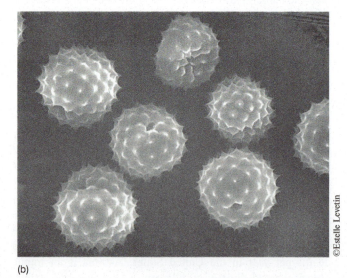

Figure 21.13 (a) Staminate inflorescence of giant ragweed bearing clusters of florets. (b) Scanning electron micrograph of ragweed pollen, ×1,130.

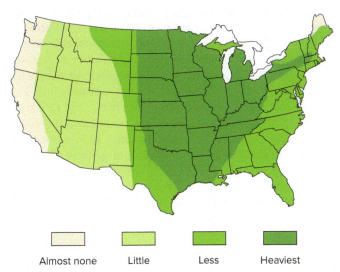

Figure 21.14 Members of the genus *Ambrosia* are the most important allergy plants in the United States, with the heaviest distribution in the central states.

Source: Data from "Advice from Your Allergist," in *Annals of Allergy*, 59(1), July 1987.

caused by trees, grasses, and weeds, correlating symptoms with the season of pollen production.

Tree pollen is normally considered the leading cause of spring hay fever. In most cases, the inconspicuous flowers of wind-pollinated trees appear early in the season, often before the leaves develop (fig. 21.15a). Depending on climate and latitude, pollen release begins as early as late January for some species in the southern United States and extends through May for others. Oak, maple, elm, birch, pecan, walnut, and mulberry trees are among the many types recognized as major producers of airborne allergens. Unfortunately, this list includes genera that are major components of the native vegetation as well as popular shade and ornamental trees. Individuals allergic to tree pollen should avoid wooded areas in early spring and use species with showy (insect-pollinated) flowers for landscaping their yards.

Although angiosperm pollen is responsible for most cases of hay fever, pollen from certain conifers, especially members

results in the release of an estimated 1 million tons of pollen each year (fig. 21.14).

Ragweed has recently become a serious invasive species problem in parts of Europe, with sensitivity to the pollen becoming a serious health issue. It is believed that ragweed seeds were first brought to Europe as early as the mid-nineteenth century, but there was little spread at that time. After World War I, seeds from several ragweed species were again accidentally introduced, mixed in with shipments of clover seed and grain. After the 1950s, the plants began spreading, and since the 1990s they have rapidly migrated into new areas. Although four species of ragweed were introduced, short ragweed has become the major problem. At the present time, the countries most seriously affected are Hungary, Croatia, and parts of France and Italy. However, ragweed's invasion of Europe continues.

Pollen dispersal normally occurs in mid-morning as the dew dries and humidity decreases. During the pollinating season, the daily atmospheric pollen concentration depends on local weather conditions, with warm, dry, windy conditions optimal for dispersal. On the other hand, heavy rainfall will wash pollen from the air and bring temporary relief to allergic individuals. Seasonal abundance of pollen often depends on growing conditions earlier in the year. Computer-generated pollen-forecasting systems (based on weather and previous pollen levels) are being developed to predict the severity of the pollen season.

Other Hay Fever Plants

In addition to the pollen from ragweed, the pollen from hundreds of seed plants are also known to trigger respiratory allergies. Allergists usually categorize pollen allergies into those

Figure 21.15 During late winter and early spring, tree pollen allergens cause suffering for millions of people. (a) Staminate inflorescence of pin oak, *Quercus palustris*, with inset showing the pollen. (b) Pollen cone of mountain cedar, *Juniperus ashei*, with inset showing the pollen. Pollen magnification × 800.

of the Cupressaceae (the cypress family) are major causes of allergic rhinitis in some regions. One of the best-studied allergens in this family is the pollen produced by *Juniperus ashei* (fig. 21.15b), which is commonly called mountain cedar or Ashe juniper. Mountain cedar is a small- to medium-sized evergreen tree native to the south-central United States, with the largest populations in central Texas, where it is considered the most serious hay fever plant in the state. Additional small populations of the species occur in southern Oklahoma and northern Arkansas. An unusual feature of mountain cedar is that pollen is released in December and January, a time of the year when other pollen types are usually absent. Other members of the Cupressaceae are also important allergens in other parts of the United States and other countries.

Grasses are the leading cause of early summer hay fever problems. Recall that the Poaceae, the grass family, is a large family with close to 12,000 species (see Chapter 12). Although ragweed is considered the most important hay fever plant in North America, grass pollen allergy is the principal cause of allergic rhinitis throughout much of Europe and Central Asia.

There are approximately 1,200 grass species in the United States; however, only 35 to 40 species are allergenic, and of those, only 11 are important allergens. Bermuda grass, timothy, Kentucky bluegrass, Johnson grass, and orchard grass are the most important contributors to grass pollen allergies. Although many native grasses pollinate in the late spring or summer, some species of lawn grass, such as Bermuda grass (*Cynodon dactylon*), will continue to flower throughout the growing season, especially when watered. Because of their proximity to people, lawn grasses present problems. Although these can be somewhat lessened by frequent mowing to prevent flowering, nearby lawns that are less well kept may produce pollen for 6 months or longer. In fact, in parts of Florida and other southern states, grass pollen may be present in the atmosphere year-round.

Weeds constitute a category of hay fever plants that include nongrass and nontree species; however, this is not a natural botanical category. As a category, "weeds" includes various types of vegetation that are not intentionally planted and are usually considered undesirable. These aggressive plants quickly invade disturbed sites associated with human activity. As such, they grow and pollinate close to populated areas. Although ragweed is the most important plant in this group, many other common weeds are also well-known allergen producers. They often pollinate in summer and fall, along with ragweed, and include marsh elder, lamb's quarter, plantain, and sage. Some of these plants cross-react with ragweed pollen; that is, they share common allergens, so symptoms can be triggered by pollen from these plants in ragweed-sensitive people. In addition to the shared molecules, distinctive allergens are also present on the pollen grains.

Climate Change and Allergy Plants

Increasing global temperatures in many parts of the world have changed the growing season of various plants, including hay fever plants. **Phenology** is the study of periodic life cycle events in plants and animals that are influenced by climate (see Chapter 26). First flowering, first leaf appearance, and autumn coloration are examples of these periodic events. Spring-flowering plants are particularly sensitive to temperature, because their flowering typically occurs as a response to accumulated temperatures above a certain threshold. This means that warmer temperatures in late winter and spring will result in earlier flowering in many species. Many phenological studies have documented this response. The International Phenological Gardens are a network of gardens in Europe containing genetically identical clones of 23 species of trees and shrubs; several of the trees are important hay fever plants, including birch, oak, willow, hazelnut, and poplar. These gardens were specifically established to study phenological phases, including first flowering, over a broad geographical range. Data from these gardens show that spring phenological events advanced 6.3 days over the period 1959 to 1996. In North America, the longest observations for first flowering dates are for lilac and honeysuckle. Data on these plants have been collected in some states since the late 1950s to study the relationship between plant development and climate. Both plants showed earlier flowering over time, with lilac flowering 7.5 days earlier over 38 years and honeysuckle 10 days earlier over 27 years. Although lilac and honeysuckle are nonallergenic, other studies of allergy plants have shown similar trends.

In addition to visual observations of flowering, airborne pollen records have been used to study the effect of climate change on allergy plants. Because knowledge of airborne pollen is important to the allergy community (see A Closer Look 5.2: Pollen Is More Than Something to Sneeze At), long-term pollen records exist in many areas. The records have shown earlier starts for the pollen season for many spring-flowering trees. In addition, some of these databases have documented increasing pollen levels for some species. A field experiment on ragweed also found that plants growing in plots that were experimentally warmed produced more pollen per plot.

The enriching effects of increased carbon dioxide (CO_2) concentration have been studied by numerous investigators in growth chamber, greenhouse, and field experiments. Increased rates of photosynthesis under elevated CO_2 result in both greater plant biomass and greater reproduction, with crop plants and wild species showing similar trends. Several researchers have shown increases in ragweed pollen production under elevated CO_2. One study even found increased levels of the actual allergenic proteins in ragweed pollen. Another study showed an increase in the length of the ragweed season since 1995 in northern latitudes due to later first frosts in the fall. All this is bad news for ragweed-sensitive individuals.

Allergy Control

The best way to control respiratory allergies is to avoid the allergen. For someone allergic to cats or dogs, this solution is possible, but for those with pollen allergies, avoidance is sometimes impossible. In the past, some cities have attempted to eradicate ragweed plants. This undertaking was very expensive and not terribly effective. Clearing a local area does not prevent the pollen from being carried from a more distant source, nor does it prevent new growth the following year from buried seeds. For many years, it was believed that moving to the desert would help allergy sufferers; however, this was only a short-term solution. The growth of Phoenix and Tucson can be partly attributed to this belief. Allergic individuals soon became sensitive to the allergy plants of the area. In addition, landscaping plants such as mulberry, which grows fast but produces allergenic pollen, were introduced. In fact, the problem became so bad in both cities that each passed an ordinance against the further planting of mulberry trees.

Avoidance of pollen may necessitate taking a long vacation to an area where the allergenic pollen does not occur (even there, long-distance pollen transport may cause problems) or remaining indoors in an air-conditioned building (where efficient filters can remove the pollen from the incoming air) for the duration of the pollen season. For most allergy sufferers, therefore, total avoidance is not achievable.

Figure 21.16 Commercial harvesting of rye grass pollen for the preparation of allergy extracts.

> ### Thinking Critically
>
> Exposure to allergenic pollen is highest near the source. Hay fever sufferers who have allergenic plants in their yards may be especially aware of the effects of pollen.
>
> *If you suffer from hay fever and are allergic to several types of spring tree pollen, what type of landscaping can you have in your yard to keep symptoms to a minimum? Prepare a list of these trees suitable for your area.*

There are many options available, however, for the treatment of respiratory allergies, including over-the-counter or prescription antihistamines (which can counteract the symptoms associated with the release of histamine and other compounds from mast cells) and decongestants (which can relieve nasal or bronchial congestion or both). Newer antihistamines no longer result in drowsiness, an unwanted side effect commonly associated with older antihistamines. Topical corticosteroids used as an aerosol or spray have also proven effective in reducing allergy symptoms. Immunotherapy or desensitization (commonly called allergy shots) is often indicated when symptoms cannot be adequately controlled by medication. Allergy patients are given a series of injections (often over several years) that contain weak extracts of the offending allergens. The extracts used in immunotherapy are prepared from actual pollen samples that have been collected from nature or from cultivated sources (fig. 21.16). In allergic individuals, these injections stimulate the production of IgG antibodies specific to the allergen. These block the binding to IgE antibodies and thus prevent the IgE-induced response. Immunotherapy is often very effective in the treatment of pollen allergies (up to 90% effective in some studies) but has less success in the treatment of other allergies. A recent development in immunotherapy replaces allergy shots with small doses of the extract applied under the tongue either as drops or tablets. This technique is known as sublingual immunotherapy (SLIT). SLIT has been in use in many countries for several years now, and the Food and Drug Administration has recently approved SLIT for use in the United States for treating grass and ragweed allergies.

Contact Dermatitis

Contact dermatitis is an allergic reaction of the skin to something touched. While hay fever and allergic asthma are examples of immediate hypersensitivity, contact dermatitis is considered delayed hypersensitivity because symptoms may take several hours to days to appear. Like other allergies, contact dermatitis is triggered by the immune system's responding to a harmless substance.

The most common plant that causes contact dermatitis is poison ivy, *Toxicodendron radicans,* a member of the Anacardiaceae, the cashew family. Poison ivy is a widespread weed native to the United States and southern Canada, where it is common in disturbed sites, on floodplains, along stream banks, on lakeshores, and in areas that border woodlands. It is extremely variable in form, growing as a woody shrub or vine that either spreads along the ground or climbs trees, fences, and poles. The leaves are alternate, each with three leaflets, but the leaf margin shows several forms (fig. 21.17a). The margin may be smooth, toothed, or lobed, adding to the variability of the species. Clusters of white to whitish yellow berries are found on all forms.

Related species include poison oak (*Toxicodendron quercifolium*), western poison oak (*T. diversilobum*), and poison sumac (*T. vernix*), which are equally potent allergens but not as widely distributed as poison ivy (figs. 21.17b and c). Poison sumac and poison oak are found in the East but occur as far west as Texas; western poison ivy occurs in the Pacific coastal

(a) Poison ivy

(b) Poison oak

(c) Poison sumac

Figure 21.17 (a) Poison ivy, (b) poison oak, and (c) poison sumac are common causes of contact dermatitis.

states. Some authorities consider all these species as the genus *Rhus* and consider poison oak to be a variant of poison ivy and not a separate species.

It is estimated that one out of every two people is allergic to poison ivy and related species, which causes a serious skin rash on sensitive people. About 10% of the population is so sensitive that they require medical care after even a brief exposure. The appearance of the rash is normally delayed for 24 to 48 hours, or even longer, after contact with the plant. The rash usually appears as a red, swollen area, followed by blisters and intense itching. The symptoms may last for a week or longer.

Urushiol, a resin that is present in all parts of the plant, is the actual allergen. Although the oily resin is produced within the plant, the slightest bruising of leaves, stems, or roots will release urushiol to the surface. It can be picked up by touching the plant or by touching clothing, animals, or gardening tools that were in contact with the plant. Urushiol rapidly bonds to proteins in the skin and may be spread to other parts of the body by rubbing, scratching, or just touching the rash or blisters. The compound is so reactive that one drop can cause dermatitis in 500 sensitive individuals. Urushiol is also long lasting; botanists have developed dermatitis after studying dried poison ivy plants that had been stored for over 100 years. Inhaling smoke from burning plants can also cause a massive skin rash and seriously affect eyes and lungs. Inhalation has caused serious problems for firefighters battling forest fires, but it can even be dangerous to the homeowner trying to eradicate plants by burning them. Climate change may also affect contact dermatitis caused by poison ivy. When poison ivy was grown under enriched CO_2 conditions, the plants produced a more allergenic form of urushiol.

Food Allergies

Food allergies are sensitivities to certain foods that can cause a great diversity of symptoms in various body systems, including the gastrointestinal tract (abdominal pain, vomiting, and diarrhea), skin (hives, eczema, skin rash), and respiratory tract (rhinitis or asthma symptoms). Recent studies indicate that approximately 4% of the U.S. population suffers from food allergies, with the greatest prevalence among children under 3 years of age. The most severe result of food sensitivity is anaphylaxis, a rare but sometimes fatal reaction with multiple symptoms, including swelling of the respiratory tissues, rapid drop in blood pressure, and cardiovascular collapse. Anaphylactic reactions are not limited to food allergies; in fact, the most common triggers of anaphylaxis are bee sting and penicillin allergies.

The diversity of symptoms often makes food allergies difficult to diagnose, with the offending food difficult to identify because hundreds of foods are capable of triggering allergic reactions. Milk, egg, and fish allergies are some of the most thoroughly studied food allergies. Among plant-based foods, peanuts, wheat, tree nuts, soy-beans, strawberries, and citrus fruits are considered leading allergens. The most common allergenic plant foods are listed in Table 21.1. The principal therapy for food allergies is avoidance, which makes individuals constantly vigilant about ingredients. Although it is relatively easy to avoid a food when its presence is obvious, soy products, peanut products, and wheat are almost omnipresent staples in prepared foods. These hidden allergens, even in

Table 21.1 Plant Foods Reported as Allergenic

Allspice	Gluten
Almond	Honey
Anise seed	Hops
Apple	Horseradish
Artichoke	Juniper berry
Baker's yeast	Lentil
Banana	Lima bean
Bay leaf	Mango
Beet	Millet
Black pepper	Mushrooms
Brazil nut	Mustard
Brewer's yeast	Nutmeg
Buckwheat	Orange
Cantaloupe	Pea
Caraway seed	Peach
Cashew nut	Peanut
Castor bean	Pine nut
Celery	Pistachio
Chamomile	Poppy seed
Chestnut	Potato
Chicory	Psyllium seed
Chili pepper	Raspberry
Chocolate	Sage
Cinnamon	Sesame seed
Clove	Soybean
Coconut	Strawberry
Corn	Sunflower seed
Cottonseed	Sweet potato
Cumin seed	Tangerine
Dill seed	Tapioca
Fennel seed	Thyme
Filbert	Tomato
Flaxseed	Turmeric
Garbanzo bean	Vanilla
Garlic	Walnut
Ginger	Wheat

small quantities, can produce severe reactions in sensitive individuals. A new U.S. food labeling law, which took effect in January 2006, requires food manufacturers to clearly indicate on the package whether products contain any of the following food allergens: milk, eggs, fish, shellfish, peanuts, tree nuts, wheat, and soy. This makes it easier for individuals with food allergies to avoid specific allergens. The peanut is the most important plant food allergen and is also the food allergen most commonly involved in anaphylaxis. Unintended exposure to peanuts or peanut butter has been reported in restaurants, schools, and commercial airliners. This exposure has resulted in peanut-free tables in schools and preschools and even peanut-free airplane flights.

Recently, there has been increased concern about hidden allergens in genetically engineered crops (see Chapter 15). An example of this danger was shown during tests of soybeans that were genetically engineered with a Brazil nut gene that transferred extra methionine. When the soybeans were tested, the Brazil nut protein caused reactions in susceptible individuals. Although this project was quickly canceled, the dangers of this type of reaction are real. Also, the mere possibility of allergic reactions led to the recall of StarLink corn during the fall of 2000.

Gluten is a complex of proteins found in wheat, rye, and barley (see Chapter 10). Gluten intolerance is characterized by adverse reactions when any of these grains are consumed. As described in Chapter 10, the most severe form of gluten intolerance is celiac disease, an autoimmune disease that damages the intestines. In addition, many people suffer from non-celiac gluten sensitivity, which causes similar symptoms such as abdominal pain, bloating, and diarrhea or constipation. Around 18 million people in the United States are believed to be gluten sensitive; this is about six times greater than the number with celiac disease. It is different from celiac disease since the IgA antibodies characteristic of celiac disease are not present, and intestinal damage is not usually severe. Gluten sensitivity also differs from wheat allergy since IgE antibodies characteristic of allergic reactions are not present. Researchers are still trying to determine the causes of non-celiac gluten sensitivity. A gluten-free diet is the standard treatment for gluten sensitivity as well as for celiac disease.

Thinking Critically

Food allergies can trigger serious, life-threatening reactions, and avoidance is often difficult to achieve with prepared foods.

Prepare a list of food products that you would have to avoid if you were allergic to corn. Be sure to check labels for all corn products.

Latex Allergy

Within the past 30 years, latex allergy has emerged as a major medical concern, especially among health care workers and patients. **Latex** is a milky exudate produced by many plants. The composition of latex varies in different species; latex may include hydrocarbons, alkaloids, resins, terpenes, proteins, and other compounds. The latex of some plants, such as *Hevea brasiliensis*, is rich in hydrocarbons with elastic properties. This tree, native to the rain forests of the Amazon basin, is the source of natural rubber (see A Closer Look 26.1: Buying Time for the Rain Forest). Although synthetics account for 70% of the rubber used, natural rubber is preferred for certain applications because of its superior properties.

Natural rubber latex is found in gloves, intravenous tubing, stethoscopes, catheters, bandages, and other medical supplies. It is also found in pacifiers, balloons, tires, and condoms. The most important single source of latex allergen exposure is latex gloves. A 2016 report indicated that about 4% of the general population (13 million in the United States) and over 9% of health care workers have latex allergies. Before 1980, allergy to natural rubber latex was a rarely reported condition. It is believed that the problem became evident because of the increased use of latex gloves and other barriers against HIV infection and hepatitis. Natural latex gloves are preferred to synthetics, which sometimes allow fluids and viruses to leak through.

Allergic reactions to latex products range from dermatitis and urticaria (hives) to allergic rhinitis, conjunctivitis (inflammation of the eye), asthma, and anaphylaxis. The type of reaction depends on the type of exposure, with airborne latex allergen causing respiratory symptoms. The allergen sticks to the cornstarch used as powder on gloves. When powdered gloves are put on, the starch particles with allergen are readily airborne, then inhaled, or they come in contact with the eyes and thereby trigger symptoms in sensitive individuals.

There is a 500-fold difference in allergen levels in various brands of gloves, which are made by different processes. This broad range indicates that the technology already exists to make low-allergen gloves. Allergists are recommending using powder-free, low-allergen latex gloves when gloves are mandated by Universal Precaution protocols.

The manufacturing community is also looking at other approaches for producing latex products with less allergen, and much interest has focused on the desert shrub guayule, *Parthenium argentatum*, native to the deserts of the U.S. Southwest. Although guayule has been known as a potential source of rubber for more than 100 years, development of the crop was never commercially practical (see Chapter 26). Guayule rubber is as strong and resistant as rubber from *Hevea* trees, and breeding programs have tripled its rubber yield. Guayule latex output is now on a par with tropical rubber. Tests have shown that latex-sensitive individuals do not respond to guayule proteins. In 2008, the U.S. Food and Drug Administration approved patient examination gloves made from guayule latex, and large-scale commercialization of guayule rubber now seems feasible.

CHAPTER SUMMARY

1. Poisonous and allergy plants are known to adversely affect the health of humans and other animals. As foraging societies identified plants according to use, knowledge of poisonous plants accumulated. Poisonous plants have been employed as arrow poisons, as "succession powders," and as a form of capital punishment. Glycosides and alkaloids are usually the toxic compounds in poisonous plants.

2. Many plants in the natural environment and in the garden and house are poisonous. Some plant toxins, such as curare, have been used for beneficial purposes in medicine. Most cases of plant poisoning in the home involve the ingestion of aroids by young children.

3. White snakeroot, locoweed, Indian hellebore, and St. John's wort are plants that have been implicated in the poisoning deaths of livestock.

4. Plants can cause mechanical injury by puncturing the skin via thorns, barbs, or spines. Stinging hairs not only penetrate the skin but also inject histamine-like irritants. Other plants injure by releasing irritating resins or latexes that damage the skin and other organs.

5. Plant toxins have become an important source in the search for natural insecticides. Pyrethrum and rotenone are two examples of plant-derived insecticides; the neem tree of India is the origin of an oil that has been shown to be effective against insect pests, bacteria, viruses, and fungi.

6. Allelopathy is a type of chemical warfare waged by plants that affects the growth and development of competing plants. The effects of allelopathy on the spacing of vegetation are especially pronounced in desert and chaparral plant communities.

7. Allergies are hypersensitivity diseases caused by the immune system's response to harmless substances, such as ragweed pollen or fungal spores. Hay fever and asthma are the most common forms of respiratory allergies, affecting 20% to 25% of the human population. Hay fever is characterized by sneezing, runny nose, nasal congestion, and watery eyes; the symptoms of asthma include wheezing, coughing, and choking.

8. Pollen from ragweed plants is the most important cause of allergic rhinitis in North America. Ragweeds are pioneer species that are well adapted to invading disturbed soils. They occur abundantly through the eastern two-thirds of the continent and in most parts of North America. Ragweed pollen generally occurs in the atmosphere from August to October. Approximately 1 million tons of pollen are produced each year in the United States.

9. In addition to pollen from ragweed, pollen from hundreds of seed plants is known to trigger respiratory allergies. Like ragweed, hay fever plants produce abundant, lightweight, airborne pollen. Pollen allergies are usually categorized into tree, grass, and weed allergies, correlating the symptoms with the season of pollen production. Tree pollen is considered the leading cause of spring hay fever, grasses the leading cause of summer hay fever, and weeds the leading cause of fall hay fever.

10. The most common cause of contact dermatitis is poison ivy, a widespread climbing shrub in North America. The actual allergen is a resin, urushiol, present in all parts of the plant. The resin can be picked up by touching the plant directly or by touching clothing, animals, or even gardening tools that were in contact with the plant.

11. Food allergies can produce gastrointestinal symptoms, dermatitis, hives, or respiratory symptoms. A wide variety of plant foods are known to cause reactions in hypersensitive individuals.

12. Within the past 30 years, latex allergy has emerged as a major concern in the health care community.

REVIEW QUESTIONS

1. How are some poisonous plants beneficial to society?
2. In terms of accidental plant poisonings, what dangers exist in the garden and home?
3. What is allelopathy? How does it account for spacing of plants in the chaparral?
4. Select one of the following poisonous plants and detail the mechanism of its action: giant hogweed, castor bean, white snakeroot.
5. Describe the source and action of the plant-derived insecticides: pyrethrum, rotenone, neem.
6. The medical term for hives (an allergic skin reaction) is *urticaria*. In what ways are plants responsible for causing urticaria?
7. In what way are allergic responses immune system "mistakes"?
8. What are the characteristics of a "hay fever" plant, and what common plants are major triggers of hay fever?
9. Avoidance is an effective treatment for what types of allergic reactions? Why is it not effective for other types of allergies?
10. What is latex? Why have latex allergies increased?

UNIT VI

CHAPTER 22

The Algae

Giant kelp, Macrocystis pyrifera, *off the coast of Australia.*

KEY CONCEPTS

1. Algae are a diverse group of photosynthetic organisms that occur in all marine and freshwater habitats, where they are the base of the food chain, providing both carbohydrates and oxygen.
2. Cultures throughout the world have used various algae as food for humans and livestock and for a variety of other economically important products.
3. The harmful effects of algae are largely due to the production of toxins by various species under bloom conditions.

©Karen Gowlett-Holmes/Photo disc/Getty Images

CHAPTER OUTLINE

Characteristics of the Algae 401
Classification of the Algae 401
    Cyanobacteria 402
    Euglenoids 403
    Dinoflagellates 403
    Diatoms 404
    Brown Algae 404
    Red Algae 405
    Green Algae 405
Algae in Our Diet 409
    Seaweeds 409
Biofuels from Algae 410
Other Economic Uses of Algae 411
Toxic and Harmful Algae 411
    Toxic Cyanobacteria 412

A CLOSER LOOK 22.1 Drugs from the Sea 413

    Red Tides 414
    Pfiesteria 414

A CLOSER LOOK 22.2 Killer Alga—Story of a Deadly Invader 415

    Other Toxic Algae 417
Chapter Summary 417
Review Questions 418

The algae are a diverse group of photosynthetic organisms that range from microscopic single cells to giant, complex, multicellular organisms. Ecologically, the algae occupy the base of food chains in both freshwater and marine ecosystems. Various algae, especially seaweeds, have been used as food for humans and livestock by cultures throughout the world. Some algae supply additives used in foods, cosmetics, and other commercial products, and other algae are the focus of biofuel research. On the negative side, algal blooms have caused damage to aquatic environments throughout the world by depleting oxygen and interfering with water purification. In addition, some algae can produce toxins that have resulted in blooms that cause massive fish kills and even human poisonings.

CHARACTERISTICS OF THE ALGAE

Some people immediately think of green pond scum when they hear the word *algae;* others think of seaweeds clinging to rocks at the seashore or huge kelps that form extensive underwater forests. Although quite diverse in appearance, these organisms have some similarities to land plants and several characteristics that distinguish them from land plants. Like land plants, the algae are photosynthetic, with chlorophyll *a* as the major light-absorbing pigment; however, many algae have various types of accessory pigments not found in land plants. Another feature in common with land plants is the presence of a cell wall, but the characteristics of the wall vary among the different groups of algae.

Unlike land plants, algae lack extensive tissue differentiation. The body of an alga, called a **thallus** (pl., **thalli**), can vary from a microscopic unicellular organism to a large macroscopic, multicellular organism. The thallus type is important for identification and classification. The most common types include (1) single cells that may be nonmotile, amoeboid, or flagellated (A **flagellum** is a whiplike organelle, composed of mircotubules, that imparts motility.); (2) colonies that may be nonmotile or flagellated; (3) filaments (chains) that may be branched or unbranched; (4) parenchymatous thalli that produce a three-dimensional structure and may be in the form of blades, sheets, or tubes.

Algae also differ from land plants in reproduction. Recall that one of the characteristics of land plants, from the bryophytes through the angiosperms, is the retention of the embryo (see Chapter 9). After fertilization in plants, the zygote develops into a multicellular embryo while still enclosed in the female gametangium. Reproduction is diverse among the various groups of algae; however, there is no retention of the embryo in any group. Some algae only reproduce asexually by simple cell division in some unicellular species or by spore formation in others; in addition, some filamentous algae reproduce by fragmentation. Other algae have complex life cycles with both sexual and asexual stages.

The life cycles are also diverse across the various algal groups. In some algae the zygote is the only diploid cell; meiosis occurs when the zygote germinates, and the vegetative stage is always haploid. By contrast, some algae are diploid, with the gametes being the only haploid cells. Many algae have distinct sporophyte and gametophyte generations. The alternation of generations may be isomorphic, in which the haploid and diploid stages appear identical, or, as with land plants, heteromorphic, in which the sporophyte and gametophyte forms are recognizably different.

Algae are photosynthetic organisms, and the amount of photosynthesis that takes place in freshwater and marine environments is enormous. About half the photosynthetic carbon fixation that takes place on the planet actually occurs through the algae. In addition to providing food, algae provide oxygen for organisms in the food chain.

CLASSIFICATION OF THE ALGAE

A recent study suggests that there are approximately 72,500 species of algae, while other estimates have proposed as few as 30,000 species to as many as 1 million species. Although the term *algae* seems to imply that they are a single group of related organisms, molecular evidence indicates that the algae are polyphyletic and represent several independent evolutionary lines. At one time the algae were considered the most primitive members of the Plant Kingdom; today they are usually classified in two domains and at least three separate kingdoms.

The domain Bacteria (kingdom Eubacteria), which includes prokaryotic organisms, contains the division (phylum) Cyanobacteria (cyanobacteria or blue-green algae). All other algae have eukaryotic cells and are in the domain Eukarya. Recall from Chapter 9, the kingdom Protista includes many diverse lineages, and a number of biologists have split the protists, which include several groups of algae, into several smaller kingdoms. Since the evolutionary relationships among various groups of algae are still uncertain, we will retain the term *protists* (kingdom Protista) only as a convenient tool to discuss these algae. The algal protists include the divisions Euglenophyta (euglenoids), Dinophyta (dinoflagellates), Bacillariophyta (diatoms), Phaeophyta (brown algae), and Rhodophyta (red algae). The diatoms and brown algae have some ultrastructural and biochemical similarities; as a result these two groups of algae are sometimes considered as classes within the division Ochrophyta. Several other groups of algal protists also are included in the Ochrophyta, but these will not be covered in this text. The green algae are the ancestors of all green plants and two divisions of green algae, Chlorophyta and Charophyta, are part of the kingdom Plantae. There are many features that are used to distinguish the various groups of algae. These include DNA sequences, ultrastructural features, pigment types, morphology, storage products, and cell wall composition (table 22.1). The term *seaweed* has no taxonomic status but refers to various macroscopic marine algae from the divisions Phaeophyta, Rhodophyta, and Chlorophyta.

Table 22.1 Major Groups of Algae

Division	Photosynthetic Pigments	Morphological Types	Storage Products	Cell Wall Structural Material	Habitat
Domain Bacteria					
Blue-green algae (Cyanobacteria)	Chlorophyll a, phycobilins, carotenoids	Unicells, filaments, colonies	Glycogen	Peptidoglycan	Marine, freshwater, terrestrial
Domain Eukarya—Protists					
Euglenoids (Euglenophyta)	Chlorophyll a and b, carotenoids	Unicells	Paramylon	Proteinaceous pellicle	Freshwater
Dinoflagellates (Dinophyta)	Chlorophyll a and c, carotenoids including peridinin and fucoxanthin	Mainly unicells	Starch	Cellulose plates	Marine, freshwater
Diatoms (Bacillariophyta/Ochrophyta)	Chlorophyll a and c, carotenoids including fucoxanthin	Mainly unicells	Oil	Silica frustule	Marine, freshwater
Brown algae (Phaeophyta/Ochrophyta)	Chlorophyll a and c, carotenoids including fucoxanthin	Branched filaments, sheets	Laminarin	Cellulose	Marine
Red algae (Rhodophyta)	Chlorophyll a, phycobilins, carotenoids	Unicells, branched filaments, sheets	Floridean starch	Cellulose	Marine, freshwater
Domain Eukarya—Plantae					
Green algae (Chlorophyta/Charophyta)	Chlorophyll a and b, carotenoids	Unicells, filaments, colonies, sheets	Starch	Cellulose	Freshwater, marine, terrestrial

Cyanobacteria

In traditional systems of classification, the cyanobacteria were called blue-green algae. Although the cells are prokaryotic, many phycologists (scientists who specialize in the algae) still study the cyanobacteria along with the eukaryotic algae. We will follow this practice here as well. These primitive organisms first appeared in the fossil record approximately 3.5 billion years ago. Today cyanobacteria are widely distributed and can be found in oceans, fresh water, and damp terrestrial habitats. Some cyanobacteria can live in alkaline hot springs, such as those at Yellowstone National Park. The temperature in these hot springs often exceeds 80°C (176°F). Cyanobacteria often form a slimy scum on greenhouse floors, clay flowerpots, and pool decks.

In structure, the cyanobacteria include species that are microscopic unicells, filaments, and colonial forms, and they are usually bluish green. The accessory pigments present in the cyanobacteria are the **phycobilins,** with phycocyanin and phycoerythrin the most important phycobilins. Phycobilins are also found in the red algae. Phycocyanin is more abundant in the cyanobacteria and provides the distinctive color to these algae. Because the cells are prokaryotic, no chloroplasts are present, and the photosynthetic pigments occur on membranes distributed within the cytoplasm (see fig. 1.4a).

Reproduction in the cyanobacteria is by simple cell division or fragmentation of filamentous forms. Molecular evidence indicates that certain cyanobacteria are the progenitors of the chloroplasts of algae and plants, further proof of the Endosymbiont Theory (see A Closer Look 2.1: Origin of Chloroplasts and Mitochondria).

Many cyanobacteria are enclosed in a mucilaginous sheath, and multiple filaments may be present in the same sheath. The mucilaginous sheath of some cyanobacteria, such as *Nostoc*, is quite firm, and in culture *Nostoc* may appear as mucilaginous balls or clumps (fig. 22.1a). Several cyanobacteria are able to fix nitrogen from the air or water; this process occurs in specialized cells called **heterocysts,** which are large, thick-walled, faintly pigmented cells found in a number of filamentous species (fig. 22.1b). Nitrogen fixation is an important component of the nitrogen cycle and provides nitrogen compounds in forms that other organisms can utilize. The cyanobacterium *Anabaena* has a symbiotic relationship with the aquatic fern *Azolla*, which is frequently found in rice paddies. The nitrogen-fixing ability of this cyanobacterium plays a prominent role in reducing the need for fertilizers during the cultivation of rice (see Chapters 12 and 13). Cyanobacteria can also be found in symbiotic associations with other organisms (see A Closer Look 23.1: Lichens).

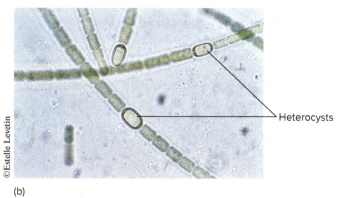

Figure 22.1 (a) *Nostoc* filaments are embedded in a firm, gelatinous sheath. (b) Filaments of *Anabaena* with heterocysts.

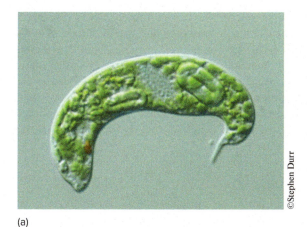
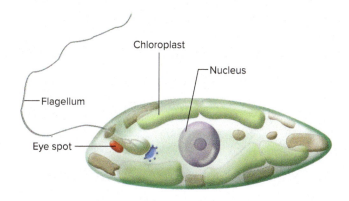

Figure 22.2 (a) *Euglena* shown in a light micrograph, and (b) drawing.

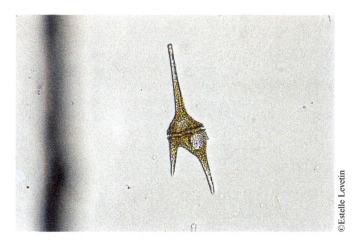

Figure 22.3 *Ceratium*, a dinoflagellate.

Even though the greenish scum on a greenhouse floor may look unappetizing, some societies use a few types of cyanobacteria as a food source. In addition, some species are currently sold as a beneficial dietary supplement. Cyanobacteria have caused environmental problems when they form harmful algal blooms.

Euglenoids

The division Euglenophyta (euglenoids) is a small group of mainly freshwater unicells that are typically grass-green in color. Euglenoids can be extremely abundant in nutrient-enriched or polluted waters, such as farm ponds and ditches. Instead of a rigid cell wall, they have a proteinaceous covering called a pellicle that allows for flexibility and shape change during swimming. In addition to their ability to change shape, euglenoids have several other distinctive features, one of which is a large red eye spot involved in sensing the direction of light (fig. 22.2). *Euglena* have two flagella, one very short and apparently nonfunctional and the second long and whiplike. Reproduction is by binary fission.

Dinoflagellates

The organisms in the division Dinophyta, more commonly known as dinoflagellates, are a group of bizarre unicells, which are abundant in both freshwater habitats and oceans.

Many dinoflagellates are covered with cellulose plates, giving them an armored appearance (fig. 22.3). Dinoflagellates typically possess two flagella, one trailing and one wrapped around the middle, that enable them to roll through the water like a top. These organisms have gained notoriety because some marine species are responsible for the phenomenon known as red tides (discussed later in this chapter). Other dinoflagellate species are known for their ability to luminesce (emit light). You may have seen tiny specks of light at night, especially as ocean waves break on the shore; these were caused by bioluminescent dinoflagellates. Dinoflagellates generally reproduce by simple cell division, although sexual reproduction does occur among some species. Ecologically, dinoflagellates along with diatoms are vitally important as the base of food chains in both freshwater and marine environments.

Diatoms

Diatoms, in the division Bacillariophyta, are mainly unicells surrounded by a silicon-based wall composed of two halves that fit together like the top and bottom of a box. The glasslike diatom wall, called a **frustule,** is often intricately marked with pits, grooves, and ridges, giving these microscopic forms the

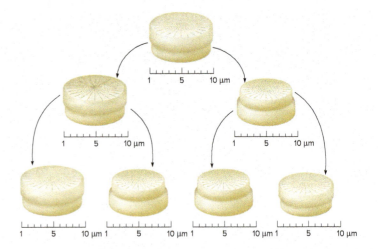

Figure 22.5 Cell division in diatoms. Each successive division forms new halves inside the parent halves. As a result, one of the two new cells is smaller than the parent cell, and one is the same size.

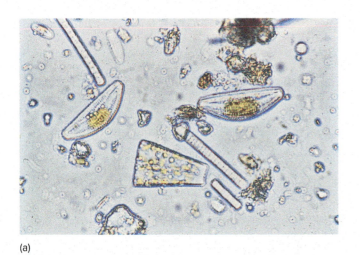

(a)

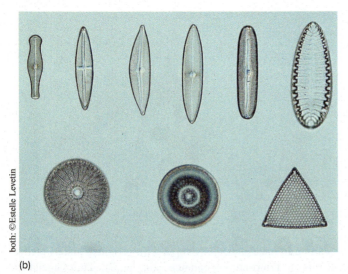

(b)

Figure 22.4 Diatoms. (a) Diatoms are abundant in both freshwater and marine habitats. (b) The ornamentation on the diatom walls has the appearance of cut crystal when viewed microscopically (pennate diatoms top row, centric bottom row).

appearance of cut crystal (fig. 22.4). The shape of diatoms can be either pennate, elongate with bilateral symmetry, or centric, often round with radial symmetry. Some pennate diatoms are motile by means of a poorly understood gliding motion that results from the secretion of mucilage; others are nonmotile.

Asexual reproduction in the diatoms is mainly through cell division; however, there are some unique features. After mitosis and cytokinesis, the two halves of the frustule separate (fig. 22.5). Each daughter cell receives one of the halves, and then a new half-wall forms in each daughter cell, fitting inside the old half. Because the bottom half of the frustule is smaller than the upper half, the daughter cells are not the same size. This process continues for several generations, with one-half of the cells continuing to get smaller. A minimum size is eventually reached, and meiosis and sexual reproduction are initiated. The full-size frustule is restored when the zygote (called an auxospore) enlarges and develops a new wall.

Diatoms are extremely abundant in both freshwater and marine ecosystems, especially in colder areas of the ocean; they can also be found in many moist terrestrial habitats. The economic value of the diatoms comes from harvesting fossilized diatom frustules (see Other Economic Uses of Algae later in this chapter).

Brown Algae

The division Phaeophyta, the brown algae, includes the huge kelps, which form extensive underwater "forests" off the California coast, and the rockweeds commonly found in the intertidal zone in coastal areas. Some species of kelp are among the largest organisms on Earth, reaching 100 meters (over 300 feet) in length. The brown seaweeds are the most complex structurally of all the algal divisions, with the simplest species consisting of highly branched filaments and the most advanced almost resembling the form of a land plant in having a rootlike **holdfast,** stemlike **stipe,** and leaflike **blade** (fig. 22.6a).

Life cycles are varied although the conspicuous phase is usually diploid. In some brown algae, there is an alternation of generations; in others, such as *Fucus*, or rockweed (fig. 22.6b), there is no free-living haploid stage. In *Fucus*, the swollen areas at the ends of the blade are reproductive regions known as **receptacles.** Just beneath the surface of the receptacles are spherical cavities called **conceptacles,** which have openings on the surface. These openings appear as dots on the surface of the receptacle. Within the conceptacles, gametangia are produced. Depending on the species, male and female gametangia form in the same conceptacle or in different individuals. Meiosis occurs in each **oogonium** (female gametangium) followed by mitosis to give rise to eight eggs; 64 sperms are produced following meiosis and several mitotic divisions in each **antheridium** (male gametangium). Both egg and sperm are released into the water, where fertilization occurs (fig. 22.7).

Red Algae

The red algae, division Rhodophyta, are most familiar as seaweeds. They are mainly large, multicellular marine forms occurring in coastal waters, often attached to rocks; however, a few freshwater species also occur. The red algae are most diverse in warm tropical waters, where they often occur at depths of as much as 200 meters (650 feet). The red color of these algae is mainly due to the presence of phycoerythrin, which is one of the phycobilin pigments. The thallus color varies from red to purple to almost black, depending on the amounts of the accessory pigments.

Most red algae have a multicellular thallus, and many consist of highly branched filaments that give them a feathery appearance (fig. 22.8). The base of the thallus is usually modified into a holdfast where the alga is attached to rocks or other substrates. Other red algae have a sheetlike thallus.

Life cycles are especially complex, and the life cycle of many red algae has three stages, two diploid phases and one haploid phase. Red algae differ from the other eukaryotic algae in the absence of any flagellated cells; therefore, they must depend on water movements for fertilization and spore dispersal.

DNA sequencing data indicate that the red algae are a sister group to the green algae–green plant lineage. As a result, some researchers have suggested including the red algae in the kingdom Plantae.

A number of species of red algae and brown algae are used directly as food or as food additives.

(a)

(b)

Figure 22.6 Brown algae. (a) *Postelsia*, brown alga from the Pacific coast. (b) *Fucus*, or rockweed, is a common brown alga along the New England coast.

Green Algae

The green algae are a large, diverse group of organisms, characterized by the presence of chlorophyll *a* and *b* in the chloroplasts and use of starch as a storage product. Green algae are classified into two divisions, the Chlorophyta and Charophyta, both of which are part of the kingdom Plantae.

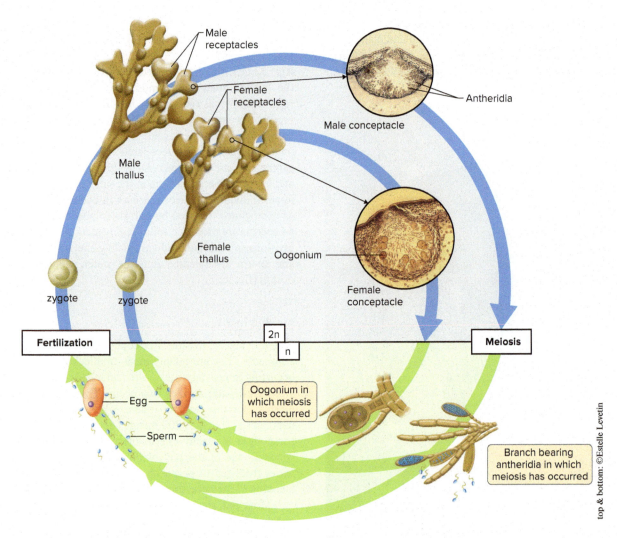

Figure 22.7 Life cycle of *Fucus*, or rockweed.

Figure 22.8 *Microcladia*, a red alga common along the northern California coast.

DNA analysis indicates that land plants (embryophytes) evolved from green algae; in fact, botanists believe that land plants evolved from the charophyte lineage more than 470 million years ago. Various charophyte species show biochemical and ultrastructural features that are shared with land plants but not with chlorophytes. These include the presence of cellulose-synthesizing proteins in the plasma membrane, the ultrastructure of flagella in sperm cells, and the formation of a phragmoplast during cytokinesis (see Chapter 2). Genetic analysis from DNA sequences supports this hypothesis that the charophytes are the closest living relatives of land plants.

A great many morphological types occur among the green algae, including motile and nonmotile unicells, colonies, branched and unbranched filaments, tubular forms, and sheets (figs. 22.9, 22.10). They are found in oceans, in fresh water, and in moist terrestrial environments. Along with the red and the brown algae, some of the larger marine species

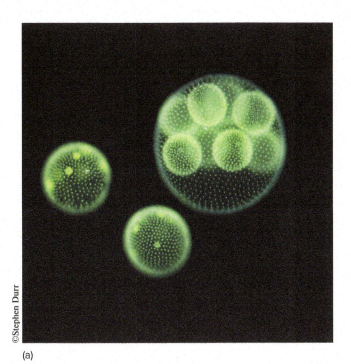

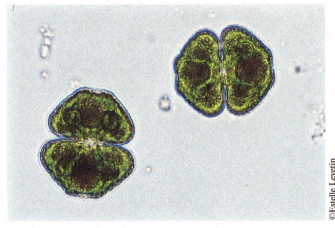

Figure 22.10 Representative Charophyta. (a) *Cosmarium*, a nonmotile unicell, (b) *Spirogyra*, a filamentous genus with ribbon-like, spirally arranged chloroplasts, and (c) *Chara*, a branching organism bearing an oogonium and antheridium.

Figure 22.9 Representative Chlorophyta. (a) *Volvox*, a motile colony with daughter colonies visible within the main colonies. (b) *Cladophora*, a branched filamentous alga.

of chlorophytes are referred to as seaweeds. However, the green algae are most abundant and diverse in fresh water, especially the charophytes. Several green algae are able to form a symbiotic relationship with fungi to make up lichens (see A Closer Look 23.1: Lichens). Unlike the brown and the red algae, only a few species of green algae are used as human food; sea lettuce (*Ulva* spp.) can be eaten raw in salads and is also used as an ingredient in soups. Although green algae play an important role in aquatic food chains, on the negative

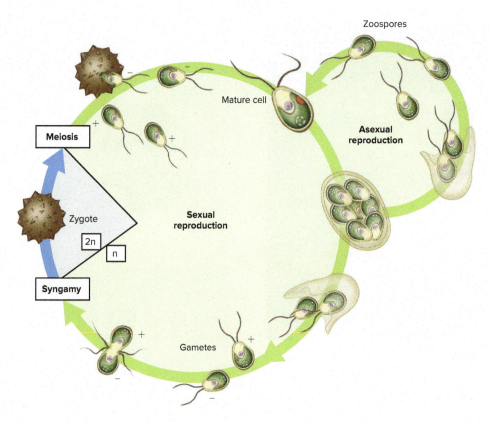

Figure 22.11 Life cycle of *Chlamydomonas*, an isogamous green alga, showing a simple life cycle with the zygote the only diploid cell.

side, seasonal blooms of green algae are often noticeable in ponds and lakes.

Life cycles are especially diverse in the Chlorophyta. Some green algae reproduce asexually by means of flagellated spores called **zoospores**. Zoospores are the products of mitosis. Sexual reproduction may be **isogamous, anisogamous,** or **oogamous**. In isogamous green algae, the small, flagellated gametes that fuse to form the zygote are identical in size and shape. In anisogamy, both types of gametes are flagellated, but they come in two sizes, large and small. Some green algae are oogamous, producing a nonmotile, larger egg and a motile, smaller sperm. The life cycle of *Chlamydomonas*, a freshwater genus, is an example of an isogamous life cycle (fig. 22.11). *Chlamydomonas* is a haploid, unicellular alga. Each cell is biflagellated. Most of the time, *Chlamydomonas* reproduces asexually. A cell loses its flagella and undergoes several mitotic divisions to produce zoospores. After release from the parent cell, each zoospore grows into a mature individual. If environmental conditions are right, *Chlamydomonas* will enter a sexual phase. An individual cell divides to produce many gametes. If compatible gametes of different mating strains meet, the isogamous gametes fuse, producing a diploid zygote surrounded by a protective wall. After a period of dormancy, the zygote undergoes meiosis and releases four zoospores, beginning the cycle once again.

The genus *Ulva*, which are seaweeds also known as sea lettuce (see fig. 1.2a), appears as a thin, wavy blade attached by an anchoring holdfast. This genus has an isomorphic alternation of generations with indistinguishable sporophytes and gametophytes (fig. 22.12). Meiosis in specialized cells called sporangia within the sporophyte produces zoospores, each with four flagella. Upon release, the haploid zoospores develop into gametophytes. The gametophytes in turn produce biflagellated gametes in gametangia. When gametes of compatible mating types meet, they fuse, forming the diploid zygote and restoring once again the sporophyte generation.

Thinking Critically

The algae are known to have a variety of photosynthetic pigments, while in all land plants the photosynthetic pigments are the same.

Can you speculate why a variety of photosynthetic pigments have evolved in the algae that live in an aquatic environment while this diversity is not seen in the land plants?

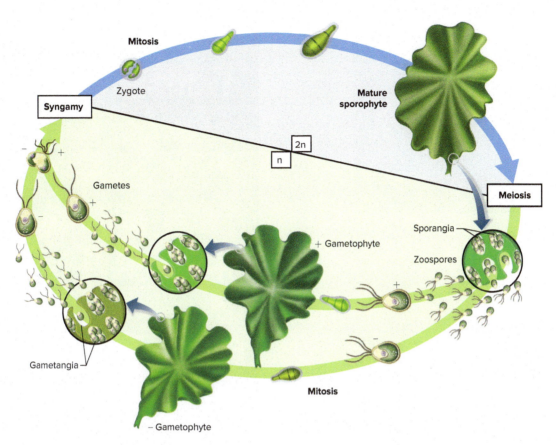

Figure 22.12 Life cycle of *Ulva*, with an isomorphic alternation of generations.

ALGAE IN OUR DIET

Many algae, especially cyanobacteria and seaweeds, are widely used as food in many parts of the world. Dried red and brown algae are available in Asian markets as well as in many grocery stores and can be found in several dishes served in Asian restaurants.

Cyanobacteria have played an important role directly in the diet of many cultures. It has even been speculated that the "manna" the Israelites consumed during their 40-year sojourn in the desert might have been the cyanobacterium *Nostoc*. Another example is *Arthrospira*, which is known commercially as spirulina. *Arthrospira* is a corkscrew-shaped filament (fig. 22.13a) that has significant quantities of protein, vitamins, and minerals and is sold in the United States as a health food and as a dietary supplement. The ancient Aztecs cultivated mats of spirulina in shallow ponds and lakes; once dried and cooked, it was eaten like a vegetable. This cyanobacterium has also been a traditional part of meals in Chad, where it is made into a sauce for beans, meat, or fish. Today spirulina is used to prepare a high-protein food additive that has been considered one solution to the malnourishment problems in developing countries. Some scientists are concerned about this current use of spirulina because of the danger of contamination with toxin-producing cyanobacteria.

Unsaturated omega-3 fish oils have been promoted as heart-healthy oils that reduce the risk of coronary heart disease (see Chapter 10). Fish cannot synthesize these oils themselves but obtain them preformed from certain algae. Omega-3 fatty acids are essential fatty acids and constituents of the gray matter of the brain; they are found in breast milk, but not in most infant formulas. There is interest in using algal cultivation to obtain these essential nutrients to make healthier formulas.

Seaweeds

In coastal areas, the red and brown seaweeds have long been used as a source of food. In Asia, this use dates back to prehistoric times. It has been estimated that over 100 species of marine algae are eaten in one form or another. Among the brown algae, the most popular is kombu (*Laminaria japonica*). The source of wakame is the brown alga *Undaria pinnatifida*. It is usually boiled and salted, transforming into a green product that is used in soups, salads, and noodles. Some favorite red seaweeds are dulse (*Rhodymenia* and *Palmaria*) and nori (*Porphyra*) (figs. 22.13b and c). These can be eaten many different ways: as a confection, as a relish, with rice, or in soups and salads. The nutritional value of seaweeds lies in their high protein content as well as their essential vitamin and mineral content, particularly iodine and potassium.

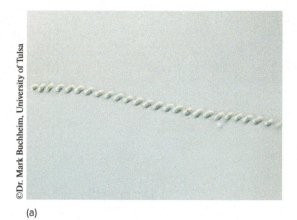

(c)

Figure 22.13 Edible algae. (a) Individual filament of *Arthrospira* (spirulina), a cyanobacterium that has been used as a source of food by various cultures and today is a well-known dietary supplement. (b) Sushi rolls with an outer wrap of nori, the red alga *Porphyra*. (c) Edible red and brown seaweed for sale in a grocery store.

Porphyra, used more extensively and by more cultures than any other seaweed, has a long history of food use dating back to the year A.D. 533. It is also the most widely cultivated seaweed and is known by many names: the Chinese call it *zicai;* the Japanese, *amanori* or *nori;* the Koreans, *kim;* the British, *laver;* and the Maoris of New Zealand, *karengo.* Nori is a primary constituent of Japanese sushi. The outer wrap of sushi is nori, which has been washed, chopped, dried, and toasted (fig. 22.13b). Nori is cultivated in shallow ocean bays on nets that have been seeded with reproductive propagules. In addition to its high protein content, nori is also considered a good source of Vitamin B12 making it a healthful choice for vegetarians.

With current interest in developing new foods, perhaps the algae will play a bigger role in the human diet in the future.

BIOFUELS FROM ALGAE

Biofuels figure prominently in the search for alternative energy sources and include land-based plants, such as members of the grass (see Chapter 12) and legume (see Chapter 13) families, as well as aquatic and marine algae.

Algal biofuels are generating a great deal of interest among researchers around the world because of the high rates of growth and photosynthetic efficiencies of the algae. The focus has been on microalgae, which are unicellular or small, multicellular algae or cyanobacteria. *Chlamydomonas* (fig. 22.11), *Haematococcus, Chlorella, Anabaena,* and *Dunaliella* are just a few of the genera being studied as biofuel sources. The biomass production of these and other microalgae is much greater than that of land plants, often doubling in less than 24 hours. These algae can be used to produce various biofuels, such as bioethanol, biomethanol, and biogasoline; however, the greater interest has been in their use for producing biodiesel.

Many microalgae produce high levels of oils (triglycerides) that can readily be converted into biodiesel fuel. In fact, the oil content of some algae is close to 80% of the dry biomass, although yields of 20% to 50% are more common. Currently, soybean, palm, and canola are being used for biodiesel production; the oil in these crops is typically less than 5% of the total crop biomass. Many researchers believe that biodiesel from microalgae is the only sustainable substitute for petroleum-based diesel fuel. In addition, the use of microalgae for biodiesel production will not compete with food production on farmland, since the algae require limited land area. The algae also require less water than crops and can utilize various water sources, including brackish water, ocean water, and wastewater. However, the cost of biodiesel production from algae needs to be dramatically reduced to become economically viable.

The microalgae are grown in either shallow, open ponds or closed photobioreactors; these can be located near the ocean or even in the desert. The open ponds are often called raceway ponds because the channels resemble the hairpin turns at a racetrack. The closed photobioreactors are composed of long, narrow glass or Plexiglas tubes that are exposed to sunlight in parallel arrays or spirals. In both systems, the culture broth for the algae continually circulates. The open pond systems are often favored due to the lower start-up costs, but photobioreactors provide for better control over contamination and growing conditions. Also, the density of the algal cell is much greater in the photobioreactors. The algae from either system must be harvested and the oil extracted from the harvested biomass. Although microalgae are potentially far more productive than land-based oil crops, at the present time harvesting the algal biomass is more expensive than growing crops and amounts to 20% to 30% of the total production costs.

Research on algal biofuels is currently focused in several directions: improving the design of photobioreactors, enhancing technologies for biomass harvesting and drying, developing uses for the residual biomass after the oil has been extracted, identifying environmental factors that favor oil accumulation, and increasing the productivity of the microalgae. Scientists are using genetic engineering to enhance photosynthesis and algal growth rates, increase oil synthesis, and improve temperature tolerance. Research in these areas may help make algal biofuels a reality in the near future.

OTHER ECONOMIC USES OF ALGAE

In addition to their direct use as food and a source of biofuels, algae provide several economically important products, three of which are **agar, carrageenan,** and **alginic acid.** These compounds are all **hydrocolloids** from the cell walls of various algae. Hydrocolloids are complex polysaccharides that are used as emulsifiers, stabilizers, and gelling agents. These products have been used for centuries in foods and more recently in industrial products. Carrageenan and agar are obtained from the red algae. Carrageenan is a common ingredient in ice cream, pudding, cottage cheese, weight-loss products, toothpaste, lotions, and paints; this natural additive imparts a creamy texture. Although agar is also used in a variety of commercial products, its most important use is as a solidifying agent in culture media used in laboratory research on bacteria, fungi, and plant tissue culture (see Chapter 15). Another algal product is **agarose,** derived from agar and used to form gels for mapping DNA in biotechnology. Several species of brown algae are the source of alginic acid, which has broad commercial use in the treatment of latex during tire manufacturing, as a binding agent for charcoal briquettes, and to give a creamy texture to confections, ice cream, and other products.

One additional algal product is **diatomaceous earth,** also known as diatomite. This substance is composed of the fossilized remains of diatom frustules. When diatoms die, their virtually indestructible walls accumulate on the ocean floor. Deposits from past geological ages are known as diatomaceous earth and can be mined (fig. 22.14). Diatomaceous earth is useful as a polishing agent in silver polish, as a filter in the wine and petroleum industries, and as the reflective material in highway paint. A recent use of diatomaceous earth is as a soil additive; it acts like a field of broken glass to discourage some microscopic garden pests.

TOXIC AND HARMFUL ALGAE

A negative aspect of the algae is related to environmental damage caused by algal blooms, which are sudden population explosions of certain algal species (fig. 22.15). In recent years, there has been an increase in the occurrence of algal blooms in both marine and freshwater environments throughout the world. Blooms are triggered by changes in environmental conditions, principally the availability of nutrients. Although blooms sometimes occur naturally, there is good evidence that this increase is related to nutrient pollution, especially nitrogen and phosphorus overload. It has been suggested that agricultural runoff, human sewage, and animal waste (such as hog and poultry manure) are all contributing to these algal population explosions in our waterways. Researchers are also investigating the connection between algal blooms and climate change. In some bloom events, scientists have found evidence to support the connection between the bloom and warmer ocean temperatures caused by global warming.

Blooms of any algae can cause problems that result from depletion of oxygen in the water. After the bloom, the nutrient levels drop, and bloom organisms begin to die. Bacteria in the water that cause the decay of the algae can use up all the available oxygen. Depletion of oxygen can even cause the death of the whole ecosystem. Blooms can also contribute to taste and odor problems when they occur in reservoirs that supply water to cities and towns, but blooms are especially harmful

Figure 22.14 Mining diatomaceous earth in northern Mexico.

Figure 22.15 Red tide bloom, Puget Sound, Washington.

when the algae are capable of synthesizing toxins. Only a small percentage of algae produce toxins, but their impact is dramatic. They can cause massive fish kills or human poisoning. Scientists are actively studying these toxins to determine their health effects and are hoping to find medical uses for these biologically active compounds (see A Closer Look 22.1: Drugs from the Sea).

Toxic Cyanobacteria

In freshwater environments, major problems have been caused by toxin-producing cyanobacteria that have been responsible for the deaths of large populations of wild and domestic animals. Migrating ducks and geese as well as cattle and sheep have been poisoned after consuming water contaminated with toxic cyanobacteria. Toxins can be found in lakes, drinking water reservoirs, and freshwater fish farms. Normally, algal cells release the toxins only when the cells die or when they age and become leaky. However, if the water is treated to break up the bloom, it is possible for even higher levels of toxins to be released into the water. Numerous genera of cyanobacteria are currently known to be toxin producers. Two of the most toxic are *Anabaena* and *Microcystis*.

Scientists have identified several groups of toxins from the cyanobacteria; the best known are the neurotoxins, which affect the nervous system, and hepatotoxins, which affect the liver. Neurotoxins interfere with the normal functioning of the neurotransmitter acetylcholine and can lead to death in minutes by causing paralysis of muscles in the respiratory system. Research on the toxins and their modes of action has had some intriguing payoffs. One of the neurotoxins (anatoxin-a) is used in research on Alzheimer's disease and other degenerative neurological diseases with the hope that a modified version of the toxin might one day be used to slow the degeneration. Another neurotoxin is saxitoxin, which is also produced by certain dinoflagellate species and is more commonly associated with red tides and paralytic shellfish poisoning. Recall from Chapter 9 that another cyanobacterial neurotoxin, BMAA, has been implicated in the neurological conditions known as Guam disease.

The hepatotoxins appear to be more widespread and have caused problems throughout the world. The majority of

A CLOSER LOOK 22.1

Drugs from the Sea

Some researchers are looking to the oceans for the next generation of wonder drugs. Although the antibiotic activity of certain algae has been known for over 50 years, there has been little development of sea-based pharmaceuticals. One reason for slow progress has been the inability to secure adequate amounts of the active compounds for screening, refining, testing, and clinical trials. Current work may change that situation. Although researchers are looking at many different types of drugs from the algae, promising areas are in the development of anticancer drugs and antiviral drugs.

In the search for bioactive compounds, researchers frequently look at organisms known to form toxins, and dinoflagellates fit this role well. One focus has been a dinoflagellate in the genus *Amphidinium* that produces a class of molecules known as amphidinolides. Approximately 35 amphidinolides have been identified, and several of these show significant anticancer activity against a mouse leukemia, human epidermoid carcinoma, and human colon tumor cell lines. Researchers from the University of Rhode Island have been growing *Amphidinium* cultures in 15,000-liter tanks; however, the yield of amphidinolide B has been very low, suggesting that mass culture may not be practical. Other researchers have been working on the chemical synthesis and modification of these compounds, and 10 amphidinolides have currently been synthesized. Also, progress is being made in determining the modes of action of these compounds. Preliminary studies indicate that amphidinolide H binds to actin molecules in the cytoskeleton and stabilizes polymerized actin microfilaments. These results suggest that the compound may disrupt cytokinesis in dividing cancer cells.

A second promising line of research is in the development of antiviral drugs from various types of algae. Compounds extracted from the algae have shown activity against a range of viruses, including herpes simplex viruses and human immunodeficiency virus (HIV). Although a variety of compounds have been tested against HIV, most of the research has been done with sulfated polysaccharides (polysaccharides with SO_4 groups attached) extracted from red and brown algae. These compounds have been able to block HIV replication in cell culture at very low concentrations with no toxicity to the host cells. Research has shown that carrageenan from the cell walls of several red algae can block replication of human rhinoviruses, which cause the common cold. This study, conducted in cultured nasal epithelial cells, suggests that treatment of nasal membranes with a carrageenan-based product may be effective in preventing and treating rhinovirus infections. Also, several lectins (carbohydrate-binding proteins) from cyanobacteria and algae have high anti-HIV activity. It is thought that these proteins may interfere with the attachment of HIV to the host cell membrane and therefore may prevent HIV infections. Based on this mechanism, it has been suggested that algal lectins can be developed into a novel and effective HIV therapy.

Sulfated polysaccharides from red, brown, and green algae have also shown anticoagulant properties. Forty species of red algae, 60 species of brown algae, and 45 species of green algae have been investigated as anticoagulants and for promoting the lysis of blood clots. Some of these compounds have been chemically characterized. Fucoidan from the brown alga *Fucus vesiculosus* has been extensively studied and is being clinically tested for therapeutic use. It is thought that algal anticoagulants may add a new dimension to the management of blood clots and vascular disorders. Many other algae also produce fucoidans, and their therapeutic uses are being investigated.

the hepatotoxins are microcystins, a group of over 80 peptide toxins that differ slightly in structure. Microcystins are produced by a number of cyanobacteria with *Microcystis aeruginosa* the most frequent species associated with toxin production. Within liver cells, these toxins inactivate certain enzymes and disrupt the cytoskeleton; this causes cell death and can ultimately cause liver failure. Microcystins are responsible for the deaths of livestock and domestic pets; however, there are no reports of human fatalities from ingestion of microcystins. Although these toxins do not directly cause cancer, studies have shown microcystins may stimulate the growth of cancer cells. Researchers have suggested that microcystins may be responsible for the high rates of liver cancer in parts of China, where blooms occur frequently.

In August 2014, news media across the United States reported on the toxic algal bloom in Lake Erie near Toledo, Ohio. For 3 days, Toledo residents were warned not to drink tap water due to contamination with microcystins. Algal blooms in Lake Erie are not new; in fact, in 2011 a massive algal bloom covered 20% of the surface of Lake Erie. While the Toledo Water Treatment Division was able to neutralize the toxin within a few days, the underlying problem of fertilizer runoff into Lake Erie continues.

Researchers have developed immunological tests to quickly find out if cyanobacterial toxins are present and genetic

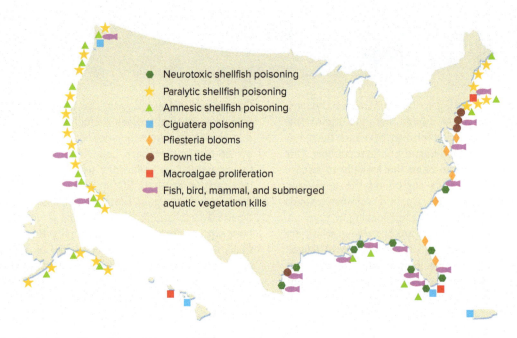

Figure 22.16 Outbreaks of harmful algal blooms in U.S. coastal areas.

tests to determine if the cyanobacteria present are genetically capable of producing toxins. These efforts should help ensure the safety of water supplies and aquaculture resources.

Less deadly cyanobacterial toxins are responsible for conditions such as "swimmer's itch." Toxins of several marine cyanobacteria cause this form of contact dermatitis, characterized by swelling and inflammation of mucous membranes and redness of the skin.

Red Tides

Although cyanobacterial toxins have caused major problems, much of the attention on harmful algal blooms has focused on dinoflagellates in the marine environment, where red tides are a growing threat (fig. 22.15 and 22.16). Under certain environmental conditions, dinoflagellates undergo a population explosion, becoming so abundant (up to 20 million organisms per liter) that they color the water red, orange, yellow, brown, or any hue in between.

Several species of dinoflagellates, especially *Karenia brevis,* produce powerful toxins that can cause massive fish kills. As schools of fish swim through a bloom, the dinoflagellates may be disrupted or killed, releasing neurotoxins into the water. The toxins damage the gills or suppress heart rate and result in asphyxiation. This massive kill has frequently occurred in the Gulf of Mexico, with hundreds or thousands of dead fish washing up on beaches.

Shellfish can accumulate, but not be affected by, these dinoflagellate toxins; this makes the shellfish poisonous to humans, other mammals, and birds. It is possible for a single clam to accumulate sufficient toxin to kill a human. The shellfish poisonings have been categorized as paralytic, diarrhetic, and neurotoxic. Each of these types is caused by a different species of dinoflagellate and different toxins. In addition, various studies indicate that aerosolized brevetoxins, produced by *K. brevis,* are associated with respiratory illness in humans. The brevetoxins in the seawater become incorporated into sea foam and bubbles in breaking waves. As the bubbles burst, the toxins become airborne and have been reported up to 1 mile inland. Symptoms include chest tightness, bronchoconstriction, and cough. A recent study found a significant increase in emergency room visits for respiratory complaints during red tides in Sarasota, Florida. People with asthma are especially susceptible; another study showed an increase in symptoms and a decrease in lung function after a 1-hour visit to the beach during red tide events. Red tides in coastal areas of the United States and other countries have caused tremendous economic losses to the shellfish industry, tourism, and fishing.

Pfiesteria

During the 1990s, a new dinoflagellate problem was recognized in the coastal waters off the mid-Atlantic and southeastern states. *Pfiesteria piscicida, P. shumwayae,* and related dinoflagellates have been implicated as the organisms responsible for major fish kills and fish disease. It has been estimated that since the early 1990s toxins produced by *Pfiesteria* may have killed a billion fish in North Carolina alone. These heterotrophic dinoflagellates normally feed on bacteria and algae but begin producing toxins from certain environmental stimuli, including the presence of excretion products from fish. Many types of fish have been affected by the toxins, which appear to stun the fish and to injure the fish skin, causing them to lose the ability to maintain their internal osmotic balance. As the

A CLOSER LOOK 22.2

Killer Alga—Story of a Deadly Invader

From zebra mussels infesting the Great Lakes to the kudzu vine smothering the U.S. South, biological invasions by exotic species are an escalating threat to biodiversity around the globe. Invasive species can displace native ones and irrevocably alter ecosystems. Most imported species either die or have limited growth in their new home, but a small percentage of the newcomers are opportunists whose growth is dangerously out of control.

Invasive species can be animals, plants, fungi, or microbes, but they often share certain characteristics. The invaders usually grow fast and reproduce rapidly and profusely. There are usually no checks to this unrestricted growth because few or no natural predators or diseases exist in the new locale. The invaders are often extremely adaptable and can live in a range of environmental conditions. Often, ecosystems that are already stressed by pollution or other types of disruption are especially subject to invasions. Some exotic species appear to be innocuous for decades and then suddenly become invasive when a favorable environmental change, such as increased rainfall or warmer temperatures, facilitates a population explosion.

Introductions of invasive species may be either accidental or deliberate. Larvae of the zebra mussel were carried along in the ballast of a European ship. When the water was later discharged into a port in the Great Lakes, the zebra mussel invasion began. The importation of infested Asian chestnuts to New York introduced the blight-causing fungus that led to the loss of the American chestnut tree from the eastern forests of the United States (see Chapter 6). Kudzu was at first heralded as an ornamental vine with lovely, fragrant flowers that could tolerate heat and provide cooling shade during the hot southern summers. Tamarisk, or saltcedar, is a tree that was introduced as a windbreak to control erosion in the western United States. It now covers 500,000 hectares (1,235,500 acres), crowding out native cottonwood and willow trees that provide food and shelter for bighorn sheep and many birds.

Consider the case of *Caulerpa taxifolia*, a green alga within a group known as the siphonous algae because they have a tubular organization with no internal cell walls or specialized cells. The entire alga is constructed like a green garden hose through which runs a continuous stream of cytoplasm. Most of the hose is taken up by a large central vacuole that pushes the cytoplasm and its contents of chloroplasts, mitochondria, and other organelles against the outer cell wall. Most cells are microscopic, but the cells of some of the siphonous algae can reach a length of 2.5 meters (8 feet). They are coenocytic, with millions of nuclei flowing through the cytoplasm.

Although *Caulerpa* lacks specialized cells, it has an organization that resembles the structure of land plants (box fig. 22.2a). There is a stolon, a horizontal axis that creeps along the substrate bottom, fronds that resemble pinnately compound leaves, and rhizoids that anchor the alga.

The story of how *C. taxifolia* became known as the killer alga began in 1989, when Dr. Alexandre Meinesz, a phycologist at the University of Nice in France, learned that this exotic alga was thriving in the Mediterranean waters off the coast of Monaco. Upon investigation, Meinesz discovered that this alga, which is native to tropical seas, was dumped into the Mediterranean with other aquarium refuse by the prestigious Oceanographic Institute of Monaco.

Eventually, the *C. taxifolia* was traced back to a stock developed at the Wilhelmina Zoologisch-Botanischer Garten in Stuttgart, Germany. This stock of *C. taxifolia* caught the attention of amateur and professional aquarists for its exceptional beauty and hardiness and became widely distributed in public and private aquaria in the 1980s.

Instead of dying in the cold winter waters of the Mediterranean Sea, as the tropical *C. taxifolia* does, this mutant variety can withstand several months at temperatures that range between 10° and 13°C (50° and 55°F). In addition, the alga grows equally well on a variety of ocean substrates—sand, mud, or rocks—and from near the surface to depths of 50 meters (150 feet). When temperatures exceed 18°C (64°F), the mutant *Caulerpa* grows at a remarkable rate, capable of elongating by 2 cm (0.8 in.) in a single day and forming a new frond every 2 days. A single square meter (square yard) of seafloor can be matted with 5,500 leafy fronds. All of this growth is asexual. Unlike the tropical form, this form of *C. taxifolia* produces only male gametes in the Mediterranean and thus replicates only by fragmentation. Just a small piece can regenerate and colonize an area. In fact, it is believed that fishing nets and boat anchors as well as the ripping action of storms have helped the alga spread to new areas.

With such a rapid rate of growth, the Mediterranean *Caulerpa* grows quickly over the sea bottom (box fig. 22.2b),

Box Figure 22.2 *Caulerpa taxifolia*. (a) Close-up showing the fronds, stolon, and rhizoids. (b) The alga covers the sea bottom as a dense mat of green.

rapidly blanketing all other sea life, such as marine angiosperms, native algae, corals, and sea fans. This living Astroturf cuts off vital essentials, such as sunlight for photosynthesis and currents for filter feeding of nutrients. The area becomes a *Caulerpa* lawn while species diversity is lost.

An additional twist is that *C. taxifolia* is inedible to most marine herbivores because of the presence of toxic compounds, primarily caulerpenyne. With no predator to inhibit its phenomenal growth, the killer alga had encompassed the Mediterranean coasts of Monaco, France, Italy, Spain, and the Croatian coast on the Adriatic Sea, covering 30,000 hectares (more than 74,000 acres) by 2002. *Caulerpa* infestations have also been found in southern Australia, in estuaries near Sydney and Adelaide.

Various methods have been tried to eradicate this deadly invader. Divers have been employed to rip the alga out by hand, but care must be taken because even a small fragment is capable of colonizing new areas. Black plastic tarps have been placed over the alga in an attempt to deprive it of life-giving sunlight. Chemicals such as copper, a conventional algicide, and toxic levels of salt have also been applied to *Caulerpa* fields. There has been limited success if the area of invasion is small.

One hopeful avenue is biological control. In biological control, a disease or predator of the invasive organism is introduced and allowed to wreak havoc. In the native range of the tropical *C. taxifolia*, two species of sea slug are immune to its poisons and feed on the alga exclusively. They suck the cytoplasm out of the fronds and incorporate the algal cholorplasts into their skin. The chloroplasts provide camouflage as the snails feed in the *Caulerpa* fields and continue to photosynthesize, providing additional nutrition for the slugs. Dr. Meinesz is suggesting the importation of these slugs to the Mediterranean as the only way to eliminate the immense area that *Caulerpa* now covers. But many experts fear that another exotic importation may lead to more environmental damage.

There are more than 70 species of *Caulerpa*, and California state legislators have recently banned the importation of *C. taxifolia* and eight other species of *Caulerpa*. Many researchers would prefer that the whole genus be banned because of the difficulty in distinguishing between invasive and harmless species of *Caulerpa*, but aquarium hobbyists lobbied against this action.

Caulerpa taxifolia was first identified in the coastal waters of California in a saltwater lagoon north of San Diego during the summer of 2000. From a single aquarium discard, *Caulerpa* had multiplied, carpeting roughly 1,500 square meters of the lagoon. Another population was later discovered further north in Orange County. To eradicate the alga, workers covered the populations with tarps and pumped in liquid chlorine.

Twenty years after the detection of *Caulerpa* in the coastal waters of southern California, there is no sign of the alga. Prompt action and routine monitoring has successfully eradicated the only known invasion of *Caulerpa* in the Western Hemisphere—for now.

skin is destroyed, open, bleeding sores develop, and even hemorrhaging can occur. This process often leads to death within hours. When the fish become incapacitated, *Pfiesteria* feed on the epidermal cells, blood, and other fluids that leak from the open sores. Since the initial descriptions from North Carolina and Maryland, these organisms have been reported in many other countries. In 2014, *Pfiesteria* and a related dinoflagellate caused extensive fish kills in two fish farms in Denmark.

Several people who worked with laboratory cultures of *Pfiesteria* reported suffering from adverse health effects. In addition, fishermen, water-skiers, and other people using waterways where fish kills occurred reported various symptoms, including rashes, skin lesions, headache, eye irritation, and memory loss. By contrast, a 4-year study of commercial fishermen in Maryland found no neurological impairment or other symptoms, even though *Pfiesteria* was detected in some of the water samples collected from Chesapeake Bay, where the fishermen worked.

Thinking Critically

Blooms by toxin-producing algae have been increasing in recent years, causing serious problems in both marine and freshwater environments.

Speculate on the evolutionary value of this toxin-producing ability.

Other Toxic Algae

Another dinoflagellate phenomenon is ciguatera poisoning, the leading cause of nonbacterial food poisoning in the United States, Canada, and Europe. Many species of tropical fish accumulate ciguatoxin, a dinoflagellate toxin, in their fatty tissues. Although the fish themselves are unaffected, people eating the tainted fish experience nausea, vomiting, and diarrhea and develop bizarre neurological symptoms that can be long lasting. People have reported reverse sensations; for example, hot objects feel cold to the touch.

Other algae that can form toxic blooms include several species of diatoms. *Pseudo-nitzschia* spp. produce domoic acid, a toxin that causes amnesic shellfish poisoning, which is characterized by both gastrointestinal and neurological symptoms and can cause fatalities. In 1987, four Canadians died after eating contaminated mussels from Prince Edward Island. Nontoxic diatom blooms have also been linked to fish kills. Salmon have been killed by blooms of *Chaetoceros*. These diatoms are characterized by long spines, which apparently lodge in the gill tissues of the fish and trigger the release of massive amounts of mucus, which cause suffocation.

Most coastal areas of the United States have been affected by one or more of these harmful algal blooms (fig. 22.16), and the situation is similar in other parts of the world. This growing threat is the focus of intensive studies as well as efforts to reduce the pollutants that may be responsible for the blooms.

Environmental problems have also developed where larger algae have become invasive. *Codium fragile* is a green seaweed native to Japan that has spread to many parts of the world. It was first noted in Holland in 1900 and has since spread throughout coastal areas in Europe. *Codium* was reported on the east coast of North America in 1957 in New York and can now be found from South Carolina to Nova Scotia. It is a weedy species that exhibits rapid growth, high rates of dispersal, and a range of tolerance to environmental conditions. *Codium* has displaced many of the kelp beds along the East Coast, with resultant changes in the ecosystems. A more recent introduction is the red alga *Grateloupia turuturu* (*G. doryphora*) that is native to the Pacific coast. This seaweed was first identified along the Rhode Island coast in 1996 and is rapidly spreading along the coast both northward and southward. A more notorious invasive species is described in A Closer Look 22.2: Killer Alga—Story of a Deadly Invader.

CHAPTER SUMMARY

1. The algae are a diverse group of photosynthetic organisms that occur in fresh water, oceans, and moist terrestrial environments where they are the base of the food chain. They range from unicells to huge, multicellular organisms. They differ from land plants in many features. In particular, they lack extensive tissue differentiation and do not retain an embryo after sexual reproduction.

2. The cyanobacteria, also known as blue-green algae, are the only group of algae with prokaryotic cells. Some cyanobacteria are able to carry out nitrogen fixation. Diatoms and dinoflagellates are typically unicellular organisms. They serve a vital role in both marine and freshwater food chains as the food for other organisms. Red and brown algae, commonly called seaweeds, include large, multicellular marine organisms, often with complex life cycles. The greatest diversity of form is found among the green algae, which also play a major role in aquatic food chains. The green algae are the most closely related to all land plants.

3. Various algae have been used as food by cultures around the world. Although some cyanobacteria have proved useful as food or food supplements, it is the red and brown seaweeds that provide products for both food and industrial applications. Some seaweeds are used directly as food; others are used as sources of alginic acid, carrageenan, and agar. Algae are also being actively investigated as a source of biofuels.

4. A negative aspect of the algae is related to environmental damage caused by algal blooms. Algal blooms are especially hazardous when the algae are capable of producing toxins. Dinoflagellates are probably the best known toxin producers, and blooms of these organisms are called red tides.

REVIEW QUESTIONS

1. What criteria are used to distinguish among the various groups of algae?
2. Describe the economic importance of the various groups of algae.
3. Describe the life cycles in green and brown algae.
4. Discuss the ecological importance of the algae.
5. How are algal blooms related to oxygen depletion?
6. The term *seaweed* refers to which groups of algae?
7. Which groups of algae are known to produce toxins?
8. What are the distinguishing features of the cell walls of diatoms?
9. Describe the industrial uses of algal polysaccharides.
10. Why are algae considered a potentially important source of biofuels?

CHAPTER

23

Fungi in the Natural Environment

©Estelle Levetin

A spectacular assortment of fleshy fungi collected during a foray in southern Rhode Island.

KEY CONCEPTS

1. Fungi are a diverse group of heterotrophic organisms with an absorptive type of nutrition; they generally have a threadlike body and reproduce by spores.
2. The majority of fungi are saprobes, actively recycling nutrients in the environment by decomposing organic matter; however, the ecological importance of the symbiotic fungi is also well known.
3. Some plant diseases caused by fungi have had devastating influences on crop plants, resulting in major impacts on human society.

CHAPTER OUTLINE

Fungi and Fungal-like Organisms 420
Fungal-like Protists 420
The Kingdom Fungi 422
    Characteristics of Fungi 422
    Classification of Fungi 424
Role of Fungi in the Environment 433
    Decomposers—Nature's Recyclers 433

A CLOSER LOOK 23.1 Lichens: Algal-Fungal Partnership 434

    Mycorrhizae: Root-Fungus Partnership 435

A CLOSER LOOK 23.2 Dry Rot and Other Wood Decay Fungi 436

Plant Diseases with Major Impact on Humans 437
    Late Blight of Potato 438
    Rusts—Threat to the World's Breadbasket 439
    Corn Smut—Blight or Delight 441
    Dutch Elm Disease—Destruction in the Urban Landscape 441
    Sudden Oak Death—Destruction in the Forest 443
Chapter Summary 445
Review Questions 445

As a group, the fungi are among the most abundant organisms in the world, yet for most people they are probably the least understood group. The term *fungus* often elicits memories of moldy basements or grocery store mushrooms without any awareness of their ecological importance or the tremendous impact these organisms have on society. Fungi are vital to the environment as decomposers or as partners in symbiotic relationships. They provide lifesaving antibiotics, fermented foods, and beverages; however, they are also responsible for devastating crop losses and debilitating human disease.

FUNGI AND FUNGAL-LIKE ORGANISMS

Organisms that were traditionally called *fungi* are now known to represent multiple lines of evolution within the domain Eukarya. The true fungi—those organisms in the main line of fungal evolution—are classified in the kingdom Fungi. In addition to the true fungi, several groups of fungal-like organisms have been included in the kingdom Protista for many years. Recall from Chapter 9 the kingdom Protista includes many diverse lineages, and a number of biologists have split the protists into several smaller kingdoms. Because the evolutionary relationships among these organisms are still uncertain, we will retain the term *protists* (kingdom Protista) only as a convenient tool to discuss these fungal-like organisms.

One feature that the true fungi and fungal-like protists share is their mode of nutrition. Unlike plants that contain chlorophyll and carry out photosynthesis, fungi and fungal-like organisms are **heterotrophic**; that is, they depend on external sources of organic material for food. Although they share this characteristic, there are many features that distinguish the true fungi from the fungal-like protists. These include molecular evidence from DNA sequences, ultrastructural features, biochemical pathways, and the presence of chitin in the cell walls of true fungi (table 23.1). Even though fungal-like organisms are no longer included in the kingdom Fungi, they are still studied by mycologists, and we will follow this practice in the current chapter. We will also follow the practice of mycologists and use the term *phylum* instead of division for the taxonomic rank below kingdom.

FUNGAL-LIKE PROTISTS

There are several groups of fungal-like protists; however, only two of these groups will be considered in this chapter, the slime molds and the water molds. These are two best-known groups of fungal-like organisms, and they are classified in the phyla Myxomycota and Oomycota, respectively.

Table 23.1
Major Groups of Fungi and Fungal-like Organisms

Phylum	Morphological Types	Sexual Spores	Asexual Spores	Cell Wall Structural Material
Fungal-like Protists				
Slime molds (Myxomycota)	Plasmodium, myxamoeba, and swarm cells	Spores produced in a fruiting body	None—myxamoeba and swarm cells may divide	Cellulose—only on spores and sclerotia
Water molds (Oomycota)	Primarily mycelium with nonseptate hyphae	Oospores produced in an oogonium	Zoospores	Cellulose
Fungi				
Zoosporic fungi (Chytridiomycota) (Blastocladiomycota)	Unicells; simple mycelium with nonseptate hyphae	Zygote only	Zoospores	Chitin
AM fungi (Glomeromycota)	Mycelium with nonseptate hyphae	None identified	Spores	Chitin
Zygomycetes (formerly Zygomycota)	Mycelium with nonseptate hyphae; yeasts	Zygospore (zygosporangium)	Sporangiospores	Chitin
Ascomycetes (Ascomycota)	Mycelium with septate hyphae; yeasts	Ascospores produced within an ascus, often in a fruiting body	Conidia	Chitin
Basidiomycetes (Basidiomycota)	Mycelium with septate hyphae; yeasts	Basidiospores produced from a basidium, often in a fruiting body	Conidia or other specialized spores	Chitin

Slime Molds

Slime molds in the Myxomycota are organisms that have affinities to both animals and fungi. The amoeboid feeding stage known as the **plasmodium** shows animal-like characteristics in actively engulfing bacteria and organic matter in the environment. The reproductive stage, however, shows fungal-like characteristics because the spores have cell walls and are produced within sporocarps (also known as fruiting bodies). Although several groups of organisms are similar, this discussion will be limited to the plasmodial slime molds, or the myxomycetes.

Slime molds produce unicellular spores that may remain dormant for long periods of time. When environmental conditions are favorable, the spores germinate and give rise to either a flagellated cell known as a swarm cell or an amoeboid cell known as a myxamoeba. The swarm cells and myxamoebae can ingest food, grow, and divide. Under unfavorable conditions, they can also encyst, forming spherical dormant cells that germinate when favorable conditions return. Eventually, the swarm cells or myxamoebae function as gametes. Compatible gametes fuse to form a zygote that is amoeboid. The diploid nucleus undergoes repeated nuclear divisions, forming a motile, multinucleate plasmodium. The plasmodium, which is frequently brightly colored, ingests food, grows, and crawls over the substrate, often reaching macroscopic size. The plasmodium stage is the main feeding stage of the organism, and its slimy appearance gives rise to the common name. Under unfavorable conditions, the plasmodium is capable of forming a hard, dry sclerotium, another resistant structure. Plasmodia eventually give rise to sporangia or similar sporocarps (fig. 23.1). Within the developing sporangia, meiosis and spore formation occur.

Slime molds frequently appear in cool, damp areas and can be found on moist soil and decaying vegetation, where they act as scavengers feeding on bacteria, other microorganisms, and organic matter. Myxomycetes are found in all terrestrial habitats, although they are most diverse in forest biomes. They have been considered a curiosity by many biology students. The sudden appearance of the miniature fruiting bodies occurring on decaying wood, or the plasmodium creeping over tree stumps, damp ground or grass, has surprised many woodland explorers unfamiliar with these fascinating organisms.

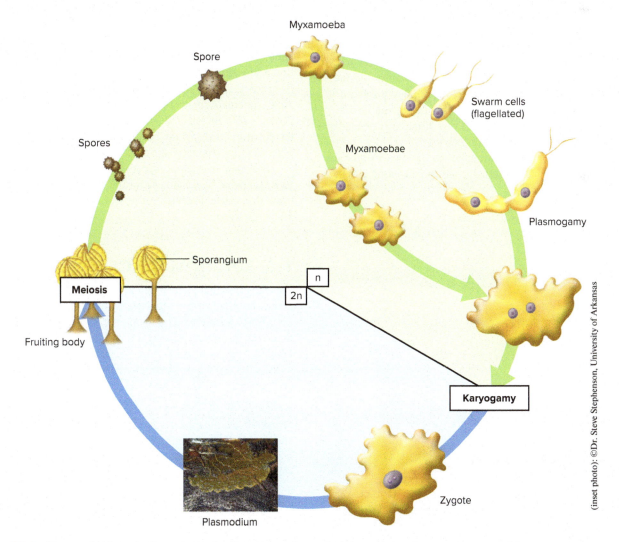

Figure 23.1 Slime mold life cycle. Inset shows the slimy plasmodium, which gives rise to the common name of these organisms.

Water molds

Organisms in the Oomycota are commonly referred to as water molds or oomycetes. Although many of these protists do occur in water, some are devastating plant pathogens and still others are soilborne **saprobes.** Saprobes are organisms that obtain nutrients from nonliving organic material or the remains and by-products of organisms. Although oomycetes and true fungi are unrelated, they share some physical traits. Like fungi, oomycetes generally have a filamentous body composed of threadlike **hyphae.** The hyphae may branch extensively, and the collective mass of hyphal filaments is referred to as a **mycelium.** In the water molds, hyphae are **nonseptate,** lacking cross-walls or **septa** (sing., **septum**). As a result multiple nuclei exist in a common cytoplasm (fig. 23.2). Septa usually occur at the base of reproductive structures.

Oomycetes also have a nutritional mode like fungi. Both groups are **absorptive heterotrophs,** producing enzymes that are secreted into the environment and break down complex organic compounds into smaller molecules. These small molecules are then absorbed by the fungus or oomycete. Despite these similarities, oomycetes have many biochemical, ultrastructural, and phylogenetic differences that separate them from fungi. The cell wall of oomycetes contains cellulose, in contrast to chitin in the cell wall of fungi.

The characteristic sexual structure of oomycetes is the **oogonium,** which contains one or more **oospores** that result from fertilization. Another feature of this group is the presence of **zoospores,** motile asexual spores that are produced in sporangia (also called zoosporangia) and dependent on water for dispersal. These zoospores have two different flagella; one flagellum is a whiplash and the other a tinsel. Under certain conditions, however, some oomycetes are known to produce nonflagellated propagules (actually one-spored sporangia) that are dispersed by wind. These fungi can be found in aquatic habitats as well as damp soil.

The life cycle of organisms in this phylum will be represented by the genus *Saprolegnia,* a common saprobic water mold; a few species of *Saprolegnia* can be pathogenic to fish. *Saprolegnia* forms a branched, nonseptate mycelium and can complete its entire life cycle submerged (fig. 23.3). Biflagellate zoospores are produced in an elongated sporangium that has a cross-wall at the base. A pore develops at the tip, through which the mature pear-shaped spores are released. After a short period of swimming, each zoospore encysts (withdraws its flagella and surrounds itself with a thin wall). The encysted state is maintained for several hours. When the zoospore reemerges, it appears bean-shaped, with laterally inserted flagella. Again, the zoospore swims about for a short time, and then reencysts. The cyst germinates immediately to produce a mycelium, thus completing the asexual phase of the life cycle.

The sexual stage of the *Saprolegnia* life cycle includes the production of male and female organs, or **gametangia** (structures housing gametes). Meiosis occurs during the development of the gametangia. The female gametangium, or oogonium, is spherical, with a cross-wall at the base. Within the oogonium, the protoplasm divides into 5 to 10 uninucleate oospheres, or eggs. Nearby hyphae develop that grow toward and press themselves on the surface of each oogonium. The ends of these hyphae become delimited by cross-walls as multinucleate **antheridia.** Specialized threadlike hyphae develop from each antheridium and penetrate the oogonium and an oosphere. A male nucleus from the antheridium passes through this thread and enters each oosphere; the ensuing fertilization results in the production of thick-walled, resistant oospores, the characteristic structure of the phylum.

Included in the Oomycota are some well-known plant pathogens that have had a major impact on society, such as the fungi that cause late blight of potato, downy mildew of grape, and white rusts.

THE KINGDOM FUNGI

The kingdom Fungi includes a large diversity of eukaryotic organisms that form a monophyletic group. Although at one time fungi were aligned with plants, molecular studies show that they are more closely related to animals. More than 120,000 species of fungi have been identified, and estimates of the total number of fungi exceed 1.5 million.

Characteristics of Fungi

Morphologically, fungi range from small, inconspicuous yeasts and molds to large, conspicuous mushrooms, puffballs, and bracket fungi. Microscopic observation of their structure shows that fungi exist as single cells, such as yeast, or, far more commonly, as threadlike hyphae. *Yeast* is a general term to describe unicellular fungi that reproduce by budding. Although many people think that the only yeasts are those involved in baking bread or brewing beer, over 1,500 different

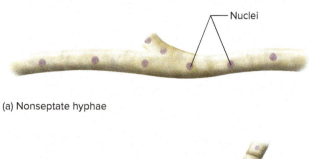

(a) Nonseptate hyphae

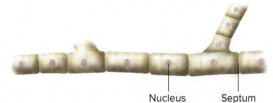

(b) Septate hyphae

Figure 23.2 Fungal hyphae can be (a) nonseptate, with multiple nuclei existing in a common cytoplasm, or (b) septate, with cross-walls (septa) dividing the hyphae into many short cells.

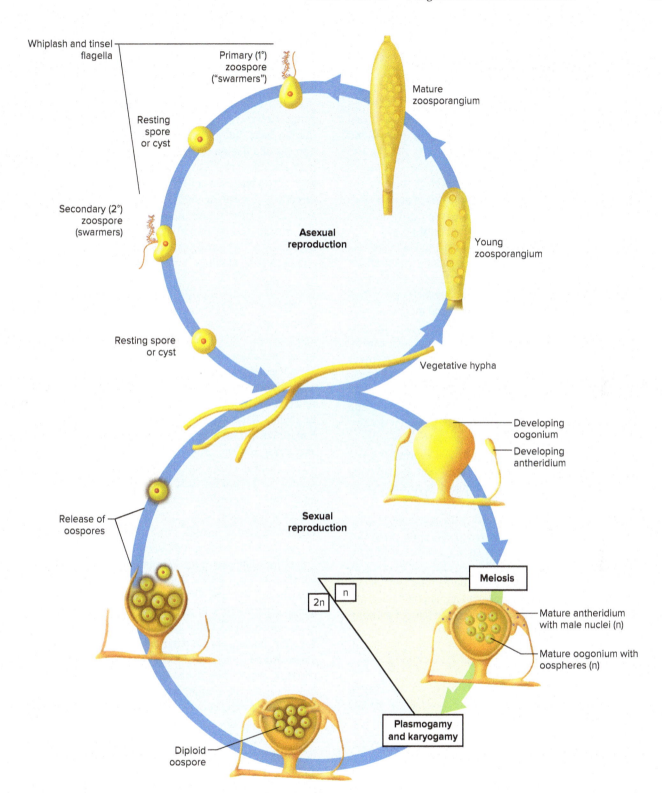

Figure 23.3 Life cycle of *Saprolegnia* in the Oomycota.

species of morphologically similar yeasts can be found in several phyla (table 23.1).

Depending on the phylum each hypha may be **septate**, divided by cross-walls into many short cells, or nonseptate (fig. 23.2). Among the fungi with septate hyphae, each cell may contain one nucleus or two genetically different nuclei, depending on the fungal group and the stage of development. Hyphae with two genetically different nuclei within each cell are called **dikaryons**. Although individual hyphae are microscopic, the whole mycelium is often visible to the naked eye. The mycelium may also develop highly specialized reproductive structures, such as mushrooms, brackets, or puffballs. These are actually compact masses of tightly interwoven hyphae and are the visible portion of an extensive mycelium within the substrate. Mycelia may be growing in countless substrates, such as underground, throughout a decaying log, or within a living tree.

Reproduction in the fungi is usually carried out through the formation of **spores,** which may result from either sexual or asexual processes. Many fungi undergo both asexual and sexual reproduction at different stages in their life cycle; others carry out just one of these reproductive processes. The sexual phase of the life cycle is known as the **perfect stage** or the teleomorph, and the asexual phase is called the **imperfect stage** or the anamorph. Life cycles in many fungi are highly complex and vary considerably among the different fungal phyla.

Spores differ in size, shape, color, and method of formation, but they are always microscopic, with unicellular to multicellular types. Spores may be formed from the breakup of hyphae or on highly specialized hyphal branches often contained within **fruiting bodies.** Mushrooms, puffballs, brackets, and morels are well-known examples of these specialized fruiting bodies. In most groups of fungi, spore characteristics are important elements for identification and classification. The majority of fungal spores are dispersed by the wind, with various types of discharge mechanisms occurring within different groups of fungi. As a result, fungal spores are a normal component of the atmosphere and are present in large numbers any time that the ground is not covered with ice and snow. From spring through fall, it is possible to have spore concentrations exceeding 100,000 spores per cubic meter of air. This airborne transport is significant in the spread of plant disease as well as in the suffering of people allergic to certain types of fungal spores.

One feature that the fungi share with plants and oomycetes is the presence of cell walls. The fungal wall, consisting of fibrils embedded in a matrix, is largely composed of polysaccharides (often more than 90%) but also contains significant amounts of protein and lipid. The fibrillar component of the cell wall is **chitin,** a nitrogen-containing polysaccharide. The matrix, on the other hand, contains a variety of carbohydrates (glucose polymers) and proteins.

Classification of Fungi

In general, classification of the fungi is based on the method of sexual reproduction. Sexual spores are the result of genetic recombination and normally follow **karyogamy** (fusion of two haploid nuclei into a diploid zygote) and meiosis. These processes occur in specialized cells, and in some groups, the sexual spores form within the fruiting bodies. In most sexually reproducing organisms, **plasmogamy** (the fusion of cytoplasm from two cells or gametes) is followed immediately by karyogamy. However, in many fungi, plasmogamy occurs early in the life cycle, but karyogamy is delayed, leading to the development of a dikaryon. Asexual spores result from mitosis and may also be characteristic of certain groups. Asexual spores may form enclosed within a sporangium or be produced directly by hyphae or modified hyphae without any enclosing wall. These latter spores are known as **conidia.**

Within the kingdom Fungi, the zoosporic fungi, AM fungi, zygomycetes, ascomycetes, and basidiomycetes (table 23.1) will be described here.

Zoosporic Fungi

Within the kingdom Fungi, the phyla Blastocladiomycota and Chytridiomycota have motile cells and are often referred to as zoosporic fungi. In older classification systems, the Blastocladiomycota were considered an order within the Chytridiomycota; however, molecular data indicated they reflect a separate lineage of zoosporic fungi and were elevated to phylum status. Zoosporic fungi are primitive organisms and include both unicellular and mycelial fungi that occur in aquatic and moist habitats. Many of these fungi exist as parasites of algae, vascular plants, animals, and even other fungi. Saprobic species are known for their ability to degrade cellulose, chitin, and keratin. Asexual reproduction is by zoospores with a single, posterior flagellum. Sexual reproduction is usually by the fusion of gametes, and in many of these fungi both gametes are motile.

A well-studied genus of Blastocladiomycota is *Allomyces*, a saprobe that shows an alternation of haploid and diploid generations (fig. 23.4). Both generations develop a sparsely branched mycelium. During the haploid stage, specialized gametangia produce motile gametes that are released into the environment. The male gamete is smaller and more active than the female gamete. After fertilization, the united gametes swim around briefly, then settle down. The nuclei fuse and the resulting zygote gives rise to the diploid generation. Zoospores are produced in sporangia on the diploid mycelium; these asexual spores repeat the diploid stage and allow for many generations of the diploid mycelium. The diploid generation also produces thick-walled, resistant sporangia that can remain dormant for many years. When environmental conditions are suitable, meiosis occurs in this structure and the resulting haploid zoospores give rise to the haploid mycelium.

Batrachochytrium dendrobatidis in the Chytridiomycota has gained considerable scientific attention over the past two decades. It is believed this fungus is responsible for the extinction or population declines of frogs and salamanders around the world. The fungus infects the epidermal cells, causing thickening of the skin and interfering with respiration, since gas exchange occurs at least partially through the skin in amphibians. In 2018 researchers identified the original source of the pathogen was the Korean peninsula. Researchers suggested

> **Thinking Critically**
>
> Although traditionally considered nonvascular plants, fungi are now classified in their own kingdom (kingdom Fungi) or as fungal-like protists. Current molecular studies even show that fungi are more closely related to animals than to plants; however, certain features of fungi seem plantlike.
>
> *Which characteristics of fungi are animal-like, and which are plantlike?*

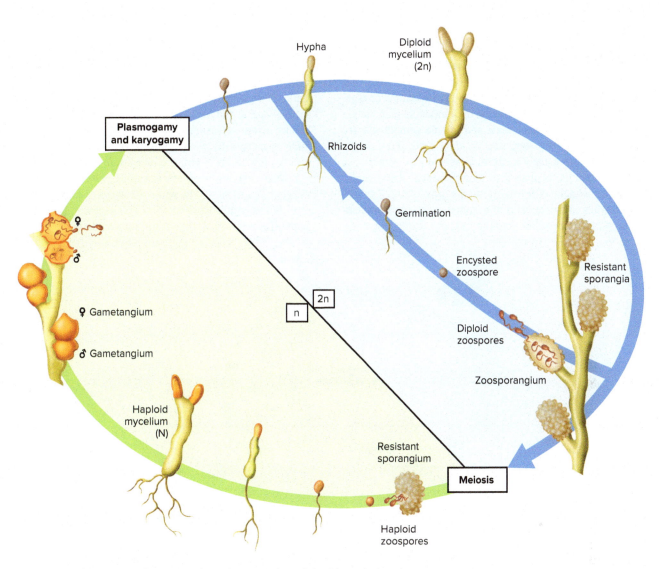

Figure 23.4 Life cycle of *Allomyces arbusculus*, a member of the Blastocladiomycota.

that early in the twentieth century the global expansion in the amphibian trade transmitted the pathogen to many areas. They also established that East Asia is a hotspot for *B. dendrobatidis* biodiversity, and new strains of the pathogen are still being spread from this area through the trade in amphibians. Also in 2018, a separate report provided some encouraging information. Another group of researchers found that populations of some frog species in Panama appeared to be recovering after near extinction 10 years earlier. The researchers determined that there was a change in the antimicrobial skin secretions in the surviving frogs, and this improved their defenses against the chytrid pathogen. Although the fungus is still present, it appears to do less damage. The researchers believe that this has caused a shift to a state where these few frog species and pathogen can co-exist.

AM Fungi

Fungi in the Glomeromycota are symbionts involved in mutualistic relationships with land plants and are known as **arbuscular mycorrhizae** (AM) fungi (see the Mycorrhizae: Root-Fungus Partnership section in this chapter). For AM fungi, the relationships with the host plants are obligate; they are unable to grow without the plants. AM fungi have nonseptate hyphae and produce large, multinucleate, asexual spores in the soil surrounding the host plant; these spores can remain dormant in the soil for long periods of time. No known sexual stage has been found in these fungi.

The Glomeromycota is the smallest fungal phylum, including only about 288 species in 29 genera. In spite of the small number of species, AM fungi are among the most abundant fungi in the environmen. In some classification systems, the AM fungi are included with the zygomycetes; additional molecular evidence should clarify the relationships of these groups.

Zygomycetes

Zygomycetes are a diverse group of fungi that were previously classified in the phylum Zygomycota. However, molecular evidence indicates that the fungi in the Zygomycota represent multiple evolutionary lines. Therefore, mycologists have recommended that the Zygomycota be split into two or more separate phyla. Because the evolutionary relationships in this group of fungi are still being clarified, we will use the common name zygomycete for the present. Zygomycetes are characterized by thick-walled sexual spores, called **zygosporangia** or **zygospores,** that develop from the fusion of compatible gametangia on hyphal tips. No distinct male and female gametangia are recognizable; however, opposite mating types are clearly present. These are usually designated as + and − strains. As compatible haploid hyphae grow toward each other, multinucleate gametangia differentiate at the tips of the hyphae. The end walls dissolve, and the contents of the gametangia mix with compatible haploid nuclei, fusing together.

The resulting zygosporangium is a multinucleate, thick-walled, resistant structure.

Asexual spores, known as **sporangiospores,** are produced within a sporangium, which develops on a specialized erect stalk, the sporangiophore, that is separated from the hyphae by a cross-wall at the base. At maturity, the thin sporangial wall breaks down, releasing thousands of spores, which are normally dispersed by wind. These spores germinate to develop into the mycelium of either + or − strains.

Many zygomycetes are saprobes, which can occur in a variety of substrates. The genus *Rhizopus,* a common zygomycete, is frequently isolated from soil samples and can be an aggressive and bothersome laboratory contaminant (fig. 23.5).

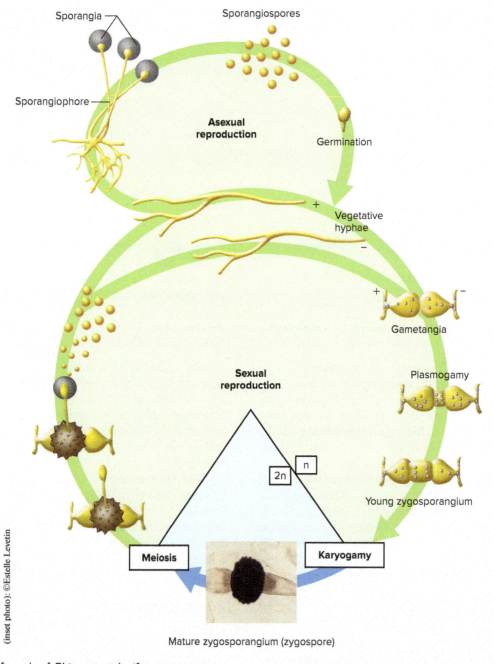

Figure 23.5   Life cycle of *Rhizopus stolonifer*, a zygomycete.

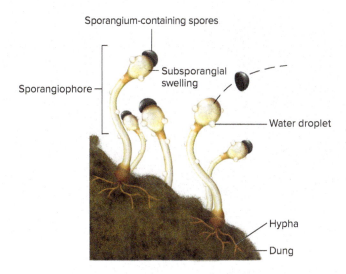

Figure 23.6 Asexual reproduction in *Pilobolus* is highly specialized. The entire sporangium is "shot" away from the sporangiophore toward the light. Discharged sporangia adhere to nearby vegetation.

Rhizopus can occur on stored fruits and vegetables, such as strawberries, tomatoes, and sweet potatoes.

One remarkable genus in this group is *Pilobolus*, commonly found on the dung of herbivores and possessing a unique dispersal mechanism for its sporangia (fig. 23.6). The sporangium is borne on top of a subsporangial swelling (an enlargement of the sporangiophore tip). When the spores are mature, turgor pressure builds up in the subsporangial swelling until the sporangium is explosively shot away from the sporangiophore at a speed of about 14 meters (45 feet) per second. The sporangium travels up to 2 meters (6 feet) and adheres to nearby vegetation. The explosive discharge shoots the sporangium away from the "ring of repugnance," which usually keeps herbivores from feeding too close to their feces. When grazing animals eat the vegetation, the attached sporangia pass unharmed through the digestive tract and emerge with the feces. There, the spores germinate, the hyphae grow, and new sporangia develop, completing an unusual life cycle.

Ascomycetes

Members of the Ascomycota are characterized by the production of sexual spores, **ascospores,** within a saclike structure called an **ascus** (pl., **asci**). Within the ascus, karyogamy and meiosis occur, followed by an additional mitotic division to produce eight haploid ascospores. Members of this phylum are commonly called ascomycetes. Both the morphology of the ascus and the type of fruiting body **(ascocarp)** are important features for identification of these fungi. The Ascomycota is the largest phylum of fungi with over 65,000 species.

Ascomycetes range from simple yeasts, in which a naked ascus is produced by fusion of two compatible cells without a fruiting body, to the cup fungi and morels, in which asci line the depressions of the fruiting body (figs. 23.7a–c).

The mycelium of these fungi is typically branched and septate, with individual cells initially uninucleate. When hyphae of different mating types become closely intertwined, male antheridia may form on one and female **ascogonia** (sing., **ascogonium**) on the other. Male nuclei pass from the antheridium into the ascogonium; however, karyogamy does not follow immediately. Ascogenous hyphae develop from the ascogonium with two compatible nuclei within each cell. The terminal cell of each dikaryotic hypha becomes an ascus, eventually giving rise to eight ascospores. In many ascomycetes, the asci form in a layer, known as the **hymenium,** that lines the base of the ascocarps (fig. 23.7d).

In most ascomycetes, the ascospores are shot from the ascus at maturity. High osmotic pressure develops within the ascus, causing it to swell. Soon, the ascus tip ruptures or opens, and the spores are explosively forced out into the atmosphere.

Asexual reproduction in mycelial ascomycetes is accomplished by the production of conidia, either singly or in chains. These spores, which are rapidly and abundantly produced, are generally dependent on airborne dispersal. In the yeasts, asexual reproduction is usually achieved by **budding:** a small outgrowth from the parent cell slowly balloons out and eventually becomes pinched off (see fig. 24.1).

Neurospora crassa, a common saprobic ascomycete, has a long history of use as a genetic tool (fig. 23.8). The asexual stage produces conidia that develop in chains (this stage is often called red bread mold, a troublesome contaminant in bakeries), and sexual reproduction results in the production of flask-shaped ascocarps known as **perithecia** (sing., **perithecium**). Genetic studies on *Neurospora* by George Beadle and Edward Tatum in the 1930s established the one gene–one enzyme hypothesis, which states that the function of a gene is to specify the production of a specific protein. In 1958, Beadle and Tatum received the Nobel Prize in Physiology for this landmark work in molecular biology.

Dutch elm disease, chestnut blight, and powdery mildews are among the devastating plant diseases caused by members of the Ascomycota. Economically useful members include the baking and brewing yeasts as well as the morels and truffles often considered among the choicest edible fungi. While morels have an aboveground ascocarp (fig. 23.7b), truffles develop underground with no visible signs on the surface. In Europe, these delicacies are collected using trained dogs and pigs that locate the scent of the underground fruiting body (fig. 23.9).

One ascomycete that has received a great deal of interest in the past few years is *Pseudogymnoascus destructans* (also known as *Geomyces destructans*). This species is associated with white nose syndrome, a condition responsible for the deaths of over 6 million hibernating bats. The syndrome was first described in February 2006 from a cave west of Albany, New York, and by July 2018 it had spread to caves in 33 states in the eastern United States through the Great Plain and the state of Washington. The fungus has also been detected in three more states, but no infected bats.

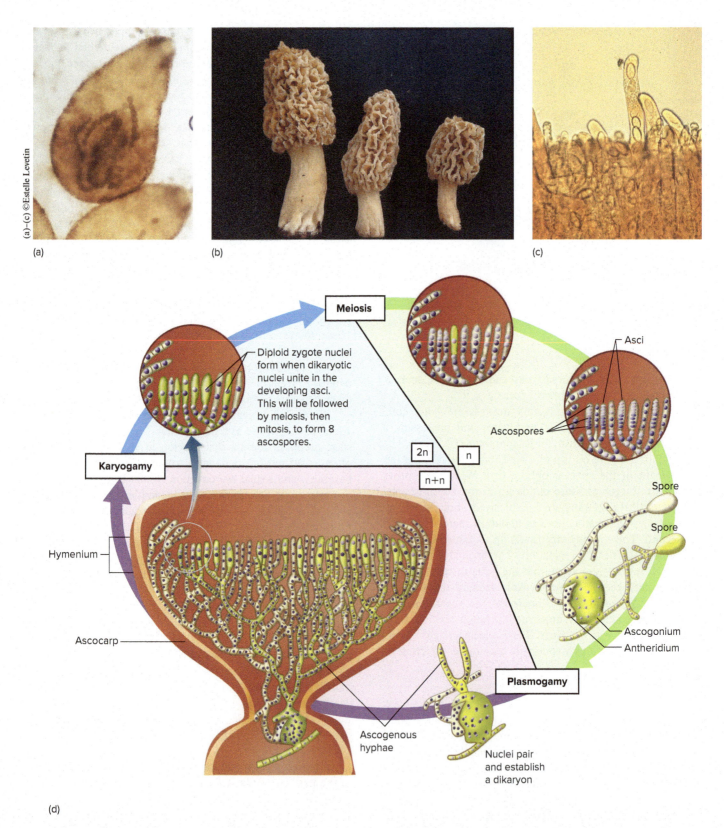

Figure 23.7 Various types of ascocarps occur in the Ascomycota. (a) Small, flask-shaped perithecium of *Sordaria*, ×150, within which ascospores and asci are visible. (b) Large apothecium of morel (*Morchella esculenta*). (c) Asci with ascospores line the depression of the morel, ×300. (d) Life cycle of a member of the Ascomycota.

Figure 23.8 *Neurospora crassa*, an ascomycete used in genetic studies. (a) Asexual stage produces chains of conidia. (b) Ascocarp showing asci and ascospores.

Figure 23.9 Truffles are the most costly fungal delicacies. (a) In France, truffle hounds are trained to sniff out the underground fruiting bodies. (b) Close-up of black truffles, *Tuber melanosporum*.

In addition, the disease been reported from caves in seven Canadian provinces. Bats with white nose syndrome have a white growth of fungal hyphae on their muzzles, wing membranes, and ears (fig. 23.10). They are affected during their hibernation and are severely weakened, although it is not known for certain how the fungus causes the death of infected bats. Some research suggests that the infection increases the bats' metabolism during hibernation, resulting in a depletion of the fat reserves. As a result the bats may wake early or starve to death. The syndrome has caused over 90% mortality in at least two bat species; several other species have been less severely affected. The ecological implications of the widespread loss of bats need to be considered, since insect-eating bats normally keep insect populations in check. There are signs that some bat species may be adapting, because mortality rates in some areas seem to decrease after the first couple of years.

The ascomycetes also include a large number of mycelial fungi we commonly call *molds*. These fungi are among the most abundant in the environment and can readily be isolated from air and soil samples and from many substrates, indoors or outdoors. At one time, the sexual stages (perfect stage) of all these fungi were unknown, and the classification was based on the morphology and development of conidia (the asexual stage). These fungi were previously called deuteromycetes or fungi imperfecti. Over the past 25 years, DNA sequencing has resulted in the correct classification within the Ascomycota even without the sexual stage. As a result, the terms *deuteromycetes* and *fungi imperfecti* have been abandoned.

Figure 23.10 Little brown bat with white muzzle, indicative of white nose syndrome. This disease is caused by the fungus *Pseudogymnoascus destructans* and has resulted in the death of over 5.5 million bats in North America.

Several well-known fungi within this group are illustrated in Figure 23.11. Among the common molds included here is the genus *Penicillium* (fig. 23.11a), a widespread genus that naturally occurs in the soil and on leaf litter. Species of *Penicillium* can also be readily isolated from moldy fruit or vegetables and often are the source of the "moldy" smell from basements, wet carpets, or old shoes. However, some species of this genus are quite beneficial and are the source of the antibiotic penicillin (see Chapter 25); other species are used in the production of cheeses. *Penicillium roquefortii* is responsible for the unique taste, color, and flavor of blue, Roquefort, Stilton, and similar cheeses. *Penicillium camembertii* is used in the production of Camembert and Brie, two additional varieties of cheese; enzymes produced by this fungus break down the curds of milk and develop their characteristic buttery consistency.

Aspergillus (fig. 23.11b) is a common fungal genus that is found worldwide. In nature, species of *Aspergillus* typically occur in the soil, growing on decaying plant material; however, other species of *Aspergillus* are human and plant pathogens. Several species of *Aspergillus* have been implicated in aspergillosis (a serious lung infection) and in sinus infections (see Chapter 25), while other species can infect important food crops before and after harvest. Also, some *Aspergillus* species produce toxic metabolites under certain environmental conditions, resulting in contaminated foods (see Chapter 25).

By contrast, some species, such as *Aspergillus oryzae*, are used in food processing to manufacture soy sauce, miso, and other fermented foods.

Cladosporium (fig. 23.11c) is a cosmopolitan mold that is readily isolated from leaf surfaces, soil, and decaying plant material. It is a weak plant pathogen, and it can also be found growing on many indoor substrates. Spores of fungi in this genus are the most common airborne spores in temperate parts of the world.

Alternaria (fig. 23.11d) is also a widespread genus that is found in soil, leaf surfaces, and decaying plant material. Species of *Alternaria* are common plant pathogens that cause leaf spot and leaf blight diseases on a number of crops and ornamental plants. *Alternaria* produces large, distinctive, multicellular conidia, which may be one of the most easily recognized fungal spores.

Basidiomycetes

The Basidiomycota includes mushrooms, puffballs, and bracket fungi (fig. 23.12), the largest and most conspicuous fungi in the environment, as well as rusts and smuts, two groups of plant pathogens that lack fruiting bodies.

The fleshy fungi in this phylum are relatively familiar to the average person and have been extensively studied since antiquity because of the edible or poisonous nature of many mushrooms (see Chapters 24 and 25). These **basidiocarps** (fruiting bodies) are the visible portion of an extensive mycelium within the substrate, often extending over large stretches. In 1992, mushrooms made headlines across the United States with the discovery of a vast area on the Michigan-Wisconsin border where a population of *Armillaria bulbosa* that covered 15 hectares (37 acres) was found to be a single organism. Studies showed that all the mushrooms in the population were genetically identical, indicating that they were all basidiocarps of a single mycelial colony. It was estimated that the colony weighed between 100 and 1,000 tons and was at least 1,500 years old based on observed growth rate, placing it among the oldest and largest living organisms. Since that discovery, larger colonies have been identified. To date, the largest has been a colony of *Armillaria ostoyae* (honey mushroom) found in eastern Oregon. The colony has been estimated at 956 hectares (2,364 acres) in size and at least 2,400 years old.

The characteristic sexual spore in the basidiomycetes (the common name for this group) is the **basidiospore,** which forms externally on a **basidium** (pl., **basidia**). Each basidium bears four spores on peglike appendages (fig. 23.13). Karyogamy and meiosis occur within the basidium, and the four haploid products of meiosis give rise to the four basidiospores.

Basidia line the gills of mushrooms and the pores of bracket fungi, with the basidiospores exposed to the atmosphere throughout development. When the spores are mature, they are actively shot away from the basidium. Although the ballistics of discharge are not as dramatic as in other fungi, this mechanism ensures that the spores are freed from the

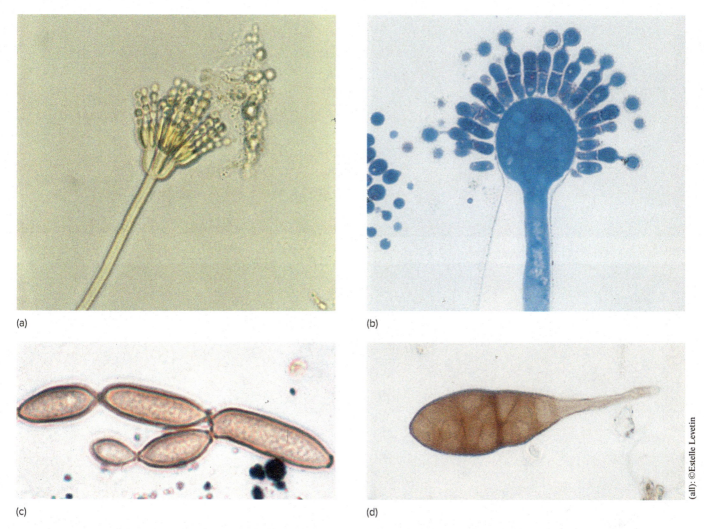

Figure 23.11 Asexual ascomycetes are some of the most common fungi in the environment, and many are easily identified by the morphology of the conidia and the conidiophore. (a) *Penicillium* has a branched, brushlike, conidiophore and small, spherical spores produced in chains, ×550. (b) *Aspergillus* conidiophore has an inflated vesicle at the top, ×650. (c) *Cladosporium* produces branching chains of oval to elongate conidia that easily separate for dispersal, ×1,200. (d) *Alternaria* conidia are large, deeply pigmented spores with a beak at the apex, ×1,200.

gills or pores and reach the surrounding air currents for wind dispersal.

If the basidiospore lands on a suitable substrate, it will germinate, giving rise to a haploid mycelium. This stage of the life cycle is of limited duration because the haploid mycelium must fuse with the haploid mycelium of a compatible mating type to establish a dikaryon before an extensive mycelium can develop. The dikaryon is the major stage in the life cycle and even includes the hyphae of the fruiting body. Fusion of the compatible nuclei occurs only in the basidium and is followed immediately by meiosis to form haploid basidiospores (fig. 23.13).

In puffballs, the basidia develop within the fruiting body, exposing the basidiospores to the atmosphere only when fully mature. These spores are passively dispersed in clouds or puffs as raindrops strike the fruiting body. Strong gusts of wind or small animals impacting the surface accomplish the same puffing action. The numbers of spores produced by puffballs, as well as by mushrooms and brackets, are astronomical (table 23.2), contributing to the atmospheric spore load.

In contrast to the airborne dispersal of spores from these fruiting bodies, basidiospores produced by stinkhorns are dispersed by flies (fig. 23.12d). The mature spores are immersed in a slimy material with a foul odor that attracts flies. While visiting the fetid fruiting body, the flies eat some spores and pick up others on their body, thus playing an important role in the fungal life cycle.

Asexual spores are also formed by basidiomycetes, especially the rusts and smuts, two groups of plant pathogens in which asexual spores are the major dispersal units. The life cycles and impact of these pathogens are discussed later in this chapter.

Figure 23.12 Basidiocarps are the most conspicuous and familiar fungal fruiting bodies and include (a) mushrooms, *Armillariella tabescens*, (b) bracket fungi, *Laetiporus sulphureus*, (c) puffballs, *Lycoperdon perlatum* releasing spores, and (d) stinkhorns, *Phallus indusiatus* (also known as *Dictyophora indusiata*).

Table 23.2
Basidiospore Numbers

Fungus	Total Number of Spores	Duration of Spore Release	Number of Spores per Day
Calvatia gigantea (giant puffball)	7×10^{12}		
Ganoderma applanatum (artist's bracket)	5.5×10^{12}	6 months	3×10^{10}
Polyporus squamosus (bracket fungus)	5×10^{11}	14 days	3.5×10^{10}
Agaricus bisporus (commercial mushroom)	1.6×10^{10}	6 days	2.6×10^{9}
Coprinus comatus (shaggy-mane mushroom)	5.2×10^{9}	2 days	2.6×10^{9}

Source: Modified from A. H. R. Buller. *Researchers on Fungi*, Vol. 2. 1922. Longmans, Green & Co., Ltd., London.

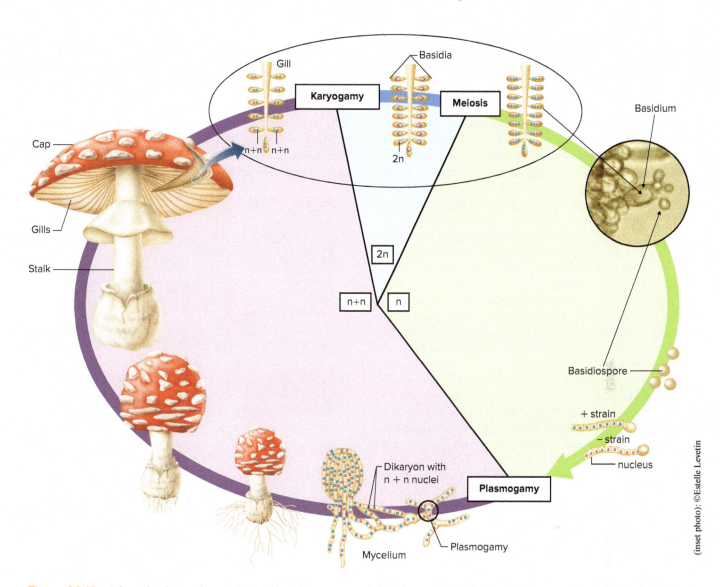

Figure 23.13 Life cycle of a mushroom. Haploid basidiospores result from karyogamy and meiosis that occur in basidia that line the gills of the mushroom. On a suitable substrate, a basidiospore will germinate and give rise to a short-lived haploid mycelium. Compatible strains will pair to establish the dikaryon. The mycelium produced by the dikaryon is often very extensive, producing mushrooms when environmental conditions are right.

ROLE OF FUNGI IN THE ENVIRONMENT

Recall that fungi are absorptive heterotrophs that obtain organic material from their environment. As such, they can exist as **parasites** (obtaining nutrients from a living host and ultimately harming that host) or **mutualistic symbionts** (obtaining nutrients from another organism while providing some benefit to that organism; see A Closer Look 23.1: Lichens), but the majority of species and the most abundant in the environment are **saprobes.**

Decomposers—Nature's Recyclers

Many saprobic species are found in the soil or as leaf-surface microorganisms, where they exist on organic matter produced by the plant. Among the fungi are species that are capable of decomposing all the carbon compounds that occur in other organisms, including cellulose, lignin, and keratin. In the environment, fungal saprobes play an essential role as decomposers, degrading organic materials and recycling nutrients.

The line between saprobes and parasites is not always clear-cut; some fungi are able to utilize both lifestyles under certain conditions. Facultative parasites (also known as opportunistic pathogens) can live as saprobes but will invade living tissue when suitable hosts are available. These fungi can be readily isolated from air and soil samples.

Soil is usually considered the most important environment for fungi; many species spend at least part of their life in or on the soil. It is known that 1 gram of soil may contain over 100,000 fungal spores.

When nutrients are available and environmental conditions are suitable, spores germinate and hyphae grow and branch within the substrate, typically producing a colony that eventually forms more spores. The types of fungi and their abundance in an area depend on the availability of nutrients,

A CLOSER LOOK 23.1

Lichens: Algal-Fungal Partnership

Lichens are dual organisms, generally consisting of an alga or a cyanobacterium (phycobiont) and a fungus (mycobiont) in a mutualistic relationship. These organisms can survive in the harshest environments: on bare rocks, in hot deserts, in the frozen Arctic. Through photosynthesis, the phycobiont produces food for both species, even though it constitutes only 5% to 10% of the total lichen body. The mycobiont obtains water and minerals, forms a complex vegetative body, and develops both sexual and asexual reproductive structures.

The overall appearance of lichens is described as foliose (leaflike), fruticose (shrubby), or crustose (crustlike) (box fig. 23.1). Many lichens produce specialized asexual propagules known as soredia, which are small groups of algal cells intertwined with fungal hyphae. Wind, rain, and animals can all act as dispersal agents for these soredia, which eventually establish a new lichen colony.

There are approximately 18,000 species of lichens, which are named and classified by the mycobiont. Interestingly, the phycobiont in more than 90% of all lichens belongs to just three genera: *Trebouxia* and *Trentepohlia*, which are green algae, and *Nostoc*, a cyanobacterium. (In fact, *Trebouxia* is the algal component in 70% of all lichens.) By contrast, thousands of species of fungi occur in lichens. These fungi are generally believed to be unable to complete their life cycle without their phycobiont. Ninety-eight percent of these fungi are ascomycetes. An obvious reproductive structure visible on many lichens is a cup-shaped ascocarp typical of certain ascomycetes.

Lichens have incredibly slow growth rates, expanding only a few millimeters each year. Existing colonies often have great longevity, with some believed to be 4,500 years old. Although tolerant to extremes of temperature and drought, lichens are very sensitive to air pollution, especially sulfur dioxide. Scientists have even used the disappearance of certain lichen species as a means of measuring the extent of pollution within an area. In fact, biomonitoring through the use of lichens is becoming mandatory in several European countries. In urban areas in which environmental regulations have led to a decrease in air pollution levels, lichen populations are slowly recovering.

In the Arctic, caribou and reindeer feed on a lichen species known as reindeer moss, which is the dominant vegetation in some areas. If other food is not available, these animals can exist on a purely lichen-based diet. Although lichens have not been used as food for humans, extracts from some lichens have been used medicinally as antibiotics.

Many lichens are brightly colored and for centuries were the source of dyes used by native peoples throughout the world (see A Closer Look 18.2: Herbs to Dye For). Although the use of lichen dyes has largely been replaced by synthetic dyes in industrial societies, lichens are still used in the manufacture of Scottish tweeds. Lichens are also used in the production of litmus paper, the acid–base indicator widely used in chemistry labs.

(a)

(b)

(c)

Box Figure 23.1 Various growth forms that are common in lichens. (a) Crustose lichen growing on a rock. (b) Fruticose lichen on a branch. (c) Foliose lichen on the bark of a tree.

water, and suitable temperatures. Saprobic fungi are especially versatile and able to utilize many different substrates for growth.

In recent years, there has been an interest in using fungi in bioremediation of toxic materials. **Bioremediation** involves the use of microorganisms to reclaim soil and water that have become contaminated with hazardous materials. This method of removing pollutants is quite inexpensive when compared with conventional physical or chemical methods of decontamination. Bioremediation is typically performed on-site and often requires only the addition of the appropriate nutrients to the soil to stimulate the growth of microorganisms in the immediate environment. Other approaches call for the addition of other microorganisms that are not typically found in the area. Much of the bioremediation currently taking place has focused on the cleanup of petroleum-contaminated soils.

Although bacteria have been considered prime players in the remediation of petroleum contamination, recently more and more attention has been given to the white-rot fungi. Some of the bracket fungi in the Basidiomycota are white-rot fungi. These wood-rotting fungi are able to degrade both lignin and cellulose; however, a few species of white-rot fungi can selectively degrade the lignin with little or no loss of cellulose. These fungi are being proposed for use in the pulp and paper industry to reduce the use of hazardous chemicals (see A Closer Look 23.2: Dry Rot and Other Wood Decay Fungi). The enzymes that degrade lignin have been the focus of intense research, and many of the genes have been identified and cloned. Studies have shown that these enzymes are also able to degrade a variety of other complex molecules, including many environmental toxins and petroleum hydrocarbons. At this time, research has focused on just a few fungi, especially on *Phanerochaete chrysosporium*, which has been described as a model system for studying degradation of petroleum compounds. Continuing research in this area is currently identifying other fungi with these abilities as well. As bioremediation becomes a more broadly used approach for environmental cleanup, fungi will be widely utilized because of their diverse enzyme systems and ability to degrade complex molecules.

Mycorrhizae: Root-Fungus Partnership

Scientists estimate that 90% of all vascular plants possess fungi in mutualistic associations with their roots. This fungus-root association is termed **mycorrhizae** (literally, "fungus root"). The plant furnishes the fungus with sugars and amino acids, while the fungus aids in the absorption of minerals, especially phosphorus, and water from the soil. The fungal hyphae grow far from the roots to utilize soil nutrients and water, which are then shared with the plant. This mutualistic association between plants and fungi has a long evolutionary history. Fossil evidence indicates that some of the earliest plants that colonized land more than 400 million years ago had mycorrhizal fungi associated with them. Although early land plants were photosynthetic, they lacked extensive root systems. The mycorrhizal relationship was clearly an asset in obtaining water and minerals.

Different types of mycorrhizae occur. In one common type, the **ectomycorrhizae (ectotrophic mycorrhizae)**, the fungus forms a sheath or mantle around the root, penetrating between the cells of the root epidermis and cortex (figs. 23.14a and b). It is estimated that over 5,000 species of fungi form ectomycorrhizal relationships, typically on forest trees in temperate areas. Some trees, such as pine, oak, and beech, absolutely require the association for normal development.

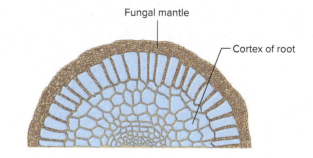

(a)

(b)

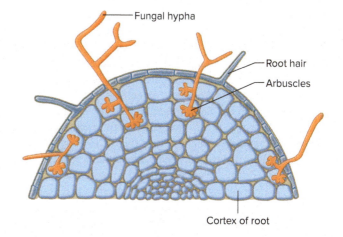

(c)

Figure 23.14 Fungi can establish mutualistic relationships with the roots of plants. (a) In ectomycorrhizae, the fungus forms a sheath around the root, penetrating between the cells of the epidermis into the cortex. (b) Close-up of ectomycorrhizae, showing the thickened, clublike roots due to the fungal sheath around the roots. (c) In arbuscular mycorrhizae, the hyphae grow through the soil and into the roots. The hyphae grow between cells in the cortex and form arbuscles within the cells.

A CLOSER LOOK 23.2

Dry Rot and Other Wood Decay Fungi

Although the components of wood, especially lignin, are known for their durability and resistance to decay, certain specialized fungi are able to decompose cellulose and lignin. Many of these produce large fruiting bodies in the form of brackets, shelves, or crusts on the trunks and branches of infected trees (box fig. 23.2). Saprobic species, which are essential for the decay of dead trees, produce similar fruiting bodies. The evident fruiting body reflects the vegetative mycelium within the tree, often extending 1 to 3 meters (3 to 10 feet) above or below the fruiting body. The mycelium of some fungi may grow within the tree for many years before the fruiting body develops on the bark or from the roots. Until the fungus is identified, it is usually not possible to know whether the fruiting body is that of a pathogen that just killed the tree or a saprobe that is merely hastening the decay.

Fungi can selectively attack either the heartwood or the sapwood. Sapwood rot and destruction eventually kill the tree. When the heartwood is attacked, the result is heart rot, which seriously weakens the tree and makes it susceptible to being blown over. The timber value of the tree is also greatly reduced. Two types of heart rot are known: brown rot and white rot. Brown rot is caused by fungi that decompose the cellulose and leave the brownish lignin relatively untouched. White rot is caused by fungi that degrade both the cellulose and the lignin, leaving the wood a whitish color. A few white-rot species are even able to selectively degrade the lignin with little or no loss of cellulose.

There are at least 1,600 species of wood-rotting fungi responsible for the destruction of valuable timber and shade trees as well as fallen and rotting trees. It is estimated that the damage caused by wood-rotting fungi amounts to $9 billion per year. In North America, approximately 10% of the annual timber harvest is destroyed by wood-rotting fungi, which cannot discriminate between rotting logs, valued timber, and stored lumber.

One economically important wood-rotting fungus is *Serpula lacrymans* (*Merulius lacrymans*), which is responsible for dry rot of timber. The name *dry rot* refers to the dry, crumbly appearance of decayed wood, but it is somewhat misleading because the fungus needs moisture to become established. In fact, dry rot occurs only in wood exposed to moisture. The species breaks down cellulose, leaving the brown, crumbling wood with many cracks. Ships and older homes are especially susceptible to this destructive fungus, which often starts in poorly ventilated areas and spreads. The fungus is difficult to eradicate once it becomes established. When a timber or support beam is infected, it must be replaced, along with adjacent beams.

Dry rot was the blight of the British navy for hundreds of years. In 1684, Samuel Pepys, who was Secretary to the Admiralty Board, reported on the poor ventilation and dampness as well as the "toadstools" growing in the ships. Although fungi were a well-recognized symptom of decay, at the time no one realized they were actually responsible for the decay. Inadequately seasoned timber and lack of

---

The fungal partner (**mycobiont**) of these ectomycorrhizae is usually a member of the Basidiomycota, and many woodland mushrooms are the reproductive structures of these fungi. In fact, some mushroom families are exclusively mycorrhizal.

Another category of mycorrhizae are **endomycorrhizae (endotrophic mycorrhizae)**. In this type, the fungus grows into the cortex of the root. There are actually several types of endomycorrhizae, of which **arbuscular mycorrhizae (AM)** are the most prevalent. These have also been called vesicular-arbuscular mycorrhizae. The mycobionts in these relationships are members of the Glomeromycota. Spores of these fungi are found in most soils but germinate only when close to roots of a compatible host. After germination, the hyphae grow through the soil and into the roots. Roots colonized with AM fungi are not apparent to the naked eye and must be stained and microscopically examined for the fungus to be seen. The hyphae grow between cells in the cortex and form branching structures called arbuscles within the cells (fig. 23.14c); exchange of nutrients occurs through these arbuscles. Arbuscular mycorrhizae occur on a wide range of herbaceous plants and on trees; it has been estimated that 300,000 plant species have AM fungi.

Mycorrhizal networks are hyphal connections between the mycorrhizal fungi of two or more plants. The plants can be individuals of the same species, but even distantly related plants can be linked by mycorrhizal networks. These networks occur in all terrestrial ecosystems and are formed by AM fungi as well as by ectomycorrhizal fungi. These networks transfer carbon compounds and mineral nutrients from plant to plant; often nutrients are moved from mature plants to seedling to aid seedling establishment. In addition, allelochemicals (see Chapter 21) as well as defensive compounds may be transferred from one plant to others. Research on mycorrhizal networks over the past 20 years has led to the understanding that these networks have important roles in ecosystem function.

Scientific interest in mycorrhizal fungi has increased in recent decades as their ecological role has become clearer. Their importance in forestry and agriculture has been recognized as attempts are made to revegetate areas abused by human activity.

ventilation created ideal conditions for fungal growth. British ships often rotted at anchor, and the designation "Rotten Row" was used for the warships at the docks. During the American Revolution, so many British ships were damaged by fungi that it has been suggested the outcome of the war may be partially attributed to dry rot. This problem continued until ironclad warships were introduced in the 1860s.

The presence of resins or tannins in the heartwood makes some trees more resistant to fungal decay than others. Tannins, which may be present in amounts as high as 30% in certain woods, are especially useful in helping trees resist fungal decay. Redwood, cedar, bald cypress, and Douglas fir are the most resistant gymnosperms; black locust, black walnut, catalpa, chestnut, and red mulberry are resistant angiosperms. Wood from these species is, therefore, useful for outdoor furniture, wooden decks, fences, and other articles exposed to the elements. Today, treatment of nonresistant wood with preservatives, such as creosote, also prevents attack by wood-rotting fungi. Creosote is obtained from the distillation of coal and blast-furnace tar and is used for railroad ties, telephone poles, pilings, and other wood products.

The biotechnical potential of wood-rotting fungi has recently been highlighted. Testing is being done on the efficacy of pretreating wood chips with fungi that selectively degrade the lignin for pulp and paper production. Results indicate that the process improves paper strength while requiring 68% less energy than standard pulping techniques. Also, the ability of these fungi to bleach wood is being considered as a substitute for the pollution-causing chemicals now used to bleach pulp for paper production. Further uses of wood-rotting fungi include the breakdown of toxic chemicals. The enzymes that help break down lignin can also degrade an assortment of environmental pollutants such as petroleum hydrocarbons, pesticides, industrial dyes, and chlorinated hydrocarbons. As knowledge of these lignin-degrading fungi increases, it is likely that other industrial applications will develop.

Box Figure 23.2 *Ganoderma applanatum,* a widely distributed wood-rotting fungus.

For over 15 years, mycorrhizal inoculants have been available commercially for use in revegetation, farming, forestry, and home gardening. Although testing can be done to determine the most suitable mycobiont for each plant species, commercial inoculants generally contain mixtures of several species of AM fungi or mixtures of AM and ectomycorrhizal fungi. Even when a plant species can grow without mycorrhizae, the same species with mycorrhizae will be more productive and may be more tolerant of pollution, need less fertilizer, or grow in marginal soils.

Thinking Critically

In the environment, saprobic fungi play a vital role as decomposers, degrading most types of organic material.

What happens to the nutrients that are released by decomposition?

PLANT DISEASES WITH MAJOR IMPACT ON HUMANS

There are many examples of plant diseases caused by fungi and oomycetes that have made a major impact on society and have even changed human history. In fact, over 70% of all major crop diseases are caused by fungi or oomycetes. Recall the discussion of late blight of potato (see Chapter 14), when 25% of the population of Ireland was lost due to the ravages of *Phytophthora infestans.* During the 1840s, over 1 million people died from starvation or famine-related diseases, and more than 1.5 million emigrated from Ireland.

Similar disasters have occurred at other times in human history. A more recent epidemic that resulted in large-scale famine was caused by *Cochliobolus miyabeanus* (also known as *Bipolaris oryzae*), the fungus responsible for brown spot of rice. Symptoms of the disease include brown spots on the leaves and glumes of the rice plant. When the infection occurs at the seedling stage, the young foliage is severely damaged, the plants develop poorly, and the grain yield is significantly reduced. The fungus can also infect the seed during development, resulting in poor germination and weak seedlings. The most severe outbreak of this disease resulted in the great Bengal famine of 1943, when 2 million people died from starvation. A related fungus, *Bipolaris maydis* (also known as *Cochliobolus heterostrophus*), which attacks corn and causes southern leaf blight, resulted in a widespread epidemic in the United States in 1970. About 15% of the total U.S. corn crop was lost, with yields in some states reduced 50% (see Chapter 12).

Although various plant diseases had been recognized since ancient times, they were not connected with pathogenic microorganisms until the nineteenth century. Early in that century, Benedict Prevost, a French scientist, reported that wheat bunt was caused by a smut fungus, but his findings were not accepted by the scientific community, which still believed in spontaneous generation. It had been generally accepted that fungi, other microorganisms, maggots, and the like appeared spontaneously on diseased and dying plants. They were considered the result, not the cause, of the condition. During the investigations of the potato blight, German botanist Anton de Bary, in 1861, proved experimentally that a fungus, *P. infestans*, was the cause of the disease. (*Phytophthora infestans* was originally described as a fungus.) After de Bary's work, other plant pathogenic fungi were described (pathogenic bacteria and viruses were not identified until later in the nineteenth century), and plant pathology (the study of plant diseases) soon became an important field of study.

Even with the many advances in twentieth-century sciences, plant diseases still pose a major threat, and the world's food supply is vulnerable. Plant pathologists wage a constant battle to stay ahead of newly evolving strains of pathogenic fungi and oomycetes that are capable of destroying crop plants.

Late Blight of Potato

The potato ranks as one of the world's major food crops, but the continued utilization of this crop may be limited by the cost of disease control measures. Potatoes are attacked by many pathogens and pests, and the total amount of agricultural chemicals applied to the potato crop is greater than for any other food crop. Late blight of potato has been the most important disease of potato since the 1840s, when it caused the destruction of the potato crop in Ireland and the resulting widespread famine (see Chapter 14).

Phytophthora infestans, the pathogen responsible for late blight of potato (and tomato), is a member of the Oomycota. It occurs wherever potatoes are grown, and all potato cultivars are susceptible, although some have moderate resistance. The organism invades host tissue, resulting in the rapid death of infected parts. Productivity of the potato plants is greatly reduced, and tuber destruction frequently occurs as well. Sporangia are produced on aerial hyphae that grow out of the stomata. The sporangia detach and are dispersed by wind, possibly carried for many kilometers. At low temperatures (10°–15°C; 50°–59°F) and high humidity, sporangia germinate by producing numerous zoospores; at higher temperatures, each sporangium gives rise to a single germ tube that develops into hyphae. As a result, during cool, wet periods, the production of zoospores leads to an astonishingly rapid spread of the disease. A blighted field can be destroyed within a couple of days. Cool, wet conditions existed during the 1840s in Ireland, resulting in widespread devastation. Similar disease progression also occurs in infected tomato fields. Even today, without fungicidal protection, this pathogen can cause havoc in these crops.

Sexual reproduction in the Oomycota results in the formation of oospores (see fig. 23.3). In some oomycetes, such as *Saprolegnia*, compatible male and female gametangia develop on a single strain. In other oomycetes, two compatible strains are needed for sexual reproduction. *Phytophthora infestans* falls in this second category and requires compatible mating types, known as A_1 and A_2.

Phytophthora infestans can overwinter as mycelium in infected tubers, but in areas where both mating types occur, it can also overwinter in the soil as dormant oospores. When the pathogen was introduced to Europe and the United States in the 1840s, only one mating type (A_1) was introduced. In 1950, the second mating type (A_2) was identified in central Mexico, the native home of the oomycete. The A_2 mating type was confined to Mexico until 1980, when it spread throughout Europe. It is speculated that this mating type was introduced in a large shipment of potatoes from Mexico to Europe in the late 1970s. Outbreaks of the A_2 type began occurring in Europe in 1980; this type has subsequently spread throughout the world. The introduction of the A_2 mating type has increased concerns that sexual reproduction and genetic recombination might occur in the field, developing new strains. This possibility has broad implications for potato breeders searching for blight resistance. These concerns were increased by the outbreaks of late blight in North America during the 1990s. The strains of *P. infestans* involved were generally resistant to metalaxyl, a widely used fungicide to control late blight. At the present time, fungicide-resistant strains have been found around the world.

The economic impact of fungicide application to control late blight (or other pathogens) must not be overlooked. Often, the cost of repeated applications may actually outweigh the value of the crop at harvest time. Rather than relying on chemicals alone, farmers can use other techniques—which include sanitation, crop rotation, biological controls, disease forecasting, and genetic resistance—to reduce the need for fungicides. This multifaceted approach is referred to as **integrated pest management** and has been widely recommended for controlling plant disease. Reducing fungicide use requires detailed information about the pathogen's life cycle, especially the dispersal phase. For pathogens with airborne dispersal, knowledge of the environmental conditions that promote dispersal can supply information on the optimum timing of fungicide applications to protect the crop. With the development of fungicide-resistant *Phytophthora* strains, farmers may need to adopt these other approaches to controlling late blight.

Rusts—Threat to the World's Breadbasket

There are about 6,000 species of rust fungi that attack a wide range of seed plants and cause destructive plant diseases. Some of the most important diseases of cereal crops are caused by these basidiomycetes, which have produced serious epidemics throughout history. Coffee, apple, and pine trees, and the economies depending on these crops, have also been devastated by rust fungi. Keep in mind that coffee rust destroyed the coffee plantations in Ceylon (Sri Lanka) in the 1870s and 1880s (see Chapter 16) and today threatens production wherever coffee is grown.

The rust fungi and the smut fungi (see Corn Smut—Blight or Delight, the next section) lack a fruiting body, unlike other members of the Basidiomycota. Although sexual reproduction does occur, these fungi are recognized by the characteristic asexual spores formed on infected plant parts.

Millennia before scientists understood that pathogenic fungi can cause plant diseases, rust epidemics were studied and the reddish lesions noted on plants. These epidemics were recognized in ancient Greece and described in the writings of Aristotle and Theophrastus, who even noted that different plants varied in their susceptibility to these diseases. The Romans held a religious ceremony or festival, the *Robigalia*, to propitiate the gods Robigo and Robigus, whom they believed responsible for the rust epidemics. In ancient times, wheat was the most important crop in these Mediterranean countries, and wheat rusts were major threats. Wherever wheat is grown today, rust fungi still remain serious pathogens, with different species of *Puccinia* causing stem rust, stripe rust, and leaf rust.

In wheat-growing regions around the world, the most important rust pathogen is *Puccinia graminis* subspecies (subsp.) *tritici*, the fungus responsible for stem rust of wheat. The organism has a complex life cycle that involves five spore stages **(basidiospores, spermatia, aeciospores, uredospores,** and **teliospores)** on two separate host plants, wheat and barberry (fig. 23.15). Basidiospores are produced in the spring when

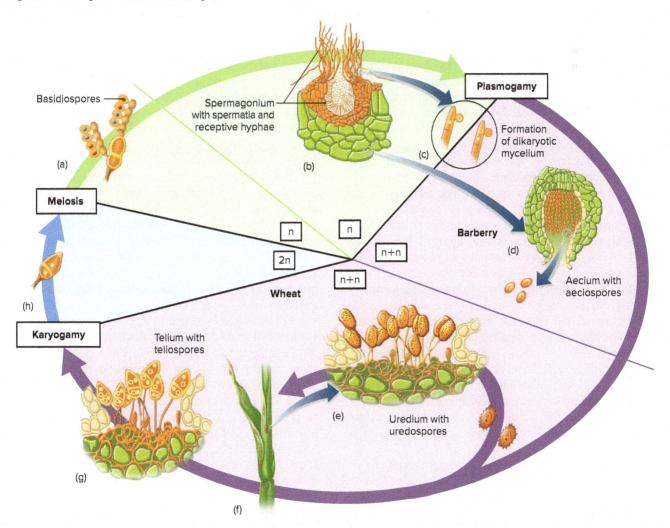

Figure 23.15 Life cycle of *Puccinia graminis*, the cause of stem rust of wheat. The two hosts are barberry and wheat. Basidiospores (a) infect barberry and result in the development of spermagonia that produce spermatia and receptive hyphae (b) and (c). Aeciospores are also produced on barberry (d). These infect wheat, transferring the infection to the economically important host. Within 14 days, uredia with uredospores (e) appear as long narrow lesions on the wheat stem (f). The uredospores can reinfect wheat throughout the growing season. Near the end of the season, uredospores are replaced by teliospores (g) and (h), which overwinter. The following spring, these give rise to basidia and basidiospores (a), thereby completing the life cycle.

the overwintering teliospores germinate. The basidiospores, which are uninucleate and haploid, belong to either + or − mating types. If the basidiospores land on a young barberry leaf, infection is initiated, eventually giving rise to small, flask-shaped structures, **spermagonia** (sing., **spermagonium**), on the upper surface of the leaf.

Each spermagonium produces masses of tiny, unicellular spermatia that function as male gametes. These spores are formed in a sugary nectar, or "honeydew," that oozes out of the neck of the spermagonium. Receptive hyphae also grow out of the spermagonium; these structures function as female gametes. Insects, which are attracted by the honeydew, unknowingly transfer the spermatia from one spermagonium to another. When spermatia are carried to the receptive hyphae of the opposite mating type, successful fertilization is initiated. The nucleus of the spermatia enters the receptive hypha and a dikaryon is established, with two nuclei of opposite mating types within each cell. Within a short time, binucleate aeciospores are produced in cuplike structures known as **aecia** (sing., **aecium**) on the lower surface of the barberry leaf.

The aeciospores transfer the infection only to wheat plants. They cannot reinfect barberry. On wheat the aeciospores germinate, and hyphae enter the plant through stomata. The dikaryotic mycelium is now established in wheat, and within 2 weeks the uredial stage develops, appearing as long, narrow lesions on the stem. Dark red, powdery masses of unicellular, binucleate uredospores occur in these lesions. Uredospores can reinfect wheat to produce several generations of **uredia** (sing., **uredium**). This stage in the rust life cycle is known as the repeating stage. The uredospores are airborne and easily carried from one plant to another, giving rise to epidemics. In fact, the uredospores can be carried by prevailing winds for hundreds of miles. In Mexico and southern Texas, this stage can continue all winter (on winter wheat) and give rise to spring infections in northern states, since the uredospores are carried by prevailing southerly winds (fig. 23.16).

Near the end of the growing season, the uredia turn black as two-celled teliospores replace the uredospores. Teliospores are the overwintering stage of the fungus and part of sexual reproduction. The haploid nuclei within each cell fuse. In the spring, the teliospore germinates, each cell giving rise to a short hypha that becomes a basidium. Meiosis produces four haploid nuclei, two of each mating type. Each nucleus migrates into each of the developing basidiospores, completing the life cycle.

Wheat plants infected by *P. graminis* are severely weakened but not destroyed, and the grain yield is significantly reduced. It is estimated that worldwide over 1 million metric tons of wheat are lost annually to stem rust.

Although the connection between wheat and barberry (as alternate hosts for the pathogen) was not made until the 1860s, farmers had long before known that wheat rust was worse whenever barberry plants were nearby. In fact, there were laws enacted in France in 1660 requiring the removal of

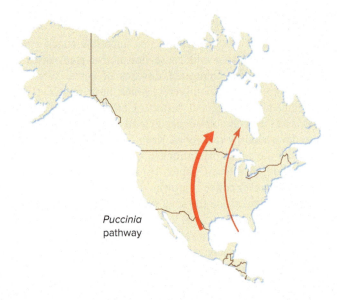

Figure 23.16 Uredospores of *Puccinia graminis* are blown northward by prevailing winds from Mexico and the southern states to the northern plains and Canada.

barberry plants from grain fields. Similar laws were enacted in the North American colonies of Connecticut in 1726 and Massachusetts in 1755. Eradication of barberry continues to be one of the major methods of control of stem rust of wheat. However, as noted, long-distance transport of uredospores can introduce wheat rust to areas where no barberry plants exist.

The most effective and cheapest control of wheat rust is through the use of resistant varieties of wheat. The ultimate goal of wheat breeding programs is the development of a wheat variety resistant to all varieties of rust fungus. However, there are over 200 physiological races of *P. graminis* subsp. *tritici*, which differ in their ability to attack the many commercial varieties of wheat. New races of the fungus are discovered each year, so wheat breeders must be diligent to stay ahead of the fungus.

Currently, the most serious threat to wheat is from Ug99, a race of *P. graminis* that was first detected in Uganda in 1998 and described in 1999. Plant pathologists estimate that approximately 80% to 90% of the wheat varieties currently being grown around the world are vulnerable to Ug99. Since it was first identified, Ug99 has spread from Uganda to ten other countries in east Africa and to Yemen and Iran in Asia. Scientists fear that spores could be carried by prevailing wind patterns into the important wheat-growing regions of India and Pakistan. Researchers around the world are expanding their breeding programs, trying to identify resistant wheat varieties. This collaborative effort is producing promising new lines, and initial plantings of some of these varieties in several countries have shown good yield. In 2016 a new strain of stem rust (unrelated to Ug99) devastated thousands of hectares of durum wheat on the Italian island of Sicily. This was the largest outbreak of stem rust in Europe since the 1950s. The following year, stem rust was again detected in Sicily as well as

Figure 23.17 Galls on eastern red cedar are produced by the cedar-apple rust fungus, *Gymnosporangium juniperi-virginianae*.

(a)

(b)

Figure 23.18 *Ustilago maydis*, the cause of corn smut, produces galls that replace kernels. Although a devastating loss for the corn crop, the infected ears are actually considered a delicacy in Mexico (where they are known as cuitlacoche [huitlacoche]) and are gaining in popularity in other areas as well. (a) Infected ear of corn. (b) Canned cuitlacoche purchased in a California market.

several regions of mainland Italy. In addition to stem rust, new aggressive strains of *Puccinia striiformis*, a related fungus causing stripe rust (also called yellow rust) of wheat, have recently been attacking wheat cultivars previously thought to be resistant. These new strains have been identified in wheat-growing areas around the world; the search for resistant wheat varieties continues.

With five spore stages, *P. graminis* is an example of a long-cycled rust. Some rust fungi have fewer spore stages, and some complete their life cycle on a single host plant. Cedar-apple rust causes yellow spots on the leaves and fruit of apple and crab apple (and related species) and produces galls on red cedar (juniper) trees. This rust fungus lacks a uredial stage. Spermatia and aeciospores are produced on the apple trees, and teliospores develop from the galls on the cedar (fig. 23.17). Because no repeating stage occurs, cedar-apple rust can be efficiently controlled by eliminating the alternate host from the immediate area.

Corn Smut—Blight or Delight

The smut fungi, like rusts, are plant pathogens that cause significant losses of grain crops. Unlike the rusts, only one host plant is involved and only two spore stages occur, teliospores and basidiospores. Incredible numbers of teliospores form from the mycelium, usually within galls. It is these masses of dark brown or black spores, resembling soot or smut, that give the common name to these plant pathogens. Basidiospores form on germination of the teliospores in a situation similar to the rust. In each infected plant, there is only one generation of teliospores. Many smuts overwinter as teliospores; however, other species overwinter as mycelium in infected grains.

Corn smut is a widespread disease, occurring wherever corn is grown, and is more common on sweet corn than on other varieties. The fungus, *Ustilago maydis*, forms galls on any aboveground plant part but is most conspicuous when the galls form on the ear (fig. 23.18a). The size and location of the galls reflect the degree of crop loss. Galls on the ear result in total loss, while those in other places reduce the yield or cause stunted growth.

The fungus overwinters as teliospores on plant debris and in the soil. In the spring, the teliospores germinate to produce a basidium that forms basidiospores. The haploid basidiospores are carried by wind or splashed by rain to the young tissues of corn plants, either young seedlings or the growing tissue of older plants. The basidiospore germinates and invades the plant tissue. For infection to continue, the dikaryon must be established by fusion of hyphae of compatible mating types. Alternatively, compatible basidiospores may fuse even before germination to establish dikaryotic mycelium directly. The mycelium grows prolifically in the infected tissue and stimulates the host cells to divide and form galls that may reach a size of 15 cm (6 in.) in diameter. The galls are initially covered by a membrane. As they mature, the interior darkens as the mycelium is converted into teliospores. Later, the membrane ruptures and releases the masses of dry spores.

One interesting feature of corn smut is the recent popularity in the United States of the young galls as a gourmet delicacy. The fungus has actually been cultivated for centuries in Mexico as a food item (fig. 23.18b). At the present time, farmers in several states are growing corn smut for the gourmet market, where it is known as "smoky maize mushroom."

Dutch Elm Disease—Destruction in the Urban Landscape

The plant diseases just discussed have led to widespread problems by devastating crop plants; other pathogenic fungi have also made an impact on society by attacking native forest and ornamental trees. During the first half of the twentieth century, the chestnut blight fungus caused the death of most of the native American chestnut trees throughout the eastern forests from Maine to Georgia (see Chapter 6). *Ophiostoma ulmi,* the cause of Dutch elm disease, has led to destruction of native American elm (*Ulmus americana*) trees and has altered urban ecology by killing ornamental elms across the country. *Ulmus americana* had long been a popular shade tree throughout much of this country, valued for its overarching growth and broad canopy.

The disease first appeared in Europe around 1919, and the fungus was first described in Holland in 1921 by two Dutch plant pathologists. Although the name of the disease reflects these early studies on the fungus, it is believed that the disease originated in Asia. The fungus was accidentally introduced to North America on contaminated logs and was first found in Ohio in 1930. After its introduction, Dutch elm disease spread across the continent; in the decades since its appearance in the United States, it is believed that over 77 million elms have died. Although the aesthetic loss cannot be estimated, the cost of removing diseased or dead elms amounts to millions of dollars each year.

The disease affects all elm species, but it is most destructive to the American elm. The fungus causes a vascular wilt, and the first symptom is a sudden and permanent wilting of individual branches. The disease usually appears first on one or two branches, then slowly spreads to other branches, gradually killing the tree over a period of a few years (fig. 23.19a). In some cases, however, the entire tree develops the symptoms at once and dies within a few weeks.

The spread of this disease results from the activities of insects. Although the fungus *O. ulmi* is responsible for the disease, elm bark beetles are the vectors that carry the fungal spores from infected elms to healthy trees (fig. 23.19b). These beetles feed on healthy trees or burrow into the bark and wood. Spores are deposited in the tree, germinate, and grow rapidly. The fungus grows in the vessels and produces spores that are carried upward in the xylem stream, thereby establishing new points of infection. Toxins are also formed by the growing fungus; these are actually responsible for the yellowing, wilting, and withering of the leaves.

Bark beetles lay their eggs in diseased or dying trees, forming breeding galleries or tunnels in the area between bark and wood (fig. 23.19b). Mycelia and spores of *O. ulmi* are often found in the breeding tunnels. When adult beetles emerge from the galleries, they pick up the spores on their body and begin the infection cycle anew. The disease also spreads through natural root grafts between adjoining trees. The effects have been seen repeatedly on elm-lined streets, as one tree after another succumbs to the disease.

Since the disease first appeared in the United States, various measures have been employed to control its spread. For years, the major approaches were sanitation to remove infected trees and the use of insecticides to destroy the bark beetle.

Figure 23.19 Dutch elm disease, caused by *Ophiostoma ulmi,* has led to the destruction of native elm trees in North America. (a) Infected elm tree leafless on the left side. (b) Infection of elm trees is spread by bark beetles and even through natural root grafts.

These measures had limited effectiveness. Attempts at biological control of the beetle have recently focused on the use of another fungus that appears to disrupt the reproductive cycle of the beetles. Various systemic fungicides have also been used for many years, but these treatments are expensive and usually are reserved for valuable or historically important trees. A promising new approach is being tested by scientists from the University of Toronto; this is not a cure but a preventive treatment that has been referred to as an "elm vaccine." Trees are injected with a glycoprotein (sugars attached to a protein) produced by *O. ulmi*. This molecule stimulates the trees' natural defense system by eliciting the production of mansonones, which are phytoalexins. **Phytoalexins** are secondary products produced by plants in response to bacterial and fungal pathogens and are known to kill or inhibit the growth of pathogens. Phytoalexins are important mechanisms of natural disease resistance in many plants. By stimulating the production of mansonones, the "vaccine" will increase the trees' ability to resist attack from the Dutch elm disease fungus. Also, intensive breeding programs at many research facilities around the world have been searching for resistant elm varieties. Scientists from the USDA have spent more than 40 years screening for varieties of elm that are resistant to *O. ulmi*. A number of resistant varieties of elms have been identified, many of which are commercially available. Since 2005, the National Elm Trial has been underway in 15 states to evaluate 17 resistant elm varieties under various growing conditions and climatic zones. In 2017, researchers published the results of this 10-year study. They found that the disease-tolerant American elm cultivars, New Harmony and Princeton, were the preferred cultivars with both showing over 80% survival. In addition, some American elm x Asian elm hybrids had over 90% survival. Meanwhile, research continues to develop additional disease-tolerant cultivars. With the use of tolerant varieties, the American elm may one day regain its former glory.

Sudden Oak Death—Destruction in the Forest

A new plant disease affecting trees in the central California coastal forests was first described in 1995. The disease, initially found in tanoak trees and several species of oak trees, was soon named *sudden oak death* because of the rapid death of trees once symptoms become apparent. Since 1995, millions of trees have died, and the epidemic has spread along the California coast and into southern Oregon. This epidemic appears to be one of the most virulent forest epidemics in recent history. In 2000, David Rizzo from the University of California-Davis and Matteo Garbelotto of UC-Berkeley identified the organism responsible for sudden oak death, an oomycete called *Phytophthora ramorum*. This species was first described in 1993 in Europe on rhododendron and viburnum shrubs and is related to *P. infestans*, the late blight of potato pathogen. The origin of *P. ramorum* is currently unknown; the California and European strains are genetically different, which rules out an introduction from Europe.

Since the initial identification of *P. ramorum*, other native trees and shrubs have also been found to be infected. Over 125 species of plants have been shown to be susceptible to the pathogen; however, two different types of diseases are caused by *P. ramorum* on different hosts. In some plant species, the pathogen causes lethal infections of the trunk and branches, while other species exhibit nonlethal infections on leaves and twigs.

Hosts for the lethal infections include tanoak, coast live oak, California black oak, Shreve's oak, canyon live oak, coast redwood, Douglas fir, and others. Tanoak is the most susceptible host identified thus far, and in some areas over 80% of the tanoaks have died. The discovery of *P. ramorum* on coast redwood and Douglas fir had researchers concerned for many reasons because these trees are highly valued ecologically as well as economically for both forestry and tourism. However, no large redwood or Douglas fir trees have been killed by the pathogen. Most of the research has focused on seedlings and understory redwood and Douglas fir; mature trees do not appear to be in danger. However, the destruction of the tanoak trees within the redwood forests have impacted the whole forest ecosystem.

Trees may be infected with the pathogen for a year or more before symptoms appear. The first visible indications are **cankers** (localized areas of dead tissue) oozing a black or dark red sap (fig. 23.20a). Trees with cankers generally survive for one to several years while the pathogen invades and kills the vascular cambium. Once crown dieback begins, the foliage turns from green to yellow to brown within a few weeks. In addition, the cankers create an ideal environment for insects and fungi to invade and possibly hasten the death of the tree. An opportunistic wood decay fungus, *Hypoxylon thouarsianum* (fig. 23.20b), is commonly found on infected trees along with bark beetles and ambrosia beetles. In fact, the beetles are often so abundant that some entomologists initially thought that an outbreak of beetles might be the cause of sudden oak death.

Infections are very different on rhododendrons, California bay laurel, big leaf maple, and other species that only develop leaf spots and show twig dieback. Although their infections are mild, these species play a major role in the spread of *P. ramorum* by acting as a reservoir of the spores that infect oaks and tanoaks. Abundant sporangia are formed on the leaves of rhododendron and other species with foliar infections. The sporangia as well as the zoospores are thought to be spread by wind-blown rain and by rain splash. Mortality among oaks is most common where California bay laurel and rhododendrons are found growing together with oaks. Infection may also be spread by contaminated soil or wood and by moving infected plants.

Although the disease has largely been confined to the central California forests, various plants in nurseries in southern California, Oregon, Washington, and British Columbia have

Figure 23.20 Sudden oak death is a relatively new forest disease. (a) Cankers oozing black sap (arrows) are the first visible symptoms of an infected tree. (b) Other fungi, such as *Hypoxylon thouarsianum*, often invade infected trees and may hasten tree death. The presence of these small *Hypoxylon* fruiting bodies (black nodules) is indicative of an advanced stage of sudden oak death.

tested positive for the pathogen. This finding raised fears cross the country, because southern California nurseries ship plants throughout the United States. In fact, infected plants subsequently showed up in nurseries in Georgia, Florida, Louisiana, Maryland, and Virginia. Many states have closed their borders to plants from California and are developing educational, survey, and detection programs to limit the spread of the pathogen. Many oak species are the dominant trees throughout the Southeast, and testing has already shown that some eastern oak species are highly susceptible to the pathogen.

In Europe, the disease was initially confined to shrubs that showed the milder foliar symptoms; however, in late fall of 2003, a few infected oak trees were identified in England and the Netherlands. Infected beech, horse chestnut, and yew trees were also found in England. Subsequently, thousands of larch (*Larix* spp.) trees in the United Kingdom and Ireland have been killed by this pathogen. In 2017 diseased larch trees were identified in France as well.

Various fungicides have been used as preventative treatments, especially for plants in nurseries. In addition, a fungicide, Agri-Fos, is being used to treat infected trees and to protect healthy trees. The fungicide can be injected into trees or paired with other chemicals and sprayed on the bark. Agri-Fos is not a cure-all and is effective only in the early stages of the disease. It is intended to treat individual trees that are at high risk of infection and is not for widespread use.

The loss of individual mature trees is an economic and aesthetic hardship for homeowners, but the destruction of a portion of a forest may be an ecological devastation. The effect on wildlife is largely unknown at this time. In California, oak forests cover 11 million acres and provide habitat for over 200 species of mammals and 100 species of birds. Also, dead trees within the forest greatly increase the fire danger; however, removal of dead trees may be cost prohibitive. Researchers are just starting to assess the potential damage to the ecosystem from sudden oak death.

Thinking Critically

Fungi are responsible for the majority of plant diseases. Although fungicides can effectively control many diseases, these chemicals are expensive and often hazardous.

What approaches can be used to control fungal diseases without the widespread use of fungicides?

CHAPTER SUMMARY

1. Organisms that have traditionally been called fungi are found in both the kingdom Protista and the kingdom Fungi. Slime molds and water molds are fungal-like organisms that are protists, whereas other groups are included in the kingdom Fungi. The presence of chitin in the cell wall is one of the major distinguishing features. Classification in the kingdom Fungi is normally based on the method of sexual reproduction, and the two largest groups are the ascomycetes and basidiomycetes.

2. Fungi are a diverse group of organisms, ranging from small, inconspicuous yeasts and molds to large, conspicuous mushrooms, puffballs, and bracket fungi. The majority of fungi consist of a mycelium composed of highly branched hyphae. Fungi reproduce by spores that result from sexual or asexual processes. Fungi have an absorptive type of nutrition, producing enzymes that are secreted into the environment and break down complex organic compounds into smaller molecules. These small molecules are then absorbed by the hyphae.

3. The majority of fungi are saprobes, actively recycling nutrients in the environment by decomposing organic matter. Symbiotic relationships among the fungi are well known, and both mycorrhizal fungi and lichen fungi play important ecological roles. Other fungi exist as pathogens, obtaining nutrients from a living host.

4. The majority of plant diseases are caused by fungi and oomycetes. Some of these have devastating influences on crop plants, and historically some have resulted in major impacts on human society. Rust fungi and smut fungi are especially damaging pathogens on cereal crops, and Dutch elm disease has resulted in the loss of millions of ornamental elm trees throughout North America. Sudden oak death is causing major devastation in California forests.

REVIEW QUESTIONS

1. What characteristics distinguish fungi from plants? What characteristics distinguish them from animals?
2. How does *Puccinia graminis* begin new infections on wheat when no barberry plants are in the area?
3. Why are scientists interested in mycorrhizal fungi?
4. Describe the life cycle of a mushroom.
5. Describe the ecological roles played by fungi.
6. Compare and contrast the rust and smut fungi.
7. Compare and contrast the sexual reproductive structures of the Ascomycota and Basidiomycota.
8. In what ways are the slime molds funguslike, and in what ways are they animal-like?
9. Describe the use of fungi in bioremediation.
10. Why has late blight of potato resurfaced to become a threat to potato growers?

CHAPTER

24

Beverages and Foods from Fungi

©Karen McMahon

In the brewing process, a grainer (top left) stores and releases wheat grains into the brew kettle (bottom left). Yeast is added to the wort in the fermentation tank (right). Marshall Brewing Company, Tulsa, Oklahoma.

KEY CONCEPTS

1. The step common to both beer and wine making is the fermentation of sugar to alcohol, a conversion made possible through the action of yeast.
2. In distillation, the alcoholic content of a beverage is raised considerably through the principle of differential boiling points.
3. Fungi have a proven nutritional value when consumed directly as food or as agents of fermentation in fermented foods.

CHAPTER OUTLINE

Making Wine 447
    The Wine Grape 448
    Harvest 449
    Red or White 449
    Fermentation 449

A CLOSER LOOK 24.1 Disaster in the French Vineyards 450

    Clarification 452
    Aging and Bottling 452
    "Drinking Stars" 453
    Fortified and Dessert Wines 454
    Climate Change and the Wine Grape 454
The Brewing of Beer 454
    Barley Malt 455
    Mash, Hops, and Wort 455
    Fermentation and Lagering 455
    Sake, a Rice "Beer" 457

Distillation 457
    The Still 458
    Distilled Spirits 458
    The Whiskey Rebellion 458
    Tequila from Agave 459
    Hard Cider 460
A Victorian Drink Revisited 461
    Absinthe 461**
    The Green Hour 461
    Active Principles 461
    Absinthism 462

A CLOSER LOOK 24.2 Alcohol and Health 463

Fungi as Food 465
    Edible Mushrooms 465
    Fermented Foods 466
    Quorn Mycoprotein 467
Chapter Summary 468
Review Questions 468

In 1810 Joseph-Louis Gay-Lussac, a French biochemist, was the first to deduce the chemistry of fermentation. He recognized that each sugar molecule is converted into two molecules of ethanol and two molecules of carbon dioxide:

$$C_6H_{12}O_6 \rightarrow 2C_2H_5OH + 2CO_2$$

It is known today that two ATPs are also formed from the chemical energy released when glucose is broken down during fermentation. The detailed steps of fermentation were not completely elucidated until the first half of the twentieth century and have been discussed in Chapter 4.

Yeast is a single-celled ascomycete (see Chapter 23). In an anaerobic environment, yeast will forgo aerobic respiration in favor of fermentation. In the modern processing of beer and wine, most of the yeast fermenters are strains of *Saccharomyces cerevisiae* (fig. 24.1).

MAKING WINE

The practice of growing grapes to make wine is an ancient one. Wine's first appearance was most likely an accident of nature when wild yeasts present on the grape skins fermented the juice. Evidence of a winelike drink was found recently in 8,000- to 9,000-year-old pottery fragments from the site of a prehistoric village in China. This is the oldest evidence for an alcoholic beverage. Analysis revealed that the beverage was made from rice, honey, and either hawthorn berries or grapes.

Whatever their exact origins, the skills in wine making were known to many ancient cultures. Five-thousand-year-old sealed wine jars have been found in Egyptian tombs. The ancient Egyptians developed fairly sophisticated practices in the cultivation of wine grapes, such as building trellises for training vines, developing precise pruning methods, and constructing irrigation canals to the vineyard. Egyptians apparently developed the techniques to make both red and white wines. An analysis of the residue in six jars from the tomb of Pharaoh Tutankhamen revealed evidence of white wine. All of the jars contained tartaric acid, which is a marker for grapes. Half of the jars had traces of syringic acid, a breakdown product of the main pigment malvidin in red grapes, and is a sure sign that the jars held red wine. The other jars stored white wine because, although there was tartaric acid in the residue, syringic acid was not found. Written records had previously dated white wines to 1,800 years ago, but this recent finding confirms that Egyptians were making white wines much earlier, sometime before the Tutankhamen dynasty from 1332 B.C. to 1322 B.C. (fig. 24.2).

Both the ancient Greeks and Romans were also skilled vintners. The Romans introduced many innovations to wine making, such as classifying grape varieties and adding spices and herbs to flavor wines. They took the wine grape and the practice of wine making to what is now France, Germany, and England. The skill was preserved and passed on in the monasteries of Europe during the Middle Ages and the Renaissance. Not until the renowned French microbiologist Louis Pasteur

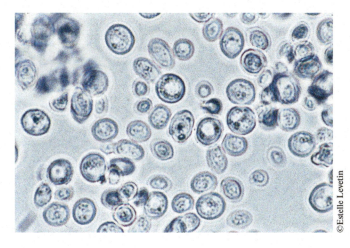

Figure 24.1 Yeast, *Saccharomyces cerevisiae*, ferments sugar to ethanol and carbon dioxide during the production of wine and beer, ×1,200.

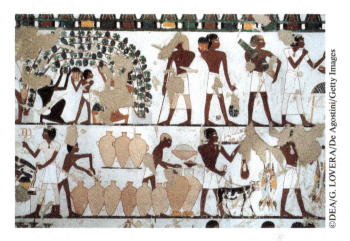

Figure 24.2 The ancient Egyptians were skilled wine makers. This tomb painting depicts workers harvesting grapes on the trellised vines.

was called in for his services by the French wine industry in the mid-nineteenth century did the ancient art of wine making become the science of **enology.** It was Pasteur who discovered that yeast, through an anaerobic process, converts grape juice into wine and that bacterial contamination by *Acetobacter,* through an aerobic reaction, can turn wine into vinegar. He reasoned correctly that eliminating air (oxygen) from the barrel (primarily by keeping it full) can prevent the conversion to vinegar. He also discovered that heating wine can reduce bacterial contaminants, a process now known as pasteurization, which is routinely used to process many substances besides wine.

In North America, the wine grape was introduced to California in 1769 by the Franciscan missionary Father Junipero Serra, but wine making did not become established until the 1800s. The industry was almost destroyed by the ratification of the Eighteenth Amendment in 1919 and the nearly two-decades-long Prohibition Era that followed. Renewed growth in the 1960s and 1970s established California as one of the major wine centers in the world. Today, California wines favorably compare with those from the finest wineries in Europe.

Next, we trace the making of wine from grape to bottle, according to the most commonly practiced methods of the California grape growers and vintners.

The Wine Grape

Although the sweet juice of other fruits, such as apricots or elderberries, can be fermented, technically speaking, wine is fermented grape juice. Wine begins with the wine grape, *Vitis vinifera,* native to the area around the Caspian Sea between Europe and Asia. Members of the Vitaceae, or grape family, are found in warm temperate or tropical areas and are economically important as the sources of wine, table grapes, raisins, and grape juice. North America has many native species of grape, but most are not suitable for making wine. One well-known exception is *V. labrusca* (fig. 24.3), native to the eastern United States and the base for several red wines.

Vitis vinifera is a woody climbing vine with five-parted flowers borne in inflorescences opposite the palmately lobed leaves. Unlike some species of grape, the wine grape bears perfect flowers and is self-pollinated. The wine grape is basically of two types, red or white, referring to the color of the grapes. Thousands of varieties of these two types exist (table 24.1). Shoots of the wine grape have been routinely grafted onto rootstocks of North American vines since the turn of the twentieth century. One North American species used for this purpose is *V. rupestris.* Grafting of *V. vinifera* to North American grapes came about because of the root aphid *Phylloxera,* an insect that destroys the grapevine by attacking its roots. In 1860, *Phylloxera* was accidentally introduced to European vineyards when grapevines native to eastern North America were sent to Europe. Native North American grapes are resistant to damage by the insect but can harbor this microscopic pest. Between 1870 and 1910, most French, German, Austrian, and even Californian (planted with the European wine grape) vineyards were devastated (fig. 24.4). In all areas of the world infested by *Phylloxera,* rootstocks of resistant American varieties are grafted to varieties of the wine grape.

Table 24.1 Some Varieties of the Wine Grape, *Vitis vinifera*

Variety	Color
Cabernet Sauvignon	Red
Chardonnay	White
Gamay Noir	Red
Grenache	Red
Muscat	White
Pinot Noir	Red
Riesling	White
Sauvignon Blanc	White
Semillon	White
Zinfandel	Red

Figure 24.4 "*Phylloxera*, a true gourmet, finds out the best vineyards and attaches itself to the best wines." From *Punch,* September 6, 1890.

Figure 24.3 Wine is made from the fermented juice of grapes.

The cultivation of grapes is known as **viticulture,** and great care and planning are required for a successful harvest. The rootstock is planted in early spring. In late July or August, the desired grape variety is grafted to the rootstock. Alternatively, some growers plant the rootstock with the specific graft already in place. The graft will determine the variety of grape to be harvested. During the second season of growth, the vine is selectively pruned and trained to trellis wires for support, maximizing photosynthesis by spacing the leaves for optimal exposure to the sun (fig. 24.5) and encouraging shoot growth and more abundant fruit production. In the third year of

Figure 24.5 The vine is pruned and trained to trellis wires to support the grapes.

growth, the vines have grown sufficiently to produce enough grapes to harvest. Fruit production continues to improve during the next 3 years until a plateau is reached when the vines are 6 to 8 years old. Vines may be productive for 40 to 50 years. Grapevines are subject to several fungal diseases as well as insect pests. Powdery mildew is a constant threat, and a dusting of sulfur powder is applied annually to prevent the mildew from taking hold. The story of downy mildew, a more serious fungal disease, is presented in A Closer Look 24.1: Disaster in the French Vineyards.

Harvest

In the fall, grapes are harvested when their sugar content reaches a certain critical level. The grape, more than most other fruits, concentrates high levels of sugars. Sucrose forms in the leaves of the grapevine as a product of photosynthesis. The sucrose is then transported to the developing fruits and broken down into glucose and fructose. Approximately 20% of the juice of ripe grapes is a mixture of these two sugars. Organic acids, mainly tartaric and malic, are also found in the grapes in concentrations around 1%. Sugar content is determined by analysis of a sample of grapes. Determining the sugar content of the juice is necessary because growers are paid according to relative concentrations of sugars to acids. In colder regions grapes tend to have higher acid levels, but in warmer areas the grapes are sweeter. Too sugary a grape produces a wine that is flat and more susceptible to bacterial spoilage. Too acidic, and the wine tastes green or hard. When the sugar concentration is optimal, the grapes are harvested either by hand or by machine into gondolas and then trucked to the winery. White grapes are harvested earlier than the red varieties.

Red or White

Grapes must be crushed, pressed, and fermented to become wine (fig. 24.6). Today many grape varieties are crushed and destemmed mechanically. *Crushing* refers to breaking the skins to start the flow of juice, and mechanical crusher-stemmers remove the stems at the same time. The resulting mix of skins, seeds, and juice is referred to as *must*. *Pressing* squeezes the skins to express the last bit of liquid in such a way as to free the juice of skins, pulp, and seeds. Leftover skins and seeds make up a cake, or pomace, usually returned to the vineyard as fertilizer. The expressed juice is transferred to stainless steel holding tanks or the traditional oak barrels in preparation for the fermentation process.

The juice of all wine grapes, whether red or white varieties, is initially white. Only through contact of the juice with broken red skins does the wine develop a red color. Red skins contain anthocyanin pigments and tannins that impart the rich color. In creating a white wine from red grapes, the skins are promptly separated from the juice in the press before fermentation. In the making of red wine, red skins are not removed from the juice until after fermentation. A rosé results when the pressing to remove the red skins takes place in the middle of the fermentation process, producing wine with a pink or pale red shade. To sum up, white wines, using either red or white grapes, are crushed, pressed, and then fermented. With red wines, the red grapes are crushed, fermented, and then pressed to remove the skins.

Fermentation

Sulfur dioxide is added to the juice to inhibit the growth of any unwanted, naturally occurring yeasts and bacteria brought in with the grapes (fig. 24.6). It also removes oxygen from the fermentation vat, a further insurance against the presence of any aerobic *Acetobacter* spp. that would turn the ethanol to vinegar. Specially developed wine yeast strains are then introduced into the grape juice. This method contrasts with the age-old practice, called spontaneous fermentation, of relying on yeasts naturally present on the grapes. The yeast ferments the grape sugars, glucose and fructose, to ethanol and carbon dioxide gas. During this process, appreciable amounts of waste heat are produced that can negatively affect the wine and must be monitored carefully.

A CLOSER LOOK 24.1

Disaster in the French Vineyards

Although plant diseases can always cause serious problems, the mid- to late nineteenth century was an especially adverse period for the wine industry in France. Starting in the 1860s, root-feeding aphids, *Phylloxera*, were attacking the grapevines. In an attempt to solve the problem, vintners introduced rootstocks of North American grape plants, and shoots of local French varieties were grafted onto the rootstocks. (North American species of grapes are resistant to damage by the aphids, even though roots are often infested with *Phylloxera*.) The grafted plants produced grapes of the desired variety while the roots tolerated the aphids. Unfortunately, the introduction of the North American plants introduced a far more lethal pathogen, *Plasmopara viticola*, the organism that causes downy mildew of grape.

Plasmopara viticola is a member of the Oomycota and produces zoospores within a sporangium. However, unlike those of *Saprolegnia* (see Chapter 23), the sporangia themselves are airborne. Zoospores are released on landing if conditions are favorable. Infection occurs when a sporangium lands on a grape leaf. The zoospores will swim around for a short time, then settle down, encyst, and soon produce a germ tube. If the zoospore is on the undersurface of a leaf, the germ tube may reach a stoma and start the infection cycle. The mycelium spreads through the leaf tissue, obtaining nutrients from mesophyll cells. Pale yellow spots typically appear on the upper surface of the leaf; the lower surface of leaves or the grapes themselves appear cottony, or "downy," due to the presence of branched sporangiophores growing out through stomata (box fig. 24.1). Sporangia developing on these stalks repeat the asexual stage of the fungus. Sexual reproduction often occurs at the end of the growing season with the production of oospores.

By the late 1870s, this pathogen threatened vineyards throughout France, with Bordeaux vineyards especially hard hit. The wine industry was suffering major new losses and had not financially recovered from the *Phylloxera* epidemic. The turnaround came with the widespread use of a fungicide promoted by Alexis Millardet, professor of botany at the University of Bordeaux. Millardet's discovery of this fungicide came by chance in 1885 as he was walking by a local vineyard. Millardet noticed that the grape plants along the path were a blue-gray color but were surprisingly healthy when compared with other plants in the vineyard. Upon investigation, Millardet learned that the farmer had sprayed the plants along the roadside with a poisonous-looking substance

Box Figure 24.1 Grapes infected by *Plasmopara viticola*, the fungus that causes downy mildew. The "downy," or cottony, appearance is due to the growth of sporangiophores on the surface of the grapes.

(a mixture of copper sulfate and lime) to discourage people from picking and eating his grapes. Millardet realized the significance of the healthy-looking plants and perfected the blend of chemicals, known as Bordeaux mixture, which solved the disaster in the vineyards. Bordeaux mixture became the first widely used fungicide and continues to be one of the most important ones in use.

Even today in the vineyards of Bordeaux, grape leaves look blue because of the application of copper fungicides used as a prophylactic to prevent infection by zoospores of *P. viticola*. Topical fungicides, such as Bordeaux mixture and other copper compounds, are surface protectants designed to prevent entry of the pathogen into the plant. Therefore, the plants require complete coverage. Unfortunately, the compounds are washed off by rain, and repeated applications are required throughout the growing season. However, once inside the plant, *Plasmopara* is not affected by the surface application of these compounds.

The constant use of copper fungicides can lead to a heavy buildup of the copper in the soil, which may eventually cause toxicity problems. Similar problems occur with the use of mercury compounds, which have also been used as fungicides. (Accumulation of toxic levels of heavy metals, such as copper or mercury, is a serious environmental problem in many areas. Heavy metals are known to denature many enzymes and are, therefore, a general biotoxic substance.)

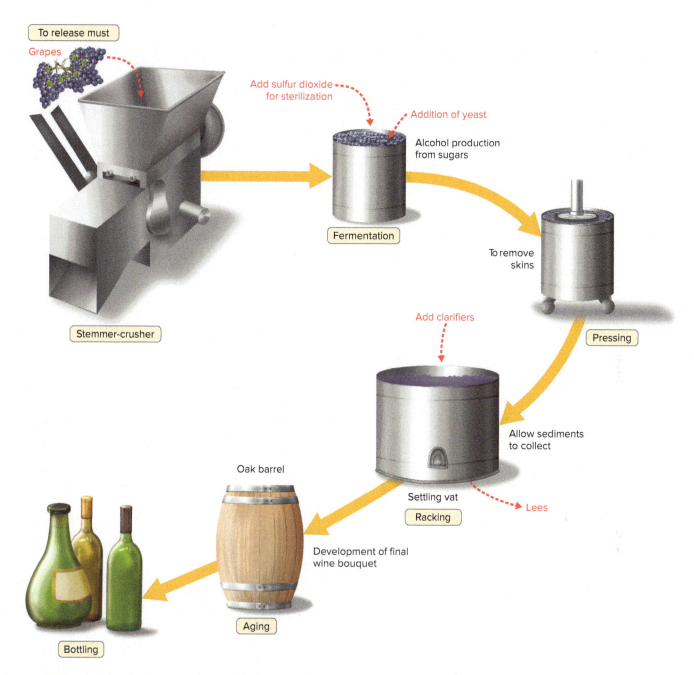

Figure 24.6 Steps in the wine-making process. Wine production begins with the crushing of grapes to release the juice. In the processing of white wines, the grapes are pressed immediately to remove the skins. In the production of red wines, shown here, pressing is delayed until after fermentation. Before the yeast is added to begin the fermentation process, sulfur dioxide is added to eliminate any unwanted microorganisms. Fermentation may continue for days or weeks. Racking allows sediments to collect, clarifying the wine. In red wines, malolactic fermentation is often induced to reduce the acidity of the wine. The aroma and flavor of fine wines are obtained by aging in oak barrels. Bottling is the last step in wine production.

White wines are fermented at temperatures from 10° to 15°C (50°-59°F); red wines are fermented at higher temperatures, ranging from 25° to 30°C (77°-86°F). Temperature control is extremely critical. If the fermentation temperature is too low, white wine develops an overly yeasty taste and red becomes too fruity and thin-bodied. If the temperature during fermentation is too high, the wine loses its fruitiness and tastes "cooked." Also, at higher temperatures, the yeast may be killed, stopping fermentation completely. In general, controlling heat production during white wine fermentation is not as difficult as in the fermentation of red wines. In red fermentation, heat production is greater and more likely to reach unacceptably high temperatures above 30°C (86°F). In small wineries, temperature control is limited to keeping the winery cool. In large-scale operations, metal cooling jackets manage the temperature in each tank or barrel.

Red fermentation is very active because pressing takes place after fermentation. Consequently, a bubbling surface cap of skins and seeds forms on the top of the fermentation tank. This cap needs to be broken up several times a day during the fermentation process. It is broken manually with a wooden paddle or mechanically by pumping up a constant stream of juice from the bottom of the tank to the surface.

Fermentation goes on for days or weeks, depending on the temperature. The cooler the temperature, the longer the fermentation. Three to 14 days is about right for a red wine, and 10 days to 6 weeks is usual for white wines. The rate of fermentation can be monitored in several ways. The most common method is to measure the decreasing sugar concentration. When there are no more fermentable sugars in the wine, fermentation is completed. In the production of a dry wine, fermentation is complete when there is no residual sweetness. If a sweet wine is desired, the vintner must interrupt fermentation while there are still sugars present by lowering the temperature, filtering the wine of yeast, or adding sulfur dioxide to inhibit the yeast. The alcohol concentration of table wines averages 12%; yeast dies at alcohol concentrations just above 14%. If higher alcohol concentrations are desired, the wine must be either fortified or distilled, options that are discussed later in the chapter. Most of the carbon dioxide produced during fermentation is allowed to escape, producing a still wine, a wine not highly carbonated.

Clarification

The newly fermented wine now undergoes the process of clarification, during which the wine loses its murky appearance and becomes crystal clear. Clarification begins with racking (fig. 24.6), as the wine is allowed to settle. Sediments, mostly used-up yeast known as *lees,* collect on the bottom, and the wine is transferred to a second barrel or tank. A clarifier, such as bentonite clay, egg whites, or gelatin, is added to speed clearing. These substances act as adsorbents, collecting suspended particles and proteins as they move through the wine. Some wineries may even centrifuge the wine to remove sediments. Next, the wine is filtered. Cellulose pads, diatomaceous earth, or membrane filters are common methods. In premium-quality wines, the wine is passed through millipore filters, with pores so tiny they exclude unwanted microbes. Pasteurization, heating the wine to 81°C (180°F), or adding a preservative, such as sorbic acid, are other methods that preserve the wine by inhibiting any undesirable microbes that may be present.

Aging and Bottling

The aroma and flavor important to fine red wines, and even a few whites, are obtained by aging the wine in oak barrels (usually white oak imported from France) or small casks for several months to a year or two (fig. 24.6). A substitute but inferior method is to age the wine in tanks with oak chips. White wines, if aged at all, are generally not aged as long as red wines. For some red wines, a second fermentation may be induced (in the barrel) during the aging process. This second fermentation may be started by introducing lactic acid bacteria cultures (*Lactobacillus*) to break down malic acid still present in the wine. This type of fermentation is known as malolactic fermentation because the malic acid is converted into lactic acid, a weaker acid, and is an effective way of reducing the acidity of a wine.

During bottling, the wine is placed in sterile bottles that are often flushed with nitrogen to minimize exposure to oxygen and the possibility of the aerobic conversion of ethanol to vinegar by *Acetobacter*. The bottles are precisely topped with wine to allow just enough room for the cork and to minimize the amount of air. Corked bottles are sealed with tinned lead and are stored bottoms up in order to wet the corks and keep air and bacteria out. Labels (fig. 24.7) reveal a great deal of information about the history of the grapes and wine. For example, the area where the grapes were cultivated and the

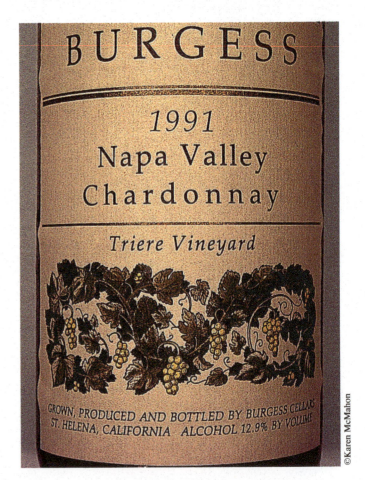

Figure 24.7 A wine label reveals important information about the bottled wine. The designation *Chardonnay* indicates that the wine contains at least 75% of that grape variety. *Napa Valley* indicates that 85% of grapes in the wine were grown and processed in this established viticultural area. The year *1991* is the vintage year, and 95% of the wine must be of that year's growth. *Grown, produced, and bottled by* discloses that Burgess Cellars fermented and finished at least 75% of the wine on its premises. *St. Helena, California* is the bottling location, and the alcohol content of the wine is 12.9%.

year the wine was made can be found by reading a label. If the vintage is given, it means that 95% of the wine was made during the indicated year.

"Drinking Stars"

Whether champagne or Asti Spumante, these wines fizz because of an excess of carbon dioxide bubbles from a secondary fermentation. The origin of champagne was accidental, the result of an unfinished fermentation, and probably occurred in the colder wine-growing regions. Most likely, the primary fermentation of the wine began as usual after the harvest but stopped before completion, as the weather cooled with the onset of winter. The wine maker assumed the fermentation was completed, and the wine was subsequently bottled. When temperatures warmed up the following spring, the fermentation process restarted in the bottle because the wine still retained some residual sugar and yeast. The wine fermented to dryness and, when eventually opened, had a noticeable fizz from the abundant carbon dioxide bubbles.

Although often ignored, champagne, strictly speaking, is a sparkling wine produced from grapes grown in the Champagne region of France, the northernmost grape-growing region in that country. The chalky soils in the area are said to influence the flavor of the grapes in a unique way. Dom Perignon, the name most associated with champagne making, was a seventeenth-century Benedictine monk from the Champagne region in France. It was Dom Perignon who aptly described the sensation of champagne as "drinking stars." He is credited with the first careful study on how to deliberately induce the bubbles. He also perfected the art of blending different wines to come up with the base for the champagne and was responsible for developing the method of pressing red grapes immediately to achieve a white wine. The methodology was further improved in the eighteenth century, and techniques developed during this time established the standards for the champagne industry today.

The base wines (cuveé) are made separately from three principal grape varieties: the white Chardonnay and the reds Pinot Noir and Pinot Meunier. The processing of these grapes into wine follows the same procedure as for any white wine: that is, crushing followed by immediate pressing. The must undergoes the primary fermentation in stainless steel tanks or, in some companies, the more traditional oak barrels.

The three young wines thus produced are then blended according to the champagne producer's specifications to achieve a characteristic taste. In the traditional method, the blended wine is bottled and undergoes a second fermentation in the bottle. The second fermentation is enabled by the addition of sugar from any natural source (cane sugar, corn syrup, honey, or grape sugar) and yeast to the bottled wine. Some companies employ the tank method to induce the second fermentation. In this method, the cuveé is placed *en masse* in large, pressurized tanks to which sugar and yeast have been added. In cheaper productions, the wine is artificially carbonated by cooling it to below freezing and pumping in carbon dioxide.

Secondary fermentation is generally finished in 2 weeks, but the wine stays in contact with the lees for several months to a year to develop aroma and flavor. Carbonation under pressure accumulates in the bottle. The champagne bottle is made of thicker glass than the typical wine bottle because the high carbon dioxide content exerts a considerable amount of pressure, 75 to 90 pounds per square inch. Even when precautions are taken, champagne bottles do sometimes explode. The bottle is sealed with a metal crown cap, the type seen on bottled soft drinks.

Riddling, the process of collecting sediments from the secondary fermentation in the bottle, is the next step. Madame Veuve Clicquot, an eighteenth-century champagne maker in France, invented riddling, which begins with placing the champagne bottles upended in a rack. Each day, the champagne is swirled and the bottle is gently uplifted and turned slightly (fig. 24.8). Although some riddling is done using a mechanical shaker, much of the industry still turns the bottles by hand. Over a period of time, sediments collect in the neck of the bottle. Formerly, the sediment collected against the cork. In modern practice, a tiny plastic bucket is inserted, during bottling, just inside the crown cap to collect the sediments.

Next, the sediments are removed from the bottle during the stage known as disgorgement. The most common method today is to immerse the neck of the inverted bottle in a subzero solution, freezing the collected sediment into a plug. When the crown cap is removed, either by hand with a bottle opener or mechanically from the now-uprighted bottle, carbonation pressure blows out the plug. The dosage step adds wine or brandy to replace any liquid lost during disgorgement. Cane sugar may also be added, depending on the amount of sweetness desired. *Brut* champagne is the driest, and the designation *doux* indicates the sweetest. Although champagne is basically a white wine, pink champagne can be made either by blending a red wine in the cuveé or by permitting some skin contact with the red grapes before pressing.

Figure 24.8 Riddling is a step in the champagne making process in which upended bottles are slowly and systematically turned to collect sediments in the neck.

Fortified and Dessert Wines

Table wines have a final alcohol content of 9% to 14%. Alcohol concentrations just above 14% kill the yeast, ending fermentation. If a higher alcohol content is desired, alcohol, usually in the form of brandy, must be added and the wine is said to be fortified. Fortified wines usually have an alcohol content between 15% and 21%. The higher alcohol content tends to act as a preservative, preventing bacterial spoilage, and preservation, in fact, might have been the original intent behind fortifying wines. In the production of a dessert wine, adding brandy brings the fermentation process to a halt and with it the conversion of sugars to alcohol. The natural sweetness of the wine is maintained, yet the brandy brings the wine up to a sufficient alcoholic content. Examples of dessert wines are port and sweet Madeiras. Brandy added to fully fermented and relatively dry wine prevents spoilage during aging. Dry sherry is an example of this type of fortified wine, also called an aperitif.

Sherry, probably the earliest fortified wine, begins as a dry white wine with an alcohol concentration between 11% and 14%. The taste and aroma of sherry are due to the oxidation of some of the ethanol to acetaldehydes. It is this oxidation that imparts the "nutty" flavor of a sherry. Oxidation can be accomplished in several ways, but the traditional method is to induce the oxidation by yeast indigenous to the Jerez region of Spain (from which the word *sherry* is derived). The yeast forms a white surface on the wine in patterns that resemble flowers and is called *flor*. After it is fortified with brandy, sherry has an alcohol content of 17% to 20%.

Climate Change and the Wine Grape

Vineyards may be a harbinger of the effects of climate change (see Chapter 26) upon agriculture. A recent study documented that the average temperature in twenty-seven prime wine producing areas has risen in the last 50 years with more than half showing an average increase of 1.26°C. Vineyards in Spain, Portugal, France, and Italy were subjected to even higher increases. The same study predicted that by 2049, these regions will show an average increase of another two degrees. Many famous wine regions will be too warm for the cultivation of grapes; the Tuscany area in Italy will shrink by 70% and the Bordeaux region of France will no longer be suitable for the cultivation of red grapes. Already vineyards are being established in northern areas such as southern England and parts of China. The expansion of vineyards into new areas may put pressure on wildlife and natural resources. One study stated that the best area for vineyards in China is an area now preserved as giant panda habitat.

The wine grape grows best in the temperature range of 12-22°C (54-72°F). Warmer temperatures in recent years have allowed the grape to remain on the vine longer, resulting in a higher sugar content. A higher sugar concentration results in more fermentable sugars for yeast and results in a higher alcoholic content in the wine. The alcoholic content of Napa Valley wines in California have risen from 12.5% in 1971 to 14.8% in 2001 as the temperature has risen. On the other hand, higher temperatures will cause photosynthesis to shut down resulting in lower sugars and acids which are the basis for flavor as well as reduced levels of anthocyanins which impart color to the wine. The prediction of more extreme events such as hailstorms and heat waves as the climate warms will result in damaged vines and greater susceptibility to fungal and other microbial diseases.

Grape growers have already begun to make modifications to offset the effects of global warming as much as possible. For example, positioning trellises to face north to keep grapes cooler and developing wine grape varieties more tolerant of warmer temperatures.

Climate change may also be involved in the decline in quality cork stoppers for wine bottles. Cork is harvested from the outer bark of the oak tree (*Quercus suber*) found in southwestern Europe and northwestern Africa; it takes 20-25 years to produce a cork harvest (see Chapter 18). Poor quality cork has thinner layers of bark with many lenticels; lenticels are passageways in the bark that allow air to enter the tree. Cork stoppers with many lenticels allow air and oxygen to enter the wine bottle and increase the probability of *Acetobacter* contamination and the subsequent conversion of alcohol into vinegar. In contrast, good quality cork has thicker layers of bark and fewer lenticels. Researchers discovered that the trees that produce good quality cork produce a greater number of heat shock proteins which prevent other proteins from denaturing (losing structure and function) even under stressful conditions such as drought and high temperatures. Also heat shock proteins promote cell division resulting in thicker layers of bark, which also shields the tree from ultraviolet radiation.

In contrast, poor quality cork had fewer heat shock proteins, but did produce twice the amount of phenolic compounds than good quality cork. Phenolic compounds pooled in the lenticular channels and provided protection by absorbing ultraviolet light. Also poor quality cork had greater activity in genes that repair damaged DNA and reduce oxidative stress but cell division was suppressed resulting in thinner layers of bark. High temperature and drought, brought on by climate change, is already becoming a problem in Portugal, one of the major cork-producing countries.

THE BREWING OF BEER

Beer is basically fermented grain and second only to tea as one of the world's most popular drinks. It has a long history that dates back at least 6,000 years. Some anthropologists have argued that the origin of beer predates that of bread and that agriculture developed to supply a ready source of cereal grains as much for the purpose of making alcoholic beverages as for food. Records and artifacts indicate that ancient peoples such as the Sumerians, Egyptians, Hebrews, Incas, and Chinese all knew and practiced the art of brewing. Recent discoveries of 5,000-year-old beer jars (pottery vessels used for fermentation or storage) verify that the Sumerians were

Figure 24.9 A Sumerian cylinder seal from the third millennium depicts beer drinkers. Note that the beer was drunk through straws; early beers were not filtered to remove the spent grains.

one of the earliest brewers (fig. 24.9). A later Sumerian clay tablet not only praises Ninkasi, the goddess of brewing, but also includes an ancient recipe for beer that uses *bappir,* a type of barley bread, as the start and honey and dates for sweetening. Recently, Anchor Brewing of San Francisco followed this 3,000-year-old recipe to recreate what might have been one of the oldest beers.

Barley Malt

Beer can be and has been made from any starchy carbohydrate source, but barley, used by the ancients, is the basis of most modern beers. Since beer making begins not with sugars but with starch, there are several preliminary steps in brewing before the yeast is introduced and fermentation can begin. The first step is the preparation of the *malt* (fig. 24.10). Barley grains are moistened with water and then allowed to germinate for a short time, spread out on a malting floor. As the barley grains germinate, they produce enzymes that catalyze the breakdown of starch. The germinated grains are then dried in a kiln (oven) and crushed to make malt powder. (Malt shakes are flavored with this powder.) The color of the finished beer is determined by how the malt was roasted; pale or standard malts are dried at low temperatures, while specialty malts are roasted at higher temperatures for a longer period of time to caramelize the malt sugars and so impart a darker color.

Mash, Hops, and Wort

The malt is now added to the grain starch and water and heated in a mash tun. This mixture is the mash, and during this step, the malt enzymes break down the starch to fermentable sugars (fig. 24.10). The starch of any cereal grain, such as wheat, corn, rice, or barley, can be the start for the brewing process. Beers made with both barley malt and grain are known as malt liquors. German *weissbier* uses wheat as both the malt and the carbohydrate source. Beers in the United States typically use rice as the grain starch. Eventually, the mash is strained, producing a clear, amber liquid called wort (fig. 24.10). The wort is boiled with hops in a brew kettle. The hop, *Humulus lupulus* (fig. 24.11), is a vine in the hemp family (Cannabaceae), which is found in northern temperate regions of Europe, Asia, and North America. It is actually the yellow-green pistillate flowers, whole or pelletized, that are added to the wort. It was discovered centuries ago by Bohemian brewmasters that the hops add a source of desirable bitterness to the beer to counteract the sweetness of the malt. With all the sugars in beer, it would be sickeningly sweet without the tartness of hops. Other plants have been added to beer for the same purpose, such as spruce, ginger, ground ivy, sweet mary tansy, sage, wormwood, and sweet gale, but hops is the most widely used. Much of the flavor and aroma of beer is attributable to the hops. Also, hopping apparently has an antibacterial action, keeping the beer from spoiling, and hopped beers are said to retain the foamy head longer.

Fermentation and Lagering

Before the yeast is introduced to begin the fermentation process, the wort is strained and cooled and placed in the fermentation tank (fig. 24.10). The main action of the yeast is to convert sugars into ethanol and carbon dioxide. Two types of yeast are used in the brewing process. Lager beers are fermented by yeast that settles to the bottom of the fermentation tank. (The German word *lager* means to store and refers to the aging process of this type of beer.) Other beers are fermented by yeast that rises to the top of the fermentation tank. (Ale is produced by a top-fermenting yeast that ferments at higher temperatures than the cold-fermenting lager yeasts. It is one of the traditional English beverages, a robust, full-bodied, high-hopped brew. Today's beer, first developed by German monks, in comparison, is lighter and more delicate tasting.) Most beers in the United States are lager beers. Fermentation continues for up to a week, at which time the wort is now a beer. This new, or green, beer is then transferred to lager tanks, where the beer is aged for 3 weeks (fig. 24.10). The flavor is often developed by adding beechwood chips to the tank. A second fermentation is started during aging by adding freshly yeasted wort, a process known as kraeusening. It produces the natural carbonation of the beer, although most mass-produced American beers are carbonated at bottling by adding carbon dioxide recovered from earlier fermentations. The lagering process lasts for several weeks. After lagering, the beer is clarified and packaged in bottles, cans, or kegs. Draft beers are sterilized by cold filtering, while most canned or bottled beers are pasteurized. The alcoholic content of beer is variable, from 3% to 12%, depending on the type of beer, but most range from 4% to 6%.

Two modern variants of brewing are light and nonalcoholic beers. Fermenting more of the carbohydrates in the mash produces a glass of light beer that has 95 kilocalories, as compared with the 150 in regular beer. Nonalcoholic beers are dealcoholized by brewing a beer normally and then reducing its alcoholic content to less than 1.1% by evaporation or distillation. A

Figure 24.10 Steps in the brewing of beer. Malting begins when barley grains are moistened in a mixing tank and spread out on the malting floor to induce germination. During germination, the grains produce enzymes that will break down starch. In the mash stage, malt and cereal grain are mixed with water, during which the malt enzymes convert grain starch to fermentable sugars. After several hours, the mash is strained, yielding a clear liquid, the wort. Wort and hops are added to the brew kettle and boiled. After the spent hops have been strained off, the wort must be cooled before the yeast can be added. With the addition of yeast, alcoholic fermentation begins. After several days, the green beer is transferred to lager tanks, where it is aged for several weeks. Finally, the beer is pasteurized or filtered before packaging in bottles, cans, or kegs.

CHAPTER 24 Beverages and Foods from Fungi

(a)

(b)

Figure 24.11 (a) The carpellate inflorescence of hops, *Humulus lupulus*, are added to the wort for flavoring. (b) Hop vines.

variant of this process is the malt beverage, which is brewed at a low temperature to reduce the fermentation rate so as not to go above 1.1% alcohol. Dilution can also be used to reduce the alcohol content of a beer to 0.5%. Nonalcoholic beers average about 55 to 82 kilocalories per glass.

Sake, a Rice "Beer"

Japanese sake, or rice wine, is produced by an intricate process that is more similar to beer making than to wine making. As in the brewing process, a cereal—in this case, rice—provides the grain starch that must be converted into fermentable sugars before yeast fermentation can begin. This conversion is accomplished by mixing steamed rice with the spores of the mold *Aspergillus oryzae*. The inoculated rice is then heated to a temperature of 35°C (95°F) for about a week. As the fungus grows on the rice, it produces enzymes that convert the starch into sugars. After several days, some of the resulting *koji* is mixed with more steamed rice and yeast is introduced to begin fermentation. The resultant *moto* serves as a starter culture and is mixed with the remaining batch of steamed rice (the *moromi*), continuing fermentation for up to 3 weeks. Sake has an alcoholic content of 20%, much higher than most beers and closer to what is found in distilled beverages.

Thinking Critically

Yeast exists in many different strains; the selection of the strain will determine to a great extent the flavor and other desirable characteristics of a beer. Lambic beers from Belgium are unique in the modern beer-making industry because they rely on spontaneous fermentation. In the lambic process, the wort is exposed to the open air to be inoculated by windborne wild yeasts.

Can you think of any difficulties that might arise from the practice of spontaneous fermentation? Does it surprise you that lambic beer is produced successfully in only a small area around Brussels and only during the months of October to April, when outside temperatures remain under 15°C (59°F)? Explain.

DISTILLATION

Distillation is the process by which a mixture of substances is separated by their differing boiling points. The process of distillation was developed by Arab alchemists in about A.D. 700, and the word *al kohl*, from which *alcohol* is derived, simply meant the distilled essence of any substance. A **distilled spirit** is made by boiling a beer or wine to leave the water behind and concentrate the alcohol. Alcohol has a lower boiling point than water, so heating an alcoholic beverage above the boiling point of alcohol, but below the boiling point of water, will vaporize all the alcohol and some of the water. The vapors are collected in a pipe; cool water running over the pipe recondenses the alcoholic vapors, creating a liquid with a higher alcohol content, or proof, than is possible through fermentation. (Proof number is equal to twice the alcoholic percentage.)

The Still

In the pot-still, or batch, method of distillation (fig. 24.12), the beer or wine is heated in what is essentially a huge, covered copper kettle until the alcohol vaporizes. Rising from the pot is a neck, or chimney, to collect the alcoholic vapors. The vapors are passed on to cooling coils, worm-shaped tubes encased in cold water. As the alcoholic vapors pass through, they recondense to liquid.

In a column, or continuous, still, the distiller has more control in separating out undesirable fractions from the alcoholic vapors. This type of still is shaped like a column, with a series of perforated plates on the inside. The beer or wine is introduced near the top of the column and percolates downward. Steam is introduced, and as the liquid moves down the column, it is repeatedly boiled and continually releases alcoholic vapors. By the time the liquid has reached the bottom of the still, only water and solids remain. The alcoholic content of the liquid has all been boiled off. The undesirable higher alcohols, essentially the extremely potent "heads" or acetaldehyde (the substance that gives moonshine its raw "kick"), vaporize at the top of the column at the lower temperatures (21°C; 70°F), while the equally unwanted "tails," the lower alcohols or fusel oils, are given off at the higher temperatures (84°C; 183°F). Vapors from ethanol are given off in the middle temperatures (78°C; 173°F). Fractioning, or separating out, the different alcoholic vapors eliminates the undesirable higher and lower alcohols from the distilled product and concentrates the favorable alcoholic vapors from ethanol. This process of rectification or purification requires the skill of an expert distiller. In contrast, with the pot still, the alcoholic beverage is heated for a certain amount of time to allow the higher alcohols to vaporize off and escape. The separation is not as fine as in a continuous still; some of the higher and lower alcohols are invariably mixed with the ethanol, producing a harsher and more potent product.

Figure 24.12 Copper pot stills are used in the distillation of beer to Scotch whisky.

Distilled Spirits

The alcohol content of distilled spirits ranges from 40% to 50%, or 80 to 100 proof. A grain spirit or neutral spirit is produced when the fermented beverage is highly distilled above 190 proof and is almost pure ethanol. It is colorless and tasteless and is used in the making of gin and vodka. Both are classified as unaged spirits because they age for only a few weeks. Vodka is the distillation of a grain or potato beer and can be either unflavored or flavored with herbs. Gin is the distillate of a grain or cane beer (made from fermenting molasses). The characteristic taste of gin is from its steeping in the modified cones, or "berries," of juniper and other herbs.

Most other distilled spirits are aged for a period of time in wooden barrels. Brandy, including cognac, is a twice-distilled grape wine that has been aged in oak casks. The color, aroma, and flavor associated with brandy are derived from the tannins, lignin, vanillin, and sugars that the alcohol picks up from the oak wood during the aging process. Different designations on the label of the bottle indicate the number of years the brandy has been aged. The most common designation is V.S.O.P., which stands for *very special old pale* and indicates that the hue of the brandy is not from adding caramel coloring but the result of aging in oak for at least 20 years. Some brandies are aged up to 50 years. The most distinctive brandies are produced in the Cognac region in southwestern France.

The term *whiskey* is derived from the Gaelic and means water of life. Any whiskey[*] is a distillate of a grain beer. The grain in Scotch whisky is malted barley, kilned over a peat fire. The smoke from the peat gives this whisky its distinctive flavor. After distillation, it is aged in oak for at least 3 years.

Bourbon is a corn whiskey whose development coincides with the colonial period and early years of the American republic. It is named after its place of origin, Bourbon County, Kentucky. The principal grain is corn, from 70% to 90%, with small amounts of rye or wheat. The color, aroma, and flavor of bourbon are distinct from those of other corn whiskeys because of charred barrel aging. To legally be classified as bourbon, the whiskey must be aged in new, charred oak barrels for at least 4 years. Some are aged for 10 to 12 years. (The oak barrels must be new and cannot be used more than once. The used barrels are often sawed in half and sold as flower planters.) It is said that the Lewis brothers, Elijah and Craig, invented this distinctive aging process in 1789.

The Whiskey Rebellion

Every student of American history knows that the Civil War divided the country into North and South camps and nearly ended the union known as the United States, but few remember that an earlier rebellion, incited by a tax on whiskey, pitted East against West and threatened the survival of the Republic during the first decade of its existence. With the support of Secretary of the Treasury Alexander Hamilton and

---

[*]*Whiskey* or *whisky* is spelled according to place of manufacture; the first spelling indicates the United States or Ireland and the latter is made in Scotland.

the Federalist Party, Congress passed the excise Tax Act of 1791. Hamilton argued that the tax on whiskey was necessary to offset a budget shortfall incurred by the expanding responsibilities of the government. Hamilton and the Federalists supported a strong central government and believed that the passage of the whiskey excise would also solidify the authority of the federal government to impose and collect taxes.

The tax incensed the people of the western frontier because whiskey was their single most important item of trade. Frontier farmers grew their fields of corn or rye and routinely converted these crops into whiskey. Barrels of whiskey were easier to transport across the Appalachian Mountains and could be expected to sell at a higher rate than grain in eastern markets. The excise was an internal tax, a tax on what the westerners made and sold and not on what they purchased. It was exactly this type of taxation that the Founding Fathers had opposed so vehemently in the years leading up to the Revolutionary War, yet the same leaders were now imposing a whiskey excise on their western brethren.

The fact that the tax was directed specifically at westerners augmented feelings of isolation and alienation in the region. Westerners felt distanced from the seat of power in the East and felt that the concerns of the frontier region were being ignored by the federal government. The government had yet to improve travel in the western counties; no major roads or canals were being built in the West. Indian attacks on settlers were rampant, and the army had recently suffered two major defeats in the military expeditions of 1790 and 1791 in Ohio country. The West demanded protection from the warring Indians that the government had as yet been unable to provide its citizens.

Not surprisingly, federal officers who were sent to collect the tax met strong resistance, even hostility. Some of the westerners simply refused to pay the excise; others took a more aggressive stance. Revenue officers were tarred and feathered, and their homes and offices were burned to the ground. Peaceful demonstrations evolved into riots as the frequency and violence of the resistance movement escalated in the summer of 1794. Radicals increasingly urged the West to separate from the Union and establish an independent country.

President Washington could no longer ignore the uprising and talk of secession. To preserve the Republic, he issued a proclamation in August 1794, calling for a militia to suppress the insurrection. Under the command of General "Light Horse Harry" Lee, governor of Virginia and father of Robert E. Lee, a force of 13,000 men was mobilized from New Jersey, Virginia, Maryland, and Pennsylvania to squelch the rebellion in the West. The "Watermelon Army," as it was derisively nicknamed, marched on to western Pennsylvania to meet the rebels.

As the militia approached Pittsburgh, several prominent leaders fled to the wilderness to avoid capture. By November, the Watermelon Army had rounded up the remaining leaders of the Whiskey Rebellion, 12 of whom were marched back east for trial in Philadelphia. Only two were found to be guilty of treason, and they were later pardoned by President Washington. In 1802, President Thomas Jefferson, who had always been a strong opponent of the tax and a firm supporter of states' rights, repealed the whiskey excise that had started a rebellion.

Tequila from Agave

Found primarily in arid regions of the American Southwest and Mexico, agaves are a monocot family (Asparagaceae) of perennial plants with large, water-storing leaves arranged in a rosette around a short stem. Leaves can be variegated in stripes of green and yellow or in solid shades of greenish to bluish gray. *Agave tequilana,* the blue agave of tequila commerce, is so called because of the definite blue tint of its leaves. Leaves of agaves look and cut like sword blades, with a large, sharp spine at the tip and regularly placed smaller spines along the margins. The common name, century plant, refers to the characteristically long wait before an agave flowers. Agaves bloom only once, near the end of their long lives, when a towering inflorescence sprouts from the center of the rosette. Soon after, the main plant dies but new shoots arise vegetatively from the underground rhizome as replacements. Indigenous peoples have used agaves for a multitude of products: soap, medicine, fiber (see Chapter 18), and a variety of drinks—aguamiel, pulque, and mezcal.

Agaumiel and pulque are agave beverages that date from the pre-Columbian era. Aguamiel—literally, "honey water"—is the fresh, sugary juice of agave. When an emerging flowering stalk of agave is cut, a milky sap starts to flow. This is the source of aguamiel, which can be collected daily from the wound for up to 2 months. If the sap is collected in barrels and fermented by wild yeast, pulque results. Pulque has an alcoholic content of 8%. Both the Mayans and the Aztecs restricted the drinking of pulque to certain religious ceremonies and other special occasions. Mayahuel, the Aztec goddess of fertility, is usually depicted seated in an agave, nursing babies with pulque. Magueys are agaves from which pulque is made, and they typically have thick, gigantic leaves, which can span the length of a human being.

Mezcal (also spelled mescal) is distilled pulque. There is some debate, but it is believed that the distillation of pulque to make mezcal came about after the Spanish took the still to Mexico. By law, tequila is a type of mezcal produced only from the sap of the blue agave and cultivated primarily in the state of Jalisco (where the town of Tequila is located) and four other areas in north-central Mexico. The mezcal from these areas was officially named tequila in 1873 to distinguish this product from other mezcals.

Two of the prime areas for growing blue agave are the lowlands that surround the town of Tequila and the highlands of Jalisco. In the lowlands, the elevation is about 1,200 m (4,000 ft), and the black soils are volcanic in origin. Agaves mature in approximately 7 years in the Tequila lowlands, but the stems of lowland agaves are smaller at harvest, 55 to 80 kilograms (125 to 175 pounds), and their sugar content averages around 25%. The elevation of the highlands is about 2,100 m (7,000 ft) and the red clay soil is rich in iron. Agaves grown in the highlands have sap with a higher sugar content (27%–28%)

and produce larger stems, averaging from 90 to 125 kilograms (200 to 275 pounds), but the maturation time is 8–10 years.

During cultivation of the blue agave, any vegetative offshoots are cut off the mother plant and transplanted to other fields. The stem of the blue agave is harvested prior to flowering, which increases the sugar content. Field workers, or jimadors, use a *coa*, a tool with a long handle and a round-edged blade, to prune off the leaves and dig up the stem. The pruned stem is called a *piña* because of its resemblance to a pineapple.

Traditionally, agave piñas were roasted in pits or ovens for several days, but in the modern process they are commonly pressure-cooked in autoclaves. During baking, the primary complex carbohydrate, found in agave, is converted into fermentable sugars. The baked piñas, now a caramelized brown, are then sent to be mashed in a mill, where the agave juice is siphoned off from the vegetable remnants. In the old method, the piñas were placed in a shallow pit and crushed by a large, heavy, round stone as it turned in a circle powered by the work of men or burros. Today the modern method is to mill the piñas between steel rollers.

The expressed agave juice is then transferred to vats, where yeast is added to begin fermentation. By law, tequila must be made from at least 51% pure agave juice. A designation of 100% indicates that the tequila is made solely from agave sugars. If cane sugar or other sugars are added, the tequila is designated *mixto*. After fermentation, two successive distillations follow. Three basic types of tequila are distinguished according to the time spent in aging. Tequila *blanco* is the color of water and bottled soon after the second distillation. Tequila *reposado* is light golden and aged in oak barrels for a few months. Tequila *añejo* is darker gold and aged for a year or more.

Thinking Critically

As the popularity of tequila spreads beyond Mexican borders, more acreage has been devoted to the cultivation of blue agave. Increasingly, blue agave cultivation has been subjected to devastating attacks by insects (especially weevils) and plagues, particularly from the bacterium *Erwinia carotovora* in combination with the fungus *Fusarium oxysporum*. These pathogens cause the stem to rot and die in a characteristic pattern known as TMA or *tristeza y muerte de agave* (wilting and death of agave). By the late 1990s, it was estimated that more than 25% of the agave crop was infected with one or more diseases. Although there was some recovery in the mid-2000s, agricultural experts expect that the blue agave crop is still at risk.

Why is the blue agave as it is cultivated for tequila production particularly susceptible to rampant insect infestation and disease? Some researchers have suggested that part of the agave crop should be allowed to flower and be naturally pollinated by moths and bats. How could this practice help reduce the spread of insect pests and disease throughout the agave growing regions?

Hard Cider

Apple juice is the source of both an unfermented and fermented drink called cider. To Americans, unfiltered apple juice is called *cider*, and fermented apple juice is *hard cider*, but to most of the world, cider is synonymous with the fermented alcoholic beverage.

The harvesting of apple fruits to make cider and other alcoholic beverages in an ancient one. As discussed in Chapter 3, the domesticated apple tree, *Malus pumila* in the Rose Family (Rosaceae) is native to Kazakhstan, a nation in Central Asia bordering China and Russia. When the Romans invaded western Europe, they found that apple trees had already spread across Europe. The peoples of southwest England, northwestern France, and Spain grew apple trees and developed methods to ferment the juice into cider. Most settlements in the American colonies produced cider. The first apple orchard in Massachusetts was established in 1623 and Peter Stuyvesant, governor of New Amsterdam, planted an apple orchard in 1647 in what is now the Bowery district of Manhattan. Cider became so important to colonial America that it served as currency; was used to pay salaries; and often, the prices for goods and services were translated into number of barrels of cider.

Originally apple orchards were seedling orchards where every tree was started from seed and each tree would have a unique genetic profile and distinctive tasting juice. Consequently, cider would be made by blending all of the fruit of the orchard not sweet enough to eat to produce a desired taste. The Ancient Greeks and Romans perfected cider making after grafting (see Chapter 3) developed around 50 B.C. where trees with fruits of sought after qualities could be propagated asexually and desirable cultivars for different qualities became available. By the mid-1500s, the Normandy area of France was home to 65 named apple cultivars and for centuries, many of the best ciders came from here. As long ago as the mid-1600s, cider was being distilled in France to boost alcoholic content.

The process of making cider begins with the harvesting of the apples. Sweating is the practice of storing the harvested apples for several days or weeks before they are milled. During the sweating process, the apples lose up to 10% of their moisture and more starch is converted into sugars at a gain of 6–8%. Both of these processes result in a sweeter juice, which is directly related to the alcoholic content of the hard cider. Next the apples are ground into a pulpy mass known as pomace. Traditionally, large circular grinding stones driven by animal or people power were used to grind the fruit into the pomace, but modern techniques use mechanical grinders. Pressing is the next step. Historically, the apple pomace was interspersed between layers of straw; modern practices have replaced the straw layers with heavy duty fabric called press cloths. A screw type press is used; as the screw is tightened, the increasing force on the large wooden press disk mounts, squeezing out the apple juice from the pomace. The juice is collected and the spent pomace is often used as mulch.

Two qualities important in selecting apple cultivars suitable for the making of cider are tannin concentration and acidity. Sweet cultivars, such as Morgan Sweet, are both low in tannin and acidity; Frederick apples are classified as sharp being low in tannin but higher acidity; apples with higher tannin and lower acidity as the White Jersey cultivar are classified as bittersweet; whereas Kingston Black and Foxwhelp, bittersharp cultivars, are higher in both tannin and acidity.

Since apples are much lower in sugar concentration than grapes, yeast have access to a lower supply of fermentable sugars resulting in a lower alcoholic content only averaging 4–6% in cider.

Cider could be transformed into the higher potency of applejack by freeze distillation, by placing barrels of cider outdoors in the freezing cold of winter. As the water freezes in the cider, the ice is removed, increasing the alcoholic concentration. Not only is applejack more potent in alcoholic content, it often contains high quantities of toxic substances, such as aldehydes and fusel oils, which if consumed regularly over time, damage the liver and result in a condition called apple palsy. Distillation of cider in a still also raises the alcoholic content but removes these undesirable heads (acetaldehyde) and tails (fusel oils).

Apple brandy is made by distilling fermented apple juice or apple mash and has a minimum of 40% alcoholic content. Some commercial producers of apple brandy will add nonfermentable sweeteners such as aspartame or saccharine to retain the sweetness of a cider. Often apple brandy is aged in oak barrels, which imparts additional flavors. In the United States, applejack is another commonly used name for apple brandy. Calvados is an apple brandy named after the particular area of northern France where it was first developed. By law, calvados must contain at least 20% of the cider made from local apple varieties; also calvados must be made from at least 70% of the bitter or bittersweet varieties and no more than 15% sharp varieties. It has a minimum alcoholic content of at least 40%.

Apple liqueur is also a distilled cider, but it is sweeter with a lower alcoholic content of 20% and may also be aged in oak barrels. Sometimes additional sugars and yeast are added to induce a second fermentation in the bottle, making the cider bubbly in a process that mimics that of making champagne. Apple wine is a cider to which sugars and yeast are added to promote greater fermentation to bring up the alcohol content to at least 7%; it is typically not carbonated.

Hard cider, the staple of colonial America that all but disappeared during Prohibition in the 1920s–1930s, was reintroduced by vintners in Vermont and California in the 1990s and continues to gain in popularity in the twenty-first century.

A VICTORIAN DRINK REVISITED

Recently, there has been a revival of the Victorian influence in architecture, home decor, and crafts. This interest has extended to Victorian-era concoctions such as laudanum (a mixture of alcohol and opium; see Chapter 20) and absinthe. Absinthe is a liqueur infused with oil of wormwood, which is extracted from the leaves, flowers, and stems of *Artemesia absinthium* or Roman wormwood, *Artemesia pontica*. Popular in the nineteenth century, absinthe had been banned by the second decade of the twentieth century in the United States and most European countries when its toxicity was finally recognized.

Absinthe

Wormwood has been valued as a medicinal herb since antiquity. Both the Roman naturalist Pliny and Dioscorides, a physician for the Roman army, recommended it to expel intestinal roundworms from humans and animals. Another wormwood species, *Artemesia annua,* has been used by the Chinese as an antimalarial (see Chapter 19). Wormwood is an herbaceous, somewhat shrubby, perennial of the sunflower family (Asteraceae). The leaves of wormwood are whitish green with a pinnately lobed shape, and when crushed, they release a pleasant aroma. The flowering heads are globe-shaped and composed of small, tubular flowers of a yellowish green hue.

Although the primary component of absinthe was oil of wormwood, several other herbs were added as flavoring agents. These included anise, fennel, hyssop, lemon balm, angelica, juniper, nutmeg, star anise, and others, depending on the specific recipe (of which there were many variations). Absinthe was made by macerating wormwood and the other herbs and steeping them in a concentrated ethanol solution for at least half a day. An equal amount of water was then added, and the mixture was distilled. To the distillate, more wormwood and some of the other herbs were added, heated, and then filtered. An alternative method of preparation was to add the essential oils of wormwood and other herbs directly to grain alcohol. The end product was an emerald-green, tart-tasting liqueur with an alcoholic content around 75%.

The Green Hour

Absinthe became so popular in France that Parisian cafés and clubs arose solely to serve the public's demand for the drink. New Orleans imported the French fascination with the liqueur, and visitors to the French Quarter can still find the Old Absinthe House on Bourbon Street. The daily ritual of unwinding with a glass of absinthe became known as *l'heure verte* (the green hour). Presentation was part of the absinthe mystique. A shot of absinthe was poured into a glass, and a sugar cube on a specially designed slotted spoon was placed atop the glass. Cold water was then poured over the sugar. The sugary water diluted the absinthe, making it less bitter, and magically transformed its color from crystal green to milky yellow.

Many famous artists and poets of the period embraced the absinthe habit, claiming that it enhanced creativity. The Irish poet and playwright Oscar Wilde was enthusiastic about the drink. The absinthe experience even became the subject of artwork by such major artists as Edouard Manet, Edgar Degas, Henri de Toulouse-Lautrec, and Pablo Picasso (fig. 24.13).

Active Principles

The bitter taste of absinthe is due to the presence of absinthin, a terpene, in oil of wormwood. Absinthin is an extremely bitter substance that still can be detected even in dilutions of 1 part to 70,000.

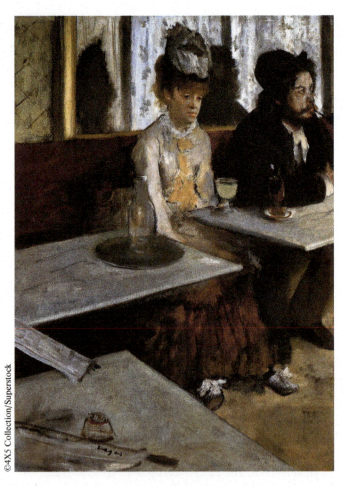

Figure 24.13 Absinthe was often the subject of artwork by major artists of the period. *L'Absinthe*, Edgar Degas, 1876.

In addition to ethanol, absinthe contains thujone, a potent psychoactive compound. Chemically, thujone is another terpene, and is the primary component of wormwood oil. Thujone is found in the essential oils of other plants, including sage (*Salvia officinalis*), tansy (*Tanacetum vulgare*), and the arborvitae group of conifers, specifically *Thuja occidentalis*, from which it was first extracted. Thujone is soluble in the high alcoholic content of the liqueur, but when water is added, the essential oil precipitates out, forming the milky yellow suspension.

Absinthism

The dangers of the absinthe habit were first reported in the 1850s. Absinthism was described in medical texts in the mid-nineteenth century as a form of alcoholism, but physicians also noted that absinthe abusers suffered from distinctive neurological damage. A syndrome of auditory and visual hallucinations, psychoses, sleeplessness, tremors, convulsions, and paralysis was observed in persons suffering absinthism. It has been suggested that the madness and eventual suicide of Vincent van Gogh were in part attributable to his addiction to absinthe. Medical researchers, beginning in the 1860s, attempted to verify absinthe's effects and identify the toxic principle in experiments with laboratory animals. Rabbits and dogs given absinthe in laboratory studies suffered hallucinations, convulsions, and impaired respiration—symptoms that had also been observed in human beings addicted to absinthe. In 1864, experiments at a Parisian hospital reported that dogs and rabbits given oil of wormwood foamed at the mouth and went into convulsions. In 1868, "Experiments and Observations on Absinth and Absinthism" appeared in the *Boston Medical and Surgical Journal* (now the *New England Journal of Medicine*). Later studies showed that the administration of a single large dose of oil of wormwood caused epileptic seizures in laboratory animals.

In 1900, the chemical nature of thujone was deciphered by a German chemist. Experiments showed that thujone stimulates the autonomic nervous system. It also causes unconsciousness and convulsions. Experimental evidence suggested that the toxic constituent in oil of wormwood and the underlying cause of absinthism are due to the activity of thujone.

The findings of the scientific community eventually reached the public, but few heeded the warnings. The lure of the green hour was too strong, despite its obvious dangers. In fact, in France, the consumption of absinthe peaked in 1913, 2 years before the sale and manufacture of the liqueur were officially banned in that country. After the ban, one of the largest French manufacturers of absinthe switched to the production of a wormwood-free substitute, Pernod, that is still available today.

Thinking Critically

Absinthe was a liqueur that inspired a devoted following in the nineteenth century. It was banned in many countries when experimental evidence confirmed its toxicity.

Recently, people are rediscovering the absinthe habit. How might the history of absinthe be useful to alcohol and drug education programs?

Despite the lessons learned from the past century, absinthe is the latest trend in the bars and coffeehouses of the Czech Republic, where its production and sale are legal.

Recently in the United States, a 31-year-old man purchased oil of wormwood through the Internet and, mistaking it for absinthe, drank about 10 milliliters (0.3 fluid ounces) of the oil. Several hours later, he was found disoriented and he began having seizures. After being hospitalized, he suffered complete kidney failure brought on by the presence of myoglobin in his urine. Myoglobin is a muscle protein that is released during muscle injury—in this case, brought on by the violent convulsions. After a week in the hospital, the patient was released. A Closer Look 24.2: Alcohol and Health continues the discussion on the many physiological effects of alcohol.

A CLOSER LOOK 24.2
Alcohol and Health

Alcohol is a psychoactive drug affecting body organs, the brain, and the peripheral nerves. When a person drinks an alcoholic beverage, some alcohol is absorbed into the bloodstream directly through the stomach wall, but most is absorbed through the wall of the small intestine. This route of absorption explains why drinking on a full stomach slows the absorption of alcohol, whereas drinking on an empty stomach speeds up its absorption and hence its effects.

Alcohol is classified as a depressant of the CNS (central nervous system), inhibiting the centers in the brain that deal with speech, vision, balance, and judgment. In the case of an alcohol overdose, the level of alcohol is so high that it depresses the respiratory control center in the brain stem, and death results.

Alcohol eventually ends up in the liver, where the ethanol is metabolized at the rate of 10 milliliters per hour (1/3 ounce per hour). The enzyme alcohol dehydrogenase degrades ethanol first to acetaldehyde, which is then converted to acetate, and eventually to carbon dioxide and water by other enzymatic reactions.

Evidence is accumulating that moderate consumption of alcohol lowers the risk of coronary heart disease (CHD) by more than 30%. (Moderate consumption is defined as drinking 20 to 30 grams of alcohol or one to two standard drinks per day. A standard drink [box fig. 24.2a] is approximately one 12-ounce beer, a 5-ounce glass of wine, or a 1.5-ounce drink of distilled spirits per day.) Despite a sedentary lifestyle, a diet high in saturated fats (14% to 15% of the daily intake), and serum cholesterol levels over 200 milligrams, the people of France have a 40% lower incidence of CHD than Americans do. Known as the French Paradox, this seeming contradiction is most likely due to the consumption of red wine on a daily basis.

Alcohol seems to afford protection against CHD in several ways, primarily by raising HDL levels and reducing the tendency to form blood clots. HDL is the "good cholesterol" that removes cholesterol from the walls of blood vessels and returns it to the liver for degradation (see Chapter 10). People who drink one to two alcoholic beverages every day have HDL levels that are 10% to 20% higher than do individuals who are similar in every way except that they abstain from alcohol. A person with an optimal level of HDL will develop fewer atherosclerotic deposits in the walls of blood vessels. Atherosclerosis is completely inhibited when the amount of alcohol is excessive. Since the early 1900s, physicians who have performed postmortems on alcoholics who had died from liver disease have noted that there were no signs of atherosclerosis. The HDL associated with moderate alcohol consumption is different from the HDL produced by regular exercise, but both types protect against atherosclerosis and CHD. It is still unclear how alcohol elevates HDL. One suggestion is that alcohol stimulates HDL-producing enzymes in the liver.

Alcohol is also known to disrupt the abnormal blood clotting that takes place in atherosclerotic regions of the coronary artery and leads to angina (pain radiating down the left arm due to decreased blood flow to the heart muscle) and heart attacks. Lowering the risk of blood clots decreases the likelihood that a clot will form and block the coronary artery and thus prevents a heart attack. In the presence of alcohol, platelets—cellular fragments in the blood—appear to be "less sticky," or less likely to aggregate, which is one of the preliminary steps in the formation of a blood clot. Additionally, alcohol raises the blood level concentration of plasminogen activator, a clot-dissolving enzyme, and lowers that of fibrinogen, a factor that promotes clots. People who are light drinkers, consuming only two to three drinks per week, also show a reduced risk of CHD that is probably attributable to these inhibitory effects on blood clotting.

Alcohol consumption may also lower the risk of CHD through other modes of action. Some research indicates that moderate alcohol consumption decreases the risk of type 2 diabetes, a disease that confers an increased risk of CHD. Alcohol may also have an anti-inflammatory effect on endothelial (blood vessel lining) tissue; some researchers have suggested that the inflammation of the lining of the blood vessels contributes to the development of CHD.

Some researchers have identified that the flavonoids from grape skins present in red wine also inhibit platelet aggregation. **Flavonoids** are a group of polyphenols that include flavonols, flavanols, and anthocyanins. They are the substances that impart color, taste, and texture to plant food crops. In addition to preventing clotting, flavonoids are powerful antioxidants. One factor in the development of atherosclerosis is the oxidation of LDLs, which makes them more easily taken up by the lining of blood vessels. Once the oxidized LDLs have been taken up by the blood vessel lining, other changes are triggered that make platelets prone to aggregate and form clots. It is suggested that flavonoids, because of their antioxidant activities, prevent the oxidation of LDLs and thus prevent the onset of changes leading to atherosclerosis. Other data suggest that polyphenols prevent atherosclerosis by inhibiting the release of the hormone endothelin. Under the influence of endothelin, blood vessels constrict. Reduced blood flow leads to platelet aggregation and thus atherosclerosis.

A recent study sought to discover which wine-making region produces wines that have the greatest inhibitory effect on the release of endothelin. Researchers obtained 100 red wines from wine-producing regions around the world. In the laboratory, they exposed cells lining the blood vessels to the wines and measured the inhibition of endothelin. They found that wines made from grapes grown in southwestern France and the Italian island of Sardinia were the most effective in inhibiting the release of endothelin. These wines also had the highest concentration of polyphenols, which are some of the tannins in red wines. When the

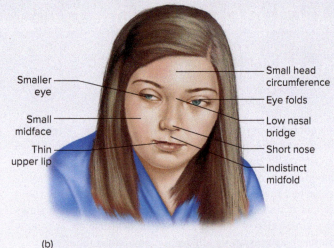

Box Figure 24.2 (a) A standard drink of beer (12 ounces), wine (5 ounces), and distilled spirits (1.5 ounces). (b and c) Fetal alcohol syndrome occurs in children whose mothers drank heavily during pregnancy. These children exhibit telltale facial abnormalities and often are mentally retarded, are learning disabled, or have behavioral problems.

researchers checked census data for France, they found that the people in the southwest lived the longest.

Resveratrol, one of the polyphenols found in red wine, has been reported to extend the longevity of a variety of test organisms, such as yeast, worms, flies, and fish. Resveratrol appears to work by increasing the production of proteins, called sirtuins, that were linked to longevity by earlier research. A study sought to examine the effects of resveratrol on mammals. Mice were divided into three groups. One group ate a healthy, nutritionally balanced diet. A second group was fed a high-fat diet. The final group ate the same high-fat diet but were dosed daily with resveratrol. Both groups on the high-caloric diet quickly became fat, but the mice not fed resveratrol soon died from obesity-related diseases, such as diabetes and heart disease. The resveratrol-dosed mice remained as healthy as those fed the nutritionally balanced diet. Resveratrol apparently protected the mice from the ill effects of obesity. A recent study found that resveratrol may have anti-cancer benefits. Mice, who were fed a high-fat diet, an associative cause of colon cancer, developed fewer tumors when administered low doses of resveratrol than those given high doses of the chemical.

Some studies show that slight differences in the incidence of CHD are dependent on the type of alcohol consumed. The risk is slightly lower for wine drinkers than for those who consume beer or hard liquor. However, these results are not conclusive because wine drinkers generally tend to have healthier lifestyles than those who regularly consume beer or hard liquor. Many people who drink wine follow a Mediterranean diet that consists of plenty of fruits, vegetables, fish, salads, and olive oil, which might also contribute to their lower risk of CHD.

Alcoholism affects 5% of drinkers. In the chronic alcoholic, the liver is one of the first of many organs to be seriously impaired. First, liver cells become fatty because alcohol disrupts the metabolic breakdown of fats. As fat accumulates in these cells, the cells enlarge and often rupture or form cysts, replacing normal cells. In the last stage, these are replaced by scarring, or cirrhosis, that can impede blood flow to the liver and disrupt normal functioning of this vital organ.

Heavy drinking is not more protective than light or moderate drinking; in fact, heavy drinking and binge drinking seem to reverse the benefits of moderate consumption. Chronic heavy drinking has adverse effects on many of the body's vital organs. The brains of alcoholics show atrophy due to the death of nerve cells. Recent studies seem to indicate that the brains of female alcoholics are more severely injured than are those of male alcoholics. An alcoholic's heart is usually larger than normal because one indirect effect of alcohol is to elevate blood pressure. Scarring of the heart muscle results in arrhythmia. The immune system is depressed, and malnutrition is common. In young women, heavy drinking is associated with a higher risk of breast cancer.

In 2018, two meta analyses were released that stated that the drinking of alcohol, even in moderate amounts, is detrimental to health. It found that those who drank more than seven drinks per week, had a lower-life expectancy and increased risk for strokes, heart attacks, aneurysms, and other health problems than non-drinkers. As the amount of alcohol imbibed increased, the risk of earlier death was also raised. In fact, the increased risk of death begins if a person consumes slightly more one drink per day. These findings have led some researchers to state that no amount of alcohol is good for health.

Perhaps the most tragic consequence of heavy drinking is fetal alcohol syndrome (box figs. 24.2b and c). First identified in 1970, fetal alcohol syndrome is the damage done to unborn children whose mothers drank during pregnancy. When a pregnant woman drinks, the fetus is also exposed to the alcohol because it can cross the placenta.

Laboratory experiments with rat pups have shown that a single exposure to high blood alcohol levels during the first 2 weeks after birth causes brain damage by killing nerve cells: this damage comes about because alcohol selectively inhibits the excitatory neurotransmitter glutamate while stimulating GABA, an inhibitory neurotransmitter. The rise in GABA triggers nerve cell death 24 hours after exposure. This study has implications for women who drink in the third trimester of pregnancy, a time of fetal brain development that corresponds to that of the 2-week-old rats. Late-pregnancy drinking is extremely harmful to fetal brain development and is strongly discouraged.

Children with fetal alcohol syndrome exhibit a range of symptoms such as facial abnormalities, mental retardation, and behavioral problems. The recommendation from the medical community is to avoid alcohol completely during pregnancy because even moderate consumption of one to three drinks per day can cause damage to brain development.

FUNGI AS FOOD

Fungi are eaten directly as a type of vegetable or are used as fermentative agents to convert foods into alternative forms.

Edible Mushrooms

Mushrooms have long been a fixture in many European and Asian dishes. As Americans have become familiar with their delicate flavors and aromas, a demand for a greater variety of these edible fungi has developed.

Most commonly eaten fungi are true mushrooms and members of the Basidiomycota, or club fungi. The basidiocarp, or fruiting body, is the part that is eaten. Two popular exceptions to this general rule are the highly prized and costly morels and truffles, fruiting bodies of the ascomycetes *Morchella* and *Tuber,* respectively (see Chapter 23).

Until recently, most edible fungi had been collected from the wild, but the number cultivated to produce a greater and more reliable supply increases with every year. Cultivation eliminates the danger of mistaking a poisonous mushroom for an edible one, no small threat because every year hundreds of deaths are reported from accidental mushroom poisonings (see Chapter 25). Contrary to folk wisdom, there is no reliable test to distinguish a toadstool from an edible mushroom. The only sure way is to correctly identify the mushroom; in some cases, that is a difficult task, even for an expert.

Two mushrooms, the button, or field, mushroom, *Agaricus bisporus,* and the shiitake mushroom, *Lentinus edodes,* have been domesticated and cultivated for some time. The field mushroom is first in worldwide production and is the common white mushroom sold in Western produce departments. Its cultivation dates back to 1650 in France. The shiitake mushroom appeared more recently in U.S. markets but is one of the most widely eaten mushrooms in East Asia, having been cultivated in China and Japan for several hundred years. The shiitake is cultivated on oak logs or, more recently, on synthetic logs created from sawdust and other organic materials and allowed to grow. In addition to its culinary uses, it has been valued in eastern Asia for its medicinal properties. In Japan, it is widely prescribed in cancer therapy for its antitumor action as well as to counteract undesirable effects from conventional chemotherapy. In the United States, shiitake tea, made from soaking the dried mushrooms in boiling water, is claimed to boost the immune system, lower blood cholesterol, and promote weight loss. Others widely cultivated are the paddy straw mushroom (*Volvariella volvacea*) and the oyster mushroom (*Pleurotus* spp.) (fig. 24.14). Some examples of the edible fungi currently in cultivation are listed in Table 24.2.

Nutritionally, fresh mushrooms, as do most fresh fruits and vegetables, have a high water content, 85% to 92% of the fresh weight. They are a source of complete protein and have appreciable amounts of vitamins C, D, and some of the Bs. They are also naturally low in kilocalories and high in fiber.

Cultivated mushrooms are grown on compost (fig. 24.15) made from agricultural and industrial wastes. The exact nature of the compost differs with the specific fungus under cultivation, but cereal straws are the main ingredient of most. An important consideration in preparing the compost is to have the correct carbon-to-nitrogen ratio. Cellulose, hemicellulose, and lignin present in the straw are the carbon sources. Nitrogen is derived from proteins and amino acids, and the compost is sometimes enriched with organic nutrients (such as horse manure) and inorganic nutrients to meet the necessary nitrogen requirements. Even so, it has been difficult to meet the specific nutrient requirements and environmental

Figure 24.14 The oyster mushroom, *Pleurotus ostreatus*, can be found on dead or dying trees. Although still sought in the wild by many collectors, it is now widely cultivated and available in many supermarkets. World production is approximately 20,000 tons annually.

Figure 24.15 Commercial cultivation of *Agaricus bisporus* produces approximately 1 million tons of mushrooms annually, outpacing all other cultivated mushroom species.

Table 24.2 Some Edible Mushrooms: Either Grown in Cultivation or Collected from the Wild

Common Name	Scientific Name
Chanterelle	*Cantharellus cibarius*
Cultivated white mushroom	*Agaricus bisporus*
Enoki	*Flammulina velutipes*
Inky cap	*Coprinus comatus*
Mu-erh	*Auricularia polytricha*
Nameko	*Pholiota nameko*
Oyster mushroom	*Pleurotus ostreatus*
Paddy straw mushroom	*Volvariella volvacea*
Périgord truffle	*Tuber melanosporum*
Shiitake	*Lentinus edodes*
Silver ear	*Tremella fuciformis*
Stropharia	*Stropharia rugoso-annulata*
Winter mushroom	*Flammulina velutipes*
Yellow morel	*Morchella esculenta*

conditions of some edible fungi. For example, deciphering the specific requirements for cultivation of the delectable morel has been possible only recently. Other choice edible fungi are still collected from the wild.

Fermented Foods

Fungi as well as bacteria have been utilized to modify foods through fermentation. Some common fermented foods well known in the United States are cheese, yogurt, sausage, pickles, sauerkraut, soy sauce, and, of course, bread. Other fermented foods are little known outside their culture of origin (table 24.3).

Fermenting foods is a widespread and time-honored practice, predating historical records, and can be found in cultures throughout the world. In eastern Asian countries, soybeans sometimes mixed with cereals have been fermented by various bacteria and fungi to yield an impressive collection of flavoring agents and protein sources, such as soy sauce, miso, and tempeh (see Chapter 13). In southern Asia, the Middle East, and Africa, the traditional fermented foods are the result of bacterial and yeast action on cereals to which a protein source has been added. In the Middle East, the protein source is usually milk; in the Indian subcontinent, it is legumes.

Fermented foods have several inherent advantages regarding nutritional and keeping qualities. Microbial enzymes released during fermentation improve flavor and appearance, increase digestibility, and destroy undesirable components of the unfermented food. Protein values are enhanced in fermented foods through the enzymatic release of amino acids, and many of the microbial agents synthesize vitamins, further improving nutritional quality. In addition, fermented foods are less subject to spoilage, an important factor in countries where refrigeration is uncommon.

Table 24.3
Examples of Traditional Fermented Foods throughout the World

Food	Origin	Base	Use
Ang-kak	China, Philippines	Rice	Coloring agent (red)
Bongkrek	Indonesia	Coconut presscake	Meat substitute; snack
Doza	India	Rice, blackgram	Pancake
Gari	West Africa	Cassava	Starchy entree with sour taste
Ogi	Dahomey, Nigeria	Maize	Fermented cake
Pozol	Mexico	Maize	Fermented dough diluted with water and drunk
Sierra rice	Ecuador	Unhusked rice	Fermented rice
Soy sauce	Asia	Soybeans	Flavoring
Sufu	China	Soybean curd	Condiment
Tarhana	Turkey	Wheat meal and yogurt	Used in soup
Tempeh	Indonesia	Soybeans	Meat substitute; snack

Thinking Critically

Many edible mushrooms are not grown in cultivation but are picked from wild stands in our national forests. For example, the pine mushroom (*Tricholoma magnivelare*) flourishes in the forest of the Pacific Northwest, growing on the undisturbed forest floor in association with 100-year-old lodgepole pines, Douglas firs, and western hemlocks. Closely related to Japanese species, top-grade "pines" may sell for up to $200 per pound or more when shipped to Japanese markets. This lucrative trade has created a type of modern-day gold rush whereby hundreds of mushroom pickers descend on forests during harvest season to "strike it rich."

Can you foresee any problems with the unregulated harvesting and sale of wild mushrooms? How would you go about regulating the harvest of wild mushrooms? (Consider that Illinois bans the sale of wild mushrooms altogether and, in Michigan, harvesters of wild mushrooms must be licensed.)

Quorn Mycoprotein

For many years, soy products have been the leading meat substitute for people wanting to reduce or eliminate animal products from their diet (see Chapter 13). Textured vegetable protein from soybeans is found in a wide range of vegetarian products, from veggie burgers to veggie dogs to imitation bacon bits. Another meatless alternative is now available in U.S. markets. The product is Quorn **mycoprotein,** a food material produced from the mycelium of the fungus *Fusarium venenatum*. Like soy protein, mycoprotein can be made into a wide variety of imitation meats. Quorn has been available in Great Britain for nearly 30 years and in other European countries for about 25 years; FDA approval for sale in the United States was given in January 2002. A month later, a nonprofit health advocacy group filed a complaint with the FDA that the mycoproteins in Quorn have not been adequately tested for allergenicity and that some people have suffered nausea, vomiting, or diarrhea after eating Quorn. Also, the group pointed out that the manufacturer has falsely advertised Quorn as a product made from a mushroom.

Quorn was developed by Marlow Foods in England after an extensive search for a high-protein meat substitute. Researchers examined 3,000 soil samples from all over the world, but the fungus selected was isolated from a sample obtained close to the Marlow laboratories. After 10 years of research, Marlow Foods developed large-scale production methods for the mycoprotein. *Fusarium venenatum* is grown in large vats called fermenters (or bioreactors), which are 45 meters (150 feet) tall and capable of producing 20,000 tons of mycoprotein annually. For growth, the fungus requires oxygen, water, glucose, a source of nitrogen, vitamins, and minerals. Glucose is usually supplied from corn syrup, but other sources of glucose are also suitable.

After a period of growth, the fungal mycelium is harvested, heated, and dried. In bulk, the harvested material looks like pastry dough; however, on closer inspection the hyphal strands of the mycelium are visible. One of the advantages of mycoprotein is the fact that the mycelium provides a texture that is similar to muscle fiber found in meats. A binder is added to hold the mycelial strands together, and then the mycoprotein is processed to resemble chicken, steaks, cutlets, patties, nuggets, crumbles, or sausages. Flavorings and coloring agents are usually added during the processing.

Mycoprotein offers many nutritional advantages as a high-protein, high-fiber, low-fat food. The high-quality protein supplies all the essential amino acids and is similar to egg whites and milk protein. The harvested mycelium is about 12% protein, but on a dry-weight basis, Quorn is about 50% protein. It is also cholesterol-free, and clinical studies have shown that the consumption of mycoprotein may lower cholesterol levels in the blood.

In addition to the value of Quorn for vegetarian diets, there is tremendous potential for using mycoprotein to supplement protein-deficient diets in developing countries. The production of mycoprotein by *F. venenatum* is an efficient and economical way to convert glucose from any surplus carbohydrate source into high-quality protein foods.

CHAPTER SUMMARY

1. The fermentation of sugar to alcohol through the action of yeast is the crucial step in the making of both wine and beer. Wine begins with the wine grape (*Vitis vinifera*). The wine grape is basically of two types, red or white, depending on the color of the grapes. The processing of a red or white wine differs in many respects. White wines, using either red or white grapes, are crushed, pressed, and then fermented. In the making of red wines, the red grapes are crushed, fermented, and then pressed. Newly fermented wines are racked to remove impurities, aged in oak barrels, and then bottled.

2. Champagne is made by the blending of three young white wines that undergo fermentation in the bottle to produce the excessive carbonation associated with champagne. Table wines have a final alcohol content ranging from 9% to 14%; in fortified wines, brandy has been added to raise the alcohol concentration to levels of 15% to 21%.

3. Beer is the product of a fermented grain. Because beer making begins not with sugars but with starch, preparation of a malt, mash, and wort are the steps necessary to provide a source of fermentable sugars. The alcohol content of beer averages around 4% to 6%.

4. Sake production uses enzymes released by the fungus *Aspergillus oryzae* to convert rice starch into fermentable sugars. The alcohol content of sake is 20%, similar to what is found in distilled spirits.

5. In the production of distilled spirits, beer or wine is boiled and the alcoholic vapors are collected and condensed to raise the alcohol concentration to the 40% to 50% range. Brandy is a twice-distilled white wine; whiskey, gin, and vodka are distilled beers; and tequila is the fermented sap from the stem of agave.

6. As a psychoactive drug, alcohol is classed as a depressant of the central nervous system; it inhibits the brain centers that control speech, vision, balance, and judgment. Moderate drinking of alcohol can significantly reduce the incidence of coronary heart disease by raising HDL levels and inhibiting blood clots. Drinking alcohol during pregnancy is not recommended because of the risk of fetal alcohol syndrome.

7. Many fungi are edible, and several cultures have a tradition of collecting wild mushrooms. Recently, it has been possible to cultivate a greater variety of edible mushrooms and other fungi, eliminating the risk of accidental poisoning by mistaking a poisonous mushroom for an edible one. Fungi have been employed to modify foods through the fermentative process. Fermented foods are often enhanced nutritionally and have improved flavor and appearance through the action of enzymes released during fermentation.

8. Quorn mycoprotein is a high-quality protein food made from the mycelium of the fungus *Fusarium venenatum*. The product has a texture similar to that of meat, and after processing, it is used to make a large variety of meat substitutes.

REVIEW QUESTIONS

1. Contrast the steps of wine making to those required in the brewing process.
2. How does the making of champagne or a sparkling wine differ from the making of a still wine?
3. Contrast the operation of a pot still to that of a continuous still. Identify the following distilled beverages: grain neutral spirits, vodka, gin, whiskey, bourbon, brandy, and tequila.
4. Describe the steps in making cider. How is applejack made? Apple brandy?
5. What is the connection between the herb wormwood and absinthe? What is absinthism?
6. Describe the effects, both positive and negative, of alcohol on health.
7. What is the nutritional value of edible mushrooms? How are mushrooms cultivated?
8. What are the advantages associated with fermented foods?

©Dariusz Majgier/Shutterstock

CHAPTER

25

Fungi That Affect Human Health

The death cap mushroom, Amanita phalloides, *is responsible for more mushroom fatalities than other fungi.*

KEY CONCEPTS

1. Like plants, fungi produce a large number of secondary products, and some have been put to use as medicinal compounds.

2. Some fungal metabolites are toxic to humans and other animals; these include mycotoxins and mushroom toxins, both of which can cause serious or fatal reactions.

3. Some fungi affect humans directly by causing allergic reactions as well as diseases ranging from mild skin conditions to fatal systemic infections.

CHAPTER OUTLINE

Antibiotics and Other Wonder Drugs 470
    Fleming's Discovery of Penicillin 470
    Manufacture of Penicillin 471
    Other Antibiotics and Antifungals 471
Fungal Poisons and Toxins 472
    Mycotoxins 472

A CLOSER LOOK 25.1 The New Wonder Drugs 473

    Ergot of Rye and Ergotism 475
    The Destroying Angel and Toadstools 477
    Soma and Hallucinogenic Fungi 478
Human Pathogens and Allergies 480
    Dermatophytes 480
    Systemic Mycoses 481
    Allergies 482
    Indoor Fungi and Toxic Molds 482
Chapter Summary 484
Review Questions 485

Fungi produce a remarkable diversity of organic compounds. In addition to all the carbohydrates, lipids, and proteins that are components of cell structure and metabolic pathways, fungi produce a vast number of **secondary products,** organic compounds that have no direct role in major metabolic pathways but may discourage predators or suppress competition. The formation of secondary compounds is quite specific. Often, a certain compound is confined to one species or sometimes one strain of a species.

Thousands of secondary metabolites from fungi have been analyzed and characterized. Many of them have widespread commercial importance; others have well-known health effects. Included in this group are alkaloids as well as other compounds that may function as **antibiotics** or **toxins.** In many respects, these secondary metabolites have made a significant impact on society. The value of the antibiotics (and even some of the fungal alkaloids) to medicine and society is clearly recognized, as are some of the harmful effects caused by toxic compounds.

ANTIBIOTICS AND OTHER WONDER DRUGS

Antibiotics are compounds that are toxic to microorganisms. In the natural environment, these substances give the producing organism advantages over competing microorganisms for available nutrients and space. Production of antibiotics is one of the mainstays of the pharmaceutical industry and the primary weapon in the physician's arsenal for fighting bacterial infections. Although it is hard to imagine modern medicine without antibiotics, large-scale production of these compounds is only about 70 years old. It all started with the accidental discovery of a contaminated bacterial culture by a British scientist.

Fleming's Discovery of Penicillin

Penicillin is a by-product of certain *Penicillium* species that inhibits the growth of Gram-positive bacteria (a group of bacteria separated on the basis of the chemical composition of the cell walls as identified with the Gram stain). The antibiotic acts by blocking wall synthesis in the bacterium and results in death of the bacterial cell by lysis.

Penicillin was first discovered in 1928 by British physician Alexander Fleming. However, its potential was not realized for a number of years, and mass production was not achieved until World War II. Although the antibacterial properties of other compounds had been seen previously, penicillin surpassed known therapeutic agents by suppressing bacterial growth without being toxic to animals or humans.

Fleming's discovery occurred while he was examining some old bacterial cultures before throwing them away. He found that a mold had contaminated his cultures while he was away on vacation. The fungus that Fleming identified as *Penicillium notatum* had killed the culture of *Staphylococcus aureus* bacteria growing in the petri dish. (This isolate has recently been reclassified as *Penicillium rubens*.) He conducted some additional experiments on the bacteria-killing substance that he called penicillin and published his findings. Fleming's paper attracted little attention at the time, and his own experiments at purifying penicillin failed. Eleven years passed before research on this compound advanced.

In 1939, Howard Florey and Ernst Chain at Oxford University began an investigation of naturally occurring antibacterial compounds and came across Fleming's report on penicillin. Within a year, the team at Oxford had chemically analyzed the compound and demonstrated that it could destroy certain types of bacteria in test tubes. With World War II intensifying, Florey and Chain realized the potential of the drug for treating war wounds. Tests on infected animals were successful, and by 1941 the first human tests were conducted. Because of the escalating war in England, much of the penicillin research was moved to various sites in the United States. Miraculous cures were reported in human tests, and mass production was finally achieved.

One of the most dramatic cures was the recovery of Anne Miller of New Haven, Connecticut. Mrs. Miller fell ill with a massive streptococcal infection on February 14, 1942. For 4 weeks she barely clung to life as her temperature spiked to 41°C (106°F) and 42°C (107°F). Doctors tried everything from sulfa drugs to transfusions, but nothing could stop the bacteria from multiplying in her bloodstream. There was virtually no likelihood of survival. By chance, her physician knew someone working on penicillin research in New Jersey, and he was able to obtain a tiny amount of the drug. It was administered during the afternoon of March 14, 1942. By 9:00 the next morning, Anne Miller's fever was gone; her vital signs were stable; and she was able to eat for the first time in 4 weeks. By afternoon, she was acting as if she had never been ill; however, she was very weak and had lost 40 pounds. The miracle of her recovery electrified the scientific community. Shortly after Anne Miller's recovery, Alexander Fleming visited her in New Haven and acknowledged her contribution to the development of the miracle drug, saying, "Thank you, Mrs. Miller, you are my most important patient."

At the North Regional Research Lab of the U.S. Department of Agriculture, one team of researchers was looking for more high-yielding sources for the drug (fig. 25.1). Moldy fruits and vegetables were routinely collected from local grocery stores; then the fungi were isolated and tested for antibiotic production. In the summer of 1943, a cantaloupe was found contaminated with *Penicillium chrysogenum.* The isolated fungus produced 200 times more penicillin than Fleming's isolate. This species was used in the industrial production of the drug and continues to be used today.

By D-Day in 1944, there was enough penicillin to treat all the British and American casualties of the European invasion. Many even credit the development of penicillin as one of the reasons behind the Allied victory. By the time World War II ended, sufficient penicillin was available for civilian use. In

Figure 25.1 USDA researchers at the North Regional Research Lab in Peoria, Illinois, identified high-yielding sources of penicillin and also developed the process for mass-producing the antibiotic.

1945, Florey, Chain, and Fleming received the Nobel Prize for their work in developing the first "miracle" drug.

Soon after World War II, the pharmaceutical industry developed chemically altered versions of the penicillin molecule. These modified penicillins provided for greater stability, broader antibacterial activity, and oral administration of the drug, which would permit home use of antibiotics.

Although other antibiotics have since been identified and developed, penicillin is the most widely used antibiotic and is still the drug of choice to treat many bacterial infections. Scientists have continued to improve the yield of the drug, and the present-day strains of *P. chrysogenum* are biochemical mutants that produce 10,000 times more penicillin than Fleming's original isolate.

Although penicillin is viewed as a miracle drug, its development has had some drawbacks. Once large-scale production of penicillin was achieved by the pharmaceutical industry and the drug was readily available, overprescribing by physicians and veterinarians commonly occurred. In addition, the antibiotics were incorporated into animal feed for use in feedlots. This widespread use led to the evolution of penicillin-resistant bacteria. (Many species of bacteria reproduce every 20 minutes, so the timetable for the evolution of new strains is much faster than evolutionary changes in other organisms.) By the early 1960s, penicillin resistance was evident among many types of bacteria, and by the early 1990s, antibiotic resistance had become a major cause for concern among the medical community.

A second drawback to penicillin is that a small percentage of the population is allergic to the drug, often resulting in severe or even fatal anaphylactic reactions. Anaphylaxis is a rapid and dramatic allergic reaction that can result in death through airway obstruction or irreversible vascular collapse. Penicillin is the most frequent cause of anaphylaxis in humans; it has been estimated that several hundred people die each year from anaphylaxis due to penicillin allergy.

Manufacture of Penicillin

Commercial production of penicillin is carried out in huge tanks with a high-yielding strain of *P. chrysogenum*. The normal culture medium used is a corn steep liquor, a by-product of corn starch production. The main carbohydrate in the corn steep liquor is lactose, a disaccharide composed of glucose and galactose. About 30,000 liters (7,500 gallons) of the culture medium is sterilized and inoculated with spores of *P. chrysogenum*. During incubation, glucose is added periodically, the temperature is kept at about 24°C (75°F), and the pH is adjusted to about 7.0. The culture is aerated with sterile air and agitated. The penicillin accumulates in the culture medium. After 5 or 6 days, the mycelium is filtered from the medium, which is then subjected to an extensive extraction and purification process to recover the penicillin. Although penicillin can be chemically synthesized today, doing so is far too expensive. As a result, penicillin is still produced by fermentation; however, a large group of antibiotics are made by chemically modifying the basic penicillin structure. Some of these semisynthetic penicillins are amoxicillin, ampicillin, and methicillin.

Other Antibiotics and Antifungals

As early as 1943, while penicillin was still being developed, Selman Waksman of Rutgers University discovered streptomycin, a new antibiotic produced by *Streptomyces griseus*, which was effective against some bacteria that were immune to penicillin. Waksman's aim was to find an antibiotic that would inhibit the growth of *Mycobacterium tuberculosis*, the bacterium that causes tuberculosis. Although not produced by fungi, this antibiotic led to the development of a whole class of well-known drugs from actinomycetes (a phylum of bacteria). In addition to streptomycin, other well-known antibacterial and antifungal drugs produced by actinomycetes are erythromycin, tetracycline, chloramphenicol, and nystatin. The majority of these antibiotics are produced by species of *Streptomyces*.

The development of penicillin led to the search for other fungi that produced antibiotic compounds. In 1948, Giuseppe Brotzu, an Italian microbiologist, had identified a compound produced by *Cephalosporium acremonium* that was an effective treatment for Gram-positive infections as well as some Gram-negative ones, such as typhoid. Although Brotzu had achieved some success with his new antibiotic, the Italian authorities showed no interest in his work. Brotzu sent a culture of this fungus to Florey. The team at Oxford once again isolated the active compound, which they named cephalosporin. Since its initial isolation, a whole group of cephalosporins has been manufactured. These antibiotics have a broader spectrum than penicillins and are effective against many penicillin-resistant strains of bacteria. They are also much more expensive to produce; many of the newer cephalosporins are reserved for hospital use.

Another fungal metabolite is griseofulvin, first isolated in 1939 from *Penicillium griseofulvum*. Unlike penicillin and the other antibiotics described in this chapter, this species produced no antibacterial properties. Griseofulvin is effective in the treatment of many fungal diseases, especially dermatophytic mycoses that affect the skin, hair, or nails. Because the compound is only **fungistatic** (stopping growth) and not **fungicidal** (killing the fungus), treatment must be continued until all the infected skin has been sloughed off.

Fusidic acid is an antibiotic produced by the fungus *Fusidium coccineum*. It was developed in the 1960s in Denmark and has been widely used in European countries, many Asian countries, Canada, Australia, and New Zealand. Fusidic acid has also been isolated from two other fungi, *Mucor ramannianus* and *Isaria kogana*. The drug is mainly effective against Gram-positive bacteria and is especially effective against *Staphylococcus* species. This includes methicillin-resistant *Staphylococcus aureus* (MRSA), a bacterium, which shows resistance to many antibiotics in addition to methicillin. MRSA can cause many illnesses ranging from skin infections to pneumonia to life-threatening blood stream infections. Although fusidic acid has been widely used in many countries, it has not been previously approved by the Food and Drug Association (FDA) in the United States. With the rise of MRSA infections worldwide, there has been renewed interest in the therapeutic use of fusidic acid. Cempra, a North Carolina pharmaceutical company, has been conducting clinic studies of this compound and has been given fast track and priority review status by the FDA based on preliminary data. In the near future, this antibiotic may be approved and available by prescription in the United States.

In 1985, a fungus isolated from a pond near Madrid, Spain, produced compounds that showed potent antifungal activity. The fungus was later found to be a new species, *Glarea lozoyensis,* and research on the active metabolites led to the development of a new class of antifungal drugs, the echinocandins. In 2001, caspofungin, marketed by Merck Pharmaceuticals as Cancidas, was the first drug in this class to be approved. This compound is currently used to treat cases of invasive aspergillosis, a severe lung infection, that do not respond to conventional antifungal drugs. Clinical trials suggest that this drug and others in the echinocandin class will also be effective for treating other serious fungal infections.

Today the search for new antibiotics and antifungals is once again in high gear as drug resistance has reached critical levels. Certain bacteria have developed resistance to the entire medical arsenal of antibiotics, including vancomycin, which was considered the antibiotic of last resort. Several new drugs are in the midst of clinical trials and are showing promising results. However, it takes approximately 10 years of trials before approval and costs hundreds of millions of dollars. Researchers recently reported success in creating new versions of current antibiotics by manipulating the enzymes used in production. The scientists predict that this approach may lead to the development of new antibiotics, antifungals, and immunosuppressants. Other researchers have been looking beyond the microbial world for compounds that have antibiotic properties; the search has extended to higher plants, insects and other invertebrates, and higher animals. While the world waits for new drugs, the proper use of current antibiotics will help minimize the further progression of drug resistance in bacteria.

Drugs derived from fungi can also be used for benefits other than their antibiotic abilities. A Closer Look 25.1: The New Wonder Drugs discusses two of them.

> **Thinking Critically**
>
> Antibiotics produced by fungi are essential weapons for fighting infectious diseases. With the increased bacterial resistance to antibiotics, infectious diseases are once again on the rise.
>
> *How has this problem developed?*

FUNGAL POISONS AND TOXINS

Although antibiotics may be toxic to microorganisms, a compound is normally referred to as a fungal toxin when it affects humans or other animals. Fungal toxins broadly fall into two groupings: **mycotoxins,** formed by hyphae of common molds growing under a variety of conditions, and mushroom toxins or poisons formed in the fleshy fruiting bodies of some fungi.

Mycotoxins

Mycotoxins are commonly produced by fungi growing in contaminated foods. These compounds can develop in grains or nuts in the field, but more commonly, they develop in storage and remain within the food after processing and cooking. These toxins have been shown to have profound chronic and acute effects on both humans and livestock when the contaminated foods are eaten. In addition to direct toxic effects, mycotoxins are among the most potent known carcinogens.

The scientific community first became aware of the mycotoxins while investigating the cause of Turkey X disease, which killed more than 100,000 young turkeys in 1960 in England. The affected turkeys stopped eating, became lethargic, suffered hemorrhages under the skin, and died. Autopsies showed that their livers had undergone extensive necrosis (localized tissue death) and their kidneys had developed lesions. It was soon found that partridges, pheasants, ducklings, and other animals were also affected. Some detective work revealed that the only factor in common with all the poisonings was the use of

A CLOSER LOOK 25.1

The New Wonder Drugs

Antibiotics produced by fungi have been a staple in modern medicine for over 60 years, and many scientists believe that fungi still contain a wealth of novel compounds, yet to be discovered, that could provide even greater benefit to humanity. Two examples of the newer wonder drugs from fungi are cyclosporin and the statins; these are among the top-selling pharmaceuticals worldwide.

Cyclosporin A, a metabolite isolated from *Tolypocladium inflatum*, was called the "wonder drug of the '70s" for its ability to suppress the immune system and prevent rejection of transplants. Cyclosporin was first isolated when workers from Sandoz Pharmaceutical Company were searching for new antifungal agents from soil fungi. Although the extracts from *T. inflatum* showed only weak fungicidal activity, there was unusually low toxicity. This finding prompted further examination and a pharmacological screening program. The screening showed that the extract has the ability to selectively suppress components of the immune system. With these results, the compound was purified and structurally analyzed. By 1980, the immunosuppressive drug cyclosporin A was ready for further clinical trials.

Cyclosporin is used for many organ transplants: liver, kidney, heart, and bone marrow. It has improved the success rate of organ transplants by preventing rejection reactions without completely suppressing the immune system. The advantages of cyclosporin A over other drugs are its level of effectiveness and its low toxicity. Cyclosporin A is also the most cost-effective immunosuppressant and is considered the prototype of a new generation of immunosuppressants.

Statins are cholesterol-lowering drugs that were originally isolated from fungi and have been widely used since the early 1990s. Seven statins are currently prescribed by physicians in the United States. Three of these (pravastatin, simvastatin, and lovastatin) are derived by fermentation of the fungus *Aspergillus terreus*. Four (fluvastatin, rosuvastatin, atorvastatin, and pitavastatin) are synthetics. Cerivastatin, another synthetic, is no longer being produced. The brand names of these drugs are listed in Box Table 25.A, and low-cost generic versions of six of these are available. Statin drugs act as inhibitors to one of the enzymes necessary for cholesterol synthesis in the body and can reduce blood cholesterol levels up to 60%. They specifically lower LDL (bad) cholesterol levels and even produce some increases in HDL (good) cholesterol levels (see Chapter 10). The cholesterol reduction significantly reduces a patient's risk of heart disease.

The benefits of statin drugs go far beyond just lowering cholesterol levels. Statins appear to be effective in preventing strokes, especially in patients with known heart disease.

Table 25.A Statin Drugs

Chemical Name	Brand Name in United States	Generic Available	Production Method
Pravastatin	Pravachol	✓	Fermentation
Simvastatin	Zocor	✓	Fermentation
Lovastatin	Mevacor	✓	Fermentation
Fluvastatin	Lescol	✓	Synthetic
Atorvastatin	Lipitor	✓	Synthetic
Rosuvastatin	Crestor	✓	Synthetic
Pitavastatin	Livalo		Synthetic
Cerivastatin	Baycol		Synthetic; no longer available

Brazilian peanut meal as a feed supplement. A toxin was isolated from the feed and found to be associated with a common fungal contaminant, *Aspergillus flavus*.

The toxin was named **aflatoxin** to designate the fungus producing it (A [*Aspergillus*] -fla [*flavus*] -toxin) although it was soon learned that four toxins were present, not just one. Within a few years, it was shown that aflatoxins are not only toxic but also carcinogenic. In experiments, it was found that, even if a food did not contain toxic levels of aflatoxins, prolonged exposure caused liver cancer in laboratory animals. It is believed that these toxins are responsible for high rates of liver cancer among some groups of people in Asia and Africa, where contaminated food is often consumed. Although the association with human cancer has not been confirmed experimentally, the toxic effects in humans were shown in India in 1974, when about 400 people were poisoned after eating corn containing aflatoxins; 106 died. Within the past 15 years multiple outbreaks of aflatoxin

Statins reduce coronary artery inflammation, which has recently been found to be a major cause of heart attacks and strokes. Studies have shown that the use of these drugs decreases the risk of fractures by increasing bone density and thereby decreases the risk of osteoporosis. Statins reduce the risk of colon cancer when combined with nonsteroidal anti-inflammatory drugs, and they reduce the symptoms of multiple sclerosis. Recent studies also show a reduced risk of Alzheimer's disease. In fact, patients taking lovastatin or pravastatin had a 60% to 73% lower risk of developing Alzheimer's than did patients taking neither drug. Researchers suggest that these drugs possibly prevent dementia by dilating blood vessels and increasing blood flow to the brain. The medical community is saying that the use of statin drugs should be expanded because many more people could benefit from their use. The expanding role of statins has already been called one of the top 10 medical advances of the new millennium.

Like all drugs, statins have side effects and are not for everyone. Side effects include muscle pain and elevated liver enzymes. In August 2001, Bayer Pharmaceuticals voluntarily withdrew Baycol after the deaths of 31 patients in the United States over a 4-year period. The deaths resulted from rhabdomyolysis, which destroys muscle cells and causes severe muscle pain. In addition to the 31 deaths, hundreds of nonfatal cases of rhabdomyolysis were reported in the United States. Although this condition is a side effect of all statin drugs, it is exceedingly rare with the use of the seven approved statins, whose health benefits clearly outweigh the slight risk. The side effects may be exacerbated by interactions with other drugs, and grapefruit juice also increases the side effects (see Chapter 6). Another fungal source of statins is the yeast *Monascus purpureus*, which is used to produce red yeast rice. When rice is inoculated and incubated with *M. purpureus*, the rice develops a reddish purple color.

Red yeast rice has been consumed in China for more than a thousand years as both food and medicine, and today it is widely available as an herbal supplement. The active ingredients are monacolins, and one of these, monacolin K, is identical to lovastatin; however, the concentrations of monacolins in red yeast rice are highly variable. Because of the possible side effects of statins, patients should consult their physicians before using red yeast rice supplements to lower cholesterol levels.

poisoning have occurred in Kenya, also from contaminated corn. Overall, 500 people have been sickened and 200 have died.

Today, in industrialized countries, foods that are most frequently contaminated by aflatoxins (such as peanut butter and grain products) are routinely screened before processing or sale. The permissible limit of mycotoxins is generally quite low (15 to 20 parts per billion); however, some scientists feel that no detectable levels of aflatoxins should be permitted because of the carcinogenic effects. Because cows and goats fed grain contaminated with aflatoxins produce milk with aflatoxins, limits also exist for livestock feed.

Aflatoxin was the first mycotoxin identified and studied; today, over 400 mycotoxins have been identified from various fungi, with new ones being discovered each year. Species of *Aspergillus*, *Penicillium*, *Fusarium*, *Alternaria*, *Cladosporium*, and *Stachybotrys* are among the common fungi known to form mycotoxins.

Aspergillus flavus is a common saprobe that occurs on many grains and legumes in storage (fig. 25.2). Aflatoxins are produced under certain conditions by some strains of this species. *Aspergillus flavus* is also used to prepare fermented foods in Asia. Fortunately, the strains of *A. flavus* that are used in fermentation do not develop toxins, or else the environmental conditions used in fermentation are not favorable for toxin production. Apparently, the ability to produce toxins may occur in one strain of a given species, while another strain of the same species is not toxigenic. Some environmental conditions also promote the development of toxins. Although much

Figure 25.2 Aflatoxins are secondary metabolites produced by *Aspergillus flavus*, ×1,500.

basic research still needs to be done to identify the precise conditions that cause common fungi to produce toxins, it is believed that warm temperatures are important environmental triggers for some species. These conditions frequently occur in tropical and subtropical countries, where storage conditions are often less than ideal.

All foods are subject to spoilage by fungi (fig. 25.3), but the mold growing on forgotten food in the back of the cupboard might also be producing deadly mycotoxins. Knowledge of these toxins suggests caution about eating moldy food!

> **Thinking Critically**
>
> Mycotoxins are potent carcinogens produced by the hyphae of certain saprobic fungi. These can develop in grains or nuts in the field or in storage.
>
> *What can be done to protect our food supply from these toxins?*

Ergot of Rye and Ergotism

In many respects, the contamination of rye grains by the plant pathogen *Claviceps purpurea* could be considered another case of mycotoxin poisoning. The long history of this plant disease, the human and animal poisonings resulting from eating contaminated grain, and the medical uses of some of the compounds warrant a fuller discussion.

Ergot, caused by the fungus *C. purpurea*, affects rye and other grains. The fungus infects the flower, producing dark **sclerotia** (hardened mycelial masses) called ergots in place of the grain (fig. 25.4a). The sclerotia contain a complex mixture of secondary metabolites (over 100 compounds), including numerous alkaloids. When infected rye plants are harvested, sclerotia can be collected and ground along with the healthy grain. The resulting flour and baked breads contain these

Figure 25.3 When foods are not properly stored, fungi can cause spoilage, as seen on this moldy bread.

(a)

(b)

Figure 25.4 Ergot of rye caused by *Claviceps purpurea* can lead to ergotism if infected flour is made into bread and eaten. (a) Infected head of rye showing ergots. (b) Sixteenth-century woodcut depicts a peasant appealing to St. Anthony for relief from ergotism. The peasant has lost his right foot; symbolic flames surround his left hand. Woodcut from Hans von Gersdorff's *Feldtbuch de Wundartzney (Fieldbook of Surgery)*, Strasburg, 1540.

alkaloids and cause the condition known today as **ergotism** in people consuming the contaminated bread. Animals grazing on infected grains in the field can also develop ergotism.

Historical records of epidemics caused by ergotism indicate that symptoms of ergotism followed two different patterns: chronic poisoning and acute poisoning. Chronic poisoning led to gangrenous ergotism, which began with cold prickling or burning sensations that started in the fingers and toes and gradually spread to the entire limb. The limbs became swollen, and burning pains alternated with icy coldness; finally the limbs became numb. The affected parts turned black with a dry gangrene that gradually spread upward and necessitated amputation of the limbs (fig. 25.4b). Pregnant women frequently miscarried, and fertility was generally lower during outbreaks of ergotism. Acute poisoning resulted in convulsive ergotism, which affected the central nervous system. Although itching, prickling under the skin, numbness, and muscle cramps or spasms were the initial symptoms, the condition quickly intensified into hallucinations and convulsions. Severe brain damage and fatalities commonly resulted from acute poisoning. Similar symptoms occurred in animals with ergotism; legs, hooves, and tails became gangrenous, and spontaneous abortions of calves often occurred.

Today it is known that the symptoms can be attributed to the abundant alkaloids contained within the fungal sclerotia. Some of the alkaloids affect smooth muscles and cause vasoconstriction (constriction of blood vessels); these led to the muscle pain, spasms, miscarriages, and gangrene. Other alkaloids affect the central nervous system, resulting in the convulsions, hallucinations, brain damage, and death.

Ergotism has been known for thousands of years, with historical records dating back to ancient Greece. During the Middle Ages, when rye bread was consumed by masses in Europe, epidemics of the plant disease, and consequently outbreaks of ergotism, were common. Many epidemics occurred in France, where rye was the staple crop of the peasants.

In France during the eleventh century, the condition was known as "sacred fire" because of the burning sensations common in the extremities of afflicted individuals. In the twelfth century, the condition began to be associated with St. Anthony, a fourth-century Christian monk who was thought to have power over fire. Many people suffering from sacred fire made pilgrimages to the church of La Motte-au-Bois in central France, where the bones of St. Anthony rested. Many "miraculous cures" occurred, and soon a hospital was built to treat the patients. Additional hospitals and houses were built for the monks to treat victims of the sacred fire, which eventually became known as St. Anthony's Fire. We know today that the "cures" brought about by the monks of St. Anthony can be attributed to the ergot-free diet that patients received at the hospitals.

The historical impact of ergotism goes beyond the misery brought to the thousands of afflicted individuals. In 1722, Russian czar Peter the Great was halted in his ambitions to invade Turkey and capture a warm-water port when both his soldiers and horses developed ergotism after eating contaminated bread and grain in the Volga Valley. The association of ergotism with other historical events has been somewhat more controversial. Several historians have implicated ergot poisoning with the hysteria that led to the Salem witch trials in 1692 (fig. 25.5). At that time, rye was grown in the Massachusetts colony, and weather conditions the preceding year were ideal for the growth of *Claviceps*. It has been suggested that stored grain was contaminated with ergot. Common complaints by persons who thought they were bewitched included muscle pain and spasms, tingling limbs, and convulsions, symptoms that are consistent with ergotism. Historian M. K. Matossian has also suggested that ergotism played a role in the panic of the French peasants, contributing to the French Revolution of 1789. Matossian claims that the panic, known to historians as the Great Fear, was caused by the consumption of contaminated bread rather than an insurrection against paying taxes.

Although the connection between the ergot (the fungal sclerotium) and ergotism was first made in 1676, it was not until the late eighteenth century that harvested grain was regularly checked for ergot. Removing the ergots from harvested grain put an end to the dreaded epidemics. Even though ergot poisoning of livestock still occurs, careful harvesting and milling make human poisonings rare. However, in the late summer of 1951 in Pont-St. Esprit, France, a number of people were stricken with ergotism after eating contaminated bread, and 136 cases were reported in 1977 in Ethiopia. The last outbreak resulted in a number of deaths following the consumption of infected oats, also susceptible to *C. purpurea*.

Figure 25.5 Salem witch trial.

Documented medical uses of ergot go back to 1582, when it was discovered that preparations of ergot can cause uterine contractions and hasten birth. For centuries, midwives employed ergot to induce abortions and as an aid in childbirth. Today the purified alkaloid ergometrine is used medicinally to reduce postpartum bleeding. The alkaloid ergotamine is an effective treatment for migraine headaches. Constricting the diameter of the cranial arteries relieves the pulsating pressure and resulting headache. Additional alkaloids are used medicinally for treating mental problems, high blood pressure, and other ailments. Fields of rye are deliberately inoculated with spores of *C. purpurea* to produce the ergot needed for the pharmaceutical industry.

Another group of ergot compounds is the lysergic acid alkaloids; these are the toxins responsible for the hallucinations associated with convulsive ergotism. **Lysergic acid diethylamide (LSD)** was first synthesized from lysergic acid by Swiss chemist Albert Hofmann in 1938. LSD is the most potent psychoactive drug known. The hallucinogenic effects were demonstrated in 1943, when Hofmann ingested 0.25 milligram of the compound and described the dreamlike state and fantastic visions that followed. In his hallucinations, furniture took on grotesque forms and people appeared as witches. Hofmann wrote,

> A demon had invaded me, had taken possession of my body, mind and soul. I jumped up and screamed, trying to free myself from him, but then sank down again and lay helpless on the sofa. ... I was seized by a dreadful fear of going insane.

LSD affects midbrain activity by interfering with the action of the neurotransmitter serotonin. In small amounts, LSD mimics the action of the neurotransmitter, but in larger amounts it is antagonistic to the action of serotonin. The hallucinations and changes in sensory perception are due to disruptions in the normal pathways of sensory stimulation.

In addition to the hallucinations, LSD induces various physiological responses, including increases in blood pressure, respiration, and perspiration, often accompanied by heart palpitations. In some cases, the effects are more serious, causing unpredictable side effects and a schizophrenia-like condition. These "bad trips" have resulted in many fatalities.

During the 1960s, Timothy Leary, a professor at Harvard University, popularized the use of this hallucinogen and urged young people to "drop out, turn on and tune in." Leary helped usher in the hippie generation, with its provocative music and psychedelic colors. One of the Beatles' popular songs of the period was "Lucy in the Sky with Diamonds," describing what many believed were LSD-induced images of tangerine trees and marmalade skies. Despite the psychedelic images described in the lyrics, both John Lennon and Paul McCartney denied the song was referring to LSD.

Even with the possibility of long-term mind-altering activity and unpredictable side effects, there has been a sustained interest in the medical uses of LSD. The drug has been used with limited success in psychotherapy for the treatment of schizophrenia.

Recreational use of LSD peaked during the late 1960s. Although the illegal use of this drug decreased during the 1970s, the street use increased again in the 1990s.

The Destroying Angel and Toadstools

The common term *toadstool* refers to poisonous mushrooms; however, there is no easy way to find out if a mushroom is poisonous. In many instances, both poisonous and edible species occur within a single genus. Although folktales suggest that cooking mushrooms with a silver spoon (or silver coin) will indicate poisonous species when the silver turns black, these fables have no scientific basis and can lead to death. No single test or rule exists to indicate the edibility of a mushroom. The only way to know for certain whether a wild mushroom is edible or poisonous is to learn mushroom identification from a trained mycologist. **NEVER** eat a wild mushroom unless the identification is certain.

Although stories of mushroom poisonings abound, only a small percentage of species are actually toxic to humans. The actual number of poisonous (or even edible) species is difficult to determine because the vast majority of mushrooms are not utilized as a food, nor have they been analyzed for the presence of poisons or toxins.

Mushroom toxins are broadly categorized by the physiological effects of the toxin on humans:

1. protoplasmic poisons that destroy cells in the liver and kidneys and are often lethal;
2. neurotoxins that affect either the central or autonomic nervous system and often lead to hallucinations; and
3. gastrointestinal irritants that generally lead to nausea, vomiting, cramps, and diarrhea but seldom cause fatalities.

The following discussion focuses on the group of toxins that are responsible for most fatalities.

Amatoxins are a group of deadly protoplasmic toxins that are widely distributed in species of several genera, including *Amanita, Conocybe, Galerina,* and *Lepiota.* The amatoxin levels in these genera are high enough to cause human fatalities; however, the toxin also occurs in minute levels in other mushrooms. Although a few species in the genus *Amanita* are actually edible, 50% of all mushroom poisonings and 95% of all fatalities are caused by members of the genus *Amanita.* Common warnings for wild mushroom collectors are geared to recognize and avoid members of this genus:

1. Never eat white-capped mushrooms not positively identified.
2. Beware of any mushroom with an **annulus** (ring) on the stalk.
3. Beware of any mushroom that bears remnants of a **universal veil,** such as a **volva** (cup or sac) at the base of the stalk, or warts, scales, or patches on the surface of the cap (fig. 25.6a).

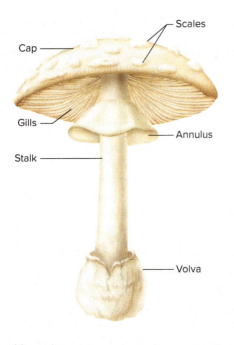

Figure 25.6 The genus *Amanita* contains many toxic species. (a) Morphological features of mushrooms often found in the genus *Amanita*. (b) *Amanita virosa*, the destroying angel.

The universal veil is a membrane that totally surrounds some young mushrooms. When the mushroom expands, the veil ruptures, often leaving a volva at the base of the stalk and remnants on the cap. The **partial veil** is a membrane that covers the gills during development. After it ruptures, it remains as an annulus on the stalk. In some members of the genus *Amanita,* both volva and annulus are prominent structures that aid in identification. One difficulty of looking for these features is that the volva and annulus may weather away, leaving the poisonous mushroom without its most distinguishable characteristics. *Amanita phalloides* (the death cap) and *Amanita virosa* (the destroying angel; fig. 25.6b) are the most notorious killers in this group.

The mechanism of amatoxin poisoning is at the level of gene expression by inhibiting mRNA synthesis. Without mRNA, the cells stop synthesizing proteins and eventually die. Usually, there is a delay, often 12 to 24 hours, between eating the mushroom and the appearance of symptoms. During this delay, considerable cellular damage is taking place. The first symptoms include intense vomiting, severe diarrhea, and fierce abdominal pain. These symptoms are caused by damage to the cells of the gastrointestinal lining. The vomiting and diarrhea often result in severe dehydration. If the victim survives these initial symptoms, there is a brief period of ostensible improvement. However, the toxins soon start affecting the liver, and severe liver damage develops. This stage is often fatal, but for persons surviving the liver damage, kidney failure follows, with secondary effects on the heart and brain.

No antidote exists for amatoxin poisoning. Treatments consist of attempts to remove the toxin from the body, to increase the rate of its excretion, and to maintain electrolyte and blood sugar levels through supportive care. In addition, various measures to support the damaged liver and kidneys have been used, including intravenous administration of high doses of penicillin G and silibinin (an extract of milk thistle). The value of some of these measures has not been established, but silibinin seems to reduce the uptake of the toxins in the liver.

Soma and Hallucinogenic Fungi

Although the genus *Amanita* is well known for its deadly members, other species of this genus are noted for their psychoactive properties. *Amanita muscaria,* the fly agaric, is a widely recognized mushroom with a long history of use as an intoxicant. The morphological features of *A. muscaria,* with its orange-to-red cap and white scales, make it one of the most commonly photographed and widely illustrated mushroom species (fig. 25.7). The common name of this mushroom arises because flies are attracted to the mushroom and then killed by its insecticidal properties.

Ibotenic acid, the major toxin in this mushroom, is converted into muscimol when consumed, and it is muscimol that produces the psychoactive effects. Several species of *Amanita* contain ibotenic acid; however, *A. muscaria* is the best known of the group. Symptoms usually develop within

20 to 90 minutes and last for up to 12 hours. Many symptoms are similar to alcohol intoxication but may progress into far more serious effects, including epilepsy-type seizures. Typical effects begin with dizziness, loss of coordination, and drowsiness, often accompanied by nausea and vomiting. These are followed by a period of sleep with vivid dreams. Later there is a period of hallucinations, with colored visions and distortions of size. In other cases, the medulla of the brain is affected, causing distortions in the motor control center and possibly resulting in seizures. Although fatalities are few from this type of mushroom toxin, they have occurred after a large dose of these mushrooms.

The properties of *A. muscaria* were discovered thousands of years ago by native peoples in various parts of the world. It is believed that this mushroom was used in ancient India to prepare the intoxicant Soma described 4,000 years ago in the *Rig Veda*, the book of sacred Hindu psalms. For centuries among some tribes in Siberia, the mushroom was dried and used as an intoxicant. The toxin is apparently excreted unaltered in the urine, which was collected and drunk for a second dose. In fact, it was known that intoxication would occur for up to four or five passages through the kidneys. *Amanita muscaria* was also used by several Native American tribes in the northern United States and Canada.

Among the native tribes of Mexico and Central America, another group of mushrooms were revered for their hallucinogenic properties. These mushrooms have been used for thousands of years in religious and healing rituals. The Aztecs called these sacred mushrooms teonanacatl, which means "flesh of god". It was believed that the mushrooms were a gift from a god who communicated with the people through the mushroom. Ancient mushroom-shaped statues found throughout the area attest to the reverence paid to these sacred mushrooms. Early Spanish explorers and missionaries described the rituals during which the mushrooms were consumed. Diego Duran described part of the ceremonies at the coronation of Montezuma in 1502:

> all the company proceeded to eat the toadstools, a food which deprived them of their reason and left them in a worse state than if they had drunk much wine. . . . Others had visions during which the future was revealed to them.

Rituals involving the ingestion of sacred mushrooms have continued to the present time in parts of Mexico, especially among the Mazatec Indians. In recent decades, scientists have observed and participated in these ceremonies and described the visual hallucinations and euphoria following the ingestion of various species of *Psilocybe* (fig. 25.8).

The major toxic compounds in these mushrooms are the alkaloids psilocybin and psilocin. In the body, psilocybin is converted into psilocin, which is the biologically active molecule. Psilocin is structurally similar to the neurotransmitter serotonin and, like LSD, interferes with the action of this substance in the brain. Hallucinations usually begin within 30 to 60 minutes of ingestion and last for several hours. In general, the results are visual hallucinations and distortions of

Figure 25.7   *Amanita muscaria*, the fly agaric.

Figure 25.8   Teonanacatl mushrooms of the genus *Psilocybe*.

time and space; however, sometimes the experience has caused vomiting, depression, and irritability. It has also been reported that excessive doses have resulted in paralysis, insanity, or suicide. These mushrooms are especially dangerous for young children; elevated temperatures, convulsions, and one fatality have been reported.

Recently, there has been interest in using psilocybin for treating various forms of mental illness, especially major depression. Pilot studies at both New York University Medical School and Johns Hopkins University Medical School used a synthetic version of psilocybin to treat patients suffering from major depression and anxiety due to cancer. The results were very promising showing that for some patients a single dose of psilocybin provided relief from anxiety and depression for 6 months.

Psilocybin and psilocin also occur in certain species of *Panaeolus*, *Conocybe*, and *Galerina* in addition to *Psilocybe*. During the 1970s, recreational use of these "magic mushrooms" gained considerable popularity in some parts of the United States. The mushrooms in this group are small and brown, often without distinctive features. It should be noted that similarity with amatoxin-containing species of *Conocybe* and *Galerina* can lead to misidentification and far more serious consequences.

Thinking Critically

Mushroom toxins are frequently poisonous to humans and other animals.

What might be the evolutionary advantage for the mushroom of producing such compounds?

HUMAN PATHOGENS AND ALLERGIES

The connection between exposure to airborne fungi and disease was first made by Blackley in the 1870s, when he described chest tightness and "bronchial catarrh" after inhalation of *Penicillium* spores. It has since been clearly demonstrated that exposure to fungal spores or even hyphal fragments can cause a variety of diseases ranging from allergic diseases (hay fever and asthma) to infectious diseases (such as histoplasmosis and aspergillosis). Infectious diseases caused by fungi are loosely categorized into

1. superficial mycoses (fungal diseases) that attack the skin, hair, and nails and
2. deep or systemic mycoses that attack internal organs.

Only about 100 species of fungi are known to be human pathogens, and most cause relatively minor superficial mycoses. More serious fungal diseases occur in individuals whose immune system is compromised by disease, the prolonged use of antibiotics or immunosuppressive drugs, or other causes. Some of the organisms are found only as human or animal pathogens; others can exist either as saprobes in the soil or as human pathogens. One final group of fungi that are responsible for human infections are opportunistic pathogens. These are common saprobic fungi, such as *Aspergillus* and *Fusarium* (species of *Fusarium* also cause plant diseases), which can cause systemic infections in immune-compromised individuals.

Dermatophytes

Among the most common superficial fungal pathogens are dermatophytes that attack the outer layers of the body. These fungi utilize keratin, a protein found in skin, hair, and nails; most dermatophytes are species of *Microsporum*, *Trichophyton*, and *Epidermophyton*. Many cause a condition known as tinea, or ringworm, which refers to the circular pattern on the skin produced by growth of the fungus (fig. 25.9). Specialization often occurs, with some species infecting only certain parts

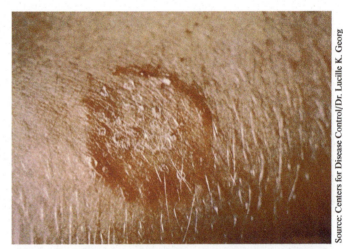

(a)

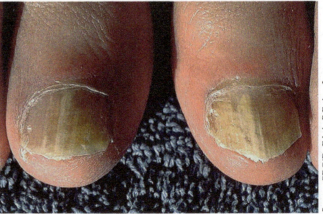

(b)

Figure 25.9 Superficial mycoses. (a) Tinea capitis, ringworm, caused by several species of *Trichophyton* and *Microsporum*. (b) Tinea unguium or onychomycosis caused by *Trichophyton rubrum*, shown on right and left big toes.

of the body. For example, tinea capitis is a ringworm found mainly on the scalp, whereas tinea corporis may infect the whole body. Probably the most common ringworm infection—and, in fact, the most common mycotic infection—is tinea pedis, better known as athlete's foot. This well-known condition may persist indefinitely, causing occasional flare-ups, especially in warm weather.

These fungi do not grow in living tissues; they invade only dead areas of the skin, hair, or nails. The symptoms caused by these infections are actually due to fungal metabolites. Transmission of the infection is generally by person-to-person contact in shared bathing facilities or locker rooms. Spores enter the skin through minor cuts or scratches to initiate the infection. Because these fungi utilize keratin, they cannot exist as soil saprobes. However, their conidia can survive in a dormant state for long periods of time in carpets, upholstery, and similar environments.

While some ringworm fungi parasitize only humans, other species infect domestic animals. A small number of dermatophytes are able to infect cats or dogs as well as people. For many ringworm infections, topical applications of nystatin or oral griseofulvin are often the recommended treatment; however, other antifungal compounds are also used.

Yeast infections, or candidiasis, are another type of common superficial mycosis. Over 95% of these infections are caused by the fungus *Candida albicans,* a normal constituent of the microorganisms present on the skin as well as in the mouth, rectum, and vagina. Various conditions (such as excessive wetness) or weakened host defenses (caused by prolonged antibiotic therapy, steroidal therapy, diabetes, and so on) can trigger an overgrowth by this yeast, resulting in soreness and itching. This overgrowth can cause diaper rash in babies or infection in the armpits, the vagina, or under the breasts in adults. Mucous membranes are particularly susceptible to yeast infections, and oral candidiasis, also known as thrush, is common in infants and young children. In adults, thrush is frequently seen as an early mycotic infection in AIDS patients. Although candidiasis is usually limited to the skin, it can become systemic and cause a life-threatening illness in severely immune-compromised individuals.

Systemic Mycoses

All fungal diseases that infect tissues deeper than the skin are regarded as systemic mycoses. Most systemic mycoses are chronic conditions that develop slowly over long periods of time. Often, months or years elapse before medical attention is sought. Again, it is immune-compromised individuals who are at risk. Many infections are seen in well-defined geographic areas, especially the most common mycoses in North America, histoplasmosis and coccidioidomycosis.

Histoplasmosis occurs in the East and Midwest and is especially common in the Ohio and Mississippi River valleys. The disease is caused by the fungus *Histoplasma capsulatum,* which is prevalent in the excrement of chickens, other birds, and bats. Anyone disturbing these droppings may inhale

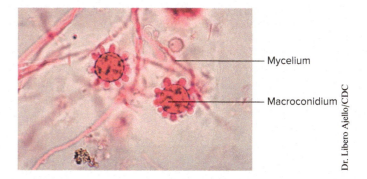

Figure 25.10 *Histoplasma capsulatum,* cause of histoplasmosis, occurs as mycelia and spores in the soil.

conidia, which can cause an infection in the lungs. In more than 95% of the cases, there are no symptoms. The individual is left with a small, calcified lesion in the lung and resistance to further infection by the fungus. It is estimated that about 90% of the population living in the endemic area has had this mild infection. However, in a small percentage of the cases, the disease does not stay localized and continues as a progressive lung disease, with symptoms similar to tuberculosis. Eventually, the infection may become systemic and spread to other organs in the body. At this stage, the disease is usually fatal. *Histoplasma* is a dimorphic fungus. When growing in the soil or droppings, or in culture at room temperature, the fungus has a mycelial form with branching hyphae (fig. 25.10). However, in the body or when cultured at 37°C (98.6°F), it has a yeastlike morphology.

Coccidioidomycosis, also known as valley fever or San Joaquin fever, is prevalent in the hot, dry deserts of the Southwest, especially southern California and Arizona, where the species *Coccidioides immitis* exists saprobically in the soil. During the short rainy season, the fungus grows rapidly, producing thousands of spores that readily become airborne during the subsequent dry season when the soil is disturbed (fig. 25.11). Infection occurs when the spores are inhaled, and like histoplasmosis, it is centered in the lungs. Millions of people in the Southwest have been infected by this fungus. In about 60% of the cases, there are no symptoms at all, and in about 40%, there are mild, flulike symptoms with fever. However, in approximately 0.2% to 0.5% of the infected individuals, the disease becomes systemic and may be fatal if it is not treated or is misdiagnosed. During the 1990s, the incidence of valley fever increased in endemic areas. In southern California, there was a dramatic rise after the January 1994 earthquake. Scientists speculate that the earthquake dislodged settled spores, which then became airborne. Another reason for the increase is believed to be the additional numbers of immunosuppressed people in the population.

Opportunistic infections are caused by a wide range of common saprobic fungi. These species are common and cause no infections when inhaled by healthy individuals. When the immune system is weakened by drug therapy or

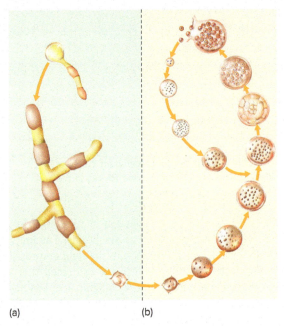

Figure 25.11 *Coccidioides immitis,* cause of valley fever. (a) Mycelial stage occurs in culture and in the soil. (b) Spherules occur in infected lung tissue.

disease, infection by these fungi can occur. It has been estimated that invasive fungal infections occur in 25% of these high-risk individuals. The AIDS epidemic and improved drug therapies for treating cancer patients and organ transplant patients have resulted in a sharp increase in the incidence of opportunistic infections. Among the most common fungi causing these infections are species of *Aspergillus,* which cause aspergillosis, a lung disease with diverse symptoms. Often, a pneumonia-like condition develops, whereas at other times pulmonary lesions occur. Formation of an aspergilloma, a fungus ball consisting of tightly interwoven hyphae within the lungs, is another possible symptom of aspergillosis. Allergic broncopulmonary aspergillosis is caused when the fungus, usually *A. fumigatus,* colonizes the mucus within the lungs, resulting in a severe allergic reaction. In many instances, aspergillosis is fatal.

Treatment of systemic mycoses is difficult because there are few effective drugs. Amphotericin B, produced by species of *Streptomyces* (one of the actinomycetes), is effective against many of the fungi when administered intravenously, but it is a very toxic compound with serious side effects. Miconazole and ketoconazole are also effective, but they, too, show side effects. These two compounds belong to a class of antifungal drugs known as azoles, which block a step in the synthesis of fungal sterols required for membrane structure in fungi. Newer azoles have been developed, and scientists are finding increased applications. Caspofungin and other echinocandins, which are the newest class of antifungal drugs, inhibit the formation of β-glucans, a fungal wall component. These are becoming increasingly useful drugs for treating systemic fungal infections with fewer side effects.

Allergies

Many fungal spores contain **allergens** that are capable of causing respiratory allergies (both hay fever and asthma) in susceptible individuals (see Chapter 21). Allergens have been reported among every major group of fungi, but the majority of attention has been focused on the asexual spores of the ascomycetes. Since the 1930s, these spores have been recognized as an important cause of both asthma and hay fever. Many of the important allergenic fungi include those whose spores are known to be most abundant in the atmosphere, *Cladosporium* and *Alternaria* (see fig. 25.12). These are cosmopolitan genera that can be isolated from the atmosphere throughout the world (fig. 25.12a). When conditions are right, atmospheric concentrations of *Cladosporium* spores can exceed 100,000 spores per cubic meter of air. Although *Alternaria* concentrations are far lower, its spores are considered much more allergenic. Unlike allergenic pollen, which occurs in definite seasons, allergenic fungi can be found in the atmosphere whenever the ground is not covered with ice or snow. This means that, in many parts of the world, fungal allergens occur year-round.

Since the 1990s, research has also focused on other groups of fungi as a source of allergens. Basidiospores from mushrooms, bracket fungi, and puffballs have been implicated as airborne allergens. These spores are produced at incredible levels and released into the atmosphere (see Chapter 23). From spring through fall, the atmosphere may also contain high levels of other spores, including ascospores, smut spores, and rust spores. Far less is known about the allergenic status of these spores, but they are also believed to be potentially important airborne allergens.

Indoor Fungi and Toxic Molds

Fungal spores also occur indoors. Whenever unfiltered outside air is introduced through either ventilation systems or open doors and windows, airborne spores will enter. Spores are also easily introduced on the clothing and shoes of people entering a room. Eventually, many of these spores settle out and, if moisture is available, will colonize indoor substrates. In buildings with well-maintained central heating, ventilation, and air conditioning (HVAC) systems, in-duct filters should remove many of the spores present. However, many instances are known in which the HVAC system itself is contaminated. In these cases, fungi have been found growing on the air filters as well as in the ducts.

Common sites for indoor fungi include carpeting, upholstered furniture, bathroom fixtures, wallboard (sheetrock), and cellulose-based ceiling tiles. All these contain abundant organic material for the growth of fungi (fig. 25.12b and c). With water damage from leaks, floods, or even high humidity,

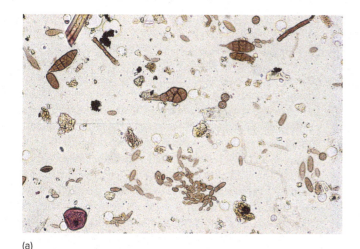

(a)

(b)

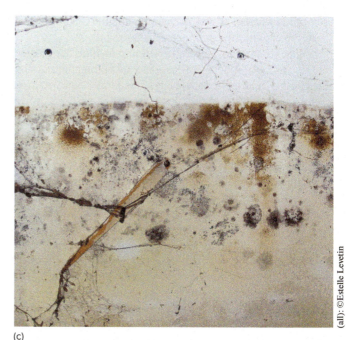

(c)

Figure 25.12 Airborne fungal spores are a major cause of allergic diseases. (a) Outdoor air sample from Tulsa, Oklahoma, showing abundant *Cladosporium* and *Alternaria* spores, ×240. (b) Extensive mold growth beneath a kitchen sink resulted from a plumbing leak. (c) Mold growth on sheetrock in a New Orleans building that was flooded following Hurricane Katrina. Upper limit of mold growth is indicative of the flooding level in this building.

spores can germinate and fungi can grow on these substrates and produce thousands of new spores. The amount of water available is the most important trigger of indoor fungal contamination. The mold growth on water-damaged building materials was one of the major problems in New Orleans homes following the flooding that occurred after Hurricane Katrina in 2005. Among the common fungi found in indoor environments are species of *Cladosporium*, *Aspergillus*, and *Penicillium*. The last two genera occur more frequently indoors than outdoors.

Over the past several decades, indoor mold contamination has become a serious environmental and health problem because more and more people have insulated their homes and businesses for energy efficiency. These measures have made buildings tighter, trapping moisture indoors. The resulting amplified mold contamination has been considered one of the possible factors contributing to the increased incidence of asthma as well as a possible cause of sick building syndrome. This condition is usually characterized by numerous reports of symptoms such as burning eyes, headache, fatigue, dizziness, and breathing difficulties by workers in large, well-insulated office buildings. Sick building syndrome is often due to poor ventilation, with the accumulation of chemical and biological contaminants in the air.

Recently, concern about fungi in indoor air has focused on those fungi known to produce mycotoxins. Studies have shown that mycotoxins can occur in the spores, leading to concern about the inhalation of mycotoxins (within spores) in contaminated indoor environments. Buildings have been closed and totally renovated when high concentrations of certain *Aspergillus* species or *Stachybotrys chartarum* were identified. *Stachybotrys* is a toxin-producing fungus that can be found growing on ceiling tiles or wallboard in water-damaged indoor areas; in the natural environment, *Stachybotrys* occurs in the soil. The toxins in *Stachybotrys* are among the most potent mycotoxins known. In 1997, the Centers for Disease Control and Prevention (CDC) reported that *Stachybotrys* contamination in Cleveland-area homes may have contributed to the deaths of several infants and sickness in many others over a 3-year period from 1993 to 1996. The infants all suffered from acute pulmonary hemorrhage, and high concentrations of *Stachybotrys* were identified in the homes of the affected infants. The CDC issued a more extensive report in 1999, stating that additional research was needed before it could determine whether or not *Stachybotrys* and the other fungi had been the cause of the pulmonary hemorrhage.

Since that time, *S. chartarum* contamination has resulted in the closing of schools, offices, and public buildings around the country, and many people have found it necessary to move from their homes. Lawsuits have been filed by homeowners against insurance companies and builders and by apartment tenants against their landlords when *Stachybotrys* contamination was found. The case that made network news shows and headlines across the country was the contamination of a $3 million mansion in Dripping Springs, Texas. The mold problems developed after leaky plumbing caused extensive damage to the hardwood floors. The homeowners filed a lawsuit against their insurance company, claiming that the company had failed to adequately repair the floors before the *Stachybotrys* growth occurred and spread throughout the house. The family experienced breathing difficulties, dizziness, and memory loss, which they blamed on the *Stachybotrys* contamination. In June 2001, a jury awarded the homeowners $32 million in their suit against the insurance company. The judgment was appealed, and in December 2002 the Texas Court of Appeals reduced the judgment to $4 million.

In response to the recent awareness of indoor air quality and mold problems, state legislatures around the country passed a wide variety of mold-related regulations. Some laws require the disclosure of prior mold problems by real estate owners, while others are aimed at landlords renting apartments or other property with mold contamination. Various laws relate to insurance coverage of mold-related claims, and other legislation is aimed at regulating the mold inspection and mold remediation industries.

In May 2004, the Institute of Medicine from the National Academy of Science released the results of an extensive study on the effects of damp buildings and indoor mold on human health. The findings indicated that a link exists between asthma symptoms and mold and other factors (biological, chemical, and particulate) found in damp buildings. Mold and other factors also have a relationship to wheezing, coughing, and other respiratory symptoms. The study indicated that available evidence does not currently support a connection between mold exposure and other health complaints that have been allegedly associated with mold. However, few well-conducted studies have examined these problems. The report recommended continued research to determine the health effects of exposure to molds and microbial toxins in damp indoor environments.

The World Health Organization conducted a similar study and in 2009 released guidelines for the protection of public health from the risks associated with damp indoor spaces and microbial contamination. The report emphasized that occupants of damp or moldy buildings are at increased risk for respiratory problems and infections and that individuals with allergies or asthma are particularly susceptible to adverse health effects. The guidelines recommend proper building techniques and sufficient ventilation to minimize or avoid dampness and prevent mold-related problems.

CHAPTER SUMMARY

1. Among the secondary products produced by some fungi are antibiotics, compounds that are toxic to microorganisms. In the natural environment, these substances provide advantages over competing microorganisms for available nutrients and space. In modern medicine, antibiotics are the primary weapons for fighting bacterial infections. Penicillin was discovered by Alexander Fleming when he examined a culture of bacteria contaminated with *Penicillium*. Within the next 15 years, purification and mass production of penicillin had been achieved. Although other antibiotics have since been identified and developed, penicillin is still the most widely used antibiotic and is the drug of choice to treat many bacterial infections. Today antibiotic resistance has been found in many common bacteria, causing major concern among the medical community.

2. Some fungal metabolites are toxic to humans and other animals and are capable of causing serious or fatal reactions. These toxins include mycotoxins, which are formed by hyphae of common molds, and mushroom toxins, formed in the fleshy fruiting bodies of basidiomycetes. Mycotoxins are commonly produced by fungi growing in contaminated foods, especially grains and nuts. Aflatoxin was the first mycotoxin identified and studied; today over 400 mycotoxins have been identified.

3. Toxic alkaloids produced by the ergot fungus, *Claviceps purpurea*, accumulate in the fungal sclerotium on contaminated rye plants. If the sclerotia are harvested and milled along with the grains of rye, contaminated bread results. Throughout the centuries, contamination has led to outbreaks of ergotism. Although some of the ergot alkaloids are known to have medicinal value, the lysergic acid alkaloids are responsible for hallucinations and other symptoms associated with convulsive ergotism. The hallucinogen LSD was first isolated from these alkaloids.

4. Mushroom toxins are categorized as protoplasmic poisons, neurotoxins, and gastrointestinal irritants. Most fatalities are caused by the protoplasmic poisons known as amatoxins that occur in various species of *Amanita* as well as other mushrooms. *Amanita muscaria*, which contains the neurotoxin ibotenic acid, has a long history of use as an intoxicant by native peoples in various parts of the world. Among the tribes of Mexico and Central America, various species of *Psilocybe* were revered for their hallucinogenic properties. These mushrooms, containing the alkaloids psilocybin and psilocin, have been used for thousands of years in religious and healing rituals.

5. Some fungi affect humans directly by causing superficial or systemic mycoses. Most fungal pathogens cause relatively minor skin infections, such as ringworm and yeast infection, but more serious fungal diseases occur in individuals whose immune system is compromised. Histoplasmosis and coccidioidomycosis are two serious systemic infections that occur in North America. Opportunistic infec-

tions are caused by a wide range of common saprobic fungi. When the immune system is weakened by drug therapy or disease, infection by these fungi can occur. Invasive fungal infections occur in 25% of these high-risk individuals. The airborne spores of many fungi are known to cause respiratory allergies, such as asthma and hay fever. Many of the important allergenic fungi include those whose spores are known to be most abundant in the atmosphere, *Cladosporium* and *Alternaria*. Recent concern has focused on the allergenic and toxin-forming fungi proliferating in indoor environments.

REVIEW QUESTIONS

1. How did Fleming's discovery change the scope of medical science?
2. Differentiate between antibiotics and fungal toxins in terms of their human impact.
3. In what ways do mycotoxins threaten food safety in both developed and developing countries?
4. Discuss the positive and negative impacts of *Claviceps purpurea* on society.
5. Why are mushrooms in the genus *Amanita* among the most feared mushrooms? Discuss the various types of toxins found in this genus.
6. Describe the use of hallucinogenic mushrooms by native peoples in various parts of the world.
7. Why has there been an increase in the incidence of mycotic infections in recent decades? What are the most common serious mycoses?
8. Describe the importance of fungi as causes of allergic diseases.

UNIT VII

CHAPTER 26

Plant Ecology

Due to global warming, melting glaciers around the world are contributing to sea level rise.

KEY CONCEPTS

1. An ecosystem is the functional unit of the environment; it includes interactions between organisms and the physical and chemical components of the environment.

2. Biomes are large terrestrial regions recognizable by their characteristic vegetation and shaped by climatic conditions; major biomes include grasslands, forests, and deserts.

3. The policy of extractive reserves is a strategy to maintain natural areas for the extraction of economically valuable, yet renewable, products with minimal environmental degradation.

©Image Source

CHAPTER OUTLINE

The Ecosystem 487
    Ecological Niche 487
    Food Chains and Food Webs 488
    Energy Flow and Ecological Pyramids 489
    Biological Magnification 491
Biogeochemical Cycling 492
    The Carbon Cycle 492
    Source or Sink? 492
    The Greenhouse Effect 494
    Species Shift 496
    Global Warming Pact 496
Ecological Succession 497
The Anthropocene Epoch? 498

The Green World: Biomes 500
    Deserts 501
    Crops from the Desert 501
    Chaparral 504
    Grasslands 504
    Forests 505
    Harvesting Trees 506

A CLOSER LOOK 26.1 Buying Time for the Rain Forest 508

Chapter Summary 511
Review Questions 511

Although ecological issues became prominent in the latter half of the twentieth century, the concept of ecology, taken from the Greek *oikos* meaning house, was first proposed by the German biologist Ernst Hackel in 1869. Ecology is the study of organisms in the environment, the surroundings in which organisms live. The areas on Earth in which organisms are found are collectively known as the **biosphere.**

THE ECOSYSTEM

An **ecosystem** is the functional unit of study in the environment. It includes the study of **biotic factors,** organisms, and the interrelationships between organisms as well as their connection to nonliving, or **abiotic,** components in the environment, such as energy flow and the cycling of minerals. The biosphere itself is an example of an ecosystem, although most ecologists study ecosystems on a smaller scale. A pond or an abandoned field is an example of an ecosystem. A terrarium or an aquarium is an ecosystem in miniature for classroom or laboratory study.

The organization of the biosphere is well structured and clearly defined, as depicted in Figure 26.1. A **population** is a collection of organisms of the same species in a given area. A **community** is all the populations of organisms occupying a given region. For example, the winged elm (*Ulmus alata*) is one of several populations of understory trees in an oak-hickory forest community in northeastern Oklahoma.

Ecological Niche

Identifying the **habitat** and **niche** is essential to understanding the role of an organism in the ecosystem. *Habitat* refers to the particular place in the ecosystem that an organism or a population occupies. It is the "address" of an organism, the specific location where the organism can be found. The habitat of cattails is a freshwater marsh; river birch is found along stream banks. The concept of niche is broader; *niche* is defined as the functional role an organism plays in the ecosystem. It encompasses not only where an organism lives but also its relationships to other organisms in the ecosystem. Niche defines such factors as the nutritional requirements, temperature tolerance, or moisture conditions of a species.

Species that fulfill similar or identical niches in different geographical regions are known as **ecological equivalents.** For example, cacti in New World deserts fulfill a niche similar to that of the euphorbs in the deserts of Africa. Species within these completely different plant groups even appear superficially similar shaped by similar selective pressure to convergent forms. Both cacti and euphorbs possess photosynthetic stems that are armed with spines and contain water-storing tissues (fig. 26.2).

Under natural conditions, no two species may occupy the same niche indefinitely. One species may eventually displace the other as a result of **competitive exclusion,** outcompeting the other for nutrients, habitat, or other resources. Competitive exclusion has been demonstrated by experimentally growing two species of clover together: white clover (*Trifolium repens*) and strawberry clover (*T. castaneum*). The faster-growing white clover quickly established a leaf canopy that seemingly shaded out the strawberry clover. In time, however, the strawberry clover sent out shoots that overtopped the white clover, and the white clover was deprived of light. Displaced species, such as the white clover in this experiment, must occupy a different niche in order to survive. In some cases, the displaced species cannot adapt and becomes extinct. Introduction of alien species often results in the displacement and eventual extinction of native species (see A Closer Look 22.2: Killer Alga). This is the case with the Catalina mahogany (*Cercocarpus traskiae*), a tree native to Santa Catalina Island off the coast of California. The introduction of herbivores, such as goats and sheep, onto the island has caused severe overgrazing to the point that the population has been reduced to just a few trees. Another case in point is the introduction of the water hyacinth (*Eichhornia crassipes*). This species, native to South America, was deliberately introduced to Florida because of its beautiful flowers. Today it is a major nuisance in southern waterways, where its population has grown to such an extent that dense mats crowd out native species.

At other times, niche competition can lead to strategies that reduce conflict, allowing the two species to coexist in a given area. In resource partitioning, a change in behavior or physiology reduces the amount of overlap between species for shared resources in the environment. For example, moths and butterflies both feed on flower nectar. But they do not directly compete for this resource because butterflies feed in the daytime

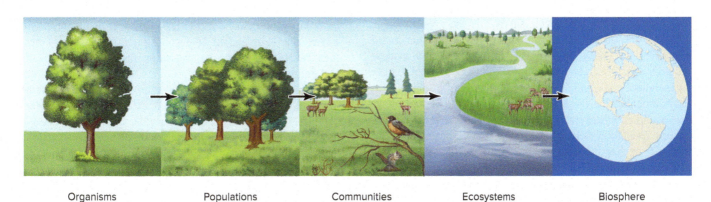

| Organisms | Populations | Communities | Ecosystems | Biosphere |

Figure 26.1 The field of ecology deals with studies of organisms, populations, communities, ecosystems, and the biosphere.

Figure 26.2 (a) Organ pipe cactus (*Lemaireocereus thurberi*) in the Sonoran Desert in Arizona and (b) a giant spurge (*Euphorbia*) in the Namib Desert, Namibia, Africa, are ecological equivalents shaped by similar selective pressures to convergent forms.

and moths at night. Likewise, day-blooming and night-blooming flowers are not competing for the same pollinator population.

Another factor in species survival is the type of niche characteristic of the species. Some species have rather broad tolerances for many environmental conditions and general requirements for survival. Species such as these are said to have a **generalized niche** and are less sensitive and more adaptable to environmental changes. Dandelions, coyotes, and human beings are examples of generalist species. In contrast, some species occupy **specialized niches;** these species have a narrow range of tolerance or exacting requirements for survival. The giant panda of China is a specialist, feeding exclusively on bamboo, and is understandably highly susceptible to extinction if its sole food source is threatened. Plants such as the swamp pink (*Helonias bullata*) of the eastern United States have become endangered because they are found only in freshwater wetlands, and many of these have been drained for development or agriculture.

Food Chains and Food Webs

Crucial to the functioning of an ecosystem is the relationship of organisms in a **food chain.** A food chain essentially asks, Who eats whom? A typical grazer food chain (fig. 26.3) begins with the **producers,** green plants or algae capable of producing complex organic compounds through the process of photosynthesis (see Chapter 4). These organisms are also classified as **autotrophs**—literally, "self-feeders." All other organisms in the grazer food chain fulfill their nutritional needs by using others as a food source and are known as **consumers** or **heterotrophs** ("other feeders"). All animals are consumers; those that feed exclusively on producers are called **herbivores.** Those that feed on other consumers are **carnivores.** Some animals are classified as **omnivores;** their diet includes a mix of plants and animals. In sequential fashion, the grazer food chain begins with the producers, which are eaten by **primary consumers;** the primary consumers, in turn, are eaten by **secondary consumers;** those eating secondary consumers are tertiary consumers, and so forth. Each step in a food chain is referred to as a **trophic level;** most food chains have maximally four or five levels. In a **detritus** food chain (fig. 26.3), consumers (detritivores) meet their nutritional needs by degrading the remains of plants and animals and their waste products (detritus). Some larger animals—vultures and hyenas for example—are scavengers of the dead, and bacteria and fungi fulfill the role of **decomposers** or **saprobes,** degrading organic material. In most situations, a food chain is a simplification of the actual trophic interactions in an ecosystem. In reality, there is usually more than one type of producer, and most consumers have several alternative food sources. A **food web** such as that found in the salt marsh depicted in Figure 26.4, with multiple interactions between several food chains, is a more accurate representation.

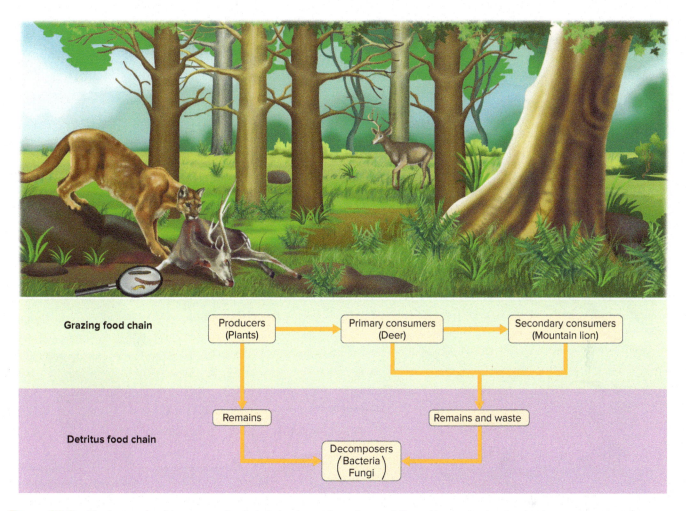

Figure 26.3 The bottom level in a grazer food chain begins with producers, followed by succeeding levels of consumers. In a detritus food chain, both macroconsumers and microconsumers degrade the remains of dead plants, animals, and waste products as a source of nutrition.

Energy Flow and Ecological Pyramids

An important characteristic of ecosystems is the transfer of energy. The ultimate source of energy for most ecosystems is the sun; solar energy is trapped by the chlorophyll present in algae and green plants for the purpose of photosynthesis. During photosynthesis, solar energy is transferred into chemical bond energy in the construction of organic compounds from the raw materials carbon dioxide and water (see Chapter 4). Producers incorporate these organic compounds or break them down to release energy during cellular respiration. Primary consumers eat the producers, obtaining the raw materials needed for the construction of their own tissues and energy. Secondary consumers eat herbivores to acquire energy, and energy is thus transferred along the food chain. Recall from Chapter 4 that energy transformations are governed by the First and Second Laws of Thermodynamics. According to the First Law, energy cannot be created or destroyed but can be transformed from one form to another. Photosynthesis transforms the radiant energy of the sun into the chemical bond energy within glucose. According to the Second Law, when energy is converted into another form, some of the energy is transformed into low-quality heat energy that is lost to the environment. With each transformation, therefore, there is less high-quality energy available and the ability to do useful work is reduced. For example, much of the chemical bond energy in fuel molecules is dispersed as heat during the transformation into mechanical energy to drive an engine.

Energy flow in ecosystems can be depicted by **ecological pyramids of numbers, biomass,** and **energy.** A pyramid of numbers (fig. 26.5) indicates the number of organisms supported at each trophic level in a food chain. Fewer organisms are supported at successively higher trophic levels. In other words, it takes many small organisms to support the nutritional needs of one large organism. **Biomass** is the organic material contained in living organisms. It is usually represented as the dry weight (fresh weight minus the water content) of living matter. As with the pyramid of numbers, the total biomass for each trophic level usually declines with each step up the food chain.

Figure 26.4 A food web in a salt marsh shows the complex interactions between food chains among organisms. The producers (1) are terrestrial and aquatic salt marsh plants. Primary consumers include herbivorous insects (2), other invertebrates (3), and fish (4). Primary consumers are in turn eaten by carnivores, such as the common egret (5) and the shrew (6). Secondary carnivores (7) include the marsh hawk and short-eared owl. Mallard ducks (8), song sparrows (9), sandpipers (10), Norway rats (11), and salt marsh harvest mice (12) are omnivores, feeding on both plants and animals.

Likewise, the pyramid of energy illustrating the total energy content in kilocalories for each trophic level follows the same pattern as the other ecological pyramids. It also shows the inefficiency of energy transfer between trophic levels and explains the shape of the ecological pyramids. Only 5% to 20% of energy (averaging 10%) is passed between trophic levels. Consequently, most food chains consist of a maximum of only four or five levels. Beyond this number, the energy transferred is minimal and incapable of supporting life. The loss of energy from lower trophic levels to higher ones also explains why more herbivores can be supported than carnivores. One suggestion to conserve food resources for the burgeoning human population (about 7.7 billion in 2019) is to eat lower on the food chain. More humans could be fed if they ate plant crops (producers) directly instead of feeding the crops to livestock and eating the meat produced. For example, 20,000 kilocalories of corn produces 2,000 kilocalories of food for cattle to yield 200 kilocalories of meat, enough to feed only one human being. In contrast, the same amount of corn eaten directly would provide 10 people 200 kilocalories apiece.

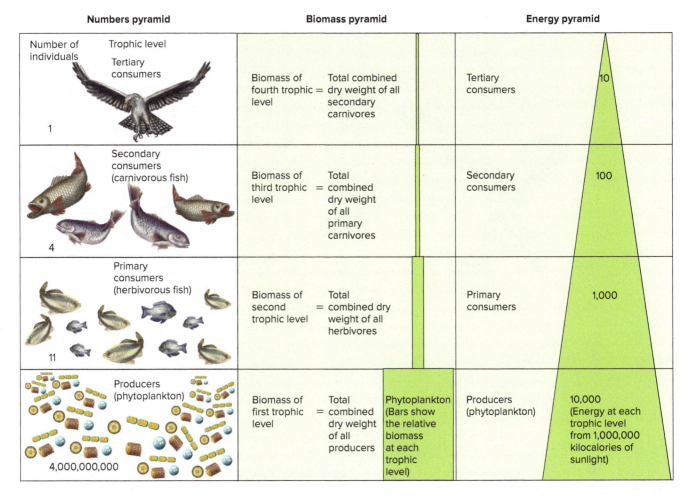

Figure 26.5 Ecological pyramids of numbers, biomass, and energy. A pyramid of numbers indicates the number of organisms in each trophic level that can be supported in the food chain. Pyramids of biomass indicate the total biomass at each trophic level. Pyramids of energy indicate the amount of energy transferred between ascending trophic levels.

Biological Magnification

In 1962, a book entitled *Silent Spring,* by Rachel Carson, brought to the attention of the scientific community and the public some disturbing observations about the use of pesticides. Since the end of World War II, numerous synthetic pesticides had been created and released in an attempt to control humanity's most serious pests. These pests, primarily insects, have spread human disease or have been destructive to crops and livestock. One of the new generation of pesticides that was widely used to control mosquitoes as well as other harmful insects was a chlorinated hydrocarbon known as DDT (dichloro-diphenyl-trichloroethane). DDT, as are many of the other synthetic pesticides produced during that time, is slow to degrade, persisting in the environment for several months or years. Also, it is known to be soluble in lipids, a major component of organisms. Stored in an organism's lipids, DDT was later shown to be passed up the food chain and, with each transfer to a higher level, became more concentrated in the fatty tissues. The transfer and concentration of a toxic material along a food chain is known as **biological magnification** (fig. 26.6). As *Silent Spring* pointed out, the biological magnification of DDT had dire consequences for many species of birds at the top level of food chains. For example, the American bald eagle, as a secondary consumer of fish-concentrated DDT (and its primary breakdown product, DDE or dichloro-diphenyl-dichloroethylene), was exposed to such toxic levels that the amount of calcium deposited in the eggshells was reduced. Shells were so thin that the fragile eggs broke during incubation, killing the developing embryo. Consequently, the numbers of bald eagles and other raptors (birds of prey), as well as some familiar songbirds, declined rapidly. DDT was also showing up in human tissue, even in the milk of lactating women. For these reasons, DDT has been banned in the United States since 1972. Unfortunately, this ban is not global; DDT is still used in many developing nations. Other toxic substances that can be biologically magnified have been identified and their use banned or restricted as well. Examples are mercury, PCBs (used as insulators in electrical systems), aldrin (a crop pesticide), and chlordane (used in termite control). (For another example of biological magnification, see Chapter 9 and the connection between cycad toxins, bats, and Lou Gehrig's disease.)

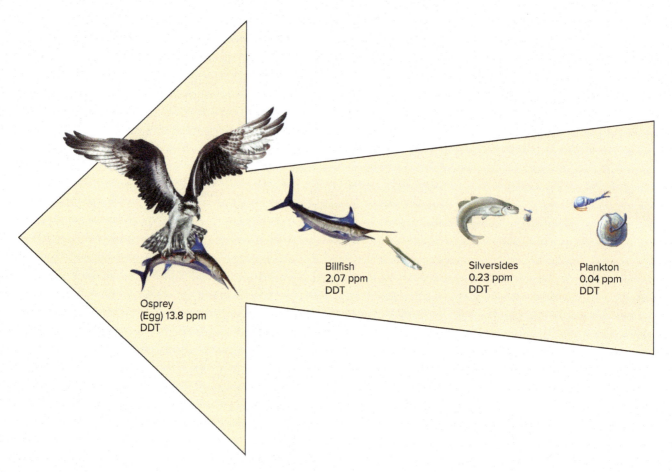

Figure 26.6 The pesticide DDT is concentrated and passed along a food chain in a process called biological magnification. (DDT values are expressed in parts per million, wet weight, whole body basis.)

BIOGEOCHEMICAL CYCLING

The natural world is the original recycler, cycling substances such as sulfur, phosphorus, and water through the abiotic and biotic components of the environment. The cycling of nitrogen was discussed in Chapter 13. Here, the carbon cycle, and how human activities have affected and upset this cycle, will be discussed.

The Carbon Cycle

All life on Earth is carbon-based; the primary organic compounds are constructed from carbon backbones. It is not surprising, therefore, that carbon makes up 40% to 60% of the dry weight of all living matter. Carbon, in the form of gaseous carbon dioxide (CO_2), is removed from the atmosphere through the process of photosynthesis as carried out by green plants, algae, and cyanobacteria (fig. 26.7). Recall that the summary equation for photosynthesis is

$$6CO_2 + 12H_2O + \text{sunlight} \rightarrow C_6H_{12}O_6 + 6O_2 + 6H_2O$$

Sunlight captured by chlorophyll provides the energy to drive the photosynthetic reaction. Carbon dioxide is fixed and reduced during photosynthesis to form glucose (see Chapter 4). Because they remove carbon dioxide from the atmosphere, the photosynthetic organisms are known as *carbon sinks.*

The carbon in glucose can be used to make numerous organic compounds in the producers and later in the herbivores and carnivores along the food chain. Glucose is also utilized as fuel to yield energy for producers, consumers, and decomposers through cellular respiration. It is through cellular respiration that carbon dioxide is released back into the atmosphere:

$$C_6H_{12}O_6 + 6O_2 \rightarrow 6CO_2 + 6H_2O + 36ATP(\text{energy})$$

Likewise, the burning of any organic material, whether firewood or fossil fuels (coal, natural gas, or oil), becomes a *carbon source,* releasing carbon dioxide to the atmosphere.

Source or Sink?

It may seem that the photosynthetic power of large tracts of vegetation, such as the tropical rain forest, would always

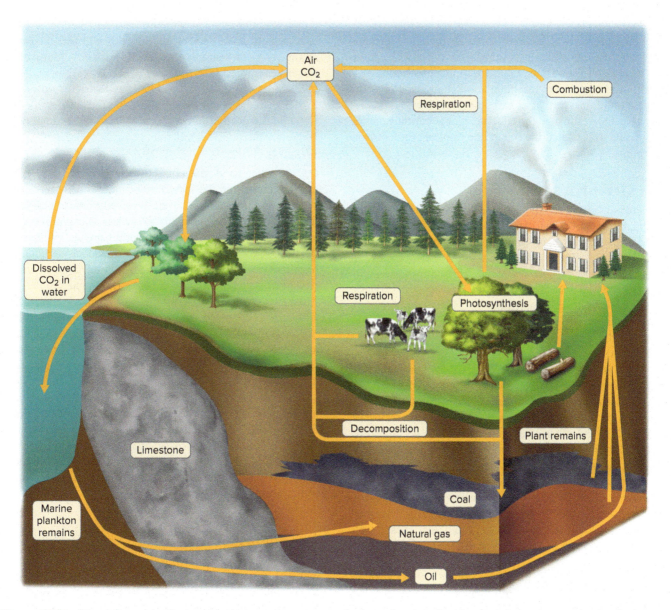

Figure 26.7 The carbon cycle. Carbon dioxide enters organisms in the form of carbon dioxide when plants photosynthesize. Carbon is released back to the atmosphere when organisms respire carbon dioxide or during the combustion of carbon-based fossil fuels. Fossil fuels (oils, natural gas, coal) were formed from the remains of ancient organisms.

act as carbon sinks, but that may not be the case. A report noted that, during unusually hot summers with little rainfall, a mature rain forest in Costa Rica became a carbon source, emitting substantial amounts of CO_2 into the atmosphere. Most researchers originally thought that the net impact of the tropical rain forest on the carbon cycle was negligible because the amount of CO_2 that the trees absorb through photosynthesis is balanced by the amount of CO_2 they continuously release through cellular respiration. But in 1995, a research report suggested that rain forests in the Amazon were now absorbing more CO_2 than they were releasing. Other studies of old-growth tropical forests followed with similar results, reinforcing the proposal that rain forests play a major role in slowing greenhouse warming.

Recently, researchers who have studied the trees of a mature tropical rain forest at La Selva, Costa Rica, disagreed. They believed that these earlier studies overestimated tree growth, a reflection of photosynthetic rate, which inflated the importance of trees as carbon sinks. When the La Selva researchers looked at the growth rates of six canopy trees for almost two decades, they noticed that in warmer years the trees grew more slowly. During the winter and drier summers of the El Niño years, tree growth declined. Tree growth was 81% higher in the two coolest years as compared to the hottest period, from 1997 to 1998.

These findings may indicate that photosynthesis rates are lower and respiration rates higher during warmer years. In fact, respiration may exceed photosynthesis, resulting in the slower growth rates of the trees. It is known that the rate of photosynthesis will increase up to an optimum temperature, then decline, while respiration rates continue to go up as temperatures get warmer. Global warming may inhibit photosynthesis but promote respiration, switching the role of the tropical rain forest to that of carbon source. Previous studies may have underestimated the effects of global warming on tree growth and exaggerated the role of the tropical rain forests as a carbon sink.

The Greenhouse Effect

Humankind has upset the balance of the carbon cycle in several ways. Global deforestation, especially of the tropical rain forest, removes vast areas of vegetation, reducing the effectiveness of what is usually considered a carbon sink. At the same time, more CO_2 has been released into the atmosphere, primarily from the accelerated burning of fossil fuels. Global carbon dioxide increased by 25% in the twentieth century (fig. 26.8), and this increase in atmospheric CO_2 has contributed to the **greenhouse effect**. Carbon dioxide, as well as some other greenhouse gases, acts as a heat trap. Much like the glass panes in a greenhouse, CO_2 is transparent to sunlight. As radiant energy strikes the surface of the Earth, some of it is converted into infrared (heat) energy and reradiated back to the atmosphere. Infrared radiation cannot pass through CO_2 and other greenhouse gases. The accumulating heat causes atmospheric temperatures to rise. Carbon dioxide, of all the **anthropogenic** (human-made) greenhouse gases, has had the greatest impact on global warming. One reason is its persistence; as much as 40% of it remains in the atmosphere for centuries. In fact, the temperature of Earth has increased by 1.0°C in 2018. It has been predicted that these levels will continue to rise as fossil fuel consumption continues to increase.

Carbon dioxide levels in the atmosphere reached a record high of 408.97 ppm (parts per million) in May 2018, an increase over the reading of 406.37 ppm recorded a year before. Before the Industrialized Age, atmospheric CO_2 was probably around 278 ppm. The four warmest yearly global temperatures on record are 2016, 2015, 2017 with 2018 appearing to end up as one of the top five warmest years recorded since records were first compiled in the mid-1800s. Atmospheric scientists have determined that the most likely cause is the accelerated burning of fossil fuels and consequent release of CO_2. The Intergovernmental Panel on Climate Change estimates that atmospheric carbon dioxide levels may reach concentrations from 650 to 970 ppm in the year 2100. With these higher levels, global temperatures may rise between 1.8°C and 40°C by 2100.

Further global warming is predicted to produce dramatic climatic changes. Agricultural regions in the United States will shift northward as greater heat stress and more frequent droughts prevail in lower latitudes. Much of the world's best agricultural lands of today will be lost as the climate becomes too dry and hot to grow the major agricultural crops. Warmer climates will shift the distribution of plants and animals; some species undoubtedly will become extinct as the range of heat-tolerant species expands. Many of the eastern hardwood trees, such as beech (*Fagus*), will die out in southeastern states and will become restricted only to the higher altitudes of a few mountains and to New England and Canada, the northernmost realm of its current range.

Disease Threats

On the other hand, the range of disease-carrying pests will expand, increasing the incidence of mosquito-borne diseases such as malaria, dengue fever, West Nile virus, and yellow fever. Currently, the possibility of malaria infection is a threat to those areas where 45% of the world's population is found. By the end of the twenty-first century, global warming will have increased the possibility of malaria infection to encompass a larger area where 60% of the world's population resides. It has been predicted that, by the year 2020, as temperatures rise, the risk of malaria transmission in many areas will have doubled. Since 1970, the elevation at which temperatures are always below freezing has ascended almost 1,640 meters (500 feet) in the tropics. Mosquitoes that carry yellow fever and dengue fever, which were once limited to lowlands because of the prohibitive cooler temperatures at higher elevations, have extended their range upward and have been found 1 mile higher in the highlands of northern India and in the Colombian Andes. Also, higher temperatures decrease the length of the breeding cycle for the pathogen *Plasmodium falciparum,* one of the protozoan species that causes malaria, from 26 days at 20°C (68°F) to just 13 days at 25°C (77°F), enhancing the possibility of infection.

Ocean Effects

The polar ice caps and glaciers will shrink, and the oceans will take up more space because of thermal expansion of water molecules. During the twentieth century, sea levels rose by 17 centimeters. A predicted 18- to 40-cm rise in sea levels by the end of the twenty-first century will flood the coastal areas of many nations, where half the world's population resides.

Ice in Antarctica is melting at an increasingly rapid pace, especially in West Antarctica. Before 2012, ice was lost from the continent at an average of 76 billion tons per year but from 2012 to 2017, the rate almost tripled to 219 billion tons each year. All of this water has raised the global sea level by an average of 7.6 millimeters (0.3 inches) with most of the increase occurring since 2012. Antarctica currently still contains enough ice to raise ocean levels by 58 meters (nearly 200 feet). It is the melting of Antarctica ice that is major threat to flooding of coastal communities and islands. Flooding may also result from greater precipitation at the higher latitudes, fueled as moisture from increased evaporation at the hotter lower latitudes moves poleward.

About one-third of the carbon dioxide gas released by the burning of fossil fuels is absorbed by ocean water. The natural

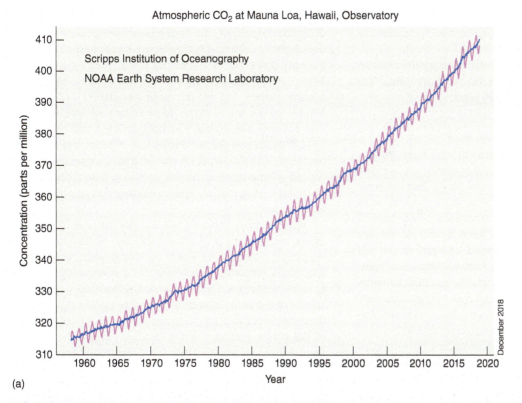

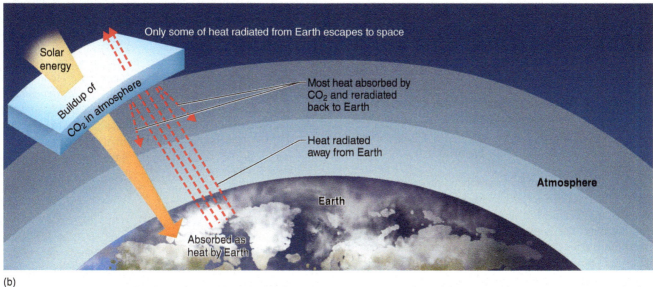

Figure 26.8 (a) A steady rise of carbon dioxide in the atmosphere due to the burning of fossil fuels and the loss of carbon sinks has been documented. (Carbon dioxide levels are lower in the summer because photosynthetic rates are high during the summer.) (b) Carbon dioxide and other greenhouse gases trap solar energy, reradiating it back to Earth as heat and, consequently, warming the atmosphere.

state of ocean water is slightly basic, but the pH of the ocean decreases as carbon dioxide combines with water to form carbonic acid. Since the start of the Industrial Revolution, the pH of ocean waters has dropped from 8.2 to 8.075, a 30% increase in acidity since the pH scale is logarithmic. The acidification of ocean water is threatening the survival of many species, such as certain plankton, corals, snails, and other organisms with calcium carbonate skeletons, because calcium carbonate shells cannot be made under acidic conditions. Ocean acidification is expected to have additional effects on sea life. One study reported that sexual reproduction and the survival of offspring in elkhorn coral (*Acropora palmata*) is impaired when, in laboratory tests, the coral was exposed to carbon dioxide levels of 500 ppm, a level expected to be reached by midcentury.

In July 2015, the global ocean surface temperature was an average 17.15°C (62.87 °F), which is the warmest average sea surface temperature on record in 136 years.

Species Shift

Two large-scale analyses reported that many diverse species from geographic locations across the world are already responding to a warming climate. In one study, data collected over several decades from more than 1,700 species were analyzed. The range boundaries of more than 100 species of birds, butterflies, and alpine herbs have moved an average of 6.1 kilometers (3.7 miles) northward, or a rate of 1 meter (3.3 feet) per decade upward. More than 400 species of trees, shrubs, herbs, lichens, mammals, insects, reptiles, amphibians, fish, marine invertebrates, and marine zooplankton were determined to have shifted in either abundance or range, and 80% of these changes follow a pattern consistent with global warming. For example, lichens and butterflies have invaded areas that were previously too cold for these species, and the range of some Arctic species has condensed as temperatures have risen.

The **phenology,** or timing of life cycle events, of about 170 species of herbs, shrubs, trees, birds, butterflies, and amphibians was discovered to take place an average of 2.3 days earlier per decade. A different statistical analysis of nearly 700 species showed that, over a time period of 16 to 132 years, 27% of the species showed no trend in phenology, 9% showed a delay, and over 60% exhibited advances in the onset of spring events, such as first flowering, tree budburst, bird nesting, frog breeding, or the arrival of migrant birds and butterflies. Again, the phenological advances are consistent with global warming.

In another major study, researchers reviewed data on almost 700 species or species groups. Eighty percent showed a temperature-related shift in phenology. Consistent with global warming, the higher latitudes nearer the poles have warmed more than the lower latitudes closer to the equator. At latitudes above 50° (about the level of Calgary, Canada, and Kiev, Russia), the phenological events are taking place an average of 5.5 days earlier. Below 50° latitude, the average advance is about 4.2 days. Interestingly, trees stand apart from other species (invertebrates, amphibians, birds, and other types of plants) in that they show an advance of only 3 days.

Shifts in the timing of life cycle events may also disrupt food chains and upset the balance in an ecosystem. Marcel Visser, an ecologist in the Netherlands, discovered that the life cycles of the winter moth (*Operophtera brumata*), the oak tree (*Quercus robur*), and the great tit (a small bird, *Parus major*) are no longer in synchrony. Oak leaves are the primary food for the caterpillar stage of the winter moth. Oak leaves now appear 10 days earlier than they did in 1985, and caterpillars now hatch 15 days earlier than the oak buds open. Back in 1995, the caterpillars would hatch just a few days before budburst. Oak leaves are edible to the caterpillars for only a few weeks before the leaves accumulate tannins. Now, caterpillars are without food for several days; and as a consequence, the winter moth population has declined. Winter moths are a major food source for the young tits that hatch in the spring. The timing of the tits' nesting and egg-laying did not change during the 20 years, but the earlier hatching of the caterpillars means that a major food source for the young birds is no longer at peak numbers. Global warming is expected to produce further disruptions in other food chains.

On the other hand, CO_2 is the raw material of photosynthesis, and the growth of some plants may be limited by its low availability (less than 0.04%) in the atmosphere. As CO_2 levels rise, some plants may be favored in a community, upsetting the natural composition of ecosystems. One study reported the effects of elevated levels of CO_2 on the growth of native and nonnative trees and shrubs in the forests of Switzerland. One of the plants studied was the invasive laurel cherry (*Prunus laurocerasus*), a woody plant from southwest Asia that was introduced as a landscape ornamental into many parts of Europe. Birds eat the fruit and disperse the seeds of the laurel cherry into the deciduous forests of Switzerland. Its leaves contain the toxin cyanide, which protects the laurel cherry from herbivores; hence, it grows unchecked as an understory tree, displacing native species.

Researchers built chambers to compare the growth of the laurel cherry to that of native plants at elevated levels of CO_2. The native species studied were holly (*Ilex aquifolium*), ivy (*Hedera helix*), hornbeam (*Carpinus betulus*), and ash (*Fraxinus excelsior*). Seeds were planted, and the seedlings were grown in pots under three treatment chambers (a control and two levels of CO_2, 365 and 500 ppm), placed in the forest, and monitored for three growing seasons.

Laurel cherry seedlings in the chambers with elevated CO_2 levels showed a 56% increase in biomass over 3 years. The ivy doubled its biomass only at the highest CO_2 level. Holly and hornbeam seedlings were unaffected by CO_2 levels. It appears that the invasive laurel cherry will be favored as CO_2 levels rise with global warming. The climbing ivy will also be favored in the enhanced CO_2 environment, and although it is a native, its climbing habit is likely to negatively impact tree growth. This study points out that community composition is likely to change in an enhanced CO_2 world in which invasives may replace native species.

Global Warming Pact

In 1997, representatives from 167 nations met in Kyoto, Japan, to discuss the problem of global warming. As a result of this conference, 38 industrialized nations agreed to reduce carbon dioxide and other greenhouse gases to 5% below 1990 levels by 2012 in order to curb the greenhouse effect. This represents a 30% reduction from projected emissions expected in the year 2010 without any controls. After 3 years of negotiations, the proposals of the Kyoto Protocol were finalized on July 2001 in Bonn, Germany, as 178 countries approved several strategies to counteract global warming. In the proposed Kyoto Treaty, 1,700 diplomats agreed to a common methodology for quantifying estimates of greenhouse gas emissions and for calculating the balancing effects of sinks in absorbing CO_2. One controversial provision of the Kyoto Treaty allows a nation to receive credits for sinks such as maintaining forests. These credits can offset mandatory reduction limits in emissions of greenhouse gases. An industrialized nation can also receive credit for aiding a developing nation with emission-reducing technologies and projects. This proviso, known as the Clean Development Mechanism, awards credits to an industrialized nation if it helps a developing nation build a nonpolluting hydroelectric dam or protect a forest to soak up CO_2. In addition, nations are allowed to trade credits. For example, one nation may buy credits from others that have cut back their greenhouse emissions to below the set limits. The Kyoto Protocol was ratified as a binding international agreement in the fall of 2004, when it was signed by 55 nations, which account for 55% of the 1990 greenhouse gas emissions released from industrialized nations. Russia was the last industrialized nation to ratify the protocol; the global climate pact took effect at the start of 2005.

Unfortunately, the United States was not one of the signers. (Australia also has not ratified the Kyoto Protocol.) Although U.S. delegates to the Kyoto Protocol had agreed to reduce greenhouse gas emissions to 7% below 1990 levels by 2012, Congress never ratified the proposal. Greenhouse gas emissions in the United States continued to increase, by more than 11% from 1991 to 2001, and the U.S. government called the proposed level of reductions unrealistic. In the fall of 2003, the U.S. Senate rejected a bill to curb emissions of carbon dioxide and other greenhouse gases from industrial sources to 2000 levels by 2012. This is in contrast to the strong commitment of the European Union to reduce greenhouse gases; their goal to reach the set limit of 8% below 1990 levels should be met by 2012. Several states, however, have created the Regional Greenhouse Gas Initiative (RGGI) in which participants have agreed to begin capping CO_2 emissions in 2009 and begin reducing emissions by 2016.

In December 2009, 193 nations met in Copenhagen, Denmark, to hammer out a successor to the Kyoto Protocol. The Copenhagen Accord established a trust fund with contributions of $10 billion per year to compensate poorer nations for local climate changes caused by the greenhouse gas emissions of wealthier countries. The document also pledged to keep global temperatures from rising less than 2°C (35.6°F) above pre-industrial levels. (This goal has been lowered from the original target of 1.4°C (34.5°F).) The Copenhagen Accord also stated that industrial nations must implement emission targets voluntarily set by 2020. Many advocates wanted industrialized nations to reduce greenhouse gas emissions from 75% to 90% below 1990s levels. The accord does not set specific emission levels (each nation will set its own reduction levels), nor does it require external examiners to verify that these levels have been met. Guidelines for emission reductions in developing nations were also not set by the Copenhagen Accord, which is not legally binding to the signatory nations.

In December 2015, representatives of 196 nations met in Paris to work out climate action proposals to address the problem of global warming caused by greenhouse gas emissions. As of July 1918, the Paris Agreement's long-term goal is to put policies into place that limit the increase in global warming to 1.5°C (34.7°F) above preindustrial temperatures. At one point, the goal was to put policies into practice that would limit the increase in warming to 2°C but further research revealed that this projected rise in average global temperature would severely damage biodiversity. In a 2018 study of more than 115,000 terrestrial species, researchers estimate that if global temperatures increase by 2°C, the geographic range would be halved for 18% of insect species, 16% of plant species, and 8% of vertebrate species. However, if the temperature increase is topped at a 1.5°C increase, the halving of geographic ranges is markedly reduced to 6% of insect species, 8% of plants, and 4% of vertebrates. Additionally many of the island nations and less developed countries that are expected to be most affected by flooding, severe drought, and scorching temperatures also campaigned for an upper limit of 1.5°C. The United States and China, two of the world's biggest carbon emitters, both signed the Paris Agreement. However, in June 2017, the United States announced that it will pull out of the Paris Agreement at the end of 2020.

ECOLOGICAL SUCCESSION

Ecosystems are not static; they experience changes over time. **Ecological succession** is the gradual progression of orderly and predictable changes that occur in an ecosystem over time. These changes result from modifications of the physical environment through the actions of the biotic community. These modifications of the abiotic component of the environment, in turn, bring about changes in the biotic community. Two types of succession are recognized: **primary** and **secondary.**

Primary succession (fig. 26.9) begins in an area bare of vegetation. The colonization of plants in a pond, a newly formed sandbar, and barren rocks are examples of primary succession. Pioneer species are the first inhabitants in the area; in the terrestrial environment, lichens are usually the first community. As pioneer species, they begin the process of soil formation by dissolving minerals from the rocks and, as they die, adding organic material. Also, the presence of vegetation tends to trap and hold water. All these modifications bring

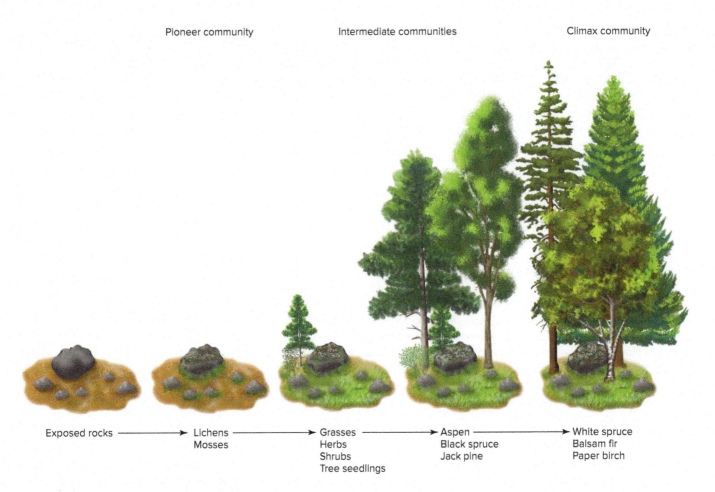

Figure 26.9 Primary succession begins as pioneer species of lichens and mosses invade bare rock. Several intermediate communities inhabit the area for a time before the self-perpetuating climax community of spruce-fir and birch forest forms.

about changes in the environment, making it more amenable to succeeding communities of invading plants. Mosses, then grasses and herbs, follow and replace the pioneer lichens. Eventually, the environment may be sufficiently modified to support a mature community of trees.

In secondary succession, natural forces or human intervention destroys the existing vegetation. Usually, the soil remains, so the soil formation associated with primary succession does not occur. One of the best-studied examples of secondary succession is old-field succession of abandoned farmland. Clearing the land of the natural vegetation is the first step in preparing the land for planting. If that field is abandoned and no longer farmed, the process of secondary succession begins (fig. 26.10), eventually returning the land to the original natural community. Other examples of secondary succession are the return of vegetation following the volcanic explosion of Mount St. Helens, and the recovery of Yellowstone Park after the devastating fires. The last successional stage in both primary and secondary succession is self-perpetuating and stable and is known as a **climax community.**

Under certain conditions, the pathway of ecological succession is not always so straightforward. Often, because of local factors, a community does not reach the predicted climax stage. For example, in the southeastern coastal plains of North America, the stable community is the longleaf pine forest. Unlike oaks, hickories, and other deciduous trees, the pines have adaptations to withstand the periodic fires that sweep through the area.

THE ANTHROPOCENE EPOCH?

A new geologic epoch, the **Anthropocene,** has been proposed to reflect the permanent imprint humanity has left upon the Earth. The current epoch is the Holocene which began as the Great Ice Age ended around 11,700 years ago. First suggested in 2002, the Anthropocene would reflect how humans' exploitation of the natural resources on earth have irrevocably changed the planet. For example, human actions have caused climate change through the accelerated burning of fossil fuel and technofossils, artifacts produced by human technology, such as plastics, pesticides, and concrete now contaminate the earth's sediments. People have also impacted wildlife on the one hand by introducing domesticated species such as the domesticated chicken worldwide while at the same time driving other species to

Figure 26.10 The secondary succession of abandoned farmland begins with the invasion of weedy species, such as crabgrass, that quickly colonize the area. Other herbaceous plants follow. Eventually, sun-loving pine species invade and establish, in time, a pine forest. Hardwood saplings develop in the shade of the pine forest and form the climax hardwood forest 70 or more years after the successional process began.

extinction through habitat destruction. All of these human-driven impacts support the finding that the Anthropocene is distinct from the Holocene. For scientists who support the creation of this new epoch, one controversy is when the Anthropocene began. Did it start when humanity first converted forest to farmland, thousands of years ago? Or with the start of the Industrial Revolution and its use of fossil fuels? Other scientists argue that although those events are significant they did not have a simultaneous worldwide effect. They propose that the beginning of the nuclear age in the late 1940s and 1950s with the development of the atomic bomb and the subsequent aboveground nuclear bomb tests is the starting point for the Anthropocene as evidenced by the global distribution of radioactive carbon and plutonium. Researchers are seeking out other evidence of human impact on the natural world in rocks, mud layers, and tree rings to gather evidence for support of this new epoch.

THE GREEN WORLD: BIOMES

Biomes are the largest terrestrial divisions of the biosphere. They are the climax communities for huge regions of the land and are recognized and defined by their distinctive vegetation and animal life. Climatic conditions, such as rainfall and temperature, act as environmental filters to largely determine the biome that develops in an area. Yearly precipitation is a key factor in whether a given latitude will develop into a desert, grassland, or forest. Deserts are found in areas with low yearly precipitation. If the yearly precipitation increases, grasslands develop. Still higher precipitation values favor the development of forest. Altitude also affects biome development (fig. 26.11). The cooler temperatures encountered at a mountain's higher altitudes promote the development of biomes normally seen at the higher latitudes. The major biomes of the world are depicted in Figure 26.12 and are discussed next.

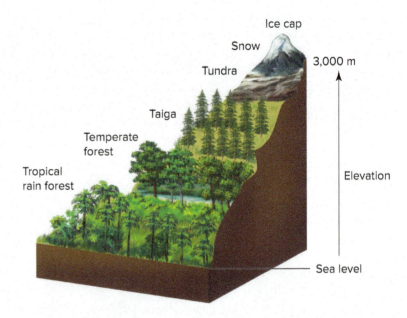

Figure 26.11 Altitude affects the distribution of biomes. The cooler temperatures at the higher elevations of a mountain, even one located on the equator, favor the development of biomes similar to those encountered when traveling a latitudinal journey from equator to the Poles.

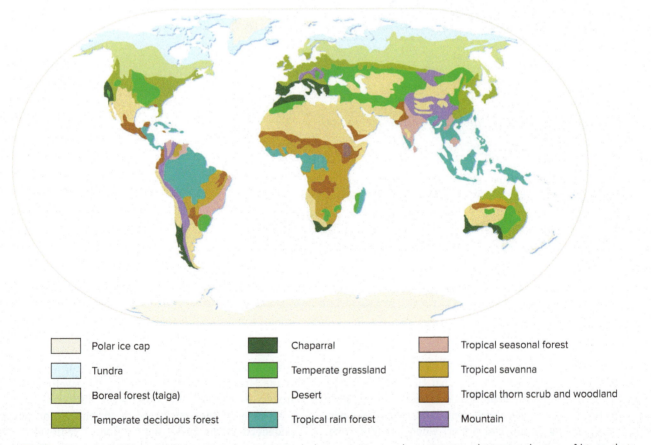

Figure 26.12 Map of world biomes. Climatic conditions, particularly temperature and precipitation, determine the type of biome that forms in a region.

Thinking Critically

Reclamation is the process of restoring to its natural state or biome land disturbed by mining, pollution, farming, grazing, logging, and other human activities. For example, to reverse the global problem of deforestation, international organizations fund tree-replanting efforts.

Typically, the first step in restoring a forest of mixed hardwoods is to plant pine seedlings. Why don't the foresters plant hardwood seedlings to restore a hardwood forest? Is there an explanation to be found in ecological theory?

Deserts

The **desert** biome develops when annual rainfall falls below 25 cm (10 in.). Hot or low deserts form when evaporation and transpiration rates exceed rainfall. The Mojave Desert in California and the Sonoran Desert, spanning New Mexico, Arizona, and southern California, are two low deserts in North America. Various cacti and drought-tolerant shrubs dominate the landscape of the low desert (fig. 26.13a). Cold deserts form when a mountain range creates a rain shadow. An example of a cold, or high, desert is the Great Basin (comprising most of Nevada and parts of Utah, California, Oregon, and Idaho), formed because few precipitation clouds can pass over the Sierra Nevada. Sagebrush and other low shrubs characterize the high desert (fig. 26.13b). Desert soils are nutrient poor because there is not enough moisture for degradation of organic material. Some are high in minerals but in levels that are toxic to most vegetation. Desert soils are also subject to wind erosion because the vegetative cover is often sparse. Both the plants and the animals that are the normal occupants of the desert exhibit adaptions to conserve water. Plants that have evolved adaptations to survive the xeric (dry) conditions of the desert are called **xerophytes.** One strategy to avoid the lack of moisture is seen in desert annuals, which germinate, grow, and reproduce only during the rainy season. Others, such as the cacti, have lost their leaves to reduce water loss from transpiration and have developed stems that are both photosynthetic and succulent (water-storing). Other desert plants have extremely long taproots that tap into underground water supplies. Native animals, such as the kangaroo rat, have also adapted to the continual drought of the desert. The kangaroo rat is a burrower, coming out only in the cool of the evening. It also never needs to drink water; its highly efficient kidney system extracts sufficient water from its food alone.

Crops from the Desert

One way to help meet the food demands of the forecasted human population expansion in the twenty-first century is to convert marginal lands into agricultural lands. Many of these marginal lands are in arid and semiarid regions. Irrigation is one method of converting desert into farmland. But as water resources are already strained to meet agricultural, industrial, and residential demands, another solution has evolved: native desert plants already adapted to the arid conditions of high temperatures and water scarcity are developed into crops.

Natural rubber is typically thought of as a product of the tropical rain forest, but the desert shrub guayule (*Parthenium argentatum* of the Asteraceae or sunflower family) might change that perception (fig. 26.14a). Native to the Chihuahuan Desert of Mexico and southwestern Texas, this plant produces a **latex** in the root that is stored in parenchyma cells of both root and stem. (Latex may be variously colored but is most often milky white. The chemical makeup of latex is also varied, but the terpenoids are among the most common constituents.) Wild stands of guayule were first used as a source of natural rubber in the late 1800s and early 1900s. During World War II, when supplies of natural rubber from the East were disrupted, the U.S. government initiated plantings of guayule as a domestic source of natural rubber. These plans were abandoned with the development of synthetic rubber and the end of the war. Although synthetic rubber is a suitable substitute in most cases, there is still a demand for the natural product. In fact, the United States imports nearly $1 billion in natural rubber each year. Considering the financial drain, there has been renewed interest to develop guayule as a natural rubber crop. After years of research, plant breeders in Arizona have improved the latex output of guayule so that cultivars now produce approximately 900 pounds per acre (809 kilograms per hectare). Guayule latex is of particular interest as a replacement for tropical rubber in the manufacture of latex gloves. Gloves made from guayule do not seem to induce the latex allergies associated with tropical rubber (see Chapter 21). Guayule bagasse, the remains of stems and leaves after the latex has been extracted, can be bonded with recycled high-density polyethylene, the type of plastic found in milk jugs, to make a composite particleboard that is both strong and resistant to termites and wood rot fungi. Alternatively, guayule resin can be extracted from the bagasse and incorporated into wood. Wood permeated with the guayule resin is protected against termite damage and wood rot.

An evergreen shrub native to the Sonoran Desert, jojoba (*Simmondsia chinensis;* fig. 26.14b) has been nicknamed the botanical whale because its seed oil is similar in many ways to the oil obtained from sperm whales. The development of jojoba as an oil crop has been encouraged since 1969 when the sperm whale was placed on the endangered species list and the import of sperm whale oil was banned. The seed oil of jojoba is unique; most seed oils are triglycerides, but jojoba oil is liquid wax, both colorless and odorless. Its unique chemical properties make it suitable for a wide range of applications, including high-pressure lubricants, antifoaming agents in the fermentation industry, protective coatings, and even an edible cooking oil.

(a)

(b)

(c)

(d)

(e)

(f)

(g)

(h)

(i)

Figure 26.13 Major biomes. (a) Hot deserts are dominated by cacti and drought-tolerant shrubs. (b) Sage bush and other low shrubs characterize cold deserts. (c) Broadleaf evergreen shrubs form thickets in the California chaparral. (d) Grasslands, such as this prairie with bison, are home to large grazing mammals. (e) The savanna, such as this one in Africa, is grassland with scattered trees. (f) The tundra is a treeless, frozen plain located in the northernmost part of North America. (g) South of the tundra lies the boreal forest, a vast band of coniferous trees. (h) Trees in the moist coniferous rain forest of the Pacific Northwest can reach gigantic proportions. (i) Much of the eastern half of North America is covered by deciduous forest, characterized by brilliant autumn color. (j) Broadleaf evergreen trees make up the dominant vegetation of the tropical rain forest.

(j)

Figure 26.14 Potential desert crops. (a) Guayule (*Parthenium argentatum*) produces a latex that is a source of natural rubber. (b) The seed oil of jojoba (*Simmondsia chinensis*) has many industrial uses. (c) The fibers of kenaf (*Hibiscus cannabinus*) can be used to make newspaper.

Another desert crop on the brink of worldwide importance is kenaf (*Hibiscus cannabinus;* fig. 26.14c), which may replace trees as a source of pulp for papermaking. This annual member of the mallow family (Malvaceae) is native to east-central Africa. Field trials indicate that kenaf would grow well in many of the U.S. southern states without irrigation. It is quick growing and can tolerate saline soils. Paper made from kenaf is said to be equal in strength to that made from wood pulp and has the added advantages of not yellowing as quickly and holding ink better. Newsprint from kenaf is now a viable alternative, with greater productivity per hectare and half the cost of wood pulp (see Chapter 18). Kenaf fiber is also used to make carpet pads, mats, fire logs, particleboard, and cardboard.

Chaparral

Chaparral is found extensively in North America in southern California. The chaparral biome has a Mediterranean climate of hot, dry summers and cool, rainy winters. Evergreen, broadleaf shrubs form dense thickets across the landscape (fig. 26.13c). Allelopathy (see Chapter 21) has been shown to influence the scattered spacing pattern of the vegetation and is the reason for the bare zones of inhibition around the shrubs. Many of the shrubs are **sclerophyllous,** having thick, leathery leaves because of an abundance of sclerenchyma cells. The leathery leaves cut down on transpiration rates during the long summers. Periodic fires sweep through the chaparral, but the shrubs regenerate from underground organs. Mule deer, chipmunks, lizards, and several bird species are examples of the animals found in the chaparral. The chaparral soil is thin, relatively infertile, and unstable. Human population pressures along the California coast have encroached on the chaparral. The ever-present threat of fire during the dry season and the potential of mudslides from heavy winter rains are the dangers of living in this biome.

Grasslands

The pampas, veld, steppes, and prairies are all examples of **grasslands** (fig. 26.13d). The average rainfall is relatively low (25 to 75 cm; 10 to 30 in.), with warm summers and cool winters. Grasses are the dominant vegetation in this biome. **Tallgrass prairies,** named for the 5-meter (15-foot) heights typical of some grasses, are found in the eastern range of this temperate biome where precipitation is greater. In the drier areas, **shortgrass prairies** are found. In past ages, these vast "seas of grass" attracted huge herds of grazing mammals. In the North American grasslands, bison and antelope thrived, and ground-nesting birds, such as the prairie chicken and quail, abounded. Periodic fires have also influenced the appearance of the grasslands, destroying trees and releasing nutrients from organic matter. Grasses, with their rhizomes belowground, can sprout back after a fire and quickly colonize a burned area. Many of the grass species are sod formers, producing compacted mats of soil, roots, and rhizomes, which were cut into blocks by the pioneers and used to build homes and barns. The topsoil layer was extensive, reaching depths of 0.7 meter

(2 feet), and of a rich, dark color because of its high organic content. Few untouched grasslands still exist in the continental United States. These rich lands now support herds of cattle and farmland where the native grasses have been replaced by domesticated grasses, grains, or pasture crops, such as wheat, corn, alfalfa, or fescue.

A variant of the grassland is the **savanna.** Trees are scattered throughout the grassy plain of the savanna. In North America, the oak savanna can be found in the transition from eastern forests to the central plains. Dominant tree species are the black oak, bur oak, and chinquapin oak. The dominant grass species is usually little bluestem. The tropical savanna has little seasonal temperature variation, but the annual rainfall of 85 to 150 cm (34 to 60 in.) is distributed unevenly into pronounced dry and wet seasons. Giraffes, elephants, zebras, wildebeests, and antelopes are indigenous to the African savanna (fig. 26.13e). The herbivores in turn attract predators and scavengers, such as the lion and hyena.

Cold shapes the **tundra,** a polar grassland that is the northernmost biome bordering the ice cap. The tundra is an icy, rolling plain where grasses, sedges, mosses, and lichens dominate the landscape (fig. 26.13f). A few shallow-rooted shrubs, such as birch and willow, may be found, but they are extremely stunted, reaching heights of only 30 cm (12 in.). The soil is thin and nutrient poor, inhibiting root growth. In the summer, only the top 10 cm (4 in.) of soil thaws. The rest is **permafrost,** permanently frozen year-round. Temperatures are cold throughout the year, averaging 10°C (50°F). The growing season in the summer is short, lasting only several weeks, but intense, for the days are 24 hours long. During the summer thaw, the tundra has a waterlogged appearance, dotted with many ponds. This appearance is deceiving, for the ponds are actually shallow puddles; the permafrost prevents drainage and the soil is saturated. Yearly precipitation is low, usually from 10 to 25 cm (4 to 10 in.). Herds of caribou are the predominant herbivores in the North American tundra. Smaller, burrowing mammals, such as lemmings, snowshoe hares, and weasels, are also tundra inhabitants. Bird life includes ptarmigans, snowy owls, and migrant waterfowl that come to breed during the summer. Insects such as deer flies, mosquitoes, and black flies are also abundant in swarms during the summer. The tundra is a fragile biome that recovers slowly from disturbances. Damage from the building of the Alaskan pipeline in the 1970s is still evident.

Global warming is accelerating the loss of permafrost, especially the permafrost found near the soil surface. Soil temperature is the result of the interaction between geothermal heat from the Earth's core and the average annual air temperature. If the average air temperature hovers around the freezing point of water, few sites in an area have permafrost, but if the average air temperature is −6°C (21°F), much of the soil is permafrost. As the average air temperature has risen during the past 100 years, soil temperature has also increased, warming the permafrost. Permafrost serves as a stable foundation for pipelines, roads, buildings, and bridges; as the extent of permafrost shrinks, the infrastructure in higher latitudes is also affected. Some computer models suggest that 90% of the world's near-surface permafrost will have disappeared by the year 2100.

Forests

Forests form when the annual precipitation is sufficient to support tree species. There are several types of forest biomes, distinguished by their latitudinal distribution. Just south of the tundra, the **boreal forest** (also known as the **taiga** or **northern coniferous forest**) occurs as a gigantic band across North America, Europe, and Asia. Although the winters are long and cold, they are less severe than those in the tundra and, combined with the 4-month growing season, allow for the complete thawing of the soil. An annual precipitation rate of 38 to 102 cm (15 to 40 in.) also contributes to the milder conditions of the boreal forest. Conifers such as spruce, pine, hemlock, and fir dominate this biome in immense stands of only one or two species (fig. 26.13g). Leaf litter accumulates beneath the tree stands and decays slowly because of the cool temperatures. The soil is relatively infertile and acidic. Many species of large and small mammals are permanent residents of or visitors to this biome. Caribou overwinter here and moose, wolverine, bear, mink, lynx, and many rodents are other inhabitants of this biome. Many migratory species of birds also temporarily inhabit the forest. The pure stands of conifers in the boreal forest have been extensively harvested as a source of softwoods for lumber and pulp for paper.

The **moist coniferous forest** biome is found in North America in the Pacific Northwest, where the northern climate is moderated by the influence of the Pacific Ocean. The winters are mild and the summers cool. This biome is also referred to as the **temperate rain forest** because of the high yearly rainfall, 200 to 380 cm (80 to 152 in.), the high humidity, and the prevalence of fog from the ocean. Conifers such as Douglas fir, northern arborvitae, and Sitka spruce are the dominant trees and often reach gigantic proportions. There is a well-developed understory of shrubs, ferns, club mosses, mosses, and lichens (fig. 26.13h). Squirrels, deer, and many species of birds, such as the endangered spotted owl, inhabit this biome. The moist coniferous forest has been heavily exploited, but tracts of old-growth forests (forests that are over 100 years old and have never been logged) still exist. The future of old-growth forests is at the center of a controversy between the timber industry and environmentalists.

Much of the eastern half of the United States was once covered by an extensive **deciduous forest** of oak, maple, beech, hickory, and other hardwoods (fig. 26.13i). **Deciduous** means to shed and refers to the fact that these trees shed their leaves annually in the autumn, often in brilliant color displays, in contrast to conifers, which retain their leaves for several seasons. The winters can be long and hard, but the summers are hot. Precipitation is evenly distributed throughout the year and relatively high, from 75 to 152 cm (30 to 60 in.) annually. The topsoil is characteristically rich in organic material. Animal life is abundant; reptiles, amphibians, insects, birds, and many familiar small and large mammals (white-tailed deer, wolves, bears, red foxes, cottontail rabbits, raccoons, and squirrels) abound. Most of the original deciduous forest is gone; less than 0.1% of the old-growth forest remains. The forest has been cleared since colonial days for urban expansion, agriculture, and timber; however, as the agricultural center moved westward during the 1800s, from the eastern states to the central United States, much of the abandoned farmland reverted to forest.

The **tropical rain forest** has the greatest species diversity of any other biome. In one-tenth of a hectare (one-quarter of an acre) of tropical rain forest in South America, there are 208 tree species. (Compare that with the approximately 25 tree species found in an area of the same size in the deciduous forest in the state of New Hampshire.) The dominant vegetation of the rain forest is broadleaf evergreen trees, which are divided into stories according to their height (fig. 26.13j). The upper story consists of the tallest trees, rising to heights of 80 meters (260 feet); beneath lies the middle layer, where most trees are found, with an average height of 50 meters (160 feet). There is little understory because of the dense shade. However, many epiphytes and lianas (woody vines) climb the trees to sunlight. Most animal life is arboreal: birds, insects, amphibians, monkeys, and reptiles. The tropical rain forest is so named because there is little seasonal variation in temperature and because of the high daily rainfall, 200 to 450 cm (80 to 180 in.) per year. The soils of the tropical rain forest are extremely poor because decay of organic material is rapid and nutrients are quickly taken up by existing vegetation. The tropical rain forest is currently being deforested at an alarming rate; it is estimated that half the world's tropical rain forests have already been destroyed. Each year, an area of forest equivalent to the state of Indiana is cleared; if this rate continues, the tropical rain forest will be gone in approximately 30 years. Several factors are contributing to this destruction.

Population pressures in many developing nations are causing governments to settle people in former wildernesses. With the introduction of people and roads, the forests are cleared in a slash-and-burn (swidden) type of agriculture. Because the forest soil is so nutrient poor, crops can be grown in an area for only a few seasons before the land must be abandoned and a new area has to be burned. Land is also cleared to make pastures for cattle. Often, the cattle are grown for shipments to overseas markets in developed nations.

Harvesting the rain forest for highly prized tropical hardwoods, such as mahogany or teak, is often done with poor forest management practices (see Chapter 18). Also, in much of the world, the primary fuel for cooking and heat is firewood, another cause of deforestation.

Rain forest destruction is also affecting the health care of people who rely on plant-based medicines. A report about the use of medicinal plants in Belén, a port city in eastern Brazil, revealed that much of the urban population still turns to fresh preparation of roots, barks, flowers, seeds, and leaves when ill. Forty-five percent of the more than 200 plants found in the marketplace and nine of the top 12 sellers are native to the mature Amazon rain forest. Over the course of nearly a decade, the sale of medicinal plants increased despite rising prices. Many of the most popular plants have no effective counterpart in pharmaceutical medication. They are used to treat a variety of ailments: urinary and vaginal infections, intestinal worms, coughs, fatigue, burns, tumors, impotence, hair loss, and others. Most of these plants have yet to be domesticated and must still be collected from the wild. Despite increasing demand, supplies of some species have declined, probably because of deforestation. For example, the shrub *Ptychopetalum olacoides,* whose root is the source of a powder used to treat tremors and impotence, is found only in mature forests. Some of the medicinal trees have also been harvested for their timber, such as *Carapa guianensis,* which resembles mahogany wood in color and grain. Its seed oil is used to treat bruises, sprains, and rheumatism and as an insect repellent. Another pressure is the developing export market for some of the medicinal plants. Medications from the bark of *Tabebuia* species are now exported because its reputation as an effective treatment for stomach ulcers and cancer has spread. Resin from the bark of copaiba (*Copaifera reticulata*), a popular remedy to heal wounds, can be extracted in a way that does not harm the plant, but many collectors just slash the bark with machetes, shortening the life of the tree. Action must be taken to preserve the diversity of medicinal plants in the rain forest because it is estimated that more than 80% of the developing world continues to rely on traditional medicines that are primarily plant-based. One strategy to save the tropical rain forest is the policy of extractive reserves, presented in A Closer Look 26.1: Buying Time for the Rain Forest.

Harvesting Trees

Exploitation has characterized many of the interactions between humanity and the forests of the world. Forests are often cut for timber without any regard for regeneration. Since the mid-eighteenth century, it has been realized that forests can be renewable resources, producing a sustainable harvest of lumber, if they are managed wisely. Several methods of forest management are currently in practice (fig. 26.15).

Clear-cutting (fig. 26.15a) is one of the most controversial methods for harvesting trees. In this method, all the trees of every species and age in a specified area of forest are cut. Clear-cuts can be strips, scattered patches, or large blocks. Regeneration of the forest may occur naturally from the surrounding intact forest, as when a few seed trees are deliberately spared in the cut area, or the cut may be prepared and planted with tree seedlings of a commercially desirable species.

Proponents of clear-cutting argue that, if an area is opened to full sunlight, regeneration of sun-loving species, such as the commercially valuable lodgepole pine or Douglas fir, is hastened (see Chapter 18). Opponents note that the scarred forest declines in recreational value and disrupts wildlife by destroying habitat. In addition, because every tree, regardless of age, is cut, young trees, which are not suitable for lumber, are wasted. Research on clear-cuts has shown that soil erosion is accelerated, especially in regions with a pronounced slope and

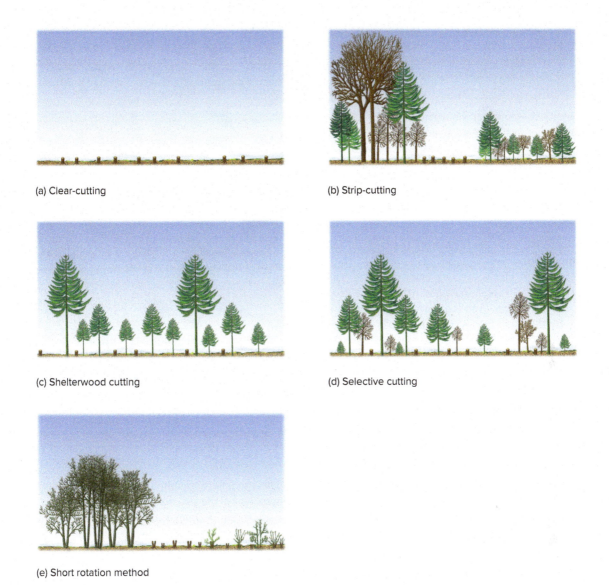

Figure 26.15 Tree harvesting methods. (a) In clear-cutting, all trees in the selected area are cut down. (b) In strip-cutting, only a narrow strip through the forest is clear-cut at any one time. (c) Shelterwood cutting removes some of the mature trees but leaves others to provide seed and shelter from wind and sun. (d) In selective cutting, trees are individually evaluated to determine the optimum time of harvest. (e) With the short rotation method, roots are left intact so that trunk sprouts can grow.

with heavy to moderate rainfall. The accompanying increase in soil runoff results in the loss of soil nutrients and beneficial microbes, which causes the decline of soil fertility. The runoff also pollutes the surrounding watershed with sediment. Tree planting replaces the original forest of many species and varying ages with a tree plantation of a single species and age.

A variant of clear-cutting, strip-cutting, limits the harvesting of trees to narrow rows through the forest (fig. 26.15b). Strips are cut sequentially after enough time has passed for the previous strip to begin regeneration. It may take several decades before a particular strip is reharvested. Most of the forest is left intact, so seed trees from wooded strips regenerate the cleared strip naturally. The wooded areas also protect developing tree seedlings in the cut from wind and sun. Wildlife is minimally disturbed because the wooded corridors still provide habitat and food. There is also less erosion and loss of soil fertility than when the entire forest is clear-cut. Because the strips are narrow, the scarring of the land is reduced, and most of the forest can still be enjoyed for recreational purposes.

In whole-tree harvesting, another variation of clear-cutting, machines uproot each tree. Every part of the tree, including roots and small branches, is harvested. The cut trees are then usually sent to a chipping machine that reduces them to wood chips for fuel or wood pulp for papermaking (see Chapter 18). The advantage is there is no waste, but nutrients to the soil are lost because there is no tree litter for decomposition and recycling that can replenish the soil and begin regeneration.

A CLOSER LOOK 26.1

Buying Time for the Rain Forest

In 1985, a plan arose to protect the tropical rain forest from timber concerns yet maintain an economic base for the indigenous peoples who live in the forest and depend on it for a livelihood. Nontimber forest products, such as latex, resins, fruits, and nuts, would be harvested and sold on the world market. These products are part of a renewable, sustainable harvest in that they could be extracted for the long term with minimal damage to the intact rain forest. More than 30 products are sold, primarily from the tropical rain forest in South America.

Natural rubber and chicle are two latex products from tropical forests. Natural rubber, obtained from several tropical plant sources, was known to the indigenous peoples of Mesoamerica. By 1600 B.C., they were making balls, figurines, and wide bands of rubber. What most astounded the early European explorers such as Columbus and Cortez, however, were the "bouncy" balls used in sport. From northern South America to present-day Arizona, a ball game was played in stone courts that in some arenas seated thousands of spectators. In one version of the game, teams of athletes would try to send solid rubber balls through hoops without using either hands or feet. Games were fast-paced and sometimes bloody as the balls ricocheted off the court walls. European observers noted that the balls were prepared by collecting white latex from *Castilla elastica*, a lowland tree of the mulberry family (Moraceae) found in Mexico and Central America. The latex was later processed with the sap of a vine (*Ipomoea alba*) of the morning glory family (Convolvulaceae) that changed the raw latex into rubber. This technique is still being practiced in parts of Mexico today.

The name *rubber* was later coined for this New World product because of its ability to rub out pencil marks. It remained a curiosity until 1823, the year Charles Macintosh discovered that rubber-treated fabric becomes waterproof. In 1839, Charles Goodyear learned that treating rubber with sulfur and heat, a process called vulcanization, stabilizes rubber's elastic properties. Vulcanization opened up numerous applications for rubber and accelerated the search for an abundant, quality source in tropical Africa and South America. *Hevea brasiliensis* (in the Euphorbiaceae), a tree native to the Amazon basin, yielded the purest and most elastic rubber of all and soon became the mainstay of the Brazilian rubber industry. Wild rubber trees are widely scattered in the tropical rain forest, usually only two to three trees per hectare. A single person collects rubber from 60 to 150 trees, and much time and effort are spent in clearing trails from tree to tree. Rubber is found in latex vessels in the inner bark, close to the vascular cambium. Rubber tapping consists of wounding the bark with shallow, spiral incisions to release the latex (box fig. 26.1a). Incisions are

Box Figure 26.1 (a) To tap a rubber tree, several shallow incisions are cut into the bark to release latex from the vessels; the runoff is then collected in small cups.

made in the morning, when latex flow is greatest. Small cups placed below the incisions catch the latex; it is collected in the afternoon. Tapping is usually done during the dry season because in the rainy season the cups tend to fill with water and the trails to the trees are often impassable. The collected latex is dripped on a type of paddle or stick, which turns over a fire. Gradually, a large ball of solid rubber forms; the balls are then shipped to factories for further processing.

The *seringueiros*, Brazilian rubber tappers, have led the environmental movement to save the tropical rain forest and protect the indigenous people and their way of life. Their cause has been met with violent political opposition; one of the most visible leaders of the rubber tappers, Franciso Alves Mendes Fihlo (Chico Mendes), was brutally murdered on December 22, 1988, by local ranchers who opposed the ban on clear-cutting the forest.

More recently, Sister Dorothy Stang was an environmental activist who died trying to protect the Amazon rain forest and the subsistence farmers who live there. Born and raised in Ohio, she went as a member of the Notre Dame de Namur community to northern Brazil in 1966. Stang soon made it her mission to help the settlers who had been moved by government-sponsored programs to farm the Amazon jungle. She organized the farmers into cooperatives, teaching them how to rotate crops in place of slash-and-burn practices. In these settlements, she sought to enforce the often ignored law that farmers could clear up to 20% of the land only if the remaining 80% were left intact. Stang encouraged the sustainable harvest of products from the preserved rain forest. She was often quoted as saying, "The death of the forest is the end of our life." She also intervened in the bloody land grabs between squatters who farmed small plots of the forest and large landowners who illegally clear-cut the forest for timber and to plant forage for cattle or soybeans for export. Before her murder at age 73, the environmental activist was campaigning to stop the illegal clearing of a part of the jungle called Lot 55. On February 12, 2005, she was on her way to help displaced families whose homes and crops had been burned to the ground by men hired by ranchers. On a jungle path, Sister Stang was challenged and then callously shot by two hired guns. A year after her murder, the Brazilian government began a crackdown on fraudulent logging permits throughout the Amazon. The president of Brazil also established a preserve of 16 million acres of rain forest in which logging is limited to environmentally conscious companies and no clear-cuts or settlements are allowed. Unfortunately, Lot 55 is not part of the preserve and is still being illegally logged for hardwoods for export to the United States, Europe, and Asia. In 2010, 5 years after Dorothy Stang's violent death, the ranchers who had ordered the murder because she blocked them from taking over land the government had given to local farmers were convicted and sentenced to 30 years in prison. Also, the man they had hired to pull the trigger confessed and was sentenced to 28 years. This final verdict came after a three-judge panel overruled the findings of trials in 2008, in which the jury voted to overturn earlier convictions and set them free. The judges' panel noted in its reversal that the jury in the 2008 trial had ignored the evidence and the defense attorneys had presented illegally obtained evidence.

Chicle, another product of the rain forest, is used in the manufacture of chewing gums. The technique used is comparable to the extraction of rubber latex. Workers called *chicleros* scar the bark of the chico tree (*Manilkara zapota*, in the Sapotaceae). The raw latex is solidified into 10-kilogram (22.5-pound) blocks, or marquetas, and stamped with the workers' initials. The latex is high-quality, but the chicle market has declined over the years since the availability of cheap synthetic gums. However, with the current interest in natural products, it is beginning to rebound.

One of the most important economic plants of the South American rain forest is the Brazil nut tree (*Bertholletia excelsa*) in the Lecythidaceae. The trees are found in lowland forests with a prolonged dry season in the Guianas, Colombia, Venezuela, Peru, Bolivia, and Brazil. In fact, flowering begins shortly before the onset of the dry season. After fertilization of the flowers, the fruits take 15 months to develop, maturing during the rainy season. The large, woody fruits fall to the ground intact with the seeds still encased. The fruits weigh from 0.5 to 2.5 kilograms (1 to 5.5 pounds) and contain 10 to 25 seeds or nuts. The long-lived trees produce from 63 to 216 fruits per season. Various rodents gnaw through the woody pericarp to remove the seeds, which also have an incredibly tough seed coat. Most seeds, shelled

(b)

Box Figure 26.1 (b) Fruit, shelled seed, and unshelled seed of the Brazil nut tree.

or unshelled, are sent to Europe and the United States (box fig. 26.1b). Over the long term, Brazil nut production appears to be more profitable than harvesting timber or clearing the forest for pasture.

A recent study examined the effects of harvesting Brazil nuts from 23 populations across the Brazilian, Peruvian, and Bolivian Amazon. The populations of Brazil nut trees differed in the intensity and regularity of the nut harvest. Some of the stands were consistently overharvested; others were only lightly to moderately exploited; and some populations were never harvested. It was discovered that the moderately to heavily harvested stands consisted mostly of older trees, but juvenile Brazil nut trees were rare or completely absent. Young trees were most common in the unharvested or lightly harvested populations. Apparently, intense harvesting of Brazil nuts is an unsustainable practice that prevents the regeneration of the Brazil nut trees. There may still be time to halt the exploitation of the Brazil nut harvest. Most of the mature trees in the overly exploited stands are still capable of producing fruits and seeds that could regenerate the population if overharvesting were stopped. Another strategy to improve age distribution would be to transplant nursery-grown seedlings into those stands with no young trees. The harvest of nuts must be managed to ensure the survival of the Brazil nut tree in the Amazon.

Another potentially profitable product from the rain forest is tagua, or vegetable ivory. Tagua is the very hard, cream-colored endosperm from the seeds of the *Phytelephas* genus of palms, with *P. macrocarpa* the most commercially important species. Tagua palms grow wild in the rain forests of northwestern South America. Tagua was once a very popular material in the manufacture of buttons and jewelry for European and North American markets until it was replaced by the availability of cheap plastics in the 1940s and 1950s. Presently, the Tagua Initiative has sought to reestablish the tagua industry in countries such as Ecuador and Colombia. Tagua buttons have already been introduced and sold in the United States on clothing by Esprit, L. L. Bean, and others, and tagua jewelry and carvings (chess pieces, animal figurines) are also available (box fig. 26.1c).

Ethnomapping is an emerging strategy to preserve the Amazon rain forest. Underwritten by ACT (Amazon Conservation Team) based in Virginia and founded by noted ethnobotanist Mark Plotkin, tribal leaders are asked to locate sites in their territories that are remarkable with respect to history, culture, or wildlife. As Mark Plotkin has said, "Indigenous peoples have repeatedly been shown to be the most effective guardians of the rain forests they inhabit." The first step is to record the memories of the elders to map out significant locations, such as ancestral cemeteries, ancient hunting grounds, battle sites, breeding territories for wildlife, and stands of medicinal plants. Next teams travel to find and verify the sites described and use GPS (Global Positioning System) to pinpoint exact locations. Already, millions of acres of rain forest have been mapped in Suriname, and Colombia. By documenting the value of the indigenous landscape, the profitability of sustainable forest products can be increased. In the state of Amazonas, a Green Free Trade Zone was established that lowered taxes on sustainable products from Brazil nuts to medicinal plants. Ethnomapping is progressing but not without risk. From 2002 to 2007 in Rondonia, a state of Brazil, 11 tribal chiefs were gunned down. Suspicion points to loggers and miners who continue to illegally plunder Indian reserves.

(c)

Box Figure 26.1 (c) Figurines made from vegetable ivory. The *Phytelephas* or tagua palm seed is shown in the upper right.

Thinking Critically

Short-term economic gain is often the underlying cause of rain forest destruction. The strategy of extractive reserves argues that a renewable harvest from an intact rain forest can yield greater economic profits over the long term.

Do you think the strategy of extractive reserves could be employed to preserve natural areas other than the tropical rain forest? What products might be harvested from other biomes, such as the prairie, the deciduous forest, the temperate rain forest, or the desert? Do you think the strategy of extractive reserves would work equally well in developed nations as in developing countries? Explain.

With the shelterwood method, mature trees in an area are targeted for removal, typically in a series of three cuts, over a period of years or decades (fig. 26.15c). The first cut removes dead, diseased, and damaged trees as well as undesirable tree species. Thinning of the forest in this way encourages the growth of younger, commercially valuable trees. Some mature trees are left uncut so that the forest can reseed in the opened areas. In a second, later cut, again some of the mature trees are harvested, but a few are left to shelter the seedlings and saplings from sun and wind. The third cut removes all the remaining mature trees because the younger trees are now of sufficient growth to replace them. Because the forest is never completely denuded of trees, there is not as much soil erosion or disruption of wildlife habitat as in clear-cutting and its variations. And because the trees are of varying ages, the forest maintains an uneven-aged natural look.

Selective cutting is the method that is least disruptive to an intact forest if it is practiced correctly. Trees are individually selected and evaluated to determine the optimum time for harvest. Because only a limited number of mature trees are cut at any one time, selective cutting maintains a diversity of tree species and ages, preserving the look of the original forest (fig. 26.15d). Harvesting mature trees reduces crowding and encourages the growth of younger trees. At the same time, because most of the forest remains intact, natural regeneration continues, sensitive species are protected from wind damage and full sunlight, and soil erosion is minimized.

Unfortunately, selective cutting has a very destructive variant: high grading, or creaming. In this method, the most valuable trees are harvested indiscriminately without regard for any damage caused to the rest of the forest. Harvesting the rain forest for highly prized tropical hardwoods, such as mahogany and teak, is often done with high grading. A road of destruction and waste leads from one desirable tree to the next as logging equipment simply demolishes every tree in its path.

The short rotation, or coppice, method is used to harvest certain fast-growing tree species that produce multiple trunks and readily sprout from stumps. In this method, the trunks are cut at ground level but the roots are left intact (fig. 26.15e). The stumps quickly sprout multiple trunks that can be harvested every 2 to 3 years. Several species, such as aspen, beech, and cottonwood, are suited to this type of harvesting. There is minimal disruption to the soil because root systems remain intact. Because the diameter of the sprouts is moderate, the wood is usually converted into chips for fuel or industrial products.

A 2015 study which sought to estimate the number of trees on Earth found out that the global number of trees is approximately 3.04 trillion with slightly less than half of the trees existing in tropical and subtropical forests. Boreal and temperate forests contained 740 billion (0.74 trillion) and 610 billion (0.61 trillion) respectively. Although these numbers seem enormous, it was also estimated that nearly half of all the trees on Earth have been removed since the beginning of human civilization with an estimated 15 billion trees cut down every year.

CHAPTER SUMMARY

1. The ecosystem is the functional unit of study of the environment. The environment is an organism's surroundings; it includes the study of biotic factors, such as the interrelationships between organisms, as well as investigations into abiotic factors, such as energy flow and the cycling of mineral nutrients.

2. Ecological niche is the functional role an organism plays in the ecosystem. Plants as producers are fundamental to most food chains, the sequential ordering of trophic relationships.

3. Biological magnification is the transfer and concentration of a toxic substance as it is passed along a food chain.

4. Human impact on the carbon cycle has resulted in the greenhouse effect, in which the buildup of CO_2 acts as a heat trap warming Earth's lower atmosphere. Global warming is likely to cause a shift in agricultural regions and in the range of many species of plants and animals.

5. Ecological succession is the gradual progression of orderly and predictable changes that may occur in ecosystems over time. The end point of ecological succession is the climax community.

6. Biomes are large, terrestrial regions recognizable by their distinctive vegetation and animal life. Climatic conditions, such as temperature and precipitation, largely determine the type of biome that develops in a region. Major biomes of the world include the desert, chaparral, grassland, savanna, tundra, boreal forest, moist coniferous forest, deciduous forest, and tropical rain forest.

7. The strategy of extractive reserves is an attempt to preserve the tropical rain forest from destructive exploitation by developing economic markets for a renewable harvest of natural products, such as natural rubber, chicle, Brazil nuts, and vegetable ivory. Ethnomapping identifies and protects locations in the rain forest that are significant to the livelihood and culture of indigenous peoples.

8. Several methods of forest management are currently in practice: clear-cutting, strip-cutting, whole-tree harvesting, shelterwood, selective cutting, and short rotation. Clear-cutting is the most controversial because this method produces significant damage to the environment: soil erosion, decline of soil fertility, sediment runoff into the watershed, and loss of habitat. Selective cutting is the least damaging to a forest. Most of the forest is left intact because desirable trees are cut for timber as they reach maturity.

REVIEW QUESTIONS

1. What are the limits of the biosphere? What factors would limit the distribution of plants? What factors would limit the distribution of animals?

2. What are the components of an ecosystem? Detail the flow of energy in a food chain.

3. How do the concepts of habitat and niche overlap? How do they differ? Give examples.

4. It is clear that the greenhouse effect caused by rising levels of CO_2 will have many dire consequences in agriculture. Would there be any benefits to plants from higher CO_2 levels in the atmosphere?

5. List examples of primary and secondary succession, and trace the transitional stages in each type of succession.

6. List and describe the major biomes found on the North American continent. Climbing up a mountain, the vegetational zones encountered are similar to the biomes passed on a journey north. Why? What is the altitudinal effect on temperature and precipitation?

7. Make a list of rain forest products obtained by extractive reserves available in local stores.

Appendix A

Atoms, Molecules, and Chemical Bonds

ATOMS

All matter in the universe is composed of a limited number of substances called **elements**; there are 92 naturally occurring elements and 23 more that have been synthesized in the laboratory. Each element is assigned a symbol consisting of one or two letters. Of all the elements, nine make up approximately 99.5% of living matter: carbon, oxygen, hydrogen, nitrogen, potassium, magnesium, calcium, phosphorus, and sulfur. These are not the only elements required by life, but the others, known as **trace elements**, or **micronutrients**, are required in extremely small amounts (table 1).

An **atom** is the smallest part of an element that still retains the chemical and physical characteristics of that element. Each atom is made up of subatomic particles; the three most important of these are **protons, neutrons,** and **electrons.** Protons and neutrons occupy the central position in atoms, a region referred to as the **atomic nucleus,** while electrons are in orbit around the nucleus. Protons and electrons are both electrically charged particles, with protons carrying a positive (+) charge and electrons a negative (−) one. Neutrons, however, have no electrical charge. Because of the presence of protons, the atomic nucleus has an overall positive charge, and it is the attraction of unlike charges that helps keep the negatively charged electrons in orbit.

Each element is assigned an **atomic number,** which refers to the number of protons in the atomic nucleus, ranging from the smallest atom, hydrogen, with an atomic number of 1, to the largest naturally occurring atom, uranium, with an atomic number of 92. All atoms of an element have the same atomic number. Atoms are also characterized by their **atomic mass,** or **mass number,** which is determined by the number of protons and neutrons in the atomic nucleus. Protons and neutrons have approximately the same mass, each equal to about 1 **dalton (atomic weight unit),** while the mass of electrons is negligible and not usually considered. Thus, the mass of an atom is concentrated in the nucleus (fig. 1).

Although the atomic number is constant for an element, the mass number may vary, since the number of neutrons may differ among atoms of the same element. **Isotopes** are atoms of an element with different mass numbers. For example, the most common isotope of carbon is ^{12}C, with 6 protons and 6 neutrons; however, another isotope of carbon is ^{14}C, with 6 protons and 8 neutrons.

Some isotopes, such as ^{14}C, are radioactive; their atomic nuclei are unstable and decay, emitting radiation. Radioactive isotopes are useful in research and have been used in dating fossils and artifacts. The time period of some early agricultural societies has been determined by this method. Radioactive isotopes have other applications in biology as tracers in metabolic pathways. For example, ^{14}C has been used to trace the path of carbon dioxide in photosynthesis.

Table 1
Essential Elements in Plants

Element	Symbol	Approximate Percent of Dry Weight
Macronutrients		
Carbon	C	45
Oxygen	O	45
Hydrogen	H	6
Nitrogen	N	1.5
Potassium	K	1.0
Calcium	Ca	0.5
Magnesium	Mg	0.2
Phosphorus	P	0.2
Sulfur	S	0.1
Micronutrients		
Chlorine	Cl	0.01
Iron	Fe	0.01
Manganese	Mn	0.005
Zinc	Zn	0.002
Boron	B	0.002
Sodium	Na	0.001
Copper	Cu	0.0006
Molybdenum	Mo	0.00001
Nickel	Ni	0.00001

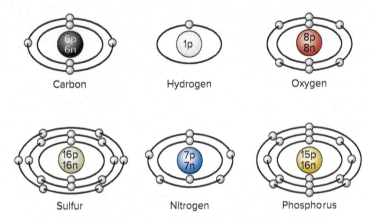

Figure 1 Atomic structures of six atoms commonly found in organic molecules.

MOLECULES AND BONDING

Most atoms do not occur singly in nature but exist combined with like or different atoms as **molecules**. A molecule of oxygen contains two atoms of oxygen chemically bonded together and represented by the symbol O_2. A molecule of water, H_2O, contains two atoms of hydrogen and one of oxygen. A **compound**, such as water, is any substance formed by two or more elements in a definite proportion.

Chemical bonds are forces that hold atoms together; three common types of bonds are **ionic, covalent,** and **hydrogen.** Atoms are naturally electrically neutral, having an equal number of positively charged protons and negatively charged electrons. When an atom gains or loses one or more electrons, it becomes a charged particle known as an **ion.** The number of electrons lost or gained is specific for each element and is known as valence. If an atom loses electrons, it becomes positively charged because the number of protons exceeds the number of electrons. Conversely, if an atom gains one or more electrons, it develops a negative charge because the number of electrons is greater. Ions are shown with the charge indicated as a superscript next to the chemical symbol; for example, sodium atoms can lose an electron, leaving 11 protons and 10 electrons, resulting in a sodium ion with a single positive charge (symbolized by Na^+). Chlorine atoms can gain an electron and thereby have an overall negative charge with 17 protons and 18 electrons (Cl^-). Calcium ions (Ca^{++}) have an overall positive charge of 2, since 2 electrons are lost, leaving 20 protons and 18 electrons.

When two oppositely charged ions are in contact, they are strongly attracted to each other. This attraction is referred to as an ionic bond. Common table salt, sodium chloride (NaCl), forms as a result of an ionic bond between oppositely charged Na^+ and Cl^- (fig. 2). In water, ionic compounds tend to dissociate, separating into the component negative and positive ions. The minerals that plants require for growth and development are absorbed as ions from soil water. In addition, many biologically important molecules occur as ions in living tissues.

Covalent bonds form when two atoms share electron pairs. Covalent bonds are strong bonds that are not easily broken. In the oxygen molecule, two electron pairs are shared by each atom, making a double covalent bond; the covalent bonds between the atoms are represented by $O=O$ (fig. 3). Most biologically important molecules are formed from various combinations of the elements carbon, hydrogen, and oxygen joined by covalent bonds. The number of covalent bonds formed by an atom is usually specific for an element; for example, each carbon atom must form four covalent bonds (fig. 3).

When electron pairs are shared between atoms of the same element, such as O_2, H_2, and N_2, they are shared equally. These covalent bonds are referred to as nonpolar covalent bonds. When electrons are shared between two atoms that differ in their ability to attract electrons, the electrons are shared unequally. This unequal sharing creates a polar covalent bond, and the atom with the greater attraction for the electrons develops a slight negative charge. Polar covalent bonds are found in water molecules (fig. 4). The electrons are pulled more closely to the oxygen atom, giving the oxygen a slight negative charge ($\delta-$) and each hydrogen atom a slight positive charge ($\delta+$).

Attractions form between each weakly charged oxygen of a water molecule and weakly charged hydrogens from adjacent molecules (fig. 4). These attractions are known as **hydrogen bonds.** Hydrogen bonds occur in other molecules also, forming between the weakly charged oxygen or nitrogen of one molecule and the weakly charged hydrogen of another. Hydrogen bonds are weak bonds. They are easily made and easily broken, but when they occur in number, they add stability to a molecule. Hydrogen bonds are important for the structure, and therefore the function, of both proteins and nucleic acids.

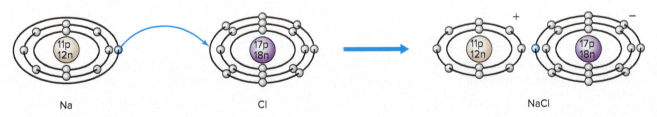

Figure 2 Ionic bonding in table salt, NaCl. The sodium atom donates a single electron to the chlorine atom. The two oppositely charged ions bond to form the compound NaCl.

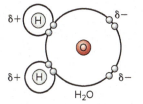

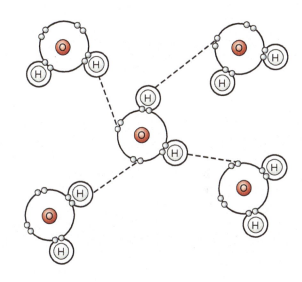

Figure 3 Covalently bonded molecules share electrons. (a) A molecule of hydrogen is formed when two atoms share one pair of electrons. (b) A molecule of oxygen is formed when two atoms share two electron pairs. (c) A molecule of methane consists of one carbon atom covalently bonded to four hydrogen atoms.

Figure 4 Water molecules. (a) The shared electrons are pulled toward the oxygen atom, creating a polar covalent bond and giving oxygen a slight negative charge ($\delta-$) and each hydrogen a slight positive charge ($\delta+$). (b) The weakly charged regions of the water molecule are attracted to the opposite charges on adjacent water molecules. These attractions are known as hydrogen bonds.

Appendix B

Classification of Plants and Those Organisms Traditionally Classified as Plants Discussed in *Plants and Society*

Domain Bacteria—unicellular or simple multicellular prokaryotes

    Cyanobacteria—blue-green bacteria (also called blue-green algae)

Domain Eukarya

Protists—unicellular or simple multicellular eukaryotes

    Division Euglenophyta-euglenoids

    Division Dinophyta-dinoflagellates

    Division Bacillariophyta/Ochrophyta-diatoms

    Division Phaeophyta/Ochrophyta-brown algae

    Division Rhodophyta-red algae

    Division Myxomycota—myxomycetes, slime molds

    Division Oomycota—oomycetes, water molds

Kingdom Fungi—eukaryotic, mainly multicellular, absorptive heterotrophs

    Phylum Chytridiomycota/Blastocladiomycota—zoosporic fungi

    Phylum Glomeromycota-AM fungi

    Former Phylum Zygomycota—zygomycetes

    Phylum Ascomycota—ascomycetes, sac fungi

    Phylum Basidiomycota—basidiomycetes, club fungi

Kingdom Plantae—eukaryotic, multicellular photosynthetic autotrophs

    Division Chlorophyta-green algae

    Division Charophyta-charophytes

    Division Hepatophyta-liverworts

    Division Bryophyta-mosses

    Division Anthocerophyta—hornworts

    Division Lycophyta—club mosses

    Division Pterophyta—ferns, whisk ferns, and horsetails

    Division Cycadophyta—cycads

    Division Ginkgophyta—ginkgo

    Division Coniferophyta—conifers

    Division Gnetophyta—gnetophytes

    Division Magnoliophyta—flowering plants

Appendix C

Metric System: Scientific Measurement

Units of Length

Unit	Abbreviation	Equivalent
Kilometer	km	1,000 m
Meter	m	1 m
Centimeter	cm	0.01 m
Millimeter	mm	0.001 m
Micrometer	μm	0.000001 m
Nanometer	nm	0.000000001 m
Angstrom	Å	0.0000000001 m

Length Conversions

1 km = 0.6 mi	1 in = 2.54 cm
1 m = 39 in	1 ft = 30 cm
1 cm = 0.39 in	1 yd = 0.9 m
1 mm = 0.039 in	1 mi = 1.6 km

To Convert:	Multiply By:	To Obtain:
inches (in)	2.54	centimeters
feet (ft)	30	centimeters
miles (mi)	1.6	kilometers

Units of Area

Unit	Abbreviation	Equivalent
square meter	m^2	m^2
hectare	ha	10,000 m^2
square kilometer	km^2	1,000,000 m2 (100 ha)

Area Conversions

1 m^2 = 10.76 ft^2	1 ft^2 = 0.09 m^2
1 ha = 2.47 ac	1 ac = 0.40 ha
1 km^2 = 0.39 mi^2	1 mi^2 = 2.59 km^2

To Convert	Multiply By:	To Obtain:
square feet (ft^2)	0.09	square meter
acre (ac)	0.40	hectares
square mile (mi^2)	2.59	square kilometers

Units of Volume

Unit	Abbreviation	Equivalent
liter	L	1 L
milliliter	ml	0.001 L

Volume Conversions

1 tsp = 5 ml	1 gal = 3.79 L
1 Tbsp = 15 ml	1 fl oz = 30 ml
1 cup = 0.24 L	1 L = 1.06 qt
1 pt = 0.47 L	1 L = 0.26 gal
1 qt = 0.95 L	

To Convert:	Multiply By:	To Obtain:
fluid ounces (fl oz)	30	milliliters
quart (qt)	0.95	liters
milliliters	0.03	fluid ounces
liters	1.06	quarts

Units of Mass

Unit	Abbreviation	Equivalent
kilogram	kg	1,000 g
gram	g	1 g
milligram	mg	0.001 g
microgram	μg	0.000001 g

Mass Conversions

1 oz = 28.3 g	1 g = 0.035 oz
1 lb = 453.6 g	1 kg = 2.2 lb
1 lb = 0.45 kg	

To Convert:	Multiply By:	To Obtain:
ounces (oz)	28.3	grams
pounds (lb)	453.6	grams
pounds	0.45	kilograms
grams	0.035	ounces
kilograms	2.2	pounds

Energy Conversions

calorie (cal) = energy required to raise the temperature of 1 g of water by 1°C
1 Calorie = 1,000 cal = 1 kilocalorie
1 kilocalorie (kcal) = 1,000 cal

Temperature Conversions

Celsius (Centigrade) °C $\quad °C = \dfrac{(°F - 32) \times 5}{9}$

Fahrenheit °F $\quad °F = \dfrac{°C \times 9 + 32}{5}$

	°F	°C
Boiling point of water	212	100
Human body temperature	98.6	37
Freezing point of water	32	0

Glossary

abiotic factor Nonliving component in the environment.
abscisic acid A plant hormone that promotes dormancy and is associated with water balance by causing stomatal closure.
absorption spectrum Graph of absorbance values for different wavelengths of light.
absorptive heterotrophs Organisms, such as fungi, whose mode of nutrition is to secrete digestive enzymes into a substrate and then absorb the products of digestion.
accessory fruit Type of fruit in which most or part of the fruit is derived from tissue other than the ovary of a flower.
achene Simple, dry, indehiscent fruit with seed attached to pericarp at only a single point.
actinomorphic Regular or radially symmetric.
active transport Movement of materials across a cell membrane and against a concentration gradient that requires the expenditure of cellular energy.
adaptation Trait that enhances the survival of an organism in its environment.
addictive drug Substance that induces an addiction by causing physiological dependence, psychological dependence, and/or tolerance.
adenine One of the purine bases found in nucleotides and nucleic acids.
adhesion Attraction between unlike molecules. The polar nature of water molecules causes them to adhere (stick) to a surface.
ADP (adenosine diphosphate) A nucleotide diphosphate that is often phosphorylated to form ATP.
adventitious Refers to an organ that forms in an unusual place; example: roots from leaves.
aeciospore Binucleate spore produced in an aecium.
aecium (pl., aecia) Cup-shaped structure in rust fungi composed of binucleate hyphae; aeciospores are formed here.
aerenchyma Type of parenchyma tissue that has many large, intercellular spaces for the passage of air and exchange of gases.
aerobic respiration The complete oxidation of glucose to CO_2 and H_2O and energy; this process requires oxygen as an electron acceptor.
aflatoxin A complex of several mycotoxins produced by *Aspergillus flavus;* often found in peanut products.
agar Gelatinous product extracted from the walls of some red algae; used in growth media for microorganisms and in tissue culture.
agarose Gelatinous product derived from agar that is a common gel base in biotechnology.
aggregate fruit A fruit derived from a single flower with several separate ovaries; example: blackberry.
aleurone layer Protein-rich, outermost layer of endosperm in a cereal grain.
alginic acid Gelatinous material extracted from the walls of certain brown algae; used in a variety of commercial and industrial products.
alkaloid Nitrogenous substance, alkaline in solution, that produces physiological and/or psychological effects on the nervous system of animals.

allele Alternate expression of a gene.
allelopathy The release of chemicals by certain plants that inhibit the growth of competing plants.
allergen Foreign substance that induces an allergic response.
alternate arrangement One leaf borne per node.
alternative medicine A wide range of therapies outside the mainstream of traditional Western medicine; includes such treatments as aromatherapy, acupuncture, biofeedback, chiropractic manipulation, herbal medicine, hypnosis, and massage therapy.
amatoxin A deadly protoplasmic toxin produced by several genera of fungi, including *Amanita, Conocybe, Galerina,* and *Lepiota.*
amino acid Nitrogen-containing building block of proteins.
amylopectin Component of starch consisting of highly branched chains of repeating glucose units; insoluble in water.
amyloplast Starch-storing plastid.
amylose Component of starch consisting of unbranched chains of repeating glucose units; soluble in water.
anabolic Chemical reactions that synthesize and require energy.
anaerobic respiration The break down of organic compounds to release energy without the use of oxygen as an electron acceptor; also known as fermentation.
analgesic Compound that reduces pain.
anaphase Stage in mitosis in which sister chromatids separate and move to opposite poles of the cell.
androecium (pl., androecia) Collective term for all the stamens in a flower.
angiosperm A flowering plant whose seeds are contained within fruits.
anisogamy Sexual reproduction involving motile gametes of two different size classes.
annual ring Ring of xylem in woody stem composed of springwood and summerwood that corresponds in temperate regions to a chronological year.
annulus Remnant of the partial veil around the stipe of some mushrooms.
anther Pollen-producing part of a stamen.
antheridium (pl., antheridia) Male gametangium in which the male gametes, sperm, are produced.
anthocyanin Plant pigment usually of red, blue, or purple and commonly found in flowers and some fruits; one of the flavonoids.
Anthropocene Epoch Proposed geologic epoch that reflects the permanent imprint that humanity has left upon Earth; would follow the Holocene Epoch that began when the Great Ice Age ended 11,700 years ago.
anthropogenic Caused by human activity.
antibiotic Substance that inhibits the growth of microorganisms, such as bacteria or fungi.
antibody Immunological protein produced by B-lymphocytes in response to a specific antigen and capable of acting against said antigen.
anticodon The sequence of three nucleotides in a transfer RNA molecule that is complementary to a specific codon.

antigen Substance, usually a protein, that is foreign to the body and elicits an immunological reaction by antibodies.
anti-inflammatory Compound that reduces inflammation or swelling.
antipodal One of several cells (usually three) of the embryo sac (female gametophyte) of angiosperms; located opposite the egg and synergids.
antipyretic Compound that reduces fever.
apical meristem Meristematic tissue located at the tips of roots and stems; responsible for primary growth of a plant.
apoplast Pathway of water entering a root that flows along intercellular spaces and cell walls.
aquifer Underground water reservoir.
arbuscular mycorrhizae The most prevalent type of endotrophic mycorrhizae, in which hyphae grow between the cells in the cortex and form branching structures called arbuscles within the cells where nutrient change occurs.
archaebacteria Prokaryotes that are biochemically distinct from eubacteria and inhabit extreme environments.
archegonium (pl., archegonia) Female gametangium in which the female gamete or egg is produced and housed.
aril Thick, fleshy seed covering around some seeds.
aromatherapy A holistic approach to healing using essential oils extracted from plants.
artificial selection Selective breeding as practiced by humans on domesticated plants and animals.
ascocarp Fruiting body of an ascomycete. Also known as ascoma (pl., ascomata).
ascogonium (pl., ascogonia) A female reproductive structure in some ascomycetes.
ascospore Sexual spore of an ascomycete fungus produced within an ascus.
ascus Saclike reproductive structure of ascomycetes in which meiosis, followed by mitosis, produces eight haploid cells called ascospores.
asexual reproduction Any type of reproduction not involving the union of gametes.
atom The smallest individual unit of an element that still retains the properties of that element.
atomic mass The total number of protons and neutrons in the atomic nucleus.
atomic nucleus Most of the mass of an atom: composed of protons and neutrons.
atomic number Number of protons in the atomic nucleus; there is a characteristic number for each type of element.
atomic weight unit The mass of one proton or neutron.
ATP (adenosine triphosphate) The energy compound of the cell.
ATP synthase A membrane-bound enzyme in mitochondria and chloroplasts that phosphorylates ADP to form ATP using energy from the passage of protons through the enzyme.
autotroph Self-feeder; an organism that can synthesize organic compounds from simple inorganic ones.

auxins A class of plant hormones that influences cell elongation and is involved in many stages of growth and development.
awn A slender bristle.
axil Upper angle between a leaf and stem.
axillary bud Bud found in the axil of a leaf; also called a lateral bud.

basidiocarp Fruiting body of a basidiomycete. Also known as **basidioma** (pl., **basidiomata**).
basidiospore Sexual spore of basidiomycetes.
basidium (pl., **basidia**) Club-shaped reproductive cell in basidiomycetes; undergoes meiosis to produce four haploid basidiospores.
bast fiber Fibers located in phloem.
beriberi Deficiency disease due to inadequate thiamine B_1, symptoms include fatigue, mental confusion, cramping, and numbness in the legs.
berry Simple, one- to many-seeded, fleshy fruit with thin exocarp; example: tomato.
beta-carotene Yellow to red pigment in plants; one of the most important of the carotenoids; converts to vitamin A in the body.
biennial Plant that completes its life cycle within two growing seasons; example: carrot.
binomial Two-word scientific name.
biodiesel A vegetable oil product used as a cleaner alternative to conventional petroleum-based diesel fuel; produces little soot or other pollutants.
biodiversity The sum variety of living organisms on Earth.
biological magnification Increase and concentration of toxins as they are passed along a food chain.
biological species concept Members of an interbreeding population reproductively isolated from any other such group of populations.
biomass Total dry weight of living organisms.
biome Large terrestrial communities recognized by characteristic vegetation.
bioremediation The use of microorganisms to reclaim contaminated soil and water.
biosphere The entire world of living organisms.
biotechnology The use of living organisms to provide products for humanity; using genetic engineering to create organisms with useful traits.
biotic factor Living component in the environment.
blade Flat, green part of a leaf; expanded or flattened portion of a brown alga.
B-lymphocytes Cells that manufacture antibodies involved in immunity.
body mass index (BMI) Body weight in kilograms per squared meter.
boreal forest Northern coniferous forest biome, also called the taiga, located south of the tundra and dominated by conifers.
bract A floral leaf.
bran The husk of a cereal grain; including the pericarp and seed coat of a cereal grain. The bran is removed in the processing of refined grains.
bud Embryonic shoot of a plant; may be composed of embryonic leaves or flowers or be mixed.
budding Type of asexual reproduction in which a small blip of the parental body develops into a new individual.
bulb Vertical, underground stem with food-storing leaves covered by papery leaves.
bulblet A small bulb or part of a larger bulb.
bundle sheath Sheath of parenchyma or sclerenchyma cells surrounding the vascular bundles in leaves.

calcium Major mineral; most abundant mineral in the body.
callus Mass of undifferentiated cells in tissue culture.
calorie The amount of heat needed to raise the temperature of 1 g of water 1° Celsius; 1,000 calories = 1 kilocalorie or 1 Calorie.
Calvin Cycle Biochemical pathway in photosynthesis in which carbon from atmospheric CO_2 is fixed and reduced into carbohydrate.
calyx Collective term for sepals of a flower.
canker Localized area of dead tissue on a plant, usually on woody stems, branches, or twigs.
capsule Simple, dry, dehiscent fruit that opens along three or more seams or pores; example: cotton.
carbohydrate Organic compound containing carbon, hydrogen, and oxygen with the general formula of $C_nH_{2n}O_n$.
carcinogen A cancer-causing agent; may be chemical, microbial, radiation, or a gene mutation.
cardioactive glycoside Sugar-containing molecules with an active compound that affects the heartbeat.
carnivore Flesh-eating organism; animal that eats other animals.
carotenes Accessory photosynthetic plant pigments; protect chlorophyll from breakdown by light or oxygen.
carotenoids Class of plant pigments that includes carotenes and xanthophylls; most are yellow, orange, or red.
carpel The ovule-bearing part of a flower.
carpellate flower A unisexual flower that contains carpels but no stamens.
carrageenan Gelatinous material extracted from the walls of some red algae and used in a variety of commercial and industrial products.
carrying capacity Maximum number of a population that can be supported by the environment over a given period of time.
caryopsis Simple, dry, indehiscent fruit with a single seed that is fused to the ovary wall; also called a grain; example: wheat.
Casparian strip Water-impermeable strip of suberin found in the transverse and radial walls of endodermal cells.
catabolic Chemical reactions that release energy by degrading complex compounds into simpler ones.
catalyst An agent that speeds up a chemical reaction but is not used up in the reaction.
catkin Inflorescence of unisexual flowers.
cell culture (tissue culture) The culture of single cells or tissues to form callus and then to develop whole plants asexually.
cell division Cell reproduction.
cell plate Double membrane across the equator of a dividing cell that develops from the phragmoplast; marks where the new cell walls will form.
Cell Theory A set of principles describing how cells are the fundamental units of life.
cellular respiration Cellular pathway for the production of ATP.
cellulose Complex carbohydrate that occurs in the cell walls of plants and oomycetes.
cell wall Rigid outer layer in the cells of plants, fungi, bacteria, and certain other organisms.
central vacuole A membrane-enclosed sac that takes up most of the volume of a mature plant cell.

centromere Constricted portion of chromosomes to which spindle fibers attach; region joining sister chromatids.
cereal Edible grains of cultivated grasses; examples: wheat, rice, corn.
chaff Bracts surrounding a cereal grain; removed during threshing.
chalazal Ovule end opposite the micropyle.
chaparral Biome characterized by dense thickets of evergreen shrubs with mild, rainy winters and dry, hot summers.
chemical bond A force that holds atoms together in a compound; some types of bonds are ionic, covalent, and hydrogen bonds.
chemiosmosis The coupling of ATP synthesis to electron transport by way of a proton gradient and ATP synthase.
chiasma (pl., **chiasmata**) The X-shaped configuration between the chromatids of homologous chromosomes that have exchanged genetic material during prophase I of meiosis.
chitin Complex polymer found in the cell walls of true fungi.
chlorophyll Green pigment found in the chloroplasts of plants, algae, and cyanobacteria that is essential to photosynthesis.
chloroplast Membrane-bound organelle in plants and algae that contains chlorophyll and in which photosynthesis takes place.
cholesterol Steroid that is an integral part of animal cell membranes and a precursor to other steroidal compounds in animals.
chromatids The identical duplicated halves of a single chromosome seen during cell division.
chromatin Dispersed form of genetic material, DNA and protein, in the nucleus of a nondividing cell.
chromoplast Membrane-bound organelle containing pigments other than chlorophyll; the carotenoid pigments may be yellow, orange, or red.
chromosome Condensed form of DNA and proteins that appear in a dividing cell.
cisgenesis *See* Precision breeding.
Citric Acid Cycle Another name for Krebs Cycle.
clade Monophyletic group of organisms, i.e., organisms that can trace descent from a common ancestor.
cladophyll Modified stem that resembles and functions as a leaf.
class Taxonomic rank consisting of related orders.
climax community Stable, self-sustaining community that is the culmination of ecological succession.
clone The production of a genetically identical individual through asexual reproduction.
cobalamin Vitamin B_{12}, not found naturally in any food of plant origin. Deficiency results in pernicious anemia.
codominance A condition in which both alleles of a heterozygous pair are expressed independently.
codon Genetic code composed of three nucleotide sequences.
coenzyme Organic molecule that is necessary for the proper functioning of an enzyme.
coevolution The interaction of species as selective forces upon each other, resulting in adaptations that enhance their interdependency.
cofactor Mineral ion which are necessary for proper functioning of enzymes.
cohesion Tendency of like molecules to stick together, usually due to hydrogen bonds.

coleoptile Sheath surrounding the embryonic shoot of monocotyledons.
coleorhiza Sheath surrounding the radicle (embryonic root) of monocotyledons.
collenchyma Ground tissue in plants with unevenly thickened primary cell walls; functions in support.
colony collapse disorder Sudden loss of honey bees from their hive, often leaving just a queen bee and a few workers; the cause is uncertain with viruses, mites, and insecticides implicated in the phenomenon.
community All the populations in a given locale.
companion cell Phloem cell associated with a sieve tube member.
competitive exclusion Principle that no two species can occupy the same niche indefinitely.
complete flower A flower with all four floral whorls (sepals, petals, stamens, and carpels).
complete protein A protein that has all the essential amino acids and in the correct proportions.
compound A substance formed by two or more elements in a definite ratio.
compound leaf A leaf with a blade divided into separate leaflets.
C_3 Pathway Another name for the Calvin Cycle.
C_4 Pathway A pathway in which CO_2 is prefixed in mesophyll cells prior to the Calvin Cycle.
concentration gradient A region along which a substance ranges from higher to lower concentrations.
conceptacles Spherical cavities where gametangia are produced and that have openings on the surface of the receptacle; found in certain genera of brown algae such as *Fucus*.
conidium (pl., conidia) Asexual reproductive spores of some ascomycetes and basidiomycetes.
consumer Organisms that cannot synthesize their food but must feed on other organisms.
contractile roots Found in plants with extremely short stems; cortical cells contract in height pulling the root down into the ground ensuring that the stem remain at ground level or slightly below.
coprolite Fossilized fecal material.
cork Suberized cells on the outer surface of woody stems and roots; produced by the cork cambium.
cork cambium Meristematic tissue that produces cork cells on its outer surface and phelloderm on its inner surface.
corm Underground, enlarged, food-storing stem covered by papery leaves.
corolla Collective term for the petals of a flower.
cortex Parenchymous tissue, the region between the epidermis and vascular tissue in an herbaceous root or stem.
cotyledon Seed leaf in the seed and seedling.
covalent bond Chemical bond formed when atoms share a pair of electrons.
Crassulacean Acid Metabolism (CAM Pathway) A variation of the C_4 Pathway that functions in a number of cacti and succulents; allows for the fixation of carbon dioxide during the night; then in the daytime the carbon dioxide is transferred to the Calvin Cycle.
crista (pl., cristae) Infoldings of the inner membrane of a mitochondrion.
CRISPR-Cas9 Clustered Regularly Interspaced Short Palindromic Repeats is composed of an enzyme to cut a segment of DNA and a RNA guide to direct the Cas9 enzyme to the specific sequence of DNA to be cut; system has the potential to alter, delete, and rearrange the genome of any organism with precision.

crossing over The exchange of genetic material between chromatids of homologous chromosomes during prophase I of meiosis.
cross-pollination Transfer of pollen from the stamen to the stigma of a flower of another plant.
culm Hollow, jointed stem of grasses.
cultivar Abbreviation for the cultivated variety of a plant.
cuticle Waterproof layer of cutin on leaves and nonwoody stems.
cutin Waxy material secreted by epidermal cells.
cyanogenic glycoside Glycoside that releases cyanide.
cyclins Cellular proteins with fluctuating concentrations that regulate the phases of the cell cycle.
Cyclin-dependent Kinase (CdK) An enzyme when attached to a specific cyclin is activated to initiate a specific stage in the cell cycle.
cytochromes Iron-containing enzymes that carry electrons in both photosynthesis and respiration.
cytokinesis The division of the cytoplasm to form daughter cells; usually accompanies mitosis and meiosis.
cytokinins A class of plant hormones that promote growth by stimulating cell division.
cytoplasm The entire contents of the cell exclusive of the nucleus.
cytoplasmic inheritance Genes found outside the nucleus in organelles (chloroplasts or mitochondria) in the cytoplasm are passed to offspring exclusively from the maternal line.
cytosine A pyrimidine base in DNA and RNA.
cytoskeleton Cellular scaffolding of microtubules and microfilaments within the cytoplasm of the cell.
cytosol The more fluid portion of the cytoplasm.

dalton Atomic weight unit; the mass of one proton or one neutron.
dark reactions Biochemical reactions that make up the Calvin Cycle of photosynthesis.
deciduous Trees and shrubs that shed their leaves during the autumn.
deciduous forest Forest biome in which the dominant trees are deciduous.
decomposer Organism that obtains its nutrition by breaking down dead plants and animals or their waste products.
decongestant An agent that relieves nasal or respiratory congestion.
decortication Removing hard fibers (vascular bundles) from leaves.
dehiscent fruit Fruit that splits open at maturity, facilitating seed dispersal.
deletion A mutation resulting from the loss of a small segment of DNA.
dendrochronology Study of the annual rings in trees to determine the timing of natural events in the past.
dendroclimatology Study of the annual rings of trees in order to interpret climatic changes in the past.
deoxyribonucleic acid (DNA) The genetic material of life.
deoxyribose Five-carbon sugar in DNA.
depressant Psychoactive drug that has a sedative effect on the central nervous system; actions include dulling mental awareness and inducing sleep.
dermal tissue Tissue that covers surfaces in plants.

desert Biome in which the annual precipitation is less than 25 cm (10 in.).
detritus Wastes and remains of dead plants and animals.
diatomaceous earth Algal product composed of the fossilized remains of diatom frustules that are mined and used in a variety of products.
dicot (dicotyledon) Member of a class of angiosperms in which the seedlings typically possess two cotyledons; dicotyledon is usually shortened to *dicot*.
differentially permeable membrane A membrane that allows the free passage of some materials but inhibits the passage of others.
diffusion The spontaneous movement of particles (ions, molecules, or atoms) from a region of higher concentration to one of lower concentration.
dihybrid cross A genetic cross between parents that differ for two traits.
dikaryon Mycelium of some fungi that have two separate haploid nuclei in each cell.
dioecious Refers to a plant species that has separate male and female plants; pollen-bearing and ovule-bearing flowers or cones are borne on different plants.
diploid Two complete sets of chromosomes in a cell; 2N.
disaccharide Sugar consisting of two monosaccharides; example: sucrose is composed of glucose and fructose.
distillation Boiling of a liquid to evaporation and subsequent condensation of the vapors for the purposes of purification and concentration.
distilled spirit Alcoholic beverage with an alcoholic content from 80 to 100 proof; obtained by distillation of a beer or wine.
division Taxonomic rank that includes related classes; synonymous with *phylum*.
Doctrine of Signatures Belief that the form of a plant reveals its medicinal properties; this was a widely held belief by many primitive cultures around the world and advanced by herbalists in the sixteenth and seventeenth centuries.
domain Most inclusive taxonomic category; above the kingdom level.
domesticated plant A plant that has been genetically changed from the wild type by artificial selection.
dominant The allele of a gene that masks or suppresses the expression of an alternate allele.
double fertilization The fusion of egg and sperm resulting in a zygote, and the simultaneous fusion of sperm with two polar nuclei resulting in the formation of endosperm that characterizes all angiosperms.
double helix The form of the DNA molecule; the complementary nucleotide strands of the DNA molecule twisted into a helix.
drupe Simple, fleshy fruit with the seed enclosed in a hard endocarp (pit); example: cherry.
dry fruit Fruit in which the cells of the pericarp are dry (dead) at maturity.
drying oils Plant oils that react with oxygen in the air to form a thin, waterproof, elastic film.

early wood The cells in the secondary xylem that are formed in the spring, usually with wide vessels (angiosperms) or wide tracheids (gymnosperms); also call spring wood.
ecological equivalent Species that fulfill similar niches in different geographical regions.

ecological pyramid of biomass Total biomass of all organisms at each trophic level in a food chain; typically, biomass declines with successively higher trophic levels.

ecological pyramid of energy Total energy content of all organisms at each trophic level in a food chain; only 5% to 20% of energy is passed between trophic levels, so energy content declines at successively higher trophic levels.

ecological pyramid of numbers Number of organisms supported at each trophic level in a food chain; typically, fewer organisms are supported at successively higher trophic levels.

ecological species concept Species defined by its role in the biological community and its unique adaptations to the environment.

ecological succession Orderly process of natural change in a community composition over time, culminating in a self-perpetuating complex community; primary and secondary types are recognized.

ecosystem Community of living organisms interacting with the abiotic factors in the environment.

ectomycorrhizae A type of mycorrhizae in which the fungus forms a sheath or mantle around the root, penetrating between the cells of the root epidermis and cortex.

egg Nonmotile female gamete.

egg apparatus Egg cell and adjacent synergids in the embryo sac (female gametophyte) of angiosperms.

elaiosome Small oil-filled outgrowth on certain seeds.

electromagnetic radiation Radiant energy released by stars such as the Sun; visible light is a component of electromagnetic radiation, which also includes radio waves, microwaves, infrared radiation, ultraviolet radiation, X rays, and gamma rays.

electron Negatively charged particle of an atom.

Electron Transport System The final stage of respiration; a series of enzymes and coenzymes on the inner membrane of mitochondria functioning in the transfer of electrons and the resulting synthesis of ATP.

element Building blocks of matter that cannot be broken down into a simpler substance; each element is composed of just one type of atom.

embryo Immature sporophyte that develops from a zygote.

embryophyte A land plant in which the zygote develops into a multicellular embryo while still enclosed in the female gametangium.

embryo sac Female gametophyte of angiosperms; retained within the ovule.

endemic (simple) goiter Mineral deficiency disease due to inadequate iodine which is required for the formation of thyroid hormone. Common symptom is a swelling of the neck due to an enlarged thyroid gland.

endergonic reaction Chemical reaction that requires energy.

endocarp Innermost layer of the pericarp (fruit wall).

endodermis Innermost layer of the root cortex surrounding the stele; many of the endodermal cells have Casparian strips.

endoplasmic reticulum (ER) Membranous network of channels throughout the cytoplasm of a cell; some regions are studded with ribosomes (rough), and others are ribosome-free (smooth).

endosperm Nutrient tissue that forms by the fusion of a sperm nucleus with two polar nuclei during double fertilization in angiosperms.

endotrophic mycorrhizae A type of mycorrhizae in which the fungus grows into the cells of the root cortex.

energy The ability to perform work.

enology Science of wine making.

enzyme Proteins that act as catalysts to chemical reactions.

epicotyl Portion of the shoot of an angiosperm embryo or seedling above the cotyledons.

epidermis Outermost tissue in all young and nonwoody plant organs.

epigenetics Inherited changes caused by chemical changes to the chromatin.

epigynous Refers to a plant whose floral parts (sepals, petals, and stamens) appear to arise from the top of an ovary; the ovary is said to be inferior.

epiphyte Plant that grows on top of another plant for support and position.

ergotism Condition caused by consumption of rye (or other grains) contaminated with sclerotia of *Claviceps purpurea*; mycotoxins within the sclerotium (or ergot) can affect the central nervous system or cause vasoconstriction.

essential amino acid Amino acid that cannot be synthesized by an organism but must be obtained ready-made in the diet for proper health.

essential nutrient Nutrient that cannot be synthesized by an organism but must be obtained ready-made in the diet for proper health.

essential oil Volatile component that contributes to the scent and flavoring of aromatic plants.

ethylene Gaseous plant hormone involved in fruit ripening and other aspects of plant growth and development.

eubacteria The majority of bacteria, including cyanobacteria, with distinctive biochemical features that distinguish them from archaebacteria.

eudicot True dicots; clade of dicotyledonous plants as indicated by phylogenetic analysis.

eukaryotic cell Refers to cells with a nucleus and distinct membrane-bound organelles.

evolution Inherited changes in populations shaped by natural selection over time.

exergonic reaction Chemical reactions that release energy.

exine The outer layer of the pollen wall.

exocarp Outermost layer in the pericarp (fruit wall).

exon An expressed segment of a gene; exons are separated from each other by introns.

F_1 generation First filial generation; offspring from a genetic cross.

F_2 generation Second filial generation of a genetic cross.

FAD (flavin adenine dinucleotide) An electron receptor in cellular respiration.

family Taxonomic rank consisting of a group of related genera.

fat Triglyceride that is solid at room temperature; usually of animal origin.

fat-soluble vitamin Vitamins that can be stored in fatty tissues of the body; vitamins A, D, E, and K.

fatty acid Long chains of carbon and hydrogen; component of phospholipids and triglycerides.

fermentation Anaerobic cellular respiration; organic compounds are broken down to release energy without the use of oxygen as an electron acceptor.

Fertile Crescent Area between the Tigris and Euphrates Rivers in the Near East; some of the earliest documented sites of agriculture.

fiber Long and narrow sclerenchymous cell; functions in support; an important dietary component that provides bulk; component of fabrics, ropes, and paper.

fibrous roots Root system with several main roots; common to many monocots.

filament Part of the stamen in a flower that supports the anther.

flagellum (pl., flagella) Whiplike cellular structure of motility; eukaryotic flagella are composed of microtubules.

flavonoids Class of plant pigments, some of which appear colorless to the human eye because their colors are activated by ultraviolet wavelengths; includes the anthocyanins, flavones, and flavonols.

fleshy fruit Fruit in which the cells of the pericarp are alive at maturity.

floret One of the small flowers that make up the inflorescences in the composite and grass families.

fluid mosaic model Model of cell membrane structure composed of a lipid bilayer with scattered proteins; often described as a sea of lipids with protein icebergs.

folic acid B vitamin.

follicle Single, dry, dehiscent fruit that splits along one seam; example: milkweed.

food chain Progression of organisms that feed on or decompose the preceding one.

food web Interrelationships between several food chains in an ecosystem.

forage crops Crops that are grown as food for domesticated herbivores.

forager Member of a hunter-gatherer group.

frame-shift mutation A mutation caused by the insertion or deletion of nucleotides (fewer than three or a number not a multiple of three) resulting in the improper grouping into codons.

free-threshing grain Grain that separates easily from enclosing bracts.

fructose Six-carbon monosaccharide; often referred to as fruit sugar.

fruit A ripened ovary of an angiosperm flower.

fruiting body Reproductive structures in fungi.

frustule The glasslike diatom wall, which is often intricately marked with pits, grooves, and ridges, giving these microscopic forms the appearance of cut crystal.

fungicidal A compound that is able to kill a fungus and stop an infection.

fungistatic A compound that slows or stops the growth of a fungus.

G_1 Part of interphase known as Gap 1; time of active metabolism in the cell cycle.

G_2 Part of interphase after the synthesis of DNA and before the start of nuclear division and known as Gap 2.

gametangium (pl., gametangia) Structure in which gametes are produced.

gamete Sex cell.

gametophyte Haploid generation of plants that produce gametes.

gene Unit of hereditary information on chromosomes.

genealogical species concept Species defined by its unique DNA sequence.
generalized niche Niche in which species have broad requirements and tolerate a range of conditions.
generative cell Cell in male gametophyte (pollen grain) that divides, producing two sperms.
genetically engineered microorganisms (GEMs) Bacteria that have been genetically engineered by the insertion or deletion of DNA segments.
genetic code The set of nucleotide triplets (codons) that code for amino acids used in protein synthesis.
genetic engineering The transfer of specific genes between organisms using techniques of molecular biology.
genetic erosion Irreversible loss of genetic diversity due to extinction of traditional varieties and wild ancestors of crop plants.
genetically modified plant (GM plant) Plant produced through genetic engineering and containing one or more modified genes.
genome The entire genetic informations, DNA, of an organism.
genotype Genetic makeup of an organism.
genus (pl., genera) Taxonomic rank consisting of a group of related species.
germ Embryo of a cereal grain.
germplasm Entire genetic makeup of an organism.
gibberellins A class of plant hormones involved in many stages of growth and development, especially stem elongation and seed germination.
ginning Removing surface fibers from a plant organ.
glucose Six-carbon monosaccharide; one of the most abundant simple sugars; the building block of both cellulose and starch and important to several metabolic pathways.
glumes Pair of bracts at the base of a spikelet in a grass flower.
gluten Protein complex in endosperm of wheat and some other cereals that is essential in making a leavened bread.
glycemic index A scale to indicate how a food affects blood glucose levels; a high GI rating indicates that the food is quickly broken down into glucose that rapidly enters the bloodstream.
glycemic load Glycemic index multiplied by the number of grams of carbohydrate present in a food.
glycogen Polysaccharide of glucose; principal carbohydrate stored in animal and fungal cells.
glycolipid A lipid with attached sugars.
glycolysis Pathway in cellular respiration in which glucose is split into pyruvate.
glycoprotein A protein with attached sugars.
glycoside Physiologically active compound in plants that always contains a sugar group, although the active part of the molecule may differ.
glyoxysome A type of microbody involved in the enzymatic conversion of stored fats to sugars in some seeds.
Golgi body (apparatus) Organelle of membranous, hollow sacs arranged in a stack: functions in modification, storage, and packaging of secretion materials; may be called dictyosome in plants.
grafting The union of a part of one plant, the scion, to the root or stock of another plant.
grain Single, dry indehiscent fruit of a single seed that is fused to the ovary wall.
granum (pl., grana) Stacked thylakoid membrane within a chloroplast.

grassland Biome in which the annual moderate precipitation is enough to support the growth of grasses but insufficient to support a forest.
greenhouse effect The warming of Earth due to the atmospheric accumulation of carbon dioxide and other gases, primarily due to the burning of fossil fuels, which trap heat and reradiate it back to Earth's surface.
Green Revolution Introduction of scientifically developed food crops that can produce high yields under conditions of high inputs of water, fertilizers, and pesticides.
ground tissue Includes primary tissues of parenchyma, sclerenchyma, and collenchyma that make up much of the bulk of the primary plant body; function in support, photosynthesis, and storage; also known as fundamental tissue.
guanine Purine base present in both DNA and RNA.
guard cell One of a pair of specialized cells in the epidermis that regulate the opening and closing of a stoma.
gymnosperm Plants that bear naked or exposed seeds.
gynoecium (pl., gynoecia) Collective term for the carpels in a flower.

habitat Place or type of place where an organism lives.
hallucinogen Psychoactive drug capable of altering moods and perceptions of time and space.
haploid One set of chromosomes in a cell; 1N.
hard fiber Commercial fiber obtained from the vascular bundles of certain monocot leaves; also known as leaf fiber.
hardwood Angiospermous trees or the wood from angiosperms.
heme iron Dietary source of iron derived from a heme complex as in hemoglobin; readily absorbed in the human digestive tract; obtained from animal sources, e.g., red meat, fish, poultry.
HDL (high-density lipoprotein) Transport molecule that removes excess cholesterol from the body's tissues to the liver for degradation and elimination.
head Horizontal inflorescence of sessile flowers.
heartwood Core in woody stems, usually darker than surrounding tissue, and no longer functioning in water conduction.
hemicellulose Polysaccharide in plant cell walls that cross-links cellulose fibrils.
herb Nonwoody plant; aromatic plant whose leaves are used in seasoning.
herbaceous Refers to nonwoody plants.
herbal A text that describes plants that are useful medicinally and in other ways.
herbarium A permanent collection of dried and pressed plants that provides information on the location and identification of the local flora.
herbivore An animal that eats plants.
hesperidium Simple, fleshy fruit with leathery exocarp; example: any citrus fruit.
heterocysts Large, thick-walled, faintly pigmentized cells where nitrogen fixation occurs in cyanobacteria.
heterotroph Other feeder; organism that is incapable of synthesizing its own food and must obtain its nutrition from other organisms.

heterozygous Having two different alleles for a given trait.
hilum Scar on a seed indicating where it was attached to the ovary.
holdfast Attachment organ or cell at the base of certain algae.
homologous chromosomes Chromosome pairs of the same size and shape that carry genes for the same traits.
homozygous Having two identical alleles for a given trait.
hormones Chemical messengers that are effective at very low concentrations.
hunter-gatherers Human social group that secures food sources from wild resources such as hunting animal prey or collecting edible plants from the wild.
hybrid Offspring of a cross between two species or between alternate homozygous conditions.
hydrocolloids Complex polysaccharides, such as agar, carrageenan, and alginic acid, found in algal cell walls and used as emulsifiers, stabilizers, and gelling agents.
hydrogenation Addition of one or more hydrogens to monounsaturated and polyunsaturated fatty acid chains.
hydrogen bond Weak chemical bond formed when the slightly positive hydrogen atom of a polar covalent bond is attracted to the slightly negative atom of a polar covalent bond of another molecule.
hydrophilic Water-attracting molecules or surfaces.
hydrophobic Water-repellent molecules or surfaces.
hydroponics Growing plants without soil in liquid nutrient solutions.
hymenium The layer of fertile cells that produces spores in a fungal fruiting body.
hyperaccumulator Any of certain plant species that concentrate specific minerals at extremely high levels.
hypertonic Solution with a greater solute concentration than that within a cell or reference solution.
hypha (pl., hyphae) Microscopic threads that make up the body of most fungi.
hypocotyl Region of stem in a plant embryo that is below the cotyledons.
hypogynous Floral whorls (sepals, petals, stamens) inserted below the ovary of a flower.
hypotonic Solution with a lesser solute concentration than that within a cell or reference solution.

IgE Immunoglobulin E; specific class of antibodies involved in allergic reactions; individuals who suffer from allergies have elevated levels of these antibodies.
immunoglobulin A category of protein known as an antibody.
imperfect flower Unisexual flower; either staminate or pistillate.
imperfect stage The asexual phase in a fungus life cycle characterized by the production of asexual spores.
incomplete dominance A type of inheritance in which the heterozygous phenotype is intermediate between the phenotypes of the dominant and recessive parents.
incomplete flower Flower lacking one or more floral whorls; typically, either the sepals, petals, or both.

incomplete protein Protein that lacks the full complement of essential amino acids in the correct proportions.

indehiscent fruit Dry fruit that does not split open on maturity.

inferior ovary Ovary that lies below the attachment of the sepals, petals, and stamens; an epigynous flower.

inflorescence A cluster of flowers.

insertion A mutation resulting from the addition of a small segment of DNA.

insoluble fiber Roughage, not digestible but moves food through the digestive tract; includes cellulose, lignin, and some hemicellulose found in plants.

integral protein A protein that spans or penetrates the lipid bilayer of a cell membrane.

integrated pest management A multifaceted approach to plant disease control that is intended to reduce the need for pesticides; includes sanitation, crop rotation, biological controls, disease forecasting, and genetic resistance in addition to pesticides.

integument Outermost layers of an ovule that typically develop into the seed coat.

intermembrane space The area between the outer and inner membranes in mitochondria.

International Unit Quantity of a biologically active substance (such as a vitamin) required to produce a standard physiological effect.

internode Region on a stem between nodes.

interphase Stage in the cell cycle when a cell is not dividing.

intine The inner layer of the pollen wall.

intron An intervening or noncoding segment of a gene; introns separate exons.

invasive species Non-native species introduced deliberately or accidentally; unrestricted growth and reproduction threaten native species.

involucre Whorl of bracts that subtend a flower or an inflorescence.

iodine Trace mineral which is a principle component of thyroid hormone.

ion An atom or a molecule that has lost or gained electrons and has either a positive or a negative charge.

ionic bond Chemical bond formed when ions of opposite charges attract.

iron Trace mineral which is a component of hemoglobin; occurs in two forms in the diet nonheme found in both plant and animal sources and heme in animal sources.

iron-deficiency anemia Improperly formed red blood cells appear pale due to a lack of iron which is necessary to make hemoglobin. Most common type of anemia usually afflicting women and children.

isogamy Sexual reproduction involving motile gametes of two different physiological types but are identical in appearance.

isotonic Solution in which the solute concentration is equal to that within the cell or reference solution.

isotope Alternate form of an element with a different number of neutrons but the same number of protons and electrons.

karyogamy In sexual reproduction, the fusion of genetically distinct nuclei.

kilocalorie One thousand simple calories; equal to one Calorie.

kinetic energy Energy of motion.

kinetochore Specialized region on the centromere that connects a chromatid to the spindle.

kingdom Taxonomic category consisting of related phyla or divisions.

Krebs Cycle The second stage of cellular respiration that occurs in the mitochondria; completes the breakdown of glucose into carbon dioxide.

kwashiorkor Diet sufficient in Calories but deficient in protein.

lacto-ovo vegetarian Vegetarian diet that includes dairy products and eggs.

lacto vegetarian Vegetarian diet that includes dairy products.

land races Traditional varieties of plant crops.

lateral bud Bud found in the axil of a leaf; also called an axillary bud.

late wood The cells in the secondary xylem that are formed in the summer, usually with narrow vessels (angiosperms) or narrow tracheids (gymnosperms); also call summer wood.

latex Milky juice exuded from some plants.

LDL (low-density lipoprotein) Transports fats and cholesterol to the body cells, including the cells lining the bloodstream.

leaflet A subdivision of a leaf blade.

legume Simple, dry, dehiscent fruit that splits along two seams, a pod; member of the Fabaceae; a type of bean or pea.

lemma Bract in a grass flower.

lenticel Raised area in the bark of woody stems that permits the exchange of gases.

leucoplast Colorless plastid typically associated with starch formation and storage.

lichen Composite organism formed by the symbiotic association of a fungus and an alga.

light-harvesting antennae A complex of several hundred chlorophyll and carotenoid molecules that form a part of each photosystem.

Light Reactions First steps in photosynthesis, in which chlorophyll traps solar energy, driving the formation of ATP and NADPH; water is also lysed, releasing oxygen.

lignin Complex organic compound that strengthens the secondary cell walls of plants.

linkage Tendency of genes located on the same chromosome to be inherited together.

lipid Organic compound that is insoluble in water; includes fats, oils, and steroids.

litter Partially decomposed plant material.

locus The location of a gene on a chromosome.

lumen Cavity bounded by secondary cell wall in dead plant cells.

lymphocyte A type of white blood cell; a component of the immune system produced by stem cells in the bone marrow; B-lymphocytes produce antibodies.

lysergic acid diethylamide (LSD) A derivative of lysergic acid alkaloids, first isolated from ergot; strongly hallucinogenic.

macromolecule Complex organic molecule formed by joining smaller molecules; example: proteins are macromolecules formed by joining amino acids.

macronutrient Nutritional requirement needed in relatively large amounts.

major minerals Inorganic nutrients needed in amounts greater than 100 milligrams per day.

malnutrition Diet insufficient in quality; lacking protein, vitamins and/or minerals.

marasmus Diet insufficient in Calories; starvation.

marker-assisted selection A molecular technique that uses specific DNA sequences as markers to detect the presence of a linked gene without waiting for the plant to mature and express the desired trait.

mass number Sum of the number of protons and neutrons in the nucleus of an atom; designated by a superscript to the upper left of the elemental symbol.

matrix Compartment of the mitochondrion enclosed by the inner membrane; site of the Krebs Cycle.

megasporangium Sporangium that contains megaspores.

megaspore Spore that develops into the female gametophyte.

megaspore mother cell Diploid cell in megasporangium that, upon undergoing meiosis, yields megaspores.

meiosis Two successive nuclear divisions during which the chromosome number is halved; in plants, meiosis results in the formation of spores.

meristem Area of actively dividing cells in plants.

mesocarp Middle layer in the pericarp (fruit wall).

mesophyll Photosynthetic middle layer in the blade of a leaf; typically composed of palisade and spongy parenchyma.

messenger RNA (mRNA) Type of RNA created from DNA template that travels to ribosomes and directs protein synthesis.

metabolism Sum total of the chemical reactions in an organism.

metaphase Stage of mitosis in which the chromosomes are aligned along the equator.

microbody Membrane-bound organelle that is the site of certain enzymatic conversions; example: peroxisomes and glyoxysomes.

microfilament Solid rod of protein and part of the cytoskeleton.

micronutrient Nutrient required in relatively small amounts; vitamins and minerals.

micropyle The opening in an ovule through which the pollen tube enters during fertilization.

microRNA (miRNA) Small, single-stranded RNA molecules that are 20 to 24 nucleotides in length and primarily involved in gene expression, often by binding to mRNA and stopping translation.

microsporangium Sporangium that contains microspores.

microspore Spore that develops into the male gametophyte.

microspore mother cell Diploid cell in microsporangium that undergoes meiosis to produce microspores.

microtubule Hollow protein rod found in cilia, flagella, the spindle, and the cytoskeleton.

middle lamella Layer of adhesive material (primarily pectins) found between adjacent cell walls.

mineral Inorganic, essential micronutrient.

mitochondrion (pl., mitochondria) Membrane-bound organelle that is the site of cellular respiration.

mitosis Nuclear division, usually accompanied by cytokinesis, in which the chromosomes are duplicated and divided to form two identical daughter cells.

moist coniferous forest Temperate rain forest in the Pacific Northwest of North America characterized by high yearly rainfall with mild winters and cool summers; conifers such as Douglas fir, Sitka spruce, and northern arborvitae and a well-developed understory of ferns, club mosses, mosses, and lichens typify the landscape.

molecular farming Growing and harvesting genetically engineered crops that are producing pharmaceuticals.

molecule Two or more atoms held together by chemical bonds that retain the properties of the compound.

monocotyledon A class of angiosperms in which the seedlings typically possess one cotyledon; commonly abbreviated to *monocot*.

monoculture The cultivation of a single crop over a large region year after year.

monoecious Refers to a plant species whose separate male and female reproductive structures are borne on the same plant; pollen-bearing and ovule-bearing flowers or cones are borne on the same plant.

monohybrid cross Genetic cross between parents that differ by a single trait.

monomer Subunit of a polymer.

monosaccharide The simplest carbohydrate, a simple sugar.

monounsaturated fat Composed of fatty acid chains in which there is only a single C-C double bond; examples: canola oil and olive oil.

multiple alleles A condition in which more than two alleles exist for a given trait.

multiple fruit A fruit derived from the fusion of the ovaries of several flowers in an inflorescence; example: pineapple.

mutualistic symbionts Two organisms in a symbiotic relationship, in which both participants benefit from the relationship.

mutation An inheritable change in genes or chromosomes.

mycelium (pl., **mycelia**) A network of fungal hyphae.

mycobiont The fungal partner in a mutualistic relationship, such as mycorrhizae or lichens.

mycoprotein A vegetarian food material produced from the mycelium of the fungus *Fusarium venenatum*.

mycorrhiza (pl., **mycorrhizae**) Symbiotic association between a fungus and a plant root.

mycotoxin A toxic compound formed by the hyphae of common molds growing under a variety of conditions, especially in contaminated foods.

NAD (nicotinamide adenine dinucleotide) A molecule capable of being reduced that acts as an electron intermediate during cellular respiration.

NADP (nicotinamide adenine dinucleotide phosphate) A molecule that acts as an electron intermediate during photosynthesis; is reduced during the Light Reactions and oxidized during the Calvin Cycle.

naked grain A grain that separates easily from the surrounding bracts.

narcotic Any psychoactive compound that is dangerously addictive; a compound that induces central nervous system depression, resulting in numbness, lethargy, and/or sleep.

natural selection A guiding force of evolution in which organisms that are most fit survive to reproduce.

nectar A sugary solution that attracts animals to plants.

nectar guide Color patterns present on petals that direct insects toward the nectar; often not visible to the human eye.

nectary A gland that secretes nectar.

Neolithic The Stone Age period following the advent of agriculture.

net venation The netlike pattern of branching of veins on a leaf blade; also known as reticulate venation; characteristic of most dicot leaves.

neurotransmitter A chemical responsible for the transmission of impulses across a neural synapse.

neutron A particle in the nucleus of an atom with no charge and a mass of approximately one atomic weight unit.

niacin Collective term for two related B vitamins, nicotinamide and nicotinic acid.

niche The particular role played by an organism in an ecosystem.

night blindness Vitamin A deficiency results in a deficit of the visual pigment found in the rod photoreceptors causing vision impairment in dim light.

nitrogen fixation The process of reducing nitrogen gas to ammonia and nitrates, forms of nitrogen that can be utilized by plants.

node The region of a stem where leaves or branches arise.

nondrying oils Plant oils that remain liquid for prolonged periods upon exposure to the air.

nonheme iron Dietary source of iron found in plant-based foods; not as readily absorbed in the human digestive tract.

nonseptate hypha Fungal hyphae that lack septa, or cross-walls.

nonshattering Refers to seeds and/or fruits that do not split off and scatter from the fruiting head; a trait associated with domesticated plants.

nucellus Tissue in the ovule within which the embryo sac develops; integuments surround the nucellus.

nuclear envelope A double membrane with pores surrounding the nucleus.

nuclear pores Small openings in the nuclear membrane.

nucleic acid A molecule consisting of joined nucleotides; the two types are deoxyribonucleic acid (DNA) and ribonucleic acid (RNA).

nucleolus Spherical structure within the nucleus consisting of RNA and protein; assembly site for ribosomal subunits.

nucleotide A single unit of nucleic acid composed of a phosphate group, a five-carbon sugar (either ribose or deoxyribose), and a purine or pyrimidine base.

nucleus Membrane-bound organelle within eukaryotic cells; contains chromosomes and the nucleolus and is essential for the regulation of all cellular functions.

nut A one-seeded, dry, indehiscent fruit with a hard pericarp.

oil A triglyceride that is liquid at room temperature.

omnivore An organism that feeds on both plants and animals.

oogamy Sexual reproduction involving a nonmotile, larger egg and a smaller, swimming sperm.

oogonium A female gametangium that occurs in some groups of algae and other organisms; gives rise to oospores.

opposite arrangement Two leaves borne per node and arranged across the stem from each other.

order Taxonomic rank consisting of a group of related families.

organelle A body within the cytoplasm of eukaryotic cells; several types of organelles occur, each with a specialized function, such as the chloroplast, which functions in photosynthesis.

osmosis Diffusion of water (or other solvents) through a differentially permeable membrane.

oospore Thick-walled sexual spore that develops within an oogonium in some algae and members of the Oomycota.

osteoporosis Weakened bones due to a deficiency of calcium.

ovary Enlarged basal portion of a single carpel or several fused carpels; contains one to many ovules.

overnutrition Excessive Caloric intake resulting in overweight and obesity.

ovule Structure that will become a seed after fertilization; an integumented megasporangium that contains the embryo sac before fertilization.

oxidative phosphorylation The formation of ATP in the mitochondria; the ATP is formed by the transfer of elections from NADH and $FADH_2$ to O_2 along a series of electron carriers.

oxidation The loss of electrons or hydrogen from an atom or a molecule.

P_{680} The reaction center for Photosystem II; a chlorophyll *a* molecule that is bound to a membrane protein and has a peak absorbance at 680 nm.

P_{700} The reaction center for Photosystem I; a chlorophyll *a* molecule that is bound to a membrane protein and has a peak absorbance at 700 nm.

paddy A flooded field used to cultivate lowland rice.

palea One of two bracts around the grass flower.

paleodicot Group of dicots from several ancient lineages excluded from the clade eudicot.

Paleolithic Old Stone Age; a cultural period during which early humans obtained food solely by foraging; ending in some areas approximately 10,000 years ago.

palisade parenchyma Parenchyma cells in the leaf mesophyll characterized by uniform rows of tightly packed cells with many chloroplasts beneath the upper epidermis.

palmately compound leaf Leaflets radiate from a common point.

panicle A branched inflorescence with the branches bearing loose flower clusters.

parallel venation Principal veins are parallel to one another; characteristic of monocot leaves.

parasite An organism that lives on or in the body of another living organism and derives nourishment from it.

parenchyma Ground tissue in plants with thin-walled cells varying in size and shape; the most abundant kind of cells in plants.

parthenocarpy Development of fruits without fertilization resulting in seedless fruits.

pathogen A disease-causing organism, such as some fungi and bacteria.

partial veil Tissue that covers the gills during growth and development of some mushrooms; when the mushroom is mature, the partial veil breaks loose from the cap edge and often remains as an annulus or ring on the stalk.

pectin A complex polysaccharide in the middle lamella and primary walls of plant cells.

pedicel An individual stalk of a flower that is part of an inflorescence.

peduncle The main stalk of an inflorescence or a single flower.

penicillin An antibiotic produced by various species in the genus *Penicillium*.

pepo A fleshy fruit with a tough outer rind that is composed of both receptacle tissue and exocarp, such as cucumber, pumpkin, and melon.

perennial A plant that continues to live for an indefinite number of years.

perfect flower A flower having both stamens and carpels.

perfect stage The phase during the life cycle of a fungus when sexual fusion occurs, producing characteristic sexual spores.

perianth The petals and sepals together.

pericarp The fruit wall that develops from the ovary wall.

pericycle Root tissue sandwiched between the endodermis and the phloem; the outermost layer of the stele; meristematic region that gives rise to branch roots.

periderm Protective tissue that replaces the epidermis after secondary growth begins; includes the cork, the cork cambium, and sometimes other cells.

perigynous Refers to a flower in which the bases of the sepals, petals, and stamens form a cup around the ovary.

peripheral protein A protein on the surface of a biological membrane.

perithecium A flask-shaped ascocarp.

permafrost Soil that is permanently frozen.

pernicious anemia Improperly formed red blood cells due to a deficiency of vitamin B_{12}.

peroxisome A microbody found in leaves and often associated with chloroplasts.

petal A floral organ that is leaflike and often brightly colored; a component of the corolla.

petiole The stalk of a leaf.

phenolics A large and diverse category of compounds, all of which contain one or more aromatic benzene rings (a ring of six carbon atoms with six hydrogen atoms attached) with one or more hydroxyl (OH) groups; they include flavonoids, tannins, and lignin.

phenology Timing of biological events (bud-burst, seed set, etc.) in the life cycle of an organism. Environmental cues, such as daylength or temperature, often trigger the event.

phenotype The physical appearance of an organism.

phloem The vascular tissue that conducts organic materials synthesized by the plant.

phospholipid A type of lipid molecule occurring in a bilayer in biological membranes; a lipid with two fatty acids and a phosphate group attached to glycerol.

photophosphorylation The formation of ATP during photosynthesis in the chloroplast; the energy of sunlight drives the phosphorylation of ADP to form ATP.

phosphorylation The addition of a phosphate group to a molecule.

photon A unit of light energy.

photosynthesis The process that results in the conversion of light energy into the chemical energy of carbohydrates.

Photosystem I A light-harvesting unit located on the thylakoid membrane of the chloroplast, with P_{700} as the reaction center.

Photosystem II A light-harvesting unit located on the thylakoid membrane of the chloroplast, with P_{680} as the reaction center.

phragmoplast A system of microtubules and vesicles that arises between two daughter nuclei at telophase and forms the cell plate.

phycobilins Accessory pigments present in the cyanobacteria and red algae and including phycocyanin and phycoerythrin.

phylogeny The evolutionary history and relationship of a species.

phytolith Crystals, composed of calcium oxalate or silica formed in plants; distinctive to varieties, species, genera or families of plants.

physiological dependence The condition in which there is a physical need for a drug to avoid withdrawal symptoms.

phytoalexins Secondary products produced by plants in response to bacterial and fungal pathogens and known to kill or inhibit the growth of pathogens; important components of natural disease resistance in many plants.

phytochemical Naturally occurring component in plants that appears to have beneficial effects on human health.

phytoestrogen Isoflavones; plant compounds that mimic estrogens in the human body.

phytoremediation Growing tolerant plants in polluted sites to decontaminate the soil or water.

pinnately compound leaf Leaflets attached on both sides of a common axis.

pistillate flower A flower having carpels but no stamens.

pit A pore in a secondary cell wall.

pith The central tissue of a dicot stem, consisting of parenchyma cells.

plasmid A small, circular DNA molecule found in bacterial cells.

plasmodesma (pl., plasmodesmata) A cytoplasmic strand that connects adjacent plant cells through pores in the cell wall.

plasmodium The vegetative stage of a slime mold.

plasmogamy The fusion of cytoplasm from two cells or gametes.

plasmolysis Shrinking of protoplasm in a cell due to loss of water in a hypertonic environment.

plastid A class of organelles that includes chloroplasts, leucoplasts, and chromoplasts.

plywood A building material consisting of two or more thin sheets of wood bonded together.

pneumatophores Aerial outgrowth of submerged root system in the bald cypress; conduct air with oxygen to root cells.

pod A dry dehiscent fruit that splits along two seams; a legume.

point mutation The smallest mutation caused by the change of a single nucleotide.

polar nuclei Two nuclei found in the embryo sac that unite with a sperm to form the primary endosperm nucleus.

pollen Immature male gametophytes of seed plants.

pollen tube A tube that develops from the pollen grain and carries the sperm to the ovule.

pollination The transfer of pollen from an anther to a stigma.

polygenic inheritance Inheritance of a trait that is controlled by more than one pair of genes; each allele has an additive effect on the same trait.

polynomial A scientific name composed of more than two words.

polypeptide A chain of amino acids connected by peptide bonds.

polyploidy Having more than two complete sets of chromosomes. More than the diploid (2N) or two sets, of chromosomes. For example, triploid (3N) or three sets; tetraploid (4N) or four sets.

polysaccharide A polymer, such as starch or cellulose, composed of thousands of monosaccharides.

polyunsaturated fat A fat having several to many double bonds between carbon atoms.

pome A simple fleshy fruit; the outer portion formed by floral parts that surrounded the ovary; examples: apple and pear.

population All the individuals of a species within a given area.

potential energy The energy stored in matter as a result of its location or chemical bonds.

precision breeding A molecular technique that uses the tools of genetic engineering to transfer genes of interest from one variety to another within the same or closely related species that are naturally able to hybridize; it differs from the creation of transgenic plants in that no foreign genes are introduced from a totally different species. Also known as cisgenesis.

Pressure Flow Hypothesis The theory that organic solutes move along a concentration gradient from source to sink through the phloem.

primary consumer An animal that feeds directly on producers.

primary endosperm nucleus The product of the fusion of a sperm and two polar nuclei in the embryo sac of angiosperms; double fertilization.

primary growth Growth in length due to the activities of the apical meristems of shoot and root.

primary wall The wall layer of a plant cell deposited during cell expansion, generally thin and elastic.

primary succession Progression of plant communities invading land bare of vegetation; pioneer species are the first colonizers beginning in the process of soil formation but are eventually replaced by succeeding communities of plants finally culminating in a self-perpetuating climax plant community.

producer An organism that manufactures food through photosynthesis.

prokaryotic cell A type of cell lacking a nucleus and membrane-bound organelles; found in the kingdoms Archaebacteria and Eubacteria.

prophase The first stage of mitosis, characterized by the condensation of chromatin into chromosomes and the formation of the spindle.

prop roots Aerial roots which sprout from the nodes of stems and then grow into the ground to anchor and support stems.

prosthetic group Nonprotein groups that are attached to an enzyme or other protein and necessary for its function.

proteasome Tunnel-shaped collection of protein-degrading enzymes that disassemble proteins tagged for destruction in cells.

protein A macromolecule composed of one or more polypeptides, each composed of many amino acids.

proton A positively charged particle in the nucleus of an atom.

protoplast All of a plant cell excluding the wall.

psychoactive drug A drug that affects the central nervous system by influencing the release of neurotransmitters or mimicking their actions.

psychological dependence A condition marked by the strong desire to repeat the use of a drug to

reexperience the feelings of well-being induced by the drug.
purine One type of nitrogen-containing base found in nucleotides; consisting of adenine and guanine.
pyrimidine One type of nitrogen-containing base found in nucleotides; consisting of cytosine, thymine, and uracil.

---

raceme A vertical inflorescence with stalked flowers.
radicle The embryonic root found in the seed.
reaction center A chlorophyll *a* molecule bound to a membrane protein.
receptacle The expanded tip of a pedicel or peduncle to which the floral organs are attached; also, swollen areas at the end of the blade that function as reproductive regions in certain members of the brown algae.
recessive Refers to an allele that is masked in the phenotype by a dominant allele.
recombinant DNA The introduction of genes from one organism into the DNA of a second organism.
redox reaction Oxidation-reduction reaction; a chemical reaction involving the transfer of electrons from one molecule to another.
reduction Gain of electrons or hydrogen from an atom or a molecule.
resin An exudate released when a tree is wounded; common in conifers but also occurring in some angiosperms.
restriction enzyme Enzyme able to cut DNA at specific base sequences; used in recombinant DNA technology.
reticulate venation Netted venation; the arrangement of veins in a leaf that resembles a net; characteristic of dicot leaves.
retting The process that frees flax fibers by allowing microbial decomposition to break down the outer part of the stem.
rhizome A horizontal, underground stem.
ribulose bisphosphate carboxylase (RUBISCO) The enzyme involved in the first step of carbon fixation during the Calvin Cycle; it catalyzes the carboxylation of ribulose-1,5-bisphosphate.
ribonucleic acid (RNA) The nucleic acid formed from DNA and involved in protein synthesis; nucleotide of chain of phosphates, ribose sugars, and purine and pyrimidines.
ribose A pentose sugar present in RNA.
ribosomal RNA (rRNA) The type of RNA that is a component of ribosomes.
ribosome A particle composed of two subunits, each containing RNA and protein; functions in protein synthesis.
root cap A thimble-shaped group of cells found at the tip of roots; protects the meristem.
root hair A root epidermal cell that functions in water absorption.
root nodule Gall-like structures on the roots of legumes that contain symbiotic nitrogen-fixing bacteria.
rootstock Root system to which a scion or upper portion of a woody plant is grafted; also known as a stock.
rough ER A portion of the endoplasmic reticulum containing ribosomes.
RNA polymerase Enzyme responsible for attaching nucleotides together in the sequence specified by DNA.
runner A horizontal stem that grows along the surface of the ground; also known as a stolon.

S Part of interphase known as Synthesis during which DNA is replicated; follows G_1 and prior to G_2 of interphase.
samara A simple, dry indehiscent fruit with the pericarp bearing winglike outgrowths; winged fruit of maple.
saponin A glycoside with a steroid molecule as the active component, such as diosgenin from yams.
saprobe An organism deriving its nutrients from non-living organic matter or the remains and by-products of organisms.
sapwood Region of secondary xylem that actively transports water; light-colored wood immediately inside the vascular cambium.
saturated fat A fat in which all the carbons in the fatty acids are connected by single bonds, thereby having the maximum number of hydrogen atoms.
savanna A tropical grassland biome with scattered trees.
schizocarp A dry indehiscent fruit that splits into two one-seeded halves at maturity.
scion A small twig or bud that is grafted to a stock.
sclereid A sclerenchyma cell with a thick, lignified secondary wall having many pits; variable in form but not usually elongated.
sclerenchyma Tissue composed of cells with thick secondary walls; functioning in support or protection.
sclerophyllous Vegetation characterized by thick, leathery leaves with abundant sclerenchyma cells; vegetation of a chaparral.
sclerotium (pl., **sclerotia**) A fungal resting body resistant to unfavorable conditions; a firm, hardened mass of hyphae (or a hardened plasmodium of a slime mold) that will germinate on the return of favorable conditions.
scutellum The single cotyledon in grass seeds.
secondary consumer An animal that feeds on other consumers.
secondary growth The increase in girth of stems and roots; produced by the activities of the vascular cambium and cork cambium.
secondary product Any chemical compound synthesized by plants or fungi but not critical for the basic metabolic functions of that organism; often functioning to deter predators or attract pollinators; a secondary metabolite.
secondary succession Natural or human forces destroy existing vegetation, reverting area to an earlier successional stage of plant community; progression of plant communities eventually culminate in a self-perpetuating climax plant community.
secondary wall The innermost layer of a cell wall formed after cell elongation has ceased; often characterized by the deposition of lignin.
seed A matured ovule containing an embryo and food supply and covered by a seed coat.
seed bank Storage facility for seeds of domesticated plants and wild relatives; facility for preserving genetic diversity.
seed coat The outer layer of a seed that is developed from the integuments of the ovule; the testa.
seed plant Common term for gymnosperms and angiosperms.
self-pollination Transfer of pollen from stamen to stigma within the same flower or plant.

---

semidrying oils Plant oils that dry slowly or at elevated temperatures; intermediate between drying and nondrying oils.
sepal A leaflike floral organ that protects the unopened flower bud.
septate Divided by cross-walls into cells.
septum (pl., **septa**) A dividing wall or partition; a cross-wall in a fungal hypha or algal filament.
sexual reproduction A type of reproduction that involves the fusion of gametes (usually egg and sperm) to form a zygote.
shattering A trait found in wild plants in which the fruiting head breaks apart to scatter the seeds over a wide area.
shortgrass prairie A grassland biome characterized by short grasses and low rainfall; also known as the plains.
sieve plate The perforated wall area in a sieve tube member.
sieve tube A long tube specialized for the conduction of food materials (products of photosynthesis) and consisting of several to many sieve tube members.
sieve tube member A phloem cell characterized by a sieve plate and enucleate condition; specialized for the conduction of products of photosynthesis.
simple fruit A fruit that develops from a single ovary.
simple leaf A leaf that is not divided into leaflets.
single nucleotide polymorphism (SNP) DNA sequence variations among individuals that occur when a single nucleotide (A, T, C, or G) in the genome sequence is altered.
sink Storage area or tissues of active metabolism using sugars that are transported in the phloem; the sink may be an underground storage organ, a developing fruit, or an actively growing meristem.
small interfering RNA (siRNA) Small, double-stranded RNA molecules that are around 20 to 24 nucleotides in length and primarily involved in gene silencing.
smooth ER The portion of endoplasmic reticulum that lacks ribosomes.
soft fiber Commercial fiber obtained from the phloem of certain dicot stems; also known as bast fiber.
softwood General term for the wood (secondary xylem) of conifers.
soluble fiber Resistant to digestion and absorption in the small intestine; includes gums, pectins, mucilages, and hemicelluloses found in plants and algal polysaccharides.
somaclonal variant A plant showing a mutation that developed asexually during the tissue culture of a single callus.
somatic mutation A mutation that occurs in cells of leaves, stems, or roots; a mutation occurring in any cells that are not involved in gamete formation.
sorus (pl., **sori**) A cluster of sporangia found on a fern leaf.
source Supply area of sugars that are transported in the phloem; the source may be a photosynthetic organ or a storage area.
specialized niche Niche in which species have a narrow range of tolerance.
species A single kind of organism; often defined as a group of interbreeding populations reproductively isolated from any other such group.
sperm A male gamete.
spermagonium (pl., **spermagonia**) A structure that produces spermatia in the rust fungi.

spermatium (pl., **spermatia**) Minute, nonmotile male gametes that occur in the rust fungi.

spice A pungent, aromatic plant product derived from plants native to tropical regions and used to flavor foods.

spike An inflorescence in which the main axis is elongated and the flowers are sessile.

spikelet A small group of grass flowers; a unit of the inflorescence in grasses.

spindle The aggregation of microtubules that is involved in the movement and separation of chromosomes during mitosis and meiosis.

spongy parenchyma Part of the leaf mesophyll; cells are loosely arranged and contain chloroplasts.

sporangiospore A spore that develops within a sporangium.

sporangium (pl., **sporangia**) A structure in which spores are produced.

spore A reproductive unit (often unicellular) that is capable of developing into a new organism without fusion with another cell.

sporophyte A diploid plant that produces spores; the diploid phase of a life cycle that has an alternation of generations.

springwood The cells in the secondary xylem that are formed early in the season, usually with wide vessels (angiosperms) or wide tracheids (gymnosperms); also called early wood.

stamen The floral organ that produces pollen; consisting of an anther and filament.

staminate flower A flower having stamens but no carpels.

starch A polysaccharide composed of a thousand or more glucose molecules; the chief food-storage material of most plants.

statins Cholesterol-lowering drugs that were originally isolated from fungi and have been widely used since the 1990s.

stele The vascular cylinder; vascular tissue making up the central cylinder of roots.

steroid A type of lipid containing four fused rings of carbon atoms with various side chains.

stigma The receptive portion of the carpel to which the pollen adheres.

stimulant A psychoactive compound that excites and enhances mental alertness and physical activity; often reduces fatigue and suppresses hunger.

stipe A supporting stalk, such as those in mushrooms and brown algae.

stipule A small appendage found in pairs at the base of leaves.

stock The rooted part of a plant to which the scion is grafted.

stolon A horizontal stem that grows along the ground surface; may form adventitious roots and plantlets; also known as a runner.

stoma (pl., **stomata**) A minute opening, bordered by guard cells, in the epidermis of leaves and stems.

stroma The ground substance of the chloroplasts where the reactions of the Calvin Cycle occur.

stroma thylakoid A thylakoid that does not occur in a granum; connects separate grana.

style The structure that connects the stigma to the ovary in the carpel.

suberin A fatty material found in the cell walls of cork cells and the Casparian strip of the endodermis.

subsidiary cells Cells in the epidermis that subtend guard cells and assist in the opening of stomata in the grasses.

substrate The substance acted on by an enzyme; the surface on which a plant or fungus grows or is attached.

sucrose A disaccharide made from a molecule of glucose linked to a molecule of fructose; table sugar.

sugar A monosaccharide; a carbohydrate with the general formula $C_nH_{2n}O_n$.

summerwood The cells in the secondary xylem that are formed late in the season, usually with few vessels (angiosperms) or narrow tracheids (gymnosperms); also called late wood.

superior ovary An ovary located above the sepals, petals, and stamens.

surface fiber Commercial fibers obtained from the surface of a plant organ.

syconium An inflorescence found in figs., consisting of an urn-shaped receptacle bearing hundreds to thousands of small unisexual flowers on its inner surface.

symbiosis A relationship in which two organisms live in intimate association with each other.

symplast The interconnected protoplasm of all cells in a plant.

synapsis The pairing of homologous chromosomes that occurs in prophase I of meiosis.

synergid One of a pair of short-lived cells that lie close to the egg in the mature embryo sac.

systemic acquired resistance Induced immune response in plants that occurs following exposure to a viral, bacterial, or fungal pathogen; salicylic acid is the signal that turns on this response.

taiga A biome in the Northern Hemisphere dominated by conifers; the northern coniferous or boreal forest.

tallgrass prairie A grassland biome characterized by many tall grasses up to 5 meters (16 feet) tall.

tannin A secondary product found in many plants that has been widely utilized as stains, dyes, inks, and tanning agents for leather; believed to function in plants by discouraging herbivores.

tannosome Plant organelle arising from the thylakoid membranes in chloroplasts and then shuttled to the central vacuole; produces tannins that impart a characteristic brown color.

taproot A relatively large primary root that gives rise to smaller, lateral roots.

taxon (pl., **taxa**) A general term for any taxonomic rank, such as species, genus, or order.

teliospore A thick-walled spore found in the rust and smut fungi; karyogamy occurs within the teliospore and it gives rise to the basidium.

telomere Repeating sequence of DNA found at the end of a chromosome; protects a chromosome from being shortened or damaged during cell divisions.

telophase The last stage of mitosis and meiosis, during which the chromosomes become reorganized into daughter nuclei.

temperate rain forest A biome dominated by coniferous trees, high rainfall, and high humidity; moist coniferous forest.

tepal Members of the perianth that are not differentiated into sepals and petals.

terpene An unsaturated hydrocarbon formed from an isoprene building block; found in many plants in the form of essential oils.

testa Seed coat.

testcross A cross involving one parent that is homozygous recessive for a given trait.

tetrad A group of four, such as the four haploid spores that form after meiosis, or the four chromatids in a bundle after homologous chromosomes pair.

thallus Body of an alga, which can vary from a microscopic unicell to a large macroscopic, multicellular organism.

thiamine Vitamin B_1.

thylakoid membrane A saclike photosynthetic membrane in chloroplasts; stacks of thylakoids form the grana.

thymine A pyrimidine base occurring in DNA but not in RNA.

Ti plasmid The tumor-inducing plasmid from the bacterium *Agrobacterium tumefaciens*; commonly used as a vector for recombinant DNA studies in plants.

tisane Herbal tea.

tissue A group of cells that perform a specific function.

tolerance Need for ever-increasing amounts of an addictive drug to produce the desired physiological effect.

toxin A poisonous substance.

trace element An inorganic element required in small amounts for plant growth.

trace minerals Inorganic nutrients needed in amounts of no more than a few milligrams per day.

tracheid An elongated, tapering xylem cell that is specialized for conducting water and support with lignified pitted walls.

trans fat Unsaturated fat subjected to hydrogenation, which changes the configuration of carbon-carbon double bonds from *cis* to *trans* and the fat from its natural liquid state to a solid form.

transcription The formation of RNA as a complementary copy of a portion of the DNA molecule.

transfer RNA (tRNA) A class of small RNA molecules that transfer amino acids to the correct position on the messenger RNA molecule at the ribosome for protein synthesis.

transgenic Refers to cells or organisms that contain genes that were inserted into them from other organisms through the techniques of genetic engineering.

translation The synthesis of a polypeptide from a specific sequence of codons on a messenger RNA molecule; occurs at the ribosomes.

transpiration The loss of water vapor from leaves; occurs mostly through the stomata.

Transpiration-Cohesion Theory The theory that explains water movement in the xylem; the driving forces are the pull of transpiration and the cohesion of water molecules.

trichome An epidermal appendage, such as a hair or a scale.

triglyceride A type of lipid formed from three fatty acids bonded to a molecule of glycerol; a fat or an oil.

triploid Refers to a cell or a nucleus that contains three sets of chromosomes; common in endosperm.

trophic level A step in the movement of energy through an ecosystem; a step in a food chain.

tropical rain forest An endangered tropical biome with high rainfall and an exceptional diversity of species.

tuber An enlarged, fleshy, underground stem tip, such as the potato.

tuberous roots Modified fibrous roots that have become fleshy and enlarged with food reserves.

tundra A treeless circumpolar biome with meadowlike vegetation above the Arctic Circle.
turgid Refers to a swollen, distended cell that is firm, owing to water uptake.

umbel A flat-lopped inflorescence in which the stalked flowers all radiate out from a common point.
undernutrition Nutritional condition in which the intake of kilocalories is insufficient to maintain daily energy requirements.
universal veil A membrane that totally encloses some young mushrooms; after it breaks, its remnants appear as a volva at the base and scales on the cap.
unsaturated fat A fat containing one or more double bonds between carbon atoms.
uracil A pyrimidine found in RNA but not DNA.
uredium (pl., uredia) The structure that produces uredospores in rust fungi; sometimes called a uredinium.
uredospore A reddish, binucleate spore formed by rust fungi; often forms the repeating stage of the rust; also called a urediniospore.

vascular bundle A strand of tissue containing primary xylem and primary phloem, often surrounded by a bundle sheath.
vascular cambium Meristematic tissue that gives rise to secondary xylem and secondary phloem.
vascular cylinder The stele; vascular tissue making up the central cylinder of roots.
vascular plant A general name for any plant that has xylem and phloem.
vascular ray Sheet of parenchyma that extends radially through the wood, across the cambium, and into the secondary phloem; rays are produced by the vascular cambium and function in lateral transport.
vascular tissue Tissue that is specialized for the long-distance transport of water or photosynthetic products; xylem and phloem.

vegan A pure vegetarian consuming no animal products at all.
vegetarian A person who does not consume animal flesh; some consume dairy products and eggs, but others are vegans.
vegetative cell The cell in the pollen grain that develops into the pollen tube.
vein A vascular bundle that forms part of the conducting and supporting tissue of a leaf.
velamen Outgrowth of epidermis of aerial orchid roots; prevents desiccation.
veneer A thin sheet of wood, often with attractive grain, used to cover less expensive wood.
vessel A tubelike column of vessel elements that are connected by open end walls and are specialized for the conduction of water and minerals.
vessel element One of the cells forming a vessel and characterized by a perforation plate.
visible light That portion of the electromagnetic spectrum that illuminates the planet and is detected by our eyes; important as the source of energy for photosynthesis in green plants and algae.
vitamin A naturally occurring organic compound that is necessary, in small amounts, for the normal metabolism of plants and animals.
viticulture Cultivation of grapes.
volva Remnant of the universal veil of certain mushrooms.

water-soluble vitamin A vitamin that is not readily stored in the body, with excess eliminated in the urine; includes B vitamins and vitamin C.
weed A plant not valued for its use or beauty and not intentionally planted; a category of hay fever plants that includes nongrass and nontree species.
whorled arrangement Three or more leaves per node.

winnowing The process that separates the grain from the fragments of chaff.
wood Secondary xylem.
wood pulp A watery suspension of pulverized wood used in the production of paper, cardboard, fiberboard, rayon, cellophane, and other products.
woody stems Stems that undergo secondary growth.

xanthophyll Any of several yellow carotenoid pigments found in chloroplasts.
xerophthalmia Severe and prolonged vitamin A deficiency causing Irreversible drying and degeneration of the cornea resulting in permanent blindness.
xerophyte A plant adapted for growth in arid conditions.
xylem The vascular tissue specialized for the conduction of water and minerals; consists of tracheids and vessel elements, fibers, and parenchyma cells.

zoospore A motile spore.
zone of cell division Region of root tip in which cell division takes place.
zone of elongation Region of root tip in which newly formed cells elongate or grow.
zone of maturation Region of root tip in which newly formed cells differentiate into the various tissues that comprise the root.
zygomorphic flower A bilaterally symmetrical flower, capable of being divided into two symmetrical halves only by a single longitudinal plane passing through the axis.
zygosporangium (zygospore) The thick-walled sexual spore formed by members of the zygomycetes.
zygote A diploid cell that is formed by the fusion of two gametes.

Index

Bold indicates use as a key term. Italics indicate material in figures and tables.

A

A (vitamin), 44-45, *162*, *162-163*, 262
Abiotic factors, **487**
Abrus precatorius (rosary pea), *221*, 384
Abscisic acid, **92**-93
Abscission, 92
Absinthe, 461-462, *462*
Absinthin, 461
Absinthism, 462
Absorption spectrum, **57**
Absorptive heterotrophs, **137, 422**
Accessory fruits, **86**
Accessory pigments, 405
Acetobacter spp., 449, 452
Acetyl-CoA, 64
Acetylsalicylic acid, 346
Achenes, **86**
Acid rain, 213
Acorns, 86
A Counterblaste to Tobacco (James I of England), 371-372
Actaea racemosa (black cohosh), *355*
Actinidia chinensis (kiwifruit), 99
Actinomorphic (regular) flowers, **74**, *74*
Actinomycetes, 471
Active transport, **20**
Adaptation, 7, **135**
Addictive psychoactive compounds, **359**. *See also* Psychoactive drugs
Adenine, **12, 111**
Adhesion, **51**
Adhesives, starch-based, 237
ADP (adenosine diphosphate), **56**
Adventitious roots, 225, 232, *233*
Adzuki bean (*Vigna angularis*), *214*
Aecia, **440**
Aeciospores, **439,** 440
Aegilops speltoides (goat grasses), 190
Aerenchyma, **35**
Aerial roots, 37
Aerobic respiration, 67-68. *See also* Cellular respiration
Aerobiologists, 80
Aeroponics, 50
Afghanistan, preservation of land races in, 250
Aflatoxins, **473**-474, *474*
African blackwood (Grenadilla), 325
African Blackwood Conservation Project, 325
Agar, **411**

Agaricus bisporus (button mushroom), 269, 465, *466*
Agaricus bisporus (commercial mushroom), *432*
Agarose, **411**
Agave fourcroydes, 309
Agave juice, 460
Agave sisalana (sisal), *309*, 318
Agave spp., 459-460
Agave tequilana, 459
Age of Cycads, 147
Age of Exploration, spice trade and, 292
Age of Herbals, 121, 338-340
Aggregate fruits, **86**
Agricultural Improvement Act, 317
Agricultural research centers, 244
Agricultural Revolution, 179-180
Agriculture, 179-183. *See also* Crops; Food plants
 biodiversity in, 131
 biological controls and, 131, 246
 characteristics of domesticated plants, 184
 crop rotation with legumes, 212
 debate over revolution or evolution of, 179-180
 dryland, 247
 early sites of, 180-183
 fertilizers and, 211, 212, 245-246, *247*
 genetic diversity and, 247
 Green Revolution and, 243-247
 importance of biodiversity to, 131
 irrigation and, 245, *246*, 247, 501
 mechanization of, 247-248
 monocultures and, 247-248
 nitrogen fertilizer use and, 212
 origins of, 79, 175-186
 selective breeding for, 242-243
 sustainable, 248-249
 tillage techniques, 246, *247*
 Vavilov's centers of plant domestication, 184-185
Agrifibers, 334
Agri-Fos, 444
Agrobacterium spp., *112, 258, 258*
Agrostis tenuis (bent grass), 134
Aguamiel, 459
AIDS, 482
Air pollution, 212, 434, 494
Alanine, *10, 157*
Alcoholic beverages, 447-462
 absinthe, 461
 beer, 454-457
 from corn, 200
 distilled spirits, 458

 health effects of, 463-465
 sake, 457
 wine, 67, 447-454
Alcoholic fermentation, 67-68
 of beer, *455, 456,* 457
 of foods, 466-467
 of wine, 449-452
Aldrin, 491
Ale, 455
Aleurites moluccana (candlenut oil tree), 320
Aleurone layer, **188**
Alexander the Great, 290
Alfalfa (*Medicago sativa*), 222, *222*
Algae, 4, 401-417
 algal blooms, 4, 212, 411, 414
 biofuels from, 410-411
 brown (Phaeophyta), 401, 404-405, *405, 406*
 characteristics of, 401
 classification of, 137, 401
 cyanobacteria, 402-403, 412-414
 diatoms (Bacillariophyta), 401, 404, *404*
 dinoflagellates (Dinophyta), 401, 403-404
 divisions of, 401, *402*
 drugs from, 413
 economic value of, 411
 euglenoids (Euglenophyta), 401, 403, *403*
 evolution of plants from, 405
 food preservatives from, 411
 foods from, 409-410
 green (Chlorophyta), 401, 405-408, *408*
 lichens and, 434
 photosynthesis by, 401
 red (Rhodophyta), 401, 414
 reproduction by, 401
 toxic, 411-417
Algal blooms, 4, 212, 411, 414
Alginic acid, 411, **411**
Al-Hasan ibn-al-Sabbah, 363
Alice in Wonderland (Carroll), *364*
Alkaloid nicotine, 372
Alkaloids, 15, 272, **343**, 359
 caffeine. *See* Caffeine
 in medicinal plants, 151, 343
 in poisonous plants, 379, 380, 381, 382-383
 psychoactive plants, 15
 tropane, 15
Alleles, **102**
 dominant, 104
 multiple, 109
 recessive, 104

Allelopathy, **383**-384, 504
Allergens, **390, 482**
 fungal, 482-484
 immune system response to, 390
 limiting exposure to, 394-395
 plant
Allergic asthma, 155, 390, 395, 414, 480, 482, 483. *See also* Asthma
Allergic broncopulmonary aspergillosis, 482
Allergic rhinitis, 390-391, 392, 394
Allergies, 390
 allergic rhinitis, 390-391, 392, 394
 asthma, 155, 390-391, 398, 414, 480, 482, 483
 controlling, 394-395
 to fungi and mold, 482-484
 to genetically modified foods, 261, 268
 hay fever, 390, 391-394
 immune system and, 390
 latex allergy, 398
 peanuts, 216, 398
 plant, 389-398
 pollen, 79
 respiratory, 390-391
Allicin, 305
Allium cepa (Onion), *302*, 304-305
Allium porrum (Leeks), *302,* 304
Allium sativum (Garlic), *302*, 304-305
Allium spp., *302*, 304, 305, 349
Allspice (*Pimenta dioica*), *294*, 300
Aloe vera (burn plant), 344, 349, 351, *351*
Aloin, 344, 351
A Long Day's Journey into Night (O'Neill), 361
Alpine pennycress (*Thlaspi caerulescens*), 51
ALS (amyotrophic lateral sclerosis), 148
Alternaria spp., 430, *431*, 474, 482, *483*
Alternate leaf arrangement, **38,** *39*
Alternation of generation, **76**, 139-142
 in angiosperms, 76
 in ferns, 139-141, *141*
 in mosses, 139-141, *140*
 in pines, 139-141, *142*
Alternative medicine, 354-357. *See also* Aromatherapy; Herbal medicine
Altitude, biome distribution and, 500, *500*
Aluminum, 50

533

Alzheimer's disease, 23, 148, 155, 412, 474
Amanita muscaria (fly agaric), 478-479, *479*
Amanita phalloides (death cap), *469*, 478
Amanita spp., 478
  morphological features, *478*
Amanita virosa (the destroying angel), 478, *478*
Amaranth (*Amaranthus* spp.), 252-253
Amaryllidaceae (Amaryllis family), 125, 304
Amaryllis, 125, 304
Amate, 320, *320*
Amatoxins, **477**
Amber, 148-149, 330
Ambrosia beetles, 443
Ambrosia spp. (ragweed), 392-393
American chestnut (*Castanea dentata*), 97, 98, 442
American elm (*Ulmus americana*), 442
American laurel (*Kalmia* spp.), 383
American Phylogeny Group (APG), 132
Amflora potato, 237
Amino acids, 9-10, *10*, **156**
  essential and nonessential, 156-157, *157*
  genetic code for, 115
  translation of, 116
Ammonia, 212, 213
Ammonium, 212, *213*
Amoxicillin, 471
Amphidinium, 413
Amphidinolides, 413
Amphotericin B, 482
Ampicillin, 471
Amsterdam Botanical Garden, 275
Amylase, 56
Amylopectin, **194**, 237
Amyloplasts, **21**, *21*
Amylose, **194**, 237
Anabaena azolla, 201, 212, *403*, 412
Anabolic reactions, **55**
Anaerobic respiration, **67-68**
Analgesics, **346**
Anaphase (mitosis), **24**, *26*, 27
Anaphase I (meiosis), 76, *77*
Anaphase II (meiosis), 76, *77*
Anaphylaxis, 398, 471
Anasazi bean (*Phaseolus vulgaris*), 214
Anathum gaveolens (Dill), *301*, 302, 303
Anatoxin-a, 412
Androecium, **70**
Andromedotoxins, 383
Angiosperm families, 359
Angiosperms, **2**, **146**, 151
  alternation of generation in, 76
  clades of and phylogenetic systems, 132
  double fertilization in, 83-84

economically important families, *128*
female gametophyte development, 79
flowers of, 2, 70
fruits of, 86, 88
germination in, 88
male gametophyte development, 77
meiosis in, 75-77
plant hormones and, 92, 93
pollination in, 79-84
reproductive cycle, *78*
seed germination and development, 88, *89*
seeds of, 88
vascular tissue in, 32
Animalia (kingdom), 137
Animals
  coevolution of with plants, 80-81
  desert, 501
  domestication of, 181
  dyes from, 321
  fats in, 160
  fibers from, 309
  forage crops for, 205, 221
  plants poisonous to, 386-387
  pollination by, 80-81, 83
  seed dispersal by, 86, 96
Anise (*Pimpinella anisum*), *302*, 303
Anisogamous reproduction, **408**
Annona cherimola (cherimoya), 99, *99*
Annual rings, **34**, 37, *38*, 327
Annulus, **477**
Anopheles mosquito, malaria and, 347, 348
Antheridia, 422
Antheridium, **139**, **405**
Anthers, 70
Anthocerophyta (hornworts), 138, *138*, 141
Anthoceros spp., *143*
Anthocyanins, 14, 50, 91, 96, 449
Anthophyta. See Angiosperms
Anthriscus cereifolium (chervil), *302*, 303
Anthropocene epoch, **498-499**
Anthropology, palynology and, 79
Antibiotic-resistant bacteria, 471-472
Antibiotics, 470-472, 473-474
Antibodies, **390**
Anticoagulants, 413
Anticodons, **116**
Antigens, **390**
Antihistamines, 395
Anti-inflammatories, **346**
Antimalarials, 348, 349
Antioxidants, 44, 163
  in cocoa, 284
Antipodals, **79**
Antipyretics, **346**
Antivirals, 413
Ants, seed dispersal by, 88
APG (Angiosperm Phylogeny Group), 132

Apiaceae, *128*, *129*, 303
Apical meristems, **29**, *29*
Apios americana (groundnut), 222
Apis mellifera (honey bee), 81
Apium graveolens (Celery), *302*, 303
Apoplast, **49**
Apple (*Malus pumila*), 86
  brandy, 461
  cider, harvesting of, 460-461
  global production of, *188*
  liqueur, 461
  plant hormones and, 92, 93
  stomata on leaves of, *40*
  symbolic meaning of, 125
  wine, 461
Appleseed, Johnny, 94
Aquifers, **245**
Arabia
  distilled spirits in, 458
  origins of coffee use in, 273-274
Arabidopsis spp., 309
Arabidopsis thaliana, 112, *112*
Araceae, 386
Arachidonic acid, 159
Arachis spp. (peanut), 216, 217, 398
Arbuscles, 436
Arbuscular mycorrhizae (AM), **425**, **436**
Archaea, **137**
Archegonia, **139**
Archeologists, 79
Archeology, 79, 178
Arecaceae, *128*, *129*
Aretaeaus
Arginine, *10*, *157*
Arils, **296**
Arisaema triphyllum (jack-in-the-pulpit), 386
Armillaria bulbosa, 430
Armillaria ostoya, 430
Aroids, 386
Aromatherapy, **291**, 341, 354
Arrow poison (curare), 379
Arsenic, bioaccumulation of, 146
Artemesia annua (wormwood), 346, 348-349
Artemesia dracunculus (tarragon), 302
Artemisinin, *337*
Arthritis, 346
Artificial selection, 102, **134**, **184**, 242, 243, *243*, 249, 257
Artificial silk. See Rayon
Artocarpus altilis (breadfruit), 255
Arundinaria amabilis (bamboo), 335
Arundo donax (reeds), 325
Asclepius spp. (milkweeds), 86, 123, *124*, 381
Ascocarps, **427**, *429*
Ascogonia, **427**
Ascomycetes (Ascomycota), **420**, **427**, *429*
Ascorbic acid, **164**
Ascospores, **427**, *428*
Ascus, **427**, *428*
-ase suffix, **56**

Asexual reproduction, **6**, 27
  budding, 427
  in diatoms, 404
  in fungi, 424
  in seed potatoes, 231
Ash (*Fraxinus excelsior*), 86, 496
Asia, early agriculture in, 182-183
Asian-Indian tree (*Strychnos nux-vomica*), 379
Asian wasps (*Tamarixia radiata*), 97
Asparagine, *10*, *157*
Asparagus (*Asparagus officinalis*), 45-46
Asparagus officinalis (Asparagus), 45-46
Aspartic acid, *10*, *157*
Aspergilloma, 482
Aspergillosis, 472, 482
Aspergillus, *423*, 430, *431*, 457, 473-474, 480, 482, 483
Aspergillus flavus, 473, 474, *474*
Aspergillus fumigatus, 482
Aspergillus niger, 56
Aspergillus terreus, 473
Aspirin, 345-346
Asteraceae (sunflower family), *128*, *129*, 306, 348, 386, 461, 501
Asthma, 155, 390-391, 398, 414, 480, 482, 483
Astragalus spp. (locoweed), 386
Atherosclerosis, 161, 372, 463
Athlete's foot, 481
Atmosphere
  carbon cycle and, 492-497
  carbon dioxide and, 61
  global warming and, 213, 496-497, 505
  nitrogen cycle, 212-213
  photosynthesis as source of oxygen, 2, 61
Atomic mass, **513**
Atomic nucleus, **513**
Atomic number, **513**
Atoms, **513**, *514*
Atorvastatin, 473, *473*
ATP (adenosine triphosphate), **55-56**
ATP synthase, **61**
Atropa belladonna (belladonna), *124*, 344, 367
Atropine, *344*, 367
Author citation, plant names and, 126
Autotrophs, **137**, **488**
Autumn crocus (*Colchicum autumnale*), 344
Auxins, **92-93**
Avena sativa (oats), 204, 248
Averrhoa carambola (carambola), 99
Avicenna, 338
Awns, **188**
Axillary bud, 38
Axils, **38**
Ayurvedic medicine, 338
Azadirachta indica (margosa), 389
Azaleas, 383

Azolla, 145, 201, *201,* 212
Aztecs, 370, 376
    amaranth ad, 252-253
    cocoa beans and, 283
    dyes used by, 321
    medicinal plants use by, 338
    psychogenic mushrooms and, 479-480
    spirullina and, 409
    vanilla use by, 300

B

B_1 (thiamine), *162,* **165**
B_2 (riboflavin), *162*
B_6 (pyroxidine), *162*
B_{12} (cobalamin), *162,* **166**
Bachelor's buttons, 125
Bacillariophyta (diatoms), 401, *402, 404, 404*
Bacillus thuringiensis (Bt), 260
Bacteria, **137**
    actinomycetes, 471
    antibiotic-resistant, 471-472
    archaebacteria, 137
    Endosymbiont Theory and, 22
    eubacteria, 137
    nitrogen-fixation by, 211, *211,* 402
    prokaryotic cells of, 8
    recombinant DNA technology and, 118
Badianus Manuscript, 338
Bagasse, 333, *334*
Bahia grass, *206*
Baker's yeast (*Saccharomyces cerevisiae*), 300, 447, *447*
Bald cypress (*Taxodium distichum*), 35, 37, 328
Bald eagles, DDT and, 491, *492*
Balick, Michael, 342
Balsa wood (*Ochroma pyramidale*), 327
Bamboo, 334-335, *335*
Banack, Sandra, 149
Banana family (Musaceae), 226
Banana Republics, 226
Bananas (*Musa* spp.), 93, 98, 226
Barbed wire, *7*
Barcode of Life Data (BOLD), 132
Barcodes, DNA, 132
Bark beetles, 443
Bark cloth, 319-320, *320*
Barley (*Hordeum vulgare*), 188, *203, 204,* 248
    beermaking and, 455
    celiac disease and, 158, 159
    domestication of, 184, *184*
    fibrous root system, *35*
    global production of, *188*
    tritordeum and, 204
Barley malt, 455
Barthlott, Wilhelm, 41
Base pairing, 111
Basidiomycota (basidiomycetes), *420,* 424, 430, 435, 465
Basidiospores, **430,** *433,* **439,** 482

Index   535

Basidium, **430**
Basil (*Ocimum basilicum*), *302,* 303
Basketmaking, 310, *311*
Basophils, 390
Bassham, James, 61
Bast fibers, **309,** 315, 316-317
Bateson, William, 109
Bats, 429, *430*
Baycol, *473,* 474
Bayer Company, 346, 361
Bayer Pharmaceuticals, 474
Bay leaves (*Laurus nobilis*), 302
B-complex vitamins, **162,** 165-167
Beadle, George W., 114, 197
Beagle, Darwin's voyage on, 133-134
Bean (Fabaceae), *128,* 214-215
    common edible, *214*
    germination of, 88, *89*
    new varieties of, 222-223
    nutritional value of, 215
    seeds of, 88, *89*
Bean curd (tofu), 218
The Beatles, 477
Beech (*Fagus*), 494
Beer, 4, 9, 454-457
    alcoholic fermentation and, 455, *456,* 457
    brewing process, 454, *456*
    fermentation of, 67, 455, 457
    lagering of, 455, 457
    origins of, 454-455
Bees, 81, *81*
Beets. *See* Sugar beet (*Beta vulgaris*)
Belladonna (*Atropa belladonna*), 124, *344,* 367
Benign prostate enlargement (BPE), 356, *356*
Benson, Andrew, 61
Bent grass (*Agrostis tenuis*), 134
Bergamots (*Citrus bergamia*), 95
Beribieri, **165**
Berlioz, Hector, 360
Bermuda grass (*Cynodon dactylis*), 205, *206,* 394
Berry, **86**
Beta-carotene, 44, **163**
Beta-glucans, 482
Beverages
    absinthe, 461
    beer, 4, 454-457
    caffeine content of, *272*
    chocolate, 283-287
    cider, 461
    coffee, 273-279
    distilled spirits, 458
    guarana, 287
    mate, 287
    sake, 457
    tea, 279-282
    tequila, 459-460
    wine, 447-454
Beyer, Peter, 262
Bhang (*Cannabis*), 363. *See also* Marijuana
Bibb lettuce, 44

Bible, plants mentioned in, 94
Biennial plants, **41**
Biennials, 92
Bile, 159
Bill and Melinda Gates Foundation, 262
Bill of Rights, 375
Bilobalide, *344*
Binomial system of nomenclature, **123,** *123,* 126
Biodiesel fuels, **220**
Biodiversity, **131.** *See also* Genetic diversity
Bioethanol, 206-208
Biofuels, 54, 206, 207, 410-411
Biofungicides, 286
Biogeochemical cycling
    greenhouse effect and, 494-496
    nitrogen cycle, 213
Biological controls, 131, 246, 286, 416
Biological magnification, 149, 150, **491,** *492*
Biological measurements, *17*
Biological mimics, 6, *7*
Biological species concept, **130**
Biomass, **489**
    ecological pyramid of, **489,** 491
Biomes, 500-511, *503*
    boreal forest, 505
    chaparral, 504
    deciduous forest, 505-506
    desert, 501-504
    distribution of, 500, *500*
    grassland, 504-505
    moist coniferous, 505
    taiga, 505
    temperate rain forests, 505
    tropical rainforests. *See* Tropical rainforests
    tundra, 505
Biomonitoring, 434
Bioremediation, 435
Biosphere, **487**
Biotechnology, **254**-257.
    See Genetically modified (GM) plants; Genetic engineering
    barcoding of species, 132
    cell and tissue culture, 256-257
    genome maps, 112
    molecular plant breeding, 257
    recombinant DNA, 118, *119*
Biotic factors, **487**
Biotin, 162
Bipolaris maydis (Southern leaf blight), 197, 438, 441
Birds, pollination by, 80
Bischofia javanica, 320
Black ash (*Fraxinus nigra*), 310
Blackberries, 86, *87*
Black cohosh (*Actaea racemosa*), 355
Black cottonwood (*Populus trichocarpa*), 114
Black-eyed peas (*Vigna unguiculata*), 214

Blackfoot tribe, 370
Black French roasts, coffee, 277
Black henbane (*Hyoscyamus niger*), *368*
Black Ivory Coffee, 278
Black locust (*Robinia pseudoacacia*), 211, 384
Black pepper (*Piper nigrum*), *294,* 294-295
Black pod rot (*Phytophthora* spp.), 286
Black Sigatoka (*Mycosphaerella fijiensis*), 227
Black teas, 281
Black turtle beans (*Phaseolus vulgaris*), 214
Black walnut (*Juglans nigra*), *322, 328,* 329, 383
Bladderpod (*Lesquerella* spp.), 219
Bladder wrack, *4*
Blade (algae), **405**
Blade (plant), **37**
BLAST search, 113
Bligh, William, 255
Blindness, preventing with golden rice, 262
Blood glucose, glycemic index and, 171, 237
Bloodroot (*Sanguinaria canadensis*), 340, *341*
Blood type, multiple alleles and, 109
Blue Agave, 459
Bluebell, *124*
Blueberries, 86
Blue-green algae. *See* Cyanobacteria
B-lymphocyte, **390**
BMAA (beta-methylamino-alanine), 148, 149, 150
Bock, Jerome, 338
Body mass index (BMI), **155**
Boehmeria nivea (ramie), *309,* 316
Bog people, 142
Bogs, 142-143
Bollgard, 315
Bollworm, 315
Bonding, 514-515
Boneset, *124*
Bordeaux mixture, 450
Boreal forests, **505**
Borlaug, Norman, 243-244
Boston lettuce, 44
Boston Tea Party, 282
Botanical insecticides, 388
Botany, *5*
    ethnobotany, *5,* 510
    father of (Theophrastus), 121
    forensic, 178
    scientific method and, 5-6
    subdisciplines of, *5*
    supermarket botany, 44
    taxonomy and. *See* Taxonomy
Bounty, HMS, 255, *255*
Bourbon, 458
Bournville, 285
Bows, string instrument, 324
Bracket fungi, 430, *432,* 435

Bracts, 72
Brake fern (*Pteris vittata*), 146
Bran, **188**
Branch roots, 35
Brandy, 458
Brassicaceae (Mustard family), 51, *128, 129*
Brassica alba, 302, 303-304
Brassica nigra, 302, 303-304
Brassica spp., 51, 243, *243, 302,* 303-304
Brazil
    bioethanol production, 206-207
    sustainable rainforest products, 508-510
Brazil nut tree (*Bertholletia excelsa*), *509, 509*
Bread, 4, 189
Breadfruit (*Artocarpus altilis*), 255
Bread making, 191
Bread wheat (*Triticum aestivum*), *187,* 191
Breakfast cereals, 199
Bristlecone pine (*Pinus aristata*), 146
Britain
    dry rot of ships of, 436-437
    Tea Act and American Revolution, 282
British East India Company, 282, 293, 361
British High Tea, 280
British opium trade, 361
Brittle bush (*Encelia farinosa*), 383
Broad bean (*Vicia faba*), 214-215
Broccoli (*Brassica oleraceae*), 172-173, 243, *243*
Bromberg, Walter, 364
Bromine, *167*
Broomcorn millet, 182
Brotzu, Giuseppe, 471
Brown algae (Phaeophyta), *4,* 401, *402,* 404-405, *405,* 409
Browning, Elizabeth Barrett, 360
Brown rice, 201
Brown, Robert, 102
Brown-rot fungi, 321, 436
Brown spot of rice, 437
Brunfels, Otto, 338
Brussels sprouts (*Brassica oleraceae*), 243, *243*
Bryology, 5
Bryophyta (mosses), 138, *138,* 139, 141
BT crop plants, 261-262
    BT corn, 261-262
    BT cotton, 267, 315
    BT potatoes, 261
BT insecticides, 232
Buckeye, 38
Buddhism, tea and, 280
Budding, **427**
Buds, **38**
Bulblet, *302,* **305**
Bulbs, **226**
Bulk flow, 51
Bundle sheath, 33

Burn plant (*Aloe vera*), 344, 349, 351, *351*
Burrs, 387
Buttercup, symbolic meaning of, 125
Butterflies, 80, *81,* 381
Button mushroom (*Agaricus bisporus*), 269, 465

C

C (vitamin), 44, 45, *162,* 299
C_3 pathway, **61**
C_4 pathway, **63**
Cabbage (*Brassica olearacea*), 40, 188, 243, *243*
Cacao tree. *See Theobroma cacao* (cacao tree)
Cactaceae, *122, 128,* 387
Cacti, 63, 387, 487
Cadbury, 283, 285, 287
Cadbury chocolate, 283
Cadbury, John, 283
Cadmium, 50, 51
Caffeine, 15, 272-273
    amount in common products, *272*
    medical benefits of, 273
    physiological effects of, 272-273
Caffeinism, 272
Caffe latte, 279
Caffe mocha, 279
Calamites, 144
Calcium, 50, *167,* **167**-168
Calcium ions (Ca^{++}), 514
Calcium oxalate, 22, 386
California black oak (*Quercus kelloggii*), 40
California chapparal, 383, 504
Callistemon citrinus (red bottlebrush), 384
Callisto, 384
Callus, **256**
Calories (C), **153**
Calvin Cycle, *58,* 61-62, *62*
Calvin, Melvin, 61
Calyx, **70**
CAM (Crassulaccean Acid Metabolism), **63**
Camellia, 124
Camellia sinensis, 279-282
Camphor, 291
Camptotheca acuminata (Chinese happy tree), 344, 353
Camptothecin, 353-354
Cancer, 155, 172-173, 352-354
Cancidas, 472
Candida albicans, 481
Candidatus liberibacter (Psyllids), 97
Candidiasis, 481
Candlenut oil tree (*Aleurites moluccana*), 320
Candy bars, 284
Canephora coffee (*Coffea canephora*), 277-278
Cankers, **443**
Cannabaceae, *128*
Cannabidiol (CBD), 365

Cannabinoids, 284, 364
Cannabis indica, 364
Cannabis sativa (marijuana, hemp), 362, 364
    chemistry of, 364
    fibers from, 309, 317, 333, 334
    in India, 363
    medicinal properties of, 363
    mind-altering effects of, 364
    paper from, 333-334
    psychoactive use of, 363
    tetrahydrocannabinol (THC) in, *344*
    three grades of, 363
    use in Africa, 363
    use in Middle East, 363
    use in Muslim world, 363
Canola oil, 160, 170, 219, *259,* 265
The Canon of Medicine, 338
Cappuccino, 279
Capsaicin, *298,* 298-299
Capsaicinoids, 298
Capsicum annuum, 294, 298
Capsicum baccatum, 294, 298
Capsicum chacoense, 298
Capsicum chinense, 294, 298
Capsicum frutescens, 294, 298
Capsicum peppers, 293, *294,* 297-299, *298*
Capsicum pubescens, 294, 298
Capsules (moss), **139**
Capsules (seed), **86**
Carambola (*Averrhoa carambola*), 99
Carapa guianensis, 506
Caraway (*Carum carvi*), *302,* 303
Carbohydrates, 8-9, **153,** 156
    as primary metabolites, 14
    structure of, *9*
Carbon cycle, 492-497
    cellular respiration and, 492
    global warming and, 496-497
    greenhouse effect and, 494-496
    photosynthesis and, 492, *493*
    sources and sinks, 492-494
Carbon dioxide
    cellular respiration and, 492
    fixation of, 63, 492
    global warming and, 251
    greenhouse effect and, 494-496, *495*
Carbon fixation, 63, 492
Carboniferous forests, *144,* 149
Carbon monoxide, 373
Carbon sinks, 492-493
Carbon sources, 493
Carboxylation, 62
Carcinogens, **172**
Cardamon (*Elettaria cardamomum*), *302*
Cardboard, 331
Cardioactive glycosides, 14-15, **343,** 381, 385
Cardiovascular disease, 155
Caribbean Islands, sugar and slave trade, 53
Carnivores, **488,** *490*

Carnivorous plants, 42, *43*
Carob (*Ceratonia siliqua*), 221
Carotenemia, 44
Carotenes, 20, **59**
Carotenoids, 14, **59**
Carpellate flowers, **73**
Carpels, **3,** 70
Carrageenan, 411
Carrion flower (*Stapelia* spp.), 80
Carroll, Lewis, 364
Carrots (*Daucus carota*), 41, 44
Carsonella ruddii, 22
Carson, Rachel, 491
Cartharanthus roseus (Madagascar periwinkle), 344, 349, 352
Carum carvi (caraway), *302,* 303
Carver, George Washington, 216, *216*
Caryopsis, **86**
Casparian strip, **35,** *36,* **50**
Caspofungin, 482
Cassava (*Manihot esculenta*), 15, 188, 234-236, *235*
Cassia (*Cinnamomum cassia*), 294
Castanea dentata (American chestnut), 97, 98, 442
Castilla elastica, 508
Castor bean (*Ricinis communis*), 385, *385*
Castor bean oil, 219, 385
Catabolic reactions, **55**
Catalina mahogany (*Cercocarpus traskiae*), 487
Catalysts, **56**
Catechins, 173, 281
Catkin, **74,** *75*
Cats, domestication of, 181
Cattails, *124*
Cattle, milk sickness in, 386-387
Caulerpa taxifolia, 415-416
Cauliflower (*Brassicea oleracea*), 172, 243, *243*
Caveman's Diet. *See* Paleo Diet
CDC (Centers for Disease Control and Prevention), 483
Cedar-apple rust, 441
Cedars (*Juniperus* spp.), 146, *147,* 341
Ceiba pentandra (kapok), *309,* 318
Celera Genomics, 112
Celery (*Apium graveolens*), *302,* 303
Celiac disease, 158-159
Cell culture, 256-257
Cell cycle, 24
Cell division, **24**-27
Cellophane, 331
Cell plate, **27**
Cells
    as basic unit of life, 8, 17
    cell wall of, **19**
    collenchyma, 31, *31,* 33
    early studies of, 17
    electron micrograph of, *18*
    Endosymbiont Theory, 22
    eukaryotic, 8, *8,* 21, 22
    movement of materials into and out of, 19-20

organelles of, 8, *8*, 20-23
organization of, *8*
parenchyma, 31, *31*, *33*
plasma membrane of, 19
prokaryotic, 8, *8*
sclerenchyma, 31, *31*, *33*
size of plant, *21*
Cell sap, 22
Cell Theory, 6, **17**
Cellular respiration, **7**, 63-68
    aerobic, 67-68
    anaerobic, 67-68
    carbon cycle and, 492
    electron transport system, 63, 67
    equation for, 63
    glycolysis, 63, 64, *65*
    in human digestion, 153
    Krebs cycle, 63, 64, *66*
    mitochondria and, 21-22, 64, *66*, *67*
    net ATP produced by, *66*
    overview of, 63
    photosynthesis *vs.*, 63
Cellulose, **8**, *9*, 19
    breakdown of, by fungi, 436
    as dietary fiber, 156
    ethanol from, 207-208
    identification of gene coding for, 309
Cellulose acetate, 331
Cellulosic ethanol, 207-208
Cell wall, **19**, *24*
    in fungi, 420
    in plants, 19
Cempra, 472
Centers for Disease Control and Prevention (CDC), 483
Centers of plant domestication, 184-185
Central America, origins of agriculture in, 183
Central Asia, origins of agriculture in, 180, 182
Central vacuole, **22**, *24*
Centric diatoms, 404
Centromeres, **25**, *25*
Cephaeline emetine
Cephalosporin, 471
Cephalosporium acremonium, 471
Ceratium, *403*
Cereals, **188**. *See also* Grasses (Poaceae)
Cereal straw, 334
Cerivastatin, 473, *473*
Cerocarpus traskiae (Catalina mahogany), 487
Chaetoceros, 417
Chaff, **188**
Chai, 280
Chain, Ernst, 470
Chamorro tribe, cycad toxins and, 148
Champagne, 453
Chapman, John (Johnny Appleseed), 94
Chapparal, 383, **504**
Charas (*Cannabis*), 363

Charcoal, 329
Chase, Martha, 111
Chaulmoogra (*Hydnocarpus* spp.), *344*
Cheeses, *124*, 430
Chemical bonds, **514**
Chemical energy, 55
Chemical reactions
    endergonic, 55
    enzymes and, 56
    exergonic, 55
    phosphorylation, 55-56
    redox, 55
Chemiosmosis, **66**, *67*
Chemotherapy drugs, 147, *344*, 353
Chenopodium quinoa (quinoa), 252, *252*
Cherimoya (*Annona cherimola*), 99, *99*
Chervil (*Anthriscus cereifolium*), 302, *303*
Chestnut (*Castanea* spp.), 86, 97-98, 442
Chestnut blight (*Cryphoectria parasitica*), 97-98, *98*, 442
Chewing gum, 509
Chewing tobacco, 372
Chiasmata, **76**
Chicha, 200
Chickpeas (*Cicer arietinum*), *214*, 215, *248*, 250
Chicle, 509
chicleros workers, 509
Chico tree (*Manikara zapota*), 509
Childhood obesity, 155
Child slave labor, chocolate cultivation and, 286-287
Chili peppers. *See* Capsicum peppers
China
    bamboo use in, 334
    marijuana and, 362-363
    medicinal plant use in, 338
    origins of agriculture in, *181*, 181-182
    origins of tea and, 279
    papermaking in, 333
    spice trade and, 290
    tea plantation, *281*
Chinese gooseberry. *See* Kiwifruit (*Actinidia chinensis*)
Chinese happy tree (*Camptotheca acuminata*), *344*, 353
Chitin, 9, **424**
Chives (*Allium schoenoprasum*), 302, 304, *305*
Chlamydomonas spp., 408, *408*, 410
Chloramphenicol, 471
Chlordane, 491
Chlorella, 410
Chlorine, *167*, 314, 331
Chlorine atoms, 514
Chlorophyll *a*, 59
Chlorophyll *b*, 59
Chlorophylls, 20, **57**, 401

Chlorophyta (green algae), *402*, 405-408
    lichens and. *See* Lichens
    life cycle of, 406, *408*
    pharmaceuticals from, 413
Chloroplasts, **20**, 22, *24*, 57-59
Chloroquine, 348
Chlorosis, 50, *51*
Chocolate, 283-287
    candy, 284
    child slave labor and, 286-287
    cultivation and processing of, 285-287
    health benefits of, 284
    history of, 283-285
    Quakers and, 283
Chocolatl, 283
Cholesterol, 12, *12*, **160**-161, 305, 463, 473
Chondrodendron tomentosum (Pareira), *344*, 379
Christian, Fletcher, 255
Christianity, 375
Chromatids, **25**, 27
Chromatin, **23**, 27
Chromium, 51, *167*
Chromoplasts, **21**, *24*
Chromosomes, **25**, *25*
    crossing over of, 76, 110, *111*
    homologous, 75, **102**
    linkages, 109-110
Chronic bronchitis, 372
Chronic obstructive pulmonary disease, 372
Church of England, 283
Chytrids (Chyrtidiomycota), *420*, 424
CIAT (Centro International de Agricultura Tropical), 244
Cicer arietinum (chickpeas), *214*, 215, *248*
Cicuta spp., 380, *380*
Cicutoxin, 380
CIFOR, 244
Cigarettes, 172. *See also* Tobacco (*Nicotiana* spp.)
Cilantro (*Coriandrum sativum*), 302, *303*
CIMMYT (Centro International de Mejoraminti de Maiz y Trigo), 243, *244*
Cinchona spp. (fever bark tree), *344*, 346, 348
Cinnamon (*Cinnamomum zeylanicum*), 293, *294*
CIP (Centro International de la Papa), 244
Cirrhosis, 464
Cisgenesis, **257**
Citrate, 64
Citric Acid Cycle, **64**. *See also* Krebs cycle
Citrons (*Citrus medica*), 95-96
Citrus aurantifolia (limes), 95, 96
Citrus aurantium (sour oranges), 95, 96
Citrus bergamia (bergamots), 95

Citrus canker, 269
Citrus family (Rutaceae), 95-96
Citrus grandis (pummelos), 95, 96
Citrus greening disease, 97
Citrus medica (citrons), 95-96
Citrus reticulata (tangerines), 95
Citrus sinensis (sweet orange), 95, 96
Citrus spp., 86
Civil War, 361
    cotton and, 313
Clades, **130**, 132
Cladophylls, **45**, *45*
Cladosporium spp., 430, 474, 482, 483, *483*
Clarification, 452
Classes (taxonomic), **127**
Classification. *See also* Taxonomy
    early history of, 121-123
    Linnaeus' sexual system of, 121-123, *122*
Claviceps purpurea, 475, *475*, 484
Clear-cutting techniques, 506-507, *507*, 511
Clifford, George, 121
Climate change, 394, 454, 494-497, 505
Climax community, **498**
Clones, **27**
Cloves (*Eugenia caryophyllata*), 294, 295, *295*
Club mosses (Lycophyta), *138*, 139, 143, 144, *144*
Clustered regularly interspaced short palindromic repeats/Cas9 (CRISPR/Cas9) system, 268-269
$C_nH_{2n}O_n$, 8
Coast rose gentian (*Sabatia arenicola*), 73
Cobalamin, *162*, **166**
Cobalt, 51, *167*
Coca (*Erythroxylum coca*), *344*. *See also* Cocaine
Coca-Cola, 272, 287, 366-368
Cocaine, 15, 343, *344*, 359, 366-370
    and Coca-Cola, 366-368
    coke and crack, 369
    deadly drug, 369-370
    Freud and, 366-368
    Holmes and, 366-368
    medical uses, 369
    South American origins, 366
Cocaine hydrochloride, 369
Coca-wine, 367
Coccidioides immitis, 481, *482*
Coccidioidomycosis, 481
Cochineal red, 321
Cockleburs, 6, *7*
Cocoa butter, 160, 284, 286
Coconuts (*Cocos nucifera*), 160, *188*, 219, *309*, 319
Cocoyam (*Xanthosoma* spp.), 238
Codeine, 15, *344*, 361
Codium fragile, 417
Codominance, **109**
Codons, **115**

Coenzyme A, 64
Coenzymes, **56**, 162
Coevolution, **80**–81
Cofactors, **56**, 167
Coffea arabica, *271*, 273, *274*, 276, 277
Coffea canephora (*robusta*), 276, 277, 278
Coffea liberica, 278
Coffee (*Caffea* spp.), 273–279
    caffeine in, *272*
    cherries, 275, *277*
    and climate change, 276
    cultivation of, 274–277
    decaffeination of, 278–279
    Fair Trade movement, 279
    germplasm ownership and, 251
    instant, 278
    origins of, 273–274
    plantations of, 274–275
    processing of, 274–277
    roasting of, 277
    shade *vs.* sun grown, 279
    specialty drinks, 279
    varieties of, 277–278
Coffee family (Rubiaceae), 128
Coffee rust (*Hemileia vastatrix*), 275, 439
Cohesion, **51**
COI gene, 132
Coiled baskets, 311, *311*
Coir, *309*, 319, *319*
Cola nitida (kola tree), 287
Colchicine, 257, *344*
Colchicum autumnale (Autumn crocus), *344*
Coleoptile, **88**
Coleorhiza, **88**
Coleridge, Samuel, 360
Collagen, 164, 167
Collenchyma, **31**, *31*, *33*
Colocasia esculenta (taro), 238, *239*
Colony collapse disorder, **81**
Colorado potato beetle, 232
Columbus, Christopher, 53, 283, 292, 297, 370
Column distillation, 458
Commercial mushroom (*Agaricus bisporus*), *432*
Common names, 123–124
Communities, **487**
Companion cell, **33**
Competition, natural selection and, 134
Competitive exclusion, **487**
Complete flowers, **72**
Complete Herbal, The, 338
Complete proteins, **157**
Complex carbohydrates, 153, 156
Composting, 206
Compound leaves, **38**
Concentration gradient, **20**
Conceptacles, **405**
Conching, 286
Confessions of an English Opium-Eater (De Quincey), 360

Conidia, *420*
Coniferophyta, *138*, 139, 146. See also Conifers
Coniferous forests, 505
Conifers, 146–147
    allergies to pollen of, 393–394
    amber from sap of, 148
    boreal forests and, 505
    economic importance of, 148
    life cycle of, *139*
    resins from, 330
Conium maculatum, 380, *380*
Conjugated linoleic acid, 161
Conocybe mushrooms, 477, 480
Consortium for the Barcode of Life (CBOL), 132
Consumers, 2, **488**, 489
Contact dermatitis, 389, 395–397
Contraceptives, 338, *339*
Contractile roots, **37**
Convallotoxin, 384–385
Convolvulaceae (morning glory family), 232, 359, 376, 508
Cook, James, 255
Copaiba (*Copaifera reticulata*), 506
Copaifera (*Copaifera officinalis*), *221*
Copenhagen Accord, 497
Copper, 50, 51, *167*
Copper fungicides, 450
Coppice (short rotation), *507*, 511
Coprolites, **176**–177
Corchorus spp., *309*, 317
Cordage fibers, 309
Cordus, Valerius, 338
Coreopsis (*Coreopsis* spp.), *124*, *322*
Coriander (*Coriandrum sativum*), *302*, *303*
Coriandrum sativum (cilantro), *302*, *303*
Coriandrum sativum (coriander), *302*, *303*
Cork, **330**
Cork cambium, **29**, *29*, 330
Cork oak (*Quercus suber*), 330, *330*
Corms, **226**
Corn (*Zea mays*), 37, 242, *243*
    ancestry of, 197–199
    bioethanol from, 206, 220
    BT corn, 261–262
    C_4 pathway and, 63
    corn smut and, 441
    cultivation requirements, 194–195
    domestication of, 183
    economic importance, 199–200, *200*
    flowers of, 194, *194*
    genome of, 114, 199
    germination and development, 88, *89*
    global production of, *188*
    grains of, 194
    hybrid, 197
    jumping genes, 196, 199
    kernel color, 195, 196
    nutritional value, 199–200
    prop roots of, 37
    scientific name, 126
    seed structure, 88, *89*
    StarLink, 262, 398
    stems anatomy, 33
    stomata on leaves of, *40*
    tassels of, 194
    transpiration by, 49
    types of, 194–197, *195*
Corn flour, 199
Corn meal, 199
Corn oil, 160, 199, 219
Corn smut (*Bipolaris maydis*), 441
Corn starch, 188, 199
Corn syrup, 199
Corolla, **70**
Coronary heart disease (CHD). See Heart disease
Cortes, Hernan, 283, 298, 300
Cortex, **33**
Cortisone, 238
Cos lettuce, 44
Cotton (*Gossypium hirsutum*, *G. barbadense*), *309*, 313
    BT cotton, 267, 315
    finishing and sizing, 314–315
    fruits of, 86
    ginning of, 310, 313–314
    global production of, *188*
    origins of, 312
    varieties of, 313
Cotton gin, 313, *314*
Cottonseed oil, 219, 314
Cotyledons, **88**, **211**
Covalent bonds, **514**, *515*
Cox, Paul, 149, 342
Cranberry (*Vaccinium macrocarpon*), 355
Crassulaccean Acid Metabolism (CAM), **63**
CREB, 359
Creosote, 437
Crepe myrtle, *124*
Crestor, 473
Crick, Francis, 111
Crimean War, 371
Criollo, 285
Crinipellis perniciosa, 286
CRISPR-Cas9 (Clustered, Regularly Interspaced Short Palindromic Repeats-Cas9) system, 268–269
Cristae, **21**, 64
Crocus sativus (saffron), 294, *297*, *297*
Crop rotation, nitrogen-fixing bacteria and, 212
Crops. See also Agriculture; Food plants; *specific crop plants*
    artificial selection of, 134, *184*, 243, *243*, 249
    disease-resistant, 244–245
    genetic erosion in, 249
    global production of, *188*
    global warming and, 251
    Green Revolution and, 244–245
    heirloom varieties, 250
    herbicide resistant, 118, 266–268
    insect resistant, 268
    monocultures of, 247–248
    world production of, *188*
Crossing over, **76**, **110**, *111*
Cross-pollination, **79**
Crotalaria, 381
Crown-of-thorns (*Euphorbia pulcherrima*), 385
Crow tribe, 370
Crude opium, *360*
Crushing, 449
Crustose lichen, 434
Cryphoectria parasitica, 97–98, *98*
Cube root (*Lonchocarpus*), 389
Cucumbers, 86, 90, 92
Cucurbitaceae, *128*
Culms, **188**
Culpeper, Nicholas, 338
Cultivars, **90**–91
Cumin (*Cuminum cyminum*), *302*, *303*
Cuminum cyminum (Cumin), *302*, *303*
Curare, 379–380
Curcuma longa (turmeric), *294*, 296–297, 320, *321*, *322*
Cuticle, **30**
Cutin, **29**
Cyamopsis tetragonoloba, 192
Cyanobacteria, 402–403
    biofuels from, 411
    classification of, 137, 150
    foods from, 408
    nitrogen-fixation by, 212, *213*
    toxic, 412–414
Cyanogenic glycosides, 15, **235**, **343**
Cyanogens, 388
Cycadophyta, *138*, 146, 147
Cycasin, 148
Cyclic photophosphorylation, **61**
Cyclins, **24**
Cyclin-dependent Kinases, 24
Cyclosporin, 473
Cynodon dactylis (Bermuda grass), 205, *206*, 394
Cyperus papyrus (papyrus), 331, *332*
Cyphomandra betacea (tamarillo), 254
Cysteine, 10, 157
Cytisine, 384
Cytochrome complex, 61
Cytochromes, **64**
Cytokinesis, **24**, 27, 76
Cytokinins, 92
Cytoplasm, **19**
Cytoplasmic inheritance, 215
Cytosine, 12–13
Cytoskeleton, **19**, 24
Cytosol, 19

D

D (vitamin), **162**, *162*, 163–164
Daffodil, 125, 384
da Gama, Vasco, 292

Daikon (*Raphanus sativa*), 45
Daily reference values (DRVs), 165n
Daily values, 165n
Daisy, 125
Daisy fleabane, *124*
Dakota burl, 334
Dalton (atomic weight unit), **513**
Dandelion (*Taraxacum officinale*), 37
Dark brown Vienna roasts, coffee, 277
Dark reactions, 61. *See also* Calvin Cycle
Dark roasts, coffee, 277
Darwin, Charles, 81, 92, *120*, 130, *133*, 134
Darwin, Erasmus, 133
Darwin, Francis, 92
Darwin, Robert, 133
Datura spp., 367
Datura stramonium, 367, *368*
Daucus carota (carrots), 41, 44
Daylily, *124*
DDT (dichloro-diphenyl-trichloroethane), 491, *492*
Deadly nightshade. *See* Belladonna (*Atropa belladonna*)
de Albuquerque, Alfonso, 293
Death cap (*Amanita phalloides*), *469*, 478
Decaffeinated coffee, 278-279
de Candolle, A. P., 383
Decarboxylation, 64, *66*
Deciduous forests, *503*, **505**-506
Decomposers, **4**, 433, 435, **488**
Decortication, **310**
Defenses. *See* Plant defenses
Deforestation, 329, 333, 506
Dehiscent fruits, 86
De Historia Stirpium Comentarii Insignes, 339
Deletion mutations, **118**
delta-9-tetrahydrocannabinol (THC), 364-365
De Materia Medica, 121, 338
De Medici, Catherine, 379
de Mestral, George, 6
Dendrochronology, 37
Dendroclimatology, 37
Dent corn, 195, *195*
Deoxyribonucleic acid (DNA). *See* DNA (deoxyribonucleic acid)
Deoxyribose sugar, **12**, **13**, 111
Department of Health and Human Services, 369
2002 National Survey on Drug Use and Health, 369
Depressants, 15, **359**
Depression, St. John's wort and, 355
Depulping, coffee berries, 275, 277
De Quincey, Thomas, 360
dermal tissues, **29**-31, *30*, 34
Dermatophytes, 480-481
Desert biome, **501**-504
Dessert wines, 454

Destroying angel (*Amanita virosa*), 478, *478*
Detritus, **488**
Detritus food chain, 488, *489*
Detrivores, 488
Deuteromycetes, 429
Developing countries
 germplasm ownership and, 250-251
 malnutrition and undernutrition in, 155
Dhatura, 367
Diabetes, 161-162, *344*, 350-352
Dias, Bartholomew, 292
Diatomaceous earth, **411**
Diatomite, 411
Diatoms (Bacillariophyta), 401, *402*, 404, *404*
Diaz, Bernal, 300
Dicke, Willem-Karel
Dicots, **3**
 floral structure, 72, *73*
 germination in, 88
 leaf venation, 38, *39*
 roots anatomy, 35-37
 seed structure, 88, *89*
 stems anatomy, 33
Dieffenbachia spp. (dumb cane), 22, 386
Diesel oil, 220. *See also* Biodiesel fuels
Diesel, Rudolf, 220
Diet. *See also* Human nutrition
 cancer prevention through, 172-173
 dietary fiber in, 156
 energy requirements, 153
 guidelines for, 169-171
 malnutrition and, 154
 overnutrition and, 154
 Paleo, 177, 179
 undernutrition, 154
 vegetarian, 172-173
Dietary fiber, 156
Dietary Goals for the United States, 169
Dietary guidelines, 169-171
Dietary Supplement Health Education Act (DSHEA), 354
Dietary supplements, 354
Differentially permeable membranes, **20**
Diffusion, **19**-20
Digitalis spp. (foxglove), 343-345, *345*
Digitoxin, 15, 344
Digoxin, 344
Dihybrid crosses, **105**-106
Dihydrotestosterone (DHT), 356
Dihydroxyacetone phosphate, 64
Dikaryons, **423**
Dill (*Anathum gaveolens*), *301*, *302*, 303
Dinoflagellates (Dinophyta), 401, *402*, 403-404
Dinophyta (dinoflagellates), 401, *402*, 403-404

Dioecious plants, **73**
Dioscorea spp. (yams), *188*, 238, *238*, 338, 343, *344*
Dioscorides, 121, 338, 355, 360
Diosgenin, 343, *344*
Diploid (2n), **75**, *76*
Diploid generation, 139
Direct dyes, 321
Disaccharides, **8**, **153**
Disease-resistant crops, 244-245, 263-264
Diseases. *See* Plant diseases
Distillation, 82
Dithiolthiones, 172
Divisions (phyla), **127**
DNA (deoxyribonucleic acid), 12-13, *13*, 102
 barcoding of species, 132
 CRISPR-Cas9, 268-269
 discovery of, 111, 114
 extraction of from fossil amber, 149
 genetic engineering and. *See* Genetic engineering
 genome maps, 112
 mapping, 112
 mutations in, 117-118
 recombinant DNA technology, 118, *119*
 semiconservative replication of, *115*
 sequencing, 112
 structure of, 115
 transcription of, 114-115
Doctrine of Signatures, **339**, 350, 367
Dogs, domestication of, 182
Dogwood flowers, 74, 125
Domesticated plants, **184**
 artificial selection of, 102, 184, 242
 characteristics of, 184
 origins of agriculture and, 184-185
Domestication syndrome, 181
Dominance
 incomplete, 106, *106*, 109
 principle of, 104
Dominant traits, **104**
Doomsday Vault, 249
Dopamine, 359
Dormancy, 93
Double fertilization, 83-**84**
Double helix, **12**, *13*, 114
Douglas, David, *124*
Douglas fir (*Pseudotsuga menziesii*), *124*, 328, *328*
Douglass, Andrew, 37
Downy mildew of grape (*Plasmopara viticola*), 450
Doyle, Arthur Conan, 368
Drake, Francis, 371
Drip irrigation, 247
Dropsy, 343-344
Drosera spp. (sundews), 42, *43*
Drosophila spp. (fruit flies), 109, 110

Drought-tolerant plants, 49
Drugs. *See also* Herbal medicine; Medicinal plants; Psychoactive plants
 from algae, 413
 antibiotics, 470-472, 473-474
 anticoagulants, 413
 antihistamines, 395
 anti-inflammatories, 346
 antimalarials, 348, 349
 caffeine in over-the-counter medications, *272*
 cancer therapies, 173, 352-353
 cyclosporin, 473
 grapefruit juice-drug interactions, 97
 molecular farming of, 264
 resistance by bacteria, 472
 statins, *473*, **473**-474
 from tropical rainforests, 506
Drupes, **86**
Dry fruits, **86**
Drying oils, **219**
Dryland agriculture, 247
Dryopteris (wood fern), *145*
Dry rot, 436-437
Dulse, 409
Dumb cane (*Dieffenbachia* spp.), 22, 386
Dump sites, origins of agriculture and, 177, 179
Dunaliella, 410
Duran, Diego, 479
Durians (*Durio zibethinus*), 99
Durio zibethinus (durians), 99
Durum wheat (*Triticum durum*), 190, 191, 204
Dusty miller, *124*
Dutch elm disease, 265-266, 427, 441-443
Dutchman's breeches, *124*
Dyes, 14, 145, 320, 321-322, 434

E

E (vitamin), **162**, *162*
Earl Grey tea, 281
Early wood, **34**
Ebers Papyrus, 290, 305, 338, 349, 352, 379
Echinacea (*Echinacea purpurea*), 340, 341, 355
Echinacea purpurea (purple coneflower), 69
Echinocandins, 472, 482
Echinopsis chamaecereus (peanut cactus), *122*
e-cigarettes, 374
Ecological equivalents, **487**
Ecological niches, **487**-488
Ecological pyramids, **489**-491, *491*
Ecological species concept, **130**
Ecological succession, **497**-498, *499*

Ecology, 487
  biogeochemical cycling, 213, 492-497
  biomes, 487-511
  ecosystems, 487-492
  levels of study, 487, *487*
Economic botany, 5
Ecosystems, **487-492**
  abiotic factors, 487
  biological magnification in, 149, 150, 491, *492*
  biomes, 487-511
  biotic factors, 487
  communities in, 487
  ecological niche, 487-488
  ecological pyramids and, 489-491
  energy flow in, 489-491
  food chains and food webs, 487, *489*, 490
  global warming and, 213, 494-497, 505
  niches, 487-488
  populations in, 487
  succession in, 487-498, *499*
Ectomycorrhizae, **436**
Edamame, 218
Edible mushrooms, 465-466, *466*
Edison, Thomas, 367
Egg, **6**, 75, **79**
Egg apparatus, 79
Egypt (ancient)
  basketmaking and, 310
  beermaking and, 454
  garlic and, 305
  leavened bread in, 192
  linen fabric in, 315
  medicinal plant use in, 338, 346
  papyrus in, 331, *332*
  spice trade and, 280
  winemaking in, 447
Eichhornia crassipes (water hyacinth), 487
Einkorn wheat (*Triticum monococcum*), 190
Elaisome, **88**
Electrical energy, 55
Electromagnetic radiation, **56**
Electromagnetic spectrum, 56
Electronic cigarettes. *See* e-cigarettes
Electron microscopes, *18*
Electrons, **513**
Electron Transport System, **63**, 67
Elements, **513**
Elettaria cardamomum (Cardamon), *302*
Elkhorn coral (*Acropora palmata*), 495-496
Elk, poisoning of by lichen, 381-382
Ellis, Havelock, 374
Elm (*Ulmus americana*), 38, 86, *87*, 125, 441-443
Elm bark beetles, 441-443
Embrophytes, **137**, *138*
Embryo, **84**, 137
Embryo sac, **79**

Emmer wheat (*Triticum turgidum*), 190
Emphysema, 372
Encelia farinosa (brittle bush), 383
Endemic goiter, **168**
Endergonic reaction, **55**
Endocarp, **86**, *95*
Endodermis, **35**, 50
Endoplasmic reticulum (ER), **20**, 23, 24
Endorphins, 361
Endosperm, **84**, 88
Endosymbionts, 22
Endosymbiont Theory, 22
Endothelin, 463
Endotrophic mycorrhizae, **436**
Energy, **55**
  ecological pyramid of, 489, 490, *491*
  endergonic reactions and, 55
  exergonic reactions and, 55
  First Law of Thermodynamics and, 55, 489
  flow of in ecosystems, 489-491
  kinetic, 55, *55*
  potential, 55, *55*
  requirements for in human diet, 153
  Second Law of Thermodynamics and, 55, 489
Energy drinks, 272-273
Enfleurage, 82
Enology, **447**
Enquiries into Plants and the Causes of Plants (Theophrastus), 121
Enriched flour, *193*
Environmental Protection Agency (EPA), 261, 266, 269, 374
Enzymes, 56
Ephedra spp., 151, *344*, 351-352, 354
Ephedrine, 151, *344*, 351-352
Epicotyl, **88**
Epidermis, **29**, *30*, *33*
Epidermophyton spp., 480
Epigallocatechin, 173, 281
Epigallocatechin gallate (EGCG), 173
Epigenetic change, 118
Epigenetics, **118**
Epigynous flowers, **73**
Epiphytes, **37**, **143**
Equisetum spp. (horsetails), 144, *144*
ER. *See* Endoplasmic reticulum (ER)
Ergometrine, 477
Ergot alkaloids, 15
Ergotamine, 477
Ergotism, *475*, **475-477**
Ergot of rye, 475, *475*
Erwinia carotovora, 460
Erythromycin, 471
Erythroxylaceae (coca family), 359
Erythroxylum (*Erythroxylon*) *coca*, 366

Erythroxylum coca (coca), *344*. *See also* Cocaine
Erythroxylum novogranatense, 366
Escherichia coli, 257
Espresso, 279
Essential amino acids, **157**, *157*
Essential elements, 50
Essential fatty acids, 159
Essential oils, 14, **80**, **281**, **290**
  aromatherapy and, 291
  perfumes and, 82
  in tea, 281
Ethanol biofuel, 54, 206-208, 220
Ethnobotany, **5**, 510
Ethnomapping, 510
Ethylene, 92
Etrogs (*Citrus medica*), 95
Eubacteria, 137
Eucalyptol, *344*
Eucalyptus globulus, *344*, 349
Eucalyptus oil, 291
Eudicots, **132**
Eugenia caryophyllata (cloves), 294, 295, *295*
Euglena, 403
Euglenophyta (euglenoids), 401, 402, 403, *403*
Eukarya, 137
Eukaryotic cells, 8, *8*, 21, 22
Euphorbiaceae (spurge family), 51, *128*, 385
Euphorbs, 487
European honey bee (*Apis mellifera*), 383
Evolution, 6, 132-135
  Darwin's voyage on HMS *Beagle* and, 133-134
  mutations and, 118
  natural selection and, 134-135
  as property of living things, 7
Exergonic reactions, 55
Exine wall, **77**
Exocarp, **86**, *95*
Exons, **115**
Experimentation, 5
Extractive reserves, 510

F

F_1 (first filial) generation, **104**
F_2 (second filial) generation, **104**
Fabaceae. *See* Legumes (Fabaceae)
Factory towns, 285
Facultative parasites, 433
FAD (flavin adenine dinucleotide), 55
$FAD/FADH_2$, 55
Fagaceae, *128*
Fair trade coffee, 279
Fairtrade Labeling Organization (FLO), 279
Family (taxonomic), **127**
  economically important angiosperm, *128*
  naming conventions for, 127
  traditional and standardized names of common, *129*

Far East, early agricultue in, 182-183
Farinha, 235, *235*, 236
Farm Bill, 317
Father of Botany (Theophrastus), 121
Father of Taxonomy (Linneaus), 123
Fats, 12, *12*, 159-162
  human nutrition and, 159-160
  saturated *vs.* unsaturated, 159-160
Fat-soluble vitamins, **162**, *162*
Fatty acids, *12*, 159, 161-162
Favism, 215
FDA. *See* U.S. Food and Drug Association
Federal Bureau of Narcotics, 364
Federal Marijuana Tax Act of 1937, 364
Fell, J. W., 341
Female gametophyte
  in angiosperms, 79
  in conifers, 139
Fennel (*Foeniculum vulgare*), 302, 303
Fenugreek (*Trigonella foenum-graecum*), 221, 302
Fermentation, **67**, **192**
  beermaking and, 455, *456*, 457
  breadmaking and, 192
  equation for, 447
  foods from, 466-467
  of rice for sake, 457
  winemaking and, 449-452
Fermented foods, 466-467, *467*
Ferns (Pterophyta), 4, *138*, 139, 143-146
Ferredoxin (Fd), 60, 61
Fertile Crescent, 181
Fertilization, 75, 137, 139
Fertilizers, 211, 212, 245-246, *247*
Fescue (*Festuca rubra*), 384
Fetal alcohol syndrome (FAS), 465
Fever bark tree (*Cinchona* spp.), *344*, 346, 348
Feverwort, 124
Fibers, **31**, *31*, **156**, 309-320
  bark, 319-320
  basketmaking and, 310, *311*
  bast, 309, 316, 317
  coir, *309*, 319
  cotton, *309*, 312-315
  extraction of, 310
  from ferns and lycophytes, 145
  flax, *309*, 315-316, *316*
  hard, 309
  hemp, *309*, 317-318
  jute and burlap, *309*, 317
  kapok, *309*, 318
  leaf, 309, 317, 318, *318*
  manila hemp, *309*, 317-318
  pineapple cloth, 318
  ramie, *309*, 316
  rayon, 319
  sisal, *309*, 318, *318*
  soft, 309, 310

Index 541

spinning of, 310-312
surface, 309, 310, 318
types of, 309-312
Fibrils, 19
Fibrous roots, **35,** *35*
Fiddleheads, 145, *145*
Field mushroom (*Agaricus bisporus*), 465
Ficus, 74
Fig, 74
Filaments, **70**
Fire, 504
Firewood, 329
First Amendment, 375
First filial (F_1) generation, **104**
First Law of Thermodynamics, **55,** 489
Fish kills, 414
Fission yeast (*Schizosaccharomyces pombe*), 300
Five-kingdom classification system, 137
Fixation, carbon, 63, 492
Flagellum, **401**
Flash powder, *Lycopodium* spores as, 144
Flavonoids, 14, 281, **463**
Flax (*Linum usitatissimum*), 219, *309,* 315-316, *316,* 355
Fleming, Alexander, 470
Fleshy fruits, **86**
Flies, 80, 431
Flint corn, 195, *195*
Flora Lapponica (Linneaus), 121
Floral organs, 70, 72
Floral symmetry, 74, *74*
Floret, **188**
Florey, Howard, 470
Flowering plants. *See* Angiosperms
Flowers, 2, 3, 70-74
carpellate, 73
coevolution of animals with, 80
complete, 72
imperfect, 73
incomplete, 73, *74*
inflorescence types, 74, *75*
irregular, 74
modified, 72-74
monocot vs. dicot, 72, *73*
organs of, 70, 72
ovary position in, 72, 73
perfect, 73
plant hormones and, 92, 93
regular (actinomorphic), 74, *74*
staminate, 73
structure of, *70*
symbolic meaning of, 125
of wind pollinated plants, 79
Flower stalk (pedicel), 70
Fluid mosaic model, **19,** 20
Fluorine, 167
Fluvastatin, 473, *473*
Fly agaric (*Amanita muscaria*), 478-479, *479*
Foeniculum vulgare (Fennel), 302, 303
Foliar spraying, 97

Folic acid, **166**
Foliose lichens, 434
Folk botany, 5
Follicles, **86**
Food additives, 409
Food allergies, 390, 397-398
Food chains, **2, 488,** *489*
Food production, increasing
genetic engineering and, 258
Green Revolution and, 246
molecular plant breeding and, 257
new food crops, 252
selective breeding for, 242
Food Pyramid, USDA, 169
Food webs, **488,** *490*
Forage crops, 205, 221-222
Foragers, **176**-177
Forastero, 285
Forensic botany, 178
Forestry, 5
Forests, 505-506
boreal, 505
coniferous, 505
deciduous, *503,* 505-506
deforestation of, 329, 333, 506
harvesting trees from, 506-507, 510-511
sustainable extraction from, 508-510
tropical rainforests, 342, 506
Fortified foods, 166
Fortified wines, 454
Fortunella japonica (kumquats), 95
Fossil fuels, burning of, 212, 494-495, *495*
Fossils
amber, 149
coprolites, 176
DNA extraction from, 149
evolution and, 134
pollen, 79
Fourdrinier, Joseph, 333
Fourdrinier paper machine, *333*
Fox, George, 283
Foxglove (*Digitalis* spp.), 15, 343-345, *345*
Frame-shift mutations, **118**
Free radicals, 163
Freese, Friedrich, *124*
Freesia, *124*
Freeze-drying, 278
French lilac, *344,* 349
French Revolution, 476
Freud, Sigmund, 366-368
Fronds, 139, 145, *145*
Frosty pod rot (*Moniliophthora roreri*), 286
Fructose, **8, 153**
Fructose-1,6-biphosphate, 62, 64
Fruit flies (*Drosophila* spp.), 109, 110
Fruiting bodies, **424**
Fruits, 3, **84,** 86. *See also specific plants*
accessory, 86

aggregate, 86
dry, 86
layers of, 86
legal definitions of, 41, 90
multiple, 86
nutritional value of, 90
plant hormones and, 92, 93
simple, 86
Fruits and Veggies—More Matters™, 173
Frustule, **404**
Fruticose lichen, 434
Fry, Joseph, 283
Fuchs, Leonhart, 338
Fucus spp., *4,* 405, *406*
Fuel
from algae, 410-411
biodiesel, 220
ethanol, 54, 206-208, 220
fossil fuels, 212, 494
wood, 329
Fundamenta Botanica (Linneaus), 122
Fungi, 4, 137, 420-444
alcoholic beverages from, 447-462
allergies to, 482-483
antibiotics from, 470-472, 473-474
ascomycetes (Ascomycota), *420,* 424, 427, 429
asexual (deuteromycetes), 429
basidiomycetes (Basidiomycota), *420,* 424, 430-432, 465
cell wall of, *420*
characteristics of, 422-424
chytrids (Chytridiomycota), *420,* 424, *425*
classification of, 137, 424
decomposition by, 433, 435
dermatophytes, 480-481
dry rot and, 436-437
ergot of rye, 475, *475*
Facultative parasites, 433, 435
foods from, 465-468, *466-467*
and fungal-like organisms, 420
glomeromycota, *420, 425*
hallucinogenic, 478-480
indoor and toxic molds, 482-484
lichens and, 321, 381-382, 407, 434
mycorrhizae and, 435-437
mycotoxins from, 472-474
plant diseases caused by, 437-444
poisonous, 475-477
reproduction in, 424
systemic mycoses, 481-482
zygomycetes (Zygomycota), *420,* 426
Fungi (kingdom), 137, 420, 422-432
Fungicidal compounds, 450, 472
Fungi Imperfecti (deuteromycetes), 429
Fungistatic, 472

Furocoumarins, 381
Fusarium oxysporum, 460
Fusarium spp., 298, 467, 474, 480
Fusidic acid, 472
Fusidium coccineum, 472

G

G_1 phase, 24, *25*
G_2 phase, 24, *25*
Galactose, **153**
Galápagos Islands, *133,* 133-134
Galen, 360
Galerina spp., 477, 480
Galitoxin, 381
Galls, 321, 441
Gallstones, 155
Gametangia, 408, **422**
Gametes, **75,** 139, *143*
Gametophytes, **75,** *76, 139,* 139-141, *140, 143*
Gamma rays, 56
Ganga. *See Cannabis sativa* (marijuana, hemp)
Gangrene, 476
Ganja (*Cannabis*), 363
Ganoderma applanatum (artists bracket), *437*
Garbonzos. *See* Chickpeas (*Cicer arietinum*)
Garcinia cambogia (*Garcinia gummi-gutta*), 355
Garden pea (*Pisum sativum*), 102-103, *103,* 214-215
Garlic (*Allium sativum*), 302, 304-305
Garrod, Archibald, 114
Gas exchange, stomata and, 30, 49
Gattefosse, Rene, 291
Gaud, William, 245
Gaultheria procumbens (wintergreen), 345
Gay-Lussac, Joseph Louis, 447
Gee, Samual
GenBank, 113
Gene, 13, **102**
Genealogical species concept, **130**
Gene editing, 268-269
Gene gun, 258, 264
Generalized niche, 488
General Sherman tree (*Sequoiadendron giganteum*), 146
Genera Plantarum (Linneaus), 122
Generative cells, **77**
Genetically modified (GM) plants, **258**
allergies to, 261, 268
bananas, 262
disease-resistant, 263-264
with enhanced nutrition, 262
environmental and safety concerns, 266-268
Golden Rice, *247,* 262
herbicide-resistant crops, 258-260

Genetically modified (GM) plants (*continued*)
  insect-resistant crops, 260-261
  molecular farming of pharmaceuticals, 264-265
  for phytoremediation, 265
  potatoes, 263
  rapeseed, 265
  regulation of, 266
  soybeans, 262
  trees, 265-266
Genetic code, 114-**115**
Genetic diversity, 247-251
  crop monocultures and, 247-248
  genetic erosion and, 249
  heirloom varieties, 250
  lack of in Green Revolution varieties, 249
  preservation of in seed banks, 249-250
  sustainable agriculture and, 248-249
Genetic engineering, 257-268
  of disease resistance, 263-264
  environmental and safety concerns, 266-268
  of Golden Rice, 247, 262
  of herbicide resistance, 258-260
  of insect reistance, 260-261
  of nonfood crops, 265
  of pharmaceuticals, 264-265
  regulatory issues, 266
  of trees, 265-266
Genetic erosion, **249**
Genetics. *See also* Molecular genetics
  codominance, 109
  crossing over, 76, 110, *111*
  dihybrid crosses, 105-106
  incomplete dominance, 106, *106*, 109
  jumping genes, 196
  linkage, 109-110
  Mendelian, 102-110
  monohybrid crosses, 103-105, *104*
  multiple alleles, 109
  polygenic inheritance, 109
  principle of dominance, 104
  principle of independent assortment, 106
  principle of segregation, 104
  solving genetics problems, 107-108
Genistein, 172
Genome mapping projects, 112, 191, 193, 199, 202, 217
Genotype, **104**
Genus (genera), 2, 127, *129*
Geomyces destructans, 427
Geranium (*Pelargonium domesticum*), *40*
Geranium pollen grain, *80*
Gerard, John, 338
Gerber daisy, *124*

Gerber, Traugott, *124*
Germ, **188**
Germination, 88, *89*
Germplasm, **249**
  ownership of, 250-251
  preservation of, 250-251
Giant fennel (*Ferula* spp.), 338
Giant Hogweed (*Heracleum mantegazzianum*), 380-381
Giant reed (*Arundo donax*), 325
Giant spurge (*Euphorbia*), 488
Gibberellins, **92**
Gills, 430
Gin, 147, 458
Ginger (*Zingiber officinale*), *291, 294, 296*, 296-297
Ginkgo (*Ginkgo biloba*), *138*, 150-151, 344, 354, 355-356
Ginkgolides, *344*
Ginkgophyta, *138*, 146, 150
Ginning, **310**, 313-314
Ginsburg, Ruth Bader, 366
Ginseng (*Panax ginseng*), *342*
Glarea lozoyensis, 472
Glaucoma, 365
Gliadin, 159
Global Crop Diversity Trust, 249
Global warming, 213, 251, 494-497, 505
Glochids, 387
Glomeromycota, *420*, 425
Glucose, **8,** 153
Glumes, **188**
Glutamic acid, *10*, 157
Glutamine, *10*, 157
Gluten, 158-159, 192, 398
Glycemic index (GI), **171**-172, 237
Glycemic load, **171**
Glyceraldehyde phosphate, *62*
Glycerol, *12*
Glycine, *10*, 157
Glycine max. *See* Soybeans (*Glycine max*)
Glycogen, 9, *9*, 153
Glycolipids, 19
Glycolysis, **63**, *64, 65*
Glycoproteins, 19, 391
Glycosides, 14-15, **343**
Glyoxysomes, **23**
Gnetophyta, *138*, 146, 151
Gnetum spp., 151
Goat grasses (*Triticum* spp.), 190
Goiter, 168
Golden chain (*Laburnum anagyroides*), 384
Golden Rice, 262
Goldenrod (*Solidago canadensis*), *124*, 251
Goldenseal (*Hydrastis canadensis*), 340
Golgi apparatus, **23,** *24*
Goobers. *See* Peanuts (*Arachis* spp.)
Goodyear, Charles, 508
Goo Goo Cluster, 284
Gourds, musical instruments from, 324
Grafting, **95,** 448

Grains, **86,** 188-189, 242, 244. *See also* Grasses (Poaceae)
Gram-positive bacteria, 470, 472
Grana (granum), **20,** *58, 59*
Grape (*Vitis* spp.), 86, 125, 448-449. *See also* Wine
  cultivation of, 448-449
  downy mildew of, 450
  global production of, *188*
  harvesting of, 448-449
  Phylloxera infestation of, 448, 450
  varieties of, *448*
Grapefruit (*Citrus paradisi*), 95-97
Grasses (Poaceae), 3, 51, *128, 129*, 188-189, *189*. *See also specific grasses*
  allergies to pollen of, 394
  flowers of, 188
  grains of, 188-189
  grasslands and, 504-505
  hyperaccumulation by, 51
  incomplete flowers of, 73
  lawn, 205-206, 394
  vegetative characteristics, 188
Grass guard cells, 31
Grassland biome, 504
Grasslands, *502,* **504**-505
Grateloupia tururura, 417
Grayanotoxins, 383
Gray, Horace, 90
Grazing food chain, *489*
Great tit (*Parus major*), 496
Greeks (ancient)
  ergotism and, 476
  linen clothing, 315
  medicinal plant use by, 338
  papyrus use, 331
  spice trade, 290
  wines, 447
Green algae (Chlorophyta), 401, *402,* 405-408, *408*
  lichens and. *See* Lichens
  pharmaceuticals from, 413
Green Belt Movement, 326
Green Hour, 461
Greenhouse effect, 213, 220, **494**-497
Greenhouse gases, 213, 220
Green peas, 215
Green Revolution, **243**-247
  disease-resistant varieties, 244-245
  high-yield varieties, 243-244
  history of, 245
  improvements on, 246-247
  problems resulting from, 245-246
Green teas, 173, 281-282, 355
Griffith, H. R., 379-380
Griseofulvin, 472
Groundnuts. *See* Peanuts (*Arachis* spp.)
Ground pine (*Lycopodium* spp.), 144
Ground tissues, **29,** 31, *31, 34*

Growth, as property of living things, 6
Guam disease, 148
Guanine, **12,** 111
Guarana (*Paullinia capana*), 287
Guard cells, **30**-31, 49, *49*
Guayale (*Parthenium argentatum*), 383, 501, *504*
Gums, 156
Gymnocladus dioicus (Kentucky coffeetree), *210*
Gymnosperms, 4, *139,* **146**-151
  divisions of, 146-151
  life cycle of, *139*
  resistance to fungal decay, 437
Gynecomastia, 303
Gynoecium, **70,** *73*

H

Habitat, **487**
Hackel, Ernst, 487
Hadza, 176, 179
Haematococcus spp., 410
Hairs, 387
Hairs (trichomes), **30,** *30*
Hairy bamboo (*Phyllostachys pubescens*), 335
Hallucinogens, 15, **359**
  absinthe, 461
  fly agaric, 478-479, *479*
  lysergic acid diethylamide (LSD), 477, 479
  nutmeg, 294
Hamamelis virginiana (witch hazel), 340
Haploid (n), **75,** *76*
Haploid generation, 139
Haplotypes, 112
Hard cider, 460-461
Hard fibers, **309**
Hard wheat, 191
Hardwoods, **326**-327
Harrison Act of 1914, 368
Harvard University, 477
Hashishins, 363
Hauptmann, Bruno, 178
Hawkins, John, 371
Hay fever, 80, 83, 390, 480, 482
  plants, 391-394
Hazelnuts, 86
Head (inflorescence), **74,** *75*
Head lettuce, 44
The Health Consequences of Smoking, 372
Healthier U.S. Initiative, 170-171
Heart disease, 159, 160, 161, 163, 166, 170, 171, 463
Heart rot, 436-437
Heartwood, **327,** *327*
Heat energy, 55
Heavy chains (antibody), 390
Heavy metals, 50, 51, 134-135
Hedge apples. *See* Osage orange (*Maclura pomifera*)
Heirloom varieites, 250

Helianthus annuus (sunflower), 7, 49, 86, 125, 220, *248*
Helianthus tuberosus (Jerusalem artichoke), 238-239, *239*
Hematoxylin, 321
Heme iron, **168**
Hemicellulose, 156
Hemileia vastatrix (coffee rust), 275, 439
Hemlock. *See* Poison hemlock (*Conium maculatum*)
Hemp fibers, *309*, 317-318. *See also Cannabis sativa* (marijuana, hemp)
Henbane (*Hyoscyamus* spp.), 367
Hennig, Willi, 130
Henry, Patrick, 371
Henry the Navigator, 292
Hepatophyta (liverworts), 138, *138*, 141
Hepatotoxins, 412
Heracleum mantegazzianum, 380-381
Herbaceous stems, **33**, *34*
Herbalism, 338
Herball or Generall Historie of Plantes (Gerard), 338
Herbal medicine, 338-342
    Age of Herbals, 121, 338-340
    dietary supplements, 354
    ginkgo, 355-356
    history of, 338-342
    Native American use of, 340
    regulation of by FDA, 352, 353, 354
    saw palmetto, *356*, 356-357
    St. John's wort, 355
    top-selling remedies, *355*
Herbals, **121**, 338
Herbicide-resistant crops, 118, 259-260, 268
Herbivores, **488**
Herbs, 290, 300-307
    allium family, 303, 304-305
    aromatherapy and, 291
    common by scientific name and family, *302*
    dyes from, 321-322
    essential oils in, 291
    mint family, 301-303
    mustard family, 303-304
    parsley family, 303
    terpenes in, 14
Heredity. *See also* Molecular genetics
    codominance, 109
    genetics problems, solving, 107-108
    incomplete dominance, 109
    jumping genes, 196
    linkages, 109-110
    Mendelian genetics, 102
    multiple alleles, 109
    polygenic inheritance, 109
Heroin, 361-362. *See also* Opium poppy (*Papaver somniferum*)
Hershey, Alfred, 111

Hershey company, 287
Hershey, Milton, 284
Hesperidium, **86**, *87*
Heterocysts, **402**
Heteromorphic, 401
Heterotrophic, **137**, 414, **420**
Heterotrophs, **488**
Heterozygous, **102**
Hevea brasiliensis (rubber tree), 14, 398, 508
Hickory (*Carya* spp.)
High-density lipoproteins (HDLs), **161**, 463
High-fructose corn syrup, 153, 237
Hilum, **88**
Hippocrates, 338, 360
Hippomane mancinella (manacheel tree), 388
Histamines, 390
Histidine, *10*, 157
Histoplasma capsulatum, *481*
Histoplasmosis, 480-481
HIV (human immunodeficiency virus), 413
HMS *Beagle*, 133-134
HMS *Bounty*, 255, *255*
Hodgkin's disease, 353
Hoffmann, Felix, 346
Hofmann, Albert, 477
Holdfast, **405**, 408
Holistic medicine, 291
Holly (*Ilex* spp.), 125, 496
Homologous chromosomes, **75**, **102**
Homozygous, **102**
Honey bees (*Apis mellifera*), 81, *81*, 251
Honeysuckle (*Lonicera japonica*), 80, 125
Hookahs (water pipes), 363
Hooke, Robert, 17
Hops (*Humulus lupulus*), 455
Hordein, celiac disease and, 158
Hordeum chilense (wild barley), 204
Hordeum vulgare. *See* Barley (*Hordeum vulgare*)
Hormones (plant). *See* Plant hormones
Hornbeam (*Carpinus betulus*), 496
Hornworts. *See* Anthocerophyta (hornworts)
Horseradish (*Armoracia rusticana*), *302*, 304, *304*
Horses, domestication of, 182
Horsetails (*Equisetum*), 144, *144*
Horticulture, 5
Hot chilies. *See Capsicum* peppers
Hot deserts, 501, *502*
Houseplants, poisonous, 386
Huanglongbing (HLA). *See* Citrus greening disease
Human diseases/illnesses. *See specific diseases*
Human genome map, 112
Human Genome Project, 112-114
Human health
    allergies to fungi, 482-484
    allergies to GM foods, 261, 268

    allergies to plants, 389-398
    alcohol consumption and, 463-465
    antibiotics and, 470-472, 473-474
    caffeine and, 272-273
    contact dermatitis, 389, 395-397
    dermatophytes, 480-481
    gluten and celiac disease, 158-159
    green tea and, 173, 281-282
    immune system and, 390
    indoor mold, 482-483
    malnutrition, 154
    medicinal plants. *See* Medicinal plants
    overnutrition, 154
    Paleo Diet, 177, 179
    poisonous fungi, 475-477
    poisonous plants, 379-386
    systemic mycoses, 481-482
    undernutrition, 154
    vitamin deficiencies, 162-167
Human immunodeficiency virus (HIV), 413
Human nutrition, 152-174
    for cancer prevention, 172-173
    carbohydrates, 153, 156
    dietary fiber, 156
    dietary guidelines, 169-171
    energy requirements, 153
    fats and cholesterol, 159-162
    macronutrients, 153-162
    micronutrients, 162-168
    minerals, 167-168
    Paleo Diet, 177, 179
    proteins and essential amino acids, 156-157
    USDA Food Pyramid, 169
    vitamins, 162-167
Human population growth, *154*, 244, 246
Human society
    agricultural revolution, 179-180
    foraging societies, 176-179
    hunter-gatherers, 176-180, 182
    impact of on biodiversity, 131
    impact of plants on, 2-6
    Neolithic, 179, 183
    Paleolithic, 177
Hunter-gatherers, **176**-180, 182
Hurricane Katrina, 483
Hyacinth, 384
Hybrid, **197**
    corn, 197
    pigeon peas, 215
Hybrid vigor, 197
Hydnocarpic acid, *344*
Hydnocarpus spp., *344*
Hydrangea, *124*
Hydrastis canadensis (goldenseal), 340
Hydrocarbons
Hydrocolloids, **411**

Hydrocyanic acid, 235
Hydrogenation, **161**
Hydrogen bonds, 51, **514**, *515*
Hydrogen cyanide, 15
Hydrophilic, **41**
Hydrophobic, **41**
Hydroponics, **50**
Hymenium, **427**
Hyoscyamine, 367
Hyoscyamus niger, 368
Hyperaccumulators, **51**, 146
Hypericum perforatum (St. John's wort), 355, 388
Hypertension, 155, 170, 171, *350*, 350-351
Hypertonic solutions, **20**, *21*
Hyphae, 422-423
Hypocotyl, **88**
Hypogenous flowers, **73**
Hypothenemus hampei, 276
Hypothesis, 5
Hypothesis testing, 5-6
Hypotonic solution, **20**, *21*
Hypoxylon thouarsianum, 443

I

Ibotenic acid, 478
ICARDA (International Center for Agricultural Research in Dry Areas), 244
Iceberg lettuce, 44
IFPRI (International food Policy Research Institute), 244
IgA antibodies, 390
IgD antibodies, 390
IgE antibodies, 390, **390**
IgG antibodies, 390
IgM antibodies, 390
IITA (International Institute of Tropical Agriculture), 244
Ilex guayusa, 287
Ilex paraguariensis, 287
Ilex spp., 125, 496
Ilex vomitoria, 287
Illicium verum (Star anise), *302*, 303
Immune system, 390
Immunoglobulins (Ig), **390**
Imperfect flowers, **73**
Imperfect stage, **424**
Imperialism, spice trade and, 293
Inca civilization, 366
Incense, 14
Incomplete dominance, **106**, *106* 109
Incomplete flowers, **73**, *74*
Incomplete proteins, **157**
Indehiscent fruits, **86**
Independent assortment, principle of, 106
India
    bamboo use in, 334-335
    marijuana and, 362-363
    medicinal plant use in, 338, 342
    spice trade and, 290
    spinning in, 311

Indian blanket (*Gallardia pulchella*), *121*
Indian corn. See also Corn (*Zea mays*)
Indian hellebore (*Veratrum* spp.), 387
Indian mustard (*Brassica juncea*), 51
Indian pipe, *124*
Indica rice, 201, 202
Indigo (*Indigofera tinctora*), *221*, 321
Indonesian civet coffee, 278
Indoor mold, 482-483
Inferior ovary, **73**
Inflorescences, **74**, *75*
Infrared radiation, 56, *57*
Ingestive heterotrophs, 137
Inheritance, 102. See also Genetics
Insecticides, 232, 388-389
Insects
    development of crops resistant to, 260-261
    dyes from, 321
    pollination by, 80
Insertion mutations, **118**
Insoluble dietary fiber, **156**
Instant coffee, 278
Institute of Medicine, National Academy of Science, 484
Institute of Plant Sciences, 262
Integral proteins, **19**
Integrated pest management, **438**
Integuments, **79**
Interational Livestock Research Institute (ILRI), *244*
Intergovernmental Panel on Climate Change (IPCC), 276
Intermembrane space, 64, *67*
Internal membrane system, cell, 22, *23*
International agricultural research centers, 243, *244*
International Barley Genome Sequencing Consortium, 204
International Code of Botanical Nomenclature, 123, 127, 129
International Coffee Organization, 275
International Crops Research Institute for Semi-Arid Tropics (ICRISAT), *244*
International Food Policy Research Institute (IFPRI), *244*
International Human Genome Sequencing Consortium, 112
International Plant Genetic Resources Institute (IPGRI), *244*
International Rice Research Institute (IRRI), 202, *244*
International Treaty of Plant Genetic Resources for Food and Agriculture, 251
International Unit, 164n
Internodes, **38**
Interphase, **24**
Intine wall, **77**

Introns, **115**
Inuits, 176, 179
Invasive species, **415**-416
    Catalina mahogany, 487
    Caulerpa taxifolia, 415-416
    Codium fragile, 417
    saltcedar, 415
    zebra mussels, 415
Iodine, *167*, **167**-168
Ionic bonds, **514**
Ions, **514**
Ipomoea violacea, 376
Iridaceae, *128*, 297
Irish potato famine, 228-229, 437
Iron, *167*, **167**-168
Iron-deficiency anemia, **168**
Irregular (zygomorphic) flowers, 74
IRRI (International Rice Research Institute), *244*
Irrigation, 245, *246*, 247, 501
Isaria kogana, 472
Isoflavones, 218
Isogamous, **408**
Isoleucine, *10*, 157
Isomorphic, 401
Isothiocyanates, 304
Isotonic solution, **20**, *21*
Isotopes, **513**
Italian Renaissance, 367
Ivory Coast, child slave labor and, 286-287
Ivy (*Hedera helix*), 125, *355*, 496

J

Jaborandi (*Papaver somniferum*), *344*
Jack-in-the-pulpit (*Arisaema triphyllum*), 386
Jamaica Blue Mountain, 278
James I, King of England, 371-372
Jamestown weed, *368*. See also Jimsonweed; thornapple
Japanese honeysuckle (*Lonicera japonica*), *124*
Japanese horseradish (*Wasabia japonica*), 304
Japanese mint (*Mentha arvensis*), 302
Japanese tea ceremony, 280
Japanese wasabi (*Wasabia japonica*), 304
Japonica rice, 201, 202
Jasmine tea, 281
Jasmonic acid, 93
Jatropha oil, 220
Java. See Coffee (*Caffea* spp.)
Jefferson, Thomas, 371
Jerusalem artichoke (*Helianthus tuberosus*), 238-239, *239*
Jimsonweed (*Datura* spp.), 367, *368*. See also Jamestown weed
The Johns Hopkins University Medical School, 480
Johnson grass (*Sorghum halepense*), 394

Johnson, Robert Gibbon, 90
Jojoba (*Simmondsia chinensis*), 501, *504*
Journal of the American Medical Association, 372
J. S. Fry & Sons, 283
Juglandaceae, *128*
Juglans nigra (black walnut), *322*, 383
Juglone, 383
Jumping genes, 196, 199
Juniperus spp., 147, 458
Jute (*Corchorus* spp.), *309*, 317

K

K (vitamin), **162**, *162*
Kale (*Brassica oleraceae*), 243, *243*
Kalmia spp., 383
Kamel, George-Josef, *124*
Kapa, 320
Kapok (*Ceiba pentandra*), *309*, 318
Karenia brevis, 414
Karyogamy, **424**
Kava (*Piper methysticum*), 274, 375-376
    active components in, 375-376
    drug of choice in the Pacific, 375-376
    preparation of the beverage, 375
Kelly, Michael, 6
Kelp, 4, 404
Kenaf (*Hibiscus cannabinus*), 334, 504, *504*
Kentucky bluegrass (*Poa pratensis*), 206, 394
Kentucky coffeetree (*Gymnocladus dioicus*), 210
Keratin, 480
Ketoconazole, 482
Kheer, 223
Kidney beans (*Phaseolus vulgaris*), 214
Kilocalorie (kcal), **153**
Kinases, **24**
Kinetic energy, **55**, *55*
Kinetochore, **25**
Kingdom Animalia, 137. See also Animals
Kingdom Fungi, 137, 420. See also Fungi
Kingdom Plantae, 137
Kingdom Protista, 137, 420, 424
Kingdoms, **127**
    five-kingdom system, 137
    six-kingdom system, 129
    in three-domain system, 137
Kiwifruit (*Actinidia chinensis*), 99
Koehler, Arthur, 178
Kohlrabi (*Brassica olearacea*), 243, *243*
Koji, 457
Kola tree (*Cola nitida*), 287
Kombu (*Laminaria*), 409
Kopi luwak. See Indonesian civet coffee
Krebs Cycle, **63**
Krebs, Hans, 64

Kudzu (*Pueraria thumbergiana*), 415
Kumquats (*Fortunella japonica*), 95
!Kung, 177
Kwashiorkor, **154**
Kyoto Treaty, 487

L

Laburnum anagyroides (golden chain), 384
Lacrimatory factor, 305
Lactase, 56
Lacto-ovo vegetarian, **172**
Lactose, **153**
Lacto vegetarians, **172**
Lactuca sativa (lettuce), 44
Lager beer, 455
Lamiaceae (mint family), *128*, *129*, 301-303
Laminaria spp., 409
Land plants, 137-138, *138*. See also specific divisions
Land races, **249**
Landscape plants, poisonous, 382-386
Language of Flowers (Greenaway), 125
La Noche de Rabanos (Night of the Radishes), 45
Late blight of potato (*Phytophthora infestans*), 228, 231, 232, 437-438
Late wood, **34**
Latex, 14, 388, **398**, 501
    allergy, 398
Latin names. See Scientific names
Laudanum, 360
Laundry starch, 237
Lauraceae, *128*, 293
Laural cherry (*Prunus laucerasus*), 496
Laurus nobilis (Bay leaves), *302*
Lavandula stoechas, 303
Lavender (*Lavendula* spp.), 125, 291, *302*, 303
Lavendula spp. (Lavender), 125, 291, *302*, 303
Laveran, Alphonse, 348
Lawn grasses, 205, *206*, 394
Laws of Thermodynamics
    first, 55, 489
    second, 55, 489
LDEF satellite, 94
Lead, 50, 51
Lead tree (*Leucaena* spp.), *221*, *221*
Leaf arrangements, 38, *39*
Leaf blight (*Xanthomonas oryzae*), 264
Leaf epidermis, 30, *30*
Leaf fibers, **309**, 317, 318, *318*
Leaflets, **38**
Leaf morphology, *39*
Leary, Timothy, 477
Leaves, 37-41
    anatomy, *40*
    arrangement of, 38, *39*
    cross-section of, *40*

fall colors of, 57
vs. leaflets, 38
morphology, *39*
photosynthesis in, 37
stomata on, *40*, 49
transpiration from, 48
venation patterns, 38, *39*
Lecithin, 218
Leeks (*Allium porrum*), *302*, 304
Legumes (Fabaceae), 51, **86**, 210-223
alkaloids in, 384
characteristics of, 211
crop-rotation using, 212
flowers of, 211, *211*, 222
fruits of, 86, *211*, 216
hyperaccumulation by, 51
nitrogen-fixation in, 212-213
rotenone from, 389
seeds of, 211, 221-223
Lemaireocerceus thurberi (organ pipe cactus), *488*
Lemma, **188**
Lemons (*Citrus limon*), 95
Lennon, John, 477
Lentils (*Lens culinaris*), *214*
Lentinus edodes (Shiitake mushroom), 465
Lepidodendron, 144
Lepiota mushrooms, 477
Leptospermone, 384
Lescol, *473*
Lettuce (*Lactuca sativa*), 44
Leucine, *10, 157*
Leucoplasts, **21,** *24*
Lewis, Walter, 342
Liberica coffee (*Coffea liberica*), 277-278
Lichens, 125, 321, 381-382, 408, 434
Licorice (*Glycyrrhiza glabra*), *221*
Life
five-kingdom classification system for, 137
fundamental properties of, 6-8
molecules of, 8-13
Life cycles
angiosperm, 69-84, 85-99
ascomycete, 427
basidiomycetes, 431, *433*
brown algae, 404-405, *407*
chytrid, 424-425, *425*
diatom, 404, *404*
Dutch elm disease, 442, *442*
fern, 139-141, *141*
foods from, 408
green algae, 405-408, *408*
moss, 139-141, *140*
myxomycete, 421
oomycete, 422
pine, 139-141, *142*
Puccinia graminis, 439
red algae, 405
zygomycete, *426*
Light, 55, 57
Light-absorbing pigments, 57-59
Light chains (antibody), 390
Light-harvesting antennae, **59**

Light-independent reactions, 61. *See also* Calvin Cycle
Light microscope, 19
Light reactions, **59**-61
Light roasts, coffee, 277
Lignin, 14, 19, 156, 330-331, 436, 437
Lignum vitae (*Guaiacum officinale*), 327
Liliaceae, *128*
Lilium lancifolium (Tiger lily), *73*
Lily-of-the-valley (*Convallaria majalis*), *125*, 384
Limes (*Citrus aurantifolia*), 95, 96
Lincoln, Abraham, 386
Lincoln, Nancy Hanks, 386
Lindbergh, Charles and Anne, 178
Lind, James, 164
Linear photophosphorylation, *60,* **61**
Linen, 219, 315-316, *316*
Linkage, 109-110
Linnaeus, Carolus, *121,* 121-123
Linoleic acid, *160,* 161
Linseed oil, 219
Linum usitatissimum (flax), 219, *309,* 315-316, *316*
Lipids, 11-12, *12,* **159,** *159*
Lipitor, *473*
Lipoproteins, 161
Livalo, *473*
Liverworts (Hepatophyta), 138, *138, 141-143. See also* Hepatophyta (liverworts)
Livestock
forage crops for, 205, 222
plants poisonous to, 386-387
"Living fossils," *136*
Lloyd, Edward, 274
Lloyd's of London, 274
Loco, 387
Locoweed (*Astragalus* spp.), 386
Locus, **102**
Lodging, 244
Logwood (*Haematoxylum* spp.), 321
Lonchocarpus spp. (cube root), 389
The London Coffee House, 275
Lonicera japonica (honeysuckle), 80, *125*
Lophophora williamsii, 374, *374*
Lost Colony of Roanoke Island, 37-38
Lou Gehrig's disease, 148
Lovastatin, *473,* 473-474
Low-density lipoproteins (LDLs), **161**
Low deserts, 501
LSD. *See* Lysergic acid diethylamide (LSD)
Lucerne. *See* Alfalfa (*Medicago sativa*)
Lumber, 327-329. *See also* Timber
Lupines (*Lupinus* spp.), 384
Lycophyta (club mosses), *138,* 139, 143, 144, 145
Lycophytes, 143-146
Lycopoidium spp., 144

Lymphocytes, **390**
Lysergic acid diethylamide (LSD), 359, 376, **477,** 479
Lysine, *10, 157*

M

Maathai, Wangari, 323, 326
Mace (*Myristica fragrans*), *294,* 295-296, *296*
Macintosh, Charles, 508
MacKill, David, 202
Maclura pomifera (osage orange), 6, 86, 123, 124, 126
Macromolecules, **8**
carbohydrates, 8-9, *9*
lipids, 8, 11-12, *12*
nucleic acids, 8, 12-13
proteins, 8, 9-10, *11*
Macronutrients (human), **153**-162
carbohydrates, 153, 156
dietary fiber, 156
fats and cholesterol, 159-162
proteins and essential amino acids, 156-157
Madagascar periwinkle (*Cartharanthus roseus*), *344,* 349, 352
Madder (*Rubia tinctorum*), 321
Magellan, Ferdinand, 292
Magnaporthe oryzae (Rice blast fungus), 269
Magnesium, *167*
Magnolia, 86, *125, 128*
Magnoliaceae, *128*
Magnoliophyta, 138, *138,* 151
Mahogany (*Swietenia* spp.), 323
Maize, 193-200. *See also* Corn (*Zea mays*)
Major minerals (human nutrition), **167,** *167*
Malaria, 346, *347,* 348-349
Male gametophyte development
in angiosperms, 77
gymnosperm, 139
Malnutrition, **154**
Malt, 204, 455
Malt beverages, 455
Maltose, 9, *9,* **153**
Malus pumila. See Apple (*Malus pumila*)
Malvaceae, *128,* 283, 504
Mandioca. *See* Cassava (*Manihot esculenta*)
Mandrake (*Mandragora officinarum*), 338, *339,* 367
Manganese, 51, *167*
Mangelsdorf, Paul C., 197, 198
Manihot esculenta (cassava), 15, *188,* 234-236, *235*
Manila hemp (*Musa textilis*), *309,* 317-318
Manilkara zapota (chico tree), 509
Mansonones, 442
Maple (*Acer* spp.), 49, 86, *87,* 325, 328

Marasmus, **154**
Marchantia spp., 143, *143*
Marchantiophyta (liverworts), 141
Marggraf, Andreas, 54
Margosa (*Azadirachta indica*), 389
Marijuana (*Cannabis sativa*), 359, 362-366
early history in China and India, 362-363
hemp fiber from, *309,* 317, 333, 334
legal issues, 365-366
medical marijuana, 365
spread to the West, 363-364
tetrahydrocannabinol (THC) in, 14
THC and its psychoactive effects, 364-365
Marjoram (*Origanum majorana*), *291,* 302, 302-303
Marker-assisted selection, **257**
Markov, Georgi, 385
Marshmallow, *124*
Mass flow, 52
Mass Flow (Pressure Flow) Hypothesis, **52**
Mass number, **513**
Master Tobacco Settlement of 1998, 373
Mate (*Ilex paraguariensis*), 287
Matossian, M.K., 476
Matrix, **22, 64**
Mayan gods, 370
Mayapple (*Podophyllum peltatum*), *344*
Mayr, Ernst, 130
Mazatec Indians, 479
McCartney, Paul, 477
McClintock, Barbara, 195, 196, 199
McElwain, Jennifer, 40
Meadowsweet (*Spirea ulmaria*), 345
Mealybug, 131
Meatless diets, 172-173
Meat substitutes, 467
Mechanical energy, 55
Mechanical injury, plants causing, 387-388
Medicago sativa (alfalfa), 222, *222*
Medical marijuana, 365
Medications. *See* Drugs
Medicinal plants. *See also* Herbal medicine
active principles in, 343
Age of Herbals, 338-340
Aloe vera (burn plant), *344,* 349, 351, *351*
alternative medicine and, 354-357
bloodroots, 340, *341*
camptothecin and, *344,* 353-354
cancer therapies, 352-354
Carapa guianensis, 506
Chinese happy tree, *344,* 353
in contemporary health care, 351
copaiba, 506
echinacea, 340, 341, 355
ephedra, *150,* 151, *344,* 351-352, 354

Medicinal plants (*continued*)
  fever bark tree, *344*, 346, 348
  foxglove, 340, 343-345, *345*
  ginkgo, *150*, 150-151, *344*, 354, 355-356
  goldenseal, 340
  history of use of, 338-342
  of known medical value, 338
  Lycopodium spp., 144
  Madagascar periwinkle, *344*, 349, 352
  modern prescription drugs, 340-342
  Native American use of, 340-341
  peat moss (*Sphagnum*), 142-143, *143*
  pharmaceuticals derived from, 340-342
  Ptychopetalum olacoidesis, 506
  saw palmetto, *356*, 356-357
  snakeroot, *344*, 350-351
  St. John's wort, 355, 388
  tropical rainforests and, 340, 342, 506
  use by early Greeks and Romans, 338
  willow, 338, 345-346
  witch hazel, 340
  wormwood (*Artemesia annua*), 346, 348-349, 461
  yews, 382
Mediterranean diet, 170
Mediterranean Sea, *Caulerpa taxifolia* in, 415-416
Megasporangium, **79**, 140
Megaspore mother cell, **79**
Megaspores, **79**
Meiosis, **74**
  in flowering plant life cycle, 77
  role in sexual reproduction, 75-79
  stages of, 75-76, *77*
Melons, 86
Mendel, Gregor, 102, *102*, 134
Mendelian genetics, 102-110
  dihybrid crosses, 105-106
  Mendel's use of garden pea in experiments, 102-103
  monohybird crosses, 103-105, *104*
  pea plant traits examined by Mendel, 102-105
  principle of dominance, 104
  principle of independent assortment, 106
  principle of segregation, 104
Mendes, Chico, 508
Mentha piperita (peppermint), 302, *302*
Mentha spicata (spearmint), 302, *302*
Menthol, 291, 302
Mercerization, 314
Mercer, John, 314
Merchant clipper ships, 361
Merck Pharmaceuticals, 472
Mercury, 50, 491

Meristems, **29**
  cell cultures from, 256
  root (zone of cell division), 35
Mescal beans, 384
"Mescal buttons," 374
Mescaline, 15
Mesocarp, **86**
Mesophyll, **41**
Mesotrione, 384
Mesozoic cycad forests, 147
Mesquite (*Prosopis glandulosa*), *221*
Messenger RNA (mRNA), 114, **114**
  transcription and, 114-115
  translation and, 116
Metabolism, **55**
  cellular respiration, 7
  energy and, 55
  enzymes and, 56
  Laws of Thermodynamics and, 55
  phosphorylation, 55-56
  photosynthesis, 7-8, 56-63
  as property of living things, 7-8
  redox reactions and, 55
Metaphase (mitosis), **24**, *25*, *26*
Metaphase I (meiosis), 76, *77*
Metaphase II (meiosis), 76, *77*
Metformin, 349-350
Methamphetamine, 352
Methicillin, 471
Methicillin-resistant *Staphylococcus aureus* (MRSA), 472
Methionine, **10**, 157
Metric system, 519-520
Mevacor, 473
Mexico, origins of agriculture in, 180, *181*, 183
Mezcal, 459
Miconazole, 482
Microalgae, biofuels from, 411
Microbiota, 156
Microbodies, **23**
Microcladia spp., *406*
Microcystis, 412, *413*
Microfilaments, **19**
Micronutrients (human), 162-168
  minerals, 167-168
  vitamins, 162-167
Micronutrients (plant), 50, **513**
Micropyle, **79**, **88**
microRNA (miRNA), **117**
Microscopes, *17*
Microsporangia, **77**, 139
Microspore mother cell, **77**
Microspores, **77**, 139
Microsporum spp., 480
Microtubules, **19**
Microwaves, 56, *57*
Middens, 177, *179*
Middle Ages, 360, 367
Middle lamella, **19**
Milk chocolate, 284
Milk sickness, 386-387
Milk vetch (*Astagalus* spp.), 386
Milkweeds (*Asclepius* spp.), 86, 123, *124*, 381
Millardet, Alexis, 450

Miller, Anne, 470
Miller, Nicole, 182
Millet, 188, *203*, 204-205
Mimics. See Biological mimics
Minerals, 50-51, 167-168
  absorption from soil by roots, 49-51
  deficiencies in plants, 50
  human nutrition and, 167-168
  hyperaccumulation of by plants, 51
  overabundance in plants, 50
  trace, **167**, *167*
Mint family (Lamiaceae), *128*, 301-303
Miracle fruit, 306, *306*
Miraculin, 306
Miscanthus giganteus, 208
Miso, 218
Mitchell, Peter, 66
Mitochondria, **21**
  cellular respiration and, 64, *66*, *67*
  origins of (Endosymbiont Theory), 22
Mitosis, **24**, *26*
Mocha, 274, 279
Moche civilization, 215, *215*
Mock orange, 123, 124, *124*, 125, 126. See also Osage orange (*Maclura pomifera*)
Moist coniferous forests, **505**
Mojave Desert, 501
Mold growth, *483*
Mold on bread, *475*
Molds, 430, 482-484
Molecular farming, **264**
Molecular genetics, 110-119
  DNA as genetic material, 111, 115
  epigenetics, 118
  gene control of proteins, 114
  genome maps, 112
  Human Genome Project, 112-114
  mutations, 117-118
  recombinant DNA, 118, *119*
  transcription, 114-114
  translation, 116
Molecular plant breeding, 257
Molecules, **514**, 514-515
  water, *515*
Molybdenum, *167*
Monarch butterflies, 381, *381*
Monascus purpureus, 474
Monilophthora roreri (frosty pod rot), 286
Monilophyta. See Pterophyta (ferns)
Monocots, **3**
  fibers from, 309
  floral structure, 72, *73*
  germination and development, 88, *89*
  leaf venation, 38, *39*
  roots anatomy, 35, *36*
  seeds of, 88
  stems anatomy, 33, *34*

Monocultures, 247-248
Monoecious plants, **73**
Monohybrid crosses, **103-105**, *104*
Monosaccharides, **8**, *9*, **153**
Monounsaturated fatty acids, **160**
Monstera spp., 386
Montezuma, 283, 479
Moonshine, 458
Moraceae (mulberry family), 508
Morales, Evo, 370
Mordant dyes, 321
Mordants, 321
Morels, 427, *428*
Moromi, 457
Morpheus (Greek god of dreams), 361
Morphine, 15, 343, 359, 361
Mosses (Bryophytes), 4, *138*, 139, 141-142, 141-143, *143*
Moto, 457
Mountain cedar (*Juniperus ashei*), *393*, 394
Mount St. Helens, 498
Mouse genome, *112*
Mucilages, 156
Mucor ramannianus, 472
Mulberry (*Morus alba*), 40
Multicellular organisms, 8
Multiple alleles, **109**
Multiple fruits, **86**
Multiple sclerosis (MS), 365
Munch, Ernst, 52
Mung beans (*Vigna radiata*), 214
Musaceae (banana family), *128*
Musa spp. (bananas), 93, 98, 188, 262-263
Musa textilis (Manila hemp), *309*, 317-318
Muscimol, 478
Mushrooms, 420, 430, *432*. See also Fungi
  cultivation of, 462
  edible, 465-466, *466*
  hallucinogenic, 479-480
  life cycle of, 431, *433*
  nutritional value of, 464
  poisonous, 477-480
Musical instruments, 324-326
Muslin, 312
Mustard, 302, 303-304
Mustard family (Brassicaceae), 51, *128*, *129*, 302, 303-304
Mustard gas, 304
Mutations, **117**-118
Mutualistic symbionts, **433**
Mycelial stage in culture, *482*
Mycelium, **4**, **422**
Mycobacterium tuberculosis, 471
Mycobionts, 434, **436**
Mycology, **5**
Mycoproteins, **467**-468
Mycorrhizae, **425**, 435-437
Mycotoxins, **472**-474, 483
Myristicaceae (nutmeg family), 295, 359
Myristica fragrans (nutmeg), *294*, 295-296, *296*, 376

Myrtaceae, *128*, 295, 300
Myxamoeba, 421
Myxomycetes (Myxomycota), 420, *420*, 421

N
NAD+, **55**
NAD (nicotinamide adenine dinucleotide), **55**
NADH, **55**
NADP (nicotinamide adenine dinucleotide phosphate), **55**
NADP+/NADPH + H+, **55**
Names of organisms
 binomial system, **123**, *123*, 126
 common names, 123-124
 meanings of, 123-124, 126
Napoleon, 363
Napoleon I, 54
Napp, Abbot Cyril, 102
Naranjilla (*Solanum quitoense*), 254
Narcissus, 384
Narcotic drug, **359**
NASA, 94
National Center for Genetic Resources Preservation, 249
National Human Genome Research Institute, 112
National Plant Germplasm System, 249
Native American Church, 375
Native Americans, 375
 basketmaking by, 310
 ceremonial use of Indian hellebore, 387
 foraging by, *176*
 hallucinogenic mushrooms used by, 479-480
 herbal remedies used by, 340-341
 symbolic meanings of pollen, 80
Native Seeds, 250
Natural dyes, 321, *322*
Natural History (Pliny), 383
Natural rubber latex, 398
Natural selection, 134-135, **184**
Navy beans (*Phaseolus vulgaris*), 214
Near East, early agricultural sites in, 181-182
Necrosis, 50
Nectar, **80**
Nectar guides, **80**
Nectaries, **80**
Neem (*Azadirachta indica*), 389
Neem oil, 388, 389
Nelumbo nucifera (sacred lotus), 41
Neolithic society, 179, **183**
Nepenthes spp., 42, *43*
Nerioside, 382
Nerium oleander, 382
Nestlé company, 287
Nestle, Henri, 284
Nettles, 388
Net (reticulate) venation, **38**
Neurons, 359
Neurospora crassa, 114, 427, *429*

Neurotoxins, 412, 414
Neurotransmitters, 359, 374, 477
Neutrons, **513**
New World, 367, 370
 early agricultural sites in, 183
New York University Medical School, 480
Niacin, *162*, 165-166
Niche, **487**-488
Nickel, 51
Nicotiana rustica, 370, 371
Nicotiana spp. (tobacco). *See* Tobacco (*Nicotiana* spp.)
Nicotiana tabacum, 370
Nicotinamide, **165**
Nicotine, 15, 370, 373
Nicotinic acid, **165**
Nicot, Jean, 371
Niemann, Albert, 369
Night blindness, **162**
Night of the Radishes (La Noche de Rabanos), 45
Nitrate, 50, 212, 213
Nitrite, 212, *213*
Nitrogen, 50, *51*
Nitrogen cycle, 212-213, *213*
Nitrogen-deficient soils, *51*
Nitrogen fixation, **212**, 247, 402
Nitrogen-fixing bacteria, 211, *211*, 247, 402
Nitrogenous base, 12-13, *13*, 111
Nitrous oxide, 212, 213
Nix, John, 90
Node, **38**
Noncyclic photophosphorylation, *60*, **61**
Nondrying oils, **219**
Non-flowering plants, 4. *See also specific divisions*
Nonheme iron, **168**
Nonseptate, **422**
Nonseptate hypha, *420*
Nonshattering fruiting heads, **184**, 242
Nori (*Poryphyra* spp.), 409, *410*
North America, early agricultural sites in, *181*, 183
Northern coniferous forests, **505**
North Regional Research Lab, 470
Norwalk virus, 265
Nostoc, 402, 403, 409, 434
No-tillage agricultural techniques, *246*, 247
Novocain, 369. *See also* Procaine
Nucellus, **79**
Nuclear pores, **23**
Nucleic acids, **8**, 12-13
Nucleolus, **23**, *24*
Nucleotides, **12**, 13
Nucleus, **8**, *8*, *24*
Nucleus accumbens (NA), 359
Numbers, ecological pyramid of, **489**, *491*
Nurses' Health Study, 161
Nutmeg (*Myristica fragrans*), *294*, 295-296, 376
Nutrition Facts panel, *152*

Nutrition, human. *See* Human nutrition
Nuts, **86**, *87*
Nystatin, 471

O
Oak galls, 321
Oaks (*Quercus* spp.)
 affect of global warming on, 496
 allergies to pollen of, 393, *393*
 cork from, 330, *330*
 corn from, 31
 sudden oak death in, 443-444
 symbolic meaning of, 127, *129*
 as timber source, *328*, 329
Oak tree (*Quercus robur*), 496
Oats (*Avena sativa*), 188, 203, 204, *248*
Obesity, 154-155, 170, 171
Observation, 5
Oca (*Oxalis tuberosa*), 254
Ochroma pyramidale (balsa), 327
Ocimum basilicum (basil), 302, 303
Ocimum basilicum (sweet basil), *301*, 302
Ogallala aquifer, 245, *246*
Oil palm, 188
Oils, 12, 14, 219-220. *See also specific types*
 essential. *See* Essential oils
 fatty acid content of, 159-160, *160*
 genetic engineering of more healthful, 262
 human nutrition and, 160, *160*
Oleaceae, 128
Oleander (*Nerium oleander*), 382
Oleandroside, 382
Oleic acid, *160*
Olive oil, 160, 219
Ololiuqui, 376
Omega-3 fatty acids, 160, 409
Omnivores, **488**
One gene-one enzyme hypothesis, 114
One gene-one polypeptide hypothesis, 114
O'Neill, Eugene, 361
Onion (*Allium cepa*), 188, 302, 304-305
On the Origin of Species (Darwin), *133*, 134
On the Tendency of Varieties to Depart Indefinitely from the Original Type (Wallace), 134
Onychomycosis. *See* Tinea unguium
Oogamous, **408**
Oogonium, **405**, **422**
Oolong teas, 281
Oomycetes, 422
Oomycota (water molds), *420*, 422
Oospheres, 422
Oospores, **422**
Ophiostoma ulmi (Dutch elm disease), 265-266, 427, 441-443
Opium, 360

Opium alkaloids, 361
Opium poppy (*Papaver somniferum*), 340, *344*, 360-362
 ancient curse, 360
 heroin, 361-362
 opium alkaloids, 361
 Opium Wars, 360-361
 withdrawal, 362
Opium trade, 361
Opium Wars, 360-361
Opportunistic infections, 481-482
Opposite leaf arrangement, **38**, *39*
Oranges (*Citrus* spp.)
 fruits of, 95-97
 global production of, *188*
 treatment with plant hormones, 92, *93*
Orchard grass, 394
Orchidaceae, *128*, 299
Orchids, 3, 37, 74
Orders (taxanomic), **127**
Oregano (*Origanum vulgare*), 302, 302-303
Organelles, 8, *8*, **19**, 20-23. *See also specific organelles*
Organic composition, as property of life, 8
Organisms, unicellular *vs.* multicellular, 8
Organization, as property of living things, 8
Organ pipe cactus (*Lemaireocerceus thurberi*), 488
Organs, plant. *See* Plant organs
Origanum majorana (marjoram), *302*, 302-303
Origanum vulgare (Oregano), 302, 302-303
Oryza glaberrima, 200
Oryza sativa. *See* Rice (*Oryza sativa*)
Osage orange (*Maclura pomifera*), 6, 86, 123, 124, 126
Osmosis, **20**, *21*
Osteomalacia, **164**
Osteoporosis, **167**-168, 171, 218
Ovary, **72**
Overnutrition, **154**
Overproduction, natural selection and, 134
Over-the-counter medications, caffeine in, *272*
Overweight, 155
Ovules, **72**
Oxaloacetate, 64, *66*
Oxford University, 470
Oxidative phosphorylation, **64**
Oxidized (redox reaction), **55**
OxO gene, 97
Oxygen, 2, 61
Oxytropis spp. (locoweed), 386
Oyster mushroom (*Pleurotus* spp.), 465, *466*

P
P$_{680}$, **60**
P$_{700}$, **60**

Pacific yew (*Taxus* spp.), 14, *124*, 147, *344*, 353, 382
Paddies, rice, 200-201
Paddy straw mushroom (*Volvariella* spp.), 465
Palea, **188**
Paleobotany, *5*
Paleodicots, **132**
Paleo Diet, 177, 179
Paleolithic societies, **177**
Palisade cells, **41**
Palmaria spp., 409
Palmately compound leaves, **38**, *39*
Palm oil, 160, 220
Palms, 3, 125, *128*, *129*
Palynology, *5*, 79
Panaeolus mushroom, 480
Panama disease (*Fusarium oxysorum*), 227
Panax ginseng (ginseng), *342*
Pandanus spp. (screw pine), 37
Pandas, bamboo and, 335
Panicle, **74**, *75*
Pansy, 125
Pantothenic acid, *162*
Papain, 56
Papaveraceae (poppy family), *128*, 359-360
Papaver somniferum, 360, *360*
Paper, 331-334
　from agrifibers, 334
　alternatives to wood pulp, 333-334
　from bamboo, 334-335
　early writing surfaces, 331
　from hemp, 333-334
　from kenaf, 334
　papermaking process, 333, *333*
　papyrus, 331, *332*
　recycled, 334
　use of fungi in producing, 437
Paper mulberry (*Broussonetia papyrifera*), 320, 333
Papua New Guinea, early agricultural sites in, 182
Papyrus (*Cyperus papyrus*), 331, *332*
Paracelsus, 339
Paraguay tea, 287
Parallel leaf venation, **38**
Parasites, **4**, **433**
Parchment, 331
Pareira (*Chondrodendron tomentosum*), *344*
Parenchyma, **31**, *31*, *33*
Parental (P) generation, 104
Paris Agreement, 497
Paris hashish clubs, 363
Parkinson's disease, 23, 148, 273
Parrot feather (*Myriophyllum aquaticum*), 51
Parsley (*Petroselinum crispum*), *302*, 303
Parthenium argentatum (guayale), 383, 398, 501, *504*
Parthenocarpy, **84**
Partial veil, **478**

Passionvine hopper (*Scolypopa australis*), 383
Pasteurization, 447
Pasteur, Louis, 447
Pathogens, **19**, 480-481
　opportunistic, 480
Pauling, Linus, 165
Paullinia capana (guarana), 287
PCBs, biological magnification of, 491
Pea (*Pisum sativum*), **40**, 102-103, *103*, 214-215
Pea aphid (*Acyrthosiphon pisum*), 21
Peaches, 86, 98
Peach-potato aphid (*Myzus persicae*), 71
Peanut cactus (*Echinopsis chamaecereus*), *122*
Peanuts (*Arachis* spp.), 216, 217, 398
Pearl millet (*Pennisetum glaucum*), *203*, 205, 248
Peat bogs, 142-143, *143*
Peat moss (*Sphagnum* spp.), 142
Pecan, 38
Pectins, **19**, 56, 156
Pedicel (peduncle), **70**
Peel, Robert, 229
Pellagra, 165, **165**-166
Pemberton, John Stith, 287, 368
Pencil tree cactus (*Eurphorbia tirucalli*), 385
Penicillin, 430, **470**-471
Penicillium chrysogenum, 470, 471
Penicillium griseofulvum, 472
Penicillium notatum, 470
Penicillium rubens, 470
Penicillium spp., 430, 470-471, 470-472, 474, 483
Pennate diatoms, 404
Pennisetum glaucum (pearl millet), *203*, 205, 248
Pepos, **86**, *87*
Peppercorns, 294-295, *295*
Peppermint (*Mentha piperita*), 302, *302*
Peppers. *See* Black pepper (*Piper nigrum*); Capsicum peppers
Peptide bond, 10, *11*
Percussion instruments, 324
Perfect flowers, **73**
Perfect stage, **424**
Perfumes, 14-15, 82
Perianth, **70**
Pericarp, **86**
Pericycle, **35**, *36*
Periderm, **30**, **31**, *33*
Perigynous flowers, **73**
Peripheral proteins, **19**
Perithecia, **427**
Permafrost, **505**
Pernicious anemia, *166*
Peroxisomes, 23
Pesticides
　from allelochemicals, 384
　biological magnification of, 491

crop monocultures and 247-248
insect-resistant crops and, 260
overreliance of Green Revolution crops on, 245
Petals, **3**, **70**, *74*
Peter, Daniel, 284
Peter the Great, 476
Petiole, **38**
Petroselinum crispum (parsley), *302*, 303
Peyote, 359, 374-375
　"mescal buttons," 374
　Native American Church, 375
Peyote cactus (*Lophophora williamsii*), *374*
Pfesteria spp., 417
PGA (phosphoglyceraldehyde acid), 62
PGAL (phosphoglyceraldehyde phosphate), 62
P (parental) generation, 104
Phaephyta (brown algae), 401, *402*, 404-405, *405*
Pharmaceuticals. *See* Drugs
Phaseolus vulgaris (navy bean), 214
Phenolics, 14-15, *343*, 383
Phenology, **394**, **496**
Phenotype, **104**
Phenylalanine, *10*, *157*
Phenylethylamine, 284
Philadelphus spp., *124*
Philip II, King of Spain, 366
Philodendron (*Philodendron* spp.), 386
Phloem, **32**, *32*, *33*, *47*, *48*, 51
　plant divisions featuring, 138
　translocation of sugars through, 52, *52*
Phosphate group, *12*, 12-13
Phosphoglyceraldehyde (PGAL), 62
Phosphoglyceric acid (PGA), 62
Phospholipids, **11**, *12*, 19, *20*
Phosphorus, 50, *167*
Phosphorylation, **55**-56
Photochemical smog, 213
Photons, 56
Photophosphorylation, **61**
Photosynthesis, **2**, 56-63
　by algae, 402
　atmospheric oxygen from, 2, 61
　C3 pathway, 61
　C4 pathway, 63
　Calvin Cycle, *58*, 61-62, *62*
　CAM pathway, 63
　cellular respiration *vs.*, 63
　chloroplasts and, 20, 57-59
　energy in ecosystems and, 489
　general equation for, 62
　leaf as major organ of, *58*
　light-absorbing pigments, 57-59
　light reactions, 59-61
　transpiration and water loss and, 49
Photosystems I and II, **59**, *60*, *61*
Phototropism, 92
Phragmoplasts, **27**
Phycobilins, **402**

Phycobionts, 434
Phycocyanin, 402
Phycoerythrin, 402, 405
Phycology, 5
Phyla (taxonomic), 127
Phyllostachys pubescens (hairy bamboo), 335
Phylloxera spp., 448, 450
PhyloCode, 130, 132
Phylogenetics, **127**, 130, *138*
Phylogeny, **130**
Physaria. *See* Bladderpod (*Lesquerella* spp.)
Physician-induced addiction, 361
Physiological dependence, **359**
Physiology, plant. *See* Plant physiology
Phytelephas macrocarpa, 509
Phytoalexins, **443**
Phytochemicals, **172**
Phytoestrogens, **218**
Phytoliths, **178**
Phytophthora infestans, 228, 231, 232, 286, 437, 438
Phytophthora ramorum, 443-444
Phytoremediation, 51
Pigeon pea (*Cajanus cajan*), 214, 215
Pigments, 20
　carotenoid, 21, 59
　chlorophylls, 20, 57
　light-absorbing, 57-59
　xanthophylls, 20, 59
Pilobolus spp., 427, *427*
Pilocarpine, *344*
Pimenta dioica (allspice), 294, 300
Pimentel, David, 206
Pimpinella anisum (Anise), *302*, 303
Piña, 460
Pineapple (*Ananus comosus*), 86, *87*, 92, 93, 318, *318*
Pine mushroom (*Tricholoma magnivelare*), 467
Pine pitch, 330
Pines (*Pinus* spp.). *See also* Conifers; Gymnosperms
　characteristics of wood of, 328, *328*
　life cycle of, 139-141, *142*
　as lumber source, 328
　resins from, 330
Pinnately compound leaves, **38**, *39*
Pin oak (*Quercus pallustris*), 393
Pinto beans (*Phaseolus vulgaris*), 214
Piperaceae, *128*, 294, 375, *376*
Piper methysticum, 375
Piper nigrum (black and white pepper), 294, *294*, 295
Pistils. *See* Carpels
Pitavastatin, 473, *473*
Pitch, 330
Pitcher plants (*Sarracenia* spp.), 42, *43*
Pith, **33**
Pits, **19**, **32**
Plain sawn timber, 327, *327*

Plaiting, 319
Plantae (kingdom), 137. See also Plants
   embryophyte divisions in, *138*
   evolutionary history, 137, *138*
   phylogenetic diagram of major groups, *138*
Plantains (*Plantago ovata, P. arenaria*), 156
Plant anatomy, subdiscipline of, 5
Plantations
   Banana Republics, 226
   cacao, 285
   coffee, 274-275
   cotton and slavery, *314*
Plant biotechnology, subdiscipline of, 5. See also Biotechnology
Plant cells. See Cells
Plant defenses
   allelopathy and, 383-384, 504
   mechanical defenses, 387-388
   secondary compounds as, 14
   systemic acquired resistance and, 346
Plant diseases, 245, 437-444. See also *specific diseases*
   breeding plants resistant to, 257
   crop monocultures and, 247
   genetic engineering of resistance to, 257-258
Plant domestication. See Domesticated plants
Plant ecology, 5. See also Ecology
Plant fibers. See Fibers
Plant for the Planet: Billion Tree Campaign, 323
Plant genetics, subdiscipline of, 5. See also Genetics
Plant hormones, 92-93
Plant morphology, subdiscipline of, 5
Plant names
   binomial, **123**, *123*, 126
   common, 123-124
   early polynomial system of, 121, 123, *123*
Plant organs, 33-46
   edible (vegetables), 41-46
   leaves, 37-41, *39*
   roots, **35**, 35-37, *36*
   stems, 33-35, *34*
Plant pathology, subdiscipline of, 5. See also Plant diseases
Plant physiology, 48-52, 55-68
   absorption of water from soil, 49-51
   cellular respiration, 63-68
   fermentation. See Fermentation
   metabolism, 55-56
   photosynthesis, 56-63
   subdiscipline of, 5
   translocation, *47*, 51-52
   transpiration, 49
   transport systems, 48-52
   water movement in plants, 51
Plants
   allergy, 389-398
   alternation of generation in, *76*, 139-142
   carnivorous, 42, *43*
   characteristics of, 137, 139-141
   classification of, 517
   domesticated. See Domesticated plants
   drought-tolerant, 49
   drugs from. See Drugs
   essential elements in, *513*
   evolution of, 137, *138*, 405
   flowering. See Angiosperms
   food from. See Agriculture; Food plants
   foods reported as allergenic, 397
   genetic engineering of. See Genetic engineering
   genome maps, 113
   gymnosperms. See Gymnosperms
   hay fever, 391-394
   human society and, 2-6
   insecticides from, 388-389
   land plants, 137-138, *138*
   medicinal. See Medicinal plants
   names of, 123-124, 126
   overview of divisions of, 137-138
   psychoactive. See Psychoactive plants
   seed producing, 138, **146**, 148
   tissues, 29-33
   vascular, 139, 143, 145
Plant sciences, 4-5
Plants of known psychoactive value, *376*
Plant systematics, subdiscipline of, 5. See also Taxonomy
Plasma membrane, 19, *20*, 22, 24
Plasmids, **118**, **258**
Plasmodesmata, **19**, *19*, 24
Plasmodial slime molds, 421
Plasmodium, *420*
Plasmodium falciparum, 494
Plasmodium spp., malaria and, *347*, 348, 349, 494
Plasmogamy, **424**, *428*
Plasmolyzed, **20**
Plasmopara viticola (downy mildew of grape), 450
Plastids, **21**
Plastocyanin (Pc), 61
Plastoquinone (Pq), 61, 384
Pliny the Elder, 121, 383
Plotkin, Mark, 342, 510
Plum pox, 263, *263*
Pluteus cervinus, *4*
Plywood, 327-**329**
Pneumatophores, **35**
Poaceae (grass family). See Grasses (Poaceae)
Pod, **211**, *211*
Pod corn, 195
Podophyllotoxin, *344*
Podophyllum peltatum (mayapple), *344*
Poi, **238**
Poinsettia (*Euphorbia pulcherrima*), *72*, 385
Point mutations, **117**
Poison arrows, 379-380
Poison hemlock (*Conium maculatum*), 380, *380*
Poison ivy (*Toxicodendron radicans*), 14, 395, *396*
Poison oak (*Toxicodendron quercifolium*), 14, 395, *396*
Poisonous fungi, 475-477
Poisonous mushrooms, 477-478
Poisonous plants, 379-387. See also *specific plants*
   allelopathy and, 383-384, 504
   in the backyard, 382-386
   in the home, 386
   insecticides from, 388-389
   livestock and, 386-387
   use of by humans, 379
   in the wild, 379-381
Poisons
   alkaloids in, 379
   amatoxins, **477**
   curare, 379-380
   fungal, 472-480
   glycosides in, 379
   mycotoxins, 472-474, 483
   ricin, 385-386
   strychnine, 379
Poison sumac (*Toxicodendron vernix*), 395, *396*
Polar ice caps, global warming and, 494-495
Polar nuclei, **79**
Pollen, **70**, **79**
   allergies caused by, 391-394
   angiosperm, 79
   gymnosperm, 139-140
   palynology and, 79
   protein quality of, 251
   symbolic meanings of, 79
Pollen chambers, 70
Pollen cones, 139, 140, 146, 147
Pollen grains, 83, 139-140, 149
Pollen tubes, **83**
Pollination, 79-84
   by animals, 80-81, 83
   cross-pollination, 79
   self-pollination, 79, 105
   by wind, 83
Polo, Marco, 291-292
Polyculture, 248
Polygenic inheritance, **109**
Polynomials, **121**, *123*, *123*
Polypeptides, **10**, *11*, 114, 115
Polyphenol oxidase, 269
Polyploid, **130**
Polysaccharides, **9**, *9*, **153**, 156
Polyterpenes, 14
Polyunsaturated fatty acid, **160**
Pomes, **86**, *87*
Popcorn, 194-195, *195*, 197-198
Poplars (*Populus* spp.), 345
Poppy. See Opium poppy (*Papaver somniferum*)
Poppy seed oil, 360
Poppy seeds, 360
Population, **487**. See also Human population growth
Porphyra spp., 409-410, *410*
Postelsia spp., 405
Post-traumatic stress disorder (PTSD), 365
Potassium, *167*
Potato beetle, 232
Potato chips, 230, 234
Potatoes. See Sweet potato (*Ipomoea batatas*); White otato (*Solanum tuberosum*)
Potato War, 229
Potential energy, **55**, *55*
Potrykus, Ingo, 262
Pot-still distillation, 458
Pravachol, *473*
Pravastatin, *473*, 473-474
Precision breeding, **257**
Prescription drugs. See Drugs
Pressing, 449
Pressure Flow (Mass Flow) Hypothesis, **52**
Prevost, Benedit, 438
Prickles, 387
Primary cell wall, **19**
Primary consumers, **2**, **488**, 489, *490*
Primary endosperm nucleus, **84**
Primary growth, **29**
Primary metabolites, 14
Primary protein structure, *11*
Primary succession, **497**-498, *498*
Primary xylem, 33
Principle of dominance, 104
Principle of independent assortment, 106
Principle of segregation, 104
Procaine, 369. See also Novocain
Producers, **2**, **488**, 489, *490*
Prohibition, 447
Prokaryotic cells, 8, *8*, 137
Proline, *10*, 157
Prophase (mitotis), **24**-25, *26*
Prophase I (meiosis), *76*, *77*
Prophase II (meiosis), *76*, *77*
Prop roots, **37**, **188**
Prostaglandins, 346
Prosthetic groups, **56**
Proteasomes, **23**, *24*
Proteins, **8**, **157**
   biological functions of, 9-10
   in cell membrane, 19, *20*
   complete, 157
   control of production of by genes, 114
   enzymes, 56
   human nutrition and, 153, 156-157, *157*
   incomplete, 157
   integral, 19
   peripheral, 19
   structure of, *11*
   synthesis of, 22-23, 116, *117*

Protista (kingdom), 137, 420, 424
Protonema, 139
Protons, **513**
Protoplast, **19**-**24**, **257**
Pseudoephedrine, 151, 352
Pseudo-nitzschia spp., 417
Pseudotsuga menziesii (Douglas fir), *124*, 328, *328*
Psilocin, 479-480
Psilocybe spp., 479
Psilocybin, 479-480
Psilotum spp., 145
Psophocarpus tetragonolobus (winged bean), 222
Psychoactive drugs, 359, **359**
    alcohol as. *See* Alcoholic beverages
    cocaine, 15, 343
    depressants, 359
    hallucinogens, 359
    lysergic acid diethylamide (LSD), 477, 479
    morphine, 15, 343
    nicotine, 15
    stimulants, 359
Psychoactive plants
    coca, *344*
    cocaine, 366-370
    kava (*Piper methysticum*), 375-376
    lesser known, 376
    marijuana, 362-366
    nutmeg, *294*, 295-296
    opium poppy, 360-362
    peyote, 374-375
    psychoactive drugs, 359
    tobacco, 370-374
    tropane alkaloids in, 15
Psychological dependence, **359**
Psyllids (*Candidatus liberibacter*), 97
Psyllium husks, 156
Pterophyta (ferns), *138*, 139, 143-146
Ptychopetalum olacoidesis, 506
Puccinia graminis (stem rust of wheat), 191, 439-441
Puffballs, 430, *432*, 482
Pulque, 459
Pummelos (*Citrus grandis*), 95, 96
Pumpkins, 52, 86
Punnett, Reginald C., 109
Punnett squares, 104
    dihybrid crosses, 105
    incomplete dominance, 106
    monohybird crosses, 105
    solving genetics problems with, 107-108
Pun-Tsao, 338
Purine, **12**, **111**
Purple coneflower (*Echinacea purpurea*), 69, 340, *341*
Pyramids, ecological, **489**-**491**
Pyrethrum, 389
Pyrimidine, **111**, 115
Pyrimidine base, **12**
Pyrixidine, *162*
Pyruvate, 64

Q

Quakers, cocoa and, 283
Quaking aspen, *124*
Quarternary protein structure, *11*
Quarter-sawn lumber cut, 327, *327*
Queen Anne's lace, 41, *41*
Quercus spp.
    affect of global warming on, 496
    allergies to pollen of, 393, *393*
    cork from, 330
    corn from, 31
    as example of genus, *129*
    Q. alba, 2, 329
    Q. coccinea, 40
    Q. kelloggii, 40
    Q. phellos, 129
    Q. rubra, 129
    Q. suber, 31, 330, *330*
    stomata on leaves of, *40*
    sudden oak death in, 443-444
    symbolic meaning of, 127, *129*
    as timber source, *328*, 329
Quetzalcoatl, Aztec god, 283
Quina-quina, 348
Quinine, 343, 348, 349
Quinoa (*Chenopodium quinoa*), 252, *252*
Quorn mycoprotein, 467-468

R

Raceme, **74**, *75*
Radial timber cut, 327, *327*
Radiant energy, 55, 56
Radicles, **88**
Radioactive isotopes, 513
Radio waves, 56, *57*
Radishes (*Raphanus sativus*), 44-45
Ragweed (*Ambrosia* spp.), 392-393
Rainforest Alliance, 279
Raleigh, Walter, 371
Ramie (*Boehmeria nivea*), 309, 316
Ranunculaceae, *128*
Rapeseed (*Brassica napus, B. campestris*), 188, 219
Raphanus sativus (radishes), 44-45
Raphides, 386
Rattlebox (*Crotalaria* spp.), 381
Rattlebox moth, 381
Rauwolfia serpentina (snakeroot), *344*, 350-351
Rayon, 319, *319*
rbcL gene, 132
Reactants, 55, 56, *63*
Reaction center, **59**
Receptacle (algal), 405
Receptacle (plant), **70**
Recessive alleles, **104**
Recognition factors, 391
Recombinant DNA molecule, 118
Recombinant DNA technology, **118**, **119**. *See also* Biotechnology; Genetic engineering
Recycled paper, wood pulp from, 334

Red algae (Rhodophyta), 401, *402*, 405, *406*, 410
Red Bean Dance, 384
Red bottlebrush (*Callistermon citrinus*), 384
Redbud, *124*
Red cabbage (*Brassica oleracea*), 322
Red fescue, 206
Red oak (*Quercus rubra*), *129*
Redox reactions, **55**
Red tides, 414. *See also* Algal blooms
Reduced (redox reaction), **55**
Reduction, 75. *See also* Meiosis
Red wine, 14, 449. *See also* Wine, 451
Red yeast rice, 474
Reeds, 325
Reference daily intakes (RDIs), 165n
Regular (actinomorphic) flowers, **74**, *74*
Reproduction, 6. *See also* Asexual reproduction; Sexual reproduction
Reserpine, *350*, 351
Resins, 14, 148, **329**-330, **362**, 437
Resource partitioning, 487-488
Respiratory allergies, 390-391
Responses, as fundamental property of life, 6-7
Restriction enzymes, **118**
Resurrection plant (*Salaginella*), 144
Resveratrol, 464
Reticulate (net) venation, **38**
Retinal, *162*, 163
Retinol, 163
Retting, **310**
Rhabdomyolysis, 474
Rhizobium spp., 211, *211*, 247
Rhizomes, **145**, **188**, **225**
Rhododendrons, 383, 443, 444
Rhodophyta (red algae), 401, *402*, 405, *406*
Rhodopsin, 162
Rhodymenia spp. (dulse), 409
Rhus, 397
Riboflavin, *162*
Ribonucleic acid (RNA). *See* RNA (ribonucleic acid)
Ribose sugar, **12**, *13*
Ribosomal RNA (rRNA), 116
Ribosomes, **22**-**23**
Ribulose-1,5-bisphosphate (RUBP), 62
Ribulose bisphosphate carboxylase (RUBISCO), **62**
Rice (*Oryza sativa*), 200-202, *201*, 248
    Azolla symbiosis and, 145, 201, *201*
    brown spot of rice, 437
    brown *vs.* white, 201
    cultivation of, 146, 200

    genetic engineering of flood tolerant, 200-201
    genome of, 114, 202
    global production of, *188*
    Golden rice, 262
    origins of, 200
    varieties of, 201-202
Rice blast fungus (*Magnaporthe oryzae*), 269
Rice wine (sake), 457
Ricin, 385-386
Rickets, **164**, *164*
Riddling, 453, *453*
Rig Veda, 479
Ringworm, 480-481
Rizzo, D., 443
RNA (ribonucleic acid), **12**-**13**
    messenger (mRNA), 114, 115
    ribosomal (rRNA), 116
    small interfering (siRNA), 117
    transcription and, 114-115
    transfer (tRNA), 116
    translation and, 116
RNA polymerase, 114-115
Robigalia, 439
Robinia pseudoacacia (black locust), 384
Rockweed, *4*, 404
Romaine lettuce, 44
Romans
    absinthe and, 461
    linen clothing, 312, 315
    medicinal plant use by, 338
    papyrus writing surfaces, 331
    spice trade and, 290-291
    winemaking in, 447
Roman wormwood (*Artemesia pontica*), 461
Ronald, Pamela, 202
Root caps, **35**
Root hairs, **35**
Root nodules, **211**
Roots, 35-37
    adaptations in, 35
    adventitious, 225, 232, *233*
    aerial, 37
    contractile, 37
    dicot, 35
    fibrous, 35, *35*
    modified as storage organs, *225*, 226-227
    monocot, *36*
    pneumatophores, 35
    prop, 37, **188**
    taproots, 35, *35*
    water uptake by, 48
Rootstock, **95**
Root tips, *36*
Rope, fiber for, 309, 318, *318*
Rosaceae, *128*
Rosary pea (*Abrus precatorius*), *221*, 384
Roselius, Ludwig, 278
Rosemary (*Rosmarinus officinalis*), *291*, 302
Roses, 80, 82, *82*, 125, 127, *128*
Rose wine, 449. *See also* Wine

Index

Rosewood (*Dalbergia* spp.), 221
Rosin, 330
Ross, Ronald, 348
Rosuvastatin, 473, *473*
Rotenone, 389
Rough ER, **22,** *23*
Roundworm (*Caenohabditis elegans*), 112
Rowntree, Joseph, 285
rRNA (ribosomal RNA), **116**
RSW1 gene, 309
Rubber, 508
   from guayle, 501
   from rubber tree, 14, 398, 508, *508*
   sustainable tapping of, 508
   synthetic, 501
Rubber tree (*Hevea brasiliensis*), 14, 398, 508
Rubiaceae (Coffee family), *128,* 359
RUBISCO, **62**
RUBP, 62
Runners, **225**
Russian tea, 280
Rusts, 430, 439-441
Rutaceae, *128*
Rutgers University, 471
Rye (*Secale cereale*), 188, 202, *203*
   celiac disease and, 158
   ergotism and, **475,** *475,* 476-477
Ryegrass (*Lolium* spp.), *206,* 383

S

Saccharomyces cerevisiae, 447
Saccharum officinarum (sugarcane), 54
   bioethanol from, 207
   C_4 pathway and, 63
   global production of, *188*
   processing of, 53-54
   slave trade and, 53-54
Sacred lotus (*Nelumbo nucifera*), 41
Safflower, 219, 321
Saffron (*Crocus sativus*), 294, 297, *297*
Sage (*Salvia officinalis*), *301,* 302
Sagebrush (*Salvia leucophylla*), 383
Sago palm, 147
Sahagun, Bernardino de, 374
Sake, 457
Salem Witch Trials, *476*
Salicaceae, *128*
Salicin, 344, *345*
Salicylic acid, 93, 345, 346
Saline soils, 50
Salix alba (white willow), 344, *345*
Saltcedar, 415
Salvia leucophylla (sagebrush), 383
Salvia officinalis (Sage), *301,* 302
Samaras, **86,** *87*
Sandoz Pharmaceutical Company, 473
Sanguinaria canadensis (bloodroot), 340, *341*
Sanguinarine, 341, *344*

San Joaquin fever, 481
Sapindaceae, *128*
Sap of a vine (*Ipomoea alba*), 508
Sapogenic glycosides, 338
Saponins, 14-15, **238,** 343, 384, 388
Saprobes, **4,** *422,* **433,** 488
Saprolegnia, 422, *423,* 438
Sap, translocation of, 52
Sapwood, **327,** *327*
Sapwood rot, 436
Sarracenia spp. (pitcher plants), 42, *43*
Sativex, 365
Saturated fatty acid, **159**-160
Savanna, *502,* **505**
Savory (*Satrueja hortensis*), 302
Saw palmetto (*Serenoa repens*), **356,** 356-357
Scarlet oak (*Quercus coccinea*), 40
Schizocarps, 303
Schizophrenia, *350,* 350-351
Schleiden, Matthias, 17
Schwann, Theodor, 17
Scientific method, 5-6
Scientific names, *124,* 126
Scion, **95**
Sclereids, **31**
Sclerenchyma, **31,** *31, 33*
Sclerophyllous, **504**
Sclerotia, **475**
Sclerotium (sclerotia), **476**
Scolypopa australis (Passionvine hopper), 383
Scopolamine, 367
Scotch whiskey, 458
Scouring rushes, 145
Scoville ratings, 298, *298*
Screw pine (*Pandanus* spp.), 37
Scurvy, **164**-165
Scutellum, **88**
Scythians (ancient nomadic Slav horsemen), 362
Sea level, global warming and, 494-495
Seaweeds, 409-410
Secale cereale (rye), 188, 202
   celiac disease and, 158
   ergotism and, **475,** *475,* 476-477
Secalin, celiac disease and, 158
Secondary cell wall, **19**
Secondary consumers, **2, 488,** 489, *490*
Secondary metabolites. *See* Secondary products (compounds)
Secondary pathways, 14
Secondary products (compounds), 14, **290,** 343, 470. *See also specific compounds*
   allelopathic compounds, 383-384
   commonly occurring, *15*
   essential oils and, 290, 291
   function of in plants, 14

   insecticides from, 388
   in medicinal plants, 343
   uses of by humans, 15
Secondary protein structure, *11*
Secondary succession, **497,** 498, *499*
Secondary xylem, 33
Second filial (F_2) generation, 104
Second Law of Thermodynamics, **55,** 489
Second Opium War, 361
Seed banks, 249-250
Seed coats, **84, 88**
Seed cones, 140-141, 146-147, *147*
Seed dispersal, 86, 96, 141
Seed plants, 4, 138, **146,** 148. *See also* Angiosperms; Gymnosperms
Seeds, **84, 88,** 138
   dicot, 88
   germination and development, 88
   monocot, 88
Seed Savers Exchange, 250
SEEDS project, 94
Segregation, principle of, 104
Selaginella spp. (resurrection plant), 144
Selective breeding, 102, 134, 242-243
Selective cutting techniques, *507,* 510, 511
Selenium, *167*
   soil concentration of, 251
Self-pollination, **79,** 105
Semiconservative DNA replication, *115*
Semidrying oils, **219**
Semolina flour, 191
Semper augustus tulip, 72, *72*
Senna pods (*Cassia fistula*), 221
Sepals, **3,** *70, 74*
Septa, *422*
Septate, **423**
Serine, *10, 157*
Seringueiros (Brazilian rubber tappers), 508
Serotonin, 477
Serpentine soils, 50
Serpula lacrymans, 436
Serturner, Friedrich, 340, 361
Sesame oil, 219
Sex hormones, from yams, 238
Sexual reproduction, **6,** 75-79
Sexual system of plant classification, *122,* 122-123
Shagbark hickory (*Carya* spp.), *124*
Shallots (*Allium ascalonicum*), *302,* 304
Shattering fruiting heads, **184**
Shellfish, 417
Shelterwood cutting technqiues, *507,* 510, 511
Shen Nung (Emperor), 362
Shiitake mushroom (*Lentinus edodes*), 465

Shortgrass prairies, **504**
Short rotation (coppice) timber management, *507,* 511
Sick building syndrome, 483-484
Siegesbeck, Johann, 122-123
Sieve plates, **33**
Sieve tube members, **33,** 51
Sieve tubes, **33**
Silent Spring (Carson), 491
Silibinin, 478
Silica, 145
Silkworm (*Bombyx mandarina*), *182*
Silphium (*Ferula* spp.), 338, *339*
Simmondsia chinensis (jojoba), *501, 504*
Simple fruits, **86**
Simple goiter, **168**
Simple leaves, **38**
Simple sugars (monosaccharides), **8,** *9,* 153
Simvastatin, 473, *473*
Sinigrin, 304
Sinks, **51**
Sinsemillas, 364
Sisal (*Agave* spp.), *309,* 318, *318*
Sizings, 237, 333
Skunk cabbage (*Symplocarpus feotidus*), *124,* 386
Slavery
   in cacao plantations, 286-287
   cotton production and, *314*
   sugar trade and, 53-54
Slime molds (myxomycetes), *420,* 421
SLIPS (slippery liquid-infused porous surfaces) technology, 41, *42*
Small interfering RNA (siRNA), **117**
Smithsonian Migratory Bird Center, 279
Smoke tree, *124*
Smooth ER, **23**
Smuts, 430, 439, 441
Snakeroot (*Rauwolfia serpentina*), 344, 350-351
Snow-on-the-mountain (*Eurphorbia marginata*), 385, 388
Snow peas, 215
Society of Friends (Quakers), 283
Socrates, poisoning of, 380
Sodium, *167*
Sodium chloride (NaCl), 514
Soft drinks, caffeine in, *272.* *See also* Coca-Cola
Soft fibers, **309,** 310
Soft wheat, 191
Softwoods, **326**-327
Soil
   heavy metals in, 50, 51
   mineral-deficient, 50
   saline, 50
   serpentine, 50
   as source of water, 48
   water uptake from, 48
Solanaceae (nightshade family), *128,* 298, 359, 367, 370

Solanum lycopersicum. *See* Tomato (*Solanum lycopersicum*)
Solanum pimpinellifolium, 91
Solanum quitoense (naranjilla), 254
Solanum tuberosum. *See* White potato (*Solanum tuberosum*)
Soldiers' disease, 361
Solidago canadensis (goldenrod), 251
Soluble dietary fiber, 156
Solvent extraction, 82
Soma, 478-479
Somaclonal variants, **256**
Somatic mutations, **118**
Sonoran Desert, 501
Sophora secundiflora, 384
Sorghum (*Sorghum bicolor*), 203, 204-205, 248
  C_4 pathway and, 63
  genome of, 114
  global production of, *188*
Sori (sorus), **139**
Source, **51**
Sour oranges (*Citrus aurantium*), 95, 96
Sousa, John Philip, 367
Souter, David, 366
Southeast Asia, early sites of agriculture in, 183
Southern leaf blight (*Bipolaris maydis*), 197, 437, 441
Soybean oil, 160, 162, 262
Soybeans (*Glycine max*), 217, 217-218
  allergies to foods from, 398
  anti-cancer diets and, 172
  crop rotation with, 212, 217
  economic importance of, 217-218
  foods from, 218
  genetic engineering of herbicide-resistant, *259*, 259-260
  genome of, 217
  global production of, *188*
  nutritional value of, 218
  oils from, 219
  phytoestrogens in, 218
Soy oil, 219
Space shuttles, studies with tomatoes on, 94
Sparkling wines, 453
Spearmint (*Mentha spicata*), 302, *302*
Specialized niche, **488**
Specialty coffee, 279
Species, **2**, 2n, 129-130
  biological species concept, 130
  DNA barcoding of, 132
  ecological species concept, 130
  genealogical species concept, 130
Species Plantarum (Linneaus), 123, *123*
Species shift, 496
Specific epithet, *127*
Sperm, **6**, 75
Spermagonia, **440**
Spermatia, **439**, 440

Spherule stage in lung tissue, *482*
S (synthesis) phase, 24, *25*
Spices, 290-300
  Age of Exploration and, 292
  allspice, 300
  aromatherapy and, 291
  black pepper, 294-295
  capsaicin peppers, 297-299
  cinnamon, 293
  cloves, 295
  commonly used, *294*
  essential oils in, 290, 291
  ginger, 296-297
  history of trade of, 290-293
  nutmeg and mace, 295-296
  saffron, 297
  terpenes in, 14
  turmeric, 296-297
  vanilla, 299-300
  white pepper, 294-295
Spice trade, 290-293
Spike, **74**, *75*
Spikelet, **188**
Spindle, **25**
Spines, 387
Spinning, 310-312
Spirulina spp., 409, *409*
Split peas, 215
Spongy, **41**
Sporangia (sporangium), **139**, 408, 424, *425*, 426
Sporangiophores, 426-427
Sporangiospores, **426**
Spores, **4**, 75, 139, 421, **424**
Sporophylls, 139
Sporophyte, **75**, *76*, 139, *143*
  dominant in gymnosperms, 139
  in lower vascular plants, 139
Spray drying, 278
Spring wheat, 191
Springwood, **34**
Spruces, 325, 328
Spurge family (Euphorbiaceae), 51, *128*, 385
Squash, 86, 90, 183
Stachybotrys chartarum, 483
Stamens, **3**, **70**
Staminate flowers, **73**
Stang, Dorothy, 508-509
St. Anthony, *475*, 476
Stapelia spp. (carion flower), 80
Staphylococcus aureus, 470
Staphylococcus spp., 472
Star anise (*Illicium verum*), *302*, 303
Starches, **8**, **153**
  amylopectin, 194, 237
  amylose, 194, 237
  commercial uses of, 237
  extraction of from plants, 237
  human nutrition and, 153
  storage of in plant organs, 225-227
  structure of, *9*
StarLink, 398
StarLink corn, 261-262
Starvation, 154
Statin drugs, *473*

Statins, 473-474
St. Augustine grass, *206*
Stearic acid, *160*
Stele, **35**, *36*
Stem rust of wheat (*Puccinia graminis*), 191, 439-441
Stems, 33-35
  dicot, 33, *34*
  herbaceous, 33, *34*
  monocot, 33, *34*
  starch storage in modified, *225*, 225-226
  woody, 33, *35*
Steroids, **11**-12, *12*, 14, *159*, 160, **343**
Stevens, John Paul, 366
Stevia, 306
Stevia rebaudiana (sweetleaf), *302*, 306
Stevioside, 306
Stigma, **72**
Stills, 458
Stimulants, **359**
  caffeine, 15
  cocaine, 15
  ephedrine, 352
  nicotine, 15
Stimulating beverages, 271-288
  caffeine content of, *272*
  caffeine in, 15
  chocolate, 283-287
  Coca-Cola, *272*, 287
  coffee, 273-279
  guarana, 287
  mate, 287
  tea, 279-282
Stinging nettles, 388
Stinkhorns, 431, *432*
Stipe, **405**
Stipules, **38**
St. John's wort (*Hypericum perforatum*), 355, 388
Stock, 95
Stolons, **188**, **225**
Stoma, **30**
Stomata, **30**
Stone, Edmund, 345
Storage organs, 225-227
Strawberries, 86, *87*, 397
Strawberry clover (*Trifolium casteaneum*), 487
Streptomyces griseus, 471
Streptomyces spp., 471, 482
Streptomycin, 471
String instruments, 324-326
Strip-cutting technique, 507, *507*, 511
Stroma, **20**, **59**
Stroma thylakoids, **59**
Strychnine, 379
Strychnos nux-vomica, 379
Strychnos toxifera, 379
Style, **72**
Subdivisions, 127
Suberin, **31**, 50, 330
Sublingual immunotherapy (SLIT), 395

Subsidiary cells, **30**-31
Substrates, **56**
Succession. *See* Ecological succession
Succulents, 30, 63
Sucrose, **8**, *9*, 53, **153**
Sudden oak death (*Phytophthora ramorum*), 443-444
Sugar, **8**. *See also* Sugar beet (*Beta vulgaris*); Sugarcane (*Saccharum officinarum*)
  disaccharides, 8, *9*, 153
  human nutrition and, 153, 156
  monosaccharides, 8, *9*, 153
  polysaccharides, 9, *9*, 153, 156
  production of and slave trade, 53-54
  translocation of, *47*, 51-52, *52*
Sugar beet (*Beta vulgaris*), 9, 54, *54*, 188
Sugarcane (*Saccharum officinarum*), 9, 54
  bioethanol from, 207
  C_4 pathway and, 63
  global production of, *188*
  processing of, 53-54
  slave trade and, 53-54
Sulforaphane, 172
Sulfur, *167*
Summerwood, **34**
Sun, 56-57
  electromagnetic radiation from, 56-57, *57*
  light reactions and, 61
Sundews (*Drosera* spp.), 42, *43*
Sunflower (*Helianthus annuus*), *7*, *40*, 49, 69, 86, 125, 220, *248*
Sunflower oil, 219, 220
Sun pitchers (*Heliamphora*), 42
Supercritical carbon dioxide process, 278
Superficial mycoses, 480, 481
Superior ovary, **73**
Supermarket botany, 44
Surface fibers, **309**, 310, 318
Surgeon General's Call to Action to Prevent and Decrease Overweight and Obesity, 155
Survival to reproduce, natural selection and, 134
Sushi, *410*
Sustainable agriculture, 248-249
Sustainable harvesting, 508-510
Svalbard Global Seed Vault, 249
Swamp pink (*Helonias bullata*), 488
Swarm cell, 421
Sweet bananas. *See* Bananas
Sweet basil (*Ocimum basilicum*), *301*, 302
Sweet corn, 195, *195*
Sweeteners, starch-based, 237
Sweetleaf (*Stevia rebaudiana*), *302*, 306
Sweet orange (*Citrus sinensis*), 95, 96
Sweet potato (*Ipomoea batatas*), 188, 232-234, *233*
  genetically modified, 263

Swimmers itch, 413
Swiss-Water decaffeination method, 278
Switchgrass (*Panicum virgatum*), *207*, 207–208
Symbiosis, **150**
Symphonie Fantastique (Berlioz), 360
Symplast, **49**
Synapsis, **76**
Synergids, **79**
Synsepalum dulcificum (miracle fruit), *302*, 306
Syconium, 74, *75*
Systema Naturae (Linneaus), 122, *122*
Systematics, 131
Systematics Agenda 2020, 131
Systemic acquired resistance, **346**
Systemic mycoses, 481–482

T

Tabebuia spp., 506
Tagua (*Phytelephas* spp.), 509–510, *510*
Taiga, **505**
Tall fescue, *206*
Tallgrass prairies, **504**
Tamarillo (*Cyphomandra betacea*), 254
Tamarind (*Tamarindus indica*), 221
Tamarisk, 415
Tamarixia radiata (Asian wasps), 97
Tangential lumber cuts, 327, *327*
Tangerines (*Citrus reticulata*), 95
Tannic acid, 330
Tannins, 14, **23**, **281**, 388, 437, 449
Tannosome, **23**, *24*
Tapioca. *See* Cassava (*Manihot esculenta*)
Taproots, **35**, *35*, **225**, 227
Taraxacum officinale (dandelion), 37
Taraxicum (dandelion) pollen grain, *80*
Tariff Act of 1893, 90
Taro (*Colocasia esculenta*), 238, *239*
Tarragon (*Artemesia dracunculus*), *302*
Tarwi (*Lupinus mutabilis*), 253
Tatum, Edward, 114
Taxine, 382
Taxodium distichum (bald cypress), 35, 37
Taxol, 14, *15*, 353
Taxon (taxa), **127**
Taxonomic hierarchy, 127–130
  five-kingdom system, 137
  major ranks in, *130*
  standard endings used in names, *129*
  three-domain system, 129, 137
Taxonomy. *See also* Classification
  early history of, 121–123
  PhyloCode, 130, 132
  plant names, 123–126
  saving species through, 132
  taxonomic hierarchy, 127–130
Taxus spp. (yews), 147, **344**, 353
Tea (*Camellia sinensis*), 279–282
  caffeine in, *272*
  cultivation of, 280–281
  flavor of, 281–282
  health benefits of, 281–282
  history of, 282
  origins of, 279
  tannins in, 14, 281
  tea ceremonies and, 280
Tea Act of 1773, 282
Tea ceremony, 280
Teliospores, **439**
Telomeres, **25**
Telophase (mitosis), **24**, *26*, 27
Telophase I (meiosis), 76, *77*
Telophase II (meiosis), 76, *77*
Tempeh, 218, 466
Temperate rain forests, **505**
Teonanacatl mushrooms, 479, *479*
Teosintes, 197–199, *199*
Tepals, *72*
Tequila, 459–460
Terpenes, 14, **290**, **343**, 383, 384
Terra Nova Forest Reserce, 342
Tertiary consumers, **488**
Tertiary protein structure, *11*
Testa, **88**
Testcross, **105**
Tetracycline, 471
Tetrads, **77**
Tetrahydrocannabinol (THC), 14, *344*
  psychoactive effects of, 364–365
Texas Court of Appeals, 484
Textured vegetable protein (TVP), 218
Thallus, **401**
Theaceae, *128*
Theaflavin, 281
Theanine, **281**
Thea sinensis, **271**
Theatrum Botanicum (Parkinson's), 338
Theobroma cacao (cacao tree), **271**, 283, *285*
  Aztecs and, 283
  child slave labor and, 286–287
  cultivation of and processing of, 285–287
  diseases affecting, 286
  Quakers and, 283
Theobromine, 284
Theol, 281
Theophrastus (Father of Botany), 121, 439
Theophylline, 282
Theory, 5
Thermal energy, 55
Thermodynamics, Laws of, 55, 489
Thiamine, *162*, 165
Thiamine pyrophosphate, 165
Thistles (*Cirsium* spp.), 383
Thlaspi caerulescens (alpine pennycress), 51
Thomas, Clarence, 366
Thornapple, *368. See also* Jamestown weed
Thorns, 387
Thousand and One Nights, 363
Three-domain classification system, 137
Threonine, *10*, 157
Thylakoid membranes, **57**, 59, *60*
Thylakoids, **20**, 23
Thyme (*Thymus vulgaris*), *302*
Thymine, **12**, 13, **111**
Thymus vulgaris (Thyme), *302*
Tiger lily (*Lilium lancifolium*), *73*
Tillage techniques, 246, *247*
Timber
  cuts of, 327
  deforestation from harvesting of, 323, 506
  management methods, 506
  sustainable forestry and, 323
Tinea, *480*, 480–481
Tinea capitis, *480*, 480–481
Tinea corporis, 481
Tinea pedis. *See* Athlete's foot
Tinea unguium, *480*
Tipiti, 235
Ti (tumor-inducing) plasmid, **258**
Tissue culture, 256–257
Tissues, **29**–33
  aerenchyma, 35
  dermal, 29–31, *30*, *34*
  ground, 29, 31, *31*, *34*
  meristems, 29
  organization of plant, 29
  vascular, 29, 32–33, *34*, 48, 138, 143
Tmesipteris spp., 145
Toadstools, 477
Tobacco (*Nicotiana* spp.), 15, 370–374
  adverse effects of smoking, 172
  cultivation practices, 371
  hazardous components of, *372*
  health risks, 371–374
  New World habit, 370–371
Tofu, 218
Tolerance, **359**
Tolypocladium inflatum, 473
Tomato (*Solanum lycopersicum*), 90–94
  annual production of, *188*
  anthocyanins in, 91
  commerical production of, 90–91
  cultivars of, 90–91
  domestication and spread of, 91
  ethylene and, 93, *94*
  fruits of, 86, 91
  genetic engineering of, 263
  heirloom varieties, 91, *91*
  legal definition of fruits and, 41, 90, *90*
  origins of, 90, 183, 184
  poisonous reputation of (false), 90, 91
  in space research, 93–94
Tonka beans (*Dipteryx odorata*), 300
Tonkin bamboo (*Arundinaria amabilis*), 335
Topoisomerase inhibitors, 353
Topotecan, 353
Toxic molds, 482–484
Toxins
  from algae. *See* Algal blooms
  amatoxins, 477
  cyanogenic glycosides, 14, 15, 235, 343
  in cycads, 148, 149
  mycotoxins, 472–474
Trace elements, **513**
Trace minerals, **167**, *167*
Tracheids, **32**, 48
Transcription, 114–115
TransFair USA, 279
Trans fats, **161**
Transfer RNA (tRNA), **116**
Transgenic plants, 257–258
Transglutaminase, celiac disease and, 158
Translation, 116
Translocation, *47*, 51–52
Transpiration, 49
Transpiration-Cohesion Theory, 48, **51**
Transport systems, 48–52
  absorption of water from soil, 48
  translocation, 51–52, *52*
  transpiration, 49
Transposable elements (transposons), 196
Travels of Marco Polo, The, 292
Trebouxia spp., 434
Tree harvest methods, 507, *507*, 511
Tree nut allergies, 397
Trees. *See also* Wood; Wood products
  allergies to pollen of, 393, *393*
  annual rings, 34, 37, *38*
  annual rings of, 327
  genetically modified, 265–266
  harvesting techniques, 506–507, *507*, 510–511
Tremetol, 386
Trentepohlia spp., 434
Triangular Trade, *53*
Trichoderma, 286
Tricholoma magnivelare (pine mushroom), 467
Trichomes (hairs), **30**, *30*
Trichophyton rubrum, 480
Trichophyton spp., 480
Trifolium spp., 487
Triglycerides, **11**–12, 159, **159**
Trigonella foenum-graecum (fenugreek), *221*, 302
Trinitario, 285
Triploid endosperm, **84**
Tripsacum spp., 198, 199
Tristeza y muerte de agave (TMA), 460
Triticale (*Tritosecale*), 202, *203*
Triticum durum (durum wheat), 204
Triticum spp., 189–193

Tritordeum, 204
tRNA (transfer RNA), **116**
Tropane alkaloids, 15, 367
Trophic levels, **488**
Tropical rainforests, **506**
    biodiversity in, 131
    carbon cycle and, 492-494
    medicinal plants found in, 340, 342, 506
    rubber tapping in, 508
    species losses in, 131, 506
    sustainable harvesting from, 508-510
    Tagua from, 509-510
True breeding plants, 103
Truffles, 427, *429*, 465
Trypsin, 217
Tryptophan, *10, 157*
Tsuga canadensis, 380
Tuba root (*Derris* spp.), *221*, 389
Tube cells, 77, 83
Tuberous roots, **226**-227
Tubers, **225**
Tubocurarine, *344*, 380
Tulip fever, 72
Tulips (*Tulipa* spp.), 71-72, *125*, 384
Tumble-weed shield lichen (*Parmelia*), 381
Tundra, *503*, **505**
Tung oil, 219
Turgid, **20**
Turkey X disease, 472
Turmeric (*Curcuma longa*), 294, 296-297, 320, 321, *322*
TVP (textured vegetable protein), 218
Twining, 310, 311
2002 National Survey on Drug Use and Health, 369
Type 2 diabetes, 155, 349
Tyrian purple, 321
Tyrosine, *10, 157*

U

Ubiquitin, 23
Ulmus americana (American elm), 441-443
Ultraviolet radiation, 56, *57*
Ulva spp., *409*
Umbel, **74**, *75*
Undaria pinnatifida, 409
Undernutrition, **154**
Unicellular organisms, 8
United Fruit Company, 226
United Nations, 370
United States Department of Agriculture (USDA), 269
Universal veil, **477**
Unsaturated fatty acids, **159**-160
Uracil, **12**, 115
Uredia, **440**, 441
Uredospores, **439**
Urokinase, 173
Urushiol, 14, *15*, 397
USDA Food Pyramid, 169
U.S. Department of Agriculture, 470

U.S. Drug Enforcement Agency, 366
U.S. Food and Drug Association (FDA), **269**, 352, 366, 373, 375, 398, 472
Usnic acid, 381
U.S. Supreme Court, 366
    legal definition of fruits, 41, 90, *90*

V

Vaccines, 264-265
Vaccinium macrocarpon (cranberry), *355*
Vacuole. *See* Central vacuole
Valine, *10, 157*
Valley fever, 481
Vancomycin, 472
Vane, John, 346
Van Gogh, Vincent, 462
van Houten, Conrad, 283
Vanilla (*Vanilla planifolia*), *124*, *294*, *299*, 299-300
Vanilla odorata, 299
Vanilla planifolia (Vanilla), *294*, 299, 299-300
Vanilla tahitensis, 299
Vanillin, 300
Variation, natural selection and, 134
Vascular bundles, **33**, *34*
Vascular cambium, **29**, *29*
Vascular cylinder, **35**, *36*
Vascular plants, 139, 143, 145. *See also specific divisions*
Vascular rays, **34**
Vascular tissues, **29**, 32-33, *34*, 138, 143. *See also* Phloem; Xylem
Vat dye, 321
Vavilov, Nikolai, 185
Vegans, **172**
Vegetable oils, 160, *219*, 219-220
Vegetables. *See also specific vegetables*
    legal definitions of, 41, 90
    supermarket botany, 44
Vegetarian, **172**
Vegetative cell, **77**
Vegetative reproduction, 27, 225
Veins, **32**
Velamen, **37**
Velcro™, **6**, *7*
Veneer, 327-**329**
Ventral tegmental area (VTA), 359
Venus flytrap (*Dionaea muscipula*), *42*, *43*
Veratrum californicum, 387
Verne, Jules, 367
Vessel elements, **32**, *48*
Vessels, **32**
Vetch (*Vicia* spp.), 384
Vicia faba (broad bean), 214
Vicks, 291
Vigna angularis (Adzuki bean), *214*
Vigna unguiculata (black-eyed pea), *214*
Vinblastine, 352

Vinca alkaloids, 352-353
Vincristine, 352-353
Vin Mariani, 367
Violins, *324*, *325*, *326*
Virchow, Rudolf, 17
Virus-resistance, genetic engineering of, 265
Viscose, 319, *319*
Visible light, **57**, *57*
Vision, vitamin A and, 162
Vitaceae, 128
Vitamins, 162-167
    A, 44, 45, *162*, 162-163, 262
    B complex, 165-167
    C, 44-45, 164-165, 299
    D, 163-164
    E, 162
    fat-soluble, 162, *162*
    K, 162
    water-soluble, 162, *162*
Viticulture, **448**-449
Vitis vinifera. *See* Wine grape (*Vitis vinifera*)
Vitus labrusca, 448
Vitus rupestris, 448
Vodka, 458
Volva, 477
Volvariella spp., 465
Vulcanization, 508

W

Wakame, 409
Waksman, Selman, 471
Wallace, Alfred Russell, 134
WARDA, 244
Wasabi (*Wasabia japonica*), 304
Washington, George, 371
Water
    absorption of from soil, 49-51
    diffusion of (osmosis), 20
    movement of through xylem, *49*, 51
    source of for land plants, 48
    transpiration and, 49
Water hemlock (*Cicuta* spp.), 380, *380*
Water hyacinth (*Eichhornia crassipes*), 487
Watermelon, *188*
Water molds (oomycetes), 420, *420*
Water molecules, *515*
Water-soluble vitamins, 162, *162*
Watson, James, 111
Wavelengths, 56-57
Waxes, 11
Weeds, 394
Welwitschia mirabilis spp., *150*, 151
Went, Frits, 92
West Africa Rice Development Association (WSRDA), 244
Western poison oak (*T. diversilobum*), 397
Western white pine (*Pinus monticola*), *147*
West Nile Virus, 494

Wheat (*Triticum* spp.), 189-193, *190*, 248
    allergies to foods containing, 397
    bread (*T. aestivum*), *187*, 191
    celiac disease and, 158-159
    cultivation of, 191
    durum (*T. durum*), 190, 191
    einkorn (*T. monococcum*), 190
    genetically modified, 260
    genome, 191, 193
    global production of, *188*
    Green Revolution varieties, 243, 244
    hard *vs.* soft, 191
    modern cultivars, 191
    monocultures of, *248*
    origin and evolution of, 190-191
    stem rust of wheat and, 191, 439-441
    stomata on leaves of, *40*
    summer, 191
    winter, 191
Wheat bran, 156
Whiskey, 458-459
Whiskey Rebellion, 458-459
Whisk ferns (*Psilotum* spp.), 145
Whisk ferns (*Tmesipteris* spp.), 145
Whitaker, Robert, 137
White blood cells, 390
White clover (*Trifolium repens*), 487
White nose syndrome, 427
White oak (*Quercus alba*), 2, *329*
White pepper (*Piper nigrum*), 294-295
White pine (*Pinus strobus*), *328*
White potato (*Solanum tuberosum*), *224*, 228-232, *230*
    annual production of, *188*
    cultivation of, *230*, 231
    genetically modified, 231, 237
    genome of, 232
    growth habit, 228
    Irish potato famine and, 228-229, 437
    late blight of potato, 228, 231, 232, 438
    modern cultivars, *230*, 232
    nutritional value of, 232
    origins of, 183, 228
White rot fungi, 321, 435
White snakeroot (*Ageratina altissima*), 386, *386*
White tea, 281
White willow (*Salix alba*), *344*, 345
White wine, 449, 451
Whitney, Eli, 313
WHO (World Health Organizaiton), 155, 373, 484
Whole-tree harvesting, 507, 511
Whorled leaf arrangement, **38**, *39*
Wild barley (*Hordeum chilense*), 204
Wild bees, 251
Wild rice (*Zizania aqautica, Z. texana*), 200
Wild tulips (*Tulipa* spp.), 71
Willow (*Salix* spp.), 338, 345-346

Willow oak (*Quercus phellos*), 129
Wind pollination, 83, 139, 147, 391
Wine, 4, 447–454
    aging and bottling, 452–453
    apple, 461
    champagne (sparkling wines), 453
    clarification of, 452
    climate change, 454
    fermentation of, 67, 449–452
    fortified and dessert, 454
    grape, 454
    health benefits of, 464–465
    origins of, 447
    red, 14, 449, 451
    rose, 449
    white, 449, 451
Wine grape (*Vitis vinifera*), 448–449
    cultivation of, 448–449
    downy mildew of grape, 448, 450
    harvesting of, 449
    Phylloxera infestation of, 448, 450
Winged bean (*Psophocarpus tetragonolobus*), 222
Winged elm (*Ulmus alata*), 487
Winnowing, **201**
Wintergreen (*Gaultheria procumbens*), 345
Winter moth (*Operophtera brumata*), 496
Winter wheat, 191
Wisteria spp., 384
Witch hazel (*Hamamelis virginiana*), 340
Witch's broom (*Crinipellis perniciosa*), 286
Withering, William, 343–344
Woad (*Isatis tinctoria*), 321
Woese, Carl, 137

Wollemia nobilis (wollemi pine), *136*
Wollemi pine (*Wollemia nobilis*), *136*, 147
Wolves, domestication of, 182
Wonder drugs, 473
Wood
    annual rings of, 327
    bark cloth from, 320
    characteristics of different species, 323, 327
    deforestation from harvesting of, 323, 506
    as fuel source, 329
    fungi that decay, 330, 436–437
    grain of, 327
    hardwood, 326–327
    harvesting from biodiversity, 131
    lignin in, 19
    musical instruments from, 324–326
    sapwood, 327, *327*
    secondary xylem, 33
    softwood, 326–327
    springwood, 34
    summerwood, 34
    timber cuts, 327, *327*
Woodcrafts, 321
Wood products, 323, 326–331
    fuel, 329
    lumber, 327–329
    plywood, 327–329
    resins, 329–330
    veneer, 327–329
    wood pulp, 330–331
Wood pulp, 330–331
Wood-rotting fungi, 321, 326, 331
Woodwind instruments, 325
Woody stem, **33**, *35*

World of Coca-Cola museum, Atlanta, 287
World Health Organization (WHO), 155, 373, 484
World Health Organization Framework Convention on Tobacco Control (WHO FCTC), 373
World War I, 371
Wormwood (*Artemesia annua*), 346, 348–349, 461

X

Xanthomonas citri, 269
Xanthomonas oryzae, 264
Xanthophylls, 20, **59**
Xathomonas campestris, 192
Xerophthalmia, **162**
Xerophytes, **49, 501**
X rays, 56, *57*
Xylem, **32**, *32, 33*
    plant divisions featuring, 138
    transpiration and, *48*
Xylocaine (lidocaine), 369

Y

Yam beans (*Pachyrhizus erosus*), 222
Yams (*Dioscorea* spp.), 15, *188*, 238, *238*, 343
Yard, 321, *322*
Yarn, spinning of, 310–312
Yautia (*Xanthosoma* spp.), 238
Yeast, 4, *420*, 427, 447
    alcoholic fermentation and, 68, 449–452, 455, 457
    anaerobic respiration in, 67
    budding by, 422
    fermented foods from, 192, 466–467
    genome of, *112*
    infections from, 481
Yeast infections, 481
Yellow chrysanthemum, 125
Yellow fever, 494
Yellow onion, *322*
Yews (*Taxus* spp.), 147, *344*, 353, 382
Yuca. *See* Cassava (*Manihot esculenta*)

Z

Zea diploperennis, 197, 199, *199*
Zea mays. *See* Corn (*Zea mays*)
Zea mexicana (teosinte), 198
Zebra mussels, 415
Zend-Avesta, 363
Zinc, 50, 51, *167*
Zingiberaceae, 296
Zingiber officinale (ginger), *294, 296*, 296–297
Zinnia, 125
Zizania aquatica (wild rice), 200
Zizania texana (wild rice), 200
Zocor, 473
Zone of cell division, **35**
Zone of elongation, **35**
Zone of maturation, **35**
Zonulin, 158
Zoospores, **408**, **422**
Zoysia grass, *206*
Zygomorphic (irregular) flowers, **74**
Zygomycetes (Zygomycota), 425–427, *426*
Zygosporangia, **426**
Zygospores, **426**
Zygote, **6, 75**